Gauge Theories

of the Strong and Electroweak Interaction

By Prof. Dr. rer. nat. Manfred Böhm, Universität Würzburg
Dr. rer. nat. Ansgar Denner, Paul Scherrer Institut Villigen
Prof. Dr. rer. nat. Hans Joos, DESY Hamburg

B. G. Teubner Stuttgart · Leipzig · Wiesbaden

Prof. Dr. rer. nat. Manfred Böhm

Born 1940 in Memmingen, studied physics in Würzburg und Munich, doctorate in physics from University of Würzburg 1966; research in elementary particle physics in Würzburg, at the Deutsches Elektronen Synchrotron (DESY) in Hamburg and at the European Organization for Nuclear Research (CERN) in Geneva; since 1977 Professor for Theoretical Physics at the University of Würzburg.

Privatdozent Dr. rer. nat. Ansgar Denner

Born 1960 in Schweinfurt, studied physics in Würzburg, doctorate in physics from University of Würzburg 1986, habilitation 1991; research in elementary particle physics in Würzburg, Karlsruhe, Leipzig, at the Max-Planck-Institut for Physics and Astrophysics in Munich and at the European Organization for Nuclear Research (CERN) in Geneva; since 1996 staff member at the Paul Scherrer Institut in Villigen/Switzerland and Privatdozent at the ETH Zurich.

Prof. Dr. rer. nat. Hans Joos

Born 1926 in Stuttgart, studied physics in Tübingen, doctorate in physics 1961, research in elementary particle in Sao Paulo/Brasil, at the University of Hamburg, at the Institute for Advanced Study in Princeton, at the University of Minnesota, at the European Organization for Nuclear Research (CERN) in Geneva, and at the Deutsches Elektronen Synchrotron (DESY) in Hamburg; from 1963 until retirement in 1990 Senior Scientist at DESY in Hamburg and since 1965 Adjoint Professor an the University of Hamburg.

Die Deutsche Bibliothek – CIP-Einheitsaufnahme
Ein Titeldatensatz für diese Publikation ist bei
Der Deutschen Bibliothek erhältlich.

Dieses Buch erschien bisher in 2 Auflagen unter dem Titel „Eichtheorien der starken und elektrostarken Wechselwirkung“ der Autoren Becher/Böhm/Joos im gleichen Verlag.

1. Auflage 1981
2., durchges. Auflage 1983
3., neubearb. Auflage April 2001

softcover reprint of the hardcover 3rd edition 2001

Der Verlag Teubner ist ein Unternehmen der Fachverlagsgruppe BertelsmannSpringer.

www.teubner.de

Umschlaggestaltung: Peter Pfitz, Stuttgart

ISBN-13: 978-3-322-80162-3 e-ISBN-13: 978-3-322-80160-9
DOI: 10.1007/978-3-322-80160-9

Preface

The gauge theories for the strong and electroweak interaction have become the Standard Model of particle physics. They realize in a consistent way the requirements of quantum theory, special relativity, and symmetry principles. For the first time, we have a consistent theory of the fundamental interactions that allows for precision calculations for many experiments. The Standard Model has, up to now, successfully passed all experimental tests. This success establishes the importance of gauge theories, despite the fact that gravity is not included and that the Standard Model is most likely an effective theory resulting from the low-energy limit of a more fundamental theory.

The aim of this book is to present the basic ideas and concepts, the technical tools, and the predictions of the gauge theories for the fundamental constituents of matter and their interactions: Quantum Chromodynamics for the strong interaction and the Electroweak Standard Model for the unified electromagnetic and weak interaction.

The first edition of this book, P. Becher, M. Böhm and H. Joos, *Eichtheorien der starken und elektroschwachen Wechselwirkung*, appeared in German in 1981. At that time, gauge theories were not yet really supported by experiment. The W and Z bosons were not discovered, the existence of the gluon was not confirmed, and the top-quark had not been found. Presently, after a lot of theoretical and experimental work, gauge theories are much better understood, many precision calculations exist, and a huge amount of experimental data is available. All this has put the gauge theories on a firm ground and required a thorough revision of the original book. This third edition is almost completely rewritten and largely extended. We tried to cover the most important developments and to guide the reader to the modern applications of gauge theories in elementary particle physics.

As before, this book is written for students who have passed the standard course in theoretical physics and an introductory course in particle physics. The book is also addressed to experimental and theoretical physicists who want to become familiar with the established treatment of the fundamental interactions. We tried to cover a large part of the applications of gauge theories, to present the material in a self-contained form, and to show in detailed calculations, as far as possible, how the results are obtained. In order to help the reader to get along with this huge amount of material we supply a detailed index and a large list of references. Evidently, the list of references cannot be exhaustive. We tried to focus on original papers, review papers, and articles that, in our opinion, should help the reader in understanding the subjects.

In the first part of the book, we review the basic concepts of the phenomenology of particle physics and relativistic quantum field theory. Then, we treat the quantum theory of gauge fields in detail. We work out the invariant perturbation theory including Feynman-diagram techniques and renormalization, which is the basis for many precision calculations. The topological properties of gauge theories are important for their non-perturbative evaluation. We discuss, in particular, instanton solutions of gauge theories and gauge theories on the lattice. This knowledge of the structure of gauge theories and of the methods for their evaluation is then applied to Quantum Chromodynamics and the Electroweak Standard Model. We conclude with short a survey of grand unified theories and supersymmetry.

Acknowledgements and thanks

We would like to thank all physicists from whom we learned and with whom we interacted, far too many to be named here. We are grateful to our colleagues at the University of Würzburg, at PSI Villigen, at DESY, and at the University of Hamburg for their help and encouragement.

For careful and critical reading of various chapters of the manuscript our special thanks go to S. Dittmaier, I. Montvay, F. Schrempp, and A. Vogt.

We are grateful to Dr. P. Spuhler of the Teubner Publishing Company for a constructive collaboration and encouragement.

Last but not least we thank our families for their patience and understanding.

Würzburg M. Böhm
Villigen A. Denner
Hamburg, January 2001 H. Joos

Contents

Corrections to this book

A list of misprints and corrections to this book is posted on the World-Wide Web at the URL `http://www.hep.psi.ch/denner/GTupdates.html`, or can be obtained by writing to the authors. We would be grateful, if you would report addtional errors in the book to Ansgar.Denner@psi.ch.

1 Phenomenological basis of gauge theories of strong, electromagnetic, and weak interactions

The gauge theories of the strong and electroweak interactions of elementary particles were developed from phenomenological models which provided first explanations of the experimental results. These models are based on the fact that the fundamental constituents of matter are leptons and quarks. Guided by the success of Quantum Electrodynamics, the fundamental forces are described by the exchange of gauge bosons between these constituents. The gluons are responsible for the strong interaction, the photon, the W and Z bosons for the electroweak interaction. The quantum numbers and the basic properties of the fundamental fermions and bosons are resumed in Sect. 1.1. In order to formulate and apply these dynamical ideas, elements of relativistic quantum theory and of group theory are needed. They are presented in Sect. 1.2. In Sect. 1.3, the quantum numbers and wave functions of the hadrons in the framework of the phenomenological quark model, especially the arguments for colour and the non-relativistic treatment of quarkonia, are reviewed. The phenomenological description of the electroweak interaction and of the parton model, together with some typical applications, are given in Sects. 1.4 and 1.5. The importance of precision calculations, radiative corrections (Sect. 1.6), and arguments for gauge theories as relativistic quantum field theories (Sect. 1.7) conclude this chapter.

1.1 Elementary particles and their interactions

In order to formulate the dynamics of elementary particles, one has to know the fundamental degrees of freedom of matter, their fundamental interactions, and how these can be described in relativistic quantum field theory. We know that hadrons are not elementary; the fundamental constituents of matter are quarks and leptons, i.e. fermions with spin 1/2. In this section we collect those properties of leptons, quarks, and gauge bosons that are

lepton	quantum number				mass
ν_l, l	L_e	L_μ	L_τ	Q	m
ν_e	1	0	0	0	$< 3\,\text{eV}$
e	1	0	0	-1	511.00 keV
ν_μ	0	1	0	0	$< 0.19\,\text{MeV}$
μ	0	1	0	-1	105.66 MeV
ν_τ	0	0	1	0	$< 18.2\,\text{MeV}$
τ	0	0	1	-1	1.777 GeV

Tab. 1.1 Quantum numbers and masses of the leptons (from Ref. [PDG00])

important for the construction of gauge theories. For more details we refer to text books on the basics of elementary particle physics [EP].

1.1.1 Leptons and quarks as fundamental constituents of matter

According to our present knowledge, leptons and quarks exist in three *families* or *generations* [LEP92]. The different leptons and quarks are characterized by quantum numbers. For each lepton and quark an antilepton or antiquark, respectively, with the same mass but with opposite inner quantum numbers exists.

1.1.1.1 Leptons

The fundamental fermions that do not interact strongly are called leptons. Every family contains a neutral and a charged lepton and is characterized by a conserved additive quantum number, the corresponding *lepton number*. The first, the electronic lepton pair, consists of the electron neutrino and the electron, (ν_e, e), with electron number $L_e = 1$, the second, the muonic lepton pair, of (ν_μ, μ) with muon number $L_\mu = 1$, and the third, the tauonic lepton pair, of (ν_τ, τ) with tau number $L_\tau = 1$. All leptons interact weakly, the charged ones also electromagnetically. Their most important properties, such as masses, lifetimes, decays etc., can be found in Ref. [PDG00]. Their quantum numbers and masses are listed in Table 1.1. Direct measurements

quark	quantum number							mass
q	I	I_3	S	C	B	T	Q	m
u	$\frac{1}{2}$	$\frac{1}{2}$	0	0	0	0	$\frac{2}{3}$	1–5 MeV
d	$\frac{1}{2}$	$-\frac{1}{2}$	0	0	0	0	$-\frac{1}{3}$	3–9 MeV
c	0	0	0	+1	0	0	$\frac{2}{3}$	1.15–1.35 GeV
s	0	0	−1	0	0	0	$-\frac{1}{3}$	75–170 MeV
t	0	0	0	0	0	+1	$\frac{2}{3}$	174.3 ± 5.1 GeV
b	0	0	0	0	−1	0	$-\frac{1}{3}$	4.0–4.4 GeV

Tab. 1.2 Quantum numbers and masses of the quarks (from Ref. [PDG00])

yield only upper limits for the neutrino masses. The detection of neutrino oscillations, however, provides evidence for non-zero neutrino masses. These violate also the conservation of the individual lepton numbers.

1.1.1.2 Quarks

In contrast to leptons, quarks interact also strongly. They have not been observed as free particles, but are the constituents of the *hadrons* which include *mesons*, i.e. quark–antiquark systems, and *baryons*, i.e. three-quark systems. The *flavour quantum numbers* of the hadrons, which are conserved in the strong interaction, strong isospin (I, I_3), strangeness S, charm C, bottomness B, and topness T, are derived from the corresponding flavour quantum numbers of the quarks (cf. Sect. 1.3). In Table 1.2 these quantum numbers are collected together with the electric charges and effective masses of the quarks. All quarks carry baryon number 1/3. The light hadrons are made up of the first quark doublet (u, d), the up and down quarks. The second quark family, (c, s), consists of the charm and the strange, the third one, (t, b), of the top and the bottom quarks. This doublet structure is important for the weak interaction of the quarks. The electromagnetic couplings are determined by the quark charge Q. For the formulation of the theory of the strong interaction, Quantum Chromodynamics (QCD), another degree of freedom, called *colour*, is necessary. The colour quantum number assumes three values: red, green, and blue $(\mathrm{r}, \mathrm{g}, \mathrm{b})$. Each quark exists in each of these three colours. The colour degrees of freedom cannot be observed directly, since hadrons, i.e. the physical particles, are colourless bound states

family	1st	2nd	3rd	
leptons	ν_e, e	ν_μ, μ	ν_τ, τ	
quarks	u_r, d_r u_g, d_g u_b, d_b	c_r, s_r c_g, s_g c_b, s_b	t_r, b_r t_g, b_g t_b, b_b	colour triplets
	doublets	doublets	doublets	

Tab. 1.3 Family structure of the fundamental fermions

of quarks. Since the quarks do not exist as free particles their mass cannot be defined in the conventional way. Instead, various definitions of effective masses are used. The values given in Table 1.2 are those recommended by the Particle Data Group [PDG00].

The flavour and colour structure of the three fermion families is displayed in Table 1.3.

1.1.2 Fundamental interactions

1.1.2.1 Standard model of strong and electroweak interaction

Presently, four fundamental interactions of elementary particles are known: the gravitational, the electromagnetic, the weak, and the strong interaction. For gravitation a consistent quantum field theory does not yet exist. This lack of our basic understanding of gravitation is not of practical importance for ordinary elementary particle physics, since the strength of gravity is by many orders of magnitude smaller than the one of the other interactions.

The three other interactions, however, are successfully described by a quantized gauge-field theory, the Standard Model (SM) of elementary particle physics. This contains as one part Quantum Chromodynamics (QCD), the gauge theory of strong interaction, which was proposed in 1973.[1] Its second part, the theory of the unified electromagnetic and weak interaction was formulated for leptons in 1967 [We67, Sa68] and extended to quarks in 1970 [Gl70]. Today this is called the electroweak standard model (EWSM). The

[1] For the historical development of QCD see Ref. [tH85].

construction, general properties, and application of these gauge theories to elementary particle physics are the main concern of this book.

1.1.2.2 One-particle-exchange model

Ideas about the dynamics of the strong and electroweak interaction were taken over from Quantum Electrodynamics. There, in lowest-order approximation, the forces between charged elementary particles are generated by the exchange of the quanta of the electromagnetic field, the *photons*, which are the gauge bosons of the electromagnetic interaction. In the static limit, the Coulomb force is reproduced in this way. The *boson-exchange model* has been generalized to the other interactions.

The exchange of the massless gauge bosons of QCD, the *gluons*, yields the simplest description of the strong force between the quarks. In contrast to the photon, the gluons are not neutral but carry colour charges. For theoretical reasons (cf. Sect. 1.2.2), eight gluons must exist for the case of three colours. Like the coloured quarks, the coloured gluons cannot be observed directly as free particles. There is, however, indirect experimental evidence for their existence from deep-inelastic scattering (cf. Sect. 1.5), the observation of the corresponding jets, and other experiments [Br79].

In contrast to the electromagnetic interaction, the weak interaction has a very short range. Therefore, the corresponding gauge bosons have to be massive. The *charged weak gauge boson*, the W boson, and the *neutral weak gauge boson*, the Z boson, were discovered in 1983 [Ar83]. Their properties have been measured with high accuracy. The W-*boson mass* is $M_{\mathrm{W}} = 80.419 \pm 0.056\,\mathrm{GeV}$ and the Z-*boson mass* is $M_{\mathrm{Z}} = 91.1882 \pm 0.0022\,\mathrm{GeV}$ [PDG00]. The W and Z bosons are unstable and decay within about $3 \times 10^{-25}\,\mathrm{s}$ almost exclusively into fermion-antifermion pairs (cf. Sect. 1.4). They can be experimentally detected only via an analysis of their decay products.

1.2 Elements of relativistic quantum field theory

In our introductory remarks we presented the quantum numbers and basic properties of the fundamental constituents of matter and their interactions, i.e. of the elementary particles. The corresponding theory must satisfy the approved physical principles of quantum mechanics and relativistic invariance. These are guaranteed by the description of the elementary particles and their interactions by quantized, relativistic, locally interacting fields. The connection between the structure of conserved charges and the symmetry groups of the fields is of great importance. Therefore, some of the relevant field- and group-theoretic concepts are collected in this section [FT, GT].

1.2.1 Basic concepts of relativistic quantum field theory

1.2.1.1 Quantum-mechanical states of a free, relativistic particle

We begin our introduction to relativistic quantum mechanics by describing the *quantum-mechanical states of a free massive particle* $|M, p, j, j_3\rangle$ [Wi39] with mass M, four-momentum $p^\mu = (E_p, \mathbf{p})$, $E_p = \sqrt{M^2 + \mathbf{p}^2}$, spin j, and third component of spin j_3, normalized as

$$\langle M, p, j, j_3 | M, p', j, j_3' \rangle = 2E_p \, (2\pi)^3 \, \delta(\mathbf{p} - \mathbf{p}') \, \delta_{j_3 j_3'}. \tag{1.2.1}$$

These free one-particle states are eigenstates of the relativistic energy–momentum operator P^μ and thus transform under space–time translations $x^\mu \to x^\mu + a^\mu$ according to

$$U(a)|M, p, j, j_3\rangle = \mathrm{e}^{\mathrm{i}pa}|M, p, j, j_3\rangle. \tag{1.2.2}$$

Homogeneous *Lorentz transformations* $x'^\mu = \Lambda^\mu{}_\nu x^\nu$, $g^{\mu\nu} = \Lambda^\mu{}_\rho \Lambda^\nu{}_\sigma g^{\rho\sigma}$, transform the one-particle state into one with four-momentum Λp and with a changed orientation of the third component j_3 of the particle spin j described by the rotation matrix $D^{(j)}$

$$U(\Lambda)|M, p, j, j_3\rangle = |M, \Lambda p, j, j_3'\rangle D^{(j)}_{j_3' j_3}(R(\Lambda, p)). \tag{1.2.3}$$

The *Wigner rotation* $R(\Lambda, p) = L^{-1}(\Lambda p)\Lambda L(p)$ is composed of the boost $L(p)$ which transforms the rest momentum $\tilde{p} = (M, \mathbf{0})$ into the actual four-momentum of the particle $p = L(p)\tilde{p}$, the Lorentz transformation Λ, and the boost $L^{-1}(\Lambda p)$ which brings Λp back into the rest frame. Equations (1.2.2) and (1.2.3) define an irreducible representation of the Poincaré group, i.e. the group of space–time translations and homogeneous Lorentz transformations.

For massless particles, no rest frame exists, and the one-particle states have to be constructed somewhat differently [Wi39, We95]. Here, the boost $L(p)$ has to be chosen such that it transforms a standard light-like momentum, e.g. $\tilde{p} = (1, 0, 0, 1)$, into the actual four-momentum of the particle. Since this vector is not invariant under SU(2) but rather under the euclidean group E(2), the states cannot be classified according to spin quantum numbers but only according to helicity quantum numbers. As a consequence, massless particles can have only two different helicities.

1.2.1.2 Creation and annihilation operators

One- and multi-particle states, which are symmetric (for bosons) or antisymmetric (for fermions) under permutation of identical particles, are described by creation operators $a^\dagger_{j_3}(p)$ and annihilation operators $a_{j_3}(p)$. These are characterized by their *(anti)commutation relations*

$$
\begin{aligned}
&[a^\dagger_{j_3}(p), a^\dagger_{j'_3}(p')]_\pm = [a_{j_3}(p), a_{j'_3}(p')]_\pm = 0,\\
&[a_{j_3}(p), a^\dagger_{j'_3}(p')]_\pm = 2E_p\,(2\pi)^3\,\delta(\mathbf{p}-\mathbf{p}')\,\delta_{j_3 j'_3},\\
&\qquad + \text{ for fermions}, \quad - \text{ for bosons}.
\end{aligned} \tag{1.2.4}
$$

The operators $a_{j_3}(p)$ annihilate the relativistically invariant ground state, the *vacuum* $|0\rangle$,

$$
a_{j_3}(p)|0\rangle = 0, \qquad \langle 0|0\rangle = 1. \tag{1.2.5}
$$

The one-particle states (1.2.1) are created by $a^\dagger_{j_3}(p)$ from the vacuum

$$
|M, p, j, j_3\rangle = a^\dagger_{j_3}(p)|0\rangle, \tag{1.2.6}
$$

and free multi-particle states by repeated application of $a^\dagger$. Creation and annihilation operators of the associated antiparticles are denoted by $b^\dagger_{j_3}(p)$ and $b_{j_3}(p)$, respectively.

1.2.1.3 Free, quantized fields

A complementary description of free particles, which emphasizes their causal propagation, uses *free quantized fields* $\psi(x)$. These are defined by field equations and commutation relations. For complex *scalar fields* these are the *Klein–Gordon equation*

$$(\partial_\mu \partial^\mu + M^2)\psi(x) = 0 \tag{1.2.7}$$

and the *commutation relations*

$$\begin{aligned} [\psi(x), \psi(x')]_- &= [\psi^\dagger(x), \psi^\dagger(x')]_- = 0, \\ [\psi(x), \psi^\dagger(x')]_- &= \mathrm{i}\Delta(x - x', M), \end{aligned} \tag{1.2.8}$$

where

$$\Delta(x, M) = -\frac{\mathrm{i}}{(2\pi)^3} \int \mathrm{d}^4 p \, \delta(p^2 - M^2) \, \mathrm{sgn}(p^0) \, \mathrm{e}^{-\mathrm{i}px}. \tag{1.2.9}$$

Commutativity of the field operators for space-like distances, i.e. $\Delta(x, M) = 0$ for $x^2 < 0$, guarantees relativistic *causality*.

The free fields $\psi_\alpha(x)$ are connected to creation and annihilation operators of particles with arbitrary spin by a Fourier transformation

$$\begin{aligned} \psi_\alpha(x) = \frac{1}{(2\pi)^3} \sum_{j_3} \int \frac{\mathrm{d}^3\mathbf{p}}{2E_p} \Big[& u_\alpha(p, j_3) \, a_{j_3}(p) \, \mathrm{e}^{-\mathrm{i}px} \\ & + v_\alpha(p, j_3) \, b^\dagger_{j_3}(p) \, \mathrm{e}^{+\mathrm{i}px} \Big], \\ px = p^0 x^0 - \mathbf{px}, \qquad & p^0 = E_p. \end{aligned} \tag{1.2.10}$$

Equation (1.2.10) incorporates the wave–particle dualism: particles are quanta of fields.

The particle and antiparticle wave functions $u_\alpha(p, j_3)$ and $v_\alpha(p, j_3)$ describe the polarization degrees of freedom. They are constructed in such a way that $\psi_\alpha(x)$ transforms locally under Lorentz transformations. For *scalar fields*,

$u(p) = v(p) = 1$. In the spin-$\frac{1}{2}$ case, $u_\alpha(p, j_3)$ and $v_\alpha(p, j_3)$ are the well-known four-component Dirac spinors. The corresponding four-component Dirac field $\psi_\alpha(x)$ satisfies the *Dirac equation* (written here with all indices)

$$\sum_{\beta=1}^{4} (\mathrm{i}\gamma^\mu_{\alpha\beta}\partial_\mu - M\delta_{\alpha\beta})\psi_\beta(x) = 0, \tag{1.2.11}$$

together with the *anticommutation relations*

$$\begin{aligned}
[\psi_\alpha(x), \psi_{\alpha'}(x')]_+ &= [\overline{\psi}_\alpha(x), \overline{\psi}_{\alpha'}(x')]_+ = 0, \\
[\psi_\alpha(x), \overline{\psi}_{\alpha'}(x')]_+ &= \mathrm{i}S_{\alpha\alpha'}(x - x', M) \\
&= (\mathrm{i}\gamma^\mu_{\alpha\alpha'}\partial_\mu + M\delta_{\alpha\alpha'})\mathrm{i}\Delta(x - x', M),
\end{aligned} \tag{1.2.12}$$

which involve the conjugate Dirac field $\overline{\psi}_\alpha = \psi^\dagger_\beta\gamma^0_{\beta\alpha}$.

The *Dirac matrices* γ^μ are defined by the *Dirac algebra*

$$[\gamma^\mu, \gamma^\nu]_+ = 2g^{\mu\nu}\,\mathbf{1} \tag{1.2.13}$$

with γ^0 hermitian and γ^i antihermitian. A useful short-hand notation for products of four-vectors with Dirac matrices is the *Feynman dagger* $\not{k} \equiv \gamma_\mu k^\mu$.

For spin-1 fields, the wave functions correspond to the polarization vectors $\varepsilon_\mu(p, j_3)$ and $\varepsilon^*_\mu(p, j_3)$.

A collection of formulas for Dirac matrices, spinors, and polarization vectors can be found in App. A.1.

1.2.1.4 Lagrangian formalism for relativistic fields

Relativistic field equations can be obtained from an invariant *Lagrangian* $\mathcal{L}(\psi, \partial_\mu\psi)$ with help of Hamilton's principle,

$$\delta\int \mathcal{L}\,\mathrm{d}^4x = 0, \tag{1.2.14}$$

as Euler–Lagrange *equations of motion*

$$\frac{\partial\mathcal{L}}{\partial\psi} - \partial_\mu\frac{\partial\mathcal{L}}{\partial(\partial_\mu\psi)} = 0. \tag{1.2.15}$$

The use of the Lagrangian formalism is important for the formulation of theories with interaction and their quantization. This can be illustrated by *Quantum Electrodynamics (QED)*, which describes—in its simplest version—the interaction of photons, electrons, and positrons [QED]. The *Lagrangian of QED* reads

$$\mathcal{L}_{\mathrm{QED}} = -\frac{1}{4} F_{\mu\nu} F^{\mu\nu} + \bar{\psi}(\mathrm{i}\partial\!\!\!/ - M)\psi - eQ\, A^{\mu} \bar{\psi} \gamma_{\mu} \psi \tag{1.2.16}$$

for an electron with charge $-Qe$ ($Q = -1$, $e > 0$), where the *electromagnetic field-strength tensor* $F_{\mu\nu}(x)$ is derived from the four-potential or *photon field* $A_{\mu}(x)$

$$F_{\mu\nu}(x) = \partial_{\mu} A_{\nu}(x) - \partial_{\nu} A_{\mu}(x). \tag{1.2.17}$$

Variation of $\mathcal{L}_{\mathrm{QED}}$ with respect to A_{ν} gives the *inhomogeneous Maxwell equations*

$$\partial^{\mu} F_{\mu\nu}(x) = eQ\bar{\psi}(x)\gamma_{\nu}\psi(x) = e j_{\nu}^{\mathrm{em}}(x). \tag{1.2.18}$$

The *homogeneous Maxwell equations* $\partial_{\mu}\varepsilon^{\mu\nu\rho\sigma} F_{\rho\sigma} = 0$ follow directly from (1.2.17). Variation of $\mathcal{L}_{\mathrm{QED}}$ with respect to $\bar{\psi}_{\alpha}(x)$ yields the Dirac equation for the electron field

$$(\mathrm{i}\partial\!\!\!/ - M)\psi(x) = eQA^{\mu}(x)\gamma_{\mu}\psi(x). \tag{1.2.19}$$

The *elementary charge* e or *Sommerfeld's fine-structure constant* $\alpha = e^2/4\pi \approx 1/137.036$ determines the strength of the coupling between the electromagnetic field and the Dirac field. For vanishing coupling, i.e. $e = 0$, (1.2.18) is the free Maxwell equation, and (1.2.19) the free Dirac equation. The electromagnetic current $j_{\mu}^{\mathrm{em}} = Q\bar{\psi}\gamma_{\mu}\psi$, as the source of $F_{\mu\nu}(x)$ in (1.2.18), as well as the potential term in the Dirac equation result from the variation of the trilinear expression, i.e. the interaction term

$$\mathcal{L}_{\mathrm{I}} = -eQA^{\mu}\bar{\psi}\gamma_{\mu}\psi = -e j_{\mu}^{\mathrm{em}} A^{\mu}. \tag{1.2.20}$$

1.2.1.5 *S* matrix

Cross sections for scattering processes are obtained from the matrix elements of the *scattering matrix* or S matrix. This transforms incoming states into outgoing states. It can be obtained as the infinite-time limit of the time-evolution operator in the interaction picture and is directly related to the interaction Lagrangian [FT]

$$S = T \exp\left[\mathrm{i} \int \mathrm{d}^4x \, \mathcal{L}_{\mathrm{I}} \right]. \tag{1.2.21}$$

This expression has to be understood as a formal power series in $\mathcal{L}_{\mathrm{I}}$, and T denotes the *time-ordered product.* Operators occurring under the T symbol are ordered from right to left with increasing times. The time-ordered product of two factors reads explicitly ($t \equiv x^0$)

$$T[\psi(x)\psi(x')] = \theta(t - t')\psi(x)\psi(x') \pm \theta(t' - t)\psi(x')\psi(x) \tag{1.2.22}$$

with + for bosonic and − for fermionic field operators. This definition is relativistically invariant owing to the commutativity of field operators at space-like distances (1.2.8).

The absolute square of the *S-matrix element* $\langle f|S|i\rangle$ is related to the probability for an incoming state $|i\rangle$ to evolve into an outgoing state $|f\rangle$. If $|i\rangle$ and $|f\rangle$ are momentum eigenstates with momenta p_i and p_f, respectively, one obtains, upon subtracting the unit matrix and extracting the δ-function of four-momentum conservation, the *invariant matrix element* $\mathcal{M}_{fi}$

$$\langle f|S|i\rangle = \langle f|i\rangle + \mathrm{i}(2\pi)^4\delta^{(4)}(p_i - p_f)\mathcal{M}_{fi}. \tag{1.2.23}$$

1.2.1.6 Differential cross section and decay width

The *differential cross section* for the scattering of two particles with momenta p_1 and p_2 and masses M_1 and M_2 into n particles with momenta $q_1, \ldots, q_n$ is calculated by squaring the invariant matrix element $\mathcal{M}$ and multiplying it with the flux of the incoming particles and the phase-space density of the outgoing particles,

$$\mathrm{d}\sigma = \frac{1}{4\sqrt{(p_1 p_2)^2 - M_1^2 M_2^2}} |\mathcal{M}_{fi}|^2 (2\pi)^4 \delta^{(4)}\left(p_1 + p_2 - \sum_{i=1}^{n} q_i\right) \times \frac{\mathrm{d}^3\mathbf{q}_1}{(2\pi)^3 2E_{q_1}} \cdots \frac{\mathrm{d}^3\mathbf{q}_n}{(2\pi)^3 2E_{q_n}}. \tag{1.2.24}$$

In this formula we implicitly assumed that the spatial parts of the incoming momenta involve no components orthogonal to their difference, i.e. $\mathbf{p}_1$ and $\mathbf{p}_2$ are antiparallel or parallel. The *decay width* of a particle with momentum $p = (E_p, \mathbf{p})$ into a final state of n particles with momenta $q_1, \ldots, q_n$ is given by

$$\mathrm{d}\Gamma = \frac{1}{2E_p} |\mathcal{M}_{fi}|^2 (2\pi)^4 \delta^{(4)}\left(p - \sum_{i=1}^{n} q_i\right) \frac{\mathrm{d}^3\mathbf{q}_1}{(2\pi)^3 2E_{q_1}} \cdots \frac{\mathrm{d}^3\mathbf{q}_n}{(2\pi)^3 2E_{q_n}}. \tag{1.2.25}$$

Degrees of freedom that are not observed have to be averaged in the initial state and summed over in the final state.

The total cross section and the total decay width are obtained by integrating over the phase space of the final state. For any set of n identical particles in the final state, a *symmetry factor* $1/n!$ has to be included in these and other inclusive quantities.

In the special case of a two-particle final state these formulas can be evaluated in the centre-of-mass (CM) system according to

$$\frac{\mathrm{d}\sigma}{\mathrm{d}\Omega_{\mathrm{CM}}} = \frac{1}{64\pi^2 E_{\mathrm{CM}}^2} \frac{|\mathbf{q}|}{|\mathbf{p}|} |\mathcal{M}_{fi}|^2, \tag{1.2.26}$$

where E_{CM} is the total energy, and $\mathbf{p}$ and $\mathbf{q}$ the three-momenta of the incoming and outgoing particles, respectively, in the CM system.

The decay width for a particle at rest with mass M becomes

$$\frac{\mathrm{d}\Gamma}{\mathrm{d}\Omega_{\mathrm{CM}}} = \frac{1}{32\pi^2 M} \frac{|\mathbf{q}|}{M} |\mathcal{M}_{fi}|^2. \tag{1.2.27}$$

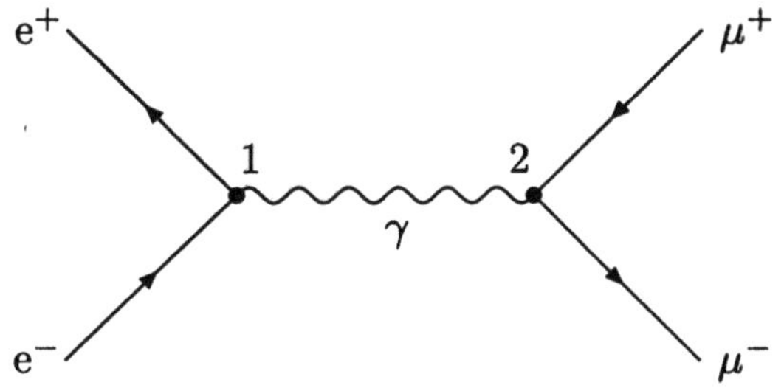

Fig. 1.1 Feynman graph for the one-photon approximation of the reaction $e^+e^- \to \mu^+\mu^-$

1.2.1.7 Feynman graphs

The quantization of the non-linear field equations (1.2.18) and (1.2.19) is considered in Sect. 2.4. For small values of the coupling constant e, the quantized theory can be evaluated by a perturbative expansion. This perturbative expansion can be expressed by Feynman graphs. These intuitively describe the interaction in terms of scattering processes between the particles (quanta) of the free fields. For example, the reaction $e^+e^- \to \mu^+\mu^-$ is described in lowest order of the electromagnetic interaction by the Feynman graph shown in Fig. 1.1. The in- and outgoing particles of the reactions are represented by external lines. The internal lines describe the field propagation between the vertices (interaction points). In this sense, the Feynman graph in Fig. 1.1 visualizes the above reaction as the annihilation of an e^+e^- pair into a photon at space–time point 1, the propagation of the *virtual* photon from point 1 to point 2, and the creation of a $\mu^+\mu^-$ pair from the photon at point 2.

1.2.1.8 Feynman rules

Together with the Feynman rules, the Feynman graphs yield analytical expressions for *S-matrix elements*, or more precisely, for i times the invariant matrix elements $\mathcal{M}_{fi}$. The momentum-space rules described here are for QED, i.e. the interaction of elementary charged fermions, namely leptons and quarks with the photon:

1. The analytical expressions for the *external lines* are given by the wave functions, i.e. spinors and polarization vectors, of the fields (1.2.10).

For scalar fields the wave functions are equal to one. Fermions with momentum p and spin j_3 contribute $u_\alpha(p, j_3)$ or $v_\alpha(p, j_3)$, photons with momentum k and helicity λ contribute $\varepsilon_\mu(k, \lambda)$ as factors:

incoming:

$$u_\alpha(p, j_3) \qquad \bar{v}_\alpha(p, j_3) \qquad \varepsilon_\mu(k, j_3) \tag{1.2.28}$$

outgoing:

$$\bar{u}_\alpha(p, j_3) \qquad v_\alpha(p, j_3) \qquad \varepsilon^*_\mu(k, j_3) \tag{1.2.29}$$

The arrows indicate the direction of fermion-number flow. The momenta flow from left to right.

2. The interaction between photons and charged fermions, corresponding to the Lagrangian term $\mathcal{L}_\mathrm{I} = -eQA^\mu \bar{\psi}\gamma_\mu \psi$ (1.2.20), is represented by a *vertex* with a fermion line passing through,

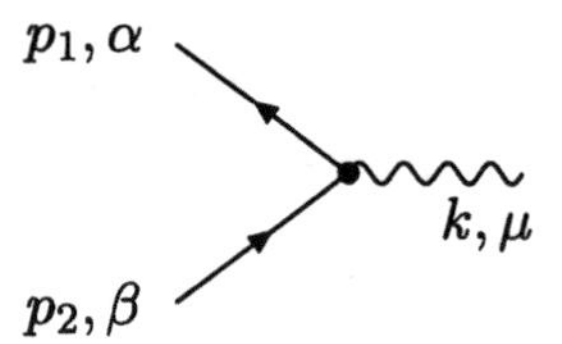

$$= -\mathrm{i}eQ\,(\gamma_\mu)_{\alpha\beta}, \tag{1.2.30}$$

All momenta are incoming and four-momentum conservation has to be respected, i.e. $p_1 + p_2 + k = 0$.

3. The propagation of the interacting fields is described by the causal Green functions of the free field equations, the *Feynman propagators* Δ_F and S_F. For scalar fields these are defined by

$$\begin{aligned} \Delta_\mathrm{F}(x, M) &= \int \frac{\mathrm{d}^4 p}{(2\pi)^4}\, \mathrm{e}^{-\mathrm{i}px} \Delta_\mathrm{F}(p, M), \\ \Delta_\mathrm{F}(p, M) &= \frac{1}{p^2 - M^2 + \mathrm{i}\varepsilon} \end{aligned} \tag{1.2.31}$$

with

$$(\partial^\mu \partial_\mu + M^2)\Delta_\mathrm{F}(x - x', M) = -\delta^{(4)}(x - x'). \tag{1.2.32}$$

In Feynman graphs, the propagation is symbolized by an *internal line*; for this, the Fourier transforms $\mathrm{i}S_\mathrm{F}(p, M)_{\alpha\beta}$ and $\mathrm{i}\Delta_\mathrm{F}^{\mu\nu}(p,0)$ are inserted as factors in the analytical expression

l :  β α

$$= \mathrm{i}S_\mathrm{F}(p, M)_{\alpha\beta} = \mathrm{i}(\not{p} + M)_{\alpha\beta}\Delta_\mathrm{F}(p, M) = \left(\frac{\mathrm{i}(\not{p}+M)}{p^2 - M^2 + \mathrm{i}\varepsilon}\right)_{\alpha\beta}, \tag{1.2.33}$$

γ :  μ ν

$$= \mathrm{i}\Delta_\mathrm{F}^{\mu\nu}(p,0) = -\mathrm{i}g^{\mu\nu}\Delta_\mathrm{F}(p,0) = \frac{-\mathrm{i}g^{\mu\nu}}{p^2 + \mathrm{i}\varepsilon}. \tag{1.2.34}$$

The arrow in the *fermion propagator* indicates the fermion-number flow, the momentum flows from left to right. The *photon propagator* (1.2.34) is given in the Feynman gauge. A systematic treatment is given in Sects. 2.3 and 2.4.

4. The composition of the elements of the graph, i.e. the factors of the analytical expression for the invariant matrix element $\mathrm{i}\mathcal{M}$, follows the structure of the reaction. This is most evident for the diagrams without loops, i.e. *tree graphs*. If the Dirac spinors and Dirac matrices are written down in the order that is obtained by following the fermion lines through the diagram oppositely to the direction of fermion-number flow, the order of the Dirac matrices is appropriate for matrix multiplication and all Dirac indices can be omitted.

   The momentum flux from the incoming to the outgoing particles can be chosen freely, provided the conservation of momentum at each vertex is respected. Free internal momenta (loop momenta) q_μ have to be integrated over using $(2\pi)^{-4}\int \mathrm{d}^4q$.

5. The relative signs between Feynman graphs are determined by the anticommutativity of fermionic operators: every closed fermion line gets a factor -1, and every interchange of two external fermion lines gives a factor -1. If the spinors and Dirac matrices are ordered oppositely to the fermion number flow, the latter sign can be easily determined as the signature of the permutation by which the actual order of all external spinors u, $\bar{u}$, v, and $\bar{v}$ in the analytic expression can be obtained from a given standard order.

6. The graph contributions have to be divided by the *symmetry factor*. It is defined by the number of possibilities of mapping the graph on itself

by a permutation of lines and vertices. (In spinor electrodynamics the symmetry factor is always 1.)

1.2.1.9 An example: e^+e^- annihilation into muon pairs

In lowest order, the one-photon exchange approximation, e^+e^- *annihilation into muon pairs* is given by the Feynman graph in Fig. 1.1. The kinematic quantities of the reaction are labelled as follows. In the centre-of-mass (CM) system, the four-momenta of the incoming $e^\pm$ ($p_\pm$) and outgoing $\mu^\pm$ ($q_\pm$) read

$$p^\mu_\pm = (E, 0, 0, \pm p), \qquad q^\mu_\pm = (E, \pm q \sin\vartheta, 0, \pm q\cos\vartheta). \tag{1.2.35}$$

The relativistically invariant *Mandelstam variables*

$$\begin{aligned} s &= (p_+ + p_-)^2 = (q_+ + q_-)^2 = 2m_e^2 + 2p_+p_- = 4E^2, \\ t &= (p_+ - q_+)^2 = (p_- - q_-)^2 = m_e^2 + m_\mu^2 - 2E^2 + 2pq\cos\vartheta, \\ u &= (p_+ - q_-)^2 = (p_- - q_+)^2 = m_e^2 + m_\mu^2 - 2E^2 - 2pq\cos\vartheta \end{aligned} \tag{1.2.36}$$

satisfy the condition $s + t + u = 2m_e^2 + 2m_\mu^2$.

According to the Feynman rules, the associated invariant matrix element is given by

$$\mathcal{M} = -e^2\bar{u}(q_-, \lambda_-)\gamma_\mu v(q_+, \lambda_+)\,\frac{-g^{\mu\nu}}{(p_+ + p_-)^2}\,\bar{v}(p_+, \sigma_+)\gamma_\nu u(p_-, \sigma_-), \tag{1.2.37}$$

where $\sigma_\pm$ and $\lambda_\pm$ denote the helicities of the in- and outgoing fermions, respectively.

It is interesting to compare this matrix element with that of the *electromagnetic current* of electrons and muons

$$j^{\text{em}}_\nu(x) = -\bar{\psi}_e(x)\gamma_\nu\psi_e(x) - \bar{\psi}_\mu(x)\gamma_\nu\psi_\mu(x) \tag{1.2.38}$$

between the vacuum $|0\rangle$ and the state $|e^+; e^-\rangle = b^\dagger(p_+)a^\dagger(p_-)|0\rangle$ of the incoming electron–positron pair. It can be calculated using (1.2.10):

$$\langle 0|j^{\text{em}}_\nu(x)|e^+; e^-\rangle = -\bar{v}(p_+, \sigma_+)\gamma_\nu u(p_-, \sigma_-)\,e^{-i(p_+ + p_-)x}. \tag{1.2.39}$$

Therefore, $\mathcal{M}$ can also be written in current–current form

$$\mathcal{M} = e^2 \langle \mu^+; \mu^- | j_\mu^{\text{em}}(0) | 0 \rangle \frac{g^{\mu\nu}}{(p_+ + p_-)^2} \langle 0 | j_\nu^{\text{em}}(0) | \text{e}^+; \text{e}^- \rangle . \qquad (1.2.40)$$

The differential cross section in the CM system for unpolarized incoming and outgoing particles is obtained from (1.2.26)

$$\begin{aligned}
\frac{\mathrm{d}\sigma}{\mathrm{d}\Omega_{\text{CM}}} &= \frac{1}{64\pi^2 s} \frac{q}{p} \frac{1}{4} \sum_{\sigma_+,\sigma_-} \sum_{\lambda_+,\lambda_-} |\mathcal{M}|^2 \\
&= \frac{e^4}{64\pi^2 s} \frac{q}{p} \frac{1}{4s^2} \\
&\quad \times \left[\sum_{\sigma_+,\sigma_-} \bar{v}(p_+,\sigma_+)\gamma_\mu u(p_-,\sigma_-)\, \bar{u}(p_-,\sigma_-)\gamma_\nu v(p_+,\sigma_+) \right] \\
&\quad \times \left[\sum_{\lambda_+,\lambda_-} \bar{u}(q_-,\lambda_-)\gamma^\mu v(q_+,\lambda_+)\, \bar{v}(q_+,\lambda_+)\gamma^\nu u(q_-,\lambda_-) \right] \\
&= \frac{\alpha^2}{16 s^3} \frac{q}{p} L_{\mu\nu}^{(\text{e})} L^{(\mu)\nu\mu} . \qquad (1.2.41)
\end{aligned}$$

Using (A.1.40) to work out the spin summations, the *leptonic tensors* $L_{\mu\nu}^{(\text{e})}$ and $L_{\mu\nu}^{(\mu)}$ are calculated with the trace formulas (A.1.30),

$$\begin{aligned}
L_{\mu\nu}^{(\text{e})} &= \sum_{\sigma_+,\sigma_-} \bar{v}(p_+,\sigma_+)\gamma_\mu u(p_-,\sigma_-)\, \bar{u}(p_-,\sigma_-)\gamma_\nu v(p_+,\sigma_+) \\
&= \text{Tr}[(\not{p}_+ - m_\text{e})\gamma_\mu(\not{p}_- + m_\text{e})\gamma_\nu] \\
&= 4\left(p_{+\mu}p_{-\nu} + p_{-\mu}p_{+\nu} - \frac{s}{2} g_{\mu\nu} \right) . \qquad (1.2.42)
\end{aligned}$$

With the kinematical variables in the CM system defined in (1.2.35) and (1.2.37) one gets for the contraction of the leptonic tensors

$$\begin{aligned}
L_{\mu\nu}^{(\text{e})} L^{(\mu)\nu\mu} &= 32 \left[(p_+q_+)^2 + (p_+q_-)^2 - \frac{s}{2}(p_+p_- + q_+q_-) + \frac{s^2}{2} \right] \\
&= 64[E^4 + p^2q^2\cos^2\vartheta + E^2(m_\text{e}^2 + m_\mu^2)] \qquad (1.2.43)
\end{aligned}$$

and for the differential cross section of $e^+e^- \to \mu^+\mu^-$

$$\frac{d\sigma}{d\Omega_{CM}} = \frac{\alpha^2}{4s}\frac{q}{p}\frac{E^4 + p^2q^2\cos^2\vartheta + E^2(m_e^2 + m_\mu^2)}{E^4} \tag{1.2.44}$$
$$= \frac{\alpha^2}{4s}\frac{q}{p}\left[1 + \cos^2\vartheta + 4\frac{m_e^2 + m_\mu^2}{s}\sin^2\vartheta + 16\frac{m_e^2 m_\mu^2}{s^2}\cos^2\vartheta\right].$$

For $E \gg m_\mu$, this simplifies to

$$\frac{d\sigma}{d\Omega_{CM}} = \frac{\alpha^2}{4s}\left(1 + \cos^2\vartheta\right). \tag{1.2.45}$$

Integration over the solid angle then gives the *total cross section*

$$\sigma(s) = \int d\Omega \frac{d\sigma}{d\Omega} = \frac{4\pi\alpha^2}{3s}, \qquad E \gg m_\mu. \tag{1.2.46}$$

Other lowest-order QED cross sections, e.g. $e^+e^- \to f\bar{f}$, can be calculated in a similar way.

1.2.2 Lie algebras and Lie groups

As we have seen in Sect. 1.1.1, the elementary particles can be characterized by charge quantum numbers. The corresponding multiplet structures can formally be described by symmetry groups [GT]. Well-known examples are the isospin doublets with the associated group SU(2) and the quark colour triplets of the SU(3) colour symmetry.

1.2.2.1 Lie algebra

In the language of group theory the *charge operators* correspond to the *generators* Q^a, $a = 1, \ldots, \dim G$, of the *Lie algebra* of a symmetry group G of dimension $\dim G$ (=number of group parameters). For instance, for the group of unitary $N \times N$ matrices with determinant one, SU(N), there are $N^2 - 1$ generators, i.e. $\dim G = N^2 - 1$. The algebraic properties of the generators can be fully specified by their commutation relations

$$[Q^a, Q^b] = i f^{abc} Q^c, \qquad a, b, c \in \{1, \ldots, \dim G\}. \tag{1.2.47}$$

Equation (1.2.47) implies the *Jacobi identity* for the *structure constants* f^{abc}

$$f^{abe}f^{cde} + f^{ace}f^{dbe} + f^{ade}f^{bce} = 0. \tag{1.2.48}$$

If $f^{abc} = 0$ for all abc, the group is abelian. For semi-simple, compact groups, the basis of the generators can be chosen such that the structure constants are totally antisymmetric. This is assumed in the following.

For SU(2), the structure constants are given by the totally antisymmetric *Levi–Civita tensor* ε^{abc} ($a, b, c \in \{1, 2, 3\}$, $\varepsilon^{123} = +1$); for the SU(3) group, up to permutations of the indices, only the following are different from zero

$$\begin{aligned} &f^{123} = +1, \qquad f^{458} = f^{678} = \frac{\sqrt{3}}{2}, \\ &f^{147} = -f^{156} = f^{246} = f^{257} = f^{345} = -f^{367} = \frac{1}{2}. \end{aligned} \tag{1.2.49}$$

In the Lie algebra of the group G, a maximum number r of commuting generators H_i, $i = 1, \ldots, r$, can be found. The number r is called the *rank of the group*. The states of a multiplet are—up to degeneracies—characterized by the eigenvalues of the H_i. The remaining generators can be combined to form ladder operators $E_{\pm\alpha}$ which transform the different multiplet states into one another.

Polynomials C in the generators Q^a that commute with all generators, $[C, Q^a] = 0$ are called *Casimir operators*. For the compact semi-simple groups the simplest Casimir operator is given by $C = \sum_a Q^a Q^a$ (in the basis where the structure constants are totally antisymmetric).

1.2.2.2 Representations of a Lie algebra

Consider an m-dimensional linear vector space V. Non-singular $m \times m$ matrices T^a fulfilling the commutation relations (1.2.47) define an m-dimensional *representation* of the Lie algebra. A representation is called *irreducible* if no subspace $U \subset V$ exists so that the T^a transform all vectors of U into U.

The algebra of the symmetry group SU(N) has a representation by $N \times N$ matrices, the *fundamental representation* of SU(N).

Examples: $N = 2$: SU(2) (isospin group), $Q^a = I^a = \tau^a/2$, $a = 1,2,3$, with the *Pauli matrices*

$$\tau^1 = \begin{pmatrix} 0 & 1 \\ 1 & 0 \end{pmatrix}, \quad \tau^2 = \begin{pmatrix} 0 & -\mathrm{i} \\ \mathrm{i} & 0 \end{pmatrix}, \quad \tau^3 = \begin{pmatrix} 1 & 0 \\ 0 & -1 \end{pmatrix}. \tag{1.2.50}$$

$N = 3$: SU(3), $Q^a = \lambda^a/2$, $a = 1, \ldots, 8$, with the *Gell-Mann matrices*

$$\lambda^1 = \begin{pmatrix} 0 & 1 & 0 \\ 1 & 0 & 0 \\ 0 & 0 & 0 \end{pmatrix}, \quad \lambda^2 = \begin{pmatrix} 0 & -\mathrm{i} & 0 \\ \mathrm{i} & 0 & 0 \\ 0 & 0 & 0 \end{pmatrix}, \quad \lambda^3 = \begin{pmatrix} 1 & 0 & 0 \\ 0 & -1 & 0 \\ 0 & 0 & 0 \end{pmatrix},$$

$$\lambda^4 = \begin{pmatrix} 0 & 0 & 1 \\ 0 & 0 & 0 \\ 1 & 0 & 0 \end{pmatrix}, \quad \lambda^5 = \begin{pmatrix} 0 & 0 & -\mathrm{i} \\ 0 & 0 & 0 \\ \mathrm{i} & 0 & 0 \end{pmatrix}, \quad \lambda^6 = \begin{pmatrix} 0 & 0 & 0 \\ 0 & 0 & 1 \\ 0 & 1 & 0 \end{pmatrix},$$

$$\lambda^7 = \begin{pmatrix} 0 & 0 & 0 \\ 0 & 0 & -\mathrm{i} \\ 0 & \mathrm{i} & 0 \end{pmatrix}, \quad \lambda^8 = \frac{1}{\sqrt{3}} \begin{pmatrix} 1 & 0 & 0 \\ 0 & 1 & 0 \\ 0 & 0 & -2 \end{pmatrix}. \tag{1.2.51}$$

Extensions to higher N are straightforward. The $N^2 - 1$ Gell-Mann matrices λ^a have the property $\mathrm{Tr}(\lambda^a) = 0$.

The *Dynkin index* T_{R} of an irreducible representation R is defined as

$$\mathrm{Tr}\left(T^a T^b\right) = \delta^{ab}\, T_{\mathrm{R}}. \tag{1.2.52}$$

For the fundamental representation of SU(N), the Gell-Mann matrices give $T_{\mathrm{F}} = 1/2$. In an irreducible representation, the Casimir operator C of SU(N) is diagonal,

$$\sum_{a=1}^{N^2-1} \sum_j \left(T^a_{ij} T^a_{jk}\right) = \delta_{ik} C_{\mathrm{R}}, \tag{1.2.53}$$

and we denote its value with C_{R}. In the fundamental representation, its value is given by

$$C_{\mathrm{F}} = \frac{N^2 - 1}{2N}. \tag{1.2.54}$$

The structure constants f^{abc} can be used to define matrices that satisfy the commutation relations (1.2.47) because of the Jacobi identity (1.2.48) according to

$$(T^a_{\mathrm{adj}})_{bc} = -\mathrm{i}f^{abc}. \tag{1.2.55}$$

This representation of the Lie algebra which has the dimension of the group is called the *adjoint representation.* In this representation, the Casimir operator C of SU(N) is given by

$$C = \sum_{a,b} f^{abc}f^{abd} = C_{\mathrm{A}}\delta^{cd} = N\delta^{cd}. \tag{1.2.56}$$

The value of the Dynkin index in the adjoint representation equals the value of the Casimir operator $T_{\mathrm{A}} = C_{\mathrm{A}} = N$.

1.2.2.3 Lie groups

Finite transformations, depending on the parameters $\boldsymbol{\theta} = (\theta^a)$, are obtained by exponentiating infinitesimal transformations, i.e. by exponential functions of the generators

$$g(\boldsymbol{\theta}) = \exp(\mathrm{i}\theta^a Q^a) = \mathbf{1} + \frac{\mathrm{i}}{1!}\theta^a Q^a + \frac{(\mathrm{i})^2}{2!}\theta^a Q^a\,\theta^b Q^b + \dots . \tag{1.2.57}$$

The finite transformations $g(\boldsymbol{\theta})$ form the *Lie group G* corresponding to the Lie algebra of the generators Q^a

$$\begin{aligned} &g(\boldsymbol{\theta})g(\boldsymbol{\theta}') = \exp(\mathrm{i}\theta^a Q^a)\exp(\mathrm{i}\theta'^b Q^b) = \exp(\mathrm{i}\theta''^c Q^c),\\ &\theta''^c = \theta''^c(\boldsymbol{\theta},\boldsymbol{\theta}'),\\ &g^{-1}(\boldsymbol{\theta}) = \exp(-\mathrm{i}\theta^a Q^a), \qquad g(0) = \mathbf{1}. \end{aligned} \tag{1.2.58}$$

Vice versa, the generators Q^a are obtained from the finite transformations $g(\boldsymbol{\theta})$ as derivatives at the origin

$$Q^a = -\,\mathrm{i}\frac{\partial g}{\partial \theta^a}\bigg|_{\boldsymbol{\theta}=0}. \tag{1.2.59}$$

Example: Fundamental representation of the group SU(2)

$$g(\theta^a) = \exp\left(\mathrm{i}\sum_{a=1}^{3}\theta^a\frac{\tau^a}{2}\right) \tag{1.2.60}$$

with

$$(\theta^1,\theta^2,\theta^3) = \theta\mathbf{e} = \theta(e^1,e^2,e^3), \qquad \mathbf{e}^2 = 1, \qquad 0 \le \theta < 2\pi, \tag{1.2.61}$$

or in polar coordinates

$$\begin{aligned} &e^1 = \sin\vartheta\cos\phi, \qquad e^2 = \sin\vartheta\sin\phi, \qquad e^3 = \cos\vartheta, \\ &(0 \le \vartheta < \pi,\ 0 \le \phi \le 2\pi). \end{aligned} \tag{1.2.62}$$

Then,

$$\begin{aligned} g &= \mathbf{1}\cos\frac{\theta}{2} + \mathrm{i}(\mathbf{e}\cdot\boldsymbol{\tau})\sin\frac{\theta}{2} \\ &= \begin{pmatrix} \cos\frac{\theta}{2} + \mathrm{i}\sin\frac{\theta}{2}\cos\vartheta & \mathrm{i}\sin\frac{\theta}{2}\sin\vartheta\exp(-\mathrm{i}\phi) \\ \mathrm{i}\sin\frac{\theta}{2}\sin\vartheta\exp(\mathrm{i}\phi) & \cos\frac{\theta}{2} - \mathrm{i}\sin\frac{\theta}{2}\cos\vartheta \end{pmatrix} \end{aligned} \tag{1.2.63}$$

is a general two-dimensional unitary matrix of determinant +1, i.e. an element of SU(2). It is related to the spin-(1/2) representation of the rotation group SO(3).

Finite group transformations that commute with the whole group form the *centre of the group*. The elements $g(\boldsymbol{\theta})$ of a Lie group depend analytically on the $\dim G$ parameters $\boldsymbol{\theta} = (\theta^a)$. For compact groups it is possible to define an integral over the group with a unique invariant measure, the *Haar measure*,

$$\mathrm{d}\mu(g) = I(\boldsymbol{\theta})\prod_{a=1}^{n}\mathrm{d}\theta^a. \tag{1.2.64}$$

The Haar measure is right- and left-invariant

$$\mathrm{d}\mu(g) = \mathrm{d}\mu(f\cdot g) = \mathrm{d}\mu(g\cdot f) = \mathrm{d}\mu(g^{-1}), \qquad g,f\in G, \tag{1.2.65}$$

and normalized in such a way that the integral over the complete group satisfies

$$\int_G \mathrm{d}\mu(g) = 1. \tag{1.2.66}$$

For a general parametrization $g(\boldsymbol{\theta})$ the Haar measure can be obtained as follows. Calculate the derivatives $\partial g/\partial\theta^a$, form the products $g^{-1}(\boldsymbol{\theta})\partial g/\partial\theta^a$, and express them as linear combinations of the generators T^a of the Lie algebra, thus defining the matrix $M(\boldsymbol{\theta})$

$$g^{-1}(\boldsymbol{\theta})\frac{\partial g}{\partial\theta^a} = \sum_b \mathrm{i}T^b M^{ba}(\boldsymbol{\theta}). \tag{1.2.67}$$

Then, the weight function $I(\boldsymbol{\theta})$ is given by

$$I(\boldsymbol{\theta}) = \frac{1}{V_G}\det M(\boldsymbol{\theta}), \tag{1.2.68}$$

with the volume of the group

$$V_G = \int_G \det M(\boldsymbol{\theta}) \prod_{a=1}^{n} \mathrm{d}\theta^a. \tag{1.2.69}$$

For SU(2) with the parametrization as in (1.2.61) and (1.2.62) and $V_G = 16\pi^2$, this measure has the form

$$\mathrm{d}\mu(g) = \frac{4}{V_G}\sin^2\frac{\theta}{2}\,\mathrm{d}\theta\sin\vartheta\,\mathrm{d}\vartheta\,\mathrm{d}\phi = \frac{1}{4\pi^2}\sin^2\frac{\theta}{2}\,\mathrm{d}\theta\sin\vartheta\,\mathrm{d}\vartheta\,\mathrm{d}\phi, \tag{1.2.70}$$

$$\int_{\mathrm{SU(2)}} \mathrm{d}\mu(g) = \frac{1}{4\pi^2}\int_0^{2\pi}\sin^2\frac{\theta}{2}\,\mathrm{d}\theta\int_0^{\pi}\sin\vartheta\,\mathrm{d}\vartheta\int_0^{2\pi}\mathrm{d}\phi = 1. \tag{1.2.71}$$

If the SU(2) group elements are parametrized in terms of Pauli matrices, like (1.2.63), the Haar measure can be written as

$$\mathrm{d}\mu(g) = \frac{1}{24\pi^2}\varepsilon^{abc}\,\mathrm{Tr}\left\{\left(g^{-1}\frac{\partial g}{\partial\theta^a}\right)\left(g^{-1}\frac{\partial g}{\partial\theta^b}\right)\left(g^{-1}\frac{\partial g}{\partial\theta^c}\right)\right\}\prod_{d=1}^{3}\mathrm{d}\theta^d. \tag{1.2.72}$$

1.2.2.4 The flavour group SU(N_F)

The group-theoretical concepts just described are widely used in elementary particle physics. The arrangement of the particles in flavour multiplets is associated with representations of the flavour group SU(N_F), i.e. the particles in these multiplets are associated with basis vectors in the representation space. Electric charge Q, *strong isospin* I, *strong hypercharge* Y, charm C, etc., are closely connected to the flavour-group generators. For details we refer to standard text books [EP].

1.2.2.5 The colour group SU(3)$_C$

The colour degrees of freedom of strongly interacting particles are described by the group SU(3)$_C$. The quarks are classified in the three-dimensional fundamental representation of SU(3)$_C$, the gluons in the eight-dimensional adjoint representation. The colour charges Q_C^a of hadrons are zero

$$Q_C^a|\text{had}\rangle = 0. \tag{1.2.73}$$

This is the group-theoretical formalization of the rule: *hadrons are colourless.*

1.2.2.6 Weak isospin SU(2)$_w$ and weak hypercharge U(1)$_Y$

In Sect. 1.4 we show how leptons and quarks are classified, and how the structure of the electromagnetic and weak interaction is group-theoretically described by the *weak-isospin* and *weak-hypercharge group* SU(2)$_w$ × U(1)$_Y$.

1.2.3 Conserved currents and charges

In this section we describe the implementation of symmetries in Lagrangian field theories.

1.2.3.1 Noether's theorem

We consider Lagrangians containing complex fields ψ_i, their hermitian adjoints $\psi_i^\dagger$, and the corresponding space–time derivatives,

$$\mathcal{L} = \mathcal{L}(\psi_i, \psi_i^\dagger, \partial_\mu \psi_i, \partial_\mu \psi_i^\dagger). \tag{1.2.74}$$

We assume that the action $S = \int \mathrm{d}^4x\, \mathcal{L}$ is invariant under infinitesimal *global symmetry transformations* of the fields

$$\begin{aligned} \psi_i(x) &\to \psi_i'(x) = \psi_i(x) + \delta\psi_i(x), \\ \psi_i^\dagger(x) &\to \psi_i'^\dagger(x) = \psi_i^\dagger(x) + \delta\psi_i^\dagger(x). \end{aligned} \tag{1.2.75}$$

The invariance of the action implies that the Lagrangian changes under the transformation (1.2.75) at most by a total divergence,

$$\begin{aligned} \delta\mathcal{L} &= \mathcal{L}(\psi_i + \delta\psi_i, \psi_i^\dagger + \delta\psi_i^\dagger, \partial_\mu\psi_i + \partial_\mu\delta\psi_i, \partial_\mu\psi_i^\dagger + \delta\partial_\mu\psi_i^\dagger) \\ &\quad - \mathcal{L}(\psi_i, \psi_i^\dagger, \partial_\mu\psi_i, \partial_\mu\psi_i^\dagger) \\ &= \partial_\mu K^\mu, \end{aligned} \tag{1.2.76}$$

where K^μ in general depends on the fields.

Then, *Noether's theorem* guarantees the existence of a *conserved current*, the Noether current. This current can be constructed by evaluating $\delta\mathcal{L}$

$$\delta\mathcal{L} = \delta\psi_i^\dagger \frac{\partial\mathcal{L}}{\partial\psi_i^\dagger} + \delta\psi_i \frac{\partial\mathcal{L}}{\partial\psi_i} + \delta(\partial_\mu\psi_i^\dagger)\frac{\partial\mathcal{L}}{\partial(\partial_\mu\psi_i^\dagger)} + \delta(\partial_\mu\psi_i)\frac{\partial\mathcal{L}}{\partial(\partial_\mu\psi_i)} \tag{1.2.77}$$

For anticommuting fields, the derivatives have to be understood as left derivatives [cf. (2.2.19)].

Using the equations of motion (1.2.15), gives

$$\begin{aligned} \delta\mathcal{L} &= \delta\psi_i^\dagger \partial_\mu \frac{\partial\mathcal{L}}{\partial(\partial_\mu\psi_i^\dagger)} + \delta\psi_i \left(\partial_\mu \frac{\partial\mathcal{L}}{\partial(\partial_\mu\psi_i)}\right) \\ &\quad + \delta(\partial_\mu\psi_i^\dagger)\frac{\partial\mathcal{L}}{\partial(\partial_\mu\psi_i^\dagger)} + \delta(\partial_\mu\psi_i)\frac{\partial\mathcal{L}}{\partial(\partial_\mu\psi_i)} \\ &= \partial_\mu \left(\delta\psi_i^\dagger \frac{\partial\mathcal{L}}{\partial(\partial_\mu\psi_i^\dagger)} + \delta\psi_i \frac{\partial\mathcal{L}}{\partial(\partial_\mu\psi_i)}\right). \end{aligned} \tag{1.2.78}$$

This together with (1.2.76) allows to define the *Noether current*

$$j^\mu(x) = K^\mu - \left(\delta\psi_i^\dagger \frac{\partial\mathcal{L}}{\partial(\partial_\mu\psi_i^\dagger)} + \delta\psi_i \frac{\partial\mathcal{L}}{\partial(\partial_\mu\psi_i)} \right) \tag{1.2.79}$$

that is conserved,

$$\partial^\mu j_\mu(x) = 0. \tag{1.2.80}$$

Current conservation, i.e. the continuity equation, implies conservation of the corresponding charge

$$Q(t) = \int j_0(t,\mathbf{x})\,\mathrm{d}^3\mathbf{x}, \tag{1.2.81}$$

namely

$$\begin{aligned} \frac{\mathrm{d}Q}{\mathrm{d}t} &= \int \frac{\partial j_0}{\partial t}(t,\mathbf{x})\,\mathrm{d}^3\mathbf{x} \\ &= \int_V \operatorname{div}\mathbf{j}(t,\mathbf{x})\,\mathrm{d}^3\mathbf{x} = \int_{\partial V} \mathbf{j}(t,\mathbf{x})\,\mathrm{d}\mathbf{S} = 0. \end{aligned} \tag{1.2.82}$$

Here we applied Gauss's law to transform the integral over the whole three-dimensional space into one over a surface at infinity, and used the boundary condition that the currents vanish sufficiently fast at infinity.

The Lagrangian of QED (1.2.16) is invariant under one-parameter phase transformations $\mathrm{U}(1)_{\mathrm{em}}$. The conserved Noether current (1.2.79) exists and is identical to the electromagnetic current (1.2.18). Thus, formally the conservation of the electric charge Q is a consequence of $\mathrm{U}(1)_{\mathrm{em}}$ global symmetry. In the interaction part of $\mathcal{L}_{\mathrm{QED}}$ the four-potential $A_\mu(x)$ is coupled directly to this conserved current.

1.2.3.2 Examples

We illustrate the application of Noether's theorem for the case of a Lagrangian for fermions which is invariant under the SU(N) transformations

$$\begin{aligned}\psi_i(x) \to \left(\mathrm{e}^{\mathrm{i}\theta^a T^a}\right)_{ij} \psi_j(x), \qquad & \psi_i^\dagger(x) \to \psi_j^\dagger(x) \left(\mathrm{e}^{-\mathrm{i}\theta^a T^a}\right)_{ji}, \\ T^a = T^{a\dagger}, & \end{aligned} \tag{1.2.83}$$

of infinitesimally

$$\delta\psi_i(x) = \mathrm{i}\delta\theta^a T^a_{ij}\psi_j(x), \qquad \delta\psi_i^\dagger(x) = \psi_j^\dagger(x)(-\mathrm{i}\delta\theta^a T^a_{ji}). \tag{1.2.84}$$

Since the $\delta\theta^a$ are arbitrary, we obtain from (1.2.79) the conserved currents

$$\begin{aligned} j^{\mu,a}(x) &= \mathrm{i}\left(\psi_i^\dagger T^a_{ij} \frac{\partial\mathcal{L}}{\partial(\partial_\mu\psi_j^\dagger)} - T^a_{ji}\psi_i \frac{\partial\mathcal{L}}{\partial(\partial_\mu\psi_j)}\right) \\ &= \mathrm{i}\left(\psi^\dagger T^a \frac{\partial\mathcal{L}}{\partial(\partial_\mu\psi^\dagger)} + \frac{\partial\mathcal{L}}{\partial(\partial_\mu\psi)} T^a\psi\right), \end{aligned} \tag{1.2.85}$$

where we used the anticommutativity of the fermionic operators. In a free fermion field theory, the currents (1.2.85) have the form

$$j^{\mu,a}(x) = \bar\psi_i(x) T^a_{ij}\gamma^\mu\psi_j(x). \tag{1.2.86}$$

1.2.3.3 Operators for the conserved charges

The operators for the *conserved charges* can be represented by the field operators

$$Q^a = \int \mathrm{d}^3\mathbf{x}\, j^{a,0}(t,\mathbf{x}) = \int \mathrm{d}^3\mathbf{x}\, \bar\psi_i(t,\mathbf{x})\gamma^0 T^a_{ij}\psi_j(t,\mathbf{x}). \tag{1.2.87}$$

As a consequence of the field *commutation relations* (1.2.12), which give at equal times ($x_0 = x_0'$)

$$[\psi_{\alpha,i}(x), \bar\psi_{\alpha',j}(x')]_+ = \gamma^0_{\alpha\alpha'}\delta^{(3)}(\mathbf{x}-\mathbf{x}')\delta_{ij}, \tag{1.2.88}$$

they satisfy the commutation relations

$$[Q^a, Q^b] = \mathrm{i} f^{abc} Q^c, \tag{1.2.89}$$

where f^{abc} are the structure constants of SU(N) according to (1.2.47). The charges Q^a generate the symmetry transformations (1.2.84)

$$-\mathrm{i}\delta\theta^a [Q^a, \psi_i(x)] = \mathrm{i} T^a_{ij} \psi_j(x) \delta\theta^a = \delta\psi_i(x). \tag{1.2.90}$$

In this way, the symmetries considered above are implemented as infinitesimal unitary transformations on the quantized fields. Besides the direct physical interpretation of currents and charges, this shows their importance in connection with symmetries in quantum field theories.

Apart from exact symmetries also broken symmetries are relevant. A symmetry can either be broken explicitly, i.e. by adding explicit symmetry-violating terms to the Lagrangian, or it can be broken spontaneously (cf. Sect. 4.1), i.e. by the very existence of degenerate vacuum states that are not invariant. Moreover, it can happen that a classical symmetry is broken in the quantized theory. This is called an *anomaly* (cf. Sect. 2.7).

If a symmetry is only valid approximately, such as the flavour symmetry, the associated currents are not exactly conserved, and the corresponding charges are time-dependent. Despite this, (1.2.89) are satisfied as equal-time commutation relation. The investigation of such structures in the framework of *current algebra* gave interesting results [Ad68]. This was developed further leading to the non-linear σ model and chiral perturbation theory. Its predictions and its relation to spontaneous breaking of flavour symmetry in QCD are discussed in Sect. 3.5.

We end this section with a general remark on the close connection between symmetry laws and field theory. In field theory, dynamics is described by means of local interaction of fields. Global changes of the fields—for example of their phase by global gauge transformations (1.2.83)—are not in the spirit of local quantum theory. This observation suggests that global symmetry transformations as described in (1.2.83) should be generalized to transformations in which the fields are transformed locally. For phase transformations (1.2.83), this would mean that the parameters θ^a become space–time dependent: $\theta^a = \theta^a(x)$. The elaboration of this idea of local symmetry leads to the concept of gauge theories [We29].

1.3 The quark model of hadrons

The *quark model* was proposed by G. Zweig [Zw64] and M. Gell-Mann [Ge64] in 1964. In spite of an intensive search, quarks have not been found as free particles. But point-like scattering centres—partons with quark properties—have been seen in deep-inelastic lepton–nucleon scattering. This apparent contradiction is explained within QCD by the confinement of quarks in hadrons. In this section, a simple explanation of the properties of hadrons derived from their quark composition is given.

1.3.1 Quantum numbers and wave functions of hadrons in the quark model

1.3.1.1 Quantum numbers

It is well-known that the many meson and baryon resonances observed can be classified by means of *quantum numbers* [EP]. The electric charge Q, the baryon number, and the geometric quantum numbers angular momentum j,j_3, parity P, and charge-conjugation parity C are conserved by the strong interaction. This is also the case for the flavour quantum numbers strong isospin I, I_3, strangeness S, charm C, bottomness B, and topness T.

The remarkable experimental fact that only certain values of these quantum numbers occur in nature can be explained within the quark model by the following hypothesis: mesons consist of one quark and one antiquark, baryons consist of three quarks.

The quantum numbers and wave functions of hadrons are formed from the quark degrees of freedom according to the rules for two- and three-particle bound states of quantum mechanics.

The quark states $|u\rangle, |d\rangle, |s\rangle, \ldots$ can be understood as basis vectors of the fundamental representation of an approximate flavour symmetry group $SU(N_F)$. The meson and the baryon flavour states are constructed by considering product representations and their reduction to irreducible components. Since $SU(N_F)$ is only an approximate symmetry, mixing between state vectors with equal quantum numbers can occur.

1.3.1.2 Wave functions of mesons as quark–antiquark systems

The composition of *meson wave functions* from flavour, spin, and orbital parts is sketched in the following [Li78, Cl79].

Flavour part $|I, I_3; S, C\rangle$: in the case of four quark flavours it is constructed from the quark state vectors $|\mathrm{u}\rangle, |\mathrm{d}\rangle, |\mathrm{s}\rangle, |\mathrm{c}\rangle$ and the antiquark states $|\bar{\mathrm{u}}\rangle, |\bar{\mathrm{d}}\rangle, |\bar{\mathrm{s}}\rangle, |\bar{\mathrm{c}}\rangle$ in the following way:

$$
\begin{aligned}
&|1,1;0,0\rangle = -|\mathrm{u}\bar{\mathrm{d}}\rangle, && |1,0;0,0\rangle = (|\mathrm{u}\bar{\mathrm{u}}\rangle - |\mathrm{d}\bar{\mathrm{d}}\rangle)/\sqrt{2},\\
&|1,-1;0,0\rangle = |\mathrm{d}\bar{\mathrm{u}}\rangle, && |0,0;0,0\rangle = (|\mathrm{u}\bar{\mathrm{u}}\rangle + |\mathrm{d}\bar{\mathrm{d}}\rangle)/\sqrt{2},\\
&|\tfrac{1}{2},\tfrac{1}{2};1,0\rangle = |\mathrm{u}\bar{\mathrm{s}}\rangle, && |\tfrac{1}{2},-\tfrac{1}{2};1,0\rangle = |\mathrm{d}\bar{\mathrm{s}}\rangle,\\
&|\tfrac{1}{2},-\tfrac{1}{2};-1,0\rangle = |\mathrm{s}\bar{\mathrm{u}}\rangle, && |\tfrac{1}{2},\tfrac{1}{2};-1,0\rangle = -|\mathrm{s}\bar{\mathrm{d}}\rangle,\\
&|0,0,0,0\rangle = |\mathrm{s}\bar{\mathrm{s}}\rangle, && \\
&|\tfrac{1}{2},\tfrac{1}{2};0,1\rangle = -|\mathrm{c}\bar{\mathrm{d}}\rangle, && |\tfrac{1}{2},-\tfrac{1}{2};0,1\rangle = |\mathrm{c}\bar{\mathrm{u}}\rangle,\\
&|\tfrac{1}{2},-\tfrac{1}{2};0,-1\rangle = |\mathrm{d}\bar{\mathrm{c}}\rangle, && |\tfrac{1}{2},\tfrac{1}{2};0,-1\rangle = |\mathrm{u}\bar{\mathrm{c}}\rangle,\\
&|0,0;1,-1\rangle = |\mathrm{c}\bar{\mathrm{s}}\rangle, && |0,0;-1,-1\rangle = |\mathrm{s}\bar{\mathrm{c}}\rangle,\\
&|0,0,0,0\rangle = |\mathrm{c}\bar{\mathrm{c}}\rangle. &&
\end{aligned}
\tag{1.3.1}
$$

The state $|\bar{\mathrm{d}}\rangle$ is always accompanied by a minus sign, because the doublet $(-|\bar{\mathrm{d}}\rangle, |\bar{\mathrm{u}}\rangle)$ transforms in the same way as the doublet $(|\mathrm{u}\rangle, |\mathrm{d}\rangle)$ under SU(2). The structure of multiplets of two (u, d), three (u, d, s) and four quark flavours (u, d, s, c) can be read off directly; an extension to five and six quarks is straightforward.

There are several states with quantum numbers $|0,0,0,0\rangle$ which can mix. For example, pseudoscalar mesons do not follow the scheme above, but show roughly SU(3) mixing

$$
|\eta\rangle \simeq -\frac{1}{\sqrt{6}}(|\mathrm{u}\bar{\mathrm{u}}\rangle + |\mathrm{d}\bar{\mathrm{d}}\rangle - 2|\mathrm{s}\bar{\mathrm{s}}\rangle), \qquad |\eta'\rangle \simeq -\frac{1}{\sqrt{3}}(|\mathrm{u}\bar{\mathrm{u}}\rangle + |\mathrm{d}\bar{\mathrm{d}}\rangle + |\mathrm{s}\bar{\mathrm{s}}\rangle).
$$

Spin part $|s, s_3\rangle$: quark and antiquark spins have two possible orientations characterized by the state vectors $|\uparrow\rangle$ and $|\downarrow\rangle$. They can be combined to form a spin triplet and a spin singlet

$$
\begin{aligned}
&|1,1\rangle = |\uparrow\uparrow\rangle, \qquad |1,0\rangle = (|\uparrow\downarrow\rangle + |\downarrow\uparrow\rangle)/\sqrt{2}, \qquad |1,-1\rangle = |\downarrow\downarrow\rangle,\\
&|0,0\rangle = (|\uparrow\downarrow\rangle - |\downarrow\uparrow\rangle)/\sqrt{2}.
\end{aligned}
\tag{1.3.2}
$$

$^{2S+1}L_J$	j^{PC}	$u\bar{d}, u\bar{u}, d\bar{d}$ $I=1$	$u\bar{u}, d\bar{d}, s\bar{s}$ $I=0$	$c\bar{c}$ $I=0$	$b\bar{b}$ $I=0$	$u\bar{s}, d\bar{s}$ $I=1/2$
$^1S_0$	0^{-+}	$\pi(140)$	$\eta(547), \eta'(958)$	η_c	????	K(494)
$^3S_1$	1^{--}	$\rho(770)$	$\omega(782), \phi(1020)$	$J/\psi(1S)$	$\Upsilon(1S)$	$K^*(892)$
$^1P_1$	1^{+-}	$b_1(1235)$	$h_1(1170), h_1(1380)$	$h_c(1P)$	????	$K_1(1270)$
$^3P_0$	0^{++}	$a_0(1450)$	$f_0(1370), f_0(1710)$	$\chi_{c0}(1P)$	$\chi_{b0}(1P)$	$K_0^*(1430)$
$^3P_1$	1^{++}	$a_1(1260)$	$f_1(1285), f_1(1420)$	$\chi_{c1}(1P)$	$\chi_{b1}(1P)$	$K_1(1400)$
$^3P_2$	2^{++}	$a_2(1320)$	$f_2(1270), f_2'(1525)$	$\chi_{c2}(1P)$	$\chi_{b2}(1P)$	$K_2^*(1430)$

Tab. 1.4 The *meson spectrum in the quark model.* The spectrum observed experimentally is taken from Ref. [PDG00]. On the left, angular momentum j, parity P, and charge-conjugation parity C quantum numbers are shown. The numbers in parentheses indicate the masses of the states in MeV. Quark-model states not experimentally established so far are marked with ????.

Orbital part $|l, m, n\rangle$: after separating off the CM motion, the relative motion between quark and antiquark is described in the non-relativistic approximation by a Schrödinger wave function $\phi(\mathbf{r}) = Y_{l,m}(\hat{\mathbf{r}})\, f_{n,l}(r)$ of the relative coordinate $\mathbf{r}$, which is made up of a spherical function $Y_{l,m}(\hat{\mathbf{r}})$ and a radial part $f_{n,l}(r)$.

Spin s and orbital angular momentum l are added to the total spin j in the usual way. This gives $j = l+1,\ l,\ l-1$ for the triplet and $j = l$ for the singlet states. The parity P, and charge-conjugation parity C of this fermion–antifermion system have the values $P = (-1)^{l+1}$ and $C = (-1)^{l+s}$. Table 1.4 shows the structure of the meson spectrum as it results from this composition of the quark degrees of freedom and a comparison with experiment.

1.3.1.3 Baryons as three-quark systems

When three identical particles are involved, the *permutation symmetry* of the wave function gives an important quantum number. Out of the 6 basic product states $|abc\rangle$, $|bca\rangle$, $|cab\rangle$, $|bac\rangle$, $|acb\rangle$ and $|cba\rangle$ one can construct the well-known symmetric $|\mathrm{Sy}\rangle$ and antisymmetric $|\mathrm{An}\rangle$ states

$$
\begin{aligned}
|\mathrm{Sy}\rangle &= \frac{1}{\sqrt{6}}(|abc\rangle + |bca\rangle + |cab\rangle + |bac\rangle + |acb\rangle + |cba\rangle), \\
|\mathrm{An}\rangle &= \frac{1}{\sqrt{6}}(|abc\rangle + |bca\rangle + |cab\rangle - |bac\rangle - |acb\rangle - |cba\rangle)
\end{aligned}
\tag{1.3.3}
$$

and the states

$$
\begin{aligned}
|\overline{\mathrm{Mi}}\rangle_\pm &= \frac{1}{\sqrt{6}}\Big[(|abc\rangle \pm |bac\rangle) + \varepsilon(|bca\rangle \pm |acb\rangle) + \varepsilon^*(|cab\rangle \pm |cba\rangle)\Big], \\
|\underline{\mathrm{Mi}}\rangle_\pm &= \frac{1}{\sqrt{6}}\Big[(|abc\rangle \pm |bac\rangle) + \varepsilon^*(|bca\rangle \pm |acb\rangle) + \varepsilon(|cab\rangle \pm |cba\rangle)\Big],
\end{aligned}
\tag{1.3.4}
$$

corresponding to the two two-dimensional mixed symmetric representations $\mathrm{Mi}_\pm$ of the permutation group of three elements. Here $\varepsilon = \exp(2\pi\mathrm{i}/3) = (-1 + \mathrm{i}\sqrt{3})/2$, $\varepsilon^* = \exp(-2\pi\mathrm{i}/3) = \varepsilon^2$. The states $|\mathrm{An}\rangle$, $|\overline{\mathrm{Mi}}_-\rangle$, and $|\underline{\mathrm{Mi}}_-\rangle$ vanish for $a = b$.

Flavour part: the states of the flavour multiplets can be obtained by combining the quark product states into those with definite symmetry according to (1.3.3) and (1.3.4). For instance for baryons made of two up and one down quarks this gives the mixed symmetric states

$$
\begin{aligned}
|\overline{\mathrm{P}}\rangle &= \frac{1}{\sqrt{3}}(|\mathrm{uud}\rangle + \varepsilon|\mathrm{udu}\rangle + \varepsilon^*|\mathrm{duu}\rangle), \\
|\underline{\mathrm{P}}\rangle &= \frac{1}{\sqrt{3}}(|\mathrm{uud}\rangle + \varepsilon^*|\mathrm{udu}\rangle + \varepsilon|\mathrm{duu}\rangle),
\end{aligned}
\tag{1.3.5}
$$

and for baryons made of three up quarks the symmetric state

$$
|\Delta^{++}\rangle = |\mathrm{uuu}\rangle. \tag{1.3.6}
$$

Spin part: the spins of the three quarks can be combined to form the total spin $(3/2)_{\mathrm{Sy}}$ and $(1/2)_{\mathrm{Mi}}$ by the same method

$$
\begin{aligned}
&|3/2,\ \ 3/2\rangle = |\uparrow\uparrow\uparrow\rangle, \qquad |3/2,\ \ 1/2\rangle = (|\uparrow\uparrow\downarrow\rangle + |\uparrow\downarrow\uparrow\rangle + |\downarrow\uparrow\uparrow\rangle)/\sqrt{3}, \\
&|3/2, -3/2\rangle = |\downarrow\downarrow\downarrow\rangle, \qquad |3/2, -1/2\rangle = (|\downarrow\downarrow\uparrow\rangle + |\downarrow\uparrow\downarrow\rangle + |\uparrow\downarrow\downarrow\rangle)/\sqrt{3};
\end{aligned}
$$

$$
\begin{aligned}
\overline{|1/2,\ \ 1/2\rangle}_+ &= (|\uparrow\uparrow\downarrow\rangle + \varepsilon|\uparrow\downarrow\uparrow\rangle + \varepsilon^*|\downarrow\uparrow\uparrow\rangle)/\sqrt{3}, \\
\overline{|1/2, -1/2\rangle}_+ &= (|\downarrow\downarrow\uparrow\rangle + \varepsilon|\downarrow\uparrow\downarrow\rangle + \varepsilon^*|\uparrow\downarrow\downarrow\rangle)/\sqrt{3};
\end{aligned}
$$

$$
\underline{|1/2, \pm 1/2\rangle}_+ = \overline{|1/2, \pm 1/2\rangle}_+^*. \tag{1.3.7}
$$

Orbital part: the relative motion of the three quarks is a function of two relative coordinates $\mathbf{z}_1 = (\mathbf{x}_2 - \mathbf{x}_1)/\sqrt{2}$, $\mathbf{z}_2 = (2\mathbf{x}_3 - \mathbf{x}_1 - \mathbf{x}_2)/\sqrt{6}$. Thus, there are many different combinations of internal angular momenta leading to the total orbital momentum l. Schrödinger wave functions of all symmetry types (Sy, An, Mi_+, Mi_-) can be systematically constructed using the mixed-symmetric, complex relative coordinates $\mathbf{z} = (\mathbf{z}_1 + \mathrm{i}\mathbf{z}_2)/\sqrt{2}$, $\mathbf{z}^* = (\mathbf{z}_1 - \mathrm{i}\mathbf{z}_2)/\sqrt{2}$. The details of these wave functions are determined by the quark–quark interaction potentials (cf. [Gr76, Bo80]).

Composition of the baryon wave functions: when flavour, spin, and orbital parts are combined to give a total wave function, the following symmetry types result

$$\begin{array}{lll} \mathrm{Sy} \otimes \mathrm{Sy} = \mathrm{Sy}, & \mathrm{An} \otimes \mathrm{An} = \mathrm{Sy}, & \mathrm{Sy} \otimes \mathrm{An} = \mathrm{An}, \\ \mathrm{Sy} \otimes \mathrm{Mi} = \mathrm{Mi}, & \mathrm{An} \otimes \mathrm{Mi} = \mathrm{Mi}, & \mathrm{Mi} \otimes \mathrm{Mi} = \mathrm{Sy} \oplus \mathrm{Mi} \oplus \mathrm{An}. \end{array} \tag{1.3.8}$$

In this way, the flavour and spin parts combine to give the multiplets with definite symmetry, as shown in Table 1.5. As an example we present the flavour-spin part of the wave function of a proton with spin up:

$$\begin{aligned} |\mathrm{P}\uparrow\rangle = \frac{1}{\sqrt{18}} & (2|\mathrm{u}\uparrow\mathrm{u}\uparrow\mathrm{d}\downarrow\rangle - |\mathrm{u}\uparrow\mathrm{u}\downarrow\mathrm{d}\uparrow\rangle - |\mathrm{u}\downarrow\mathrm{u}\uparrow\mathrm{d}\uparrow\rangle \\ & + 2|\mathrm{u}\uparrow\mathrm{d}\downarrow\mathrm{u}\uparrow\rangle - |\mathrm{u}\downarrow\mathrm{d}\uparrow\mathrm{u}\uparrow\rangle - |\mathrm{u}\uparrow\mathrm{d}\uparrow\mathrm{u}\downarrow\rangle \\ & + 2|\mathrm{d}\downarrow\mathrm{u}\uparrow\mathrm{u}\uparrow\rangle - |\mathrm{d}\uparrow\mathrm{u}\uparrow\mathrm{u}\downarrow\rangle - |\mathrm{d}\uparrow\mathrm{u}\downarrow\mathrm{u}\uparrow\rangle). \end{aligned} \tag{1.3.9}$$

If the interaction shows only a slight flavour and spin dependence, it is advantageous to combine multiplets of equal symmetry to form super-multiplets of the higher approximate symmetry group $\mathrm{SU}(2N_\mathrm{F})$ [Cl79]. The final stage consists in the combination with the orbital wave function to give states with definite total angular momentum. If the total wave function is to be symmetric under permutations, then the classification shown in Table 1.6 follows. This is a good description of the experimental spectrum for baryon resonances with strangeness 0 and -1. The many blanks which occur in the case of resonances with strangeness -2 or -3 are due to experimental difficulties in the production of these resonances. Baryons with open charm or open bottom, i.e. with non-vanishing charm or bottom number, have been observed at the expected masses.

The fact that a symmetric total wave function must be chosen to explain the experimental baryon spectrum was a problem for the naïve, phenomenological quark model: quarks with spin 1/2 should obey Fermi statistics; therefore

	$N_F = 2$	$N_F = 3$	$N_F = 4$
Sy	$20 = (4, \frac{3}{2}) \oplus (2, \frac{1}{2})$	$56 = (10, \frac{3}{2}) \oplus (8, \frac{1}{2})$	$120 = (20, \frac{3}{2}) \oplus (20, \frac{1}{2})$
Mi	$20 = (4, \frac{1}{2}) \oplus (2, \frac{3}{2})$ $\oplus (2, \frac{1}{2})$	$70 = (10, \frac{1}{2}) \oplus (8, \frac{3}{2})$ $\oplus (8, \frac{1}{2}) \oplus (1, \frac{1}{2})$	$168 = (20, \frac{3}{2}) \oplus (20, \frac{1}{2})$ $\oplus (20, \frac{1}{2}) \oplus (4, \frac{1}{2})$
An	$4 = (2, \frac{1}{2})$	$20 = (8, \frac{1}{2}) \oplus (1, \frac{3}{2})$	$56 = (20, \frac{1}{2}) \oplus (4, \frac{3}{2})$

Tab. 1.5 Symmetry types and dimensions of (flavour, spin) multiplets of baryons made from two ($N_F = 2$), three ($N_F = 3$), and four ($N_F = 4$) quarks. We show the total number of the states and their (SU(N_F), spin) content.

SU(6), l^P	j^P	octet				decuplet				singlet
		N	Λ	Σ	Ξ	Δ	Σ	Ξ	Ω	Λ
$56, 0^+$	$1/2^+$	939	1116	1193	1318					
	$3/2^+$					1232	1385	1530	1672	
$70, 1^-$	$1/2^-$	1535	1670	1620	????	1620	????	????	????	1405
		1650	1800	1750	????					
	$3/2^-$	1520	1690	1670	1820	1700	????	????	????	1520
		1700	????	????	????					
	$5/2^-$	1675	1830	1775	????					

Tab. 1.6 The *baryon spectrum in the quark model* with three flavour degrees of freedom. On the left the predicted low-lying particle multiplets are shown, arranged in order of total angular momentum values j^P. The experimentally observed spectrum is taken from Ref. [PDG00]. Quark-model states not experimentally established so far are marked with ????.

the total wave function of a three-quark system must be antisymmetric. Accordingly, the orbital part of the wave function of the Δ^{++} resonance with $I = 3/2$, $I_3 = +3/2$ and $j = 3/2$ should be antisymmetric, since the flavour and spin components are necessarily symmetric. On the other hand, the Δ^{++} is the ground state of all particles made up of three u quarks. In general, the orbital wave function of the ground state has no nodes and is symmetric. This is in contradiction to the required antisymmetry. This problem of quark statistics was solved by introducing an additional degree of freedom for the quarks, *colour* [Ge72]. This was an important step on the way to QCD.

1.3.2 Quark model with colour

The *colour* degree of freedom of the quarks can assume three values (e.g. red, green, and blue). It spans a three-dimensional complex space with basis vectors $|\mathrm{r}\rangle$, $|\mathrm{g}\rangle$, and $|\mathrm{b}\rangle$. Hadrons are constructed according to the rule "hadrons are colourless". This means that the mesons and baryons are colour singlets and have the following colour wave functions

$$
\begin{aligned}
|\text{meson}\rangle &= \frac{1}{\sqrt{3}}(|q'_{\mathrm{r}}\bar{q}_{\mathrm{r}}\rangle + |q'_{\mathrm{g}}\bar{q}_{\mathrm{g}}\rangle + |q'_{\mathrm{b}}\bar{q}_{\mathrm{b}}\rangle) = \frac{1}{\sqrt{3}}\sum_{c,c'} |q'_c\bar{q}_{c'}\rangle\delta_{cc'}, \\
|\text{baryon}\rangle &= \frac{1}{\sqrt{6}}(|q_{\mathrm{r}}q'_{\mathrm{g}}q''_{\mathrm{b}}\rangle + |q_{\mathrm{g}}q'_{\mathrm{b}}q''_{\mathrm{r}}\rangle + |q_{\mathrm{b}}q'_{\mathrm{r}}q''_{\mathrm{g}}\rangle \\
&\quad -|q_{\mathrm{g}}q'_{\mathrm{r}}q''_{\mathrm{b}}\rangle - |q_{\mathrm{r}}q'_{\mathrm{b}}q''_{\mathrm{g}}\rangle - |q_{\mathrm{b}}q'_{\mathrm{g}}q''_{\mathrm{r}}\rangle) \\
&= \frac{1}{\sqrt{6}}\sum_{c,c',c''} |q_c q'_{c'} q''_{c''}\rangle\varepsilon_{cc'c''}.
\end{aligned} \tag{1.3.10}
$$

Here, $\varepsilon_{cc'c''}$ is the Levi–Civita tensor, and $r = 1, g = 2, b = 3$.

For illustration we give two simple examples of wave functions including the colour degree of freedom. We choose the ρ^+ meson and the Δ^{++} resonance since these have vanishing orbital angular momenta. The wave function can be approximated by a gaussian

$$
\begin{aligned}
\rho^+ &= \frac{1}{\sqrt{3}}(-|\mathrm{u_r\bar{d}_r}\rangle - |\mathrm{u_g\bar{d}_g}\rangle - |\mathrm{u_b\bar{d}_b}\rangle)\;|\uparrow\uparrow\rangle \\
&\quad \times \left(\frac{\alpha_{\mathrm{M}}}{\pi}\right)^{3/4} \exp\left(-\alpha_{\mathrm{M}}\frac{\mathbf{r}^2}{2}\right), \\
\Delta^{++} &= \frac{1}{\sqrt{6}}(|\mathrm{u_r u_g u_b}\rangle + |\mathrm{u_g u_b u_r}\rangle + |\mathrm{u_b u_r u_g}\rangle \\
&\quad - |\mathrm{u_g u_r u_b}\rangle - |\mathrm{u_r u_b u_g}\rangle - |\mathrm{u_b u_g u_r}\rangle) \\
&\quad \times|\uparrow\uparrow\uparrow\rangle\;\left(\frac{\alpha_{\mathrm{B}}}{\pi}\right)^{3/2} \exp\left(-\alpha_{\mathrm{B}}\frac{\mathbf{z}_1^2 + \mathbf{z}_2^2}{2}\right)
\end{aligned} \tag{1.3.11}
$$

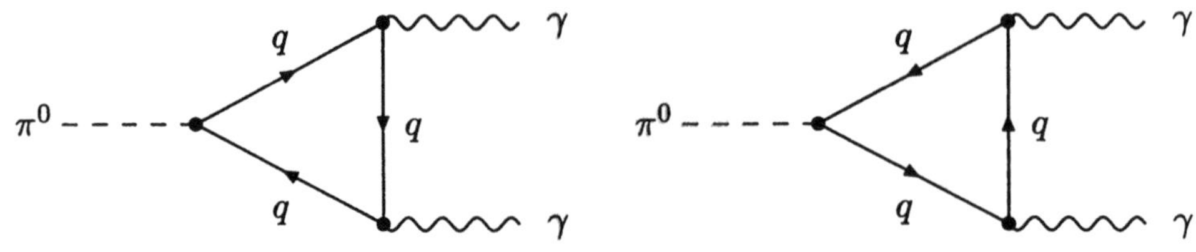

Fig. 1.2 Feynman diagrams for the π^0 decay

with range parameters α_{M} and α_{B}.

The introduction of colour is of decisive importance for the formulation of QCD. Therefore, further phenomenological confirmation of the colour degree of freedom is essential. Besides the statistics problem, the classical examples are the size of the total cross section for e^+e^- annihilation into hadrons (cf. Sect. 1.5) and the decay rate for $\pi^0 \to \gamma\gamma$. Both require three colour degrees of freedom for a quantitative agreement between the measured numbers and the quark-model results.

The *decay* $\pi^0 \to 2\gamma$ can be described by the Feynman graphs in Fig. 1.2. The decay width can be calculated herefrom using the hypothesis of the partially conserved axial-vector current (PCAC) [Ad70] giving

$$\Gamma(\pi^0 \to 2\gamma) = \frac{\alpha^2}{32\pi^3} \frac{M_\pi^3}{f_\pi^2} \frac{N_{\mathrm{C}}^2}{9}. \tag{1.3.12}$$

Experimentally, $\tau(\pi^0 \to 2\gamma) = (8.4 \pm 0.6) \times 10^{-17}\mathrm{s}$. With the π decay constant $f_\pi = 130\,\mathrm{MeV}$, $N_{\mathrm{C}} = 3$ colours are required in order to have agreement between theory and experiment. For a more detailed discussion of the pion decay we refer to Sect. 2.7.1.4.

1.3.3 The concept of quark dynamics—quarkonia

1.3.3.1 Quarkonia

In Sect. 1.3.1 the quantum numbers of mesons and baryons and their wave functions were constructed in the naïve quark model. Dynamical questions were merely touched upon when assessing the significance of colour on the strong interaction. In this section, some concepts of quark dynamics are discussed within the framework of a *non-relativistic potential model* for the

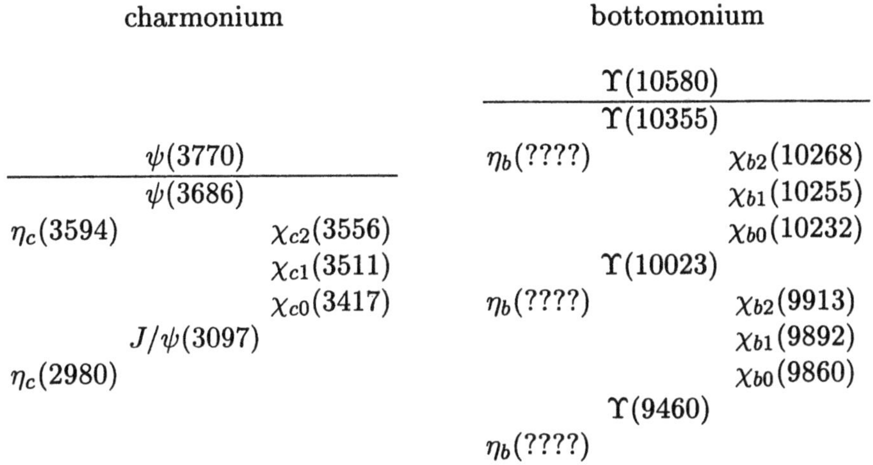

Fig. 1.3 The low-lying states of the ψ and Υ family with their masses (states not experimentally established are marked with ????) (from Ref. [PDG00])

meson and baryon spectrum. This does, in fact, give a realistic picture for bound heavy quark–antiquark systems. Experimental examples for such systems, called *quarkonia*, are charmonium ($J/\psi, \psi', \ldots$) and bottomonium ($\Upsilon, \Upsilon', \ldots$). The top is not included because its lifetime is too short to allow for the formation of bound states. Figure 1.3 shows the experimental spectrum of the ψ and Υ family. The states are characterized by their masses in MeV. The horizontal line indicates the $c\bar{c}$ and $b\bar{b}$ thresholds.

1.3.3.2 Potential model

Ideas for a potential model for quarkonia were taken from the analogy with positronium, i.e. the bound states of the e^+e^- system. In this case the binding potential is generated by photon exchange. For quarkonia it is due to the exchange of massless gluons which couple to the colour degree of freedom of the quarks. The consistent field-theoretical treatment of these ideas in QCD is discussed in Chapt. 3. For systems of heavy quarks it can be expected that the quarks move non-relativistically in the bound state. Accordingly, the quark dynamics can be described by a Hamilton operator

$$H = \sum_i \left(m_i + \frac{\mathbf{p}_i^2}{2m_i} \right) + \sum_{i>j} (V_{\mathrm{C}}(r_{ij}) + V_{\mathrm{G}}(r_{ij})). \qquad (1.3.13)$$

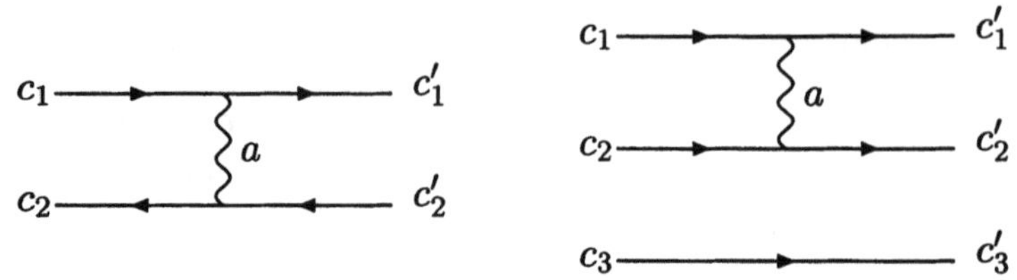

Fig. 1.4 Diagrams that determine the one-gluon exchange potential

The first term describes the rest energy of the quarks with effective constituent masses m_i. It gives an additive contribution to the hadron mass and generates the main effects of the mass splittings in the flavour multiplets. The one-gluon exchange potential V_C and the *confinement potential* V_G of the quarks are approximately flavour-independent.

An attractive *Coulomb potential* is produced in QCD by gluon exchange as shown in Fig. 1.4 [Ap75, dR75]. The colour part (1.3.10) of the wave function gives rise to colour factors. From the Feynman graphs in Fig. 1.4, we find for mesons

$$\begin{aligned}&\sum_a \frac{1}{\sqrt{3}}\delta_{c_1c_2}\, g_s \left(\frac{\lambda^a}{2}\right)_{c_1c'_1} g_s \left(\frac{\lambda^a}{2}\right)_{c_2c'_2} \frac{1}{\sqrt{3}}\delta_{c'_1c'_2}\\ &= \sum_a \frac{g_s^2}{12}\,\mathrm{Tr}\{\lambda^a\lambda^a\} = \frac{4}{3}g_s^2\end{aligned} \tag{1.3.14}$$

and for baryons

$$\begin{aligned}&\frac{1}{\sqrt{6}}\varepsilon_{c_1c_2c_3}\, g_s \left(\frac{\lambda^a}{2}\right)_{c_1c'_1} g_s \left(\frac{\lambda^a}{2}\right)_{c_2c'_2} \delta_{c_3c'_3} \frac{1}{\sqrt{6}}\varepsilon_{c'_1c'_2c'_3}\\ &= \frac{g_s^2}{24}\varepsilon_{c_1c_2c_3}\varepsilon_{c'_1c'_2c_3}\lambda^a_{c_1c'_1}\lambda^a_{c_2c'_2}\\ &= \frac{g_s^2}{24}(\delta_{c_1c'_1}\delta_{c_2c'_2} - \delta_{c_1c'_2}\delta_{c_2c'_1})\lambda^a_{c_1c'_1}\lambda^a_{c_2c'_2}\\ &= \frac{g_s^2}{24}(\mathrm{Tr}\{\lambda^a\}\,\mathrm{Tr}\{\lambda^a\} - \mathrm{Tr}\{\lambda^a\lambda^a\}) = -\frac{2}{3}g_s^2,\end{aligned} \tag{1.3.15}$$

i.e. factors 4/3 for mesons and −2/3 for baryons, respectively. In analogy to QED, the one-gluon exchange potential reads

$$V_G(r) = -\frac{\alpha_s}{r}\begin{Bmatrix} 4/3 \\ 2/3 \end{Bmatrix} \quad \begin{matrix}\text{for quark–antiquark in the meson,}\\ \text{for quark–quark in the baryon,}\end{matrix} \tag{1.3.16}$$

where the minus sign for mesons results from the presence of an antiparticle.

In a potential model, *quark confinement* is enforced by introducing an infinitely rising potential. A simple linear ansatz

$$V_{\mathrm{C}}(r) = \kappa r \tag{1.3.17}$$

gives already a good description.

The derivation of the form of V_{C} is an important problem of QCD. The value for the strong fine-structure constant α_{s} and the string constant κ have to be determined by comparing the calculated spectrum with experiment. Figure 1.5 gives a comparison between the results thus obtained for the charmonium system and experiment. The model parameters were determined from the masses of $J/\psi(3097)$ and $\psi'(3686)$ and the leptonic width of J/ψ, $\Gamma(J/\psi \to \mathrm{e}^+\mathrm{e}^-) = 6.02 \pm 0.19\,\mathrm{keV}$. Typical values for these parameters are

$$\alpha_{\mathrm{s}} \approx 0.25\text{–}0.35, \qquad \kappa \approx 0.8\text{–}1.0\,\frac{\mathrm{GeV}}{\mathrm{fm}}, \qquad m_{\mathrm{c}} \approx 1.2\text{–}1.6\,\mathrm{GeV}. \tag{1.3.18}$$

The agreement between the general structure of the experimental spectrum and the calculated masses is reasonable. However, the spectrum can be well reproduced up to masses of 4 GeV with other potentials, too. Therefore, these results alone cannot be regarded as an experimental confirmation of the potentials (1.3.16) and (1.3.17).

As non-relativistic approximations are the basis of the quarkonium model, a short remark on the quark velocity in the bound state is appropriate. The expectation value of the velocity in the charmonium ground state is calculated as $\langle v \rangle \simeq 0.4\,c$. This roughly justifies the non-relativistic calculations. However, relativistic effects, like spin–spin and spin–orbit interactions, are of importance also for heavy-quark bound states (cf. Sect. 3.4).

1.4 Basics of the electroweak interaction

Electromagnetic and weak processes between leptons and quarks are successfully described phenomenologically by photon, $\mathrm{W}^{\pm}$-boson, and Z-boson exchange [EW]. At low energies ($E \ll M_{\mathrm{W}}$) or small momentum transfers ($|q^2| \ll M_{\mathrm{W}}^2$) the W- and Z-exchange diagrams can be approximated by a *four-fermion contact interaction*, i.e. by the *Fermi model.* As the example of $\mathrm{e}^+\mathrm{e}^-$ annihilation into μ pairs in Sect. 1.2.1.9 shows, S-matrix elements of

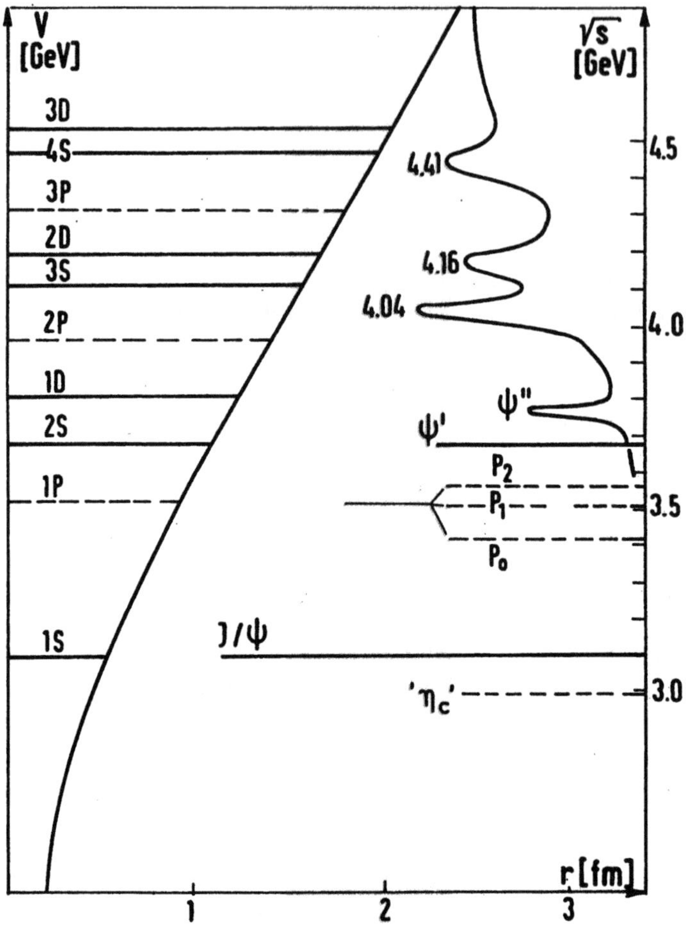

Fig. 1.5 The standard charmonium potential, the spectrum, and the charmonium states found experimentally ($\alpha_s = 0.41$, $\kappa = 0.8665\,\text{GeV/fm}$ and $m_c = 1.6\,\text{GeV}$) (from Ref. [Kr79])

the electromagnetic interaction can be built up from the matrix elements of the electromagnetic current [cf. (1.2.40)]. In a similar way, the phenomenological description of the weak interaction of elementary particles is based on vector and axial-vector current matrix elements. This section describes the structure of the electromagnetic and weak currents and compares the corresponding theoretical predictions with experiment.

The full theory of the electroweak interaction involves more than the weak currents. It has to provide a consistent treatment of the massive W and Z bosons. This can only be formulated if electromagnetic and weak interactions are combined in the form of a non-abelian gauge theory with spontaneous symmetry breaking. This is discussed in detail in Chapt. 4.

1.4.1 Electroweak interaction of leptons

1.4.1.1 Fermion–gauge-boson couplings

Experiment tells us that the couplings between the elementary fermions and the electroweak gauge bosons, i.e. the γ, the $W^{\pm}$ boson, and the Z boson are of vector and axial-vector type. No experimental indications for scalar, pseudoscalar, tensor, as well as derivative couplings have been found. The γ couplings are purely vector, the W-boson couplings are vector minus axial vector, i.e. left-handed, and the Z-boson couplings are a mixture of vector and axial vector. We write the interaction Lagrangian in the general form

$$\mathcal{L}_{\mathrm{I}} = e \sum_{V} \sum_{i,j} \bar{\psi}_i(x)(\gamma_\mu v_{ij}^V - \gamma_\mu \gamma_5 a_{ij}^V)\psi_j(x) A_V^\mu(x), \tag{1.4.1}$$

where $V = \{\gamma, W^+, W^-, Z\}$. The sum over i, j extends over all elementary fermions. The vector and axial-vector couplings are denoted by v_{ij}^V and a_{ij}^V, respectively. The couplings are specified later. In particular, they include no mixing between leptons and quarks.

An important property of the electromagnetic and weak interaction is their *universality*. On a naïve level it means that all fermion families interact in a similar way with the gauge bosons. More precisely, the electromagnetic and weak interactions of all fermions are described by equally structured contributions to the interaction Lagrangian with universal coupling constants.

We begin our discussion with the leptons. Because of universality their couplings are of the same type and strengths. They can be expressed by two parameters: the elementary charge e and the weak mixing angle θ_{w}. We explain in a first step the coupling structure for the lightest lepton doublet, $(\nu_{\mathrm{e}}, \mathrm{e})$, made of the electron neutrino ν_{e} and the electron e. Restricting the coupling matrices to this case of two fermions, we have

$$v^\gamma = -\begin{pmatrix} 0 & 0 \\ 0 & -1 \end{pmatrix} = -Q, \qquad a^\gamma = 0,$$

$$
\begin{aligned}
v^{\mathrm{W}^+} &= a^{\mathrm{W}^+} = \frac{1}{2\sqrt{2}s_{\mathrm{W}}}\begin{pmatrix} 0 & 1 \\ 0 & 0 \end{pmatrix} = \frac{1}{2\sqrt{2}s_{\mathrm{W}}}\tau^+ , \\
v^{\mathrm{W}^-} &= a^{\mathrm{W}^-} = \frac{1}{2\sqrt{2}s_{\mathrm{W}}}\begin{pmatrix} 0 & 0 \\ 1 & 0 \end{pmatrix} = \frac{1}{2\sqrt{2}s_{\mathrm{W}}}\tau^- , \\
v^{\mathrm{Z}} &= \frac{1}{2s_{\mathrm{W}}c_{\mathrm{W}}}\begin{pmatrix} \frac{1}{2} & 0 \\ 0 & -\frac{1}{2}+2s_{\mathrm{W}}^2 \end{pmatrix} = \frac{1}{2s_{\mathrm{W}}c_{\mathrm{W}}}\left(\frac{\tau^3}{2} - 2s_{\mathrm{W}}^2 Q\right) , \\
a^{\mathrm{Z}} &= \frac{1}{2s_{\mathrm{W}}c_{\mathrm{W}}}\begin{pmatrix} \frac{1}{2} & 0 \\ 0 & -\frac{1}{2} \end{pmatrix} = \frac{1}{2s_{\mathrm{W}}c_{\mathrm{W}}}\frac{\tau^3}{2} ,
\end{aligned} \tag{1.4.2}
$$

where $s_{\mathrm{W}} = \sin\theta_{\mathrm{W}}$, $c_{\mathrm{W}} = \cos\theta_{\mathrm{W}}$, and τ^+, τ^-, and τ^3 are combinations of the Pauli matrices (A.1.11).

Introducing the doublet of fields $\Psi = (\psi_{\nu_{\mathrm{e}}}, \psi_{\mathrm{e}})$ and defining the *electromagnetic current* l_μ^γ, the *charged weak current* $l_\mu^\pm$, and the *neutral weak current* l_μ^{Z},

$$
\begin{aligned}
l_\mu^\gamma &= \overline{\Psi}\gamma_\mu Q \Psi , \\
l_\mu^\pm &= \overline{\Psi}\gamma_{\mu\mathrm{L}}\tau^\pm \Psi , \\
l_\mu^{\mathrm{Z}} &= \overline{\Psi}\left[\gamma_{\mu\mathrm{L}}\frac{\tau^3}{2} - s_{\mathrm{W}}^2\gamma_\mu Q\right]\Psi = l_\mu^3 - s_{\mathrm{W}}^2 l_\mu^\gamma ,
\end{aligned} \tag{1.4.3}
$$

with

$$
\gamma_{\mu\mathrm{L}} = \gamma_\mu \frac{1-\gamma_5}{2} , \tag{1.4.4}
$$

the interaction Lagrangian can be written as[2]

$$
\mathcal{L}_{\mathrm{I}} = -e l_\mu^\gamma A_\gamma^\mu + \frac{e}{\sqrt{2}s_{\mathrm{W}}}(l_\mu^+ A_{\mathrm{W}^+}^\mu + l_\mu^- A_{\mathrm{W}^-}^\mu) + \frac{e}{s_{\mathrm{W}}c_{\mathrm{W}}} l_\mu^{\mathrm{Z}} A_{\mathrm{Z}}^\mu . \tag{1.4.5}
$$

The coupling matrices and consequently the currents of the leptons of the second and third lepton family are constructed in the same way. Altogether we have the three lepton doublets

$$
\begin{pmatrix} \nu_{\mathrm{e}} \\ \mathrm{e} \end{pmatrix} , \begin{pmatrix} \nu_\mu \\ \mu \end{pmatrix} , \begin{pmatrix} \nu_\tau \\ \tau \end{pmatrix} \begin{matrix} \leftarrow Q = 0 \\ \leftarrow Q = -1 \end{matrix} , \tag{1.4.6}
$$

and the total lepton current is the sum of the corresponding three contributions.

[2] The factor $\sqrt{2}$ in the denominator is a consequence of the different normalization of neutral and charged fields.

1.4.1.2 Charges and $SU(2)_W \times U(1)_Y$ symmetry

The charges of the left-handed currents l_μ^+, l_μ^-, and l_μ^3

$$I_W^\pm(t) = \int d^3x\, l_0^\pm(t,\mathbf{x}), \qquad I_W^3(t) = \int d^3x\, l_0^3(t,\mathbf{x}) \tag{1.4.7}$$

have the commutators

$$[I_W^+, I_W^-] = 2I_W^3, \qquad [I_W^3, I_W^\pm] = \pm I_W^\pm \tag{1.4.8}$$

owing to the equal-time commutation relations of the Dirac fields (1.2.88). This is the Lie algebra of the *weak isospin group* $SU(2)_W$.

The *weak hypercharge*, which is defined via the *Gell-Mann–Nishijima relation*

$$\frac{1}{2}Y_W = Q - I_W^3 = \int d^3x\,(l_0^\gamma(t,\mathbf{x}) - l_0^3(t,\mathbf{x})) \tag{1.4.9}$$

commutes with all components of the weak isospin. Together, they generate the group $SU(2)_W \times U(1)_Y$. This group plays an important rôle for the construction of a gauge theory of the unified electroweak interaction, thus leading to a theoretical explanation of universality. The $SU(2)_W \times U(1)_Y$ symmetry of the currents and charges seems to be broken by the prefactors $e/\sqrt{2}s_W$ and $e/s_W c_W$ in the interaction Lagrangian (1.4.5). Introducing orthogonal linear combinations of the neutral gauge fields A_γ^μ and A_Z^μ,

$$\begin{aligned} A_3^\mu &= -s_W A_\gamma^\mu + c_W A_Z^\mu, \\ B^\mu &= c_W A_\gamma^\mu + s_W A_Z^\mu, \end{aligned} \tag{1.4.10}$$

gives

$$\mathcal{L}_I = \frac{e}{s_W}\left(\frac{1}{\sqrt{2}}l_\mu^+ A_{W+}^\mu + \frac{1}{\sqrt{2}}l_\mu^- A_{W-}^\mu + l_\mu^3 A_3^\mu\right) - \frac{e}{c_W}l_\mu^B B^\mu, \tag{1.4.11}$$

where $l_\mu^B = l_\mu^\gamma - l_\mu^3$. Now the $SU(2)_W \times U(1)_Y$ symmetry of the interaction is manifest. Introducing *left- and right-handed Dirac fields*

$$\Psi_L = \frac{1-\gamma_5}{2}\Psi, \qquad \Psi_R = \frac{1+\gamma_5}{2}\Psi, \tag{1.4.12}$$

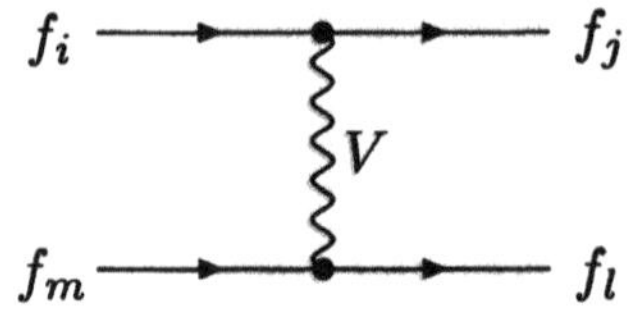

Fig. 1.6 Boson-exchange diagram for four-fermion processes

it is possible to arrange the left-handed ν_{e} and e into a doublet Ψ_{eL} and the right-handed e into a singlet ψ_{eR} (there are no right-handed couplings for the neutrino) and similarly for the other lepton families. The left-handed doublet is coupled to the weak isospin triplet (W^+, W^-, W^3) with strength e/s_{w} and to the singlet B with e/c_{w} times the weak hypercharge $Y^{\mathrm{L}} = -1$. The right-handed electron is only coupled to B with e/c_{w} times its hypercharge $Y^{\mathrm{R}} = -2$.

1.4.1.3 Four-fermion processes

We consider the invariant matrix element $\mathcal{M}$ of an elementary four-fermion process (Fig. 1.6). The Feynman rules together with the *propagator for a massive vector boson*,

V :  $$= \mathrm{i}\Delta_{\mathrm{F}}^{\mu\nu}(p, M_V) = \mathrm{i}\frac{-g_{\mu\nu} + k_\mu k_\nu/M_V^2}{k^2 - M_V^2 + \mathrm{i}\varepsilon}, \tag{1.4.13}$$

give[3]

$$\begin{aligned}\mathcal{M} = -e^2 \sum_{V=\gamma,\mathrm{W},\mathrm{Z}} & \bar{u}(p_j)\left(\gamma^\mu v_{ji}^V - \gamma^\mu\gamma_5 a_{ji}^V\right) u(p_i) \frac{-g_{\mu\nu} + k_\mu k_\nu/M_V^2}{k^2 - M_V^2} \\ & \times \bar{u}(p_l)\left(\gamma^\nu v_{lm}^V - \gamma^\nu\gamma_5 a_{lm}^V\right) u(p_m) \\ = {} & \mathcal{M}^\gamma + \mathcal{M}^{\mathrm{W}} + \mathcal{M}^{\mathrm{Z}},\end{aligned} \tag{1.4.14}$$

where $k = p_i - p_j$. This matrix element can be rewritten using the fermionic current matrix elements (1.2.40)

$$\mathcal{M}^\gamma = -e^2 \langle j|l_\mu^\gamma(0)|i\rangle\langle l|l_\nu^\gamma(0)|m\rangle \frac{-g^{\mu\nu}}{k^2},$$

[3]In the photon propagator the $k_\mu k_\nu$ term is absent.

$$\begin{aligned}
\mathcal{M}^{\mathrm{W}} &= -\frac{e^2}{2s_{\mathrm{w}}^2}\langle j|l_\mu^+(0)|i\rangle\langle l|l_\nu^-(0)|m\rangle\frac{-g^{\mu\nu}+k^\mu k^\nu/M_{\mathrm{W}}^2}{k^2-M_{\mathrm{W}}^2},\\
\mathcal{M}^{\mathrm{Z}} &= -\frac{e^2}{s_{\mathrm{w}}^2c_{\mathrm{w}}^2}\langle j|l_\mu^{\mathrm{Z}}(0)|i\rangle\langle l|l_\nu^{\mathrm{Z}}(0)|m\rangle\frac{-g^{\mu\nu}+k^\mu k^\nu/M_{\mathrm{Z}}^2}{k^2-M_{\mathrm{Z}}^2}.
\end{aligned} \tag{1.4.15}$$

The matrix element for photon exchange is identical to (1.2.40). For $|k^2| \ll M_{\mathrm{W}}^2, M_{\mathrm{Z}}^2$ and for light fermions, where the $k_\mu k_\nu$ term may be neglected since its contribution is proportional to the product of the fermion masses, the expressions for $\mathcal{M}^{\mathrm{W}}$ and $\mathcal{M}^{\mathrm{Z}}$ simplify to

$$\begin{aligned}
\mathcal{M}^{\mathrm{W}} &= -\frac{e^2}{2s_{\mathrm{w}}^2M_{\mathrm{W}}^2}\langle j|l_\mu^+|i\rangle\langle l|l^{-,\mu}|m\rangle,\\
\mathcal{M}^{\mathrm{Z}} &= -\frac{e^2}{s_{\mathrm{w}}^2c_{\mathrm{w}}^2M_{\mathrm{Z}}^2}\langle j|l_\mu^{\mathrm{Z}}|i\rangle\langle l|l^{\mathrm{Z},\mu}|m\rangle.
\end{aligned} \tag{1.4.16}$$

These matrix elements are of current–current form and can be directly compared with those in the *Fermi model* of the charged and neutral weak interaction.

In the Fermi model the interaction Lagrangian reads

$$\mathcal{L}_{\mathrm{I}}^{\mathrm{Fermi}} = \frac{4G_\mu}{\sqrt{2}}\left[l_\mu^+ l^{-\mu} + l_\mu^{\mathrm{Z}} l^{\mathrm{Z},\mu}\right]. \tag{1.4.17}$$

The corresponding matrix elements for four-fermion processes are identical to (1.4.16) if we identify

$$\frac{4G_\mu}{\sqrt{2}} = \frac{e^2}{2s_{\mathrm{w}}^2M_{\mathrm{W}}^2}, \qquad M_{\mathrm{W}}^2 = c_{\mathrm{w}}^2M_{\mathrm{Z}}^2. \tag{1.4.18}$$

These are the lowest-order relations between the *gauge-boson masses* M_{W} and M_{Z}, the *muon decay constant (Fermi constant)* G_μ, and the *weak mixing angle* θ_{w}.

1.4.1.4 Comparison with experiment

As an application, and in order to determine the parameters in the ansatz for the couplings (1.4.1) and (1.4.2), we apply (1.4.15) to a charged-current process, the *μ decay*. In this case we have indeed $|k^2| = |(p_\mu - p_e)^2| \ll M_W^2$ and $m_e, m_\mu \ll M_W$. Therefore, the Fermi matrix element (1.4.16) for $\mathcal{M}^W$ is appropriate. The decay probability per unit time, Γ_μ, is obtained from $\mathcal{M}^W$ using (1.2.25). The phase-space integration gives

$$\Gamma_\mu = \tau_\mu^{-1} = \frac{G_\mu^2 m_\mu^5}{192\pi^3}\left(1 - 8\frac{m_e^2}{m_\mu^2}\right) \tag{1.4.19}$$

for the width Γ_μ or the lifetime τ_μ of the muon. From the experimental value of the muon lifetime $\tau_{\mu,\text{exp}} = 2.19703(4) \times 10^{-6}\,\text{s}$ the following value for the *muon decay constant* is obtained[4] [PDG00]

$$G_\mu = 1.16639(1) \times 10^{-5}\,\text{GeV}^{-2} = (292.8\,\text{GeV})^{-2}. \tag{1.4.20}$$

The measured decay spectrum as well as polarization correlations are in excellent agreement with the results obtained from the left-handed matrix element (1.4.16). This is also true for τ decays, thus supporting the universality of the weak interaction.

The interaction Lagrangian $\mathcal{L}_I$ (1.4.1) describes not only the weak decays of leptons but also their scattering processes. Since only electrons are available as target, the reactions shown in Table 1.7 are possible. The experimental data can be used to determine the *weak mixing angle* from purely leptonic reactions. The result is $s_w^2 = 0.221 \pm 0.008$ [PDG00]. Together with the fine-structure constant $\alpha = e^2/4\pi = 1/137.036$, this allows for the determination of the W-boson and Z-boson masses in lowest order from (1.4.18) and (1.4.20). The result is $M_W \approx 79\,\text{GeV}$, and $M_Z \approx 89\,\text{GeV}$. These values are roughly 1–2 GeV lower than the experimental ones. This indicates that higher-order effects are not small and needed for an adequate comparison between theory and experiment. In fact, after taking into account the radiative corrections within the Electroweak Standard Model, which is described in Chapt. 4, the predicted values agree with the measured values within the experimental errors, which are at the per-mill level.

[4]including QED corrections as given in (4.6.7)

reaction	Contributions from	total cross section
$\nu_\mu e^- \to \nu_\mu e^-$	$l_\mu^{(Z)}$	$\left(\frac{1}{4} - s_W^2 + \frac{4}{3}s_W^4\right)\sigma_0$
$\overline{\nu}_\mu e^- \to \overline{\nu}_\mu e^-$	$l_\mu^{(Z)}$	$\frac{1}{3}\left(\frac{1}{4} - s_W^2 + 4s_W^4\right)\sigma_0$
$\nu_\mu e^- \to \nu_e \mu^-$	$l_\mu^{(\pm)}$	σ_0
$\overline{\nu}_e e^- \to \overline{\nu}_\mu \mu^-$	$l_\mu^{(\pm)}$	$\frac{1}{3}\sigma_0$
$\nu_e e^- \to \nu_e e^-$	$l_\mu^{(Z)}, l_\mu^{(\pm)}$	$\left(\frac{1}{4} + s_W^2 + \frac{4}{3}s_W^4\right)\sigma_0$
$\overline{\nu}_e e^- \to \overline{\nu}_e e^-$	$l_\mu^{(Z)}, l_\mu^{(\pm)}$	$\frac{1}{3}\left(\frac{1}{4} + s_W^2 + 4s_W^4\right)\sigma_0$

Tab. 1.7 Cross sections for neutrino–electron scattering ($\sigma_0 = G_\mu^2 E_{CM}^2/\pi = G_\mu^2 2m_e E_\nu/\pi = 1.72 \times 10^{-41}(E_\nu/\mathrm{GeV})\,\mathrm{cm}^2$)

1.4.2 Electroweak interaction of hadrons

1.4.2.1 Hadronic currents

Since the quarks are the fundamental constituents of matter rather than the hadrons, the hadronic currents are expressed in terms of the quark fields. In analogy to the leptonic current (1.4.3), the electromagnetic hadronic current for three generations is given by (suppressing the colour degree of freedom)

$$h_\mu^\gamma(x) = \bar{\psi}(x)\gamma_\mu Q\psi(x) \quad \text{with} \quad Q = \mathrm{diag}\left(\tfrac{2}{3}, -\tfrac{1}{3}, \tfrac{2}{3}, -\tfrac{1}{3}, \tfrac{2}{3}, -\tfrac{1}{3}\right). \tag{1.4.21}$$

Experiments on the weak interaction show that the $SU(2)_W \times U(1)_Y$ structure formulated for leptons applies to hadronic weak processes as well. Thus, one extends the definition (1.4.7) of the generators of weak isospin with commutation relations (1.4.8) by adding the hadronic current component

$$\begin{aligned} I_W^\pm(t) &= \int d^3x\,(l_0^\pm(t,\mathbf{x}) + h_0^\pm(t,\mathbf{x})), \\ I_W^3(t) &= \int d^3x\,(l_0^3(t,\mathbf{x}) + h_0^3(t,\mathbf{x})). \end{aligned} \tag{1.4.22}$$

As with leptons, the left-handed quarks are arranged into doublets

$$\begin{pmatrix} u \\ d \end{pmatrix}_L, \begin{pmatrix} c \\ s \end{pmatrix}_L, \begin{pmatrix} t \\ b \end{pmatrix}_L \begin{matrix} \leftarrow Q = 2/3 \\ \leftarrow Q = -1/3 \end{matrix}, \tag{1.4.23}$$

and the right-handed ones into singlets u_R, d_R, c_R, s_R, t_R, and b_R. The composition of the doublets is based on the quark mass: (u, d) are the light quarks, (c, s) the moderately heavy ones, and (t, b) the heavy ones.

The quarks u, ..., b are defined as flavour eigenstates, i.e. eigenstates of the conserved quantum numbers of the strong interaction and therefore mass eigenstates. Since the electromagnetic and weak interactions do not conserve these quantum numbers, the eigenstates of the weak interactions are different. As a consequence the *charged current* for the weak interaction of hadrons involves a 3×3 *quark-mixing matrix* $\mathbf{V}$ and has the form [Ko73, Ja79]

$$h_\mu^+(x) = \left(\bar{u}(x), \bar{c}(x), \bar{t}(x)\right) \gamma_{\mu\mathrm{L}} \begin{pmatrix} V_{\mathrm{ud}} & V_{\mathrm{us}} & V_{\mathrm{ub}} \\ V_{\mathrm{cd}} & V_{\mathrm{cs}} & V_{\mathrm{cb}} \\ V_{\mathrm{td}} & V_{\mathrm{ts}} & V_{\mathrm{tb}} \end{pmatrix} \begin{pmatrix} d(x) \\ s(x) \\ b(x) \end{pmatrix} . \tag{1.4.24}$$

The matrix $\mathbf{V}$ describes the transformation from the mass eigenstates to the weak-interaction eigenstates of the quarks. In the corresponding construction for leptons, $\mathbf{V}$ was chosen diagonal because of the separate conservation of lepton numbers and equal to the unit matrix because of universality [cf. (1.4.3)]. The non-electromagnetic component of the *neutral current* $h_\mu^3(x)$ is constructed as commutator of I_W^+ and $h_\mu^-(x)$ as the zeroth component of weak isospin (1.4.22)

$$\begin{aligned} 2h_\mu^3(x) &= [I_\mathrm{W}^+, h_\mu^-(x)] \qquad (1.4.25) \\ &= \left(\bar{u}(x), \bar{c}(x), \bar{t}(x) \mid \bar{d}(x), \bar{s}(x), \bar{b}(x)\right) \gamma_{\mu\mathrm{L}} \left(\begin{array}{c|c} \mathbf{V}\mathbf{V}^\dagger & 0 \\ \hline 0 & -\mathbf{V}^\dagger\mathbf{V} \end{array}\right) \left(\begin{array}{c} u(x) \\ c(x) \\ t(x) \\ \hline d(x) \\ s(x) \\ b(x) \end{array}\right) . \end{aligned}$$

The commutator is calculated using the equal-time commutation relations (1.2.88). Inserting h_μ^3 from (1.4.25) into the relation

$$[I_\mathrm{W}^\pm, h_\mu^3(x)] = \mp h_\mu^\pm(x) \tag{1.4.26}$$

implies $\mathbf{V}\mathbf{V}^\dagger\mathbf{V} = \mathbf{V}$ and thus $\mathbf{V}\mathbf{V}^\dagger = \mathbf{V}^\dagger\mathbf{V} = \mathbf{1}$, i.e. the quark-mixing matrix is unitary and the neutral current $h_\mu^3(x)$ is flavour-diagonal. This is called *GIM mechanism* [Gl70]. Indeed the neutral hadronic current is parametrized by the same *weak mixing angle* as in the leptonic case. The explicit expression is

$$\begin{aligned} h_\mu^Z(x) &= h_\mu^3(x) - s_W^2 h_\mu^\gamma(x) \\ &= \bar{u}(x)\left[\left(\frac{1}{2}-\frac{2}{3}s_W^2\right)\gamma_{\mu L} - \frac{2}{3}s_W^2\gamma_{\mu R}\right]u(x) \\ &+ \bar{c}(x)\left[\left(\frac{1}{2}-\frac{2}{3}s_W^2\right)\gamma_{\mu L} - \frac{2}{3}s_W^2\gamma_{\mu R}\right]c(x) \\ &+ \bar{t}(x)\left[\left(\frac{1}{2}-\frac{2}{3}s_W^2\right)\gamma_{\mu L} - \frac{2}{3}s_W^2\gamma_{\mu R}\right]t(x) \\ &+ \bar{d}(x)\left[\left(-\frac{1}{2}+\frac{1}{3}s_W^2\right)\gamma_{\mu L} + \frac{1}{3}s_W^2\gamma_{\mu R}\right]d(x) \\ &+ \bar{s}(x)\left[\left(-\frac{1}{2}+\frac{1}{3}s_W^2\right)\gamma_{\mu L} + \frac{1}{3}s_W^2\gamma_{\mu R}\right]s(x) \\ &+ \bar{b}(x)\left[\left(-\frac{1}{2}+\frac{1}{3}s_W^2\right)\gamma_{\mu L} + \frac{1}{3}s_W^2\gamma_{\mu R}\right]b(x), \end{aligned} \tag{1.4.27}$$

where $\gamma_{\mu R} = \gamma_\mu(1+\gamma_5)/2$.

Like h_μ^3, h_μ^Z is flavour diagonal, i.e. neutral weak processes obey the flavour selection rules

$$\Delta Q = \Delta S = \Delta C = \Delta B = \Delta T = 0. \tag{1.4.28}$$

A typical example for the validity of these selection rules are the leptonic flavour-changing K_L^0 decays. Experimentally, $\Gamma(K_L^0 \to \mu^+\mu^-)/\Gamma_{K_L^0} = (7.15 \pm 0.16) \times 10^{-9}$ and $\Gamma(K_L^0 \to e^+e^-)/\Gamma_{K_L^0} = 9 + 6 - 4 \times 10^{-12}$ [PDG00]. These branching ratios have the order of magnitude of higher-order effects of the electroweak interaction at the relevant energy scale.

1.4.2.2 Quark mixing matrix

The unitarity of the quark-mixing matrix $\mathbf{V}$ expresses the universality of the weak interaction in the quark sector. Being a general unitary $N_\mathcal{F} \times N_\mathcal{F}$ matrix for $N_\mathcal{F}$ families, $\mathbf{V}$ depends on $N_\mathcal{F}^2$ real parameters, $N_\mathcal{F}(N_\mathcal{F}-1)/2$ rotation angles and $N_\mathcal{F}(N_\mathcal{F}+1)/2$ phases. Some of these phases can be absorbed by redefining the quark fields $q_i(x) \to \exp(\mathrm{i}\phi_i)q_i(x)$, $i = 1, \ldots, 2N_\mathcal{F}$. Since this redefinition affects the quark-mixing matrix by differences of the $2N_{\mathcal{F}}$ phases, only $2N_{\mathcal{F}} - 1$ phases can be eliminated in $\mathbf{V}$. Thus, the $N_{\mathcal{F}} \times N_{\mathcal{F}}$ quark mixing matrix $\mathbf{V}$ can be parametrized by $N_{\mathcal{F}}(N_{\mathcal{F}} - 1)/2$ rotation angles and $(N_{\mathcal{F}} - 1)(N_{\mathcal{F}} - 2)/2$ phases. The presence of these phases in $\mathbf{V}$ is related to CP violation.

For $N_{\mathcal{F}} = 2$, i.e. only u, d, c and s quarks, $\mathbf{V}$ is a 2×2 matrix. According to the considerations above, it can be parametrized by a single rotation angle, the *Cabibbo angle* [Ca63],

$$\mathbf{V} = \begin{pmatrix} \cos\theta_{\mathrm{C}} & \sin\theta_{\mathrm{C}} \\ -\sin\theta_{\mathrm{C}} & \cos\theta_{\mathrm{C}} \end{pmatrix}. \tag{1.4.29}$$

There is no remaining phase. The hadronic charged current (1.4.24) is explicitly given by

$$\begin{aligned} h_\mu^+(x) &= \bar{u}(x)\gamma_{\mu\mathrm{L}}[\cos\theta_{\mathrm{C}}\, d(x) + \sin\theta_{\mathrm{C}}\, s(x)] \\ &\quad + \bar{c}(x)\gamma_{\mu\mathrm{L}}[-\sin\theta_{\mathrm{C}}\, d(x) + \cos\theta_{\mathrm{C}}\, s(x)]. \end{aligned} \tag{1.4.30}$$

The Cabibbo angle θ_{C} measures the strength of the $\Delta I_3 = 1/2$ transitions and has the value $\sin\theta_{\mathrm{C}} \simeq 0.22$, $\cos\theta_{\mathrm{C}} \simeq 0.98$.

For three flavours, $N_{\mathcal{F}} = 3$, the quark-mixing matrix (1.4.24) can be parametrized by three rotation angles and one phase. There is no unique parametrization. In accordance with [PDG00] we use ($c_{ij} \equiv \cos\theta_{ij}$, $s_{ij} \equiv \sin\theta_{ij}$)

$$\mathbf{V} = \begin{pmatrix} c_{12}c_{13} & s_{12}c_{13} & s_{13}\mathrm{e}^{-\mathrm{i}\delta_{13}} \\ -s_{12}c_{23} - c_{12}s_{23}s_{13}\mathrm{e}^{\mathrm{i}\delta_{13}} & c_{12}c_{23} - s_{12}s_{23}s_{13}\mathrm{e}^{\mathrm{i}\delta_{13}} & s_{23}c_{13} \\ s_{12}s_{23} - c_{12}c_{23}s_{13}\mathrm{e}^{\mathrm{i}\delta_{13}} & -c_{12}s_{23} - s_{12}c_{23}s_{13}\mathrm{e}^{\mathrm{i}\delta_{13}} & c_{23}c_{13} \end{pmatrix}. \tag{1.4.31}$$

The 90% confidence level limits on the angles are [PDG00]

$$s_{12} = 0.219\text{–}0.226, \qquad s_{23} = 0.037\text{–}0.043, \qquad s_{13} = 0.002\text{–}0.005. \tag{1.4.32}$$

In the limit $s_{13} = s_{23} = 0$, the third generation decouples. The remaining mixing of the first two generations is described by the Cabibbo angle.

1.4.2.3 CP violation

CP violation through the quark-mixing matrix can only occur if at least one of its non-trivial phases is non-zero. Consequently, at least three generations are needed. Parametrization-independent measures of CP violation in terms of the elements of $\mathbf{V}$ are the quantities [Ja85]

$$\chi_{il} \equiv \mathrm{Im}\{V^{\dagger}_{mk} V_{kn} V^{\dagger}_{nj} V_{jm}\}, \qquad (i,j,k) \text{ and } (l,m,n) \text{ cyclic, no summation.} \tag{1.4.33}$$

In the three-generation case there are nine of these invariants, which are all equal. In the parametrization (1.4.31) they read

$$\chi_{il} = \chi = s_{12} s_{23} s_{13} c_{12} c_{23} c_{13}^2 \sin\delta_{13}. \tag{1.4.34}$$

Furthermore, the existence of CP violation requires that the masses of the quarks with the same charge must not be degenerate. Otherwise the remaining phase δ_{13} could be eliminated from $\mathbf{V}$ by a redefinition of the quark fields through a unitary transformation in the degenerate quark fields. Consequently, the condition for CP violation in the three-generation case reads

$$\begin{aligned}&\chi(m_{\mathrm{t}}^2 - m_{\mathrm{c}}^2)(m_{\mathrm{t}}^2 - m_{\mathrm{u}}^2)(m_{\mathrm{c}}^2 - m_{\mathrm{u}}^2)\\ &\quad\times(m_{\mathrm{b}}^2 - m_{\mathrm{s}}^2)(m_{\mathrm{b}}^2 - m_{\mathrm{d}}^2)(m_{\mathrm{s}}^2 - m_{\mathrm{d}}^2) \neq 0.\end{aligned} \tag{1.4.35}$$

CP violation is found experimentally in the K^0 system where it leads to transitions between the eigenstates with different CP quantum numbers. In the quark model, the neutral K mesons are composed of quarks as

$$|\mathrm{K}^0\rangle \sim |\mathrm{d\bar{s}}\rangle, \qquad |\bar{\mathrm{K}}^0\rangle \sim -|\mathrm{s\bar{d}}\rangle.$$

The lowest-order transitions between $|\mathrm{d\bar{s}}\rangle \leftrightarrow -|\mathrm{s\bar{d}}\rangle$ are caused by the box diagrams shown in Fig. 1.7 [Ga74]. The transition amplitude resulting from these diagrams has the structure

$$\mathcal{M} = \sum_{i,j} V^{\dagger}_{\mathrm{s}i} V_{i\mathrm{d}} V^{\dagger}_{\mathrm{s}j} V_{j\mathrm{d}} F(M_{\mathrm{W}}, m_{\mathrm{u}_i}, m_{\mathrm{u}_j}), \tag{1.4.36}$$

where i, j run over the quarks with charge $+2/3$ and the function F results from the calculation of the loop integral. Note that the combination of elements of the quark mixing matrix appearing here is exactly the one in χ_{il}.

Recently CP violation has also been found in the B^0 system in the decay $\mathrm{B} \to \mathrm{J}/\psi\, \mathrm{K}^0_{\mathrm{s}}$ [Af00].

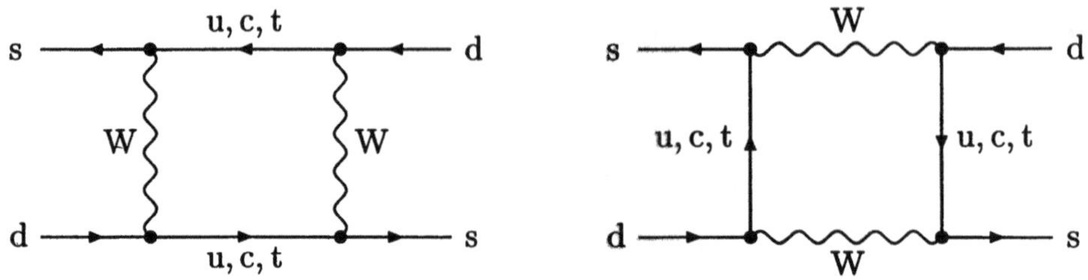

Fig. 1.7 CP-violating box diagrams

1.4.2.4 Applications and comparison with experiment

The predictions of the Lagrangian $\mathcal{L}_{\mathrm{I}}$ were evaluated for many electromagnetic and weak decays and scattering processes. We give only some typical examples. For a more complete discussion we refer to other textbooks [EP].

Z *and* W *decays*: in lowest order, the Z and W bosons decay into fermion–antifermion pairs. The corresponding decay matrix elements are obtained directly as matrix elements of $\mathcal{L}_{\mathrm{I}}$ between the in- and outgoing states. The resulting formula for the partial widths for the Z decay reads

$$\begin{aligned}\Gamma(\mathrm{Z}\to f\bar{f}) &= N_{\mathrm{C}}^{f}\frac{\alpha M_{\mathrm{Z}}}{3}\sqrt{1-\frac{4m_f^2}{M_{\mathrm{Z}}^2}}\left[v_f^2\left(1+\frac{2m_f^2}{M_{\mathrm{Z}}^2}\right)+a_f^2\left(1-\frac{4m_f^2}{M_{\mathrm{Z}}^2}\right)\right]\\ &\approx N_{\mathrm{C}}^{f}\frac{\alpha M_{\mathrm{Z}}}{3}\left[v_f^2+a_f^2\right] \quad \text{for} \quad m_f \ll M_{\mathrm{Z}}\\ &= N_{\mathrm{C}}^{f}\frac{G_\mu M_{\mathrm{Z}}^3}{12\pi\sqrt{2}}(1-4s_{\mathrm{w}}^2|Q_f|+8s_{\mathrm{w}}^4|Q_f|^2), \end{aligned} \tag{1.4.37}$$

with the colour factor $N_{\mathrm{C}}^{f}=1$ for leptons and $N_{\mathrm{C}}^{f}=3$ for quarks. The couplings v_f and a_f are the diagonal elements of v^{Z} and a^{Z} in (1.4.1),

$$v_f = \frac{1}{2s_{\mathrm{w}}c_{\mathrm{w}}}(I_{\mathrm{w},f}^3-2Q_f s_{\mathrm{w}}^2), \qquad a_f = \frac{1}{2s_{\mathrm{w}}c_{\mathrm{w}}}I_{\mathrm{w},f}^3, \tag{1.4.38}$$

and $I_{\mathrm{w},f}^3$ and Q_f stand for the third component of the weak isospin and the relative charge of the fermion f. The total width Γ_{Z} is obtained from (1.4.37) by summing over all contributing partial widths, i.e. fermions with $2m_f \le M_{\mathrm{Z}}$,

$$\Gamma_{\mathrm{Z}} = \sum_f \Gamma(\mathrm{Z}\to f\bar{f}) \approx 2.5\,\mathrm{GeV}. \tag{1.4.39}$$

In a similar way, the decay widths of the W boson into fermion–antifermion pairs can be calculated

$$\begin{aligned}\Gamma(\mathrm{W}\to f_i\bar{f}_j) &= N_{\mathrm{C}}^f\frac{\alpha M_{\mathrm{W}}}{12s_{\mathrm{W}}^2}|V_{ij}|^2\left[1-\frac{m_i^2+m_j^2}{2M_{\mathrm{W}}^2}-\frac{(m_i^2-m_j^2)^2}{2M_{\mathrm{W}}^4}\right]\\ &\quad\times\frac{\sqrt{[M_{\mathrm{W}}^2-(m_i+m_j)^2][M_{\mathrm{W}}^2-(m_i-m_j)^2]}}{M_{\mathrm{W}}^2}\\ &\approx N_{\mathrm{C}}^f\frac{\alpha M_{\mathrm{W}}}{12s_{\mathrm{W}}^2}|V_{ij}|^2 \quad \text{for} \quad m_f\ll M_{\mathrm{W}}\\ &= N_{\mathrm{C}}^f\frac{G_\mu M_{\mathrm{W}}^3}{6\pi\sqrt{2}}|V_{ij}|^2. \end{aligned} \tag{1.4.40}$$

Again, the total width is the sum over all possible decay channels

$$\Gamma_{\mathrm{W}} = \sum_{ij}\Gamma(\mathrm{W}\to f_i\bar{f}_j)\approx 2.0\,\mathrm{GeV}. \tag{1.4.41}$$

The values for the partial and total widths of the Z and W bosons obtained from the expressions involving G_μ are in good agreement with the experimental results [PDG00]

$$\Gamma_{\mathrm{Z}} = 2.4952\pm 0.0026\,\mathrm{GeV}, \qquad \Gamma_{\mathrm{W}} = 2.12\pm 0.05\,\mathrm{GeV}. \tag{1.4.42}$$

The radiative corrections to the decay widths (if parametrized with G_μ), which are discussed in Sect. 4.6.3, are small.

Z-boson production in $\mathrm{e^+e^-}$ *annihilation*: purely electromagnetic $\mathrm{e^+e^-}$ annihilation into fermion pairs was calculated in Sect. 1.2.1. In the complete electroweak model, the Z-exchange diagram has to be taken into account besides the photon-exchange diagram. Neglecting terms proportional to the electron mass, which arise e.g. from the $k_\mu k_\nu$ terms in (1.4.13), the invariant matrix element reads

$$\begin{aligned}\mathcal{M} &= \frac{e^2}{s}Q_{\mathrm{e}}Q_f\,\bar{v}(p_+)\gamma^\mu u(p_-)\,\bar{u}(q_-)\gamma_\mu v(q_+)\\ &\quad+\frac{e^2}{s-M_{\mathrm{Z}}^2+\mathrm{i}M_{\mathrm{Z}}\Gamma_{\mathrm{Z}}}\,\bar{v}(p_+)(\gamma^\mu v_{\mathrm{e}}-\gamma^\mu\gamma_5 a_{\mathrm{e}})u(p_-)\\ &\quad\times\bar{u}(q_-)(\gamma_\mu v_f-\gamma_\mu\gamma_5 a_f)v(q_+), \end{aligned} \tag{1.4.43}$$

where we have taken into account the Z-boson width in the propagator.

Squaring the matrix element, summing/averaging over the fermion spins, and multiplying with the flux and phase-space factors gives the differential cross section ($m_{\mathrm{e}}^2 \ll s$)

$$\frac{\mathrm{d}\sigma}{\mathrm{d}\Omega_{\mathrm{CM}}} = \frac{\alpha^2}{4s}\sqrt{1-\frac{4m_f^2}{s}}\left[\left(1+\cos^2\vartheta+\frac{4m_f^2}{s}\sin^2\vartheta\right)G_1(s) - \frac{8m_f^2}{s}G_2(s)+\sqrt{1-\frac{4m_f^2}{s}}\,2\cos\vartheta\, G_3(s)\right] \tag{1.4.44}$$

with

$$\begin{aligned} G_1(s) &= Q_{\mathrm{e}}^2Q_f^2 + 2Q_{\mathrm{e}}Q_f v_{\mathrm{e}}v_f \operatorname{Re}\chi_{\mathrm{Z}}(s) + (v_{\mathrm{e}}^2+a_{\mathrm{e}}^2)(v_f^2+a_f^2)|\chi_{\mathrm{Z}}(s)|^2,\\ G_2(s) &= (v_{\mathrm{e}}^2+a_{\mathrm{e}}^2)a_f^2|\chi_{\mathrm{Z}}(s)|^2,\\ G_3(s) &= 2Q_{\mathrm{e}}Q_f a_{\mathrm{e}}a_f \operatorname{Re}\chi_{\mathrm{Z}}(s) + 4v_{\mathrm{e}}v_f a_{\mathrm{e}}a_f|\chi_{\mathrm{Z}}(s)|^2, \end{aligned} \tag{1.4.45}$$

and the reduced propagator

$$\chi_{\mathrm{Z}}(s) = \frac{s}{s-M_{\mathrm{Z}}^2+\mathrm{i}M_{\mathrm{Z}}\Gamma_{\mathrm{Z}}}. \tag{1.4.46}$$

For light fermions we may neglect the m_f^2/s terms getting the simpler result

$$\frac{\mathrm{d}\sigma}{\mathrm{d}\Omega_{\mathrm{CM}}} = \frac{\alpha^2}{4s}\left[(1+\cos^2\vartheta)G_1(s)+2\cos\vartheta\, G_3(s)\right]. \tag{1.4.47}$$

For small energies ($s \ll M_{\mathrm{Z}}^2$), $|\chi|$ and $\operatorname{Re}\chi$ are small, and we recover the QED result (1.2.44). At energies around M_{Z}, the Z-boson contribution dominates compared with the γ contribution since then $|\chi_{\mathrm{Z}}|^2 \approx M_{\mathrm{Z}}^2/\Gamma_{\mathrm{Z}}^2 \gg 1$. As an example, the total cross section for $\mathrm{e}^+\mathrm{e}^- \to \mu^+\mu^-$ is shown in Fig. 1.8 in QED and in the Electroweak Standard Model (EWSM). While the large cross section at small energies is due to the photon matrix element, the resonance near 90 GeV originates from the Z-boson matrix element.

The $\cos\vartheta$ part with G_3 is forward–backward-asymmetric, i.e. charge-asymmetric, and typical for the vector–axial-vector interference of the weak interaction. While the asymmetry is negligible at small energies, it becomes experimentally significant at several 10 GeV. For illustration we show the differential cross section for $\mathrm{e}^+\mathrm{e}^- \to \mu^+\mu^-$ at $E_{\mathrm{CM}} = 40\,\mathrm{GeV}$ in Fig. 1.9.

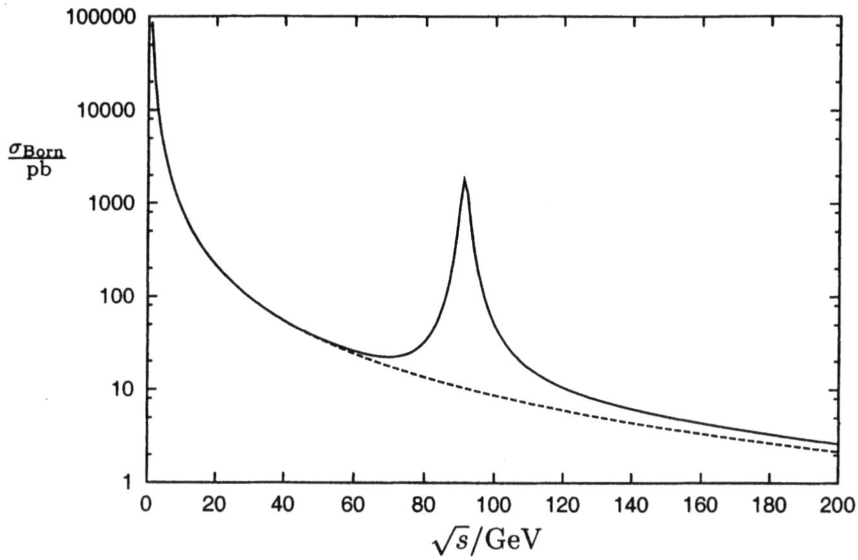

Fig. 1.8 Total lowest-order cross section for $e^+e^- \to \mu^+\mu^-$ in the EWSM (full) and in QED (dashed).

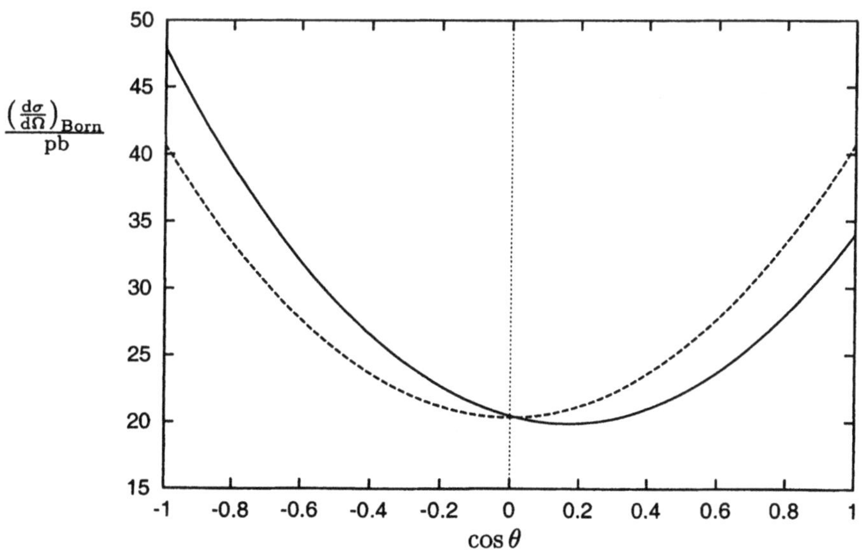

Fig. 1.9 Differential lowest-order cross section for for $e^+e^- \to \mu^+\mu^-$ at $E_{\text{CM}} = 40\,\text{GeV}$ in the SM (full) and in QED (dashed).

1.4.2.5 Concluding remarks

The results of the phenomenological discussion of the weak interaction can be summarized as follows. The concept of universality of the vector- and axial-vector couplings forms the structural basis of the theory of the electroweak interaction of leptons and hadrons. The $\mathrm{SU}(2)_\mathrm{w} \times \mathrm{U}(1)_Y$ symmetry is the quintessence of the phenomenological analysis of electroweak processes of leptons and hadrons. It is comprised in the interaction Lagrangian

$$\begin{aligned}\mathcal{L}_\mathrm{I} = e &\sum_i \overline{\Psi}_i(x) v_i^\gamma \gamma_\mu \Psi_i(x) A_\gamma^\mu(x) \\ + e &\sum_i \overline{\Psi}_i(x)(\gamma_\mu v_i^\mathrm{Z} - \gamma_\mu\gamma_5 a_i^\mathrm{Z})\Psi_i(x) A_\mathrm{Z}^\mu(x) \\ + e &\Big(\sum_{i,j} \overline{\Psi}_i(x) V_{ij}(\gamma_\mu v^{\mathrm{W}+} - \gamma_\mu\gamma_5 a^{\mathrm{W}+})\Psi_j(x) A_{\mathrm{W}+}^\mu(x) + \text{h.c.}\Big).\end{aligned} \tag{1.4.48}$$

Here the sums extend over all fermion doublets (in the quark case including the colour degree of freedom). The 2×2 coupling matrices are given in (1.4.2) for leptons. For quarks they can be read off from (1.4.21), (1.4.24), and (1.4.27).

In terms of the explicit fermion fields, the interaction Lagrangian reads

$$\begin{aligned}\mathcal{L}_\mathrm{I} = -e &\sum_l Q_l \bar{\psi}_l \gamma_\mu \psi_l A_\gamma^\mu(x) - e \sum_{q,c} Q_q \bar{\psi}_{q,c}\gamma_\mu \psi_{q,c} A_\gamma^\mu(x) \\ + e &\sum_l \bar{\psi}_l \gamma_\mu (v_l - a_l\gamma_5)\psi_l A_\mathrm{Z}^\mu(x) \\ + e &\sum_{q,c} \bar{\psi}_{q,c}\gamma_\mu (v_q - a_q\gamma_5)\psi_{q,c} A_\mathrm{Z}^\mu(x) \\ + &\frac{e}{\sqrt{2}s_\mathrm{W}} \sum_i (\bar{\psi}_{l_i}\gamma_{\mu\mathrm{L}}\psi_{\nu_{l_i}} A_{\mathrm{W}+}^\mu(x) + \text{h.c.}) \\ + &\frac{e}{\sqrt{2}s_\mathrm{W}} \sum_{i,j,c} (\bar{\psi}_{u_i,c}\gamma_{\mu\mathrm{L}} \mathbf{V}_{ij}\psi_{d_j,c} A_{\mathrm{W}+}^\mu(x) + \text{h.c.}),\end{aligned} \tag{1.4.49}$$

where l runs over all leptons, q over all quarks, c over the quark colours, and i and j over the three generations. The couplings v_l, a_l, v_q, and a_q are given in (1.4.38).

If the neutrinos are massive, then also in the lepton sector a mixing matrix analogously to $\mathbf{V}$ has to be included in the Lagrangian.

The Lagrangian (1.4.49) describes electroweak four-fermion processes in lowest order. In the low-energy limit it reproduces the Fermi model and QED. The Lagrangian $\mathcal{L}_{\mathrm{I}}$ is part of the complete electroweak gauge-theory Lagrangian which is constructed in Chapt. 4. For more detailed applications of (1.4.49) we refer to the literature [EW]. The $\mathrm{SU}(2)_{\mathrm{W}} \times \mathrm{U}(1)_Y$ structure inherent in (1.4.49) is the starting point for the construction of the corresponding non-abelian gauge theory (cf. Sect. 4.2).

1.5 The quark–parton model

The hypothesis that mesons and baryons are made up of quarks is the basis of a successful description of hadron physics. This concept has not been directly confirmed, because quarks have not been observed as free particles experimentally. However, the structure of the hadrons can be investigated by means of probes such as electrons, muons, or neutrinos which have no strong interaction. The discovery of scale invariance in *deep-inelastic lepton–nucleon scattering* showed that the nucleons contain point-like constituents, the *partons*. Tests on the quantum numbers of these constituents confirm that they are identical to quarks. Moreover, the contribution of the gluons, can be measured in these experiments.

1.5.1 Scaling in deep-inelastic lepton–nucleon scattering

The differential cross section for leptoproduction, i.e. for the reaction $l + \mathcal{N} \to l' + X$ can be calculated in the boson-exchange model. Here l and l' denote the incoming and outgoing leptons with momenta k and k', respectively, $\mathcal{N}$ the incoming nucleon with momentum p, and X an arbitrary hadronic final state. The corresponding graph is displayed in Fig. 1.10. The kinematics in the laboratory frame, i.e. in the rest frame of $\mathcal{N}$, reads

$$
\begin{aligned}
&k^\mu = (E, \mathbf{k}), \qquad k'^\mu = (E', \mathbf{k}'), \\
&p^\mu = (M, \mathbf{0}), \qquad q^\mu = k^\mu - k'^\mu, \\
&Q^2 = -q^2 = 4EE' \sin^2\theta/2, \qquad \nu = pq/M = E - E'.
\end{aligned} \tag{1.5.1}
$$

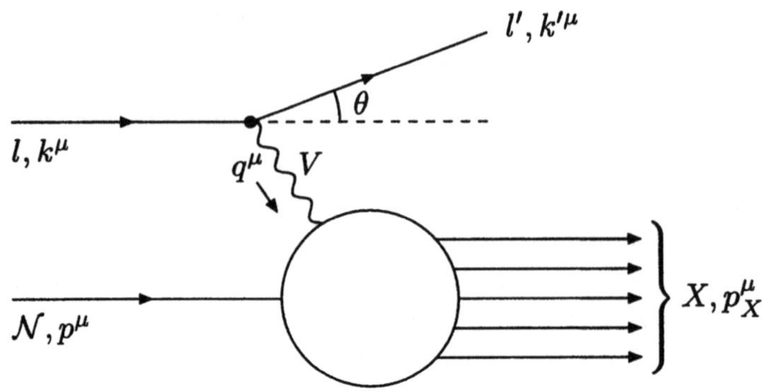

Fig. 1.10 Kinematics of deep-inelastic lepton–nucleon scattering

Since we are interested in the high-energy region, the lepton and the nucleon masses can be neglected, i.e. $k^2 = k'^2 = p^2 = 0$. The calculation is carried out using the *Mandelstam variables*

$$\begin{aligned} s &= (k+p)^2 = (k'+p_X)^2 = 2kp = 2k'p_X + M_X^2 = 2ME, \\ t &= (k-k')^2 = (p-p_X)^2 = -2kk' = -2pp_X + M_X^2 = -Q^2, \\ u &= (k-p_X)^2 = (k'-p)^2 = -2kp_X + M_X^2 = -2k'p = -2ME'. \end{aligned} \tag{1.5.2}$$

Furthermore the *scaling variables*

$$\begin{aligned} x &= \frac{Q^2}{2pq} = \frac{Q^2}{2M\nu} = -\frac{t}{s+u}, \qquad 0 \le x \le 1, \\ y &= \frac{2pq}{s} = \frac{2M\nu}{s} = \frac{s+u}{s}, \qquad 0 \le y \le 1 \end{aligned} \tag{1.5.3}$$

are used. In the rest frame of the nucleon, their values are

$$x = \frac{2EE'}{M(E-E')}\sin^2\frac{\theta}{2}, \qquad y = \frac{E-E'}{E}. \tag{1.5.4}$$

We are interested in experiments in which only the lepton is detected in the final state. The corresponding *inclusive cross section* is obtained by summing over all hadronic final states $|X\rangle$,

$$\mathrm{d}\sigma^{l\mathcal{N}} = \frac{1}{4ME}\frac{1}{4}\sum_X |\mathcal{M}|^2(2\pi)^4\delta^{(4)}(k+p-k'-p_X)\frac{\mathrm{d}^3k'}{(2\pi)^3 2E'}. \tag{1.5.5}$$

Here the sum over X includes the corresponding phase space. The invariant matrix element for exchange of the vector boson $V = \gamma, \mathrm{W}^{\pm}, \mathrm{Z}$ between leptons with *vector and axial-vector couplings*

$$\begin{aligned} v_l^{\gamma} &= -Q_l, \qquad a_l^{\gamma} = 0, \qquad v_l^{W} = a_l^{W} = \frac{1}{2\sqrt{2}s_{\mathrm{w}}}, \\ v_l^{Z} &= \frac{I^3_{\mathrm{w},l} - 2Q_l s_{\mathrm{w}}^2}{2s_{\mathrm{w}}c_{\mathrm{w}}}, \qquad a_l^{Z} = \frac{I^3_{\mathrm{w},l}}{2s_{\mathrm{w}}c_{\mathrm{w}}}, \end{aligned} \tag{1.5.6}$$

and hadrons can be written using the hadronic current operators[5] (1.4.21), (1.4.24), and (1.4.27),

$$\begin{aligned} \mathcal{M} = -e^2 &\sum_{V=\gamma,\mathrm{W}^{\pm},\mathrm{Z}} \bar{u}(k',\sigma')\gamma^{\mu}(v_l^V - \gamma_5 a_l^V)u(k,\sigma)\frac{1}{Q^2+M_V^2} \\ &\times \zeta_V \langle X|h_{\mu}^V(0)|p,\lambda\rangle, \end{aligned} \tag{1.5.7}$$

where

$$\zeta_{\gamma} = 1, \qquad \zeta_Z = \frac{1}{s_{\mathrm{w}}c_{\mathrm{w}}}, \qquad \zeta_W = \frac{1}{\sqrt{2}s_{\mathrm{w}}}. \tag{1.5.8}$$

With (1.5.5) and (1.5.7), we have

$$\begin{aligned} \mathrm{d}\sigma^{l\mathcal{N}} &= \frac{\alpha^2\mathrm{d}^3k'}{MEE'}\sum_{V,V'}\frac{1}{Q^2+M_V^2}\frac{1}{Q^2+M_{V'}^2}\zeta_{V'}^*\zeta_V \\ &\quad\times\frac{1}{2}\sum_{\mathrm{pol}}\bar{u}(k')\gamma^{\mu}(v_l^V - \gamma_5 a_l^V)u(k)\,\bar{u}(k)\gamma^{\nu}(v_l^{V'} - \gamma_5 a_l^{V'})u(k') \\ &\quad\times\frac{1}{2}\sum_X(2\pi)^3\delta^{(4)}(p+q-p_X)\frac{1}{2}\sum_{\mathrm{pol}}\langle p|h_{\nu}^{V,\dagger}(0)|X\rangle\langle X|h_{\mu}^{V'}(0)|p\rangle \\ &= \frac{\alpha^2\mathrm{d}^3k'}{MEE'}\sum_{V,V'}\frac{\zeta_V}{Q^2+M_V^2}\frac{\zeta_{V'}^*}{Q^2+M_{V'}^2}\times\frac{1}{2}L^{VV',\mu\nu}\times W_{\nu\mu}^{VV'\mathcal{N}} \end{aligned} \tag{1.5.9}$$

[5]Note that our normalization of the weak currents differs by a factor 2 from the one commonly used in the parton model.

for the unpolarized differential cross section, where we have introduced the *leptonic tensor* $L_{\mu\nu}^{VV'}$ [second line of (1.5.9)] and the *hadronic tensor* $W_{\nu\mu}^{VV'\mathcal{N}}$ [third line of (1.5.9)].

The leptonic tensor $L_{\mu\nu}^{VV'}$ can be evaluated with help of the trace formulas for Dirac matrices (A.1.30),

$$\begin{aligned} L^{VV',\mu\nu} &= \sum_{\text{pol}} \bar{u}(k')\gamma^\mu(v_l^V - \gamma_5 a^V)u(k)\,\bar{u}(k)\gamma^\nu(v_l^{V'} - \gamma_5 a_l^{V'})u(k') \\ &= \mathrm{Tr}\left[\not{k}'\gamma^\mu(v_l^V - \gamma_5 a_l^V)\not{k}\gamma^\nu(v_l^{V'} - \gamma_5 a_l^{V'})\right] \\ &= 4\Big((k'^\mu k^\nu + k'^\nu k^\mu - g^{\mu\nu} kk')G_v^{l,VV'} - \mathrm{i}\varepsilon^{\mu\nu\rho\sigma}k'_\rho k_\sigma G_a^{l,VV'}\Big). \end{aligned} \tag{1.5.10}$$

In the case of antilepton scattering the sign of the term involving $\varepsilon^{\mu\nu\rho\sigma}$ has to be reversed. The constants $G_v^{l,VV'}$ and $G_a^{l,VV'}$ represent combinations of the lepton couplings,

$$G_v^{l,VV'} = v_l^V v_l^{V'} + a_l^V a_l^{V'}, \qquad G_a^{l,VV'} = v_l^V a_l^{V'} + a_l^V v_l^{V'}. \tag{1.5.11}$$

The hadronic tensor $W_{\mu\nu}^{VV'\mathcal{N}}$ cannot be calculated from first principles. Because of Lorentz covariance it can be decomposed into tensors and invariant *structure functions* which only depend on the invariant variables Q^2 and ν defined in (1.5.1). Using current conservation and neglecting terms of the order of the lepton masses yields for the hadronic tensor the structure

$$\begin{aligned} W_{\mu\nu}^{VV'\mathcal{N}} &= -g_{\mu\nu}^{\mathrm{T}} W_1^{VV'\mathcal{N}}(Q^2,\nu) + \frac{p_\mu^{\mathrm{T}} p_\nu^{\mathrm{T}}}{pq}\frac{pq}{M^2} W_2^{VV'\mathcal{N}}(Q^2,\nu) \\ &\quad - \mathrm{i}\varepsilon_{\mu\nu\alpha\beta}\frac{p^\alpha q^\beta}{pq}\frac{pq}{M^2} W_3^{VV'\mathcal{N}}(Q^2,\nu), \end{aligned} \tag{1.5.12}$$

where $g_{\mu\nu}^{\mathrm{T}} = g_{\mu\nu} - q^\mu q^\mu/q^2$ and $p_\mu^{\mathrm{T}} = p_\mu - q_\mu pq/q^2$. Instead of the functions W_i, often the slightly different ones

$$F_1 = W_1, \qquad F_2 = \frac{pq}{M^2}W_2, \qquad F_3 = \frac{pq}{M^2}W_3 \tag{1.5.13}$$

are used.

Contracting the leptonic and hadronic tensors gives

$$\frac{\mathrm{d}^2\sigma^{l\mathcal{N}}}{\mathrm{d}E'\mathrm{d}\Omega} = 4\alpha^2\frac{E'^2}{M}\sum_{V,V'}\frac{\zeta_V}{Q^2+M_V^2}\frac{\zeta^*_{V'}}{Q^2+M_{V'}^2}$$
$$\times\left[\left(2\sin^2\frac{\theta}{2}W_1^{VV'\mathcal{N}}+\cos^2\frac{\theta}{2}W_2^{VV'\mathcal{N}}\right)G_v^{l,VV'}\right.$$
$$\left.\pm\sin^2\frac{\theta}{2}\frac{E+E'}{M}W_3^{VV'\mathcal{N}}G_a^{l,VV'}\right] \tag{1.5.14}$$

for the cross section in the laboratory system ($\mathrm{d}^3k' = E'^2\mathrm{d}E'\mathrm{d}\Omega$), where the last term is positive for lepton and negative for antilepton scattering.

At low energies, where the weak neutral current is negligible, the special cases of photon exchange

$$\frac{\mathrm{d}^2\sigma^{e\mathcal{N}}}{\mathrm{d}E'\mathrm{d}\Omega} = \frac{\alpha^2}{4ME^2\sin^4\frac{\theta}{2}}\left[2\sin^2\frac{\theta}{2}W_1^{\gamma\gamma\mathcal{N}}+\cos^2\frac{\theta}{2}W_2^{\gamma\gamma\mathcal{N}}\right], \tag{1.5.15}$$

and of charged weak processes

$$\frac{\mathrm{d}^2\sigma^{\nu\mathcal{N}}}{\mathrm{d}E'\mathrm{d}\Omega} = \frac{2G_\mu^2E'^2}{\pi^2M(1+Q^2/M_\mathrm{W}^2)^2}\left[2\sin^2\frac{\theta}{2}W_1^{\mathrm{WW}\mathcal{N}}+\cos^2\frac{\theta}{2}W_2^{\mathrm{WW}\mathcal{N}}\right.$$
$$\left.\pm\sin^2\frac{\theta}{2}\frac{E+E'}{M}W_3^{\mathrm{WW}\mathcal{N}}\right] \tag{1.5.16}$$

are relevant. Note that in the charged-current process (1.5.16) only left-handed neutrinos contribute so that no spin average is required. Equation (1.5.14) provides the general formula for the differential cross section in the boson-exchange model. Specifying the coupling terms $G_v^{l,VV'}$ and $G_a^{l,VV'}$ yields explicit expressions for electromagnetic and weak deep-inelastic scattering. The W_3 term contributes only to parity-violating processes.

The differential cross section is independent of the azimuthal angle. Therefore, the corresponding expression in terms of the Mandelstam variables is obtained using the Jacobian and performing the azimuthal integration

$$\frac{\mathrm{d}^2\sigma}{\mathrm{d}u\mathrm{d}t} = 2\pi\left|\frac{\partial(E',\cos\theta)}{\partial(u,t)}\right|\frac{\mathrm{d}^2\sigma}{\mathrm{d}E'\mathrm{d}\Omega} = \frac{\pi}{2MEE'}\frac{\mathrm{d}^2\sigma}{\mathrm{d}E'\mathrm{d}\Omega}, \tag{1.5.17}$$

and similarly for other pairs of variables like (Q^2,ν) and (x,y).

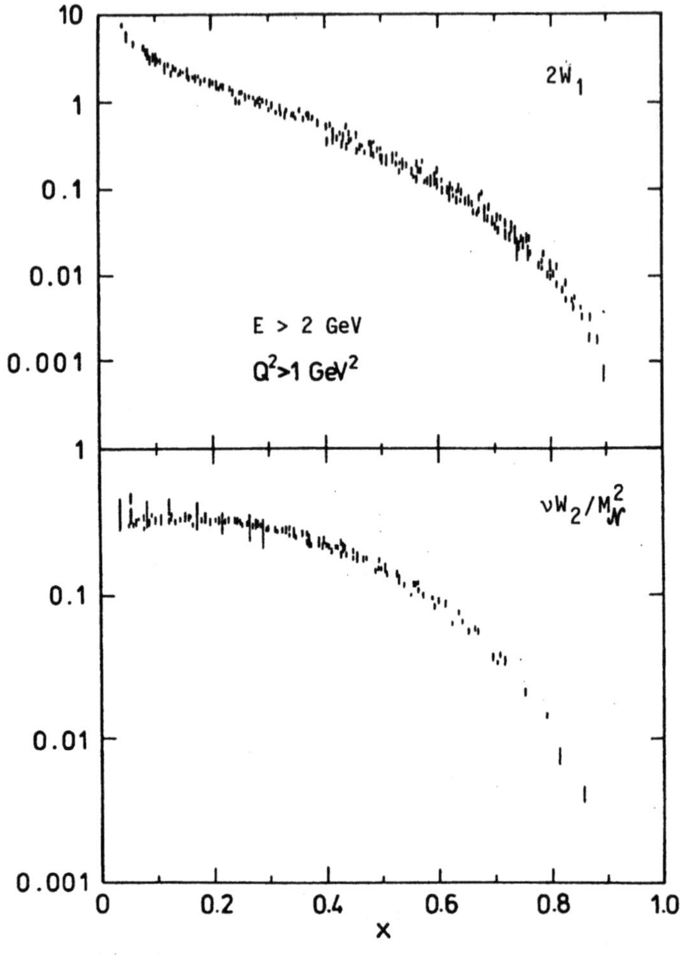

Fig. 1.11 The structure functions $W_{1,2}^{\gamma\gamma P}(Q^2, \nu)$ of deep-inelastic eP scattering according to Ref. [Ta75]

1.5.1.1 Bjorken scaling

Figure 1.11 shows the classical experimental results for the structure functions $W_1^{\gamma\gamma P}$ and $W_2^{\gamma\gamma P}$. The data are plotted as functions of the variable x for different values of Q^2. One observes that for $Q^2 > 1\,\text{GeV}^2$ and $\nu > 1\,\text{GeV}$ the structure functions do not depend on the two variables Q^2 and ν but approximately only on the dimensionless *scaling variable* $x = Q^2/2M\nu$. They are said to show *scaling behaviour* [Bj69]. To take into account this fact we define the *Bjorken limit* of the structure functions F_l,

$$\lim_{\substack{Q^2,\nu\to\infty \\ x \text{ fixed}}} F_l(Q^2,\nu) = F_l(x), \quad x = \frac{Q^2}{2\nu}, \quad l = 1,2,3. \tag{1.5.18}$$

In this Bjorken limit the cross section becomes

$$\begin{aligned}\frac{\mathrm{d}^2\sigma^{l\mathcal{N}}}{\mathrm{d}t\mathrm{d}u} = \frac{2\pi\alpha^2}{s^2}\frac{1}{(s+u)}\sum_{V,V'}\frac{\zeta_V}{t-M_V^2}\frac{\zeta_{V'}^*}{t-M_{V'}^2} \\ \times\Big[\Big((s+u)^2 2xF_1^{VV'\mathcal{N}}(x) - 2usF_2^{VV'\mathcal{N}}(x)\Big)G_v^{l,VV'} \\ \pm (s^2-u^2)xF_3^{VV'\mathcal{N}}(x)G_a^{l,VV'}\Big].\end{aligned} \tag{1.5.19}$$

The parton model, described in the next section, provides a simple explanation of this scaling behaviour and gives information on the structure functions $F_i(x)$.

A detailed analysis of the data reveals small systematic deviations from scaling behaviour. This can be explained within QCD by radiative corrections (cf. Sect. 3.3).

1.5.2 The parton model

The fact that in the Bjorken limit the structure functions of deep-inelastic lepton–nucleon scattering depend almost exclusively on the dimensionless scaling variable x shows that no particular mass scale plays a rôle in this reaction. Feynman and Bjorken [Fe69, Bj69a] described this behaviour as elastic lepton scattering off massless point-like nucleon constituents, the *partons*.

The concept of the *parton model* can be expressed by the following assumptions on the dynamics of deep-inelastic scattering processes:

(i) In hard-scattering processes the nucleon acts like an ensemble of free, massless particles, the partons. The nucleon momentum $\mathbf{p}$ is distributed over the parton momenta $\hat{\mathbf{p}}_i$ (the index i labels the partons). The parton momentum $\hat{\mathbf{p}}_i$ itself is decomposed into a longitudinal and a transverse part $\hat{\mathbf{p}}_i^{\mathrm{T}}$,

$$\hat{\mathbf{p}}_i = \xi_i \mathbf{p} + \hat{\mathbf{p}}_i^{\mathrm{T}}, \quad \mathbf{p}\hat{\mathbf{p}}_i^{\mathrm{T}} = 0, \quad \xi_i \geq 0, \quad \sum_i \xi_i = 1,$$

where ξ_i is the part of the nucleon momentum that is carried by the parton i as longitudinal momentum.

In the parton model the scattering is considered in the infinite-momentum frame defined by

$$q^\mu = (0,0,0,\sqrt{Q^2}), \qquad p^\mu = (\sqrt{M^2+p^2},0,0,p) \tag{1.5.20}$$

with the z component $p = \nu M/\sqrt{Q^2} = \sqrt{Q^2}/2x$. Hence, in the Bjorken limit p goes to infinity, which means that the nucleon can be regarded as moving at high speed and so do the partons. Furthermore, it is assumed that the transverse parton momenta remain finite.

Under these conditions, and for high initial energy, $p^0 \gg M$, the parton energy can be written as

$$\begin{aligned}\hat{p}_i^0 &= \sqrt{\hat{\mathbf{p}}_i^2} = \sqrt{\xi_i^2 \mathbf{p}^2 + (\hat{\mathbf{p}}_i^{\mathrm{T}})^2} = \sqrt{\xi_i^2 (p^0)^2 - \xi_i^2 M^2 + (\hat{\mathbf{p}}_i^{\mathrm{T}})^2} \\ &\sim \xi_i p^0 + \frac{(\hat{\mathbf{p}}_i^{\mathrm{T}})^2 - \xi_i^2 M^2}{2\xi_i p^0},\end{aligned}$$

and thus the parton four-momentum becomes, neglecting the finite $\mathbf{p}^{\mathrm{T}}$,

$$\hat{p}_i^\mu \sim (\xi_i p^0, 0, 0, \xi_i p) = \xi_i p^\mu. \tag{1.5.21}$$

Consequently, the parton i carries the fraction ξ_i of the four-momentum of the nucleon.

(ii) Neglecting final-state interactions and interference terms, the cross section can be calculated as the incoherent sum of the lepton–parton cross sections multiplied by the *parton distribution functions* $f_{i,\mathcal{N}}(\xi_i)$, where $f_{i,\mathcal{N}}(\xi_i)\mathrm{d}\xi_i$ is the probability to find a parton i with momentum fraction between ξ_i and $\xi_i + \mathrm{d}\xi_i$ in the nucleon $\mathcal{N}$. Consequently, $\int_0^1 \mathrm{d}\xi_i\, f_{i,\mathcal{N}}(\xi_i) = N_i$ is the number of partons of type i in the nucleon.

The lepton–nucleon cross section is obtained by summing over the parton flavours i and integrating over all possible momentum fractions ξ_i.

The calculation of the *lepton–parton cross section* in the boson-exchange model proceeds according to that of elementary particles (compare e.g. the treatment of e^+e^- annihilation into μ pairs in Sect. 1.2.1.9). The Mandelstam variables for the parton kinematics are related to the hadronic variables as

$$\begin{aligned} \hat{s}_i &= (k+\hat{p}_i)^2 = (k+\xi_i p)^2 \approx 2\xi_i kp = \xi_i s, \\ \hat{u}_i &= (k'-\hat{p}_i)^2 = (k'-\xi_i p)^2 \approx -2\xi_i kp' = \xi_i u, \\ \hat{t}_i &= (k-k')^2 = t. \end{aligned} \tag{1.5.22}$$

Assuming that the partons are point-like objects with spin 1/2, the unpolarized parton cross section becomes

$$\frac{\mathrm{d}\hat{\sigma}_i}{\mathrm{d}\Omega} = \frac{1}{64\pi^2\hat{s}_i}\frac{1}{4}\sum_{\mathrm{pol}}|\mathcal{M}|^2 = \frac{\alpha^2}{4\hat{s}_i}\sum_{V,V'}\frac{\frac{1}{2}L^{VV',\mu\nu}\frac{1}{2}\hat{L}^{VV'}_{\nu\mu,i}}{(Q^2+M_V^2)(Q^2+M_{V'}^2)}, \tag{1.5.23}$$

where $L^{VV'}_{\mu\nu}$ is the leptonic tensor (1.5.10), and $\hat{L}^{VV'}_{\nu\mu,i}$ a similar tensor for the parton i. Contraction of the tensors and use of Mandelstam variables leads to

$$\frac{\mathrm{d}\hat{\sigma}_i}{\mathrm{d}t} = \frac{2\pi\alpha^2}{\hat{s}_i^2}\sum_{V,V'}\frac{(\hat{s}_i^2+\hat{u}_i^2)G_v^{l,VV'}G_v^{i,VV'} \pm (\hat{s}_i^2-\hat{u}_i^2)G_a^{l,VV'}G_a^{i,VV'}}{(t-M_V^2)(t-M_{V'}^2)} \tag{1.5.24}$$

with + for lepton–parton or antilepton–antiparton scattering and − for antilepton–parton or lepton–antiparton scattering, respectively. The products of quark couplings $G_v^{i,VV'}$ and $G_a^{i,VV'}$ are defined as in (1.5.11). The Mandelstam condition for vanishing parton masses, $\hat{s}_i + \hat{u}_i + t = 0$, is made explicit in (1.5.24) by introducing a δ-function

$$\begin{aligned} \frac{\mathrm{d}^2\hat{\sigma}_i}{\mathrm{d}t\mathrm{d}\hat{u}_i}(\hat{s}_i,t,\hat{u}_i) &= \frac{2\pi\alpha^2}{\hat{s}_i^2}\sum_{V,V'}\frac{1}{(t-M_V^2)(t-M_{V'}^2)} \\ &\times\left[(\hat{s}_i^2+\hat{u}_i^2)G_v^{l,VV'}G_v^{i,VV'} \pm (\hat{s}_i^2-\hat{u}_i^2)G_a^{l,VV'}G_a^{i,VV'}\right]\delta(\hat{s}_i+\hat{u}_i+t). \end{aligned} \tag{1.5.25}$$

Use of the kinematical relations (1.5.22) allows to transform to the hadronic variables s and u.

The deep-inelastic lepton–nucleon scattering cross section can now be built up from the parton distribution functions and these parton cross sections

$$\begin{aligned}\frac{\mathrm{d}^2\sigma^{l\mathcal{N}}}{\mathrm{d}t\mathrm{d}u}(s,t,u) &= \sum_i \int_0^1 \mathrm{d}\xi_i\, f_{i,\mathcal{N}}(\xi_i)\, \xi_i \frac{\mathrm{d}^2\hat{\sigma}}{\mathrm{d}t\mathrm{d}\hat{u}_i}(\xi_i s, t, \xi_i u) \\ &= \frac{2\pi\alpha^2}{s^2}\sum_{V,V'}\sum_i \frac{(s^2+u^2)G_v^{l,VV'}G_v^{i,VV'} \pm (s^2-u^2)G_a^{l,VV'}G_a^{i,VV'}}{(t-M_V^2)(t-M_{V'}^2)} \\ &\quad\times \int_0^1 \mathrm{d}\xi_i\, f_{i,\mathcal{N}}(\xi_i)\,\xi_i\,\delta(\xi_i(s+u)+t). \end{aligned} \tag{1.5.26}$$

Because of the δ-function ξ_i is identical to $x = -t/(s+u) = Q^2/2M\nu$, i.e. the Bjorken scaling variable. Hence x determines the longitudinal part of the parton momentum.

Carrying out the integration yields

$$\begin{aligned}\frac{\mathrm{d}^2\sigma^{l\mathcal{N}}}{\mathrm{d}t\mathrm{d}u} &= \frac{2\pi\alpha^2}{s^2}\frac{1}{s+u}\sum_{V,V'}\frac{1}{t-M_V^2}\frac{1}{t-M_{V'}^2} \\ &\quad\times\sum_i x f_{i,\mathcal{N}}(x)\left[(s^2+u^2)G_v^{l,VV'}G_v^{i,VV'} \pm (s^2-u^2)G_a^{l,VV'}G_a^{i,VV'}\right]\end{aligned} \tag{1.5.27}$$

as final result for the lepton–nucleon cross section. Indeed the lepton–nucleon cross section calculated in the parton model shows scaling behaviour.

By comparing the cross section formulas for lepton–nucleon (1.5.19) and lepton–parton scattering (1.5.27), the connection between the structure functions F and the parton distribution functions f_i can be derived,

$$\begin{aligned}\zeta_{V'}^*\zeta_V F_1^{VV'\mathcal{N}}(x) &= \frac{1}{2}\sum_i f_{i,\mathcal{N}}(x)G_v^{i,VV'}, \\ \zeta_{V'}^*\zeta_V F_2^{VV'\mathcal{N}}(x) &= \sum_i x f_{i,\mathcal{N}}(x)G_v^{i,VV'}, \\ \zeta_{V'}^*\zeta_V F_3^{VV'\mathcal{N}}(x) &= \pm\sum_i f_{i,\mathcal{N}}(x)G_a^{i,VV'}.\end{aligned} \tag{1.5.28}$$

The coefficients ζ_V are given in (1.5.8). From (1.5.28), an important result of the parton model, namely the *Callan–Gross relation* [Ca69],

$$F_2(x) = 2xF_1(x) \tag{1.5.29}$$

follows. The Callan–Gross relation holds for spin-1/2 particles. For spin-zero partons one would find $F_1(x) = 0$ instead.

Since (1.5.28) was obtained from (1.5.26), a relation in the form of a multiplicative convolution relating the *nucleon structure functions* $F^{VV'\mathcal{N}}$ with the *parton distributions* $f_{i,\mathcal{N}}$ and the *parton structure functions* $\hat{F}^{VV'i}$, where

$$\begin{aligned}
\zeta^*_{V'}\zeta_V \hat{F}_1^{VV'i}(y) &= \frac{1}{2}\delta(1-y)G_v^{iVV'}, \\
\zeta^*_{V'}\zeta_V \hat{F}_2^{VV'i}(y) &= y\delta(1-y)G_v^{iVV'}, \\
\zeta^*_{V'}\zeta_V \hat{F}_3^{VV'i}(y) &= \pm\delta(1-y)G_a^{iVV'},
\end{aligned} \tag{1.5.30}$$

can be derived,

$$\begin{aligned}
F_1^{VV'\mathcal{N}}(x) &= \sum_i \int_0^1 \frac{dy}{y} \hat{F}_1^{VV'i}\left(\frac{x}{y}\right) f_{i,\mathcal{N}}(y), \\
F_2^{VV'\mathcal{N}}(x) &= \sum_i \int_0^1 dy \hat{F}_2^{VV'i}\left(\frac{x}{y}\right) f_{i,\mathcal{N}}(y), \\
F_3^{VV'\mathcal{N}}(x) &= \sum_i \int_0^1 \frac{dy}{y} \hat{F}_3^{VV'i}\left(\frac{x}{y}\right) f_{i,\mathcal{N}}(y).
\end{aligned} \tag{1.5.31}$$

This representation is the starting point for the inclusion of QCD corrections (cf. Sect. 3.3.1).

In the *quark–parton model*, the hadronic constituents are identified with the quarks. Then, the coupling terms G_v^i and G_a^i for electron–nucleon scattering are the squares of the quark and antiquark charges, $G_v = Q_q^2$, $G_a = 0$, and

$$F_2^{\gamma\gamma\mathcal{N}}(x) = x\sum_q \left[f_{q,\mathcal{N}}(x) + f_{\bar{q},\mathcal{N}}(x)\right] Q_q^2, \qquad F_3^{\gamma\gamma\mathcal{N}}(x) = 0. \tag{1.5.32}$$

Here we are considering only the photon exchange. The full relation including Z-boson exchange can be found in Ref. [PDG00].

When defining *distribution functions for the individual quark flavours* the isospin symmetry of proton (index P) and neutron (index N) can be utilized yielding

$$\begin{aligned}
f_{\mathrm{u,P}}(x) &= f_{\mathrm{d,N}}(x) = u(x), &\quad f_{\mathrm{d,P}}(x) &= f_{\mathrm{u,N}}(x) = d(x), \\
f_{\mathrm{s,P}}(x) &= f_{\mathrm{s,N}}(x) = s(x), &\quad f_{\mathrm{c,P}}(x) &= f_{\mathrm{c,N}}(x) = c(x),
\end{aligned} \tag{1.5.33}$$

and similarly for antiquarks. According to (1.5.32), the structure function for electron–proton scattering for four quark flavours becomes[6]

$$F_2^{\gamma\gamma \mathrm{P}}(x) = \frac{4}{9} x \left[u(x) + \bar{u}(x) + c(x) + \bar{c}(x)\right] + \frac{1}{9} x \left[d(x) + \bar{d}(x) + s(x) + \bar{s}(x)\right] . \tag{1.5.34}$$

In the case of charged-current scattering, the quark-mixing matrix must be taken into account. At energies high enough to create the various quark doublets with nearly equal probability, however, the effects of the quark-mixing matrix drop out. Consequently, beyond the charm threshold we have

$$\begin{aligned} F_2^{W^+W^+\mathrm{P}}(x) &= \frac{1}{2} x \left[d(x) + s(x) + \bar{u}(x) + \bar{c}(x)\right] , \\ F_3^{W^+W^+\mathrm{P}}(x) &= \frac{1}{2} \left[d(x) + s(x) - \bar{u}(x) - \bar{c}(x)\right] , \\ F_2^{W^-W^-\mathrm{P}}(x) &= \frac{1}{2} x \left[u(x) + c(x) + \bar{d}(x) + \bar{s}(x)\right] , \\ F_3^{W^-W^-\mathrm{P}}(x) &= \frac{1}{2} \left[u(x) + c(x) - \bar{d}(x) - \bar{s}(x)\right] \end{aligned} \tag{1.5.35}$$

for the neutrino structure functions. These results correspond to our normalization of the weak currents. In the literature it is common to use a different normalization such that the factor 1/2 in (1.5.35) is replaced by 2. The latter convention is used in the following. Then the sum rules take their usual form.

Explicit expressions for the different structure functions of e$\mathcal{N}$ and $\nu\mathcal{N}$ scattering via neutral currents (photon- and Z-boson exchange) and charged currents (W-boson exchange) can be found in Ref. [PDG00].

1.5.3 Applications of the simple parton model

In this section the theoretical predictions of the quark–parton model are compared with experimental results. Special emphasis is put on the question as to what extent the partons carry the quark quantum numbers.

[6]We do not include the heavy quarks b and t since their contribution is small at presently available energies.

1.5.3.1 Parton spin

If partons are identical to quarks they must be spin-1/2 particles. This demands the validity of the *Callan–Gross relation* (1.5.29). As a consequence, the ratio R of the cross sections resulting from the exchange of longitudinally polarized virtual photons σ_L and transversely polarized ones σ_T is expected to disappear in the Bjorken-limit (up to QCD corrections),

$$R = \frac{\sigma_\mathrm{L}}{\sigma_\mathrm{T}} = \frac{(1+\nu^2/Q^2)W_2 - W_1}{W_1} \underset{\substack{Q^2,\nu\to\infty \\ x \text{ fixed}}}{\longrightarrow} \frac{F_2(x) - 2xF_1(x)}{2xF_1(x)} = 0. \tag{1.5.36}$$

In contrast to this, R should diverge for spin-0 partons where $F_1 = 0$. Thus, R is a sensitive measure of the spin-1/2 property of the partons. The result of the CDHS νP-scattering experiment [Ab83], $R = 0.039 \pm 0.014(\text{stat}) \pm 0.025(\text{syst})$ for $0.4 \le x \le 0.7$, $Q^2 = 38\,\mathrm{GeV}^2$, strongly confirms that the charged partons are spin-1/2 particles. However, the nucleons contain also spin-1 particles, the gluons. These give a contribution to the longitudinal structure functions which is discussed in Sect. 3.2.4.

1.5.3.2 Sum rules for internal quantum numbers

The electric charges of proton and neutron can be calculated using (1.5.33),

$$1 = \int_0^1 \mathrm{d}x \left[\frac{2}{3}(u(x) - \bar{u}(x)) - \frac{1}{3}(d(x) - \bar{d}(x))\right], \tag{1.5.37}$$

$$0 = \int_0^1 \mathrm{d}x \left[\frac{2}{3}(d(x) - \bar{d}(x)) - \frac{1}{3}(u(x) - \bar{u}(x))\right]. \tag{1.5.38}$$

Strangeness and charm of the nucleon are zero, consequently

$$0 = \int_0^1 \mathrm{d}x\,[s(x) - \bar{s}(x)] = \int_0^1 \mathrm{d}x\,[c(x) - \bar{c}(x)]. \tag{1.5.39}$$

In the context of the parton model, those quarks that determine the quantum numbers of the relevant hadron, e.g. u and d for proton and neutron, are called *valence quarks*, the others, which appear only as quark–antiquark pairs, are called *sea quarks*.

From (1.5.37), (1.5.38), and (1.5.39) sum rules for the structure functions can be derived. Subtracting (1.5.37) and (1.5.38) leads to the *Adler sum rule* [Ad66]

$$1 = \int_0^1 \mathrm{d}x \, \left[\bar{d}(x) - d(x) + u(x) - \bar{u}(x)\right] \qquad (1.5.40)$$
$$= \int_0^1 \frac{\mathrm{d}x}{2x} \left[F_2^{W^+W^+\mathrm{N}}(x) - F_2^{W^+W^+\mathrm{P}}(x)\right].$$

This was confirmed from neutrino-scattering data using hydrogen and deuterium targets, where a value of 1.01 ± 0.20 was measured [Al85].

Similarly, a weighted sum of (1.5.37), (1.5.38), and (1.5.39) yields the *Gross–Llewellyn Smith sum rule* [Gr69]. It measures the number of valence quarks in the nucleon, since the parity-violating structure function F_3 is equal to the difference between quark and antiquark densities, which is just the valence quark density,

$$\frac{1}{2}\int_0^1 \mathrm{d}x \, \left[F_3^{W^-W^-\mathrm{P}}(x) + F_3^{W^+W^+\mathrm{P}}(x)\right] = 3. \qquad (1.5.41)$$

This parton-model result gets perturbative and non-perturbative QCD corrections [Br87, Sh79]. These corrections shift the 3 on the right-hand side of (1.5.41) to 2.55 ± 0.03 at $Q^2 = 3\,\mathrm{GeV}^2$. The experimental value of $2.50 \pm 0.018 \pm 0.078$ [Le93] confirms that the proton consists of three valence quarks, and hence the partons carry baryon number 1/3.

Neglecting the sea-quark contributions, the values of the electric charges of the quarks can be derived from the *Gottfried sum rule* [Go67]

$$\int_0^1 \frac{\mathrm{d}x}{x} \left[F_2^{\gamma\gamma\mathrm{P}}(x) - F_2^{\gamma\gamma\mathrm{N}}(x)\right] = Q_\mathrm{u}^2 - Q_\mathrm{d}^2 = \frac{1}{3}. \qquad (1.5.42)$$

The experimental result including values in the region $0.004 < x < 0.8$ at $Q^2 = 4\,\mathrm{GeV}^2$ is smaller, namely 0.2281 ± 0.0065 [Ar95], indicating a flavour asymmetry in the quark-antiquark sea.

Averaging the experimental results over neutrons and protons and using the ratio of electron and neutrino structure functions for large x, where the sea contribution is small, gives

$$\frac{F_2^{\gamma\gamma\mathcal{N}}(x)}{F_2^{WW\mathcal{N}}(x)} = \frac{Q_\mathrm{u}^2 + Q_\mathrm{d}^2}{2} = 0.29 \pm 0.02, \qquad (1.5.43)$$

and the electric charges of the u and d quarks are measured as [Au87]

$$|Q_\mathrm{u}| = 0.64 \pm 0.05, \qquad |Q_\mathrm{d}| = 0.41 \pm 0.07,$$

in agreement with the quark-model values $Q_\mathrm{u} = 2/3, \quad Q_\mathrm{d} = -1/3$.

1.5.3.3 Momentum of the partons

The parton distribution functions $f_{i,\mathcal{N}}(x)$ are normalized to the number of partons of type i in the nucleon. Therefore,

$$\int_0^1 \mathrm{d}x\, x\, [u(x) + \bar{u}(x) + d(x) + \ldots + \bar{c}(x)] = 1 - \varepsilon$$

measures the fraction of the momentum carried by quarks and antiquarks in the nucleon. Assuming isospin invariance, it can be measured, for instance, from the integral

$$\begin{aligned}\frac{9}{5}\int_0^1 \mathrm{d}x \left[F_2^{\gamma\gamma\mathrm{P}}(x) + F_2^{\gamma\gamma\mathrm{N}(x)}\right] \\ = \int_0^1 \mathrm{d}x\, x\, [u(x) + \bar{u}(x) + d(x) + \ldots + \bar{c}(x)] \\ + \frac{3}{5}\int_0^1 \mathrm{d}x\, x\, [(c(x) + \bar{c}(x)) - (s(x) + \bar{s}(x))]\,. \end{aligned} \tag{1.5.44}$$

The extra contributions of strange quarks and charm quarks in the third line are small and can be extracted from experiments and subtracted. In this way one finds [Au87] $\varepsilon = 0.53 \pm 0.02$ (for $Q^2 = 15\,\mathrm{GeV}^2$). This shows that about half the momentum in the nucleon must be carried by other (flavour-neutral) partons, i.e. partons that do not couple to photons, W bosons, and Z bosons. According to QCD these are the gluons, the quanta mediating the interaction of the quarks. This momentum sum rule presented the first hint to the existence of gluons. More direct evidence follows from other experiments.

1.5.3.4 Experimental parton distribution functions

The experiments on deep-inelastic electron–nucleon and (anti)neutrino–nucleon scattering contain enough information to determine the distribution functions of the valence quarks $u(x)$ and $d(x)$, the sea quarks $s(x)$, $c(x)$, $\bar{u}(x)$, $\bar{d}(x)$, $\bar{s}(x)$, and $\bar{c}(x)$, and the gluons from the measured values of the structure functions (cf. Sect. 3.2.4). They are presented in Fig. 3.5.

1.5.4 Universality of the parton model

The naïve parton model can also be used for a phenomenological description of other inclusive hard-scattering processes, such as the e^+e^- annihilation into hadrons and inclusive hadron production in electron–nucleon scattering. Just as in lepton–nucleon scattering, the cross section is calculated from the following ingredients:

a) the *parton cross sections* (for free, point-like particles),

b) the probabilities of detecting the partons i with momentum fraction x in the hadron h, the *parton distribution functions* $f_{i,h}(x)$,

c) the probabilities that a parton i is converted to a hadron h, the *fragmentation functions* $D_{h,i}(x)$.

We give several examples to illustrate the applications of the parton model.

1.5.4.1 Electron–positron annihilation into hadrons

The total cross section for the *electron–positron annihilation into hadrons* is calculated according to the above concepts from the point-like $e^+e^- \to q\bar{q}$ cross section similar to (1.5.23) and from the probability that quarks are converted to hadronic states. According to the confinement hypothesis the latter is equal to one. For energies that are small with respect to M_Z only the one-photon-exchange diagram has to be considered, and the cross section is given by

$$\sigma(e^+e^- \to \gamma^* \to \text{hadrons}) = \frac{4\pi\alpha^2}{3s} \sum_q N_C Q_q^2, \qquad (1.5.45)$$

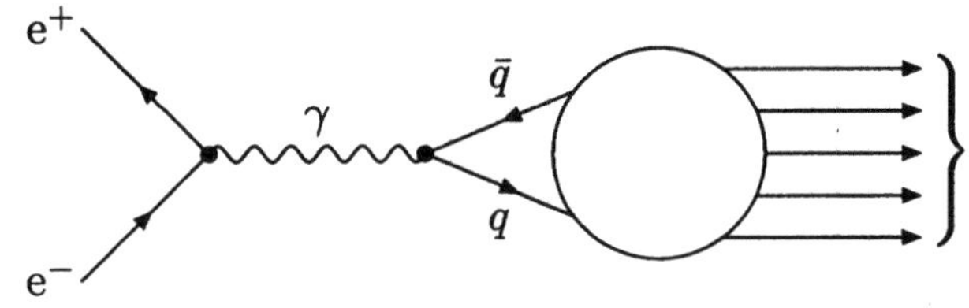

Fig. 1.12 Electron–positron annihilation into hadrons

in complete analogy to (1.2.46). In the quark–parton model, the sum runs over all quarks q with different flavours and N_C colours. The cross section (1.5.45) shows scaling behaviour. The ratio

$$R = \frac{\sigma(e^+e^- \to \gamma^* \to \text{hadrons})}{\sigma(e^+e^- \to \gamma^* \to \mu^+\mu^-)} = \sum_q N_C Q_q^2 \tag{1.5.46}$$

measures the hadron production relative to μ-pair production. For different numbers of quark flavours one obtains

$$\sum_q N_C Q_q^2 = \begin{cases} 2/3 & \text{for} \quad q = \text{u, d, s}, \quad N_C = 1, \\ 2 & \text{for} \quad q = \text{u, d, s}, \quad N_C = 3, \\ 10/3 & \text{for} \quad q = \text{u, d, s, c}, \quad N_C = 3, \\ 11/3 & \text{for} \quad q = \text{u, d, s, c, b}, \quad N_C = 3. \end{cases} \tag{1.5.47}$$

Comparison with experiment shows that the factor $N_C = 3$, supplied by the colour degree of freedom, is needed (cf. Fig. 1.13). This can be regarded as one experimental confirmation of the hypothesis of *coloured quarks.* The rise of R for $E_{CM} \gtrsim 40\,\text{GeV}$ is due to the contribution of the Z boson.

The fact that quarks carrying high energies and momenta do not occur as free particles is closely related with the quark-confinement problem. In deep-inelastic lepton–nucleon scattering, quarks do not appear as free particles but in form of partons with quark quantum numbers. In e^+e^- annihilation, quarks manifest themselves as *hadron jets.* The jet-angular distribution behaves like $(1+\cos^2\theta)$, i.e. like the distribution of elementary spin-1/2 particles. Hence one may conclude that the primary process consists of the production of two high-energy spin-1/2 particles which then hadronize in subsequent stages.

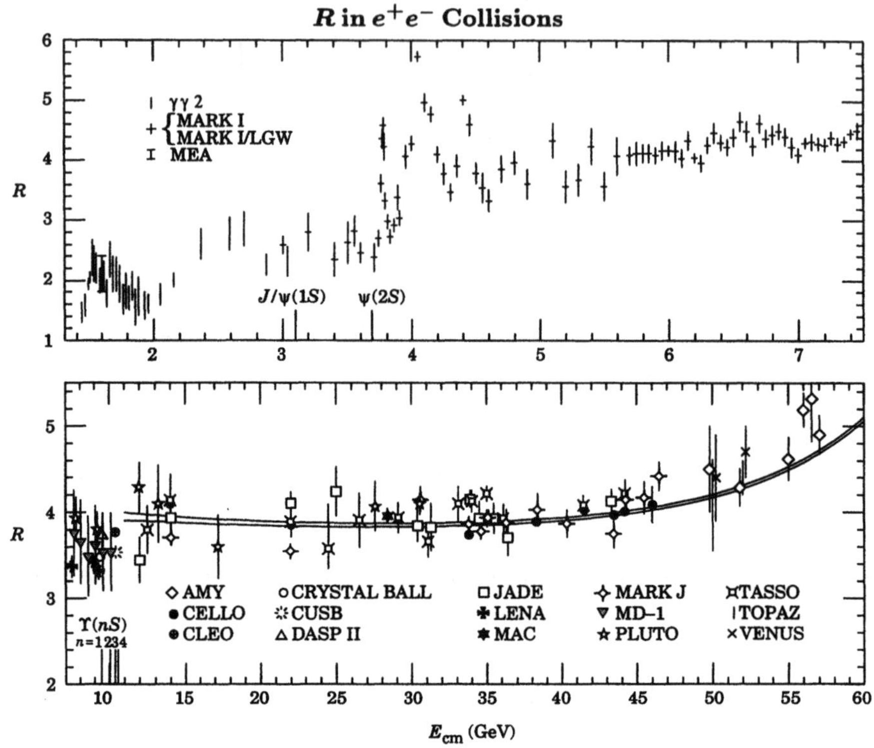

Fig. 1.13 Measurements of R (from Ref. [PDG00])

1.5.4.2 Inclusive hadron production in e^+e^- annihilation

If a specific hadron (e.g. $\pi^+, K^+, K^0, \ldots$) is detected in the final state of inclusive e^+e^- annihilation, the cross section in the parton model is again calculated from the photon-exchange cross section (photon momentum p^γ) for $e^+e^- \to q\bar{q}$ and the *fragmentation function* $D_{h,q}(x)$ [Be71, Gr73, Fi77] of the quark q with $x = 2p_h p_\gamma / p_\gamma^2 = E_h/E_{\text{beam}} = E_h/E_q$ (cf. Fig. 1.14). The fragmentation function $D_{h,q}(x)\mathrm{d}x$ is just the probability that a quark q converts into a hadron h with momentum fraction between x and $x+\mathrm{d}x$. The

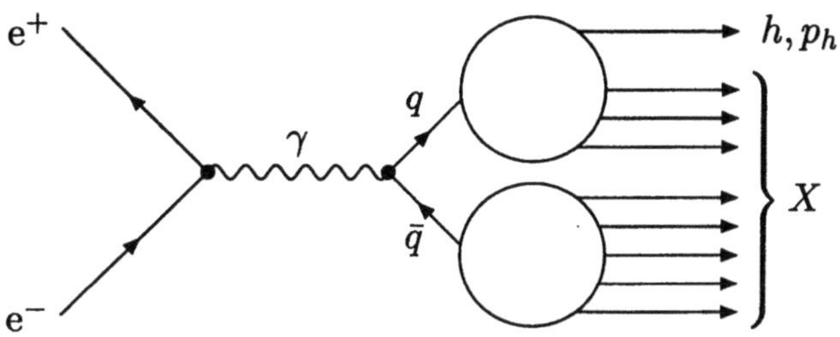

Fig. 1.14 Inclusive hadron production in e^+e^- annihilation

quantity measured is the distribution $\mathrm{d}N^h(x)/\mathrm{d}x$ which is the differential cross section normalized by the total cross section,

$$\frac{\mathrm{d}N^h(x)}{\mathrm{d}x} = \frac{\frac{\mathrm{d}\sigma}{\mathrm{d}x}(\mathrm{e}^+\mathrm{e}^- \to h + X)}{\sigma_{\mathrm{tot}}(\mathrm{e}^+\mathrm{e}^- \to h + X)}. \tag{1.5.48}$$

In the parton model $\mathrm{d}N^h/\mathrm{d}x$ shows scaling behaviour and is calculated from the quark charges and fragmentation functions according to

$$\frac{\mathrm{d}N^h(x)}{\mathrm{d}x} = \frac{\sum_q N_{\mathrm{C}} Q_q^2 D_{h,q}(x)}{\sum_q N_{\mathrm{C}} Q_q^2}, \tag{1.5.49}$$

where q indicates quark and antiquark flavours.

From isospin and charge-conjugation invariance the following relations between fragmentation functions emerge,

$$\begin{aligned} D_{\pi^+,\mathrm{u}} &= D_{\pi^-,\mathrm{d}} = D_{\pi^-,\bar{\mathrm{u}}} = D_{\pi^+,\bar{\mathrm{d}}} \equiv D^+, \\ D_{\mathrm{K}^+,\mathrm{u}} &= D_{\mathrm{K}^0,\mathrm{d}} = D_{\mathrm{K}^-,\bar{\mathrm{u}}} = D_{\bar{\mathrm{K}}^0,\bar{\mathrm{d}}} \equiv K^+, \\ D_{\bar{\mathrm{K}}^0,\mathrm{s}} &= D_{\mathrm{K}^-,\mathrm{s}} = D_{\mathrm{K}^0,\bar{\mathrm{s}}} = D_{\bar{\mathrm{K}}^0,\bar{\mathrm{s}}} \equiv K^+ + \delta. \end{aligned}$$

Moreover, assuming that all sea quarks fragment in a similar way, we obtain

$$\begin{aligned} D_{\pi^+,\mathrm{s}} &= D_{\pi^+,\bar{\mathrm{s}}} = D_{\pi^+,\mathrm{d}} = D_{\pi^+,\bar{\mathrm{u}}} \equiv D^-, \\ D_{\mathrm{K}^+,\bar{\mathrm{u}}} &= D_{\mathrm{K}^+,\mathrm{d}} = D_{\mathrm{K}^+,\bar{\mathrm{d}}} = D_{\mathrm{K}^+,\mathrm{s}} \equiv K^-. \end{aligned}$$

The *fragmentation functions* D^+, D^-, K^+, K^-, and δ can be extracted from the experimental data. The results show that the valence quarks dominate the fragmentation, but at low values of x a significant amount of sea quarks is present.

The parton model has been successfully applied to many other hard-scattering processes, e.g. inclusive $\mu^+\mu^-$ creation in nucleon–nucleon or hadron–hadron scattering with high transverse momentum, jet production, etc. The validity of the simple parton model confirms that—investigated with high resolution—hadrons contain point-like, in good approximation free scattering centres carrying quark-flavour and colour quantum numbers. Moreover, flavour-neutral partons, the gluons, exist inside hadrons. Details of this are discussed in the framework of perturbative QCD in Sect. 3.3.

1.6 Higher-order field-theoretical effects in QED

We based our phenomenological discussion of dynamics in elementary particle physics essentially on the one-particle exchange model (Sect. 1.1.2). This model is a generalization of the simplest approximation of QED. The experimental verification of higher-order effects with great precision further strengthens the rôle of QED as a prototype for theories of the other interactions. The following discussion of the anomalous magnetic moment of the muon in Sect. 1.6.2 illustrates this point.

1.6.1 QED as a quantum field theory

In Sect. 1.2.1 we already got some insight into the physical content of QED as described by the Lagrangian (1.2.16). The most important points are:

- The quanta of the free basic fields ψ and A_μ describe the observed fundamental fermions (electrons, muons, ...) and photons, respectively.
- The non-linear terms in the field equations are responsible for the interaction between these particles and can be described by Feynman graphs representing a perturbation theory with expansion parameter $\alpha = e^2/4\pi$.
- In the simplest case, a Feynman graph describes a reaction by an exchange of particles (tree graph).

Tree graphs describe elementary particle reactions within the framework of a simple, relativistic collision theory. However, a complete field theory contains more subtle effects owing to the mutual interaction of the fields. These are generally called *radiative corrections.* In Feynman's perturbation theory [Fe49] they are represented by *loop diagrams.* In the following we give some examples.

Self-energies: the one-loop *electron self-energy diagram*

describes the modification of the electron–positron field propagation by the interaction between electrons and the electromagnetic field in first-order perturbation theory. This leads among other things to a change in the electron mass which must be compensated by renormalization.

The situation is similar for the modification of the photon propagation by the *vacuum polarization*, i.e. the production and annihilation of a *virtual fermion–antifermion pair*:

Vertex corrections: in first-order perturbation theory, the interaction between photons and electrons is modified by the graphs in Fig. 1.15.

As a consequence of the radiative corrections the electron (muon, or tauon) receives, besides the Dirac coupling $e\gamma_\mu$, a form factor and a Pauli coupling $e\sigma_{\mu\nu}k^\nu/m$ [Sc49]. To $\mathcal{O}(\alpha)$ one finds for the effective photon–lepton vertex at small momentum transfer the expression

$$\mathrm{i}e\left\{\gamma_\mu + \frac{\alpha}{2\pi}\frac{\mathrm{i}\sigma_{\mu\nu}k^\nu}{2m} + \mathcal{O}\left(\frac{k^2}{m^2}\right)\right\}, \tag{1.6.1}$$

where k denotes the momentum of the incoming photon. The second term, calculated first by Schwinger, leads to an *anomalous magnetic moment for the electron.*

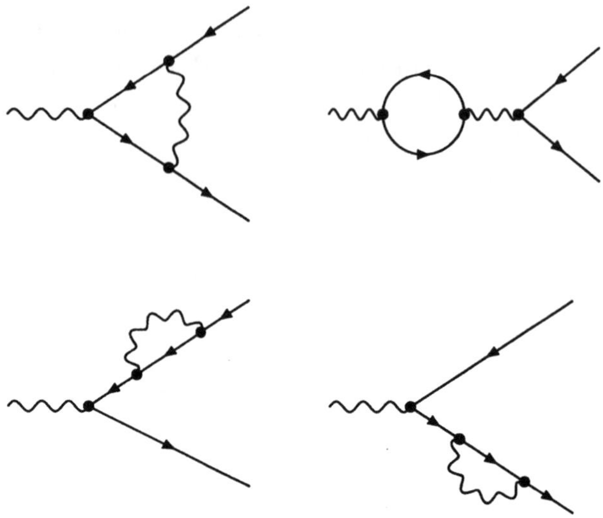

Fig. 1.15 Corrections to the photon–electron vertex

The experimental evidence of radiative corrections is the decisive test for the quantum-field-theoretical nature of a dynamics. The next section shows that this has been achieved for QED with extremely high precision.

1.6.2 A test of QED: the magnetic moment of the muon

We consider one of the precision experiments carried out to test the validity of QED, the measurement of the *magnetic moment of the muon.*

1.6.2.1 Description of the experiment

If the magnetic moment of the muon

$$\boldsymbol{\mu} = \frac{e}{2m} g\,\mathbf{s} \tag{1.6.2}$$

deviates from its Dirac value ($g = 2$), i.e. if it has a Pauli coupling, then its spin $\mathbf{s}$ rotates with the frequency

$$\omega_a = a\frac{eB}{m}, \qquad a = \frac{1}{2}(g-2) \tag{1.6.3}$$

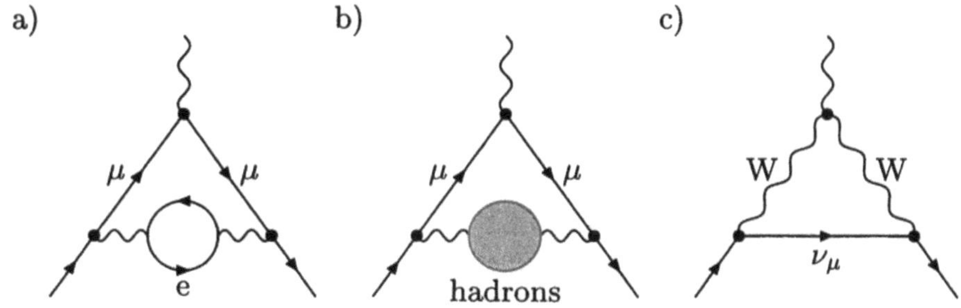

Fig. 1.16 Contributions of electromagnetic (a), hadronic (b), and weak (c) interactions to the muon magnetic moment.

relative to its momentum in a homogeneous magnetic field B. This frequency can be measured in a muon storage ring. The muons produced from the decay processes $\pi \rightarrow \mu + \nu_\mu$ are polarized. The precession of the muon spin with frequency ω_a is measured as a time modulation of the frequency with which the electrons from muon decay are detected in the storage-ring plane [Ba79, Br00].

The experiment gives the value [Br00]

$$a_\mu^{\mathrm{exp}} = 116592050(460) \times 10^{-11} \tag{1.6.4}$$

for the anomalous magnetic moment of the muon.

1.6.2.2 Theoretical calculation

The theoretical calculation up to the fifth order in α and including hadronic and weak corrections yields [Cz00]

$$a_\mu^{\mathrm{th}} = 116591597(67) \times 10^{-11}. \tag{1.6.5}$$

Thus, the theoretical and the experimental result are in agreement within 6 significant digits for a_μ and within 9 significant digits for the g factor.

The importance of this result can be assessed by listing the different contributions to the theoretical result. According to the fact that ultimately everything interacts with everything, one has to take into account the electromagnetic [Ki84, Ki85], the hadronic [Ei95], and the weak [Cz95, Ja72, Ba72]

		contribution to a_μ $(\times 10^{11})$
QED	coefficient of $\left(\frac{\alpha}{\pi}\right)^n$	
$n = 1$	0.5	116140972.9(0.4)
$n = 2$	0.765857376(27)	413217.6
$n = 3$	24.05050898(44)	30142.4
$n = 4$	126.07(41)	366.9(1.2)
$n = 5$	930(170)	6.3(1.1)
QED total		116584705.7(1.7)
hadronic		6739(67)
weak		
one-loop		195
two-loop		−43(4)
gravitation [Be75]		4×10^{-30}
total result		$a_\mu^{\text{th}}(\times 10^{11}) = 116591597(67)$

Tab. 1.8 List of the individual contributions to the anomalous magnetic moment of the muon according to Ref. [Cz00] ($\alpha^{-1} = 137.03599958(52)$ is used)

interaction. Figure 1.16 shows examples of Feynman graphs describing radiative corrections of these interactions. The numerical contributions of the different graphs are listed in Tab. 1.8.

Pure QED supplies the largest part. The contribution of the one-loop order ($n = 1$) is the Pauli term in (1.6.1) with the coefficient $\alpha/2\pi$. One surprising feature is the small contribution of the strong interaction to the g factor of only 60 ppm.

According to Table 1.8, the coefficients of the powers of α/π in the QED perturbation series increase. The increase in the number of higher-order graphs renders the perturbative expansion a divergent, asymptotic series [Hu52, Th53].

1.7 Towards gauge theories of strong and electroweak interactions

Today, the Standard Model of elementary particles describes physics by quarks and leptons interacting through fields of vector particles (Sect. 1.1). In the preceding sections we collected the basic experimental facts which led to the following picture.

- Hadrons consist of constituents, the *quarks*. Because of *confinement*, these do not exist as free particles but as *partons* which in interactions hadronize into *particle jets*.
- The charges of the strong interaction, called *colour*, are related to an $SU(3)_C$ symmetry. The strong interaction is mediated by *gluons*. Like the quarks, the gluons do not exist as free particles but only as partons. The gluons have been verified experimentally as jets.
- The disentanglement of the involved current–current structure of the electroweak interaction revealed a global $SU(2)_W \times U(1)_Y$ symmetry. The associated charges are *weak isospin* and *weak hypercharge*.
- The carriers of the weak interaction, the *weak vector bosons* have been found experimentally with masses of 80 GeV and 91 GeV.
- The discovery of the *top quark*, which was required by the $SU(2)_W$ symmetry, has completed the spectrum of fundamental fermions.
- For the successful precision tests of QED the inclusion of *radiative corrections* into the theoretical predictions is crucial.

Relativistic quantum field theory and, in particular, well-established QED provided guidance for the phenomenological discussion. Symmetry considerations structured the different experimental results. As in QED, currents and vector particles play an important rôle in the description of all interactions. It turns out that these concepts are naturally embedded in quantized gauge theories, Yang–Mills theories, in which interactions are essentially determined by local symmetries. The description of such theories of the strong and electroweak interactions is the main theme of the following chapters.

References to Chapter 1

[Ab83] H. Abramowicz et al., *Z. Phys.* **C17** (1983) 283.

[Ad66] S.L. Adler, *Phys. Rev.* **143** (1966) 1144.

[Ad68] S.L. Adler and R.F. Dashen, *Current Algebras and Applications to Particle Physics*, Frontiers in physics Vol. 30, (W.A. Benjamin, Inc., New York, 1968).

[Ad70] S.L. Adler, *Perturbation Theory Anomalies* in *Lectures on Elementary Particles and Quantum Field Theory*, eds. S. Deser et al. (Cambridge, Massachusetts, 1970) p. 1.

[Af00] T. Affolder et al. (CDF Coll.), *Phys. Rev.* **D61** (2000) 072005.

[Al85] D. Allasia et al., *Z. Phys.* **C28** (1985) 321.

[Ap75] T. Appelquist and H.D. Politzer, *Phys. Rev.* **D12** (1975) 1404 and *Phys. Rev. Lett.* **34** (1975) 43.

[Ar95] M. Arneodo et al. (NMC Coll.), *Phys. Lett.* **B364** (1995) 107.

[Ar83] G. Arnison et al. (UA1 Coll.), *Phys. Lett.* **122B** (1983) 103 and **126B** (1983) 398;
M. Banner et al. (UA2 Coll.), *Phys. Lett.* **122B** (1983) 476;
P. Bagnaia et al. (UA2 Coll.), *Phys. Lett.* **129B** (1983) 130.

[Au87] J.J. Aubert et al. (EMC Coll.), *Nucl. Phys.* **B293** (1987) 740.

[Ba72] W.A. Bardeen, R. Gastmanns and B. Lautrup, *Nucl. Phys.* **B46** (1972) 319.

[Ba79] J. Bailey et al., *Nucl. Phys.* **B150** (1979) 1.

[Be71] S.M. Berman, J.D. Bjorken and J.B. Kogut, *Phys. Rev.* **D4** (1971) 3388.

[Be75] F.A. Berends and R. Gastmans, *Phys. Lett.* **55B** (1975) 311.

[Bj69] J.D. Bjorken, *Phys. Rev.* **179** (1969) 1547.

[Bj69a] J.D. Bjorken and E.A. Paschos, *Phys. Rev.* **185** (1969) 1975.

[Bo80] M. Böhm, *Z. Phys.* **C3** (1980) 321.

[Br79] R. Brandelik et al. (TASSO Coll.), *Phys. Lett.* **86B** (1979) 243; Ch. Berger et al. (PLUTO Coll.), *Phys. Lett.* **86B** (1979) 418; D.P. Barber et al. (MARK-J Coll.), *Phys. Rev. Lett.* **43** (1979) 830; W. Bartel et al. (JADE Coll.), *Phys. Lett.* **91B** (1980) 142.

[Br87] V.M. Braun and A.V. Kolesnichenko, *Nucl. Phys.* **B283** (1987) 723.

[Br00] H.N. Brown et al. (Muon $g-2$ Coll.), *Phys. Rev.* **D62** (2000) 091101.

[Ca63] N. Cabibbo, *Phys. Rev. Lett.* **10** (1963) 531.

[Ca69] C.G. Callan, Jr. and D.J. Gross, *Phys. Rev. Lett.* **22** (1969) 156.

[Cl79] F.E. Close, *An Introduction to Quarks and Partons*, (Academic Press, London, New York, 1979).

[Cz95] A. Czarnecki, B. Krause and W.J. Marciano, *Phys. Rev.* **D52** (1995) R2619 and *Phys. Rev. Lett.* **76** (1996) 3267.

[Cz00] A. Czarnecki and W.J. Marciano, hep-ph/0010194.

[dR75] A. de Rújula, H. Georgi and S.L. Glashow, *Phys. Rev.* **D12** (1975) 147.

[Ei95] S. Eidelman and F. Jegerlehner, *Z. Phys.* **C67** (1995) 585.

[Fe49] R.P. Feynman, *Phys. Rev.* **76** (1949) 769.

[Fe69] R.P. Feynman, *Phys. Rev. Lett.* **23** (1969) 1415.

[Fi77] R.D. Field and R.P. Feynman, *Phys. Rev.* **D15** (1977) 2590.

[Ga74] M.K. Gaillard and B.W. Lee, *Phys. Rev. Lett.* **33** (1974) 108 and *Phys. Rev.* **D10** (1974) 897.

[Ge64] M. Gell-Mann, *Phys. Lett.* **8** (1964) 214.

[Ge72] M. Gell-Mann, *Quarks* in *Elementary Particle Physics, Multiparticle Aspects,* Acta Physica Austriaca, Suppl. IX, ed. P. Urban (Wien, New York, 1972) p. 733.

[Gl70] S.L. Glashow, J. Iliopoulos and L. Maiani, *Phys. Rev.* **D2** (1970) 1285.

[Go67] K. Gottfried, *Phys. Rev. Lett.* **18** (1967) 1174.

[Gr69] D.J. Gross and C.H. Llewellyn Smith, *Nucl. Phys.* **B14** (1969) 337.

[Gr73] M. Gronau, F. Ravndal and Y. Zarmi, *Nucl. Phys.* **B51** (1973) 611.

[Gr76] D. Gromes and I.O. Stamatescu, *Nucl. Phys.* **B112** (1976) 213.

[Hu52] C. Hurst, *Proc. Camb. Soc.* **48** (1952) 625.

[Ja72] R. Jackiw and S. Weinberg, *Phys. Rev.* **D5** (1972) 2396.

[Ja79] C. Jarlskog, *Gauge Theories* in *New Phenomena in Lepton Hadron Physics*, eds. D.E.C. Fries and J. Wess (New York, London, 1979) p. 1.

[Ja85] C. Jarlskog, *Phys. Rev. Lett.* **55** (1985) 1039 and *Z. Phys.* **C29** (1985) 491.

[Ki84] T. Kinoshita, B. Nižić and Y. Okamoto, *Phys. Rev. Lett.* **52** (1984) 717.

[Ki85] T. Kinoshita and V.W. Hughes, *Comments Nucl. Part. Phys.* **14** (1985) 341.

[Ko73] M. Kobayshi and K. Maskawa, *Prog. Theor. Phys.* **49** (1973) 652.

[Kr79] M. Krammer and H. Krasemann, *Quarkonia* in *Quarks and Leptons,* Acta Physica Austriaca, Suppl. XXI, ed. P. Urban (Wien, New York, 1979) p. 259.

[LEP92] The LEP collaborations, *Phys. Lett.* **B276** (1992) 247.

[Le93] W.C. Leung et al., *Phys. Lett.* **B317** (1993) 655.

[Li78] D.B. Lichtenberg, *Unitary Symmetry and Elementary Particles*, (Academic Press, New York, 1978).

[PDG00] D.E. Groom et al. (Particle Data Group), *Eur. Phys. J.* **C15** (2000) 1.

[Sa68] A. Salam, *Weak and Electromagnetic Interactions* in *Elementary Particle Theory*, ed. N. Svartholm (Stockholm, 1968) p. 367.

[Sc49] J. Schwinger, *Phys. Rev.* **76** (1949) 790.

[Sh79] M.A. Shifman, A.I. Vainshtein and V.I.Zakharov, *Nucl. Phys.* **B147** (1979) 385, 448, and 519.

[Ta75] R.E. Taylor, *Inelastic Electron–Nucleon Scattering Experiments at SLAC* in *Proceedings 1977 International Symposium on Lepton and Photon Interactions at High Energies*, ed. W.T. Kirk (Stanford, 1975) p. 679.

[Th53] W. Thirring, *Helv. Phys. Acta* **26** (1953) 33.

[tH85] G. 't Hooft, *Nucl. Phys.* **B254** (1985) 11.

[We29] H. Weyl, *Z. Phys.* **56** (1929) 330.

[We67] S. Weinberg, *Phys. Rev. Lett.* **19** (1967) 1264.

[We95] S. Weinberg, *The Quantum Theory of Fields, Vol. 1 and 2*, (Cambridge University Press, Cambridge, 1995).

[Wi39] E. Wigner, *Ann. Math.* **40** (1939) 51.

[Zw64] G. Zweig, CERN-TH 401 and 412 (1964) (unpublished).

2 Quantum theory of Yang–Mills fields

The phenomenological discussion of Chapt. 1 suggests that the theory of the strong, the electromagnetic, and the weak interactions takes the form of a quantized Yang–Mills theory. Therefore, the structure and properties of relativistic quantum field theories in general and Yang–Mills theories in particular are investigated in this chapter.

In Sect. 1.2, we have treated the quantum aspect of fields with help of the wave–particle dualism for linear field equations and with the recipe-like use of Feynman rules (cf. Sect. 1.2.1). Although many applications can be discussed on this basis, this procedure is insufficient for the general treatment of gauge theories. In the following, we need the general connection between fields and physical particles. In Sect. 2.1 we show how the expectation values of field operators, in particular Green functions, describe particles and their interactions in field theory. This is of particular importance for QCD since quarks and gluons do not appear as free particles.

One possibility to describe the full content of a gauge theory, especially the consequences of gauge invariance, is the path-integral representation (Sect. 2.2) of quantum field theory. It provides a closed expression for all Green functions of the theory directly in terms of the Lagrangian. This formulation is the starting point for a deductive treatment of both QCD and the theory of the electroweak interaction.

The principle of local gauge invariance and the construction of Yang–Mills theories are discussed in Sect. 2.3. Section 2.4 is devoted to the quantization of gauge theories in the path-integral representation. Perturbative evaluation of the path integral results in the familiar Feynman rules for calculating S-matrix elements. The path-integral representation also allows for a simple derivation of the Ward identities resulting from gauge invariance.

In order to evaluate quantum field theories in higher orders, renormalization has to be performed. The principles and the technical details of renormalization are discussed in Sect. 2.5 and applied to QED and non-abelian gauge theories at one-loop order (Sect. 2.5.3). In Sect. 2.5.2 the methods for the calculation of higher-order corrections are presented. In the case of gauge theories, the Ward identities play the decisive rôle to proof that the renormalized theory gives a gauge-invariant, unitary S matrix in the space of the physical state vectors. This is discussed in Sect. 2.5.4. The relations between different renormalization prescriptions are related by the renormalization group (Sect. 2.6).

The renormalizability can be spoiled by anomalies. The origin and the physical consequences of anomalies are discussed in Sect. 2.7.

Besides the ultraviolet divergences, quantum field theories contain also infrared divergences. Their origin, and the methods to extract and to treat these divergences are explained in Sect. 2.8.

Besides perturbative techniques, the Feynman path integral allows to apply also non-perturbative techniques to derive results for physical observables (Sect. 2.9). We consider the classical solutions of the field equations and discuss the evaluation of the path integral in the semi-classical approximation. A different non-perturbative treatment of quantum field theories consists in the lattice approximation which is discussed in Sect. 2.10. This allows, in particular, to approach the problem of confinement (Sect. 3.7) and other non-perturbative phenomena.

2.1 Green functions and S-matrix elements

This section extends the field-theoretical vocabulary which was summarized in its simplest form in Sect. 1.2.1. We show how in a general field theory the S matrix for relativistic particle reactions can be obtained from quantum-mechanical expectation values of field operators. At the same time, some general properties of the S matrix are discussed. Our presentation is predominantly descriptive; for more detailed treatments we refer to the literature [FT].

2.1.1 The principles of quantum field theory

The main concepts in quantum mechanics are states and observables. In field theory we met already the vacuum state $|0\rangle$, the free one-particle states $|M, p, j, j_3\rangle$, and those of many free particles (cf. Sect. 1.2.1.2). The latter play an important rôle in theories with interaction for the description of incoming and outgoing particle configurations.

Quantum-mechanical operators occurred in the form of quantized basic fields, e.g. free fields in Sect. 1.2.1.3, or as physical observables, e.g. currents and charges composed of them as in (1.2.79) and (1.2.81).

The *basic physical principles* can be formulated with help of these concepts in a rather general way. They are summarized in the *Wightman axioms* [Wi56]:

1. The principle of special relativity for closed systems requires a representation of the Poincaré group (the group of translations by a vector a and *Lorentz transformations* Λ) by unitary operators $U(a)$ and $U(\Lambda)$ that leaves the *vacuum state* invariant,
$$U(a)|0\rangle = |0\rangle, \qquad U(\Lambda)|0\rangle = |0\rangle, \tag{2.1.1}$$
and transforms the particle states according to (1.2.2) and (1.2.3). Accordingly, *field operators* $\psi_\alpha(x)$ must transform covariantly,
$$\begin{aligned} U(a)\psi_\alpha(x)U^{-1}(a) &= \psi_\alpha(x+a), \\ U(\Lambda)\psi_\alpha(x)U^{-1}(\Lambda) &= S^{-1}_{\alpha\alpha'}(\Lambda)\psi_{\alpha'}(\Lambda x). \end{aligned} \tag{2.1.2}$$
Here $S^{-1}_{\alpha\alpha'}(\Lambda)$ is the representation matrix of an irreducible representation of the homogeneous Lorentz group spanned by the ψ_α [Ca77]. The translation operators $U(a) = \mathrm{e}^{\mathrm{i}P^\mu a_\mu}$ and the Lorentz-transformation operators $U(\Lambda) = \exp(-\mathrm{i}M^{\mu\nu}\beta_{\mu\nu})$ are generated infinitesimally by the four-momentum operator P^μ and the generalized angular-momentum operator $M^{\mu\nu}$. The antisymmetric tensor $\beta_{\mu\nu}$ is composed of the rotation and boost parameters.

2. The eigenvalues of the energy and mass-square operator must be non-negative,
$$P^0 \geq 0, \qquad P_\mu P^\mu \geq 0. \tag{2.1.3}$$
The physical vacuum $|0\rangle$ is the only state with the lowest energy $P^0|0\rangle = 0$ and is *non-degenerate.* All other states have positive energy $P^0 > 0$ and mass squared $P^2 > 0$.

3. *Causality* is implemented by the requirement of locality. Observables must be quantum-mechanically independent at space-like distances, i.e. they have to commute. Consequently, the field operators of bosons (fermions) have to commute (anticommute) for space-like distances (observables must be even in anticommuting fermionic operators),
$$[\psi_\alpha(x), \psi_{\alpha'}(x')]_\mp = 0 \quad \text{for} \quad (x-x')^2 < 0. \tag{2.1.4}$$

4. General *charge-conservation* laws follow from the existence of charge operators Q. These generate infinitesimal symmetry transformations of the fields according to (1.2.89) and (1.2.90) and annihilate the vacuum,

$$Q|0\rangle = 0. \tag{2.1.5}$$

These Wightman axioms represent the general framework of relativistic quantum field theory.

We note that the Wightman axioms 2 and 4 are not fulfilled for realistic theories, because these contain massless particles ($P^2 = 0$) and degenerate ground states. Massless gauge bosons are typical for gauge theories. The construction of gauge theories is discussed in Sects. 2.3 and 2.4.

2.1.2 Green functions

The physical content of a quantum field theory is contained in the quantum-mechanical vacuum expectation values (vevs) of products of field operators. Consider a multiplet of fields $\psi_\alpha(x)$ that transforms according to an irreducible representation of the Lorentz group as in (2.1.2). The vacuum expectation values of products of field operators are called *Wightman functions* [Wi56],

$$\begin{aligned} W_{\alpha_1\ldots\alpha_n}(x_1,\ldots,x_n) &= \langle 0|\psi_{\alpha_1}(x_1)\cdots\psi_{\alpha_n}(x_n)|0\rangle \\ &= \langle \psi_{\alpha_1}(x_1)\cdots\psi_{\alpha_n}(x_n)\rangle. \end{aligned} \tag{2.1.6}$$

Relativistic invariance, i.e. (2.1.1) and (2.1.2), implies that the vevs are Lorentz-invariant functions of the coordinate differences

$$\begin{aligned} &\langle \psi_{\alpha_1}(x_1+a)\cdots\psi_{\alpha_n}(x_n+a)\rangle \\ &= \langle 0|U^{-1}(a)U(a)\psi_{\alpha_1}(x_1)U^{-1}(a)\cdots U(a)\psi_{\alpha_n}(x_n)U^{-1}(a)U(a)|0\rangle \\ &= \langle \psi_{\alpha_1}(x_1)\cdots\psi_{\alpha_n}(x_n)\rangle, \end{aligned} \tag{2.1.7}$$

and similarly

$$\begin{aligned} &\langle \psi_{\alpha_1}(\Lambda x_1)\cdots\psi_{\alpha_n}(\Lambda x_n)\rangle = S_{\alpha_1\alpha_1'}(\Lambda)\cdots S_{\alpha_n\alpha_n'}(\Lambda) \\ &\quad\langle 0|U^{-1}(\Lambda)U(\Lambda)\psi_{\alpha_1'}(x_1)U^{-1}(\Lambda)\cdots U(\Lambda)\psi_{\alpha_n'}(x_n)U^{-1}(\Lambda)U(\Lambda)|0\rangle \\ &= S_{\alpha_1\alpha_1'}(\Lambda)\cdots S_{\alpha_n\alpha_n'}(\Lambda)\langle \psi_{\alpha_1'}(x_1)\cdots\psi_{\alpha_n'}(x_n)\rangle. \end{aligned} \tag{2.1.8}$$

This argument illustrates how properties of Wightman functions are derived from the general principles. The analyticity properties which follow from (2.1.3) and (2.1.4) are discussed in the literature [FT].

Besides the Wightman functions another type of vevs, the *Green functions*, plays an important rôle in field theory. Green functions are the vevs of time-ordered products of field operators [cf. (1.2.22)]

$$G_{\alpha_1\ldots\alpha_n}(x_1,\ldots,x_n) = \langle T\psi_{\alpha_1}(x_1)\cdots\psi_{\alpha_n}(x_n)\rangle. \tag{2.1.9}$$

Just like the Wightman functions these are relativistically invariant functions of the coordinate differences and (anti)symmetric in the coordinates because of the (anti)symmetry of the T product for (fermions) bosons.

2.1.2.1 Green functions of free fields

The representation of the free fields by creation and annihilation operators allows for a simple calculation of the vevs. Starting from the representation of scalar hermitian (neutral) fields $\psi(x)$ by creation and annihilation operators (1.2.10),

$$\psi(x) = \frac{1}{(2\pi)^3}\int \frac{\mathrm{d}^3p}{2E_p}\Big[a(p)\mathrm{e}^{-\mathrm{i}px} + a^\dagger(p)\mathrm{e}^{+\mathrm{i}px}\Big], \tag{2.1.10}$$

and using $a(p)|0\rangle = 0$, $\langle 0|a^\dagger(p) = 0$, and $[a(p), a^\dagger(p')] = 2E_p(2\pi)^3\delta(\mathbf{p}-\mathbf{p}')$ gives

$$\begin{aligned} W(x_1,x_2) &= \langle\psi(x_1)\psi(x_2)\rangle \\ &= \frac{1}{(2\pi)^6}\int \frac{\mathrm{d}^3p}{2E_p}\frac{\mathrm{d}^3p'}{2E_{p'}}\,\langle a(p)a^\dagger(p')\rangle\mathrm{e}^{-\mathrm{i}px_1}\mathrm{e}^{\mathrm{i}p'x_2} \\ &= \frac{1}{(2\pi)^3}\int \frac{\mathrm{d}^3p}{2E_p}\,\mathrm{e}^{-\mathrm{i}p(x_1-x_2)} \\ &= \frac{1}{(2\pi)^3}\int \mathrm{d}^4p\,\delta(p^2-M^2)\theta(p^0)\mathrm{e}^{-\mathrm{i}p(x_1-x_2)} = \Delta_+(x_1-x_2,M). \end{aligned} \tag{2.1.11}$$

This and (1.2.31) imply

$$\begin{aligned} G(x_1,x_2) &= \langle T\psi(x_1)\psi(x_2)\rangle \\ &= \theta(x_1^0-x_2^0)\langle\psi(x_1)\psi(x_2)\rangle + \theta(x_2^0-x_1^0)\langle\psi(x_2)\psi(x_1)\rangle \end{aligned} \tag{2.1.12}$$

for the two-point Green function. Using the Fourier transform of the θ-function

$$\theta(x) = \lim_{\varepsilon \to 0+} \frac{\mathrm{i}}{2\pi} \int_{-\infty}^{+\infty} \mathrm{d}y \, \frac{\mathrm{e}^{-\mathrm{i}xy}}{y + \mathrm{i}\varepsilon}, \tag{2.1.13}$$

this gives

$$\begin{aligned} G(x_1, x_2) &= \frac{\mathrm{i}}{(2\pi)^4} \int \mathrm{d}^4 p \, \frac{1}{p^2 - M^2 + \mathrm{i}\varepsilon} \, \mathrm{e}^{-\mathrm{i}p(x_1 - x_2)} \\ &= \mathrm{i}\Delta_{\mathrm{F}}(x_1 - x_2, M). \end{aligned} \tag{2.1.14}$$

Thus, the two-point Green function of the free field is i times the *Feynman propagator* Δ_{F}.

In coordinate space, the Feynman propagator can be written in terms of the modified Bessel function of first order K_1 [Ab70]

$$\Delta_{\mathrm{F}}(x) = -\frac{\delta(x^2)}{4\pi} - \mathrm{i}\,\mathrm{sgn}(x^2) \frac{M^2}{4\pi^2} \frac{K_1(M\sqrt{-x^2 + \mathrm{i}\varepsilon})}{M\sqrt{-x^2 + \mathrm{i}\varepsilon}}. \tag{2.1.15}$$

For large distances, $M\sqrt{|x^2|} \gg 1$, this behaves as

$$\Delta_{\mathrm{F}}(x) \underset{M\sqrt{|x|^2} \gg 1}{\sim} -\mathrm{i}\,\mathrm{sgn}(x^2) \frac{1}{4\pi^2 x^2} \frac{\sqrt{\pi M \sqrt{-x^2 + \mathrm{i}\varepsilon}}}{\sqrt{2}} \, \mathrm{e}^{-M\sqrt{-x^2 + \mathrm{i}\varepsilon}}, \tag{2.1.16}$$

i.e. for large space-like distances it drops exponentially. The behaviour close to the *light cone*, $M\sqrt{|x^2|} \ll 1$, is

$$\begin{aligned} \Delta_{\mathrm{F}}(x) = &-\frac{\delta(x^2)}{4\pi} + \mathrm{i}\,\mathrm{sgn}(x^2) \frac{1}{4\pi^2} \frac{1}{x^2 + \mathrm{i}\varepsilon} \\ &- \mathrm{i}\,\mathrm{sgn}(x^2) \frac{M^2}{8\pi^2} \left(\log \frac{M\sqrt{-x^2 + \mathrm{i}\varepsilon}}{2} + \frac{1}{2} \right) \\ &+ \mathcal{O}\left(\sqrt{|x^2|} \log |x^2| \right). \end{aligned} \tag{2.1.17}$$

The four-point Green function for the free scalar field can be calculated in the same way,

$$G(x_1,\dots,x_4) = \mathrm{i}^2\Big(\Delta_\mathrm{F}(x_1-x_2)\Delta_\mathrm{F}(x_3-x_4) + \Delta_\mathrm{F}(x_1-x_3)\Delta_\mathrm{F}(x_2-x_4) + \Delta_\mathrm{F}(x_1-x_4)\Delta_\mathrm{F}(x_2-x_3)\Big). \tag{2.1.18}$$

Since $\Delta_\mathrm{F}(x-y) = \Delta_\mathrm{F}(y-x)$, the four-point Green function is in fact symmetric in the four space–time points x_i. The general expression for the symmetric n-point Green function of the free scalar field vanishes for n odd and is given by a sum of products of two-point functions for n even,

$$\begin{aligned} G(x_1,\dots,x_n) &= \langle T\psi(x_1)\dots\psi(x_n)\rangle \\ &= \mathrm{i}^n \sum_{\text{part.}} \Delta_\mathrm{F}(x_{i_1}-x_{i_2})\Delta_\mathrm{F}(x_{i_3}-x_{i_4})\cdots\Delta_\mathrm{F}(x_{i_{n-1}}-x_{i_n}), \end{aligned} \tag{2.1.19}$$

where the sum extends over all non-equivalent partitions of the x_i. Similarly, the Green functions of the free Dirac field are sums of products of the *Dirac propagator* $\langle T\psi_\alpha(x)\overline{\psi}_\beta(y)\rangle = \mathrm{i}S_\mathrm{F}(x-y)_{\alpha\beta} = (\mathrm{i}\partial\!\!\!/ + M)_{\alpha\beta}\mathrm{i}\Delta_\mathrm{F}(x-y)$.

In view of the importance of spin-1 particles in gauge theories, the two-point function of massive *vector bosons* is considered explicitly. Inserting the field operator

$$A_\mu(x) = \frac{1}{(2\pi)^3}\sum_{j_3=0,\pm1}\int\frac{\mathrm{d}^3p}{2E_p}\Big[\varepsilon_\mu(p,j_3)a_{j_3}(p)\mathrm{e}^{-\mathrm{i}px} + \text{h.c.}\Big], \tag{2.1.20}$$

where $\varepsilon_\mu(p,j_3)$ is the polarization vector defined in App. A.1.1.6, into the vacuum expectation value leads to

$$\begin{aligned} W_{\mu\nu}(x_1,x_2) &= \langle A_\mu(x_1)A_\nu(x_2)\rangle = \frac{1}{(2\pi)^6}\sum_{j_3,j_3'}\int\frac{\mathrm{d}^3p}{2E_p}\frac{\mathrm{d}^3p'}{2E_{p'}} \\ &\quad\times\langle M,p,1,j_3|M,p',1,j_3'\rangle\varepsilon_\mu(p,j_3)\varepsilon^*_\nu(p',j_3')\mathrm{e}^{-\mathrm{i}p(x_1-x_2)} \\ &= \frac{1}{(2\pi)^3}\int\frac{\mathrm{d}^3p}{2E_p}\frac{\mathrm{d}^3p'}{2E_{p'}}2E_{p'}\delta(\mathbf{p}-\mathbf{p}')\sum_{j_3}\varepsilon_\mu(p,j_3)\varepsilon^*_\nu(p,j_3)\mathrm{e}^{-\mathrm{i}p(x_1-x_2)} \\ &= \frac{1}{(2\pi)^3}\int\frac{\mathrm{d}^3p}{2E_p}\Big(-g_{\mu\nu}+\frac{p_\mu p_\nu}{M^2}\Big)\mathrm{e}^{-\mathrm{i}p(x_1-x_2)} \\ &= \Big(-g_{\mu\nu}-\frac{\partial_\mu\partial_\nu}{M^2}\Big)\Delta_+(x_1-x_2,M), \end{aligned} \tag{2.1.21}$$

where we have used the polarization sum (A.1.45). By time ordering in (2.1.21), the Green function of a massive vector field

$$\begin{aligned} G_{\mu\nu}(x_1, x_2) &= \frac{\mathrm{i}}{(2\pi)^4} \int \mathrm{d}^4 p \, \frac{-g_{\mu\nu} + p_\mu p_\nu / M^2}{p^2 - M^2 + \mathrm{i}\varepsilon} \mathrm{e}^{-\mathrm{i}p(x_1 - x_2)} \\ &= \mathrm{i} \left(-g_{\mu\nu} - \frac{\partial_\mu \partial_\nu}{M^2} \right) \Delta_\mathrm{F}(x_1 - x_2, M) \end{aligned} \tag{2.1.22}$$

is obtained.

For $M^2 \to 0$, (2.1.22) gives a divergent result. In order to derive the vector-boson propagator in the case of zero mass, gauge invariance is crucial. As shown in Sect. 2.4, the Fourier transform of a massless vector-boson propagator in gauge theories is given by $G_{\mu\nu}(p) = -\mathrm{i}g_{\mu\nu}/(p^2 + \mathrm{i}\varepsilon)$ in the *Feynman gauge*.

Then, the Wightman function of the free massless vector field reads

$$\langle A_\mu(x_1) A_\nu(x_2) \rangle = -g_{\mu\nu} \frac{1}{(2\pi)^3} \int \frac{\mathrm{d}^3 p}{2|\mathbf{p}|} \mathrm{e}^{-\mathrm{i}p(x_1 - x_2)}. \tag{2.1.23}$$

Now consider a state created from the vacuum by application of $A_0(x)$

$$|\Psi\rangle = \int \mathrm{d}^4 x \, f(x) A_0(x) |0\rangle. \tag{2.1.24}$$

Then, with the Fourier representation of $f(x)$

$$\tilde{f}(\mathbf{p}) = \frac{1}{(2\pi)^{3/2}} \int \mathrm{d}^4 x \, \mathrm{e}^{\mathrm{i}(|\mathbf{p}|x^0 - \mathbf{p}\mathbf{x})} f(x), \tag{2.1.25}$$

the norm of this state is

$$\begin{aligned} \|\Psi\|^2 &= \int \mathrm{d}^4 x_1 \int \mathrm{d}^4 x_2 \, \langle A_0(x_1) A_0(x_2) \rangle f^*(x_1) f(x_2) \\ &= -\int \frac{\mathrm{d}^3 p}{|2\mathbf{p}|} \tilde{f}^*(\mathbf{p}) \tilde{f}(\mathbf{p}) < 0, \end{aligned} \tag{2.1.26}$$

i.e. the vector-boson propagator of the form used above in QED implies the appearance of unphysical states with negative norm. This is not the case for a propagator of the form (2.1.22). The possible occurrence of negative-norm states is a frequent phenomenon in the relativistically covariant treatment of quantized gauge-field theories. Care must be taken to prevent these contributions from having any physical significance and violating the unitarity of the S matrix (cf. Sects. 2.5.4.3 and 2.5.4.5).

2.1.2.2 Feynman rules for Green functions

The relation between the Feynman propagator and the two-point Green function of free fields (2.1.12) suggests the extension of the *Feynman rules* in such a way that they allow for the calculation of Green functions. To this end, graphs with *field points* $\overset{x}{\bullet}$, which correspond to the arguments of the Green functions, are introduced [tH73]. The analytical expression

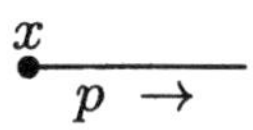

$$= \int \frac{d^4p}{(2\pi)^4}\, e^{ipx} \tag{2.1.27}$$

is assigned to a field point with momentum p flowing out of the field point. Lines that connect field points with other field points or with a vertex of the graph are internal lines. Otherwise, the Feynman rules as already described in Sect. 1.2.1.8 apply. Note, however, that the Feynman rules for Green functions involve in addition an explicit δ-function for global four-momentum conservation, $(2\pi)^4\delta^{(4)}(\sum_i p_i)$. This δ-function had been split off in the Feynman rules for the invariant matrix element as given in Sect. 1.2.1.8 [cf. (1.2.23)].

Accordingly, the two-point Green function (2.1.12) can be represented graphically as

$$\langle T\psi(x_1)\psi(x_2)\rangle =$$

$$\begin{aligned} &= \int \frac{d^4p_1}{(2\pi)^4}\, e^{ip_1x_1} \int \frac{d^4p_2}{(2\pi)^4}\, e^{ip_2x_2} \frac{i}{p_1^2 - M^2 + i\varepsilon} (2\pi)^4\delta^{(4)}(p_1 + p_2) \\ &= \frac{i}{(2\pi)^4} \int d^4p\, e^{ip(x_1-x_2)} \frac{1}{p^2 - M^2 + i\varepsilon} = i\Delta_F(x_1 - x_2, M). \end{aligned} \tag{2.1.28}$$

The four-point function can be depicted as

$$\langle T\psi(x_1)\cdots\psi(x_4)\rangle =$$

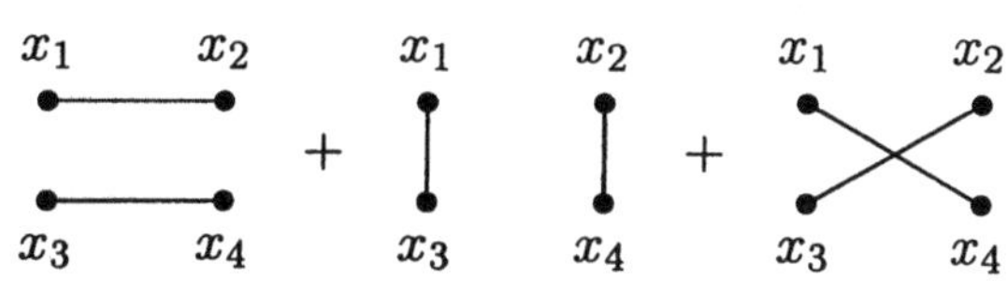

(2.1.29)

The general Green function of free fields (2.1.19) can be represented accordingly.

It is essential that the extension of the Feynman rules to include field points allows for the calculation of Green functions in theories with interaction. As an example, we consider the field theory for charged scalar particles ψ with mass M interacting by the exchange of neutral scalars ϕ with mass μ. The interaction term of the Lagrangian $\mathcal{L}_{\mathrm{I}} = -g\psi^*(x)\psi(x)\phi(x)$ corresponds to the vertex in momentum space

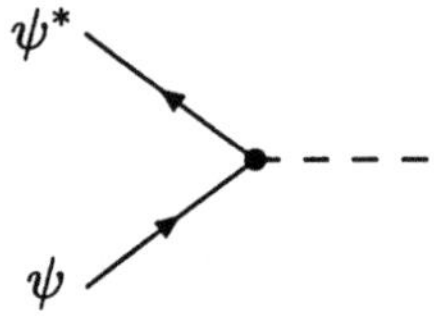

$$= -\mathrm{i}g. \qquad (2.1.30)$$

Therewith, the connected part of the lowest-order four-point Green function is calculated as (the momenta p_1, p_3 are incoming, p_2, p_4 are outgoing)

$$\langle T\psi(x_1)\psi^*(x_2)\psi^*(x_3)\psi(x_4)\rangle\Big|_{\text{conn.}} =$$

=

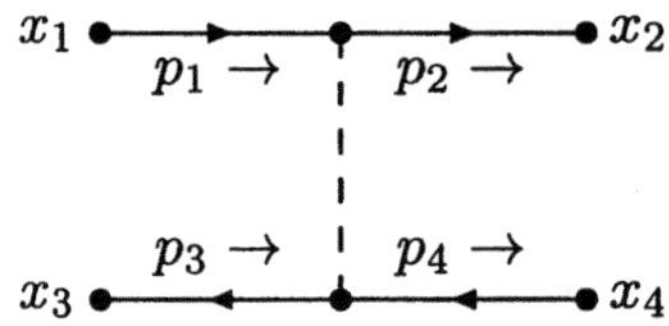

\+

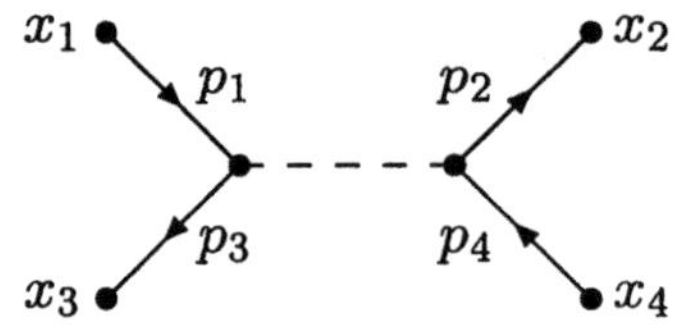

$$= -\mathrm{i}g^2 \frac{1}{(2\pi)^{12}} \int \mathrm{d}^4p_1\, \mathrm{d}^4p_2\, \mathrm{d}^4p_3\, \mathrm{d}^4p_4\, \delta^{(4)}(p_1 + p_3 - p_2 - p_4)$$
$$\times\, \mathrm{e}^{-\mathrm{i}(p_1x_1 - p_2x_2 + p_3x_3 - p_4x_4)} \frac{1}{(p_1^2 - M^2)(p_3^2 - M^2)}$$
$$\times \left(\frac{1}{(p_1 - p_2)^2 - \mu^2} + \frac{1}{(p_1 + p_3)^2 - \mu^2} \right) \frac{1}{(p_2^2 - M^2)(p_4^2 - M^2)}. \qquad (2.1.31)$$

The extended Feynman rules yield Green functions that have all the properties following from the general principles of quantum field theory outlined in Sect. 2.1.1. The ultimate justification for these rules is that they form a systematic approximation of a closed representation of field theory (cf. Sect. 2.2).

2.1.3 S-matrix elements and the LSZ formula

As already mentioned, Green functions determine S-matrix elements. This can be seen by comparing the Feynman rules for S-matrix elements and Green functions. The S-matrix elements are obtained in the following way: first, take the Fourier transform $\widetilde{G}(p_1, \ldots, p_n)$ of the corresponding Green function $G(x_1, \ldots, x_n)$,

$$\begin{aligned}\widetilde{G}(p_1, \ldots, p_n) &= \int \mathrm{d}^4x_1 \cdots \mathrm{d}^4x_n \, \mathrm{e}^{-\mathrm{i}(p_1x_1+\ldots+p_nx_n)} G(x_1, \ldots, x_n) \\ &= (2\pi)^4 \delta^{(4)}(p_1 + \ldots + p_n) G(p_1, \ldots, p_n). \qquad (2.1.32)\end{aligned}$$

Then, remove the poles of those internal lines that end in field points by multiplying by $-\mathrm{i}(p_i^2 - M_i^2)$. This operation is called *truncation.* The resulting expressions at $p_i^2 = M_i^2$ and $p_i^0 > 0$ for $i = 1, \ldots, s$ and $-p_i^0 > 0$ for $i = s+1, \ldots, n$ ("on-shell") give the *S-matrix elements* for s incoming particles with the momenta p_i, $i = 1, \ldots, s$ and $n-s$ outgoing particles with the momenta $p_i' = -p_i$, $i = s+1, \ldots, n$,

$$\begin{aligned}&\langle -p_{s+1}, \ldots, -p_n | S | p_1, \ldots, p_s \rangle \qquad (2.1.33) \\ &= R^{-n/2}(-\mathrm{i})^n (p_1^2 - M_1^2) \cdots (p_n^2 - M_n^2) \tilde{G}(p_1, \ldots, p_n) \Big|_{p_i^2 = M_i^2}.\end{aligned}$$

This is the *reduction formula for scalar fields,* also called *LSZ formula.* It expresses the S-matrix element as the multiple residue of the Green function with respect to the particle poles corresponding to the external lines. For particles with spin, the corresponding wave functions have to be attached. An extended version of (2.1.33) incorporating the description of bound states is given in Sect. 2.1.5.

The *wave-function renormalization constants* $R^{-1/2}$, also called *LSZ factors,* are required for a correct normalization of the S-matrix elements. They are determined from the one-particle parts of the field operators or equivalently by the residues of the propagators at the physical mass

$$R = |\langle M, p | \psi(x) | 0 \rangle|^2 = -\mathrm{i}(p^2 - M^2) G(p, -p)|_{p^2 = M^2}. \qquad (2.1.34)$$

Replacing the factors $p_i^2 - M_i^2$ in (2.1.33) using (2.1.34), results in

$$\begin{aligned}
&\langle -p_{s+1}, \dots, -p_n | S | p_1, \dots, p_s \rangle \\
&\quad = R^{n/2} G^{-1}(p_1, -p_1) \cdots G^{-1}(p_n, -p_n) \tilde{G}(p_1, \dots, p_n) \Big|_{p_i^2 = M_i^2} \\
&\quad = R^{n/2} \tilde{G}_{\text{trunc}}(p_1, \dots, p_n) \Big|_{p_i^2 = M_i^2} .
\end{aligned} \tag{2.1.35}$$

Thus, the S-matrix elements are given by the *truncated Green functions* [Zi59, Sy60, Sy67],

$$\tilde{G}_{\text{trunc}}(p_1, \dots, p_n) = G^{-1}(p_1, -p_1) \cdots G^{-1}(p_n, -p_n) \tilde{G}(p_1, \dots, p_n), \tag{2.1.36}$$

taken on-shell. Conversely, the truncated Green functions represent S-matrix elements which are continued to off-shell momenta p_i. A particle or an external line is called *on shell* if its momentum squared equals its mass ($p^2 = m^2$) and *off shell* otherwise.

The truncation can also be performed in configuration space. Here, a Green function is truncated in a field point (indicated by underlining the space coordinate),

$$G(\underline{x}_1, x_2, \dots, x_n) = \int \mathrm{d}^4 z \, G^{-1}(x_1, z) G(z, x_2, \dots, x_n), \tag{2.1.37}$$

by convolution with the inverse two-point function $G^{-1}(x_1, x_2)$ defined by

$$\int \mathrm{d}^4 z \, G(x_1, z) G^{-1}(z, x_2) = \delta^{(4)}(x_1 - x_2). \tag{2.1.38}$$

Truncated Green functions can be represented perturbatively by *truncated Feynman graphs.* These are obtained from the usual Feynman graphs by removing all parts that are connected to a field point and can be separated from the other field points by cutting a single internal line. This is illustrated in Fig. 2.1, where the graph is truncated at the field point x_1 by removing the subgraph C. In the following we visualize truncated graphs by omitting the field points at the external legs.

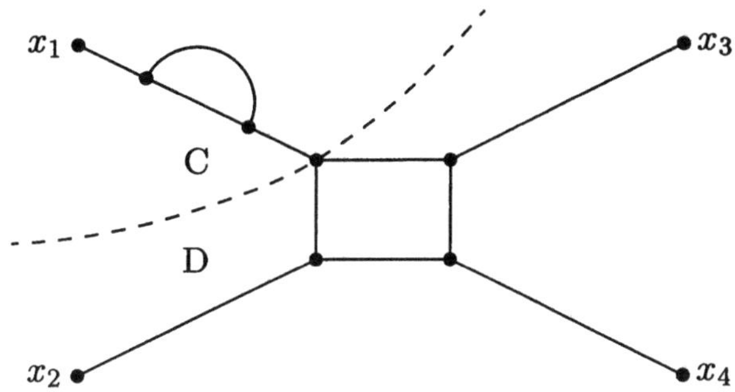

Fig. 2.1 Graphical representation of the truncation procedure

2.1.4 Connected Green functions and vertex functions

The representation of Green functions and S-matrix elements by graphs describes also some structural properties. The vacuum and one-particle structure is physically of great importance beyond perturbation theory and leads to the concepts of connected Green functions and vertex functions.

2.1.4.1 Connected Green functions

The graphs for a Green function consist of *connected subgraphs.* A graph is said to be connected if any two of its points are connected to each other via internal lines. In a free theory only the graph of the two-point function (2.1.28) is connected owing to the lack of interaction vertices.

The contributions of disconnected graphs to a Green function factorize according to the Feynman rules. Trivial examples are the Green functions of free fields (2.1.18) and (2.1.19). As another example consider the four-point function with interaction. The graphs which contribute[1] can be classified as

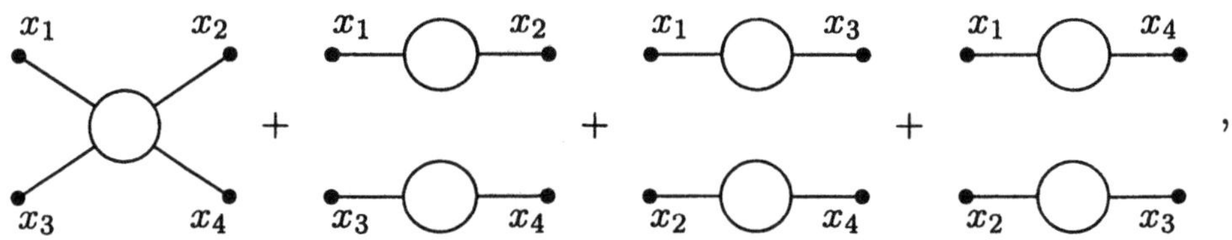

[1]We omit graphs containing vacuum subgraphs and tadpoles, which do not contribute to the S matrix.

where the open circles represent the sum of all connected graphs with two or four field points, i.e. the connected two- and four-point functions. If we denote the corresponding analytic expressions by $G_c(x_1, x_2)$ and $G_c(x_1, x_2, x_3, x_4)$, the decomposition of the general two- and four-point functions into connected parts reads (assuming that the one-point function vanishes, i.e. $G_c(x) = 0$)

$$\begin{aligned} G(x_1, x_2) &= G_c(x_1, x_2), \\ G(x_1, x_2, x_3, x_4) &= G_c(x_1, x_2, x_3, x_4) + G_c(x_1, x_2)G_c(x_3, x_4) \\ &\quad + G_c(x_1, x_3)G_c(x_2, x_4) + G_c(x_1, x_4)G_c(x_2, x_3). \end{aligned} \tag{2.1.39}$$

In general, if there is a unique vacuum [Ha92] we get the following decomposition of n-point Green functions into connected parts, the *cluster decomposition*

$$\begin{aligned} G(x_1, \ldots, x_n) &= G_c(x_1, \ldots, x_n) \\ &+ \sum G_c(x_{i_1}, \ldots, x_{i_\mu})G_c(x_{i_{\mu+1}}, \ldots, x_{i_n}) \\ &+ \sum G_c(x_{i_1}, \ldots, x_{i_\mu})G_c(x_{i_{\mu+1}}, \ldots, x_{i_\nu})G_c(x_{i_{\nu+1}}, \ldots, x_{i_n}) \\ &+ \ldots . \end{aligned} \tag{2.1.40}$$

The sums extend over all partitions of the indices $1, \ldots, n$. This equation can be inverted iteratively to give the general definition of *connected Green functions*

$$\begin{aligned} G_c(x_1, \ldots, x_n) &= G(x_1, \ldots, x_n) - 1! \sum G(\ldots)G(\ldots) \\ &\quad + 2! \sum G(\ldots)G(\ldots)G(\ldots) \mp \ldots . \end{aligned} \tag{2.1.41}$$

The decomposition of the Green function $G(x_1, \ldots, x_n)$ into connected parts, called *vacuum structure of Green functions* [Wa53, Sy54, Zi59], describes an important physical point. Independently of the representation by Feynman diagrams, the following properties of Green functions can be derived in the framework of general field theory [Jo65]

$$\begin{aligned} &\langle T\psi(x_1) \cdots \psi(x_n)\psi(x_{n+1} + a) \cdots \psi(x_{n+m} + a)\rangle \\ &\underset{|a|\to\infty}{\sim} \langle T\psi(x_1) \cdots \psi(x_n)\rangle\langle T\psi(x_{n+1} + a) \cdots \psi(x_{n+m} + a)\rangle, \end{aligned} \tag{2.1.42}$$

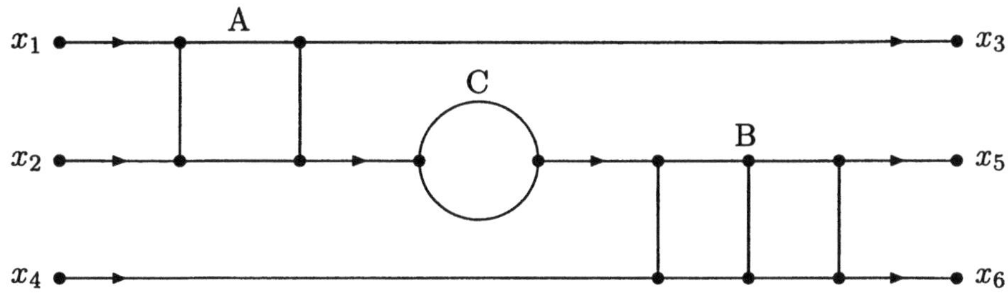

Fig. 2.2 Example of a one-particle reducible Feynman graph

or

$$\lim_{|a|\to\infty} G_c(x_1, \ldots, x_n, x_{n+1} + a, \ldots, x_{n+m} + a) = 0, \tag{2.1.43}$$

i.e. processes that take place far away from each other are quantum-mechanically independent. According to (2.1.33), the part of the S matrix derived from the connected Green functions, G_c, describes connected processes.

Two processes can be connected simply by the fact that an outgoing particle from one reaction is an incoming particle in the next reaction. The possibility of these multiple reactions is reflected by the structure of graphs and Green functions, too. Figure 2.2, in which the graph describes the processes A, B, and C, gives an example.

2.1.4.2 Vertex functions

A graph is called *one-particle reducible* if it decomposes into two parts by cutting one internal line. In graphs for Green functions, the field point x_i is separated from the rest of the graph by one internal line. This type of reducibility is treated by truncation (cf. Sect. 2.1.3).

If a connected, completely truncated graph is *one-particle irreducible*, it is called a *proper vertex graph*. The subgraphs A, B and C in Fig. 2.2 are of this type. Inclusion of the propagators attached to the two-point vertex C gives a propagator graph $\overline{\mathrm{C}}$. If $F_\mathrm{A}(x_1, \ldots, x_\nu, z)$, $F_\mathrm{B}(z, x_{\nu+1}, \ldots, x_n)$, and $F_{\overline{\mathrm{C}}}(z, z')$ are the analytical expressions for the graphs, then according to the

Feynman rules the analytical expression for the connected truncated graph is

$$F_{\mathrm{G}}(x_1,\ldots,x_6) = \int \mathrm{d}^4z\,\mathrm{d}^4z'\,F_{\mathrm{A}}(x_1,x_2,x_3,z)\,F_{\overline{\mathrm{C}}}(z,z')F_{\mathrm{B}}(z',x_4,x_5,x_6). \quad (2.1.44)$$

As in this example, also general, one-particle reducible, truncated graphs consist of several proper vertex graphs that are connected by propagator graphs.

This structure of Feynman graphs can be transferred to Green functions [Zi59, Sy60]. The sum of all contributions of proper vertex graphs gives the n-point *vertex function* Γ symbolized by

$$\mathrm{i}\Gamma(x_1,\ldots,x_n) =$$

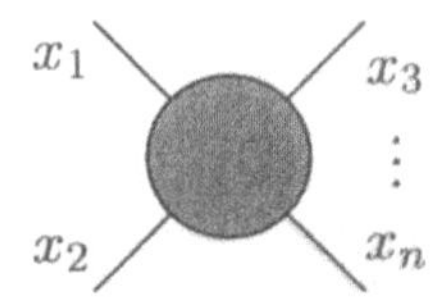

$$. \quad (2.1.45)$$

The connected Green functions are made up of vertex functions and propagators. This can be illustrated by means of *structural graphs.* The three-point Green function can be obtained by reversing the truncation,

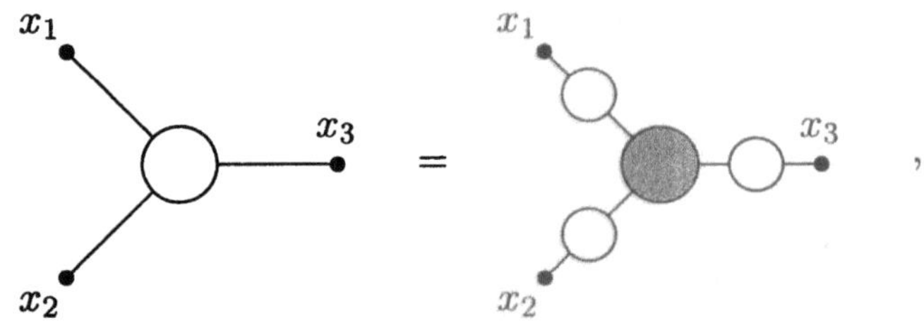

$$G_{\mathrm{c}}(x_1,x_2,x_3) = \int \mathrm{d}^4z_1\mathrm{d}^4z_2\mathrm{d}^4z_3\,G_{\mathrm{c}}(x_1,z_1)G_{\mathrm{c}}(x_2,z_2)G_{\mathrm{c}}(x_3,z_3)\mathrm{i}\Gamma(z_1,z_2,z_3). \quad (2.1.46)$$

With the help of the combinatorics of Feynman graphs, the relation of Green functions with n field points to those with $(n-1)$ field points is the following: the additional field point x_n is connected via a propagator to either an m-vertex $(n-1 \geq m \geq 3)$ or to a propagator (via a three-vertex) of a structural graph. This procedure has to be carried out in all possible ways on the different $(n-1)$ structural graphs. In the case of the connected four-point function this leads to the result

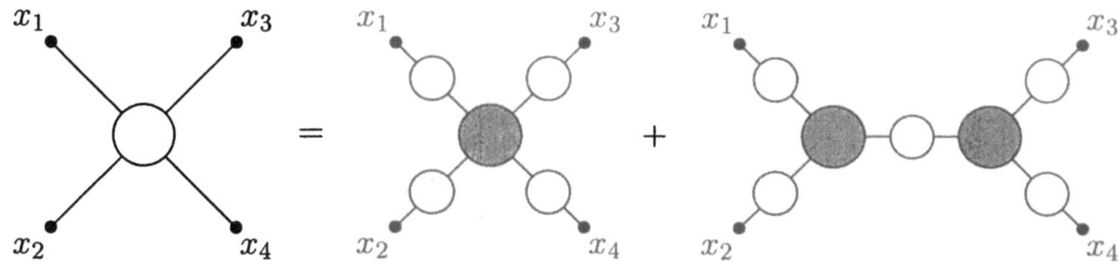

$$+ (x_3 \leftrightarrow x_1) + (x_3 \leftrightarrow x_2).$$

This recursive construction can be inverted to a recursive definition of the vertex functions $\Gamma(x_1, \ldots, x_n)$ from the connected Green functions $G_c(x_1, \ldots x_n)$ (cf. Sect. 2.2).

The representation of Green functions by general propagators and vertex functions describes a general collision process as made up of individual processes connected by propagating particles. The individual processes are described by the vertex functions, and this is the reason for their great physical importance. Important physical parameters, such as effective coupling constants, can be expressed in terms of the vertex functions as described later. Because of energy-momentum conservation, the particle propagation between individual processes might be only virtual. In any case, the mass pole of the Fourier transform of the two-point function determines the long-range component $\propto \exp(-Mx)$ of the particle propagation [cf. (2.1.16)].

2.1.5 Scattering of composite particles

Poles in the invariant-mass variables of Green functions point to the existence of free particles and the corresponding one-particle structure of their reactions. This was illustrated with help of a simple example of a field theory in which the field quanta describe directly the free particles.

However, even particles that are bound states and their reactions can be described by Green functions. In order to formulate this, the field-theoretical amplitudes for one-particle states composed of field quanta, the *Bethe–Salpeter (BS) amplitudes* χ, are introduced. For a $\psi^*\psi$ bound state $|M, k, j, j_3\rangle$, i.e. a one-particle state with mass M, four-momentum k, and spin j, j_3, these are defined by

$$\chi_k(p_1, p_2)\delta^{(4)}(p_1 + p_2 - k) = \frac{1}{(2\pi)^4}\int \mathrm{d}^4x_1\mathrm{d}^4x_2\,\langle 0|T\psi(x_1)\psi^*(x_2)|M, k, j, j_3\rangle\,\mathrm{e}^{\mathrm{i}(p_1x_1+p_2x_2)}. \tag{2.1.47}$$

The BS amplitude χ has a certain relationship to the Schrödinger wave function of a bound state in non-relativistic quantum mechanics. Like the Schrödinger equation, a field-theoretical BS equation determines both mass and amplitude of the bound state. Therefore, the BS equation provides a relativistic generalization of the phenomenological quark model for mesons and baryons. Further details on the BS equation should be gleaned from the literature [Be51, Ge51, Sy60, Bo73].

The general relation between S matrix and Green functions can finally be given by two theorems of general field theory.

1. The Fourier transformed Green functions $\tilde{G}(p_1, \ldots, p_n)$ have *poles* when a partial sum of the momenta $k = p_{i_1} + \ldots + p_{i_\nu}$ lies on the mass shell of a particle, e.g. $p_i^2 = M^2$, $(p_1 + p_3)^2 = \mu^2$, or $(p_1 - p_2)^2 = \mu^2$ in (2.1.31). Thus, in field theory the mass spectrum is found from the position of the poles in the Green functions.

2. If the momenta of a Green function are divided up into r partial sums k_l,

$$k_l = p_{i_1} + \ldots + p_{i_{\nu_l}}, \qquad \sum_{l=1}^{r} k_l = 0,$$

and if the k_l lie on mass shells, then the corresponding residue of the Green function factorizes into the S-matrix element and BS amplitudes

$$(R_1 \cdots R_r)^{-1/2}(-\mathrm{i})^r(k_1^2 - M_1^2)\cdots(k_r^2 - M_r^2)\tilde{G}(p_1, \ldots, p_n)\Big|_{k_l^2=M_l^2} = \bar{\chi}_{k_{s+1}}\cdots\bar{\chi}_{k_r}\langle -k_{s+1}, \ldots, -k_r|S|k_1, \ldots, k_s\rangle\chi_{k_1}\cdots\chi_{k_s}, \tag{2.1.48}$$

where R_i are wave-function renormalization constants for the BS amplitudes, which are defined in analogy to (2.1.34). This is the *general LSZ formula* [Le55, Zi59] of field theoretical scattering theory. It provides the general relation between Green functions and S-matrix elements. It is a generalization of (2.1.33). Notice that the BS amplitude for one field is

$$\chi_k(p_1)\delta^{(4)}(p_1 - k) = \frac{1}{(2\pi)^4}\int \mathrm{d}^4x \langle 0|\psi(x)|k\rangle \mathrm{e}^{\mathrm{i}p_1 x}, \tag{2.1.49}$$

i.e. $\chi_k(p_1) = u(k)$.

On the left-hand side of (2.1.48), $\tilde{G}(p_1, \ldots, p_n)$ is truncated by the free inverse propagator $-\mathrm{i}(p^2 - M^2)$. According to theorem 1, the full propagator $G(p, -p)$ has the same position of the pole. So we can introduce the full inverse propagator and truncated Green functions in the same way as in Sect. 2.1.3.

We illustrate the fundamental significance of the general formula (2.1,48) for the quark model of hadrons by the example of $\pi + \pi \to \pi + \pi$ scattering. As pions are described in the quark model as $q\bar{q}$ bound states, the pion BS amplitude in configuration space takes the form

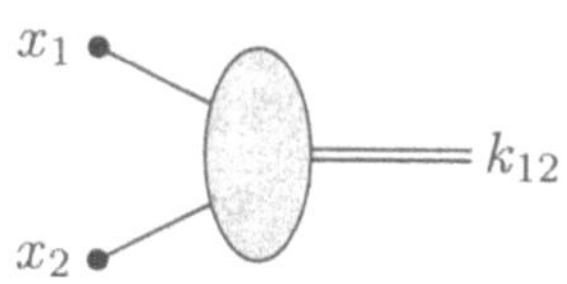

$$= \langle 0|T\psi_{q_1}(x_1)\bar{\psi}_{q_2}(x_2)|k_{12}\rangle. \tag{2.1.50}$$

The S-matrix element is contained in the pole structure of the eight-point quark function, e.g. according to (2.1.48) as a factor of the residue of the simultaneous poles $[(k_{12}^2 - M_\pi^2)(k_{34}^2 - M_\pi^2)(k_{56}^2 - M_\pi^2)(k_{78}^2 - M_\pi^2)]^{-1}$, $k_{12} = p_1 + p_2, \ldots$, etc. This can be represented symbolically by the graph

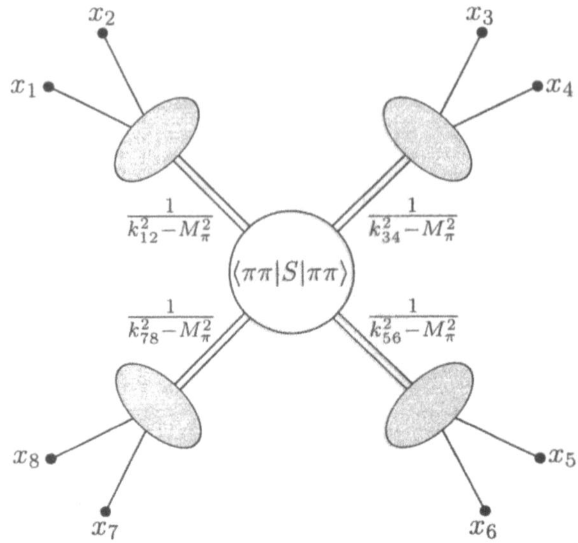

2.2 Path-integral representation of quantum field theory

The set of Green functions describes the physical content of a field theory. In the same way as in combinatorics, where sequences of numbers are combined into generating functions, in field theory it is useful to collect the information contained in the set of Green functions in a *generating functional.* The *Feynman path integral* provides a closed expression for this generating functional. Quite generally it is the starting point for the studies of the formal and structural properties of quantized field theories and for approximative methods which allow for the evaluation of physical processes.

2.2.1 Functional calculus

Summing up the Green functions $G(x_1, \ldots, x_n)$ of a field theory with powers of a source function $J(x)$,

$$T\{J\} = \sum_{n=0}^{\infty} \frac{\mathrm{i}^n}{n!} \int \mathrm{d}^4x_1 \cdots \mathrm{d}^4x_n \, G(x_1, \ldots, x_n) J(x_1) \cdots J(x_n) \quad (2.2.1)$$

defines the *generating functional*[2] $T\{J\}$ from which—as we will see—the Green functions can be reconstructed by functional differentiation.

In order to profit from this approach, several rules for the manipulation of functionals are needed. Therefore in the following, concepts from the analysis of functions of several variables, such as partial differentiation, power series expansion, and integration, are extended to functionals, i.e. to functions of an innumerable infinite number of variables [Be66, Fr72]. A systematic treatment of these methods is beyond the scope of this book; we only give a pragmatic orientation and refer to Ref. [Ro94] for more details. Another approach to functional calculus is by discretization, e.g. by the lattice approximation (Sect. 2.10).

2.2.1.1 Functional differentiation

If we restrict our analysis of functions of several variables $\mathbf{y} = (y_1, \ldots, y_n)$ to polynomials and power series, then the partial differentiation operator $\partial/\partial y_i$ can be formally defined by

$$\frac{\partial}{\partial y_i} 1 = 0, \qquad \frac{\partial}{\partial y_i} y_k = \delta_{ik}, \tag{2.2.2}$$

and the product rule

$$\frac{\partial}{\partial y_i}\Big(F(\mathbf{y})G(\mathbf{y})\Big) = \left(\frac{\partial F(\mathbf{y})}{\partial y_i}\right) G(\mathbf{y}) + F(\mathbf{y}) \left(\frac{\partial G(\mathbf{y})}{\partial y_i}\right). \tag{2.2.3}$$

Analogously, the *functional differentiation* $\delta/\delta y(x)$ is a linear operation with the rules

$$\begin{aligned} &\frac{\delta}{\delta y(x)} 1 = 0, \qquad \frac{\delta}{\delta y(x)} y(x') = \delta(x - x'), \\ &\frac{\delta}{\delta y(x)}\Big(F\{y\}G\{y\}\Big) = \left(\frac{\delta F\{y\}}{\delta y(x)}\right) G\{y\} + F\{y\} \left(\frac{\delta G\{y\}}{\delta y(x)}\right). \end{aligned} \tag{2.2.4}$$

[2] We enclose the arguments of functionals in curly brackets.

2.2.1.2 Volterra series

The generalization of the power-series expansion of a function $F(\mathbf{y})$ of k variables y_i,

$$F(\mathbf{y}) = \sum_{n=0}^{\infty} \sum_{i_1=1}^{k} \cdots \sum_{i_n=1}^{k} \frac{1}{n!} T(i_1, \ldots, i_n)\, y_{i_1} \cdots y_{i_n}, \tag{2.2.5}$$

leads via the formal transition to innumerable infinitely (continuously) many variables, $i \to x$, $y_i \to y(x)$, $\sum_i \to \int \mathrm{d}x$, to the representation of a functional $F\{y\}$ by the *Volterra series*

$$F\{y\} = \sum_{n=0}^{\infty} \int \mathrm{d}x_1 \cdots \mathrm{d}x_n \frac{1}{n!} T(x_1, \ldots, x_n)\, y(x_1) \cdots y(x_n). \tag{2.2.6}$$

The symmetric coefficients $T(x_1, \ldots, x_n)$ of the Volterra series (2.2.6) can be obtained as functional derivatives at the origin,

$$T(x_1, \ldots, x_n) = \frac{\delta^n}{\delta y(x_1) \cdots \delta y(x_n)} F\{y\}\Big|_{y\equiv 0}. \tag{2.2.7}$$

Using this formula it can be easily seen that the results (2.1.19) for the Green functions $G(x_1, \ldots, x_n)$ of the free scalar field can be reproduced from the functional

$$T_{\text{free}}\{J\} = \exp\left(-\frac{\mathrm{i}}{2} \int \mathrm{d}^4x \int \mathrm{d}^4y\, J(x) \Delta_{\mathrm{F}}(x-y) J(y)\right). \tag{2.2.8}$$

The *generating functional of free Green functions* is determined by the *Feynman propagator.*

2.2.1.3 Functional integration

The methods for *functional integration*, as used in the following, are based on the gaussian integral

$$\int_{-\infty}^{+\infty} \frac{\mathrm{d}y}{\sqrt{\pi}} \exp(-ay^2) = \frac{1}{\sqrt{a}}. \tag{2.2.9}$$

If $(\mathbf{y}, A\mathbf{y}) = \sum_{i,j} y_i A_{ij} y_j$ is a positive-definite, quadratic form in k variables, then

$$\int_{-\infty}^{+\infty} \cdots \int_{-\infty}^{+\infty} \prod_{i=1}^{k} \frac{\mathrm{d}y_i}{\sqrt{\pi}} \,\mathrm{e}^{-(\mathbf{y},A\mathbf{y})} = (\det A)^{-1/2}, \tag{2.2.10}$$

as can be proven by diagonalization of A by an orthogonal transformation under which the volume element $\prod \mathrm{d}y_i$ is invariant. This formula can be generalized to the space of functions $y(x)$

$$(y, Ay) = \int \mathrm{d}x\mathrm{d}x'\, y(x)A(x,x')y(x'),$$

$$\int \mathcal{D}[y]\, \mathrm{e}^{-(y,Ay)} = (\det A)^{-1/2}, \tag{2.2.11}$$

where the volume element $\mathcal{D}[y]$ is an infinite product of ordinary volume elements $\mathrm{d}y(x)$ at all space–time points, $\mathcal{D}[y] = \prod_x \mathrm{d}y(x)/\sqrt{\pi}$. Similarly, the determinant $\det A$ is defined as the (infinite) product of the eigenvalues of the operator $A(x, x')$. The formula

$$\det A = \exp[\mathrm{Tr}(\log A)] \tag{2.2.12}$$

often allows for an approximate calculation of $\det A$. For the precise mathematical definition of the functional integral in quantum field theory we refer to Ref. [Ro94].

Just like in the case of conventional integrals, expressions of the type $\int \mathcal{D}[y] F\{y\} \exp(-(y, Ay))$ with a power series $F\{y\}$ can be evaluated by algebraic manipulations of the gaussian integral. This is extensively used in the next section. Thereby, the (infinite-dimensional) differential $\mathcal{D}[y]$ is regarded as translationally invariant

$$\int \mathcal{D}[y] F\{y - y'\} \mathrm{e}^{-((y-y'),A(y-y'))} = \int \mathcal{D}[y] F\{y\} \mathrm{e}^{-(y,Ay)}. \tag{2.2.13}$$

In general, the substitution rule is valid with an infinite-dimensional *Jacobi determinant*

$$\int \mathcal{D}[y] \det \left(\frac{\delta z\{y\}}{\delta y(x)} \right) F\{z\{y\}\} \, \mathrm{e}^{-(z\{y\}, A\, z\{y\})} = \int \mathcal{D}[z] \, F\{z\} \, \mathrm{e}^{-(z, A\, z)}. \tag{2.2.14}$$

For a linear transformation $z\{y\}(x) = \int \mathrm{d}x' z(x,x') y(x')$, the Jacobi determinant is the determinant of the linear operator $z(x,x')$

$$\det \frac{\delta z\{y\}(x)}{\delta y(x')} = \det(z(x,x')). \tag{2.2.15}$$

2.2.1.4 Calculus with functionals of Grassmann variables

In order to treat anticommuting Fermi fields, the functional methods must be extended to the analysis of functionals of *Grassmann-algebra*-valued functions [Be66]. Again we present only the formal algebraic aspects for the corresponding definitions and relationships.

The starting point is the introduction of $2n$ *generators of a Grassmann algebra*, i.e. anticommuting objects y_i, $i = 1, \ldots, n$, and their conjugates y_i^*, $[(y_i^*)^* = y_i]$

$$y_i y_j + y_j y_i = y_i^* y_j + y_j y_i^* = y_i^* y_j^* + y_j^* y_i^* = 0. \tag{2.2.16}$$

This implies, in particular, $y_i^2 = y_i^{*2} = 0$.

The elements of the Grassmann algebra are the formal polynomials in y_i and y_j^* with coefficients $T(l_1, \ldots, l_r | k_1, \ldots, k_s)$ that are antisymmetric in k_i and l_i,

$$F(y^*, y) = \sum_{r,s=1}^{n} \sum_{l_i,k_j=1}^{n} y_{k_s} \cdots y_{k_1} y_{l_r}^* \cdots y_{l_1}^* T(l_1, \ldots, l_r | k_1, \ldots, k_s). \tag{2.2.17}$$

Owing to their anticommutativity a product of more than n Grassmann variables y_i vanishes.

Differentiation with respect to Grassmann variables is a linear algebraic operation defined by

$$\frac{\partial}{\partial y_k} 1 = 0, \qquad \frac{\partial}{\partial y_k} y_i = \delta_{ik}, \qquad \frac{\partial}{\partial y_k^*} y_i = 0, \tag{2.2.18}$$

and the product rule

$$\frac{\partial}{\partial y_i}\Big(F_1 F_2\Big) = \left(\frac{\partial}{\partial y_i} F_1\right) F_2 \pm F_1 \left(\frac{\partial}{\partial y_i} F_2\right), \tag{2.2.19}$$

with $\pm$ if F_1 is even/odd in the number of anticommuting variables. These rules actually define the *left derivative* which we use throughout this book. Similarly, integration is defined as a linear algebraic operation by

$$\int \mathrm{d}y_i = 0, \qquad \int \mathrm{d}y_i \, y_j = \delta_{ij}, \qquad \text{etc.} \tag{2.2.20}$$

The differentials obey the anticommutation relations

$$[\mathrm{d}y_i, \mathrm{d}y_k]_+ = [\mathrm{d}y_i, \mathrm{d}y_k^*]_+ = [\mathrm{d}y_i, y_k]_+ = [\mathrm{d}y_i, y_k^*]_+ = 0. \tag{2.2.21}$$

The conjugate quantities y_i^* are treated in the same way as y_i in all these calculations.

The above definitions can be used to derive the usual rules of calculus. However, one has to take care of the various signs resulting from the anticommutativity, and of the fact that $y_i^2 = 0$. For example, for $z_i = \sum_k A_{ik} y_k$ the substitution rule of integration reads

$$\int \mathrm{d}z_n \cdots \mathrm{d}z_1 \, F(z) = (\det A)^{-1} \int \mathrm{d}y_n \cdots \mathrm{d}y_1 \, F(z(y)). \tag{2.2.22}$$

In contrast to the ordinary integration this rule corresponds to a transformation rule for the differentials $\mathrm{d}z_i = \sum_k (A^{-1})_{ki} \mathrm{d}y_k$ with the inverse matrix A^{-1} instead of the matrix A. The substitution rule (2.2.22) can be derived by expanding F into a power series in the y_i and using the anticommutation relations together with the basic formulas (2.2.20).

A formula similar to the gaussian integral (2.2.9) plays an important rôle in field theory, namely the *Fubini formula*

$$\int \mathrm{d}y_1 \cdots \mathrm{d}y_n \mathrm{d}y_n^* \cdots \mathrm{d}y_1^* \exp\left(\sum_{i,k} y_i^* A_{ik} y_k\right) = \det A. \tag{2.2.23}$$

It follows directly by applying the substitution rule (2.2.22) to the integration with respect to the y_i.

The formulas of the analysis of a finite number of variables (2.2.16)–(2.2.23) for the definition of power-series expansion, differentiation, and integration are extended to Grassmann functionals:

1. Generating elements of Grassmann algebra $y(x)$, $y^*(x')$:

$$[y(x_1), y(x_2)]_+ = [y^*(x_1), y(x_2)]_+ = \ldots = 0. \tag{2.2.24}$$

2. Power series:

$$\begin{aligned} T\{y^*, y\} = \sum_{r,s} \int \mathrm{d}x_r' \cdots \mathrm{d}x_1' \mathrm{d}x_1 \cdots \mathrm{d}x_s\, y(x_s) \cdots y(x_1) \\ \times y^*(x_r') \cdots y^*(x_1')\, T_{r,s}(x_1', \ldots, x_r' | x_1, \ldots, x_s) \end{aligned} \tag{2.2.25}$$

with $T(x_1', \ldots, x_r' | x_1, \ldots, x_s)$ antisymmetric in x_i' and x_k.

3. Differentiation:

$$\begin{aligned} \frac{\delta}{\delta y(x)} y(x_1) \cdots y(x_n) = \delta(x - x_1) y(x_2) \cdots y(x_n) \\ -\delta(x - x_2) y(x_1) y(x_3) \cdots y(x_n) \\ \pm \ldots + (-1)^{n-1} \delta(x - x_n) y(x_1) \cdots y(x_{n-1}). \end{aligned} \tag{2.2.26}$$

4. Gaussian integral:

$$\int \mathcal{D}[y]\mathcal{D}[y^*] \exp\left(\int \mathrm{d}x' \mathrm{d}x\, y^*(x') A(x', x) y(x)\right) = \det A. \tag{2.2.27}$$

The formula for the gaussian integral can be easily obtained by expansion of the exponential and use of the anticommutativity of the Grassmann variables.

These formulas can be used to write the *generating functional of the Green functions of a free Dirac field* $\psi(x)$ and its adjoint $\bar{\psi}(x)$ in analogy to (2.2.1) and (2.2.8)

$$\begin{aligned} T_{\text{free}}\{\bar{\eta},\eta\} &= \sum_{n=0}^{\infty} \frac{(-1)^n}{(n!)^2} \int \mathrm{d}^4x_n \cdots \mathrm{d}^4x_1\, \mathrm{d}^4x'_1 \cdots \mathrm{d}^4x'_n\, \eta(x'_n) \cdots \eta(x'_1) \\ &\quad \times \bar{\eta}(x_n) \cdots \bar{\eta}(x_1) \langle 0|T\psi(x_1) \cdots \psi(x_n)\, \bar{\psi}(x'_n) \cdots \bar{\psi}(x'_1)|0\rangle \\ &= \exp\left(\mathrm{i} \int \mathrm{d}^4x\, \mathrm{d}^4x'\, \bar{\eta}(x) S_{\mathrm{F}}(x-x')\eta(x)\right). \end{aligned} \qquad (2.2.28)$$

The sources $\bar{\eta}(x)$, $\eta(x)$ are anticommuting, i.e. elements of an infinite-dimensional Grassmann algebra.

2.2.2 Generating functionals of Green functions

The generating functional of Green functions $T\{J\}$ was defined in (2.2.1)

$$T\{J\} = \sum_{n=0}^{\infty} \frac{\mathrm{i}^n}{n!} \int \mathrm{d}^4x_1 \cdots \mathrm{d}^4x_n\, G(x_1,\ldots,x_n) J(x_1) \cdots J(x_n).$$

The Green functions $G(x_1,\ldots,x_n)$ can be obtained from $T\{J\}$ by functional differentiation,

$$G(x_1,\ldots,x_n) = (-\mathrm{i})^n \frac{\delta^n}{\delta J(x_1) \cdots \delta J(x_n)} T\{J\}\bigg|_{J(x)\equiv 0}. \qquad (2.2.29)$$

In Sect. 2.1 we studied, besides the Green functions, also the connected Green functions, the vertex functions, and the relations between them and to S-matrix elements. In the following we extend these results to the corresponding functionals.

2.2.2.1 Generating functional of connected Green functions

The *generating functional of connected Green functions* $T_{\mathrm{c}}\{J\}$ is defined in an analogous way as $T\{J\}$,

$$T_{\mathrm{c}}\{J\} = \sum_{n=0}^{\infty} \frac{\mathrm{i}^n}{n!} \int \mathrm{d}^4x_1 \cdots \mathrm{d}^4x_n\, G_{\mathrm{c}}(x_1,\ldots,x_n) J(x_1) \cdots J(x_n). \qquad (2.2.30)$$

The relation between $T\{J\}$ and $T_c\{J\}$ is obtained from the relation between $G(x_1,\dots,x_n)$ and $G_c(x_1,\dots,x_n)$ which was given in (2.1.41),

$$T\{J\} = \exp[T_c\{J\}], \qquad T_c\{J\} = \log[T\{J\}]. \tag{2.2.31}$$

For the free field we obtain from (2.2.8)

$$T_{\text{free,c}}\{J\} = -\frac{\mathrm{i}}{2}\int \mathrm{d}^4x\,\mathrm{d}^4y\,J(x)\Delta_{\mathrm{F}}(x-y)J(y). \tag{2.2.32}$$

For an illustration of (2.2.31) we consider the four-point function. Applying the rules (2.2.4) and (2.2.29) yields

$$\begin{aligned}
G(x_1,x_2,x_3,x_4) &= \frac{\delta^4}{\delta J(x_1)\cdots\delta J(x_4)}\exp[T_c\{J\}]\Big|_{J(x)\equiv 0}\\
&= \left[\frac{\delta^4 T_c}{\delta J(x_1)\cdots\delta J(x_4)} + \left(\frac{\delta^3 T_c}{\delta J(x_1)\delta J(x_2)\delta J(x_3)}\frac{\delta T_c}{\delta J(x_4)} + \text{perm.}\right)\right.\\
&\quad + \left(\frac{\delta^2 T_c}{\delta J(x_1)\delta J(x_2)}\frac{\delta^2 T_c}{\delta J(x_3)\delta J(x_4)} + \text{perm.}\right)\\
&\quad + \left(\frac{\delta^2 T_c}{\delta J(x_1)\delta J(x_2)}\frac{\delta T_c}{\delta J(x_3)}\frac{\delta T_c}{\delta J(x_4)} + \text{perm.}\right)\\
&\quad \left. + \frac{\delta T_c}{\delta J(x_1)}\frac{\delta T_c}{\delta J(x_2)}\frac{\delta T_c}{\delta J(x_3)}\frac{\delta T_c}{\delta J(x_4)}\right]\exp[T_c\{J\}]\Big|_{J(x)\equiv 0}.
\end{aligned} \tag{2.2.33}$$

With $\delta T_c/\delta J(x)|_{J\equiv 0} = G_c(x) = 0$, this simplifies to

$$\begin{aligned}
G(x_1,x_2,x_3,x_4) &= G_c(x_1,x_2,x_3,x_4) + G_c(x_1,x_2)G_c(x_3,x_4)\\
&\quad + G_c(x_1,x_3)G_c(x_2,x_4) + G_c(x_1,x_4)G_c(x_2,x_3),
\end{aligned} \tag{2.2.34}$$

in agreement with (2.1.39).

2.2.2.2 The vertex functional

The relation between the *generating functional of vertex functions*

$$\Gamma\{a\} = \sum_{n=0}^{\infty} \frac{1}{n!} \int \mathrm{d}^4x_1 \cdots \mathrm{d}^4x_n \, \Gamma_n(x_1, \ldots, x_n) a(x_1) \cdots a(x_n) \quad (2.2.35)$$

and that of connected Green functions $T_{\mathrm{c}}\{J\}$ takes the form of a Legendre transformation

$$a(x) = \frac{\delta T_{\mathrm{c}}\{J\}}{\mathrm{i}\delta J(x)}, \quad (2.2.36)$$

$$\mathrm{i}\Gamma\{a\} = -\mathrm{i} \int \mathrm{d}^4x \, J(x)a(x) + T_{\mathrm{c}}\{J\}. \quad (2.2.37)$$

The factor i has been introduced for later convenience. Note that the variable of Γ is a and differs from that of T_{c}, namely J.

For the free scalar field we obtain from (2.2.32)

$$a(x) = -\int \mathrm{d}^4y \, \Delta_{\mathrm{F}}(x-y)J(y) \quad (2.2.38)$$

and thus

$$\begin{aligned} J(x) &= (\Box + M^2)a(x), \\ \Gamma_{\mathrm{free}}\{a\} &= -\frac{1}{2}\int \mathrm{d}^4x \, a(x)(\Box + M^2)a(x) = S_{\mathrm{free}}\{a\}, \end{aligned} \quad (2.2.39)$$

i.e. the generating functional of vertex functions for the free theory is just the action (cf. also Sect. 2.2.5.3). For this reason, the vertex functional is also called *effective action*.

The inversion of (2.2.36) is obtained by differentiation of (2.2.37) using the chain rule

$$\begin{aligned} \frac{\delta\Gamma\{a\}}{\delta a(x)} &= -J(x) - \int \mathrm{d}^4y \, \frac{\delta J(y)}{\delta a(x)} a(y) + \int \mathrm{d}^4y \, \frac{\delta T_{\mathrm{c}}\{J\}}{\mathrm{i}\delta J(y)} \frac{\delta J(y)}{\delta a(x)} \\ &= -J(x). \end{aligned} \quad (2.2.40)$$

As an example we calculate the two-point vertex function. For this we differentiate (2.2.36) using again the chain rule

$$\delta^{(4)}(x-y) = \frac{\delta a(x)}{\delta a(y)} = \frac{\delta}{\delta a(y)}\frac{\delta T_{\mathrm{c}}\{J\}}{\mathrm{i}\delta J(x)} = \int \mathrm{d}^4 z\, \frac{\delta^2 T_{\mathrm{c}}\{J\}}{\mathrm{i}\delta J(x)\mathrm{i}\delta J(z)}\frac{\mathrm{i}\delta J(z)}{\delta a(y)}$$
$$= \int \mathrm{d}^4 z\, \frac{\delta^2 T_{\mathrm{c}}\{J\}}{\mathrm{i}\delta J(x)\mathrm{i}\delta J(z)}\frac{-\mathrm{i}\delta^2 \Gamma\{a\}}{\delta a(z)\delta a(y)}. \tag{2.2.41}$$

If in this expression the integration with respect to z is formally considered as multiplication of the kernels $\delta^2 T_{\mathrm{c}}/\mathrm{i}\delta J(x)\mathrm{i}\delta J(z)$ and $\mathrm{i}\delta^2\Gamma/\delta a(z)\delta a(y)$, then one is the negative inverse of the other

$$\frac{\delta^2 T_{\mathrm{c}}\{J\}}{\mathrm{i}\delta J(x)\mathrm{i}\delta J(z)} = X(x,z), \qquad \frac{\mathrm{i}\delta^2\Gamma\{a\}}{\delta a(z)\delta a(y)} = -X^{-1}(z,y). \tag{2.2.42}$$

As $X(x,y)|_{J\equiv 0} = \langle T\psi(x)\psi(y)\rangle$ is the usual propagator, $X(x,y)$ is generally called the propagator in the presence of an external source [see the explanation after (2.2.55)]. In this sense the two-point vertex function is the inverse propagator.

We continue by considering the three-point vertex function. According to (2.2.41) and (2.2.42),

$$\frac{\delta^2 T_{\mathrm{c}}\{J\}}{\mathrm{i}\delta J(x_1)\mathrm{i}\delta J(x_2)} = -\int \mathrm{d}^4 z_1\, \mathrm{d}^4 z_2\, X(x_1,z_1)X(x_2,z_2)\frac{\mathrm{i}\delta^2\Gamma\{a\}}{\delta a(z_1)\delta a(z_2)} \tag{2.2.43}$$

is an identity. Differentiation with respect to $J(x_3)$ using the chain rule, combined with

$$\frac{\delta}{\mathrm{i}\delta J(x)} = \int \mathrm{d}^4 z\, \frac{\delta a(z)}{\mathrm{i}\delta J(x)}\frac{\delta}{\delta a(z)} = \int \mathrm{d}^4 z\, X(x,z)\frac{\delta}{\delta a(z)}, \tag{2.2.44}$$

gives, with the help of (2.2.41) and (2.2.42),

$$\frac{\delta^3 T_{\mathrm{c}}\{J\}}{\mathrm{i}\delta J(x_1)\mathrm{i}\delta J(x_2)\mathrm{i}\delta J(x_3)} = \int \mathrm{d}^4 z_1\, \mathrm{d}^4 z_2\, \mathrm{d}^4 z_3\, \frac{\mathrm{i}\delta^3\Gamma\{a\}}{\delta a(z_1)\delta a(z_2)\delta a(z_3)}$$
$$\times X(x_1,z_1)X(x_2,z_2)X(x_3,z_3). \tag{2.2.45}$$

For $J = a = 0$ (2.2.45) agrees with (2.1.46). Further differentiation of (2.2.45) with respect to $J(x)$ gives, with help of (2.2.44), the inductive definition of vertex functions as discussed after (2.1.46). This shows that $\Gamma\{a\}$ is its generating functional.

2.2.3 Functional-integral representation of the S matrix

The LSZ formula (2.1.35) relates S-matrix elements to Green functions. This relation can also be cast into a functional form. For scalar fields the S matrix can be concisely written as

$$S = :\exp\left[\int \mathrm{d}^4x\,\mathrm{d}^4y\,\psi(x)G^{-1}(x,y)R^{1/2}\frac{\delta}{\mathrm{i}\delta J(y)}\right]: T\{J\}\Big|_{J\equiv 0}, \tag{2.2.46}$$

where $\psi(x)$ is the free asymptotic field, $G^{-1}(x,y)$ its inverse propagator, and $T\{J\}$ the corresponding generating functional of Green functions. The symbol $:\ldots:$ denotes *normal ordering*, i.e. the creation operators are left of the annihilation operators, e.g.

$$:a^\dagger(p)a(q): = :a(q)a^\dagger(p): = a^\dagger(p)a(q). \tag{2.2.47}$$

For general fields (2.2.46) is generalized in an obvious way.

In order to show that (2.2.46) is equivalent to the reduction formulas (2.1.35) we expand the exponential and replace the derivatives of the generating functional by the Green functions,

$$\begin{aligned} S = \sum_{n=0}^{\infty}\frac{1}{n!}\int \mathrm{d}^4x_1\,\mathrm{d}^4y_1\cdots\mathrm{d}^4x_n\,\mathrm{d}^4y_n\,G(y_1,\ldots,y_n)\\ \times R^{1/2}G^{-1}(x_1,y_1)\cdots R^{1/2}G^{-1}(x_n,y_n)\,:\psi(x_1)\cdots\psi(x_n):. \end{aligned} \tag{2.2.48}$$

It remains to calculate the matrix element between general incoming and outgoing states

$$\begin{aligned} |p_1,\ldots,p_k\rangle &= a^\dagger(p_1)\cdots a^\dagger(p_k)|0\rangle,\\ \langle p_1',\ldots,p_l'| &= \langle 0|a(p_1')\cdots a(p_l'). \end{aligned} \tag{2.2.49}$$

If we assume that all final-state momenta differ from the initial-state momenta, as it is usually the case in experiments, only the nth term in (2.2.48) contributes, i.e. there are no disconnected terms, and the corresponding matrix element is easily evaluated with help of (1.2.10) and (1.2.4),

$$\begin{aligned}
\langle p'_1,\ldots,p'_l|S|p_1,\ldots,p_k\rangle = &\\
& R^{(k+l)/2}\int \mathrm{d}^4x_1\cdots\mathrm{d}^4x_{k+l}\,\mathrm{e}^{-\mathrm{i}p_1x_1}\ldots\mathrm{e}^{-\mathrm{i}p_kx_k}\mathrm{e}^{\mathrm{i}p'_1x_{k+1}}\ldots\mathrm{e}^{\mathrm{i}p'_lx_{k+l}}\\
& \times\int \mathrm{d}^4y_1\cdots\mathrm{d}^4y_{k+l}\,G^{-1}(x_1,y_1)\cdots G^{-1}(x_{k+l},y_{k+l})\\
& \times G(y_1,\ldots,y_{k+l})\Big|_{p_i^2=p_j'^2=m^2}\\
= & G^{-1}(p_1,-p_1)\cdots G^{-1}(p_k,-p_k)G^{-1}(p'_1,-p'_1)\cdots G^{-1}(p'_l,-p'_l)\\
& \times R^{(k+l)/2}\widetilde{G}(p_1,\cdots,p_k,-p'_1,\ldots,-p'_l)\Big|_{p_i^2=p_j'^2=m^2},
\end{aligned} \tag{2.2.50}$$

where the factor $1/n!$ is cancelled by the $n!$ pairings of field operators and one-particle states. The result is just the *reduction formula* (2.1.33).

2.2.4 The field-theoretical path integral

The representation of Green functions by the *Feynman path integral* is by far the most important application of functional calculus in field theory [Fe48, Fe65, Sc51, Ro94]. Despite the fact that the field-theoretical path integral lacks a mathematically rigorous definition, such as a measure theory, it is of great heuristic value. It is a very convenient tool for studying perturbation theory and using lattice discretization; it serves as the starting point for the investigation of non-perturbative phenomena.

2.2.4.1 Path integral for free fields

Let us first consider the free field with the Lagrangian

$$\mathcal{L}_{\text{free}} = \frac{1}{2}[\partial^\mu\psi(x)\partial_\mu\psi(x) - M^2\psi^2(x)]. \tag{2.2.51}$$

We define

$$Z_{\text{free}}\{J\} = \int \mathcal{D}[\psi]\exp\Big(\mathrm{i}\int \mathrm{d}^4x\Big[\mathcal{L}_{\text{free}} + J(x)\psi(x)\Big]\Big), \tag{2.2.52}$$

where we implicitly assume that the mass involves an infinitesimal negative imaginary part, $\mathrm{Im}\, M^2 = -\mathrm{i}\varepsilon$, in order to guarantee the convergence of the gaussian integral. We want to show that

$$\mathcal{T}_{\text{free}}\{J\} = \frac{Z_{\text{free}}\{J\}}{Z_{\text{free}}\{0\}} \tag{2.2.53}$$

is the generating functional of Green functions of the free scalar field theory.

To this end, we formally manipulate $Z_{\text{free}}\{J\}$ as a gaussian functional integral and use

$$(\partial^\mu \partial_\mu + M^2)\Delta_{\mathrm{F}}(x - x') = -\delta^{(4)}(x - x'). \tag{2.2.54}$$

The choice of the *Feynman propagators* in (2.2.54) results from the infinitesimal negative imaginary part of M^2 which is needed in (2.2.52). Performing a quadratic completion of the exponent,

$$\begin{aligned}
&\int \mathrm{d}^4x \left[\frac{1}{2}[\partial^\mu \psi(x)\partial_\mu \psi(x) - M^2\psi^2(x)] + J(x)\psi(x)\right] \\
&= \int \mathrm{d}^4x \left[-\frac{1}{2}[\psi(x)(\Box + M^2)\psi(x)] + J(x)\psi(x)\right] \\
&= \int \mathrm{d}^4x \,\frac{1}{2} \left(\psi(x) + \int \mathrm{d}^4x' \Delta_{\mathrm{F}}(x - x')J(x')\right) \\
&\qquad \times (-\partial^\mu \partial_\mu - M^2) \left(\psi(x) + \int \mathrm{d}^4x'\, \Delta_{\mathrm{F}}(x - x')J(x')\right) \\
&\qquad - \frac{1}{2} \int \mathrm{d}^4x\, \mathrm{d}^4x'\, J(x)\Delta_{\mathrm{F}}(x - x')J(x'),
\end{aligned}$$

leads with the linear substitution $\psi(x) + \int \mathrm{d}^4x' \Delta_{\mathrm{F}}(x - x')J(x') \to \psi'(x)$ in the gaussian functional integral (up to surface terms) to

$$\begin{aligned}
Z_{\text{free}}\{J\} &= \exp\left(-\frac{\mathrm{i}}{2} \int \mathrm{d}^4x\, \mathrm{d}^4x'\, J(x)\Delta_{\mathrm{F}}(x - x')J(x')\right) Z_{\text{free}}\{0\}, \\
Z_{\text{free}}\{0\} &= \int \mathcal{D}[\psi'] \exp\left(\mathrm{i} \int \mathrm{d}^4x\, \frac{1}{2}\left[\partial^\mu \psi'(x)\partial_\mu \psi'(x) - M^2\psi'^2(x)\right]\right).
\end{aligned} \tag{2.2.55}$$

The resulting expression for $Z_{\text{free}}\{J\}/Z_{\text{free}}\{0\}$ agrees with the generating functional $\mathcal{T}_{\text{free}}\{J\}$ for the free field (2.2.8).

The field equation resulting from $\mathcal{L}_{\text{free}}+J\psi$ is $(\partial^\mu\partial_\mu+M^2)\psi(x)=J(x)$. This is the reason why the argument of the functionals $T_{\text{free}}\{J\}$ and $T_{\text{free,c}}\{J\}$ is called a *source*.

2.2.4.2 Interacting fields

The representation of the generating functional of Green functions can be extended to interacting fields,

$$T\{J\}=\frac{Z\{J\}}{Z\{0\}},$$
$$Z\{J\}=\int\mathcal{D}[\psi]\exp\left(\mathrm{i}\int\mathrm{d}^4x\,[\mathcal{L}+J(x)\psi(x)]\right), \tag{2.2.56}$$

where the Lagrangian $\mathcal{L}$ is given by

$$\mathcal{L}=\mathcal{L}_{\text{free}}+g\mathcal{L}_{\mathrm{I}}. \tag{2.2.57}$$

Typical examples for the interaction Lagrangian are $\mathcal{L}_{\mathrm{I}}=-{:}\psi^3(x)/3!{:}$ or $\mathcal{L}_{\mathrm{I}}=-{:}\psi^4(x)/4!{:}$. These formulas are the basic equations of the functional-integral representation of quantum field theory with arbitrary Lagrangian. The Z functional (2.2.56) can also be written as

$$Z\{J\}=\int\mathcal{D}[\psi]\,\mathrm{e}^{\mathrm{i}S\{\psi\}}\exp\left(\mathrm{i}\int\mathrm{d}^4x\,J(x)\psi(x)\right)=\int\mathcal{D}[\psi]\,\mathrm{e}^{\mathrm{i}S_J\{\psi,J\}}, \tag{2.2.58}$$

i.e. as a functional Fourier transform of the exponential of the action $S\{\psi\}$ of the field, or as a functional average over the exponential of the action with external source $S_J\{\psi,J\}=S\{\psi\}+\int\mathrm{d}^4x\,J(x)\psi(x)$.

By expanding $T\{J\}$ one obtains the *path-integral representation of Green functions*

$$\langle T\psi(x_1)\cdots\psi(x_n)\rangle=\left.\frac{\int\mathcal{D}[\psi]\,\psi(x_1)\cdots\psi(x_n)\mathrm{e}^{\mathrm{i}S\{\psi,J\}}}{\int\mathcal{D}[\psi]\,\mathrm{e}^{\mathrm{i}S\{\psi,J\}}}\right|_{J(x)\equiv 0}. \tag{2.2.59}$$

The formulation of a quantum field theory with help of this Feynman path-integral formula is an equivalent alternative to the *quantization* with help of the canonical formalism [Ab73], i.e. use of the Hamiltonian and Heisenberg commutation relations. It forms the starting point for the non-perturbative calculation of Green functions (Sects. 2.9 and 2.10.1). In Sect. 2.4, gauge theories are defined in this compact and complete way.

2.2.4.3 Integration of fermionic path integrals

In renormalizable theories, the Lagrangian is bilinear in the fermionic fields. Consider a Lagrangian of the form

$$\mathcal{L} = \bar{\psi} Q \psi, \tag{2.2.60}$$

where Q is an arbitrary operator that may depend on other fields, as for instance on the photon field in QED (1.2.16). After adding fermion sources $J_{\bar{\psi}}$ and J_ψ, we can write

$$\begin{aligned}\mathcal{L} - \bar{\psi} J_{\bar{\psi}} + J_\psi \psi &= \bar{\psi} Q \psi - \bar{\psi} J_{\bar{\psi}} + J_\psi \psi \\ &= (\bar{\psi} + J_\psi Q^{-1}) Q (\psi - Q^{-1} J_{\bar{\psi}}) + J_\psi Q^{-1} J_{\bar{\psi}}.\end{aligned} \tag{2.2.61}$$

We can integrate the *fermionic fields* in the path integral with help of the gaussian integral (2.2.27):

$$\begin{aligned}Z\{J_\psi, J_{\bar{\psi}}\} &= \int \mathcal{D}[\psi]\mathcal{D}[\bar{\psi}] \exp\left(\mathrm{i}\int \mathrm{d}^4x \left[\mathcal{L} - \bar{\psi} J_{\bar{\psi}} + J_\psi \psi\right]\right) \\ &= \det(\mathrm{i}Q) \exp\left(\mathrm{i}\int \mathrm{d}^4x\, J_\psi Q^{-1} J_{\bar{\psi}}\right).\end{aligned} \tag{2.2.62}$$

The operator Q is related to the generalized fermion propagator $S(x, x')$ by

$$QS(x, x') = \delta^{(4)}(x - x'), \tag{2.2.63}$$

and therefore

$$Q^{-1}\psi(x) = \int \mathrm{d}^4x'\, S(x, x')\psi(x'). \tag{2.2.64}$$

Differentiating (2.2.62) with respect to the fermion sources and putting those to zero we find

$$\begin{aligned}&\int \mathcal{D}[\psi]\mathcal{D}[\bar{\psi}]\, \psi(x_1)\bar{\psi}(y_1)\cdots\psi(x_n)\bar{\psi}(y_n) \exp\left(\mathrm{i}\int \mathrm{d}^4x\, \bar{\psi} Q \psi\right) \\ &\quad = \det(\mathrm{i}Q) \sum_{\mathrm{perm}} \sigma_{\mathrm{perm}}\, \mathrm{i}S(x_1, y_{i_1})\cdots \mathrm{i}S(x_n, y_{i_n}),\end{aligned} \tag{2.2.65}$$

where the sum runs over all permutations of the y_i, and σ_{perm} is the sign corresponding to this permutation.

2.2.4.4 Euclidean field theory and statistical mechanics

The formulation of quantum field theory with help of the Feynman path-integral formula brings field theory close to statistical mechanics [Sy66, Wi75]. We mentioned already that the gaussian functional (2.2.55) requires a damping factor. From a mathematical point of view, one particularly satisfactory solution of this problem is obtained by considering the fields $\psi(x)$ at imaginary times: $x_0 = -\mathrm{i}x_4$ with x_4 real.[3] This makes the action in (2.2.58) imaginary and the exponent in the path integral negative-definite

$$\begin{aligned} \mathrm{i}S\{\psi\} &= \mathrm{i}\int \mathrm{d}x_0\,\mathrm{d}^3x \left\{\frac{1}{2}\left[\left(\frac{\partial}{\partial x_0}\psi\right)^2 - (\nabla\psi)^2 - M^2\psi^2\right] + g\mathcal{L}_\mathrm{I}\right\} \\ &\overset{x_0=-\mathrm{i}x_4}{\mapsto} -\int \mathrm{d}x_4\,\mathrm{d}^3x \left\{\frac{1}{2}\left[\left(\frac{\partial}{\partial x_4}\psi\right)^2 + (\nabla\psi)^2 + M^2\psi^2\right] - g\mathcal{L}_\mathrm{I}\right\} \\ &= \quad -S_\mathrm{E}\{\psi\} \leq 0. \end{aligned} \tag{2.2.66}$$

Fields in space–time points with imaginary time are called euclidean fields; accordingly $S_\mathrm{E}\{\psi\}$ is the euclidean action. The generating functional of *euclidean Green functions* can now be obtained by analytic continuation of (2.2.58)

$$\begin{aligned} T_\mathrm{E}\{J\} &= \frac{Z_\mathrm{E}\{J\}}{Z_\mathrm{E}\{0\}}, \\ Z_\mathrm{E}\{J\} &= \int \mathcal{D}[\psi]\,\exp\left(-S_\mathrm{E}\{\psi\} + \int \mathrm{d}^4x\,J(x)\psi(x)\right). \end{aligned} \tag{2.2.67}$$

As $S_\mathrm{E}\{\psi\} > 0$, $\mathrm{d}\mu\{\psi\} \equiv \mathcal{D}[\psi]\exp(-S_\mathrm{E}\{\psi\})$ is approximately a gaussian measure. From a physical point of view, the analytic continuation of Green functions to imaginary times may seem somewhat artificial. However, the exact conditions under which euclidean Green functions can be analytically continued back to physical times are known [Os75]. The pole structure which is important for the physical interpretation of Green functions remains essentially preserved in the euclidean region.

The formulas of euclidean field theory for the calculation of Green functions formally correspond to the basic equations of statistical mechanics. In order to show this, put $S_\mathrm{E}\{\psi\} = \beta H_\mathrm{E}\{\psi\}$. If the functional differentiation $\delta/\delta J(x)$

[3] Formulas for the euclidean Dirac matrices are collected in App. A.1.2.

for the calculation of the Green functions is carried out under the integral, it follows that[4]

$$\langle\psi(x_1)\cdots\psi(x_n)\rangle = \frac{\int\mathcal{D}[\psi]\,\psi(x_1)\cdots\psi(x_n)\exp\left(-\beta H_{\mathrm{E}}\{\psi\}\right)}{\int\mathcal{D}[\psi]\exp\left(-\beta H_{\mathrm{E}}\{\psi\}\right)}. \tag{2.2.68}$$

Hence, euclidean Green functions are statistical expectation values of products of field operators with the canonical distribution $\exp(-\beta H_{\mathrm{E}}\{\psi\})$ in four dimensions [Hu87]. The action is interpreted as Hamilton function of a four-dimensional field theory. The coupling constant $g \approx 1/\beta^{1/2}$ (cf. Sect. 2.10.1.2) gets the meaning of temperature, $T = 1/\beta$, and $Z\{0\} = Z(\beta) = \int\mathcal{D}[\psi]\exp(-\beta H_{\mathrm{E}}\{\psi\})$ that of a *partition function.* The notions *free energy* $F \sim -\log Z(\beta)/\beta$, *internal energy* $U \sim \delta F/\delta\beta$, etc., are used accordingly. The close formal relation between relativistic quantum field theory and classical statistical mechanics is of great practical importance since concepts and techniques from one of them can be taken over to the other and vice versa.

This formal analogy of euclidean field theory with statistical mechanics should be strictly distinguished from a quantum theory at finite physical temperature. It is well-known [StM] that the partition function of a finite-temperature system $Z(\beta_T)$ and the thermal expectation value $\langle A\rangle$ can be represented in terms of a functional integral over a finite euclidean time extent,

$$\begin{aligned} Z(\beta_T) &= \int\mathcal{D}[\psi]\,\mathrm{e}^{-\mathcal{S}_{\mathrm{E},\beta_T}\{\psi\}}, \\ \langle A\rangle_{\beta_T} &= \int\mathcal{D}[\psi]\,A\{\psi\}\mathrm{e}^{-\mathcal{S}_{\mathrm{E},\beta_T}\{\psi\}}, \end{aligned} \tag{2.2.69}$$

where

$$\mathcal{S}_{\mathrm{E},\beta_T}\{\psi\} = \int_0^{\beta_T}\mathrm{d}x_4\int_{\mathbb{R}^3}\mathrm{d}^3x\,\mathcal{L}_{\mathrm{E}}(\psi). \tag{2.2.70}$$

Here the fields have to satisfy periodic boundary conditions in euclidean time directions $\psi(x+\beta_T e_T) = \psi(x)$, and β_T is related to the physical temperature $\beta_T = 1/(kT)$ with the Boltzmann constant k. In this sense, one considers euclidean field theory as the zero-temperature limit.

[4]Euclidean Green functions involve no time ordering.

2.2.5 Feynman rules and path integral

In order to establish the connection between the perturbative treatment of theories with interaction, as discussed in the preceding sections, and the path-integral representation of quantum field theory, we indicate the derivation of the *Feynman rules* from the path-integral formula (2.2.56). The general procedure works as follows. First, the field variable in the interaction term $\mathcal{L}_\mathrm{I}$ is replaced by a differentiation with respect to the source $J(x)$. Then, the result of gaussian integration over free fields (2.2.55) is used. Finally, the exponential of $\mathcal{L}_\mathrm{I}$ is expanded in powers of the coupling constant. This leads in the example of scalar fields to the following sequence of equations

$$\begin{aligned} Z\{J\} &= \int \mathcal{D}[\psi] \exp\left[\mathrm{i}g\int \mathrm{d}^4x\, \mathcal{L}_\mathrm{I}(\psi(x))\right] \exp\left(\mathrm{i}\int \mathrm{d}^4x \left[\mathcal{L}_\mathrm{free} + J\psi\right]\right) \\ &= \exp\left[\mathrm{i}g\int \mathrm{d}^4x\, \mathcal{L}_\mathrm{I}\left(\frac{\delta}{\mathrm{i}\delta J(x)}\right)\right] \int \mathcal{D}[\psi] \exp\left(\mathrm{i}\int \mathrm{d}^4x \left[\mathcal{L}_\mathrm{free} + J\psi\right]\right) \\ &= \left[1 + \mathrm{i}g\int \mathrm{d}^4x_1\, \mathcal{L}_\mathrm{I}\left(\frac{\delta}{\mathrm{i}\delta J(x_1)}\right) + \right. \\ &\quad \left. + \frac{(\mathrm{i}g)^2}{2!}\int \mathrm{d}^4x_1\, \mathrm{d}^4x_2\, \mathcal{L}_\mathrm{I}\left(\frac{\delta}{\mathrm{i}\delta J(x_1)}\right) \mathcal{L}_\mathrm{I}\left(\frac{\delta}{\mathrm{i}\delta J(x_2)}\right) + \dots\right] \\ &\quad \times \exp\left(-\frac{\mathrm{i}}{2}\int \mathrm{d}^4x\, \mathrm{d}^4x'\, J(x)\Delta_\mathrm{F}(x-x')J(x')\right) Z_\mathrm{free}\{0\}. \end{aligned} \tag{2.2.71}$$

Here $Z_\mathrm{free}\{0\}$ is the Z functional of the free field for vanishing sources. The evaluation of the functional differentiations leads to the perturbative expansion of the Green functions which can be formulated with the Feynman rules.

2.2.5.1 Feynman rules in configuration space

We illustrate this procedure by a simple example. For this purpose we consider $\mathcal{L}_\mathrm{I} = -\psi^4(x)/4!$ as interaction Lagrangian of the scalar field and calculate the four-point Green function up to first order in g. Because of

$$\mathcal{L}_\mathrm{I}\left(\frac{\delta}{\mathrm{i}\delta J(x)}\right) = -\frac{1}{4!}\frac{\delta^4}{\delta J(x)^4},$$

the corresponding expansion (2.2.71) reads

$$
\begin{aligned}
&\frac{1}{Z_{\text{free}}\{0\}} \frac{\delta^4 Z\{J\}}{\delta J(x_1)\cdots\delta J(x_4)} \\
&= \frac{\delta^4}{\delta J(x_1)\cdots\delta J(x_4)} \left[\frac{1}{2!}\left(-\frac{\mathrm{i}}{2}\int \mathrm{d}^4z\,\mathrm{d}^4z'\,J(z)\Delta_{\mathrm{F}}(z-z')J(z')\right)^2 \right. \\
&\quad - \frac{\mathrm{i}g}{4!}\int \mathrm{d}^4x\,\frac{1}{4!}\frac{\delta^4}{\delta J(x)^4}\left(-\frac{\mathrm{i}}{2}\int \mathrm{d}^4z\,\mathrm{d}^4z'\,J(z)\Delta_{\mathrm{F}}(z-z')J(z')\right)^4 \\
&\quad \left. + \dots \right] \\
&= (a) + (b) + (c) + (d) + \dots \qquad (2.2.72)
\end{aligned}
$$

up to the contributing order in g and $J(x)$, where

$$
\begin{aligned}
(a) &= -\sum_{\text{part.}} \Delta_{\mathrm{F}}(x_i - x_j)\Delta_{\mathrm{F}}(x_k - x_l), \\
(b) &= -\mathrm{i}g \int \mathrm{d}^4x\,\Delta_{\mathrm{F}}(x-x_1)\Delta_{\mathrm{F}}(x-x_2)\Delta_{\mathrm{F}}(x-x_3)\Delta_{\mathrm{F}}(x-x_4), \\
(c) &= -\frac{\mathrm{i}g}{2}\sum_{\text{part.}} \Delta_{\mathrm{F}}(x_i - x_j)\int \mathrm{d}^4x\,\Delta_{\mathrm{F}}(x-x)\Delta_{\mathrm{F}}(x-x_k)\Delta_{\mathrm{F}}(x-x_l), \\
(d) &= -\frac{\mathrm{i}g}{8}\sum_{\text{part.}} \Delta_{\mathrm{F}}(x_i - x_j)\Delta_{\mathrm{F}}(x_k - x_l)\int \mathrm{d}^4x\,\Delta_{\mathrm{F}}(x-x)\Delta_{\mathrm{F}}(x-x),
\end{aligned}
\qquad (2.2.73)
$$

and

$$
Z\{0\}/Z_{\text{free}}\{0\} = 1 + \frac{\mathrm{i}g}{2^3}\int \mathrm{d}^4x\,\Delta_{\mathrm{F}}(x-x)\Delta_{\mathrm{F}}(x-x) + \dots . \qquad (2.2.74)
$$

The sums in (2.2.73) extend over all non-equivalent partitions of the x_i. It follows

$$
\begin{aligned}
G^{(1)}(x_1,\dots,x_4) &= Z\{0\}^{-1}\frac{\delta^4 Z\{J\}}{\delta J(x_1)\cdots\delta J(x_4)} \\
&= (a) + (b) + (c) + \dots . \qquad (2.2.75)
\end{aligned}
$$

The term (d) in (2.2.73) is cancelled by the product of $\mathrm{i}g/2^3 \int \mathrm{d}^4x\,\Delta_{\mathrm{F}}(0)^2$ in the expansion of $1/Z\{0\}$ in powers of g with the term (a). Thus, the vacuum loops yield no contribution. The general rôle of the factor $1/Z\{0\}$

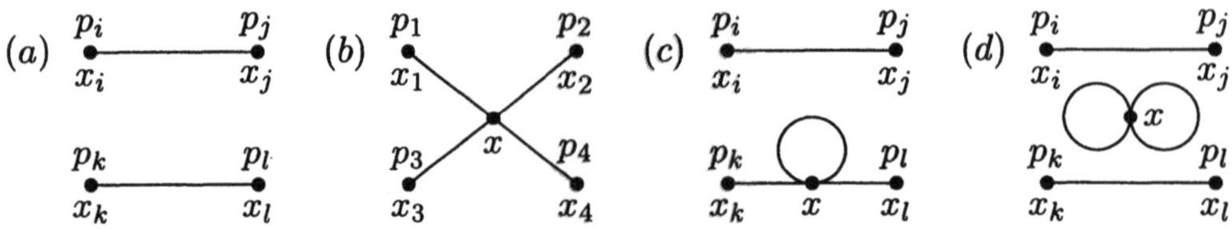

Fig. 2.3 Feynman diagrams for the four-point Green function up to first order in g

in the definition of $T\{J\}$ is to cancel all vacuum subgraphs in $Z\{J\}$. On the level of Feynman diagrams this is easily taken into account by omitting all graphs containing vacuum subdiagrams, i.e. subdiagrams without external lines (as in Fig. 2.3d).

The results (2.2.73) can be obtained from Feynman rules in configuration space. The corresponding graphs are shown in Fig. 2.3. The analytic expressions result from these graphs by attaching to each internal line joining the points x_i and x_j a space–time propagator $\mathrm{i}\Delta_{\mathrm{F}}(x_i - x_j)$, to each vertex associated with coordinate x a factor $-\mathrm{i}g \int \mathrm{d}^4x$, and to each field point a factor 1. In addition the *symmetry factors*, e.g. 1/2 for (*c*) and 1/8 for (*d*), have to be multiplied and, in case of fermions, also the factors -1 as discussed in Sect. 1.2.1.8. Note that the correct symmetry factors are obtained by simply differentiating $Z\{J\}$ with respect to its sources.

2.2.5.2 Feynman rules in momentum space

In order to derive the Feynman rules in momentum space we have to perform a Fourier transformation. To this end, we substitute the Fourier representations for the propagators (1.2.31) into (2.2.73) and obtain

$$
\begin{aligned}
(a) &= \frac{1}{(2\pi)^8} \sum_{\text{part.}} \int \mathrm{d}^4p_i\, \mathrm{d}^4p_k\, \mathrm{e}^{-\mathrm{i}p_i(x_i - x_j) - \mathrm{i}p_k(x_k - x_l)} \frac{\mathrm{i}}{p_i^2 - m^2} \frac{\mathrm{i}}{p_k^2 - m^2} \\
&= \int \frac{\mathrm{d}^4p_1}{(2\pi)^4} \frac{\mathrm{d}^2p_2}{(2\pi)^4} \frac{\mathrm{d}^3p_3}{(2\pi)^4} \frac{\mathrm{d}^4p_4}{(2\pi)^4}\, \mathrm{e}^{-\mathrm{i}(p_1x_1 + p_2x_2 + p_3x_3 + p_4x_4)} \\
&\quad \times \sum_{\text{part.}} (2\pi)^4 \delta^{(4)}(p_i + p_j)(2\pi)^4 \delta^{(4)}(p_k + p_l) \frac{\mathrm{i}}{p_i^2 - m^2} \frac{\mathrm{i}}{p_k^2 - m^2}, \\
(b) &= -\mathrm{i}g \frac{1}{(2\pi)^{12}} \int \mathrm{d}^4p_1\, \mathrm{d}^2p_2\, \mathrm{d}^3p_3\, \mathrm{d}^4p_4\, \delta^{(4)}(p_1 + p_2 + p_3 + p_4)
\end{aligned}
$$

$$\times e^{-i(p_1x_1+p_2x_2+p_3x_3+p_4x_4)} \frac{i}{p_1^2-m^2} \frac{i}{p_2^2-m^2} \frac{i}{p_3^2-m^2} \frac{i}{p_4^2-m^2}$$

$$= \int \frac{d^4p_1}{(2\pi)^4} \frac{d^2p_2}{(2\pi)^4} \frac{d^3p_3}{(2\pi)^4} \frac{d^4p_4}{(2\pi)^4} e^{-i(p_1x_1+p_2x_2+p_3x_3+p_4x_4)}$$

$$\times(-i)g(2\pi)^4\delta^{(4)}(p_1+p_2+p_3+p_4)$$

$$\times \frac{i}{p_1^2-m^2} \frac{i}{p_2^2-m^2} \frac{i}{p_3^2-m^2} \frac{i}{p_4^2-m^2},$$

$$(c) = -ig\frac{1}{2}\frac{1}{(2\pi)^{12}} \sum_{\text{part.}} \int d^4p_i\, d^4p_k\, d^4p_l\, \delta^{(4)}(p_k+p_l) \int d^4q\, \frac{i}{q^2-m^2}$$

$$\times e^{-i(p_i(x_i-x_j)+p_kx_k+p_lx_l)} \frac{i}{p_i^2-m^2} \frac{i}{p_k^2-m^2} \frac{i}{p_l^2-m^2}$$

$$= \int \frac{d^4p_1}{(2\pi)^4} \frac{d^2p_2}{(2\pi)^4} \frac{d^3p_3}{(2\pi)^4} \frac{d^4p_4}{(2\pi)^4} e^{-i(p_1x_1+p_2x_2+p_3x_3+p_4x_4)}$$

$$\times(-i)\frac{1}{2}g \sum_{\text{part.}} (2\pi)^4\delta^{(4)}(p_i+p_j)(2\pi)^4\delta^{(4)}(p_k+p_l)$$

$$\times \frac{i}{p_i^2-m^2} \frac{i}{p_k^2-m^2} \frac{i}{p_l^2-m^2} \int \frac{d^4q}{(2\pi)^4} \frac{i}{q^2-m^2}. \tag{2.2.76}$$

These results can be directly obtained from the momentum space Feynman rules introduced in Sects. 1.2.1.8 and 2.1.2.2, and the vertex

$$= -ig. \tag{2.2.77}$$

Since $\Delta_F(x)$ is singular at $x = 0$, our treatment is completely formal. Regularization methods which make this procedure more rigorous are discussed in Sect. 2.5.

2.2.5.3 Loop expansion and quantum effects

We conclude the discussion of the derivation of the Feynman rules from the path-integral representation with a short remark [Na66]. For this, the

Planck constant $\hbar$, which has been used so far as a unit, is written explicitly in (2.2.71)

$$\begin{aligned} Z\{J\} &= \int \mathcal{D}[\psi] \exp\left(\frac{\mathrm{i}}{\hbar}\int \mathrm{d}^4x \,[\mathcal{L} + \hbar J(x)\psi(x)]\right) \\ &= \exp\left[\frac{\mathrm{i}}{\hbar} g \int \mathrm{d}^4x \,\mathcal{L}_\mathrm{I}\left(\frac{\delta}{\mathrm{i}\delta J(x)}\right)\right] \\ &\quad \times \exp\left(-\frac{\mathrm{i}\hbar}{2}\int \mathrm{d}^4x \mathrm{d}^4x' \,J(x)\Delta_\mathrm{F}(x-x')J(x')\right) Z_\text{free}\{0\}. \end{aligned} \tag{2.2.78}$$

This shows that each vertex is linked with a factor $\hbar^{-1}$ and each propagator with a factor $\hbar$ in the perturbation series. The number of loops for any connected vertex graph consisting of I internal lines and V vertex points is

$$L = I - (V-1) = I - V + 1, \tag{2.2.79}$$

as $V-1$ lines are required to connect the V points simply by a tree graph and any further line generates a loop. Thus, the above counting of powers of $\hbar$ gives the following result for the perturbative expansion of the generating functional of vertex functions

$$\begin{aligned} \mathrm{i}\Gamma\{a\} &= \hbar \log(Z\{J\}/Z\{0\}) - \mathrm{i}\hbar \int \mathrm{d}^4x \,J(x)a(x) \\ &= \mathrm{i}\sum_L \hbar^L \Gamma_L, \end{aligned} \tag{2.2.80}$$

where Γ_L is the contribution of the graphs with L loops. This result indicates a close relation between the order of quantum corrections and the number of loops in Feynman diagrams. In particular, one expects that the *tree-graph approximation* without loops is closely related to the perturbative solutions of the classical field equations. Actually, for the most simple, *semi-classical approximation*, where the path-integral is evaluated using the method of stationary phase [De61], we have

$$T_\mathrm{c}\{J\} = \hbar \log\frac{Z\{J\}}{Z\{0\}} \approx \mathrm{i}\int \mathrm{d}^4x \left[\mathcal{L}(a_\mathrm{cl}(x), \partial_\mu a_\mathrm{cl}(x)) + \hbar J(x) a_\mathrm{cl}(x)\right], \tag{2.2.81}$$

where $a_{\rm cl}(x)$ is a solution of the classical equation of motion

$$\frac{\delta}{\delta a_{\rm cl}(x)} \int \mathrm{d}^4 y \, \mathcal{L}(a_{\rm cl}(y), \partial_\mu a_{\rm cl}(x)) = -\hbar J(x). \tag{2.2.82}$$

If both these equations are compared with the introduction of the functional of vertex functions (2.2.37) and (2.2.40) one concludes that in this approximation the vertices are generated by

$$\Gamma\{a_{\rm cl}\} = \int \mathrm{d}^4 x \, \mathcal{L}(a_{\rm cl}(x), \partial_\mu a_{\rm cl}(x)), \tag{2.2.83}$$

i.e. by the action. Differentiation of (2.2.83) with respect to $a_{\rm cl}(x)$ gives the vertices of the tree-graph approximation.

2.2.6 Ward identities and equations of motion for Green functions

In Sect. 1.2.3 we treated the relations between a symmetry of the classical Lagrangian, currents, and charge conservation. These considerations are now extended to the path-integral representation of quantum field theory. The results are Ward identities for Green functions, i.e. relations between Green functions resulting from a symmetry of the action [Wa50, Ta57]. They are of great importance for the treatment of gauge-field theories.

The equations of motion for the classical fields can be generalized to equations of motion for Green functions. The derivation of these equations and of the Ward identities proceeds along similar lines. The starting point is the invariance of the path integral under field transformations, which simply corresponds to the invariance of a usual integral with respect to transformations of the integration variables.

2.2.6.1 Invariance of the path integral under field transformations

We consider a theory with fields $\psi_i(x)$, $i = 1, \ldots, n$, which are collectively denoted by a vector $\psi = (\psi_1, \ldots, \psi_n)$, a Lagrangian $\mathcal{L}(\psi, \partial_\mu \psi)$, the vector of sources $\mathbf{J} = (J_1, \ldots, J_n)$, and the generating functional

$$Z\{\mathbf{J}\} = \int \mathcal{D}[\psi] \exp\left(\mathrm{i} \int \mathrm{d}^4x \Big[\mathcal{L}(\boldsymbol{\psi}, \partial_\mu \boldsymbol{\psi}) + \mathbf{J}(x)\boldsymbol{\psi}(x) \Big] \right)$$
$$= \int \mathcal{D}[\psi] \exp\left(\mathrm{i}S\{\boldsymbol{\psi}\} + \mathrm{i} \int \mathrm{d}^4x \, \mathbf{J}(x)\boldsymbol{\psi}(x) \right). \tag{2.2.84}$$

For simplicity we assume that none of the variables or fields is Grassmann-valued. In this case, extra signs would appear in the results.

Let us perform an infinitesimal transformation of the fields,

$$\psi_i \to \psi_i + \delta\psi_i, \qquad \delta\psi_i(x) = \varepsilon f_i\{\boldsymbol{\psi}; x\}, \tag{2.2.85}$$

where ε is an infinitesimal parameter, and f_i is an arbitrary functional of the fields and the space–time point x. Under this transformation the action varies as

$$S\{\boldsymbol{\psi}\} \to S\{\boldsymbol{\psi} + \delta\boldsymbol{\psi}\} = S\{\boldsymbol{\psi}\} + \delta S\{\boldsymbol{\psi}\}, \tag{2.2.86}$$

with

$$\delta S\{\boldsymbol{\psi}\} = \int \mathrm{d}^4x \frac{\delta S\{\boldsymbol{\psi}\}}{\delta\psi_k(x)} \delta\psi_k(x)$$
$$= \int \mathrm{d}^4x \left[\frac{\partial\mathcal{L}}{\partial\psi_k(x)} \delta\psi_k(x) + \frac{\partial\mathcal{L}}{\partial(\partial_\mu\psi_k(x))} \partial_\mu \delta\psi_k(x) \right]$$
$$= \int \mathrm{d}^4x \left[\frac{\partial\mathcal{L}}{\partial\psi_k(x)} - \left(\partial_\mu \frac{\partial\mathcal{L}}{\partial(\partial_\mu\psi_k(x))} \right) \right] \delta\psi_k(x), \tag{2.2.87}$$

where in the last step we performed a partial integration omitting the surface terms. From (2.2.87) we can read off the well-known *variational derivative*

$$\frac{\delta S}{\delta\psi_k(x)} = \frac{\partial\mathcal{L}}{\partial\psi_k(x)} - \partial_\mu \frac{\partial\mathcal{L}}{\partial(\partial_\mu\psi_k(x))}. \tag{2.2.88}$$

As a change of integration variables cannot change the integral $Z\{\mathbf{J}\}$, the variation $\delta Z\{\mathbf{J}\}$ must vanish

$$0 = \delta Z\{\mathbf{J}\} = \int \mathcal{D}[\psi] \exp\left(\mathrm{i}S\{\boldsymbol{\psi}\} + \mathrm{i} \int \mathrm{d}^4y \, \mathbf{J}(y)\boldsymbol{\psi}(y) \right)$$
$$\times \int \mathrm{d}^4x \left[\frac{\delta f_k\{\boldsymbol{\psi}; x\}}{\delta\psi_k(x)} + \mathrm{i} \left(\frac{\delta S\{\boldsymbol{\psi}\}}{\delta\psi_k(x)} + J_k(x) \right) f_k\{\boldsymbol{\psi}; x\} \right]. \tag{2.2.89}$$

The "trace" term $\delta f_k/\delta\psi_k$ originates from the Jacobian of the infinitesimal transformation (2.2.85).

Differentiating (2.2.89) with respect to the sources and putting these equal to zero gives the desired relations between Green functions

$$0 = \left\langle T \prod_l \psi_{i_l}(x_l) \frac{\delta f_k\{\psi; x\}}{\delta\psi_k(x)} \right\rangle + \mathrm{i} \left\langle T \prod_l \psi_{i_l}(x_l) \frac{\delta S\{\psi\}}{\delta\psi_k(x)} f_k\{\psi; x\} \right\rangle$$
$$+ \sum_k \left\langle T \prod_{l<k} \psi_{i_l}(x_l) f_{i_k}\{\psi, x_k\} \prod_{m>k} \psi_{i_m}(x_m) \right\rangle. \qquad (2.2.90)$$

2.2.6.2 Equations of motion for Green functions

Let us consider the case where $f_i\{\psi; x\}$ are ordinary functions of x

$$\delta\psi_i(x) = \varepsilon f_i\{\psi; x\} = \varepsilon f_i(x). \qquad (2.2.91)$$

The functional measure is obviously invariant under this transformation. Since $f_i(x)$ are arbitrary functions of x we find from (2.2.89)

$$0 = \int \mathcal{D}[\psi] \left(\frac{\delta S}{\delta\psi_i(x)} + J_i(x) \right) \exp\left(\mathrm{i}S\{\psi\} + \mathrm{i} \int \mathrm{d}^4y\, \mathbf{J}(y)\boldsymbol{\psi}(y) \right). \qquad (2.2.92)$$

As $\delta S/\delta\psi_i(x) = 0$ are just the classical equations of motion, the identities for Green functions resulting from (2.2.92) and corresponding to (2.2.90) are called *equations of motion for Green functions*

$$-\mathrm{i}\left\langle T \prod_l \psi_{i_l}(x_l) \frac{\delta S}{\delta\psi_i(x)} \right\rangle = \sum_k \left\langle T \prod_{l\neq k} \psi_{i_l}(x_l) \right\rangle \delta_{i i_k} \delta^{(4)}(x - x_k)$$
$$= \left\langle T \frac{\delta}{\delta\psi_i(x)} \prod_l \psi_{i_l}(x_l) \right\rangle. \qquad (2.2.93)$$

The δ-functions in (2.2.93) result from the differentiation of the explicit sources $J_i(x)$ in (2.2.92).

If we restrict ourselves to one field type and one explicit field in the Green function, this simplifies to

$$-\mathrm{i}\left\langle T\psi(y) \frac{\delta S}{\delta\psi(x)} \right\rangle = \delta^{(4)}(x - y). \qquad (2.2.94)$$

This provides a relation between the full ψ propagator and the higher Green functions, as can for instance be seen by considering a scalar field theory, where

$$\frac{\delta S}{\delta\psi(x)} = -(\Box + m^2)\psi - V'(\psi), \tag{2.2.95}$$

and V' is the derivative of the scalar potential with respect to its argument.

2.2.6.3 Ward identities from local symmetries

In order to study the consequences of symmetries of the action we consider transformations of the type

$$\delta\psi_i(x) = \varepsilon f_i\{\psi; x\} = \varepsilon g_i^a\{\psi; x\}\theta^a(x), \tag{2.2.96}$$

where $\theta^a(x)$ are the parameters of the local symmetry transformation. The invariance of the action under the local transformation (2.2.96) implies

$$\begin{aligned} 0 &= \frac{\delta S\{\psi + \varepsilon \mathbf{g}^b\theta^b\}}{\delta\theta^a(x)}\theta^a(x) = \frac{\delta S\{\psi\}}{\delta\psi_k(x)} g_k^a\{\psi; x\}\theta^a(x)\varepsilon \\ &= \frac{\delta S\{\psi\}}{\delta\psi_k(x)} f_k\{\psi; x\}\varepsilon. \end{aligned} \tag{2.2.97}$$

If in addition the functional measure is invariant under (2.2.96), as it is the case form many symmetries, the first term in (2.2.89) drops out, and we obtain, using that the $\theta^a(x)$ are arbitrary functions of x,

$$0 = \int \mathcal{D}[\psi]\mathbf{J}(x)\mathbf{g}^a\{\psi; x\} \exp\left(\mathrm{i}S\{\psi\} + \mathrm{i}\int \mathrm{d}^4y\, \mathbf{J}(y)\psi(y)\right) \tag{2.2.98}$$

from (2.2.89). Since we are only interested in $T\{\mathbf{J}\}$, it is in fact sufficient for (2.2.98) that the Jacobian of the transformation (2.2.85) is non-zero and field-independent. If there exists no regularization where this is the case, the contribution of the Jacobian gives rise to anomalies [Fu80] (cf. Sect. 2.9.3.8).

2.2.6.4 Ward identities from global symmetries

Invariance of the action under global transformations,

$$\delta\psi_i(x) = \varepsilon f_i\{\psi;x\} = \varepsilon g_i^a\{\psi;x\}\theta^a, \qquad \partial_\mu\theta^a(x) = 0, \tag{2.2.99}$$

means

$$0 = \frac{\partial S\{\psi + \varepsilon \mathbf{g}^b\theta^b\}}{\partial\theta^a} = \int \mathrm{d}^4x\, \frac{\delta S\{\psi\}}{\delta\psi_k(x)} g_k^a\{\psi;x\}\varepsilon. \tag{2.2.100}$$

Similarly to the reasoning in the preceding section, (2.2.100) implies directly the integrated relation

$$0 = \int \mathcal{D}[\psi] \int \mathrm{d}^4x\, \mathbf{J}(x)\mathbf{g}^a\{\psi;x\} \exp\left(\mathrm{i}S\{\psi\} + \mathrm{i}\int \mathrm{d}^4y\, \mathbf{J}(y)\psi(y)\right). \tag{2.2.101}$$

Equation (2.2.101) is equivalent to the following *Ward identities for Green functions*

$$0 = \sum_k \Big\langle T\Big[\prod_{l<k} \psi_{i_l}(x_l)\Big] g_{i_k}^a\{\psi, x_k\} \prod_{m>k} \psi_{i_m}(x_m)\Big\rangle = \delta^a \Big\langle T\prod_l \psi_{i_l}(x_l)\Big\rangle, \tag{2.2.102}$$

with

$$\delta^a\psi_i(x) = g_i^a\{\psi;x\}. \tag{2.2.103}$$

Note that (2.2.101) is exactly the relation (2.2.98) integrated over the Minkowski space. While the local symmetry yields a relation for all x, the global symmetry yields only a relation for the integral over x.

In order to derive a stronger relation also for the global symmetry, we consider the change of the action under the local transformation corresponding to the global symmetry

$$\begin{aligned}
&\int \mathrm{d}^4x\, \frac{\delta S\{\psi\}}{\delta\psi_k(x)} f_k(\psi,x) = \\
&= \int \mathrm{d}^4x \left[\frac{\partial\mathcal{L}}{\partial\psi_k(x)} - \partial_\mu\left(\frac{\partial\mathcal{L}}{\partial(\partial_\mu\psi_k(x))}\right)\right] g_k^a\{\psi;x\}\theta^a(x) \\
&= \int \mathrm{d}^4x \Bigg[\frac{\partial\mathcal{L}(\psi + \varepsilon\mathbf{g}^b\theta^b, \partial_\mu(\psi + \varepsilon\mathbf{g}^c\theta^c))}{\varepsilon\partial\theta^a} \\
&\qquad - \partial_\mu\left(\frac{\partial\mathcal{L}}{\partial(\partial_\mu\psi_k(x))} g_k^a(x)\right)\Bigg]\theta^a(x),
\end{aligned} \tag{2.2.104}$$

where

$$\frac{\partial \mathcal{L}(\psi + \varepsilon \mathbf{g}^b \theta^b, \partial_\mu(\psi + \varepsilon \mathbf{g}^c \theta^c))}{\varepsilon \partial \theta^a} \tag{2.2.105}$$

$$= \left(\frac{\partial \mathcal{L}}{\partial \psi_k(x)} g_k^a\{\psi; x\} + \frac{\partial \mathcal{L}}{\partial(\partial_\mu \psi_k(x))} (\partial_\mu g_k^a\{\psi; x\}) \right) \tag{2.2.106}$$

is the change of the Lagrangian under the global transformations (2.2.99). Invariance of the action under these transformations implies that this change is at most a total derivative, i.e.

$$\frac{\partial \mathcal{L}(\psi + \varepsilon \mathbf{g}^b \theta^b, \partial_\mu(\psi + \varepsilon \mathbf{g}^c \theta^c))}{\partial \theta^a} = \partial^\mu K_\mu^a\{\psi; x\} \varepsilon \tag{2.2.107}$$

with some vectors K_μ^a. Inserting this into (2.2.104) yields

$$\int \mathrm{d}^4x \, \frac{\delta S\{\psi\}}{\delta \psi_k(x)} f_k(\psi, x) = \int \mathrm{d}^4x \, (\partial^\mu j_\mu^a\{\psi; x\}) \, \theta^a(x), \tag{2.2.108}$$

with the classical *Noether currents*

$$j_\mu^a\{\psi; x\} = K_\mu^a\{\psi; x\} - \left(\frac{\partial \mathcal{L}}{\partial(\partial_\mu \psi_k(x))} g_k^a\{\psi; x\} \right), \tag{2.2.109}$$

which can be considered as a local functional of the fields $\psi_i(x)$. This definition of the Noether currents constitutes a generalization of (1.2.79).

The conservation of these currents for classical on-shell fields follows directly from (2.2.108). The classical equations of motion imply $\delta S\{\psi\}/\delta \psi_k(x) = 0$ and thus

$$\partial^\mu j_\mu^a = 0, \tag{2.2.110}$$

since $\theta^a(x)$ is an arbitrary function of x. Inserting (2.2.108) into (2.2.89) and using the invariance of the functional measure and the arbitrariness of the $\theta^a(x)$ gives the result

$$0 = \int \mathcal{D}[\psi] \left(\partial^\mu j_\mu^a\{\psi; x\} + J_k(x) g_k^a\{\psi; x\}\right)$$
$$\times \exp\left(\mathrm{i} S\{\psi\} + \mathrm{i} \int \mathrm{d}^4x \, \mathbf{J}(x) \psi(x) \right). \tag{2.2.111}$$

By replacing the fields appearing in $j^a\{\psi;x\}$ and $g_i^a\{\psi;x\}$ by derivatives with respect to the sources, we obtain the Ward identities in functional form

$$\left[\partial_x^\mu j_\mu^a\left\{\frac{\delta}{\mathrm{i}\delta \mathbf{J}},x\right\}+J_k(x)g_k^a\left\{\frac{\delta}{\mathrm{i}\delta \mathbf{J}},x\right\}\right]T\{\mathbf{J}\}=0. \tag{2.2.112}$$

Equation (2.2.111) gives rise to the following *Ward identities for Green functions*

$$\partial_x^\mu\left\langle Tj_\mu^a\{\psi;x\}\prod_l \psi_{i_l}(x_l)\right\rangle \tag{2.2.113}$$
$$=\mathrm{i}\sum_k \delta^{(4)}(x-x_k)\left\langle T\left(\prod_{l<k}\psi_{i_l}(x_l)\right)g_{i_k}^a\{\psi,x_k\}\prod_{m>k}\psi_{i_m}(x_m)\right\rangle.$$

These identities can be visualized in terms of diagrams as

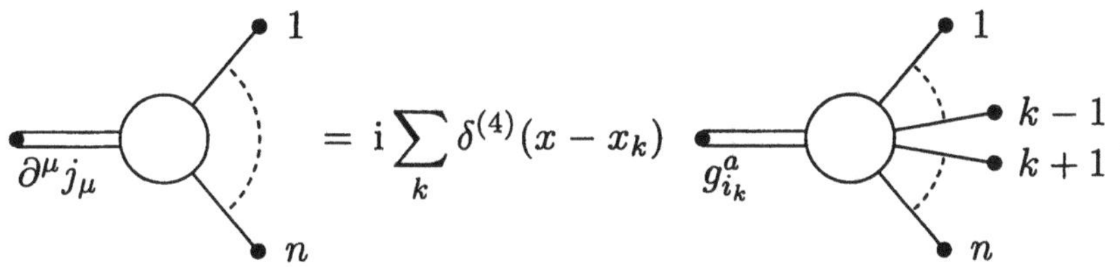

where  represents *composite operators*, like $\partial^\mu j_\mu$ or $g_{i_k}^a$. In general, composite operators are products of elementary fields with the same space–time arguments. Integration of (2.2.111) and (2.2.113) with respect to space–time reproduces (2.2.101) and (2.2.102), respectively.

Examples of Ward identities for gauge theories are derived in Sects. 2.4.3.3 and 2.4.3.5.

2.3 Local gauge invariance

The importance of symmetry groups was repeatedly stressed during our phenomenological discussion of elementary particle physics. In connection with a field theory, a symmetry has important consequences. In Sect. 1.2.3 we showed the relation between global symmetries and conserved currents. In accordance with the principle of local action, one should require that the symmetry transformations of fields are rather local than global. This is the *principle of local gauge invariance.* Gauge theories are constructed according to this principle.

At first we show in Sect. 2.3.1 how QED, the theory for the electromagnetic interaction, is obtained from the principle of local gauge invariance. Then we introduce the main concepts necessary for its generalization to non-abelian symmetries in Sect. 2.3.2. This framework is used in Sect. 2.3.3 to formulate general gauge theories.

The appropriate mathematical setting for the discussion of classical gauge fields is the theory of fibre bundles. Such a treatment would go beyond the general mathematical scope set for this book. However, we want to give at least the geometrical interpretation of the basic concepts of gauge theories. This knowledge is important for the understanding of the more subtle aspects of gauge theories. In particular, the non-perturbative structure of gauge theories is closely connected to topological properties of gauge-field configurations (Sect. 2.9). Finally, the lattice approximation of gauge theories (Sect. 2.10.1) is based on the understanding of their geometrical meaning.

2.3.1 Local gauge invariance in QED

Our starting point is the free *Dirac equation* for a fermion field with charge $Q_f e$,

$$(\mathrm{i}\partial\!\!\!/ - m)\psi(x) = 0. \tag{2.3.1}$$

It is the Euler–Lagrange equation of the Lagrangian

$$\mathcal{L} = \bar{\psi}(x)(\mathrm{i}\partial\!\!\!/ - m)\psi(x). \tag{2.3.2}$$

Both the Dirac equation (2.3.1) and the Lagrangian (2.3.2) are invariant under the *global abelian transformations* ($\partial_\mu\theta = 0$),

$$\psi(x) \to \psi'(x) = \mathrm{e}^{-\mathrm{i}Q_f e\theta}\psi(x), \quad \bar{\psi}(x) \to \bar{\psi}'(x) = \bar{\psi}(x)\mathrm{e}^{\mathrm{i}Q_f e\theta}. \tag{2.3.3}$$

They are, however, not invariant under *local gauge transformations* where θ depends on x, $\theta = \theta(x)$, because $\partial_\mu\theta(x)$ does not drop out,

$$\begin{aligned}(\mathrm{i}\partial\!\!\!/ - m)\psi'(x) &= (\mathrm{i}\partial\!\!\!/ - m)\mathrm{e}^{-\mathrm{i}Q_f e\theta(x)}\psi(x) \\ &= \mathrm{e}^{-\mathrm{i}Q_f e\theta(x)}[\mathrm{i}\partial\!\!\!/ - m + Q_f e(\partial\!\!\!/\theta(x))]\psi(x).\end{aligned} \tag{2.3.4}$$

Local gauge invariance can be achieved by introducing a *gauge field* $A_\mu(x)$ which transforms in such a way that the term $\partial\!\!\!/\theta(x)$ is cancelled. This is done by replacing ∂_μ by the *covariant derivative*

$$D_\mu = \partial_\mu + \mathrm{i}Q_f e A_\mu(x). \tag{2.3.5}$$

The resulting Dirac equation behaves under local gauge transformations as

$$\begin{aligned}&[\mathrm{i}\partial\!\!\!/ - Q_f e A\!\!\!/'(x) - m]\psi'(x)\\ &\quad = \mathrm{e}^{-\mathrm{i}Q_f e\theta(x)}[\mathrm{i}\partial\!\!\!/ - Q_f e A\!\!\!/'(x) - m + Q_f e(\partial\!\!\!/\theta(x))]\psi(x).\end{aligned} \tag{2.3.6}$$

It becomes invariant if the gauge field transforms as

$$A'_\mu(x) = A_\mu(x) + \partial_\mu\theta(x). \tag{2.3.7}$$

The essential property of the covariant derivative is

$$D'_\mu\psi'(x) = D'_\mu \mathrm{e}^{-\mathrm{i}Q_f e\theta(x)}\psi(x) = \mathrm{e}^{-\mathrm{i}Q_f e\theta(x)} D_\mu\psi(x), \tag{2.3.8}$$

or

$$D'_\mu = \mathrm{e}^{-\mathrm{i}Q_f e\theta(x)} D_\mu \mathrm{e}^{\mathrm{i}Q_f e\theta(x)}. \tag{2.3.9}$$

The *electromagnetic field-strength tensor*

$$F_{\mu\nu}(x) = \partial_\mu A_\nu(x) - \partial_\nu A_\mu(x) \tag{2.3.10}$$

is gauge-invariant.

The gauge field itself becomes a dynamical quantity by adding a gauge-field term to the Lagrangian. The simplest Lorentz- and gauge-invariant choice is[5]

$$\mathcal{L}_{\text{gauge}} = -\frac{1}{4}F_{\mu\nu}(x)F^{\mu\nu}(x). \tag{2.3.11}$$

Thus, we obtain the *Lagrangian of QED*

$$\mathcal{L}_{\text{QED}} = -\frac{1}{4}F_{\mu\nu}(x)F^{\mu\nu}(x) + \bar\psi(x)(\mathrm{i}D\!\!\!\!/ - m)\psi(x). \tag{2.3.12}$$

[5]The term $\varepsilon_{\mu\nu\rho\sigma}F^{\mu\nu}(x)F^{\rho\sigma}$ is a total derivative and therefore does not enter the equation of motion. All other gauge-invariant terms involving only the gauge field are non-renormalizable (cf. Sect. 2.5.1.3).

The corresponding field equations read

$$(\mathrm{i}\not{D} - m)\psi(x) = (\mathrm{i}\not{\partial} - Q_f e \not{A}(x) - m)\psi(x) = 0, \tag{2.3.13}$$
$$\partial^\mu F_{\mu\nu}(x) = Q_f e \bar{\psi}(x)\gamma_\nu \psi(x). \tag{2.3.14}$$

By construction these equations and the Lagrangian (2.3.12) are invariant under the *abelian gauge transformations*

$$\begin{aligned} &A_\mu(x) \to A_\mu(x) + \partial_\mu \theta(x), \\ &\psi(x) \to \mathrm{e}^{-\mathrm{i}Q_f e\theta(x)}\psi(x), \qquad \bar{\psi}(x) \to \bar{\psi}(x)\mathrm{e}^{\mathrm{i}Q_f e\theta(x)}. \end{aligned} \tag{2.3.15}$$

The equations above are valid for QED with one fermion with charge $Q_f e$. In general, one has to sum over all elementary charged fermions f with masses m_f and charges Q_f.

Thus, the principle of *minimal gauge-invariant coupling to a dynamical gauge field* introduces the interaction into a free theory with charge symmetry (global gauge invariance). This principle is highly successful for the electromagnetic interaction described in QED.

2.3.2 Geometry of non-abelian gauge symmetry

In this section we define gauge transformations, the covariant derivative, gauge fields, and the field-strength tensor of non-abelian gauge fields, and give their geometric interpretation.

2.3.2.1 Non-abelian symmetry transformations

We consider a multiplet of fields, $\psi(x) = (\psi_i(x))$, $i = 1, \ldots, n$, e.g. Dirac fields, which transforms according to a representation $\mathcal{D}_g$ of a symmetry group G with elements $g \in G$. The infinitesimal generators T^a, $a = 1, \ldots, \dim G$, the charges of G, are represented by $n \times n$ matrices, e.g. $\tau^a/2$ for SU(2) or $\lambda^a/2$ for SU(3) (cf. Sect. 1.2.2.2). Then, the local *gauge transformations* read

$$\psi(x) \to \psi'(x) = \mathcal{D}_g \psi(x), \qquad \mathcal{D}_g = \mathrm{e}^{\mathrm{i}\theta^a T^a}, \tag{2.3.16}$$

or in components

$$\psi'_i(x) = \mathcal{D}_{g,ij}\psi_j(x).$$

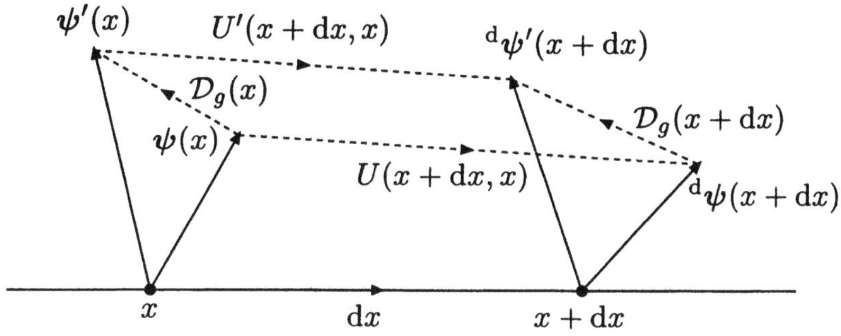

Fig. 2.4 Geometrical interpretation of an (infinitesimal) parallel displacement and the covariant derivative

The extension of global internal symmetry transformations, θ^a = constant, to local gauge transformations $\theta^a(x)$, i.e. space–time-dependent group elements $g(\theta(x)) = g(x)$ and representations $\mathcal{D}_g(\theta(x)) = \mathcal{D}_g(x)$ are discussed in the following using the concepts of differential geometry.

2.3.2.2 Gauge field as connection

The field equations connect the field vectors $\psi(x)$ at space–time points x and $x + \mathrm{d}x$. The corresponding infinitesimal *parallel displacement* $U(x + \mathrm{d}x, x)$ of the field $\psi(x)$,

$$ {}^{\mathrm{d}}\psi(x + \mathrm{d}x) = U(x + \mathrm{d}x, x)\psi(x), \tag{2.3.17}$$

is described by a *connection*, in physics called *gauge field* or *gauge potential* $\mathcal{A}^\mu(x)$, an element of the Lie algebra of G,

$$U(x + \mathrm{d}x, x) = \mathbf{1} + \mathrm{i}\mathcal{A}_\mu(x)\mathrm{d}x^\mu, \tag{2.3.18}$$

where $U, \mathbf{1}$, and $\mathcal{A}_\mu$ are $n \times n$ matrices. The compatibility of gauge transformations $\psi'(x) = \mathcal{D}_g\psi(x)$ and parallel displacement U requires (cf. Fig. 2.4)

$$\begin{aligned} {}^{\mathrm{d}}\psi'(x + \mathrm{d}x) &= U'(x + \mathrm{d}x, x)\psi'(x) = U'(x + \mathrm{d}x, x)\mathcal{D}_g(x)\psi(x) \\ &\stackrel{!}{=} \mathcal{D}_g(x + \mathrm{d}x){}^{\mathrm{d}}\psi(x + \mathrm{d}x) = \mathcal{D}_g(x + \mathrm{d}x)U(x + \mathrm{d}x, x)\psi(x), \end{aligned} \tag{2.3.19}$$

i.e. the parallel displacement behaves under gauge transformations as

$$U'(x + \mathrm{d}x, x) = \mathcal{D}_g(x + \mathrm{d}x)U(x + \mathrm{d}x, x)\mathcal{D}_g^{-1}(x). \tag{2.3.20}$$

Inserting (2.3.18) and expanding to first order in $\mathrm{d}x$, we obtain for the connection

$$\begin{aligned} \mathrm{i}\mathcal{A}'_\mu(x) &= \mathrm{i}\mathcal{D}_g(x)\mathcal{A}_\mu(x)\mathcal{D}_g^{-1}(x) + \frac{\partial \mathcal{D}_g(x)}{\partial x^\mu}\mathcal{D}_g^{-1}(x) \\ &= \mathcal{D}_g(x)\left(\mathrm{i}\mathcal{A}_\mu(x) - \partial_\mu\right)\mathcal{D}_g^{-1}(x), \end{aligned} \tag{2.3.21}$$

where we used $\partial_\mu[\mathcal{D}_g(x)\mathcal{D}_g^{-1}(x)] = \partial_\mu \mathbf{1} = 0$ in the second equation. The gauge potential transforms inhomogeneously under the local gauge transformations.

2.3.2.3 Covariant derivative

The *covariant derivative*

$$D_\mu = \partial_\mu - \mathrm{i}\mathcal{A}_\mu(x), \tag{2.3.22}$$

behaves under gauge transformations as

$$\begin{aligned} \mathcal{D}_g D_\mu \mathcal{D}_g^{-1} &= \mathcal{D}_g(\partial_\mu \mathcal{D}_g^{-1}) - \mathrm{i}\mathcal{D}_g\mathcal{A}_\mu\mathcal{D}_g^{-1} \\ &= \mathcal{D}_g(\partial_\mu \mathcal{D}_g^{-1}) - \mathrm{i}\mathcal{A}'_\mu - \mathcal{D}_g(\partial_\mu \mathcal{D}_g^{-1}) + \partial_\mu \\ &= \partial_\mu - \mathrm{i}\mathcal{A}'_\mu = D'_\mu. \end{aligned} \tag{2.3.23}$$

As a consequence, the *Dirac or Klein–Gordon equations with covariant derivatives*

$$\begin{aligned} (\mathrm{i}\not{D} - m)\psi(x) &= 0, \\ (D^\mu D_\mu + m)\phi(x) &= 0, \end{aligned} \tag{2.3.24}$$

are gauge-invariant field equations if the mass matrix m is invariant under global transformations. Since D^μ contains the gauge potential $\mathcal{A}^\mu(x)$, these are equations for fermions and scalars which are coupled minimally to the non-abelian gauge bosons.

2.3.2.4 Field-strength tensor

The parallel displacement $U(x + \mathrm{d}x, x)$ with a non-trivial gauge potential leads in the language of differential geometry to a *curvature*. It can be determined from the difference of the parallel displacements along the two paths $x \to x + \mathrm{d}x \to x + \mathrm{d}x + \mathrm{d}y$ and $x \to x + \mathrm{d}y \to x + \mathrm{d}y + \mathrm{d}x$, keeping terms up to $\mathcal{O}(\mathrm{d}x\mathrm{d}y)$,

$$\begin{aligned} &U(x + \mathrm{d}x + \mathrm{d}y, x + \mathrm{d}x)U(x + \mathrm{d}x, x) \\ &\qquad - U(x + \mathrm{d}y + \mathrm{d}x, x + \mathrm{d}y)U(x + \mathrm{d}y, x) \\ &\quad = [1 + \mathrm{i}\mathcal{A}_\mu(x + \mathrm{d}x)\mathrm{d}y^\mu][1 + \mathrm{i}\mathcal{A}_\nu(x)\mathrm{d}x^\nu] \\ &\qquad - [1 + \mathrm{i}\mathcal{A}_\mu(x + \mathrm{d}y)\mathrm{d}x^\mu][1 + \mathrm{i}\mathcal{A}_\nu(x)\mathrm{d}y^\nu] \\ &\quad = \mathrm{i}(\partial_\mu \mathcal{A}_\nu(x) - \partial_\nu \mathcal{A}_\mu(x) - \mathrm{i}[\mathcal{A}_\mu(x), \mathcal{A}_\nu(x)])\mathrm{d}x^\mu \mathrm{d}y^\nu + \dots \\ &\quad = \mathrm{i}\mathcal{F}_{\mu\nu}(x)\mathrm{d}x^\mu \mathrm{d}y^\nu + \dots . \end{aligned} \tag{2.3.25}$$

The antisymmetric *field-strength tensor* $\mathcal{F}_{\mu\nu}(x)$

$$\mathcal{F}_{\mu\nu}(x) = \partial_\mu \mathcal{A}_\nu(x) - \partial_\nu \mathcal{A}_\mu(x) - \mathrm{i}[\mathcal{A}_\mu(x), \mathcal{A}_\nu(x)] \tag{2.3.26}$$

is the non-abelian generalization of the electromagnetic one from which it differs by the bilinear term in the gauge field.

A simple calculation allows us to express $\mathcal{F}_{\mu\nu}$ as commutator of covariant derivatives

$$\mathcal{F}_{\mu\nu}(x) = \mathrm{i}[D_\mu, D_\nu]. \tag{2.3.27}$$

This gives for the behaviour under gauge transformations

$$\mathcal{F}'_{\mu\nu}(x) = \mathcal{D}_g(x)\mathcal{F}_{\mu\nu}(x)\mathcal{D}_g^{-1}(x), \tag{2.3.28}$$

i.e. in the non-abelian case the field strength is no more gauge-invariant but transforms homogeneously. From (2.3.21) and (2.3.28) the *infinitesimal gauge transformations* of the potential and the field strength follow:

$$\begin{aligned} \delta\mathcal{A}_\mu(x) &= (\mathrm{i}[T^a, \mathcal{A}_\mu(x)] + T^a\partial_\mu)\,\delta\theta^a(x), \\ \delta\mathcal{F}_{\mu\nu}(x) &= \mathrm{i}[T^a, \mathcal{F}_{\mu\nu}(x)]\delta\theta^a(x). \end{aligned} \tag{2.3.29}$$

Using the Jacobi identity for commutators, we obtain the *Bianchi identity* for the field-strength tensor

$$[D_\mu, \mathcal{F}_{\nu\rho}] + [D_\nu, \mathcal{F}_{\rho\mu}] + [D_\rho, \mathcal{F}_{\mu\nu}] = 0 \tag{2.3.30}$$

from (2.3.27). This equation generalizes the homogeneous Maxwell equations to the non-abelian case.

2.3.2.5 Component fields

Since $\mathcal{A}_\mu$, D_μ and $\mathcal{F}_{\mu\nu}$ are elements of the same representation of the Lie algebra, they can be expanded in terms of the representation matrices T^a_{ij},

$$\begin{aligned}
\mathcal{A}_{\mu,ij}(x) &= gA^a_\mu(x)T^a_{ij}, \\
D_{\mu,ij}(x) &= \partial_\mu \delta_{ij} - \mathrm{i}gA^a_\mu(x)T^a_{ij}, \\
\mathcal{F}_{\mu\nu,ij}(x) &= gF^a_{\mu\nu}(x)T^a_{ij}.
\end{aligned} \tag{2.3.31}$$

The proportionality constant g is fixed by physical considerations. The infinitesimal *gauge transformations* of the components $A^a_\mu(x)$ and $F^a_{\mu\nu}(x)$,

$$\begin{aligned}
\delta A^a_\mu(x) &= f^{abc}A^b_\mu(x)\delta\theta^c(x) + \frac{1}{g}\partial_\mu\delta\theta^a(x) = \frac{1}{g}D^{ab}_\mu\delta\theta^b(x), \\
\delta F^a_{\mu\nu}(x) &= f^{abc}F^b_{\mu\nu}(x)\delta\theta^c(x),
\end{aligned} \tag{2.3.32}$$

show explicitly that the gauge potential and the field strength are elements of the Lie algebra of the symmetry group G. The appearance of the structure constants of the Lie algebra f^{abc} [cf. (1.2.47) and (1.2.55)] indicates that the components transform according to the adjoint representation of the gauge group, and D^{ab}_μ is the corresponding covariant derivative.

In components the relation between the potential and the field strength reads

$$F^a_{\mu\nu} = \partial_\mu A^a_\nu(x) - \partial_\nu A^a_\mu(x) + gf^{abc}A^b_\mu(x)A^c_\nu(x). \tag{2.3.33}$$

For the non-abelian case, the field strength contains bilinear terms in the potentials.

2.3.2.6 Finite parallel transport

So far, the symmetry connection has been considered in an infinitesimal neighbourhood. Now we deal with the *parallel transporter* $U_C(y,x)$ for fields between points x and y of finite distance. For the displacement of the field $\psi(x)$ from x to y along the path C we write

$$^{\mathrm{d}}\psi(y) = U_C(y,x)\psi(x). \tag{2.3.34}$$

If the symmetry is conserved, then, as for infinitesimal transformations, $U_C(y,x)$ is a representation matrix of the symmetry group G. Under a gauge transformation $\mathcal{D}_g(x)$ the finite parallel displacement must be transformed according to [cf. (2.3.20)],

$$U'_C(y,x) = \mathcal{D}_g(y)U_C(y,x)\mathcal{D}_g^{-1}(x). \tag{2.3.35}$$

The parallel displacement along a curve C can be generated infinitesimally, i.e. the parallel transporter $U_C(y,x)$ can be calculated from $U(x+\mathrm{d}x,x)$ and thus from the gauge field $\mathcal{A}_\mu(x)$. For this purpose, the path C is described with help of a parameter s,

$$C = \Big\{\xi_\mu(s) \mid 0 \le s \le 1,\ \xi_\mu(0) = x_\mu,\ \xi_\mu(1) = y_\mu\Big\}.$$

Then it is divided up into a large number L of segments along which infinitesimal parallel displacement can be done according to (2.3.17),

$$\begin{aligned}
U_C(y,x) &= \lim_{L\to\infty}\prod_{l=0}^{L-1} U\left(\xi(\tfrac{l+1}{L}),\xi(\tfrac{l}{L})\right)\\
&= \lim_{L\to\infty}\prod_{l=0}^{L-1}\left[1+\mathrm{i}\mathcal{A}_\mu\Big(\xi(\tfrac{l}{L})\Big)\Big(\xi^\mu(\tfrac{l+1}{L})-\xi^\mu(\tfrac{l}{L})\Big)\right]\\
&\equiv \mathrm{P}\exp\left(\mathrm{i}\int_C \mathrm{d}\xi^\mu\mathcal{A}_\mu(\xi)\right).
\end{aligned} \tag{2.3.36}$$

For a general non-abelian potential, the contributions $\mathcal{A}_\mu(\xi)\Delta\xi^\mu$ of different path segments do not commute. Therefore, the product in (2.3.36) can only be formally summed to an exponential function by using the *path-ordering* operator P [Fe51]. From the last line of (2.3.36), one concludes that U_C is a solution of the differential equation

$$-\mathrm{i}\frac{\mathrm{d}}{\mathrm{d}s}\overline{U}(s) = \mathcal{A}_\mu(\xi(s))\frac{\mathrm{d}\xi^\mu}{\mathrm{d}s}\overline{U}(s), \qquad \overline{U}(0) = 1, \tag{2.3.37}$$

with $U_C(y, x) = \overline{U}(1)$.

The parallel transporter U_C can be used to perform gauge transformations so that $\mathcal{A}_\mu(x)$ has a simpler form. As an example we show the existence of a transformation for which $\mathcal{A}_0(x) \equiv 0$ (*temporal gauge*). For this purpose we choose a parallel displacement U_0 along a straight line from $y = (t, \mathbf{x})$ to $x = (0, \mathbf{x})$, i.e. parallel to the time axis,

$$U_0 = \mathrm{P}\exp\left(\mathrm{i}\int_t^0 \mathrm{d}t'\,\mathcal{A}_0(t', \mathbf{x})\right). \tag{2.3.38}$$

This parallel displacement U_0 can be interpreted as a gauge transformation of $\psi(t, \mathbf{x})$ and $\mathcal{A}_\mu(t, \mathbf{x})$

$$\begin{aligned} \psi(x) &\to \psi'(x) = U_0\,\psi(x), \\ \mathcal{A}_\mu(x) &\to \mathcal{A}'_\mu(x) = U_0(\mathcal{A}_\mu(x) + \mathrm{i}\partial_\mu)U_0^{-1}. \end{aligned} \tag{2.3.39}$$

It describes the transformation to the temporal gauge. This is confirmed by the following calculation:

$$\begin{aligned} \mathcal{A}'_0(x) &= U_0(\mathcal{A}_0 + \mathrm{i}\partial_0)U_0^{-1} \\ &= U_0\left(\mathcal{A}_0 - \mathrm{i}U_0^{-1}\frac{\mathrm{d}U_0}{\mathrm{d}t}\right)U_0^{-1} \\ &= U_0(\mathcal{A}_0 - \mathcal{A}_0)U_0^{-1} = 0. \end{aligned} \tag{2.3.40}$$

The last step follows from $\mathrm{d}U_0/\mathrm{d}t = -\mathrm{i}U_0\mathcal{A}_0(t, \mathbf{x})$, the analogue[6] of (2.3.37) for U_0.

[6] The interchange of $\mathcal{A}$ and U and the different sign with respect to (2.3.37) result from the fact that t is the lower limit of the integral in (2.3.38).

2.3.2.7 Remarks on gauge theories and fibre bundles

We complete this section by extending our discussion of the relations between the concepts of differential geometry and physics with some remarks on *fibre bundles* [Ch82, Go87]. In this framework, one considers the mapping of points x of a base space $\mathcal{M}$ on fields, i.e. vectors $\psi(x)$ of a local Hilbert space ${}^x\mathcal{H}$ at each point x, together with local transformations g of a gauge group G which map ${}^x\mathcal{H}$ into itself. The set of all elements $\{x, \psi, g\} = \{x \in \mathcal{M}, \psi \in {}^x\mathcal{H}, g \in G\}$ is a vector fibre bundle $\mathcal{F}$ with the fibre ${}^x\mathcal{H}$. The base space is, for instance, a Minkowski space or a euclidean space, and the fields can be matter fields or gauge fields. The symmetry transformations at different points are connected by finite parallel transports, or infinitesimally by covariant derivatives defined by the gauge potentials $\mathcal{A}_\mu(x)$. The different spaces ${}^x\mathcal{H}$ are isomorphic. If and only if the vector bundle is equivalent to the direct product of the space–time with a universal representation space $\mathcal{H}$ of the symmetry group, $\mathcal{M} \times \mathcal{H}$, it is called trivial.

There are non-trivial fibre bundles that have such a product structure only on covering subsets $\mathcal{M}_i \subset \mathcal{M}$. The mappings, $\mathcal{M} \to {}^x\mathcal{H}$, are continuous only locally but not globally. On the overlap of $\mathcal{M}_i$ with $\mathcal{M}_j$, the fields are connected by gauge transformations. For three overlapping subsets, these gauge transformations have to fulfill the cocycle condition (2.9.36). Such field configurations play an important rôle for the non-perturbative structure of gauge theories. We elaborate more on these aspects of gauge theories in Sects. 2.9 and 2.10.1.

2.3.3 Yang–Mills field theories

The extension of local gauge transformations to non-abelian symmetry groups makes it possible to transfer the principle of minimal gauge-invariant coupling to fields with non-abelian charge algebra. These field theories are called *Yang–Mills field theories* [Ya54, Ut56].

2.3.3.1 The Yang–Mills Lagrangian

Above we already mentioned the gauge-covariant Dirac equation (2.3.24) with the Lagrangian

$$\mathcal{L} = \overline{\psi}(x)(\mathrm{i}\not{D} - m)\psi(x) \tag{2.3.41}$$

as an example of a gauge-invariant coupling of a gauge field to a relativistic spin-1/2 field. However, in a self-contained physical theory, $\mathcal{A}_\mu$ cannot be an external gauge field but must itself be a dynamical quantity. This is achieved by adding a gauge-invariant term depending only on the gauge field to the Dirac part of the Lagrangian,

$$\mathcal{L}_{\text{gauge}} = -\frac{1}{4}F^a_{\mu\nu}(x)F^{a,\mu\nu}(x) = -\frac{1}{2g^2}\,\mathrm{Tr}\Big(\mathcal{F}_{\mu\nu}(x)\mathcal{F}^{\mu\nu}(x)\Big). \tag{2.3.42}$$

This form is motivated, apart from gauge invariance, by simplicity and similarity to QED [cf. (2.3.12)].[7] Altogether we obtain the following *Yang–Mills Lagrangian* for the gauge-invariant interactions of fermions with non-abelian *gauge bosons*

$$\mathcal{L} = -\frac{1}{4}F^a_{\mu\nu}(x)F^{a,\mu\nu}(x) + \overline{\psi}(x)(\mathrm{i}\not{D} - m)\psi(x). \tag{2.3.43}$$

Variation with respect to $\overline{\psi}(x)$ yields the Dirac equation (2.3.24). The variation with respect to the gauge field $A^a_\mu(x)$ gives the generalized Maxwell equation

$$D^\mu F^a_{\mu\nu}(x) = -g\overline{\psi}(x)T^a\gamma_\nu\psi(x). \tag{2.3.44}$$

In these equations all indices associated with the fermion multiplets of the group G have been suppressed in order to emphasize the formal similarity to QED.

2.3.3.2 Non-abelian gauge interactions of scalars

The restriction to Dirac fields is not necessary. By similar considerations an interaction mediated by the dynamical gauge field $\mathcal{A}_\mu(x)$ is introduced for a multiplet of free scalar fields $\phi = (\phi_i)$. To this end, the free Lagrangian

$$\mathcal{L}_{\text{free}} = (\partial_\mu\phi(x))^\dagger\partial^\mu\phi(x) - M^2\phi^\dagger(x)\phi(x) \tag{2.3.45}$$

[7] An $\varepsilon_{\mu\nu\rho\sigma}F^{\mu\nu}F^{\rho\sigma}$ term is Lorentz- and gauge-invariant but violates CP invariance. Moreover, it is a total derivative and does not contribute in perturbation theory. Therefore, it is not taken into account here. For non-perturbative effects see Sect. 2.9.3.4.

is extended in gauge-invariant form and is supplemented by the gauge-field part resulting in

$$\mathcal{L} = -\frac{1}{4}F^a_{\mu\nu}(x)F^{a,\mu\nu}(x) + \left(D_\mu\phi(x)\right)^\dagger D^\mu\phi(x) - M^2\phi^\dagger(x)\phi(x). \tag{2.3.46}$$

From this Lagrangian the field equations

$$D_\mu D^\mu\phi(x) + M^2\phi(x) = 0, \tag{2.3.47}$$

$$D^\mu F^a_{\mu\nu}(x) = -ig[\phi^\dagger(x)T^a D_\nu\phi(x) - (D_\nu\phi(x))^\dagger T^a\phi(x)] \tag{2.3.48}$$

follow by variation with respect to $\phi^\dagger(x)$ or $\phi(x)$ and $A^a_\mu(x)$. While (2.3.47) is the *gauge-covariant Klein–Gordon equation*, (2.3.48) shows the conserved current of the scalar fields that couples to the gauge field.

The *gauge couplings* of fermions and scalars exhaust all possible renormalizable gauge couplings. In addition, *general gauge theories* contain *Yukawa couplings* (scalar–fermion couplings) and *scalar self-couplings* up to the fourth order. These additional couplings do not contain derivatives and must respect the global gauge invariance. This fixes the general framework of renormalizable gauge theories (cf. Sect. 2.5.1.3).

2.3.3.3 The generalized Maxwell equation

A more detailed analysis of the *generalized Maxwell equation* (2.3.44) for a non-abelian gauge theory with fermions and its comparison with electrodynamics is useful. For this purpose it is written with the linear field-strength tensor

$$G^a_{\mu\nu}(x) = \partial_\mu A^a_\nu(x) - \partial_\nu A^a_\mu(x) \tag{2.3.49}$$

resulting in

$$\begin{aligned}\partial^\mu G^a_{\mu\nu} &= -gf^{abc}(2A^b_\mu\partial^\mu A^c_\nu - A^b_\nu\partial^\mu A^c_\mu - A^{b,\mu}\partial_\nu A^c_\mu) \\ &\quad -g^2 f^{abc}f^{cde}A^{b,\mu}A^d_\mu A^e_\nu - g\bar{\psi}T^a\gamma_\nu\psi \\ &\equiv ({}^AJ^a_\nu(x) + {}^\psi J^a_\nu(x)) = J^a_\nu(x).\end{aligned} \tag{2.3.50}$$

Equation (2.3.49) implies $\partial^\nu \partial^\mu G^a_{\mu\nu}(x) = 0$ and therefore conservation of the currents $J^a_\nu(x)$

$$\partial^\nu J^a_\nu(x) = 0. \tag{2.3.51}$$

Two aspects are important here: in non-abelian gauge theories, not only the matter fields but also the gauge fields are charged. Accordingly, there is a self-coupling of the gauge fields $A^a_\mu(x)$ expressed by the non-linear terms in the Maxwell equations (2.3.50), i.e. by the charged gauge-field current ${}^A J^a_\nu(x)$. The conserved currents $J^a_\nu(x)$ (2.3.50), which form the sources of the gauge field, are proportional to the Noether currents which are derived from the global symmetry of the Lagrangian (cf. Sect. 1.2.3).

The proportionality constant g which was introduced in (2.3.31), represents the *gauge coupling* constant and determines the strength of the universal interaction of all particles which bear the charge of the local symmetry group G. This is known from QED ($T^a = -Q, f^{abc} = 0$), where $e = g > 0$ is the elementary charge.

Several coupling constants can appear in the case of a non-semi-simple symmetry group G which consists of a direct product of simple factors, e.g. $G = G_2 \times G_1$ with generators T^a and S^a for G_2 and G_1, respectively. In this case, the Lie-algebra-valued potential of G can be written as

$$\mathcal{A}_\mu(x) = g_2 A^a_\mu(x) T^a + g_1 B^b_\mu(x) S^b \tag{2.3.52}$$

with different coupling constants g_2 and g_1. Because $[T^a, S^b] = 0$, the gauge potentials $A^a_\mu(x)$ and $B^b_\mu(x)$ are independent of each other. This case is of importance in the Glashow–Salam–Weinberg theory of the electroweak interaction, where $G = \mathrm{SU}(2)_\mathrm{w} \times \mathrm{U}(1)_Y$.

2.3.3.4 QCD, the Yang–Mills theory of the strong interaction

For the foundation of QCD [Fr73, We73, Gr73b], the field-theoretical formulation of hadron dynamics, several aspects played a rôle:

(1) the result of the phenomenological discussion;

(2) the success of QED; and

(3) the geometric beauty of Yang–Mills gauge theories.

Assuming that (2) and (3) provide sufficient motivation for the formulation of hadron dynamics as a Yang–Mills theory, the phenomenological discussion suggests to start with coloured *quark fields* as matter fields. As colour charges play an important part in the strong interaction of quarks (Sect. 1.3.1), they should be the source of the gauge fields, which mediate the interaction. Consequently, QCD is based on the colour symmetry group $\mathrm{SU}(3)_\mathrm{C}$. According to the dimension of $\mathrm{SU}(3)_\mathrm{C}$, eight gauge fields $A_\mu^a(x)$, $a = 1, \ldots, 8$, the *gluon fields*, mediate the interaction between the quarks. Deep-inelastic scattering and jet production in high-energy collisions provide direct evidence for these types of flavour-neutral constituents.

According to (2.3.43), the Lagrangian of QCD takes the form

$$\begin{aligned} \mathcal{L}_\mathrm{QCD} = &-\frac{1}{4}\sum_a F_{\mu\nu}^a(x)F^{a,\mu\nu}(x) \\ &+ \sum_{q,c,c'} \bar{\psi}_{q,c}(x)(\mathrm{i}\not{D}_{cc'} - m_q\delta_{cc'})\psi_{q,c'}(x), \end{aligned} \tag{2.3.53}$$

where $\psi_{q,c}(x)$ are the quark fields with masses m_q, flavour index q, and colour index c, and $A_\mu^a(x)$ are the gluon fields. The covariant derivative $D_{\mu,cc'} = \partial_\mu\delta_{cc'} - \mathrm{i}g_\mathrm{s}(\lambda^a)_{cc'}A_\mu^a(x)/2$ induces the minimal gauge-invariant interaction of strength g_s, and $F_{\mu\nu}^a(x) = \partial_\mu A_\nu^a(x) - \partial_\nu A_\mu^a(x) + g_\mathrm{s} f^{abc} A_\mu^b(x) A_\nu^c(x)$ is the *QCD field-strength tensor*. The definitions of f^{abc} and λ^a are given in Sect. 1.2.2; colour and flavour indices are summed up explicitly in $\mathcal{L}_\mathrm{QCD}$.

The Lagrangian (2.3.53) allows for the formulation of QCD as a quantum field theory. The non-abelian nature of the $\mathrm{SU}(3)_\mathrm{C}$ gauge transformations makes it necessary to treat gauge invariance and quantization in more detail. This is done in Sect. 2.4.

QCD has three important general properties:

(1) QCD is *renormalizable*. This guarantees its consistency, the validity of general physical principles, and its predictive power.

(2) QCD is asymptotically free. This establishes *perturbative QCD*, i.e. the parton model with many experimentally tested applications.

(3) In the non-perturbative regime, *lattice QCD* leads to an understanding of quark confinement and the properties of the hadrons.

These properties and the impressive agreement with experiment establish QCD as the theory of the strong interaction.

2.3.3.5 Yang–Mills theory for the electroweak interaction

There are similar arguments in favour of a Yang–Mills theory of the electroweak interaction. In Sect. 1.4 we saw that the electroweak interaction can be described by the exchange of the vector bosons γ, Z, and $W^{\pm}$. The charges corresponding to the electroweak currents generate the group $SU(2)_w \times U(1)_Y$. This symmetry turned out to be manifest after the neutral vector fields were replaced by suitable linear combinations (1.4.10). This suggests to formulate the theory of the electroweak interaction as a non-abelian gauge theory with the gauge group $SU(2)_w \times U(1)_Y$.

The resulting Electroweak Standard Model (EWSM) was constructed for leptons by S.L. Glashow [Gl61], S. Weinberg [We67], and A. Salam [Sa68] and extended to include the hadronic degrees of freedom by S.L. Glashow, J. Iliopoulos and L. Maiani [Gl70]. It is a minimal model as far as the number of required fields is concerned [GSW]. However, the non-zero masses of the weak gauge bosons seem to be in conflict with exact gauge invariance. In order to incorporate massive gauge bosons in a gauge theory one needs to introduce scalar fields that break the gauge invariance spontaneously and provide masses for the gauge bosons via the so-called Higgs mechanism. Therefore, we postpone a detailed formulation of this gauge theory to Chapt. 4.

2.4 Path-integral formulation of gauge theories

The quantization of systems with finite degrees of freedom usually is performed by constructing the Hamiltonian and postulating the Heisenberg commutation relations between canonically conjugate observables. Alternatively, an equivalent quantization using Feynman's path integral is possible. The latter is the preferred method for relativistic field theories, especially gauge theories. The reasons are two-fold.

- Lorentz invariance is not manifest in the Hamiltonian formulation since time is used as a distinguished variable.
- The four-potential of a gauge field has unphysical degrees of freedom which have to be eliminated before canonical quantization.

2.4.1 Path-integral quantization of gauge theories

Since the path-integral quantization is based on the (relativistically invariant) Lagrangian, or the action, it is manifestly Lorentz-invariant. Moreover, the treatment of the unphysical degrees of freedom is performed in a very elegant way.

2.4.1.1 Problems in the quantization of gauge theories

The straightforward application of the Feynman path-integral formulation of quantum field theory to non-abelian gauge theories leads to the ansatz

$$\begin{aligned} T\{\mathbf{J}\} &= Z\{\mathbf{J}\}/Z\{0\}, \\ Z\{\mathbf{J}\} &= \int \mathcal{D}[\mathbf{A}] \exp\Big(\mathrm{i}S\{\mathbf{A}\} + \mathrm{i}\int \mathrm{d}^4x J^{\mu,a}(x) A^a_\mu(x)\Big), \\ S\{\mathbf{A}\} &= \int \mathrm{d}^4x \Big[-\frac{1}{4} F^a_{\mu\nu}(x) F^{a,\mu\nu}(x)\Big] \end{aligned} \tag{2.4.1}$$

for the generating functional of Green functions, where $\mathbf{A} = (A^a_\mu)$ and $\mathbf{J} = (J^a_\mu)$. The measure $\mathcal{D}[\mathbf{A}] = \prod_{x,\mu,a} \mathrm{d}A^a_\mu(x)$ involves at each space–time point a product over all group and vector components of the field $A^a_\mu(x)$. The ansatz (2.4.1) using the four-potential with its unphysical degrees of freedom contains mathematical inconsistencies that make even a perturbative evaluation of $Z\{\mathbf{J}\}$ impossible. This problem can already be seen in the case of the free photon field. The evaluation of the corresponding functional integral

$$Z\{\mathbf{J}\} = \int \mathcal{D}[\mathbf{A}] \exp\Big(\mathrm{i}\int \mathrm{d}^4x \Big[\frac{1}{2} A_\mu (g^{\mu\nu}\Box - \partial^\mu\partial^\nu) A_\nu + J^\mu A_\mu\Big]\Big),$$

in analogy to (2.2.55), fails, because the operator $K^{\mu\nu} = (g^{\mu\nu}\Box - \partial^\mu\partial^\nu)$ has eigenvalues zero. Thus, the necessary condition for the existence of the gaussian integral, $(A_\mu, K^{\mu\nu}A_\nu) > 0$, is not satisfied and the result diverges, as $K^{-1}_{\mu\nu}$ does not exist.

For the correct quantization of any gauge theory it is necessary to solve this problem, which is caused by the gauge invariance of the Lagrangian $\mathcal{L}$. The singularity in (2.4.1) results from the integration over physically equivalent configurations ${}^g\mathbf{A}(x)$. These are related to $\mathbf{A}(x)$ by a gauge transformation $g(x)$ according to (2.3.21). As $\mathcal{L}$ and the measure $\mathcal{D}[\mathbf{A}]$ are gauge-invariant, the integration over the physically equivalent configurations results for each space–time point in the volume of the gauge group and thus in total in an infinite unique gauge-invariant factor. When forming the ratio $Z\{\mathbf{J}\}/Z\{0\}$ this infinite factor drops out, and a meaningful result emerges. In the following we describe the method proposed by L.D. Faddeev and V.N. Popov [Fa67] to separate off this factor. The Faddeev–Popov method can also be applied in the presence of zero eigenvalues resulting from other symmetries (cf. Sect. 2.9.3.2).

2.4.1.2 The Faddeev–Popov procedure

We consider the path integral without source terms,

$$Z_f = \int \mathcal{D}[\mathbf{A}]\, f\{\mathbf{A}\} \exp(\mathrm{i}S\{\mathbf{A}\}), \tag{2.4.2}$$

where $f\{\mathbf{A}\}$ is an arbitrary gauge-invariant functional.

Let $\mathbf{C} = (C^a\{\mathbf{A};x\})$ be a set of *local functionals of the gauge fields* that fixes the gauge in the sense that the condition $C^a\{{}^g\mathbf{A};x\} = 0$ selects for any initial field $\mathbf{A}$ one[8] solution ${}^{g_0}\mathbf{A}$ out of the set of gauge-transformed, physically equivalent potentials ${}^g\mathbf{A}$. In order to enforce the gauge-fixing condition in the functional integral we use the δ-functional $\delta\{\mathbf{C}\{{}^g\mathbf{A}\}\} = \prod_{x,a} \delta(C^a\{{}^g\mathbf{A};x\})$, which is an infinite product of Dirac δ-functions. Now consider the functional $\Delta^{-1}\{\mathbf{A}\}$, defined by

$$\Delta^{-1}\{\mathbf{A}\} = \int \mathcal{D}[\mu(g)]\, \delta\{\mathbf{C}\{{}^g\mathbf{A}\}\},$$

where $\mathcal{D}[\mu(g)] = \prod_x \mathrm{d}\mu(g(x))$ stands for the infinite product of invariant measures $\mathrm{d}\mu(g(x))$ (1.2.64) of the gauge group G at every space–time point x. The invariance property of the group measure (1.2.65) implies $\mathcal{D}[\mu(g)] = \mathcal{D}[\mu(gg')] = \mathcal{D}[\mu(g'g)]$, and thus the gauge invariance of the functional $\Delta^{-1}\{\mathbf{A}\}$,

$$\begin{aligned}\Delta^{-1}\{{}^{g'}\mathbf{A}\} &= \int \mathcal{D}[\mu(g)]\, \delta\{\mathbf{C}\{{}^{gg'}\mathbf{A}\}\} = \int \mathcal{D}[\mu(gg')]\, \delta\{\mathbf{C}\{{}^{gg'}\mathbf{A}\}\} \\ &= \int \mathcal{D}[\mu(g'')]\, \delta\{\mathbf{C}\{{}^{g''}\mathbf{A}\}\} = \Delta^{-1}\{\mathbf{A}\}.\end{aligned} \tag{2.4.3}$$

[8] In general, there may be more than one solution. These are called Gribov copies [Gr78]. For perturbation theory, however, these are irrelevant, and the method described here applies.

We insert

$$1 = \Delta\{\mathbf{A}\} \int \mathcal{D}[\mu(g)]\, \delta\{\mathbf{C}\{{}^g\mathbf{A}\}\}, \tag{2.4.4}$$

into the path integral (2.4.2), interchange the order of integrations, and use the gauge invariance of the functional $\Delta\{\mathbf{A}\}$, of the measure $\mathcal{D}[\mathbf{A}]$, of $f\{\mathbf{A}\}$, and of the action $S\{\mathbf{A}\}$,

$$\begin{aligned} Z_f &= \int \mathcal{D}[\mathbf{A}]\, \Delta\{\mathbf{A}\} \int \mathcal{D}[\mu(g)]\, \delta\{\mathbf{C}\{{}^g\mathbf{A}\}\} f\{\mathbf{A}\} \exp(\mathrm{i}S\{\mathbf{A}\}) \\ &= \int \mathcal{D}[\mu(g)] \int \mathcal{D}[{}^g\mathbf{A}]\, \Delta\{{}^g\mathbf{A}\}\delta\{\mathbf{C}\{{}^g\mathbf{A}\}\} f\{{}^g\mathbf{A}\} \exp(\mathrm{i}S\{{}^g\mathbf{A}\}) \\ &= \int \mathcal{D}[\mu(g)] \int \mathcal{D}[\mathbf{A}]\, \Delta\{\mathbf{A}\}\delta\{\mathbf{C}\{\mathbf{A}\}\} f\{\mathbf{A}\} \exp(\mathrm{i}S\{\mathbf{A}\}). \end{aligned} \tag{2.4.5}$$

The last line is obtained by the substitution ${}^g\mathbf{A} \to \mathbf{A}$ in the functional integral. The expression under the $\mathcal{D}[\mu(g)]$ integral is independent of g, and the integration over the gauge group gives a multiplicative divergent factor. This factor contributes only to the irrelevant normalization of Z_f and we can write

$$Z_f \propto \int \mathcal{D}[\mathbf{A}]\, \Delta\{\mathbf{A}\}\delta\{\mathbf{C}\{\mathbf{A}\}\} f\{\mathbf{A}\} \exp(\mathrm{i}S\{\mathbf{A}\}). \tag{2.4.6}$$

This expression contains the δ-functional and therefore defines Z_f in the $C^a\{\mathbf{A};x\} = 0$ gauge.

We still have to evaluate $\Delta\{\mathbf{A}\}$. We use the gauge invariance of $\Delta\{\mathbf{A}\}$ to choose $\mathbf{A}$ in such a way that it satisfies the gauge condition $C^a\{\mathbf{A};x\} = 0$. Then because of the δ-functional, we need to consider only infinitesimal gauge transformations $g = 1 + \mathrm{i}\theta^a T^a$, $\theta^a \ll 1$, and the invariant group measure takes the simple form $\mathcal{D}[\mu(g)] \propto \mathcal{D}[\boldsymbol{\theta}]$. A change of integration variables yields

$$\begin{aligned} \Delta^{-1}\{\mathbf{A}\} &\propto \int \mathcal{D}[\boldsymbol{\theta}]\delta\{\mathbf{C}\{{}^\theta\mathbf{A}\}\} \\ &= \int \mathcal{D}[\mathbf{C}] \left(\det \frac{\delta C^b\{{}^\theta\mathbf{A};x\}}{\delta\theta^c(y)}\right)^{-1} \delta\{\mathbf{C}\} \\ &= \left(\det \frac{\delta C^b\{{}^\theta\mathbf{A};x\}}{\delta\theta^c(y)}\right)^{-1}\Bigg|_{C^a\{{}^\theta\mathbf{A};x\}=0}, \end{aligned} \tag{2.4.7}$$

where ${}^{\theta}\mathbf{A}$ stands for ${}^{g}\mathbf{A}$ with $g = 1+\mathrm{i}\theta^a T^a$. Since the condition $C^a\{{}^{\theta}\mathbf{A}_\mu; x\} = 0$ is fulfilled for $\theta = 0$, we finally have

$$\Delta\{\mathbf{A}\} = \det \left.\frac{\delta C^b\{{}^{\theta}\mathbf{A}; x\}}{\delta\theta^c(y)}\right|_{\theta=0} \equiv \det M^{bc}(x,y). \tag{2.4.8}$$

We have to calculate the determinant of a matrix $M^{bc}(x,y)$, the *Faddeev–Popov determinant*, in both space–time and group indices,

$$M^{bc}(x,y) = \left.\frac{\delta C^b\{{}^{\theta}\mathbf{A}; x\}}{\delta\theta^c(y)}\right|_{\theta=0} = \int \mathrm{d}^4 z \left.\frac{\delta C^b\{{}^{\theta}\mathbf{A}; x\}}{\delta A^d_\nu(z)} \frac{\delta A^d_\nu(z)}{\delta\theta^c(y)}\right|_{\theta=0}. \tag{2.4.9}$$

As $\delta\mathbf{A}/\delta\boldsymbol{\theta}$ is proportional to the covariant derivative [cf. (2.3.32)], we find

$$\begin{aligned} M^{bc}(x,y) &= \int \mathrm{d}^4 z \frac{\delta C^b\{\mathbf{A}; x\}}{\delta A^d_\nu(z)} \frac{1}{g} D^{dc}_\nu(z) \delta^{(4)}(y-z) \\ &= \frac{\delta C^b\{\mathbf{A}; x\}}{\delta A^d_\nu(y)} \frac{1}{g} D^{dc}_\nu(y). \end{aligned} \tag{2.4.10}$$

Choosing for instance as linear local *gauge-fixing functional*

$$C^a\{\mathbf{A}; x\} = \partial^\mu A^a_\mu(x), \tag{2.4.11}$$

we obtain

$$\begin{aligned} M^{bc}(x,y) &= \int \mathrm{d}^4 z\, \partial^\mu_x \delta^{(4)}(x-z) \frac{1}{g} D^{bc}_\mu(z) \delta^{(4)}(y-z) \\ &= \frac{1}{g} \partial^\mu_x D^{bc}_\mu(x) \delta^{(4)}(y-x). \end{aligned} \tag{2.4.12}$$

Note that (2.4.8) implicitly assumes that $C^a\{\mathbf{A}; x\} = 0$. Since this is enforced by the δ-functional, $\Delta\{\mathbf{A}\}$ can be replaced by $\det M$ under the path integral (2.4.6).

2.4.1.3 The effective Lagrangian for a gauge theory

It is convenient to rewrite the path integral (2.4.6) in such a way that all contributions appear in the form of an effective Lagrangian.

To this end, we first note that the above construction works in the same way for gauge-fixing conditions of the form

$$C^a\{\mathbf{A};x\} - c^a(x) = 0, \tag{2.4.13}$$

where $\mathbf{c}(x) = (c^a(x))$ are arbitrary functions of x, resulting in

$$Z_f^{\mathbf{c}} = \int \mathcal{D}[\mathbf{A}]\,\Delta\{\mathbf{A}\}\delta\{\mathbf{C}\{\mathbf{A}\} - \mathbf{c}\} f\{\mathbf{A}\} \exp(\mathrm{i}S\{\mathbf{A}\}). \tag{2.4.14}$$

Because $c^a(x)$ is not affected by gauge transformations, $M^{ab}(x,y)$ is the same for all gauge conditions of the class (2.4.13). Moreover, by construction $Z_f^{\mathbf{c}}$ is gauge-invariant and consequently the same for different gauge-fixing conditions, i.e. independent of $c^a(x)$. Carrying out the functional integral over $\mathcal{D}\{\mathbf{c}\}$ with any weight function will thus only change the (irrelevant) normalization of Z_f. It is common to use a gaussian weight function resulting in

$$\begin{aligned} Z_f \propto & \int \mathcal{D}[\mathbf{c}] \exp\left(-\frac{\mathrm{i}}{2\xi}\int \mathrm{d}^4x\, c^a(x)c^a(x)\right) \\ & \times \int \mathcal{D}[\mathbf{A}] \det M\, \delta\{\mathbf{C}\{\mathbf{A}\} - \mathbf{c}\} f\{\mathbf{A}\} \exp(\mathrm{i}S\{\mathbf{A}\}) \\ = & \int \mathcal{D}[\mathbf{A}] \det M\, f\{\mathbf{A}\} \exp(\mathrm{i}S\{\mathbf{A}\}) \\ & \times \exp\left(-\frac{\mathrm{i}}{2\xi}\int \mathrm{d}^4x\, C^a\{\mathbf{A};x\}C^a\{\mathbf{A};x\}\right). \end{aligned} \tag{2.4.15}$$

Finally, according to the Fubini formula (2.2.27), the functional determinant $\det M$ can be expressed as a functional integral over anticommuting scalar fields $\mathbf{u}(x) = (u^a(x))$ and $\bar{\mathbf{u}}(x) = (\bar{u}^a(x))$,

$$\det M \propto \int \mathcal{D}[\mathbf{u}]\mathcal{D}[\bar{\mathbf{u}}] \exp\left(-\mathrm{i}g \int \mathrm{d}^4x\, \mathrm{d}^4y\, \bar{u}^a(x) M^{ab}(x,y) u^b(y)\right),$$

where a factor $-g$ has been introduced for later convenience. Note that the fields $u^a(x)$ and $\bar{u}^a(x)$ are independent of each other and, in particular, not related by hermitian conjugation. The auxiliary Grassmann-valued fields $u^a(x)$ and $\bar{u}^a(x)$ are called *Faddeev–Popov ghost fields.* These scalar anticommuting fields violate the spin-statistics theorem. This is unproblematic as they are unphysical degrees of freedom which occur in perturbative calculations only inside loops but never as external fields or particles.

Thus, we finally arrive at the following representation for the path integral

$$
\begin{aligned}
Z_f &= \int \mathcal{D}[\mathbf{A}]\mathcal{D}[\mathbf{u}]\mathcal{D}[\bar{\mathbf{u}}]\, f\{\mathbf{A}\} \exp\left(\mathrm{i}S\{\mathbf{A}\}\right) \\
&\quad \times \exp\left(-\frac{\mathrm{i}}{2\xi}\int \mathrm{d}^4x\, C^a\{\mathbf{A};x\}C^a\{\mathbf{A};x\}\right) \\
&\quad \times \exp\left(-\mathrm{i}g\int \mathrm{d}^4x\,\mathrm{d}^4y\,\bar{u}^a(x)M^{ab}(x,y)u^b(y)\right) \\
&= \int \mathcal{D}[\mathbf{A}]\mathcal{D}[\mathbf{u}]\mathcal{D}[\bar{\mathbf{u}}] \exp\left(\mathrm{i}\int \mathrm{d}^4x\, \mathcal{L}_{\text{eff}}\right)
\end{aligned}
\tag{2.4.16}
$$

with the *effective Lagrangian for a quantized gauge theory*

$$
\begin{aligned}
\mathcal{L}_{\text{eff}} &= \mathcal{L}_{\text{gauge}} + \mathcal{L}_{\text{fix}} + \mathcal{L}_{\text{ghost}}\,, \\
\mathcal{L}_{\text{gauge}} &= -\frac{1}{4}F^a_{\mu\nu}(x)F^{a,\mu\nu}(x), \\
\mathcal{L}_{\text{fix}} &= -\frac{1}{2\xi}C^a\{\mathbf{A};x\}C^a\{\mathbf{A};x\}, \\
\mathcal{L}_{\text{ghost}} &= -g\int \mathrm{d}^4y\,\bar{u}^a(x)M^{ab}(x,y)u^b(y) \\
&= -\int \mathrm{d}^4z\,\bar{u}^a(x)\frac{\delta C^a\{\mathbf{A};x\}}{\delta A^c_\nu(z)}D^{cb}_\nu u^b(z).
\end{aligned}
\tag{2.4.17}
$$

In the Lorentz gauge (2.4.11), the Faddeev–Popov-ghost Lagrangian reads

$$
\begin{aligned}
\mathcal{L}_{\text{ghost}} &= -\int \mathrm{d}^4z\,\bar{u}^a(x)\partial^\mu_x D^{ab}_{x,\mu}\delta^{(4)}(x-z)u^b(z) \\
&= -\bar{u}^a(x)\partial^\mu D^{ab}_\mu u^b(x) \\
&= -\bar{u}^a(x)\partial^\mu(\partial_\mu\delta^{ab} - gf^{abc}A^c_\mu(x))u^b(x).
\end{aligned}
\tag{2.4.18}
$$

As in this case, the Faddeev–Popov Lagrangian is local as long as the gauge-fixing functional $C^a\{\mathbf{A};x\}$ is local in the fields.

Note that (2.4.16) differs from (2.4.2) only by a constant normalization factor which drops out in $T\{\mathbf{J}\}$.

The path integral (2.4.16) is suitable for perturbative evaluation. The Feynman rules can be derived from the effective Lagrangian (2.4.17) as usual (cf. Sect. 2.2.5). They depend on the gauge via $\mathcal{L}_{\mathrm{fix}}$ and $\mathcal{L}_{\mathrm{ghost}}$, but the final physical results, such as S-matrix elements, do not because the path integral is gauge-invariant by construction.

The result (2.4.16) suggests to define the *generating functional of Green functions for a non-abelian gauge theory* as follows

$$Z\{\mathbf{J},\mathbf{J}_u,\mathbf{J}_{\bar{u}}\} = \int \mathcal{D}[\mathbf{A}]\mathcal{D}[\mathbf{u}]\mathcal{D}[\bar{\mathbf{u}}] \times \exp\left(\mathrm{i}\int \mathrm{d}^4x\left[\mathcal{L}_{\mathrm{eff}} + J^{\mu,a}A^a_\mu + J^a_u u^a - \bar{u}^a J^a_{\bar{u}}\right]\right), \qquad (2.4.19)$$

where we have introduced besides the sources for the gauge fields also anticommuting sources $J^a_{\bar{u}}$, J^a_u for the ghost fields. Since the source terms are not gauge-invariant, (2.4.19) differs from (2.4.1) and depends on the gauge-fixing functional. The same applies to the Green functions that are obtained as functional derivatives. Nevertheless, physical observables, as for instance S-matrix elements, that are calculated via these Green functions are unique and independent of $C^a\{\mathbf{A};x\}$.

The generalization to gauge theories including matter fields is straightforward. One uses the appropriate classical Lagrangian $\mathcal{L}$ and introduces sources for the matter fields. Gauge-fixing and ghost terms remain unchanged.

In the following we consider a gauge theory with fermions $\psi = (\psi_i)$ and $\bar{\psi} = (\bar{\psi}_i)$ that transform according to the representation T^a of the gauge group. If we denote the corresponding anticommuting sources by $\mathbf{J}_\psi$ and $\mathbf{J}_{\bar{\psi}}$ the generating functional reads

$$Z\{\mathbf{J},\mathbf{J}_u,\mathbf{J}_{\bar{u}},\mathbf{J}_\psi,\mathbf{J}_{\bar{\psi}}\} = \int \mathcal{D}[\mathbf{A}]\mathcal{D}[\mathbf{u}]\mathcal{D}[\bar{\mathbf{u}}]\mathcal{D}[\psi]\mathcal{D}[\bar{\psi}] \times \exp\left(\mathrm{i}\int \mathrm{d}^4x\left[\mathcal{L}_{\mathrm{eff}} + J^{\mu,a}A^a_\mu + J^a_u u^a - \bar{u}^a J^a_{\bar{u}} + J^i_\psi \psi_i - \bar{\psi}_i J^i_{\bar{\psi}}\right]\right) \qquad (2.4.20)$$

with

$$\mathcal{L}_{\text{eff}} = \mathcal{L}_{\text{gauge}} + \mathcal{L}_{\text{ferm}} + \mathcal{L}_{\text{fix}} + \mathcal{L}_{\text{ghost}},$$
$$\mathcal{L}_{\text{ferm}} = \bar{\psi}_i(x)(\not{D}_{ij} - m_i\delta_{ij})\psi_j(x), \tag{2.4.21}$$

and $\mathcal{L}_{\text{gauge}}$, $\mathcal{L}_{\text{fix}}$, and $\mathcal{L}_{\text{ghost}}$ as in (2.4.17).

2.4.1.4 The effective Lagrangian of QED

In QED the situation is simplified by the fact that all structure constants f^{abc} vanish and the representation matrices T^a are trivial: $T^a = -Q$, $g = e$. (In the following we put $Q = -1$.) As a consequence the Faddeev–Popov-ghost fields are free fields and the generating functional factorizes into a part containing the ghost fields and a ghost-less part. The ghost part gives only an irrelevant normalization factor and can be omitted. Considering only one fermion field ψ, we can write for the *generating functional of QED* in the gauge (2.4.11)

$$Z\{\mathbf{J}\} = \int \mathcal{D}[\mathbf{A}]\mathcal{D}[\psi]\mathcal{D}[\bar{\psi}] \exp\left(\mathrm{i}\int \mathrm{d}^4x\, \mathcal{L}_{\text{tot}}\right) \tag{2.4.22}$$

with

$$\begin{aligned} \mathcal{L}_{\text{tot}} &= \mathcal{L}_{\text{gauge}} + \mathcal{L}_{\text{fix}} + \mathcal{L}_{\text{ferm}} \\ &\quad + J^\mu(x)A_\mu(x) - \bar{\psi}(x)J_{\bar{\psi}}(x) + J_\psi(x)\psi(x), \\ \mathcal{L}_{\text{gauge}} &= -\frac{1}{4}F_{\mu\nu}(x)F^{\mu\nu}(x), \qquad \mathcal{L}_{\text{fix}} = -\frac{1}{2\xi}\left[\partial^\mu A_\mu(x)\right]^2, \\ \mathcal{L}_{\text{ferm}} &= \bar{\psi}(x)\left[\mathrm{i}\gamma_\mu(\partial^\mu - \mathrm{i}eA^\mu) - M\right]\psi(x). \end{aligned} \tag{2.4.23}$$

2.4.2 Feynman rules for gauge theories

2.4.2.1 Derivation

If the gauge coupling is small, gauge theories can be evaluated perturbatively. The Feynman rules can be derived directly from the effective Lagrangian. For a gauge theory with a multiplet of fermionic matter fields $\psi = (\psi_i)$ in a representation T^a of the gauge group and in the gauge (2.4.11) the effective Lagrangian reads

$$\begin{aligned}\mathcal{L}_{\text{eff}} = &-\frac{1}{4}F^a_{\mu\nu}(x)F^{a,\mu\nu}(x) - \frac{1}{2\xi}(\partial^\mu A^a_\mu(x))(\partial^\nu A^a_\nu(x))\\ &- \bar{u}^a(x)\partial^\mu D^{ab}_\mu u^b(x) + \bar{\psi}_i(x)(\not{D}_{ij} - m_i\delta_{ij})\psi_j(x)\end{aligned}$$

or more explicitly

$$\begin{aligned}\mathcal{L}_{\text{eff}} = &-\frac{1}{2}\partial^\mu A^a_\nu(x)\left(\partial_\mu A^{a,\nu}(x) - \partial^\nu A^a_\mu(x)\right)\\ &- \frac{1}{2\xi}(\partial^\mu A^a_\mu(x))(\partial^\nu A^a_\nu(x))\\ &- gf^{abc}A^a_\mu(x)A^b_\nu(x)\partial^\mu A^{c,\nu}(x)\\ &- \frac{1}{4}g^2 f^{abc}f^{ade}A^b_\mu(x)A^c_\nu(x)A^{d,\mu}(x)A^{e,\nu}(x)\\ &- \bar{u}^a(x)\left(\partial^\mu\partial_\mu u^a(x) - gf^{abc}\partial^\mu\left(A^c_\mu(x)u^b(x)\right)\right)\\ &+ \bar{\psi}_i(x)(\delta_{ij}\mathrm{i}\not{\partial} + gT^a_{ij}\not{A}^a - m_i\delta_{ij})\psi_j(x). \end{aligned} \qquad (2.4.24)$$

For the gauge group SU(3) and the fermions in the fundamental representation ($T^a = \lambda^a/2$) this is the *effective Lagrangian of QCD.*

The Feynman rules can be derived most easily by using the fact that the vertices are the tree-level vertex functions. Their generating functional is, according to (2.2.83), directly given by the space–time integral of the Lagrangian.

As an example, we derive the three-gauge-boson vertex. From the tree-level vertex functional

$$\Gamma_0\{\mathbf{A}, \mathbf{u}, \bar{\mathbf{u}}, \psi, \bar{\psi}\} = \int \mathrm{d}^4x\, \mathcal{L}_{\text{eff}}, \qquad (2.4.25)$$

the corresponding vertex function in configuration space is obtained as

$$
\begin{aligned}
\mathrm{i}\Gamma^{abc}_{0,\mu\nu\rho}(x,y,z) &= \left.\frac{\mathrm{i}\delta^3\Gamma_0}{\delta A^{a,\mu}(x)\delta A^{b,\nu}(y)\delta A^{c,\rho}(z)}\right|_{\mathbf{A}=\mathbf{u}=\bar{\mathbf{u}}=\psi=\overline{\psi}=0} \\
&= \frac{\delta^3}{\delta A^{a,\mu}(x)\delta A^{b,\nu}(y)\delta A^{c,\rho}(z)} \\
&\quad \left(-\mathrm{i}\int \mathrm{d}^4x'\, g f^{a'b'c'} A^{a'}_{\mu'}(x') A^{b'}_{\nu'}(x')\partial^{\mu'} A^{c',\nu'}(x')\right) \\
&= -\mathrm{i}g f^{a'b'c'} \int \mathrm{d}^4x' \Big[\delta^{aa'} g_{\mu\mu'}\delta^{(4)}(x-x')\delta^{bb'} g_{\nu\nu'}\delta^{(4)}(y-x') \\
&\qquad \partial^{\mu'}_{x'}\left(\delta^{cc'} g^{\nu'}_{\rho}\delta^{(4)}(z-x')\right) + 5 \text{ permutations}\Big] \\
&= -\mathrm{i}g f^{abc} \\
&\quad \times\Big[g_{\nu\rho}\delta^{(4)}(y-x)\partial^x_\mu\delta^{(4)}(z-x) - g_{\rho\nu}\delta^{(4)}(z-x)\partial^x_\mu\delta^{(4)}(y-x) \\
&\quad + g_{\rho\mu}\delta^{(4)}(z-y)\partial^y_\nu\delta^{(4)}(x-y) - g_{\mu\rho}\delta^{(4)}(x-y)\partial^y_\nu\delta^{(4)}(z-y) \\
&\quad + g_{\mu\nu}\delta^{(4)}(x-z)\partial^z_\rho\delta^{(4)}(y-z) - g_{\nu\mu}\delta^{(4)}(y-z)\partial^z_\rho\delta^{(4)}(x-z)\Big].
\end{aligned}
\tag{2.4.26}
$$

Fourier transformation to momentum space yields

$$
\begin{aligned}
\mathrm{i}\tilde{\Gamma}^{abc}_{0,\mu\nu\rho}(p,q,r) &= \int \mathrm{d}^4x\,\mathrm{d}^4y\,\mathrm{d}^4z\,\mathrm{i}\Gamma^{abc}_{0,\mu\nu\rho}(x,y,z)\exp(-\mathrm{i}(xp+yq+zr)) \\
&= g f^{abc}\int \mathrm{d}^4x \Big[g_{\nu\rho}(p+q)_\mu - g_{\rho\nu}(p+r)_\mu + g_{\rho\mu}(q+r)_\nu \\
&\quad - g_{\mu\rho}(p+q)_\nu + g_{\mu\nu}(p+r)_\rho - g_{\nu\mu}(q+r)_\rho\Big]\exp\left(-\mathrm{i}x(p+q+r)\right) \\
&= g f^{abc}\Big[g_{\nu\rho}(q-r)_\mu + g_{\mu\nu}(r-p)_\nu + g_{\mu\nu}(p-q)_\rho\Big] \\
&\quad \times(2\pi)^4\delta^{(4)}(p+q+r),
\end{aligned}
\tag{2.4.27}
$$

where p, q, r denote the incoming momenta. In order to integrate over the δ-functions we performed a partial integration and dropped the surface terms. The momentum-conservation δ-function is a consequence of translation invariance and can be split off by defining

$$
\tilde{\Gamma}^{abc}_{0,\mu\nu\rho}(p,q,r) = (2\pi)^4\delta^{(4)}(p+q+r)\Gamma^{abc}_{0,\mu\nu\rho}(p,q,r). \tag{2.4.28}
$$

Now $\mathrm{i}\Gamma^{abc}_{0,\mu\nu\rho}(p,q,r)$ is the vertex factor to be used in the Feynman rules. The Feynman rules for all other vertices can be obtained in the same way.

The propagators can be directly evaluated by performing the path integral over the free part of the effective Lagrangian as in Sect. 2.2.4.1. Alternatively they are given by the inverse of the corresponding two-point functions [cf. (2.2.42)]. For the gauge-boson propagator we first determine the two-gauge-boson vertex function

$$\begin{aligned}
\mathrm{i}\Gamma^{ab}_{0,\mu\nu}(x,y) &= \frac{\delta}{\delta A^a_\mu(x)\delta A^b_\nu(y)}\mathrm{i}\int \mathrm{d}^4x'\Big[-\frac{1}{2\xi}(\partial^{\mu'}A^{a'}_{\mu'}(x'))(\partial^{\nu'}A^{a'}_{\nu'}(x')) \\
&\quad -\frac{1}{2}\left(\partial^{\mu'}A^{a'}_{\nu'}(x')\right)\left(\partial_{\mu'}A^{a',\nu'}(x')-\partial^{\nu'}A^{a'}_{\mu'}(x')\right)\Big] \\
&= \mathrm{i}\int \mathrm{d}^4x'\Big[-\left(\partial^{\mu'}\delta^{aa'}g^{\mu}_{\nu'}\delta^{(4)}(x-x')\right) \\
&\quad \times\left(\partial_{\mu'}\delta^{ba'}g^{\nu\nu'}\delta^{(4)}(y-x')-\partial^{\nu'}\delta^{ba'}g^{\nu}_{\mu'}\delta^{(4)}(y-x')\right) \\
&\quad -\frac{1}{\xi}\left(\partial^{\mu'}\delta^{aa'}g^{\mu}_{\mu'}\delta^{(4)}(x-x')\right)\left(\partial^{\nu'}\delta^{ba'}g^{\nu}_{\nu'}\delta^{(4)}(y-x')\right)\Big] \\
&= \mathrm{i}\delta^{ab}\left[\partial^{\mu'}_x\partial^x_{\mu'}g^{\mu\nu}-\partial^\nu_x\partial^\mu_x\left(1-\frac{1}{\xi}\right)\right]\delta^{(4)}(y-x), \qquad (2.4.29)
\end{aligned}$$

where we again performed a partial integration. In momentum space this yields

$$\begin{aligned}
\mathrm{i}\tilde{\Gamma}^{ab}_{0,\mu\nu}(p,q) &= \int \mathrm{d}^4x\,\mathrm{d}^4y\,\mathrm{i}\Gamma^{ab}_{0,\mu\nu}(x,y)\exp\left(-\mathrm{i}(xp+yq)\right) \\
&= -\mathrm{i}\delta^{ab}\left[p^2g^{\mu\nu}-p^\mu p^\nu\left(1-\frac{1}{\xi}\right)\right](2\pi)^4\delta^{(4)}(p+q). \qquad (2.4.30)
\end{aligned}$$

According to (2.2.42), the propagator is given by the negative inverse of the two-point function. Splitting off the δ-functions, the gauge-boson Feynman propagator in momentum space is obtained as

$$\mathrm{i}\Delta^{ab}_{\mathrm{F},\mu\nu}(p) = -[\mathrm{i}\Gamma^{ab}_{0,\mu\nu}(p,-p)]^{-1} = \mathrm{i}\delta^{ab}\frac{-g_{\mu\nu}+(1-\xi)\frac{p_\mu p_\nu}{p^2}}{p^2+\mathrm{i}\varepsilon}. \qquad (2.4.31)$$

The $\mathrm{i}\varepsilon$ results from an infinitesimal imaginary bilinear term added to the Lagrangian that is required for the convergence of the path integral (cf. Sect. 2.2.4.1). As a consequence a causal propagator results. The case $\xi = 1$ is called the *Feynman gauge* and the case $\xi = 1$ the *Landau gauge.*

The propagators for the fermions and ghost fields can be calculated in the same way.

2.4.2.2 Feynman rules

Altogether, the Lagrangian (2.4.24) gives rise to the following Feynman rules for a non-abelian gauge theory (in particular QCD):

Propagators

fermion: (quark)

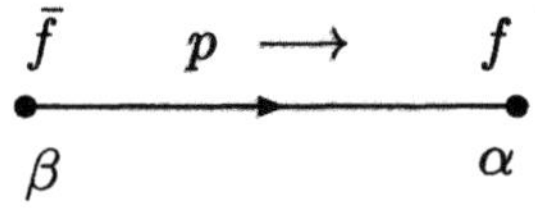

$$= \frac{\mathrm{i}\delta_{\alpha\beta}}{\not{p} - m + \mathrm{i}\varepsilon},$$

gauge boson: (gluon)

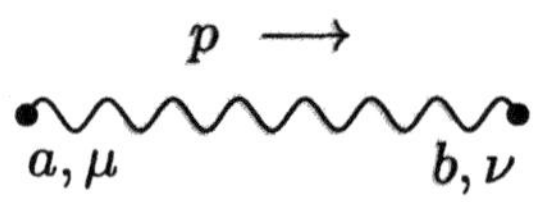

$$= \mathrm{i}\Delta^{ab}_{\mathrm{F},\mu\nu}(p)$$

$$= \mathrm{i}\left[-\left(g_{\mu\nu} - \frac{p_\mu p_\nu}{p^2}\right) - \xi\frac{p_\mu p_\nu}{p^2}\right]\frac{\delta^{ab}}{p^2 + \mathrm{i}\varepsilon},$$

ghost:

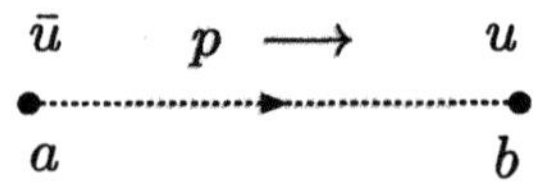

$$= \frac{\mathrm{i}\delta^{ab}}{p^2 + \mathrm{i}\varepsilon}.$$

Vertices (all momenta incoming)

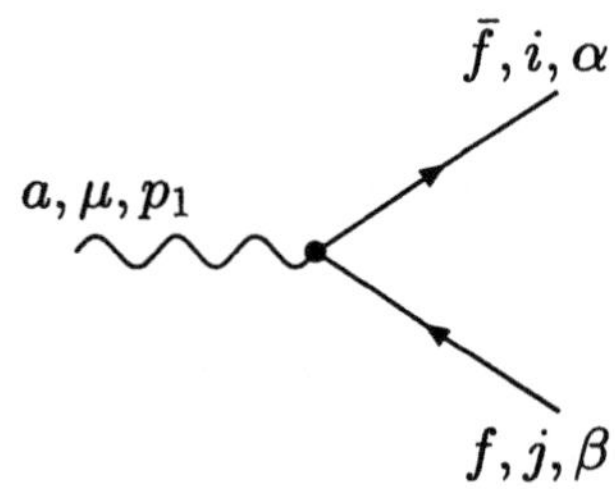

$$= \mathrm{i}g(\gamma_\mu)_{\alpha\beta}T^a_{ij},$$

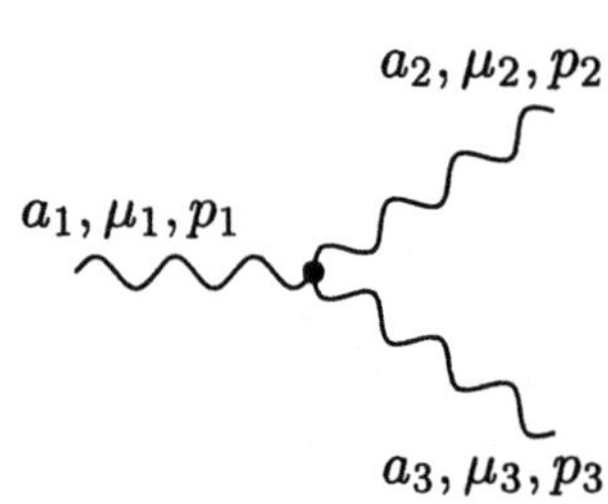

$$\begin{aligned} = g f^{a_1 a_2 a_3}[&g_{\mu_1\mu_2}(p_1 - p_2)_{\mu_3} \\ &+ g_{\mu_2\mu_3}(p_2 - p_3)_{\mu_1} \\ &+ g_{\mu_3\mu_1}(p_3 - p_1)_{\mu_2}], \end{aligned}$$

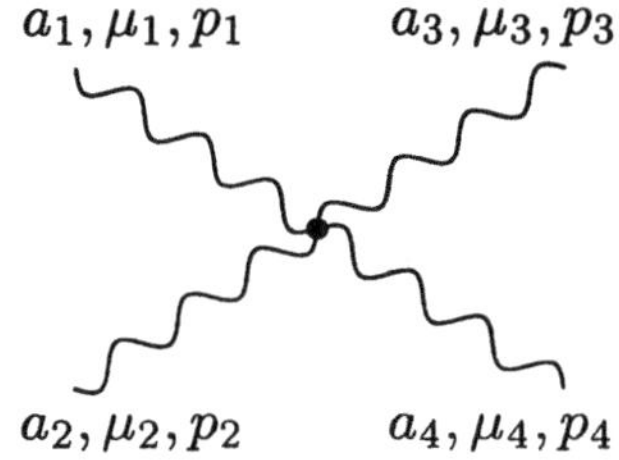

$$
\begin{aligned}
&= -ig^2 \\
&\times[f^{a_1a_2b}f^{a_3a_4b}(g_{\mu_1\mu_3}g_{\mu_2\mu_4} - g_{\mu_1\mu_4}g_{\mu_2\mu_3}) \\
&+ f^{a_1a_3b}f^{a_4a_2b}(g_{\mu_1\mu_4}g_{\mu_3\mu_2} - g_{\mu_1\mu_2}g_{\mu_3\mu_4}) \\
&+ f^{a_1a_4b}f^{a_2a_3b}(g_{\mu_1\mu_2}g_{\mu_4\mu_3} - g_{\mu_1\mu_3}g_{\mu_4\mu_2})],
\end{aligned}
$$

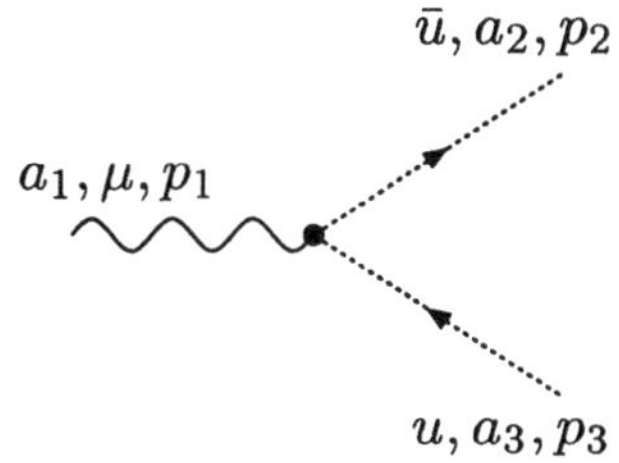

$$= -gf^{a_1a_2a_3}p_{2,\mu}. \tag{2.4.32}$$

The use of the Feynman rules is not very different from that in QED. The self-interaction of gauge bosons gives graphs with closed gauge-boson loops which have to be weighted with appropriate statistical factors. In the composition rules, Faddeev–Popov ghosts are treated as fermions, in particular each closed ghost loop gets a factor -1.

The gauge-fixing part of the Lagrangian gives rise to the gauge parameter ξ in the gauge-boson propagator. It cancels out in physical matrix elements. The Faddeev–Popov ghost fields $u^a(x)$ and $\bar{u}^a(x)$ lead to the appearance of ghost propagators and ghost–gauge-boson vertices. However, the ghost states have no physical significance (cf. Sect. 2.5.4.3). Their purpose is to cancel the unphysical contributions caused by the gauge-fixing term. External fermion and gauge-boson lines are treated as in QED (cf. Sect. 1.2.1.8).

2.4.3 BRS invariance and Slavnov–Taylor identities

The manifest gauge invariance of the non-abelian gauge theory is lost by the introduction of the gauge-fixing term $C^a\{\mathbf{A}; x\}$. However, a symmetry of the effective Lagrangian $\mathcal{L}_{\text{eff}}$ with all the consequences of gauge invariance for all physical results can be defined again by an extension of the gauge transformation to the ghost fields. As generalizations of the Ward identities of gauge symmetry the *Slavnov–Taylor identities* are obtained.

2.4.3.1 BRS invariance

The extended gauge transformation is the *Becchi–Rouet–Stora* or *BRS-transformation* [Be74]. For gauge-boson and fermion fields, the BRS transformation is a gauge transformation (2.3.32) with $\delta\theta^a(x) = g\delta\lambda u^a(x)$,

$$\begin{aligned} \delta_{\rm BRS} A^a_\mu(x) &= \delta\lambda D^{ab}_\mu u^b(x) = \delta\lambda(\partial_\mu \delta^{ab} - g f^{abc} A^c_\mu(x))u^b(x) \\ &\equiv \delta\lambda\, s A^a_\mu(x), \\ \delta_{\rm BRS}\psi_i(x) &= \delta\lambda\, u^a(x) \mathrm{i} g(T^a\psi(x))_i \equiv \delta\lambda\, s\psi_i(x), \\ \delta_{\rm BRS}\overline{\psi}_i(x) &= -\delta\lambda\, u^a(x)\mathrm{i} g(\overline{\psi}(x)T^a)_i \equiv \delta\lambda\, s\overline{\psi}_i(x), \end{aligned} \tag{2.4.33}$$

where $\delta\lambda$ is an infinitesimal, Grassmann-valued constant which anticommutes with the ghost fields $u^a(x)$ and $\bar{u}^a(x)$.[9] The *BRS operator s* is defined as the left derivative with respect to $\delta\lambda$ of the BRS transformed fields and consequently the product rule reads

$$s(FG) = (sF)G \pm FsG, \tag{2.4.34}$$

where the minus sign occurs, if F contains an odd number of Grassmann variables. Note that the BRS symmetry is a global symmetry in contrast to the local gauge symmetry.

The Lagrangian $\mathcal{L}_{\rm gauge}$ (2.4.17) and the fermionic Lagrangian $\mathcal{L}_{\rm ferm}$ are evidently BRS-invariant, $\delta_{\rm BRS}\mathcal{L}_{\rm gauge} = \delta_{\rm BRS}\mathcal{L}_{\rm ferm} = 0$. The transformation behaviour of the ghost fields,

$$\begin{aligned} \delta_{\rm BRS} u^a(x) &= -\delta\lambda \frac{1}{2} g f^{abc} u^b(x) u^c(x) \equiv \delta\lambda\, s u^a(x), \\ \delta_{\rm BRS}\bar{u}^a(x) &= -\delta\lambda\frac{1}{\xi} C^a\{\mathbf{A}; x\} \equiv \delta\lambda\, s\bar{u}^a(x), \end{aligned} \tag{2.4.35}$$

is chosen so that $\mathcal{L}_{\rm fix} + \mathcal{L}_{\rm ghost}$ is also BRS-invariant although both terms are transformed non-trivially,

$$\delta_{\rm BRS}(\mathcal{L}_{\rm fix} + \mathcal{L}_{\rm ghost}) = 0. \tag{2.4.36}$$

[9] If a *ghost number* is introduced in analogy to fermion number, then $u^a(x)$, $\bar{u}^a(x)$, and $\delta\lambda$ have ghost numbers $+1$, -1, and -1 respectively. Consequently, $\delta\theta^a(x) = gu^a(x)\delta\lambda$ has ghost and fermion number zero.

This shows the importance of the Faddeev–Popov ghost fields which re-establish the gauge invariance of a theory destroyed by $\mathcal{L}_{\text{fix}}$.

In order to prove (2.4.36), we notice that in view of the anticommutativity of the ghost fields

$$\begin{aligned}\int \mathrm{d}^4y \left(\delta_{\mathrm{BRS}} \frac{\delta C^a\{\mathbf{A};x\}}{\delta A_\nu^b(y)}\right) D_\nu^{bc} u^c(y) \\ = \int \mathrm{d}^4y\, \mathrm{d}^4z\, \frac{\delta C^a\{\mathbf{A};x\}}{\delta A_\nu^b(y)\delta A_\rho^d(z)} D_\rho^{de} u^e(z) D_\nu^{bc} u^c(y) = 0,\end{aligned} \tag{2.4.37}$$

and because of the Jacobi identity of the structure constants f^{abc} (1.2.48)

$$\delta_{\mathrm{BRS}} D_\mu^{ab} u^b(x) = \delta_{\mathrm{BRS}}(\partial_\mu u^a(x) - g f^{abc} A_\mu^c(x) u^b(x)) = 0. \tag{2.4.38}$$

This implies

$$\begin{aligned}\delta_{\mathrm{BRS}} \mathcal{L}_{\text{ghost}} &= -\delta_{\mathrm{BRS}} \int \mathrm{d}^4y\, \bar{u}^a(x) \frac{\delta C^a\{\mathbf{A};x\}}{\delta A_\nu^b(y)} D_\nu^{bc} u^c(y) \\ &= -\int \mathrm{d}^4y \left(\delta_{\mathrm{BRS}} \bar{u}^a(x)\right) \frac{\delta C^a\{\mathbf{A};x\}}{\delta A_\nu^b(y)} D_\nu^{bc} u^c(y) \\ &= \delta\lambda \frac{1}{\xi} C^a\{\mathbf{A};x\} \int \mathrm{d}^4y\, \frac{\delta C^a\{\mathbf{A};x\}}{\delta A_\nu^b(y)} D_\nu^{bc} u^c(y).\end{aligned} \tag{2.4.39}$$

On the other hand, we obtain

$$\begin{aligned}\delta_{\mathrm{BRS}} \mathcal{L}_{\text{fix}} &= -\frac{1}{\xi} C^a\{\mathbf{A};x\} \int \mathrm{d}^4y\, \frac{\delta C^a\{\mathbf{A};x\}}{\delta A_\nu^b(y)} \delta_{\mathrm{BRS}} A_\nu^b(y) \\ &= -\frac{1}{\xi} C^a\{\mathbf{A};x\} \int \mathrm{d}^4y\, \frac{\delta C^a\{\mathbf{A};x\}}{\delta A_\nu^b(y)} \delta\lambda D_\nu^{bc} u^c(y).\end{aligned} \tag{2.4.40}$$

This proves (2.4.36).

The BRS operator is nilpotent, i.e.

$$s^2 A_\mu^a = s^2 \psi = s^2 \bar{\psi} = s^2 u^a = s^2 C^a = 0, \tag{2.4.41}$$

except for the antighost

$$s^2 \bar{u}^a = -\frac{1}{\xi} s C^a\{\mathbf{A};x\} = -\frac{1}{\xi} \int \mathrm{d}^4y\, \frac{\delta C^a\{\mathbf{A};x\}}{\delta A_\nu^b(y)} D_\mu^{bc} u^c(y) \neq 0, \tag{2.4.42}$$

but

$$s^3\bar{u}^a = 0. \tag{2.4.43}$$

Equation (2.4.42) entails that the *Faddeev–Popov-ghost Lagrangian* can be written as

$$\mathcal{L}_{\text{ghost}} = -\bar{u}^a(x) sC^a\{\mathbf{A};x\}. \tag{2.4.44}$$

Finally we note that the BRS transformation leaves the measure of the functional integral invariant.

The BRS symmetry can be formulated in a more symmetric form by introducing an auxiliary (commuting) scalar field B^a, the *Nakanishi–Lautrup field*, together with the corresponding path integral over B^a, and the gauge-fixing Lagrangian [Na66a, La67]

$$\mathcal{L}_{\text{fix}} = \frac{\xi}{2} B^a B^a + B^a C^a\{\mathbf{A};x\}. \tag{2.4.45}$$

Because the field B^a appears only in quadratic terms without derivatives, it is an auxiliary field without an own dynamics. Upon performing the functional integral over B^a by completing the square one recovers the original Lagrangian and generating functional.

The *BRS transformations* are given by (2.4.33) and (2.4.35) with the antighost transformation replaced by

$$\begin{aligned} \delta_{\text{BRS}}\bar{u}^a(x) &= \delta\lambda B^a \equiv \delta\lambda\, s\bar{u}^a(x), \\ \delta_{\text{BRS}} B^a(x) &= 0 \equiv \delta\lambda\, sB^a(x). \end{aligned} \tag{2.4.46}$$

In this formulation the BRS transformations are independent of the gauge-fixing term and thus independent of the effective Lagrangian and strictly nilpotent

$$s^2\bar{u}^a = sB^a = 0. \tag{2.4.47}$$

Moreover, the sum of the gauge-fixing and ghost Lagrangian can be written in a concise form

$$\mathcal{L}_{\text{fix}} + \mathcal{L}_{\text{ghost}} = s\left(\bar{u}^a \left(\frac{\xi}{2} B^a + C^a\{\mathbf{A};x\}\right)\right). \tag{2.4.48}$$

Thus, these terms can be written as a pure BRS transformation. In this formalism, other gauges can be obtained as

$$\mathcal{L}_{\text{fix}} + \mathcal{L}_{\text{ghost}} = s\left(\bar{u}^a F^a\right), \tag{2.4.49}$$

where $F^a\{\mathbf{A}, \mathbf{u}, \bar{\mathbf{u}}, \mathbf{B}; x\}$ is an arbitrary functional of the fields with vanishing ghost number.

According to the Noether theorem (cf. Sect. 1.2.3.1), the invariance of $\mathcal{L}_{\text{eff}}$ under the BRS transformations implies the existence of a conserved current

$$\begin{aligned} j_\mu^{\text{BRS}} &= (D^\nu u)^a \frac{\partial \mathcal{L}}{\partial(\partial^\mu A^{a\nu})} - \frac{1}{2} g f^{abc} u^b u^c \frac{\partial \mathcal{L}}{\partial(\partial^\mu u^a)} + B^a \frac{\partial \mathcal{L}}{\partial(\partial^\mu \bar{u}^a)} \\ &\quad + \mathrm{i} g u^a (T^a \psi)_i \frac{\partial \mathcal{L}}{\partial(\partial^\mu \psi_i)} - \mathrm{i} g u^a (\bar{\psi} T^a)_i \frac{\partial \mathcal{L}}{\partial(\partial^\mu \bar{\psi}_i)} . \end{aligned} \tag{2.4.50}$$

Using the equations of motion and the fact that the gauge field appears in the matter Lagrangian only via the covariant derivative, the matter contribution can be eliminated. The explicit form of the conserved *BRS current* (2.4.50) depends on the gauge-fixing term. In the gauge $C^a\{\mathbf{A}; x\} = \partial^\mu A_\mu^a(x)$, it can be brought into the form [Ku78]

$$j_\mu^{\text{BRS}} = B^a (D_\mu u)^a - (\partial_\mu B^a) u^a + \frac{1}{2} g f^{abc} (\partial_\mu \bar{u}^a) u^b u^c - \partial^\nu (F_{\mu\nu}^a u^a). \tag{2.4.51}$$

The corresponding conserved charge, the *BRS charge*, reads

$$Q_{\text{BRS}} = \int \mathrm{d}^3 x \left[B^a (D_0 u)^a - (\partial_0 B^a) u^a + \frac{1}{2} g f^{abc} (\partial_0 \bar{u}^a) u^b u^c \right]. \tag{2.4.52}$$

2.4.3.2 Slavnov–Taylor identities

In Sect. 2.2.6, a functional differential equation for the generating functional of the Green functions (2.2.112) was derived from the global symmetry of the action $S\{\psi\}$ of a multiplet of fields $\psi(x)$ and the invariance of the measure $\mathcal{D}[\psi]$ in the functional integral (2.2.84). Carrying these considerations over to BRS symmetry gives the Ward identities for the generating functional of the Green functions of a non-abelian gauge theory, the *Slavnov–Taylor identities* [Ta71, Sl72, tH71, tH72, Le72]. These summarize all relationships between the Green functions of the theory resulting from local gauge symmetry. The Slavnov–Taylor identities are essential for the proof of renormalizability of gauge theories (cf. Sect. 2.5.4).

For definiteness we consider the generating functional of Green functions for a gauge theory with fermions (2.4.20). In order to get linear identities we introduce sources $\mathbf{K}, \mathbf{K}_u, \mathbf{K}_\psi$, and $\mathbf{K}_{\overline{\psi}}$ for the BRS-transformed fields, too,

$$
\begin{aligned}
&Z\{\mathbf{J},\mathbf{J}_u,\mathbf{J}_{\bar{u}},\mathbf{J}_\psi,\mathbf{J}_{\overline{\psi}},\mathbf{K},\mathbf{K}_u,\mathbf{K}_\psi,\mathbf{K}_{\overline{\psi}}\} = \\
&\qquad \int \mathcal{D}[\mathbf{A}]\mathcal{D}[\mathbf{u}]\mathcal{D}[\bar{\mathbf{u}}]\mathcal{D}[\psi]\mathcal{D}[\overline{\psi}]\exp\left(\mathrm{i}\int \mathrm{d}^4x\,\mathcal{L}_{\mathrm{tot}}\right), \\
&\mathcal{L}_{\mathrm{tot}} = \mathcal{L}_{\mathrm{eff}} + J^{a,\mu}A^a_\mu + J^a_u u^a - \bar{u}^a J^a_{\bar{u}} + J^i_\psi \psi_i - \overline{\psi}_i J^i_{\overline{\psi}} \\
&\qquad + K^{a,\mu} sA^a_\mu + K^a_u su^a + K^i_\psi s\psi_i + (s\overline{\psi}_i)K^i_{\overline{\psi}}.
\end{aligned}
\tag{2.4.53}
$$

All sources are assumed to be invariant under a BRS transformation, and $J^a_{\bar{u}}$, J^a_u, $J^i_{\overline{\psi}}$, J^i_ψ, and K^a are Grassmann variables. The vanishing of the ghost number for $\mathcal{L}_{\mathrm{tot}}$ implies ghost number -1 for K^a, K^i_ψ, and $K^i_{\overline{\psi}}$ and -2 for K^a_u.

In the following we assume that the gauge-fixing functional $C^a\{\mathbf{A};x\}$ is linear in the field A_μ, i.e.

$$C^a\{\mathbf{A};x\} = \Phi^{ab}_\mu A^{b,\mu}(x), \tag{2.4.54}$$

where Φ^{ab}_μ is field-independent but may contain a derivative as in (2.4.11). The corresponding ghost Lagrangian reads

$$\mathcal{L}_{\mathrm{ghost}} = -\bar{u}^a(x)\Phi^{ab}_\mu D^{bc,\mu}u^c(x) = -\bar{u}^a(x)\Phi^{ab}_\mu sA^{b,\mu}(x). \tag{2.4.55}$$

The use of a non-linear functional $C^a\{\mathbf{A};x\}$ would require the introduction of an extra source coupled to it.

Changing the variables of the functional integration according to the BRS transformation (2.4.33) and (2.4.35), using BRS invariance of the effective Lagrangian and the functional measure together with the relations (2.4.41), one obtains (cf. Sect. 2.2.6.4)

$$0 = \int \mathcal{D}[\mathbf{A}]\mathcal{D}[\mathbf{u}]\mathcal{D}[\bar{\mathbf{u}}]\mathcal{D}[\psi]\mathcal{D}[\bar{\psi}] \exp\left(\mathrm{i}\int \mathrm{d}^4x\, \mathcal{L}_{\mathrm{tot}}\right)$$
$$\times \int \mathrm{d}^4x \left[J^{a,\mu} sA^a_\mu - J^a_u su^a - (s\bar{u}^a)J^a_{\bar{u}} - J^i_\psi s\psi_i - (s\bar{\psi}_i)J^i_{\bar{\psi}}\right]. \tag{2.4.56}$$

Expressing the fields in the last integral by derivatives with respect to the corresponding sources, we obtain the *Slavnov–Taylor identities for the generating functional T*,

$$0 = \int \mathrm{d}^4x \left[J^a_\mu \frac{\delta}{\mathrm{i}\delta K^a_\mu} - J^a_u \frac{\delta}{\mathrm{i}\delta K^a_u} + \frac{1}{\xi} C^a \left\{\frac{\delta}{\mathrm{i}\delta \mathbf{J}}; x\right\} J^a_{\bar{u}} - J^i_\psi \frac{\delta}{\mathrm{i}\delta K^i_\psi}\right.$$
$$\left. - \frac{\delta}{\mathrm{i}\delta K^i_{\bar{\psi}}} J^i_{\bar{\psi}}\right] T\{\mathbf{J}, \mathbf{J}_u, \mathbf{J}_{\bar{u}}, \mathbf{J}_\psi, \mathbf{J}_{\bar{\psi}}, \mathbf{K}, \mathbf{K}_u, \mathbf{K}_\psi, \mathbf{K}_{\bar{\psi}}]. \tag{2.4.57}$$

In terms of the generating functional $T_{\mathrm{c}} = \log T$ of connected Green functions this reads

$$0 = \int \mathrm{d}^4x \left[J^a_\mu \frac{\delta}{\delta K^a_\mu} - J^a_u \frac{\delta}{\delta K^a_u} + \frac{1}{\xi} C^a \left\{\frac{\delta}{\delta \mathbf{J}}; x\right\} J^a_{\bar{u}} - J^i_\psi \frac{\delta}{\delta K^i_\psi}\right.$$
$$\left. - \frac{\delta}{\delta K^i_{\bar{\psi}}} J^i_{\bar{\psi}}\right] T_{\mathrm{c}}\{\mathbf{J}, \mathbf{J}_u, \mathbf{J}_{\bar{u}}, \mathbf{J}_\psi, \mathbf{J}_{\bar{\psi}}, \mathbf{K}, \mathbf{K}_u, \mathbf{K}_\psi, \mathbf{K}_{\bar{\psi}}\}, \tag{2.4.58}$$

where we used the linearity of $C^a\{\mathbf{A}; x\}$ in $\mathbf{A}$ to derive (2.4.58) from (2.4.57).

Similarly, the invariance of T under the transformation $\bar{u}^a \to \bar{u}^a + \delta\bar{u}^a$ gives rise to the relation

$$0 = \int \mathcal{D}[\mathbf{A}]\mathcal{D}[\mathbf{u}]\mathcal{D}[\bar{\mathbf{u}}]\mathcal{D}[\psi]\mathcal{D}[\bar{\psi}] \exp\left(\mathrm{i}\int \mathrm{d}^4x\, \mathcal{L}_{\mathrm{tot}}\right)$$
$$\times \int \mathrm{d}^4x \left[sC^a\{\mathbf{A}; x\} + J^a_{\bar{u}}\right], \tag{2.4.59}$$

where we used (2.4.44). This is nothing but the equation of motion for the antighost field (cf. Sect. 2.2.6.2). It results in

$$\left(J^a_{\bar{u}} + \Phi^{ab}_\mu \frac{\delta}{\mathrm{i}\delta K^{b,\mu}}\right) T\{\mathbf{J}, \mathbf{J}_u, \mathbf{J}_{\bar{u}}, \mathbf{J}_\psi, \mathbf{J}_{\bar{\psi}}, \mathbf{K}, \mathbf{K}_u, \mathbf{K}_\psi, \mathbf{K}_{\bar{\psi}}\} = 0, \tag{2.4.60}$$

or for connected functionals

$$J^a_{\bar{u}} + \Phi^{ab}_\mu \frac{\delta T_\mathrm{c}}{\mathrm{i}\delta K^{b,\mu}} = 0. \tag{2.4.61}$$

2.4.3.3 Identities for Green functions

As discussed in Sect. 2.2.6, *Slavnov–Taylor identities between Green functions* can be directly obtained by differentiating (2.4.57) with respect to the external sources and putting the sources to zero afterwards. We write the Green functions generically as $\langle T \prod_l \Psi_{I_l}\rangle$, where Ψ_I represents anyone of the fields A^a_μ, u^a, $\bar{u}^a$, ψ_i, or $\bar{\psi}_i$, and the index I comprises all indices of the actual field including the space–time coordinates. The resulting identities can be written as

$$\begin{aligned} 0 &= \frac{\delta_{\mathrm{BRS}}}{\delta\lambda}\Big\langle T \prod_l \Psi_{I_l}\Big\rangle = s\Big\langle T \prod_l \Psi_{I_l}\Big\rangle \\ &= \sum_k \sigma_k \Big\langle T\Big(\prod_{l<k} \Psi_{I_l}\Big)(s\Psi_{I_k}) \prod_{m>k} \Psi_{I_m}\Big\rangle, \end{aligned} \tag{2.4.62}$$

where σ_k stands for the sign that results from anticommuting the BRS operator with Grassmann-valued fields. The sum $\sum_k$ extends over all fields in $\prod_l \Psi_{I_l}$. These relations can also be obtained directly from the invariance of the action under BRS transformations by performing a BRS variation in the generating functional of Green functions [cf. (2.2.102)].

Similarly (2.4.60) gives rise to

$$\begin{aligned} \mathrm{i}\Big\langle T \frac{\delta}{\delta \bar{u}^a(x)} \prod_l \Psi_{I_l}\Big\rangle &= -\Phi^{ab}_\mu \Big\langle T(sA^{b,\mu}(x)) \prod_l \Psi_{I_l}\Big\rangle \\ &= -\Big\langle T(sC^a\{\mathbf{A};x\}) \prod_l \Psi_{I_l}\Big\rangle = \Big\langle T \frac{\delta S}{\delta \bar{u}^a(x)} \prod_l \Psi_{I_l}\Big\rangle, \end{aligned} \tag{2.4.63}$$

which is nothing but a consequence of the equation of motion (2.2.93) of the antighost field $\bar{u}^a$.

As an example let us consider the Green function $\langle T\bar{u}^a(x)C^b\{\mathbf{A};y\}\rangle$. Equation (2.4.62) yields

$$\begin{aligned} 0 &= s\left\langle T\bar{u}^a(x)C^b\{\mathbf{A};y\}\right\rangle \\ &= \left\langle T(s\bar{u}^a(x))C^b\{\mathbf{A};y\}\right\rangle - \left\langle T\bar{u}^a(x)sC^b\{\mathbf{A};y\}\right\rangle \\ &= -\frac{1}{\xi}\left\langle TC^a\{\mathbf{A};x\}C^b\{\mathbf{A};y\}\right\rangle + \left\langle T\left(sC^b\{\mathbf{A};y\}\right)\bar{u}^a(x)\right\rangle . \end{aligned} \tag{2.4.64}$$

If the gauge-fixing term involving the field B^a (2.4.45) is used, the same identity is obtained with the help of the equation of motion [cf. (2.2.93)] for the field B^a,

$$\xi\left\langle T\left(B^a(x) + \frac{1}{\xi}C^a\{\mathbf{A};x\}\right)\prod_l \Psi_{I_l}\right\rangle = \mathrm{i}\left\langle T\frac{\delta}{\delta B^a(x)}\prod_l \Psi_{I_l}\right\rangle . \tag{2.4.65}$$

Using (2.4.63) we obtain from (2.4.64)

$$-\frac{1}{\xi}\left\langle TC^a\{\mathbf{A};x\}C^b\{\mathbf{A};y\}\right\rangle = \mathrm{i}\left\langle T\frac{\delta}{\delta\bar{u}^b(y)}\bar{u}^a(x)\right\rangle = \mathrm{i}\delta^{ab}\delta^{(4)}(x-y). \tag{2.4.66}$$

In the covariant gauge (2.4.11) this yields

$$\partial_x^\mu\partial_y^\nu\left\langle TA_\mu^a(x)A_\nu^b(y)\right\rangle = \partial_x^\mu\partial_y^\nu G_{\mu\nu}^{ab}(x-y) = -\mathrm{i}\xi\delta^{ab}\delta^{(4)}(x-y), \tag{2.4.67}$$

or, in momentum space

$$k^\mu k^\nu G_{\mu\nu}^{ab}(k) = -\mathrm{i}\xi\delta^{ab}. \tag{2.4.68}$$

This is the *Slavnov identity* [Sl72, Ta71], the Ward identity for the gauge-boson propagator of a non-abelian gauge theory. It implies that the longitudinal part of the gauge-boson propagator is not changed by interactions and that this propagator is transverse in the Landau gauge $\xi = 0$. It is evidently fulfilled in the tree-graph approximation, i.e. for the free propagator given in (2.4.32). Its validity in higher orders implies that the gauge-fixing term need not be renormalized.

As a further example we apply a BRS transformation to $\langle T\bar{u}^a(x)\psi(y)\bar{\psi}(z)\rangle$ and obtain

$$
\begin{aligned}
0 &= s\langle T\bar{u}^a(x)\psi_i(y)\bar{\psi}_j(z)\rangle \\
&= \langle T(s\bar{u}^a(x))\psi_i(y)\bar{\psi}_j(z)\rangle - \langle T\bar{u}^a(x)(s\psi_i(y))\bar{\psi}_j(z)\rangle \\
&\quad + \langle T\bar{u}^a(x)\psi_i(y)s\bar{\psi}_j(z)\rangle \\
&= -\frac{1}{\xi}\langle TC^a\{\mathbf{A};x\}\psi_i(y)\bar{\psi}_j(z)\rangle - ig\langle T\bar{u}^a(x)u^b(y)(T^b\psi(y))_i\bar{\psi}_j(z)\rangle \\
&\quad - ig\langle T\bar{u}^a(x)\psi_i(y)u^b(z)(\bar{\psi}(z)T^b)_j\rangle. \qquad (2.4.69)
\end{aligned}
$$

In diagrams this can be written as

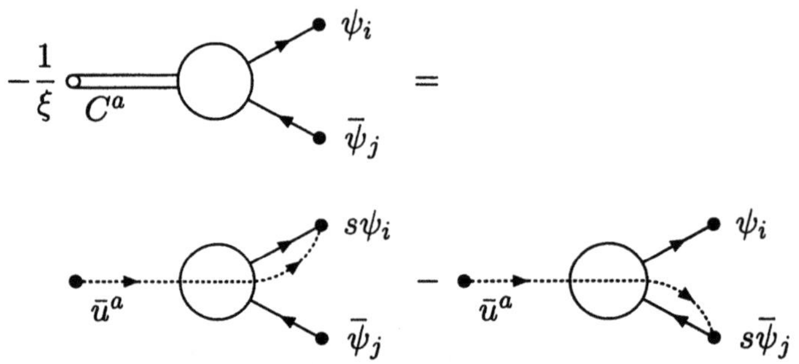

, (2.4.70)

where we use a double line with a circle to indicate the gauge-fixing term. The ghost lines go through the diagrams owing to ghost-number conservation.

Equation (2.4.69) is the generalization of the QED Ward identity (2.4.86) that relates the electron–photon vertex function to the electron propagator to non-abelian gauge theories.

Ward identities for other Green functions are obtained in a similar way.

2.4.3.4 Lee identities

The Slavnov–Taylor identities (2.4.58) and (2.4.61) can be converted into identities for the generating functional of vertex functions Γ, the *Lee identities*. The functional Γ is defined as follows:

$$
\begin{aligned}
&\mathrm{i}\Gamma\{\mathbf{A},\mathbf{u},\bar{\mathbf{u}},\psi,\bar{\psi},\mathbf{K},\mathbf{K}_u,\mathbf{K}_\psi,\mathbf{K}_{\bar{\psi}}\} \\
&\quad = T_{\mathrm{c}}\{\mathbf{J},\mathbf{J}_u,\mathbf{J}_{\bar{u}},\mathbf{J}_\psi,\mathbf{J}_{\bar{\psi}},\mathbf{K},\mathbf{K}_u,\mathbf{K}_\psi,\mathbf{K}_{\bar{\psi}}\} \\
&\qquad -\mathrm{i}\int \mathrm{d}^4x\left(J^{a,\mu}A^a_\mu + J^a_u u^a - \bar{u}^a J^a_{\bar{u}} + J^i_\psi \psi_i - \bar{\psi}_i J^i_{\bar{\psi}}\right). \quad (2.4.71)
\end{aligned}
$$

Note that the sources K^a, K^a_u, $K^i_{\bar{\psi}}$, and K^i_ψ are not transformed, and that we denote the other sources of Γ with the same symbols as the integration variables of the functional integral. These sources are given by

$$
\begin{aligned}
A^a_\mu &= \frac{\delta T_{\mathrm{c}}}{\mathrm{i}\delta J^a_\mu}, \qquad u^a = \frac{\delta T_{\mathrm{c}}}{\mathrm{i}\delta J^a_u}, \qquad \bar{u}^a = \frac{\delta T_{\mathrm{c}}}{\mathrm{i}\delta J^a_{\bar{u}}}, \\
\psi_i &= \frac{\delta T_{\mathrm{c}}}{\mathrm{i}\delta J^i_\psi}, \qquad \bar{\psi}_i = \frac{\delta T_{\mathrm{c}}}{\mathrm{i}\delta J^i_{\bar{\psi}}}, \qquad (2.4.72)
\end{aligned}
$$

or

$$
\begin{aligned}
J^a_\mu &= -\frac{\delta\Gamma}{\delta A^a_\mu}, \qquad J^a_{\bar{u}} = \frac{\delta\Gamma}{\delta\bar{u}^a}, \qquad J^a_u = \frac{\delta\Gamma}{\delta u^a}, \\
J^i_{\bar{\psi}} &= \frac{\delta\Gamma}{\delta\bar{\psi}_i}, \qquad J^i_\psi = \frac{\delta\Gamma}{\delta\psi_i}. \qquad (2.4.73)
\end{aligned}
$$

In addition we have

$$
\frac{\delta T_{\mathrm{c}}}{\delta K^a_\mu} = \frac{\mathrm{i}\delta\Gamma}{\delta K^a_\mu}, \qquad \frac{\delta T_{\mathrm{c}}}{\delta K^a_u} = \frac{\mathrm{i}\delta\Gamma}{\delta K^a_u}, \qquad \frac{\delta T_{\mathrm{c}}}{\delta K^i_{\bar{\psi}}} = \frac{\mathrm{i}\delta\Gamma}{\delta K^i_{\bar{\psi}}}, \qquad \frac{\delta T_{\mathrm{c}}}{\delta K^i_\psi} = \frac{\mathrm{i}\delta\Gamma}{\delta K^i_\psi}. \qquad (2.4.74)
$$

Inserting this into (2.4.58) and (2.4.61), we find the Lee identities

$$
\begin{aligned}
0 &= \int \mathrm{d}^4x\left[\frac{\delta\Gamma}{\delta A^a_\mu}\frac{\delta\Gamma}{\delta K^{a,\mu}} + \frac{\delta\Gamma}{\delta u^a}\frac{\delta\Gamma}{\delta K^a_u} - \frac{1}{\xi}C^a\{\mathbf{A};x\}\frac{\delta\Gamma}{\delta\bar{u}^a}\right. \\
&\qquad \left. + \frac{\delta\Gamma}{\delta K^i_{\bar{\psi}}}\frac{\delta\Gamma}{\delta\bar{\psi}_i} + \frac{\delta\Gamma}{\delta\psi_i}\frac{\delta\Gamma}{\delta K^i_\psi}\right], \\
0 &= \Phi^{ab}_\mu\frac{\delta\Gamma}{\delta K^b_\mu} + \frac{\delta\Gamma}{\delta\bar{u}^a}. \qquad (2.4.75)
\end{aligned}
$$

If we extract the gauge-fixing term from the vertex functional and define

$$\tilde{\Gamma} = \Gamma + \frac{1}{2\xi}\int d^4x\, C^a\{\mathbf{A};x\}C^a\{\mathbf{A};x\}, \tag{2.4.76}$$

the Lee identities read

$$0 = \int d^4x \left[\frac{\delta\tilde{\Gamma}}{\delta A^a_\mu}\frac{\delta\tilde{\Gamma}}{\delta K^{a,\mu}} + \frac{\delta\tilde{\Gamma}}{\delta u^a}\frac{\delta\tilde{\Gamma}}{\delta K^a_u} + \frac{\delta\tilde{\Gamma}}{\delta K^i_{\bar{\psi}}}\frac{\delta\tilde{\Gamma}}{\delta\bar{\psi}_i} + \frac{\delta\tilde{\Gamma}}{\delta\psi_i}\frac{\delta\tilde{\Gamma}}{\delta K^i_\psi}\right], \tag{2.4.77}$$

$$0 = \Phi^{ab}_\mu \frac{\delta\tilde{\Gamma}}{\delta K^b_\mu} + \frac{\delta\tilde{\Gamma}}{\delta\bar{u}^a}. \tag{2.4.78}$$

In this form the Lee identities are universal, i.e. independent of the parameters of the theory (masses, coupling constants, and gauge parameters). Among other things they are useful to proof renormalizability of gauge theories.

2.4.3.5 Ward identities in QED

The Slavnov–Taylor identities are valid in any gauge theory. However, in an abelian gauge theory like QED the situation is simplified (cf. Sect. 2.4.1.4).

The infinitesimal gauge transformations

$$\delta A_\mu = \partial_\mu\delta\theta, \qquad \delta\psi = \mathrm{i}e\psi\delta\theta, \qquad \delta\bar{\psi} = -\mathrm{i}e\bar{\psi}\delta\theta \tag{2.4.79}$$

leave $\mathcal{L}_{\text{gauge}}$, $\mathcal{L}_{\text{ferm}}$, and the measure of the functional integral invariant. Thus, the invariance of the functional integral under a change of variables according to (2.4.79) yields (cf. Sect. 2.2.6)

$$\begin{aligned} 0 = &\int \mathcal{D}[\mathbf{A}]\mathcal{D}[\psi]\mathcal{D}[\bar{\psi}] \exp\left(\mathrm{i}\int d^4x\mathcal{L}_{\text{tot}}\right) \\ &\times \int d^4x \left[-\frac{1}{\xi}[\partial^\mu A_\mu(x)]\Box\delta\theta(x) + J^\mu(x)\partial_\mu\delta\theta(x)\right. \\ &\left. + \mathrm{i}e\bar{\psi}(x)\delta\theta(x)J_{\bar{\psi}}(x) + \mathrm{i}eJ_\psi(x)\psi(x)\delta\theta(x)\right]. \end{aligned} \tag{2.4.80}$$

After integrating by parts, cancelling $\delta\theta(x)$, and expressing the explicit fields by derivatives with respect to their sources we find the *Ward identities*

$$\begin{aligned} 0 = & \frac{1}{\xi}\Box\partial^{\mu}\frac{\delta T}{\mathrm{i}\delta J^{\mu}(x)} + (\partial_{\mu}J^{\mu}(x))T \\ & - \mathrm{ie}\frac{\delta T}{\mathrm{i}\delta J_{\bar{\psi}}(x)}J_{\bar{\psi}}(x) - \mathrm{ie}J_{\psi}(x)\frac{\delta T}{\mathrm{i}\delta J_{\psi}(x)}. \end{aligned} \tag{2.4.81}$$

For connected Green functions these read

$$0 = \frac{1}{\xi}\Box\partial^{\mu}\frac{\delta T_{\mathrm{c}}}{\mathrm{i}\delta J^{\mu}(x)} + \partial_{\mu}J^{\mu}(x) - \mathrm{ie}\frac{\delta T_{\mathrm{c}}}{\mathrm{i}\delta J_{\bar{\psi}}(x)}J_{\bar{\psi}}(x) - \mathrm{ie}J_{\psi}(x)\frac{\delta T_{\mathrm{c}}}{\mathrm{i}\delta J_{\psi}(x)}, \tag{2.4.82}$$

and in terms of vertex functions

$$0 = \frac{1}{\xi}\Box\partial^{\mu}A_{\mu}(x) - \partial_{\mu}\frac{\delta\Gamma}{\delta A_{\mu}(x)} - \mathrm{ie}\bar{\psi}(x)\frac{\delta\Gamma}{\delta\bar{\psi}(x)} - \mathrm{ie}\frac{\delta\Gamma}{\delta\psi(x)}\psi(x). \tag{2.4.83}$$

As applications, we derive some relations for special Green functions. Differentiating (2.4.82) with respect to $J_{\nu}(y)$ and putting all sources to zero yields

$$\begin{aligned} \frac{1}{\xi}\Box\partial^{\mu}G^{AA}_{\mathrm{c},\mu\nu}(x,y) &= \frac{1}{\xi}\Box\partial^{\mu}\left.\frac{\delta T_{\mathrm{c}}}{\mathrm{i}\delta J^{\mu}(x)\mathrm{i}\delta J^{\nu}(y)}\right|_{\mathbf{J}=J_{\bar{\psi}}=J_{\psi}=0} \\ &= \mathrm{i}\partial_{\nu}\delta^{(4)}(x-y) \end{aligned} \tag{2.4.84}$$

or in momentum space

$$-\frac{1}{\xi}k^2k^{\mu}G^{AA}_{\mathrm{c},\mu\nu}(k,-k) = \mathrm{i}k_{\nu}. \tag{2.4.85}$$

This is the QED version of the Ward identity (2.4.68). It implies that the longitudinal part of the *photon propagator* gets no higher-order corrections.

Differentiating (2.4.82) with respect to $J_{\bar{\psi}}(z)$ and $J_\psi(y)$ results in

$$\frac{1}{\xi}\Box\partial^\mu \frac{\delta T_{\mathrm{c}}}{\mathrm{i}\delta J_\mu(x)\mathrm{i}\delta J_\psi(y)\mathrm{i}\delta J_{\bar{\psi}}(z)} = e\delta^{(4)}(x-y)\frac{\delta T_{\mathrm{c}}}{\mathrm{i}\delta J_\psi(x)\mathrm{i}\delta J_{\bar{\psi}}(z)} - e\delta^{(4)}(x-z)\frac{\delta T_{\mathrm{c}}}{\mathrm{i}\delta J_\psi(y)\mathrm{i}\delta J_{\bar{\psi}}(x)}, \tag{2.4.86}$$

or diagrammatically after putting the sources to zero

$\frac{1}{\xi}\Box\partial^\mu$ 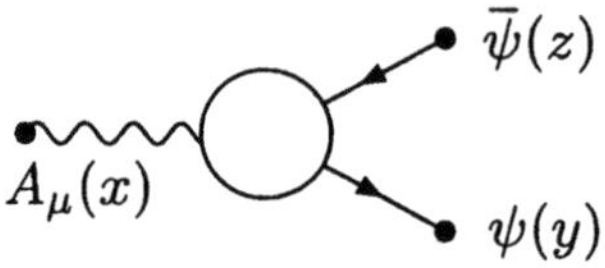

$= e\delta^{(4)}(x-y)$ 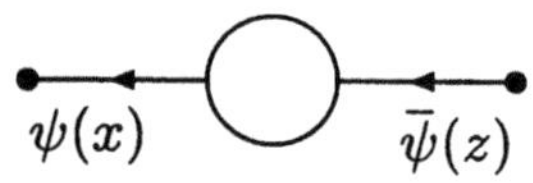

$-\,e$ 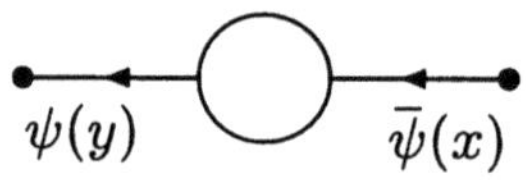

$\delta^{(4)}(x-z)$. (2.4.87)

This Ward identity can be also obtained by applying $\Box_x$ to (2.4.69) with $C\{\mathbf{A};x\} = \partial^\mu A_\mu(x)$ and using the equation of motion for the (free) ghost field.

Transforming to momentum space yields the *Ward–Takahashi identity*

$$-\frac{\mathrm{i}}{\xi}k^2 k^\mu G^{A\psi\bar{\psi}}_{\mathrm{c},\mu}(k,p,q) = e\left[G^{\psi\bar{\psi}}_{\mathrm{c}}(-q,q)) - G^{\psi\bar{\psi}}_{\mathrm{c}}(p,-p)\right], \tag{2.4.88}$$

where k,p,q denote the incoming momenta corresponding to the (outgoing) fields $A,\psi,\bar{\psi}$. In diagrams this reads

$-\frac{\mathrm{i}}{\xi}k^2 k_\mu$ 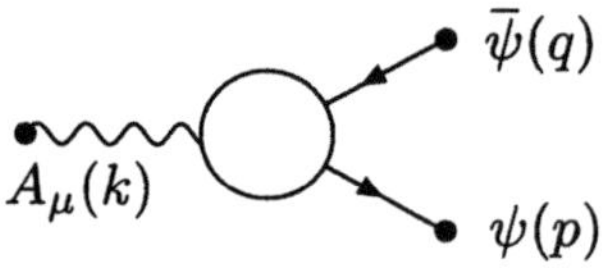

$= e$ 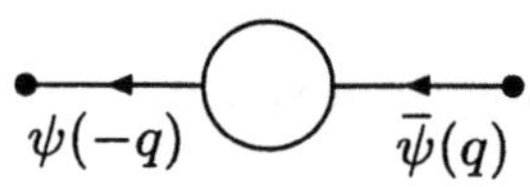

$-\,e$ 

. (2.4.89)

Equation (2.4.88) provides a relation between the electron–photon Green function and the electron propagator.

Splitting the photon propagator off in the vertex function

$$G^{A\psi\bar{\psi}}_{\mathrm{c},\mu}(k,p,q) = G^{AA}_{\mathrm{c},\mu\nu}(k,-k)G^{\underline{A}\psi\bar{\psi},\nu}_{\mathrm{c}}(k,p,q), \tag{2.4.90}$$

and using (2.4.85) yields

$$-k^{\mu}G^{\underline{A}\psi\bar{\psi}}_{\mathrm{c},\mu}(k,p,q) = e\left[G^{\psi\bar{\psi}}_{\mathrm{c}}(-q,q) - G^{\psi\bar{\psi}}_{\mathrm{c}}(p,-p)\right]. \tag{2.4.91}$$

By differentiating (2.4.82) with respect to $\{J^{\mu_i}(x_i)\}$, $i = 1,\dots,N_A$ as well as $\{J_{\bar{\psi}}(z_j)\}$ and $\{J_{\psi}(y_j)\}$, $j = 1,\dots,N_{\psi}$, we can derive in the same way the identity

$$\begin{aligned}-k^{\mu}&G^{\underline{A}A\dots\psi\dots\bar{\psi}\dots}_{\mathrm{c},\mu\mu_i}(k,k_i,p_j,q_j) = \\ &e\sum_l\left[G^{A\dots\psi\dots\bar{\psi}\dots}_{\mathrm{c},\mu_i}(k_i,p_j+\delta_{jl}k,q_j) - G^{A\dots\psi\dots\bar{\psi}\dots}_{\mathrm{c},\mu_i}(k_i,p_j,q_j+\delta_{jl}k)\right].\end{aligned} \tag{2.4.92}$$

When truncating all external lines on the left-hand side and putting them on mass shell, all terms on the right-hand side vanish because they have no pole in p_l^2 or q_l^2, and we find the identity

$$-k^{\mu}G^{\underline{AA}\dots\underline{\bar{\psi}}\dots\underline{\psi}\dots}_{\mathrm{c},\mu\mu_i}(k,k_i,p_j,q_j)\bigg|_{k_i^2=0,p_i^2=q_i^2=m_{\mathrm{e}}^2} = 0, \tag{2.4.93}$$

which holds also for S-matrix elements and states that these vanish if one of the photon polarization vectors is replaced by the photon momentum.

If we truncate all external lines apart from the one corresponding to $\bar{\psi}_1$, we obtain for $k_i^2 = 0$, $p_j^2 = q_{j\neq 1}^2 = m_{\mathrm{e}}^2$:

$$\begin{aligned}-k^{\mu}&G^{\underline{AA}\dots\underline{\bar{\psi}}\dots\underline{\psi}\dots}_{\mathrm{c},\mu\mu_i}(k,k_i,p_j,q_j)\; G^{\psi\bar{\psi}}_{\mathrm{c}}(-q_1,q_1) \\ &= -eG^{\underline{A}\dots\underline{\bar{\psi}}\dots\underline{\psi}\dots}_{\mathrm{c},\mu_i}(k_i,p_j,q_j+\delta_{j1}k)G^{\psi\bar{\psi}}_{\mathrm{c}}(-q_1-k,q_1+k).\end{aligned} \tag{2.4.94}$$

By substituting the fermion propagator on the right-hand side using (2.4.91), this can be written as

$$
\begin{aligned}
&k^\mu G^{AA\ldots\bar\psi\ldots\psi\ldots}_{\mathrm{c},\mu\mu_i}(k,k_i,p_j,q_j)\; G^{\psi\bar\psi}_{\mathrm{c}}(-q_1,q_1) = \\
&\quad G^{A\ldots\bar\psi\ldots\psi\ldots}_{\mathrm{c},\mu_i}(k_i,p_j,q_j+\delta_{j1}k) \\
&\qquad \times \left[e G^{\psi\bar\psi}_{\mathrm{c}}(-q_1,q_1) + k^\mu G^{A\psi\bar\psi}_{\mathrm{c},\mu}(k,-q_1-k,q_1)\right]
\end{aligned}
\tag{2.4.95}
$$

and after multiplication with $(G^{\psi\bar\psi}(-q_1,q_1))^{-1}$

$$
\begin{aligned}
&k^\mu G^{AA\ldots\bar\psi\ldots\psi\ldots}_{\mathrm{c},\mu\mu_i}(k,k_i,p_j,q_j) = \\
&\quad G^{A\ldots\bar\psi\ldots\psi\ldots}_{\mathrm{c},\mu_i}(k_i,p_j,q_j+\delta_{j1}k)\left[e + k^\mu G^{A\psi\psi}_{\mathrm{c},\mu}(k,-q_1-k,q_1)\right].
\end{aligned}
\tag{2.4.96}
$$

Substituting $q_1 \to q_1 - k_1$, this results in

$$
\begin{aligned}
&k^\mu \Big(G^{AA\ldots\bar\psi\ldots\psi\ldots}_{\mathrm{c},\mu\mu_i}(k,k_i,p_j,q_j-\delta_{j1}k) \\
&\qquad - G^{A\ldots\bar\psi\ldots\psi\ldots}_{\mathrm{c},\mu_i}(k_i,p_j,q_j) G^{A\psi\psi}_{\mathrm{c},\mu}(k,-q_1-k,q_1)\Big) \\
&= e\, G^{A\ldots\bar\psi\ldots\psi\ldots}_{\mathrm{c},\mu_i}(k_i,p_j,q_j)\,,
\end{aligned}
\tag{2.4.97}
$$

or in diagrams (suppressing all on-shell lines)

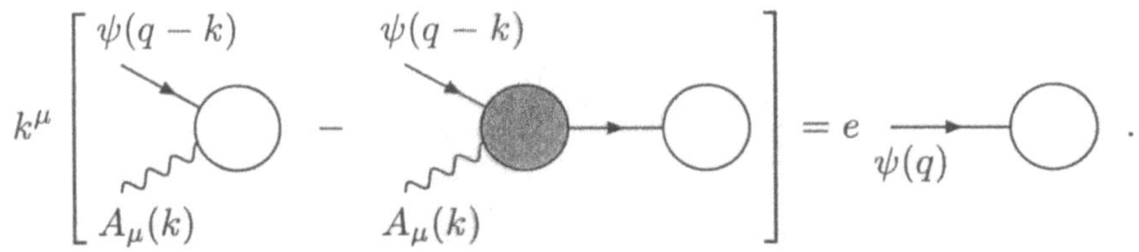

(2.4.98)

Similar relations can be derived for vertex functions. Differentiating (2.4.83) with respect to $A_\nu(y)$ and putting $A_\mu = \psi = \bar\psi = 0$ we obtain, for example,

$$
\partial^\mu \Gamma^{AA}_{\mu\nu}(x,y) = \partial^\mu \left.\frac{\delta\Gamma}{\delta A^\mu(x)\delta A^\nu(y)}\right|_{A_\mu=\psi=\bar\psi=0} = \frac{1}{\xi}\Box\partial_\nu\delta^{(4)}(x-y).
\tag{2.4.99}
$$

This is a condition on the photon–two-point function. After transformation to momentum space it reads

$$k^\mu \Gamma^{AA}_{\mu\nu}(k,-k) = -\frac{1}{\xi} k^2 k_\nu . \tag{2.4.100}$$

This identity is equivalent to (2.4.85) and ensures that the higher-order contributions to the photon–two-point function are transverse.

Differentiating (2.4.83) with respect to $\psi(y)$ and $\bar{\psi}(z)$ gives rise to

$$\begin{aligned}\partial^\mu \frac{\delta\Gamma}{\delta A^\mu(x)\delta\psi(z)\delta\bar{\psi}(y)} &= \mathrm{i}e\delta^{(4)}(x-z)\frac{\delta\Gamma}{\delta\psi(x)\delta\bar{\psi}(y)} \\ &\quad - \mathrm{i}e\delta^{(4)}(x-y)\frac{\delta\Gamma}{\delta\psi(z)\delta\bar{\psi}(x)} .\end{aligned} \tag{2.4.101}$$

Putting $A_\mu = \psi = \bar{\psi} = 0$ and transforming to momentum space yields

$$k^\mu \Gamma^{A\psi\bar{\psi}}_\mu(k,q,p) = e\left[\Gamma^{\psi\bar{\psi}}(-p,p) - \Gamma^{\psi\bar{\psi}}(q,-q)\right] , \tag{2.4.102}$$

where now k, p, q denote the incoming momenta corresponding to the (incoming) fields $A, \bar{\psi}, \psi$, or in diagrams

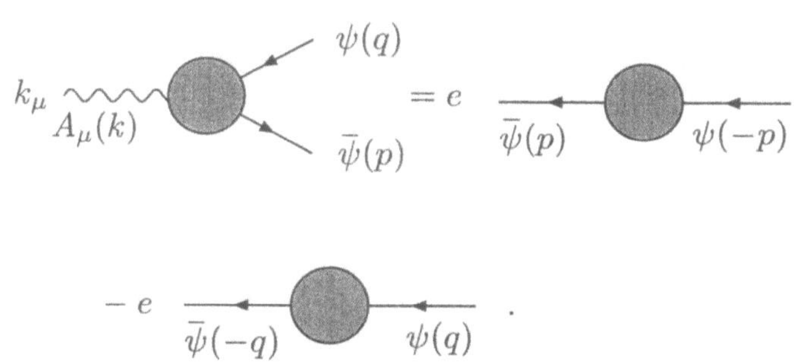

(2.4.103)

This is the *Ward–Takahashi identity* for the electron–photon vertex [Ta57]. Differentiating with respect to k_μ for fixed q, using $p = -q - k$, and putting $k = 0$, results in

$$\Gamma^{A\psi\bar{\psi}}_\mu(0,q,-q) = e\frac{\partial}{\partial q^\mu}\Gamma^{\psi\bar{\psi}}(q,-q) . \tag{2.4.104}$$

2.4.4 The background-field method

The classical Lagrangian of a gauge theory is *gauge-invariant.* The same holds for any physical quantity. But quantities with no direct physical meaning such as off-shell Green functions or counterterms are not gauge-invariant. Moreover, the quantization of gauge theories, as described in Sect. 2.4.1, requires the introduction of a gauge-fixing and a Faddeev–Popov ghost term. The resulting effective Lagrangian and the effective action are not gauge-invariant but only invariant under the non-linear BRS transformations. As a consequence, the Green functions in the conventional formalism do not directly reflect the underlying gauge invariance, but rather obey complicated Slavnov–Taylor identities resulting from BRS invariance. In addition, the Green functions depend on the particular gauge fixing chosen, and only physical quantities are gauge-independent order by order in perturbation theory.

In the background-field approach [De67, tH76a, Kl75, Bo81, Ha83, Ab81], the quantization is performed in such a way that an explicit gauge invariance is still present once gauge-fixing and ghost terms have been added. As a result, Green functions obey the naïve Ward identities, i.e. the Ward identities corresponding to the classical action (without gauge-fixing and ghost terms), and even unphysical quantities such as divergent counterterms take a gauge-invariant form. The background-field approach does not only make the structure of gauge theories more transparent and easier to understand but can also greatly simplify practical calculations.

2.4.4.1 The gauge-invariant effective action

We follow the treatment of Ref. [Ab81] and consider a gauge theory with a multiplet of fermion fields ψ_i as in Sect. 2.4.1. The classical Lagrangian reads

$$\mathcal{L}_{\text{cl}} = -\frac{1}{4}F^a_{\mu\nu}F^{a,\mu\nu} + \bar{\psi}_i(x)(\mathrm{i}\not{D}_{ij} - m\delta_{ij})\psi_j(x) \tag{2.4.105}$$

with

$$\begin{aligned} F^a_{\mu\nu} &= \partial_\mu A^a_\nu - \partial_\nu A^a_\mu + gf^{abc}A^b_\mu A^c_\nu, \\ D_\mu &= \partial_\mu - \mathrm{i}gT^a A^a_\mu. \end{aligned} \tag{2.4.106}$$

We split the gauge field into an arbitrary background field $\hat{A}^a_\mu$ and a fluctuating quantum field A^a_μ (here and in the following we denote all background fields with a caret),

$$A^a_\mu \to \hat{A}^a_\mu + A^a_\mu. \tag{2.4.107}$$

The background field is treated as an external source and the quantum field as the integration variable of the functional integral. After this split the classical Lagrangian can be written as

$$\begin{aligned}\mathcal{L}_{\mathrm{cl,BFM}} = & -\frac{1}{4}(\hat{F}^a_{\mu\nu} + \hat{D}^{ac}_\mu A^c_\nu - \hat{D}^{ac}_\nu A^c_\mu + g f^{abc} A^b_\mu A^c_\nu)^2 \\ & + \bar{\psi}_i(x)(\mathrm{i}\hat{\not{D}} + g \not{A}^a T^a - m)_{ij}\psi_j(x),\end{aligned} \tag{2.4.108}$$

where

$$\begin{aligned}\hat{F}^a_{\mu\nu} &= \partial_\mu \hat{A}^a_\nu - \partial_\nu \hat{A}^a_\mu + g f^{abc} \hat{A}^b_\mu \hat{A}^c_\nu, \\ \hat{D}^{ac}_\mu &= \partial_\mu \delta^{ac} + g f^{abc} \hat{A}^b_\mu, \\ \hat{D}_\mu &= \partial_\mu - \mathrm{i} g T^a \hat{A}^a_\mu\end{aligned} \tag{2.4.109}$$

are the field strength of and the covariant derivative with respect to the background field. For fixed background field, the Lagrangian (2.4.108) is invariant under the infinitesimal gauge transformations

$$\begin{aligned}A^a_\mu \to A'^a_\mu &= A^a_\mu + \frac{1}{g}\hat{D}^{ac}_\mu \delta\theta^c + f^{abc} A^b_\mu \delta\theta^c \\ &= A^a_\mu + \frac{1}{g}\delta^{ac}\partial_\mu \delta\theta^c + f^{abc}(A^b_\mu + \hat{A}^b_\mu)\delta\theta^c, \\ \psi_i \to \psi'_i &= \psi_i + \mathrm{i}(T^a\psi)_i \delta\theta^a.\end{aligned} \tag{2.4.110}$$

In order to define the functional integral we must fix a gauge for A^a_μ using the Faddeev–Popov procedure. We choose a gauge-fixing functional that is covariant with respect to the background gauge field

$$C^a_{\mathrm{BFM}}\{\mathbf{A}, \hat{\mathbf{A}}; x\} = \hat{D}^{ab,\mu} A^b_\mu(x). \tag{2.4.111}$$

The corresponding *Faddeev–Popov Lagrangian* is determined as described in Sect. 2.4.1.3 from the variation of the gauge-fixing operator (2.4.111) with respect to the gauge transformations (2.4.110),

$$\begin{aligned}\mathcal{L}_{\text{ghost,BFM}} &= -\int \mathrm{d}^4z\, \bar{u}^a(x)\frac{\delta C^a\{\mathbf{A},\hat{\mathbf{A}};x\}}{\delta A^c_\mu(z)}(\hat{D}^{cb}_\mu + gf^{cdb}A^d_\mu)u^b(z)\\ &= -\int \mathrm{d}^4z\, \bar{u}^a(x)\hat{D}^{ac,\mu}(\hat{D}^{cb}_\mu + gf^{cdb}A^d_\mu)u^b(z). \qquad (2.4.112)\end{aligned}$$

The complete Lagrangian

$$\begin{aligned}\mathcal{L}_{\text{BFM}} = \mathcal{L}_{\text{cl,BFM}} &- \frac{1}{2\xi}C^a_{\text{BFM}}\{\mathbf{A},\hat{\mathbf{A}};x\}C^a_{\text{BFM}}\{\mathbf{A},\hat{\mathbf{A}};x\}\\ &+ \mathcal{L}_{\text{ghost,BFM}} \qquad (2.4.113)\end{aligned}$$

is gauge-fixed but invariant under the local transformations

$$\begin{aligned}\hat{A}^a_\mu &\to \hat{A}'^a_\mu = \hat{A}^a_\mu + \frac{1}{g}\hat{D}^{ab}\delta\hat{\theta}^b, & A^a_\mu &\to A'^a_\mu = A^a_\mu + f^{abc}A^b_\mu\delta\hat{\theta}^c,\\ u^a &\to u'^a = u^a + f^{abc}u^b\delta\hat{\theta}^c, & \bar{u}^a &\to \bar{u}'^a = \bar{u}^a + f^{abc}\bar{u}^b\delta\hat{\theta}^c,\\ \psi_i &\to \psi'_i = \psi_i + \mathrm{i}(T^a\psi)_i\delta\hat{\theta}^a, & \bar{\psi}_i &\to \bar{\psi}'_i = \bar{\psi}_i - \mathrm{i}(\bar{\psi}T^a)_i\delta\hat{\theta}^a.\end{aligned} \qquad (2.4.114)$$

Under this transformation the background field transforms inhomogeneously as a gauge field, while the quantum field and the Faddeev–Popov ghost field transform (homogeneously) as a matter field in the adjoint representation. This background-field gauge invariance follows directly from the global invariance of the Lagrangian (2.4.113) and from the fact that the background field $\hat{A}^a_\mu$ appears in (2.4.113) only within the covariant derivative and the field strength.

The *generating functional of Green functions* is defined as

$$\begin{aligned}T_{\text{BFM}}\{\mathbf{J},\hat{\mathbf{A}},\mathbf{J}_u,\mathbf{J}_{\bar{u}},\mathbf{J}_\psi,\mathbf{J}_{\bar{\psi}}\} &= \frac{1}{Z}\int \mathcal{D}[\mathbf{A}]\mathcal{D}[\mathbf{u}]\mathcal{D}[\bar{\mathbf{u}}]\mathcal{D}[\psi]\mathcal{D}[\bar{\psi}]\\ &\times \exp\left(\mathrm{i}\int \mathrm{d}^4x\left[\mathcal{L}_{\text{BFM}} + J^{\mu,a}A^a_\mu + J^a_u u^a - \bar{u}^a J^a_{\bar{u}} + J^i_\psi\psi_i - \bar{\psi}_i J^i_{\bar{\psi}}\right]\right). \qquad (2.4.115)\end{aligned}$$

It depends both on the conventional sources and on the background field $\hat{\mathbf{A}}$. The functional measure is invariant under the transformation (2.4.114).

By performing a change of integration variables of the form (2.4.114) it is easy to see that the generating functional T_{BFM} is invariant under the transformation

$$\begin{aligned}
\hat{A}^a_\mu &\to \hat{A}'^a_\mu = \hat{A}^a_\mu + \frac{1}{g}\hat{D}^{ab}\delta\hat{\theta}^b, & J^a_\mu &\to J'^a_\mu = J^a_\mu + f^{abc}J^b_\mu\delta\hat{\theta}^c,\\
J^a_u &\to J'^a_u = J^a_u + f^{abc}J^b_u\delta\hat{\theta}^c, & J^a_{\bar{u}} &\to J'^a_{\bar{u}} = J^a_{\bar{u}} + f^{abc}J^b_{\bar{u}}\delta\hat{\theta}^c,\\
J^i_\psi &\to J'^i_\psi = J^i_\psi - \mathrm{i}(J_\psi T^a)^i\delta\hat{\theta}^a, & J^i_{\bar{\psi}} &\to J'^i_{\bar{\psi}} = J^i_{\bar{\psi}} + \mathrm{i}(T^a J_{\bar{\psi}})^i\delta\hat{\theta}^a.
\end{aligned} \tag{2.4.116}$$

As in the conventional formalism the *generating functional of connected Green functions* is given by

$$T_{\text{c,BFM}}\{\mathbf{J},\hat{\mathbf{A}},\mathbf{J}_u,\mathbf{J}_{\bar{u}},\mathbf{J}_\psi,\mathbf{J}_{\bar{\psi}}\} = \log T_{\text{BFM}}\{\mathbf{J},\hat{\mathbf{A}},\mathbf{J}_u,\mathbf{J}_{\bar{u}},\mathbf{J}_\psi,\mathbf{J}_{\bar{\psi}}\}. \tag{2.4.117}$$

The corresponding *generating functional of vertex functions* is obtained after a Legendre transform as

$$\begin{aligned}
\mathrm{i}\Gamma_{\text{BFM}}\{\mathbf{A},\hat{\mathbf{A}},\mathbf{u},\bar{\mathbf{u}},\boldsymbol{\psi},\bar{\boldsymbol{\psi}}\} &= T_{\text{c,BFM}}\{\mathbf{J},\hat{\mathbf{A}},\mathbf{J}_u,\mathbf{J}_{\bar{u}},\mathbf{J}_\psi,\mathbf{J}_{\bar{\psi}}\}\\
&\quad - \mathrm{i}\int \mathrm{d}^4x \left(J^{a,\mu}A^a_\mu + J^a_u u^a - \bar{u}^a J^a_{\bar{u}} + J^i_\psi\psi_i - \bar{\psi}_i J^i_{\bar{\psi}}\right),
\end{aligned} \tag{2.4.118}$$

where the sources of Γ_{BFM}

$$\begin{aligned}
A^a_\mu &= \frac{\delta}{\mathrm{i}\delta J^a_\mu}T_{\text{c,BFM}}, & u^a &= \frac{\delta}{\mathrm{i}\delta J^a_u}T_{\text{c,BFM}}, & \bar{u}^a &= \frac{\delta}{\mathrm{i}\delta J^a_{\bar{u}}}T_{\text{c,BFM}},\\
\psi_i &= \frac{\delta}{\mathrm{i}\delta J^i_\psi}T_{\text{c,BFM}}, & \bar{\psi}_i &= \frac{\delta}{\mathrm{i}\delta J^i_{\bar{\psi}}}T_{\text{c,BFM}}
\end{aligned} \tag{2.4.119}$$

are denoted with the same symbols as the integration variables of the functional integral. Note that the background field is not transformed. Since the sources of Γ_{BFM} are just the conjugate variables to the sources of T_{BFM} it follows immediately from the invariance of T_{BFM} that the effective action (2.4.118) is invariant under the transformations (2.4.114). In contrast to the BRS transformations these transformations are linear in the fields that correspond to the sources of the effective action.

We can define a *gauge-invariant effective action* $\hat{\Gamma}$ via

$$\hat{\Gamma}\{\hat{\mathbf{A}},\psi,\bar{\psi}\} = \Gamma_{\mathrm{BFM}}\{0,\hat{\mathbf{A}},0,0,\psi,\bar{\psi}\}. \tag{2.4.120}$$

It is invariant under ordinary gauge transformations of the background gauge field and the fermion fields, i.e.

$$0 = \frac{\delta\hat{\Gamma}}{\delta\hat{\theta}^a(x)} = D_\mu^{ab}\frac{\delta\hat{\Gamma}}{\delta\hat{A}_\mu^b(x)} + \mathrm{i}g\bar{\psi}(x)T^a\frac{\delta\hat{\Gamma}}{\delta\bar{\psi}(x)} + \mathrm{i}g\frac{\delta\hat{\Gamma}}{\delta\psi(x)}T^a\psi(x). \tag{2.4.121}$$

As a result, the vertex functions generated by differentiating $\hat{\Gamma}$ with respect to $\hat{\mathbf{A}}$, ψ, and $\bar{\psi}$ obey the naïve Ward identities of gauge invariance. These Ward identities are similar to those of QED (2.4.83) and lead to similar conclusions. As an example we mention that the Ward identities (2.4.121) imply that the product $g\hat{A}_\mu^b(x)$ needs not to be renormalized (cf. Sect. 2.5.1.8). Besides the background-field Ward identities (2.4.121), also Slavnov–Taylor identities related to the quantum gauge transformations (2.4.110) are valid.

2.4.4.2 Relation to the conventional effective action

In order to establish the usefulness of the background-field effective action we have to relate it to the conventional effective action. Performing a shift $A_\mu^a \to A_\mu^a - \hat{A}_\mu^a$ in T_{BFM} (2.4.115), one finds

$$\begin{aligned} &T_{\mathrm{BFM}}\{\mathbf{J},\hat{\mathbf{A}},\mathbf{J}_u,\mathbf{J}_{\bar{u}},\mathbf{J}_\psi,\mathbf{J}_{\bar{\psi}}\} = \\ &\qquad T\{\mathbf{J},\mathbf{J}_u,\mathbf{J}_{\bar{u}},\mathbf{J}_\psi,\mathbf{J}_{\bar{\psi}};\hat{\mathbf{A}}\}\exp\left(-\mathrm{i}\int \mathrm{d}^4x\, J^{\mu,a}\hat{A}_\mu^a\right), \end{aligned} \tag{2.4.122}$$

where T is the (normalized) generating functional of the conventional formalism (2.4.20) evaluated with the unconventional gauge-fixing term

$$\tilde{C}^a\{\mathbf{A},\hat{\mathbf{A}};x\} = \partial^\mu A_\mu^a(x) - \partial^\mu\hat{A}_\mu^a(x) + gf^{abc}\hat{A}^{b,\mu}(x)A_\mu^c(x). \tag{2.4.123}$$

The ghost Lagrangian of T_{BFM} in the background-field gauge goes over into the correct ghost Lagrangian of T in the gauge (2.4.123). Note that T is still a functional of $\hat{\mathbf{A}}$ because of the presence of this field in the gauge-fixing term (2.4.123). As a consequence of (2.4.122) the generating functionals of connected Green functions are related by

$$T_{\mathrm{c,BFM}}\{\mathbf{J},\hat{\mathbf{A}},\mathbf{J}_u,\mathbf{J}_{\bar{u}},\mathbf{J}_\psi,\mathbf{J}_{\overline{\psi}}\} = \\ T_{\mathrm{c}}\{\mathbf{J},\mathbf{J}_u,\mathbf{J}_{\bar{u}},\mathbf{J}_\psi,\mathbf{J}_{\overline{\psi}};\hat{\mathbf{A}}\} - \mathrm{i}\int \mathrm{d}^4x\, J^{\mu,a}\hat{A}^a_\mu. \tag{2.4.124}$$

Differentiating with respect to J^a_μ yields

$$A^a_\mu = \frac{\delta}{\mathrm{i}\delta J^a_\mu}T_{\mathrm{c,BFM}} = \frac{\delta}{\mathrm{i}\delta J^a_\mu}T - \hat{A}^a_\mu = \tilde{A}^a_\mu - \hat{A}^a_\mu, \tag{2.4.125}$$

where $\tilde{A}^a_\mu$ denotes the gauge-field source of the conventional vertex functional. Performing a Legendre transformation we find the relation between the effective action in the background-field formalism and the conventional formalism

$$\Gamma_{\mathrm{BFM}}\{\mathbf{A},\hat{\mathbf{A}},\mathbf{u},\bar{\mathbf{u}},\psi,\overline{\psi}\} = \Gamma\{\tilde{\mathbf{A}},\mathbf{u},\bar{\mathbf{u}},\psi,\overline{\psi};\hat{\mathbf{A}}\}\Big|_{\tilde{\mathbf{A}}=\mathbf{A}+\hat{\mathbf{A}}}. \tag{2.4.126}$$

Finally, the gauge-invariant effective action (2.4.120) is given by

$$\hat{\Gamma}\{\hat{\mathbf{A}},\psi,\overline{\psi}\} = \Gamma\{\hat{\mathbf{A}},0,0,\psi,\overline{\psi};\hat{\mathbf{A}}\}. \tag{2.4.127}$$

The background field $\hat{A}^a_\mu$ enters $\hat{\Gamma}$ both through its gauge-fixing term and through its field argument.

It is instructive to consider the background-field method for the case of a non-gauge field theory with field ψ. Splitting this field into a quantum part ψ and background part $\hat{\psi}$ and performing the same manipulations as above we find

$$\Gamma_{\mathrm{BFM}}\{\psi,\hat{\psi}\} = \Gamma\{\tilde{\psi}\}\Big|_{\tilde{\psi}=\psi+\hat{\psi}}, \tag{2.4.128}$$

or

$$\hat{\Gamma}\{\hat{\psi}\} \equiv \Gamma_{\mathrm{BFM}}\{0,\hat{\psi}\} = \Gamma\{\hat{\psi}\}. \tag{2.4.129}$$

The background-field effective action $\Gamma_{\mathrm{BFM}}\{\psi,\hat{\psi}\}$ is just the conventional effective action computed in the presence of the background field $\hat{\psi}$. In particular, $\Gamma_{\mathrm{BFM}}\{0,\hat{\psi}\}$ is the vacuum-to-vacuum amplitude in the presence of the background field. Thus, according to (2.4.129) the conventional effective action is equal to the vacuum-to-vacuum amplitude in the background-field method. This statement can be generalized and implies that for a non-gauge theory the background-field effective action is equivalent to the conventional effective action.

If we split in (2.4.105) also the fermion fields into a background and quantum part

$$\psi \to \hat{\psi} + \psi, \qquad \bar{\psi} \to \hat{\bar{\psi}} + \bar{\psi} \tag{2.4.130}$$

we obtain

$$\begin{aligned} &\Gamma_{\mathrm{BFM}}\{\mathbf{A}, \hat{\mathbf{A}}, \mathbf{u}, \bar{\mathbf{u}}, \psi, \hat{\psi}, \bar{\psi}, \hat{\bar{\psi}}\} = \\ &\quad \Gamma\{\tilde{\mathbf{A}}, \mathbf{u}, \bar{\mathbf{u}}, \tilde{\psi}, \tilde{\bar{\psi}}; \hat{\mathbf{A}}\}\Big|_{\tilde{\mathbf{A}}=\mathbf{A}+\hat{\mathbf{A}},\, \tilde{\psi}=\hat{\psi}+\psi,\, \tilde{\bar{\psi}}=\hat{\bar{\psi}}+\bar{\psi}} \end{aligned} \tag{2.4.131}$$

instead of (2.4.126) and (2.4.127), and

$$\begin{aligned} \hat{\Gamma}\{\hat{\mathbf{A}}, \hat{\psi}, \hat{\bar{\psi}}\} &\equiv \Gamma_{\mathrm{BFM}}\{0, \hat{\mathbf{A}}, 0, 0, 0, \hat{\psi}, 0, \hat{\bar{\psi}}\} \\ &= \Gamma\{\hat{\mathbf{A}}, 0, 0, \hat{\psi}, \hat{\bar{\psi}}; \hat{\mathbf{A}}\}. \end{aligned} \tag{2.4.132}$$

The difference between the gauge-invariant effective action and the conventional effective action is only due to the gauge-fixing term. Thus, for fields that do not appear in the gauge-fixing term, as for example the fermion fields, the split-up into background and quantum fields is often not relevant.

2.4.4.3 The S matrix in the background-field method

Once the background-field effective $\hat{\Gamma}$ action has been determined, the connected background-field Green functions and the S matrix are constructed in the same way as in the conventional formalism by forming trees of vertex functions. To this end one has to introduce propagators for the background fields. In this respect, we note that in this application of the background-field method, the background fields are purely formal objects and have no physical meaning.

In order to define the background-field propagators that connect the vertex functions one has to add a gauge-fixing term for the background fields to $\hat{\Gamma}$ resulting in

$$\hat{\Gamma}^{\mathrm{full}} = \hat{\Gamma} + \int \mathrm{d}^4x \, \mathcal{L}_{\mathrm{fix}}^{\mathrm{BF}}. \tag{2.4.133}$$

The gauge-fixing term $\mathcal{L}_{\mathrm{fix}}^{\mathrm{BF}}$ is independent of the gauge-fixing term (2.4.111) for the quantum fields and can be chosen suitably.

The generating functional of connected Green functions $\hat{T}_{\mathrm{c}}$ can be obtained from $\hat{\Gamma}^{\mathrm{full}}$ via a Legendre transformation

$$\hat{T}_{\mathrm{c}}\{\hat{\mathbf{J}}, \mathbf{J}_{\psi}, \mathbf{J}_{\overline{\psi}}\} = \mathrm{i}\hat{\Gamma}^{\mathrm{full}}\{\hat{A}, \psi, \overline{\psi}\} + \mathrm{i}\int \mathrm{d}^4x \left(\hat{J}^{a,\mu}\hat{A}^a_\mu - J^i_{\psi}\psi_i + \overline{\psi}_i J^i_{\overline{\psi}}\right), \tag{2.4.134}$$

with

$$\hat{J}^{a,\mu} = -\frac{\delta}{\delta \hat{A}^a_\mu}\hat{\Gamma}^{\mathrm{full}}, \qquad J^i_{\psi} = \frac{\delta}{\delta \psi_i}\hat{\Gamma}^{\mathrm{full}}, \qquad J^i_{\overline{\psi}} = \frac{\delta}{\delta \overline{\psi}_i}\hat{\Gamma}^{\mathrm{full}}. \tag{2.4.135}$$

We note that the background-field gauge invariance gives rise to Ward identities for $\hat{\Gamma}^{\mathrm{full}}$ and $\hat{T}_{\mathrm{c}}$ that resemble those of QED [De96].

From the generating functional of connected Green functions $\hat{T}_{\mathrm{c}}$ the S matrix is constructed as usual (cf. Sect. 2.2.3). In order to justify the background-field method, one has to show that the resulting S-matrix is equal to the conventional one [Ab83a, Re85]. In non-gauge theories this follows trivially from (2.4.129). In gauge theories the proof from the relation (2.4.127) is complicated by the extra dependence of $\Gamma\{\hat{\mathbf{A}}, 0, 0, \psi, \overline{\psi}; \hat{\mathbf{A}}\}$ on $\hat{\mathbf{A}}$ via the gauge-fixing term. While the conventional vertex functions are obtained by first differentiating $\Gamma\{\mathbf{A}, 0, 0, \psi, \overline{\psi}; \hat{\mathbf{A}}\}$ with respect to $\mathbf{A}$ for fixed $\hat{\mathbf{A}}$ and then putting $\mathbf{A} = \hat{\mathbf{A}}$, the background-field vertex functions are obtained by first putting $\mathbf{A} = \hat{\mathbf{A}}$ and then differentiating $\Gamma\{\hat{\mathbf{A}}, 0, 0, \psi, \overline{\psi}; \hat{\mathbf{A}}\}$ with respect to $\hat{\mathbf{A}}$. This is equivalent to taking derivatives with respect to $\mathbf{A}$ *and* $\hat{\mathbf{A}}$ and then putting $\mathbf{A} = \hat{\mathbf{A}}$. In order to prove that the background-field method yields the correct S matrix one therefore has to proof that the extra derivatives with respect to $\hat{\mathbf{A}}$ do not affect the S matrix. This is plausible because the dependence of $\Gamma\{\mathbf{A}, 0, 0, \psi, \overline{\psi}; \hat{\mathbf{A}}\}$ on $\hat{\mathbf{A}}$ is entirely due to the gauge-fixing term. In fact, the proof of this statement resembles closely the proof of the gauge independence of the S matrix.

2.4.4.4 Feynman rules

From the background-field Lagrangian $\mathcal{L}_{\mathrm{BFM}}$, Feynman rules can be derived as usual; one merely has to distinguish between quantum and background fields.

The gauge-invariant effective action $\hat{\Gamma}$ can be computed by summing all one-particle irreducible Feynman graphs. Since we have put $\mathbf{A} = 0 = \mathbf{u} = \bar{\mathbf{u}}$ in the gauge-invariant effective action (2.4.127), no quantum fields or ghosts appear on the external lines. Likewise no background fields appear inside loops since the functional integration is performed only over the quantum fields, i.e. no backgound-field propagators appear within the loops. The fermions and other fields that are not split into a background and quantum part appear both inside loops and on external lines.

Once the background-field vertex functions have been calculated, the connected Green functions are composed by joining the one-particle irreducible graphs with background-field propagators. At this point a gauge for the background field must be chosen. Note that in particular the lowest-order diagrams contain only background fields.

The Feynman rules can be obtained from the Lagrangian (2.4.113) in the usual way. All vertices that result only from the classical Lagrangian are equal to those in the conventional formalism, no matter whether they contain background fields and/or quantum fields. The difference between both formalisms results only from the gauge-fixing term. Since the gauge-fixing term is strictly quadratic in the quantum fields, the only vertices that are different from the conventional ones are those involving exactly two quantum fields or Faddeev–Popov ghosts. Taking all momenta as incoming and indicating the background fields by hats over the group indices, these differing vertices are listed in Fig. 2.5. Note that in contrast to the conventional formalism vertices involving two Faddeev–Popov ghosts and two gauge-bosons appear. All vertices involving fermion fields coincide with those of the conventional formalism. Therefore, a split of the fermion fields into background and quantum fields would not matter. Similarly, in non-gauge theories the Feynman rules in the background-field formalism and the conventional formalism are identical.

For connected Green functions one needs in addition to the above Feynman rules the propagator of the background fields. If $\mathcal{L}_{\mathrm{fix}}^{\mathrm{BF}}$ is chosen using a gauge-fixing condition as in (2.4.11), the background-field propagator takes the usual form given in (2.4.32) but with a new gauge parameter $\hat{\xi}$. An important

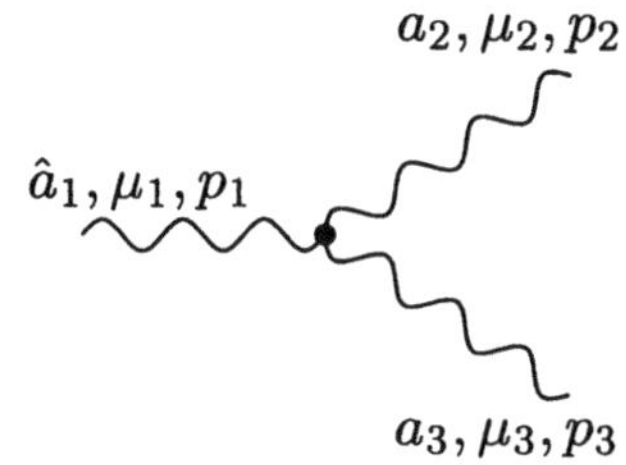

$$= g f^{\hat a_1 a_2 a_3}[g_{\mu_1\mu_2}(2p_1 + p_3(1 - 1/\xi))_{\mu_3} + g_{\mu_2\mu_3}(p_2 - p_3)_{\mu_1} - g_{\mu_3\mu_1}(2p_1 + p_2(1 - 1/\xi))_{\mu_2}]$$

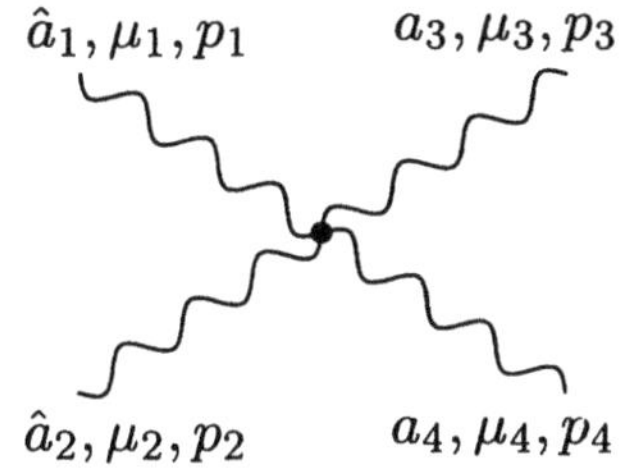

$$\begin{aligned} = -\mathrm{i}g^2 &\times [f^{\hat a_1 \hat a_2 b} f^{a_3 a_4 b}(g_{\mu_1\mu_3} g_{\mu_2\mu_4} - g_{\mu_1\mu_4} g_{\mu_2\mu_3}) \\ &+ f^{\hat a_1 a_3 b} f^{a_4 \hat a_2 b}(g_{\mu_1\mu_4} g_{\mu_3\mu_2} - g_{\mu_1\mu_2} g_{\mu_3\mu_4} - g_{\mu_1\mu_3} g_{\mu_2\mu_4}/\xi) \\ &+ f^{\hat a_1 a_4 b} f^{\hat a_2 a_3 b}(g_{\mu_1\mu_2} g_{\mu_4\mu_3} - g_{\mu_1\mu_3} g_{\mu_4\mu_2} + g_{\mu_1\mu_4} g_{\mu_3\mu_2}/\xi)] \end{aligned}$$

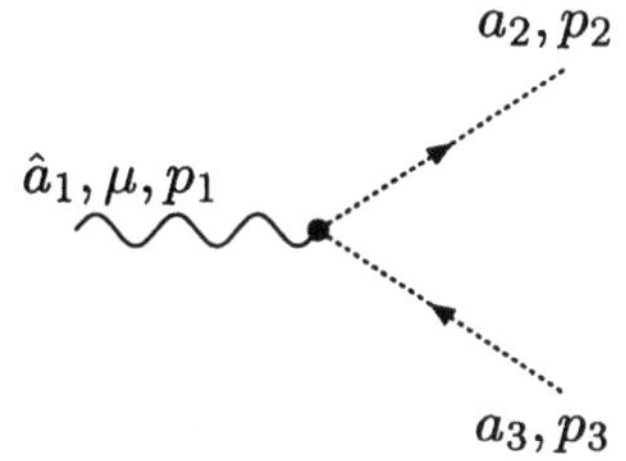

$$= -g f^{\hat a_1 a_2 a_3}(p_2 - p_3)_\mu$$

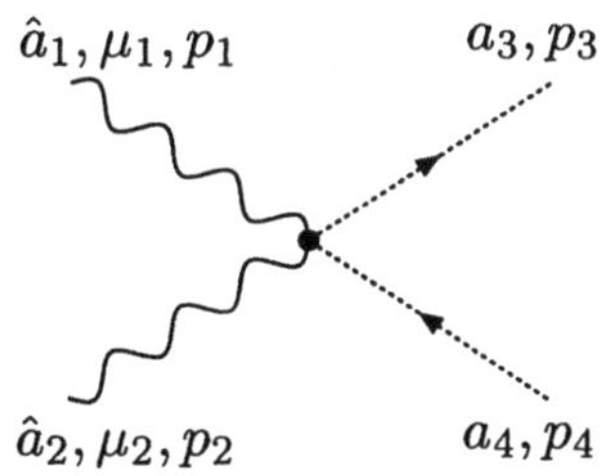

$$= \mathrm{i}g^2 (f^{\hat a_1 a_3 b} f^{\hat a_2 a_4 b} + f^{\hat a_1 a_4 b} f^{\hat a_2 a_3 b}) g_{\mu_1\mu_2}$$

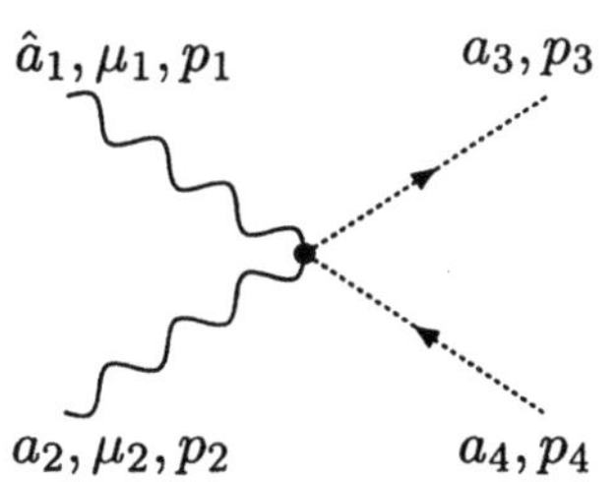

$$= \mathrm{i}g^2 f^{\hat a_1 a_3 b} f^{a_2 a_4 b} g_{\mu_1\mu_2} \qquad (2.4.136)$$

Fig. 2.5 Extra vertices in the background-field formalism involving gauge-boson and Faddeev–Popov fields

advantage of the background-field method is that $\mathcal{L}_{\text{fix}}^{\text{BF}}$ can be chosen in such a way that actual calculations simplify.

At first sight the use of the background-field method seems to be more complicated than conventional calculations. However, once the Feynman rules have been determined, calculations within the background-field method are in fact often simpler. In addition, the background-field Green functions have many desirable properties that the conventional Green functions are lacking.

In this section, the background field was used as a technical device to simplify calculations. The background-field formalism can also be used for background fields with physical significance. An example for this can be found in the discussion of the one-instanton calculation in Sect. 2.9.3.2.

2.5 Renormalization of quantum field theories

The precision of many experiments requires the evaluation of the theories for the strong and electroweak interaction beyond the tree-graph approximation. The higher-order corrections to Green functions and S-matrix elements, i.e. the quantum effects, result from Feynman graphs containing loops. Since these corrections change the relations between the parameters in the Lagrangian and the observables, the original parameters of the theory, the *bare parameters*, are no longer directly related to physical quantities. Moreover, the bare parameters become divergent. To allow for a proper treatment of divergent quantities, a *regularization* procedure is needed. This amounts to a modification of the theory so that the possibly divergent expressions become well-defined, and that in a suitable limit the original (divergent) theory is recovered. Consequently, a redefinition of the basic parameters, a *renormalization* of the theory, is needed. As is well-known from second-order quantum-mechanical perturbation theory also the normalization of the state vectors is modified. This has to be corrected for by a renormalization of the fields and/or external wave-functions (states).

2.5.1 Divergences and renormalization

A technical and conceptual complication arises in relativistic quantum field theories because *divergences* occur in the evaluation of loop diagrams.

Therefore, regularization prescriptions [Pa49] that are compatible with the structure of gauge theories are required. The divergences in all Green functions and S-matrix elements must be eliminated in the renormalization procedure by absorbing them in renormalization constants for parameters and fields. A theory in which this is possible in all orders of perturbation theory is called renormalizable. The proof of the *renormalizability* of gauge theories [Le72, tH72] was one of the most important steps in the development of QCD and the Glashow–Salam–Weinberg (GSW) theory. A detailed and complete treatment of the theory of renormalization is beyond the scope of this book (cf. [Co84]). We restrict our survey to the principles of the renormalization program, address the crucial points, and illustrate the renormalization procedure by presenting the techniques needed for the evaluation and renormalization of the one-loop corrections in gauge theories.

The requirement that divergences are compensated does not fix the finite parts of the renormalization constants. Indeed, these can be defined in many different ways leading to different parametrizations of the theory. As a consequence, calculations in finite orders of perturbation theory performed in different renormalization schemes differ by higher-order contributions (*renormalization-scheme dependence*). In an exact (all-order) calculation all different renormalization schemes would lead to equivalent relations between physical quantities. Their independence of the choice of the renormalization scheme and the consequences of this fact are studied with the help of the renormalization-group equation (Sect. 2.6).

2.5.1.1 Concept of renormalization

In order to clarify the meaning of *renormalization* we begin with a simple example: a theory described by a field ψ_0, some parameter g_0, and a Lagrangian $\mathcal{L}(\psi_0, \partial_\mu\psi_0, g_0)$. For the moment we assume that one can exactly calculate n measurable quantities $\sigma_1(g_0), \dots, \sigma_n(g_0)$. From an exact measurement E_1 of $\sigma_1(g_0)$ it is possible to determine g_0 and thus to obtain predictions for $\sigma_2, \dots, \sigma_n$. These can be compared to the measured values $E_2, \dots, E_n$ yielding $n-1$ tests of the theory.

Let us introduce a new parameter g by substituting $g_0 = g_0(g)$. Then, $\sigma_i = \sigma_i(g_0(g)) = \tilde{\sigma}_i(g)$. Under this reparametrization which may be extended to act also on the field ψ_0 we have exact equivalence of the relations between the physical results. In this sense the exact results are strictly parametrization-independent. This is of importance since by this procedure the original parameters in the Lagrangian can get replaced by others which have a more direct physical meaning. For example, it is possible to introduce well-defined quantities that have been measured with high precision as parameters of the Lagrangian.

Relativistic quantum field theories involve additional complications. Since higher-order calculations lead to divergent expressions, a *regularization scheme* is required. This introduces an effective *cut-off parameter* Λ, where $\Lambda \to \Lambda_0$ corresponds to the absence of regularization, i.e. to the original but possibly divergent expression. Consequently, we have $\sigma_i = \sigma_i(g_0, \Lambda)$. If we determine the bare coupling g_0 via σ_1 from E_1, $g_0 = \sigma_1^{(-1)}(E_1, \Lambda)$, we obtain a value for the bare coupling that is cut-off-dependent, possibly ultraviolet-divergent, and therefore evidently has no direct physical meaning. However, in renormalizable theories, elimination of g_0 in $\sigma_2(g_0, \Lambda) = \sigma_2(\sigma_1^{(-1)}(E_1, \Lambda), \Lambda)$ gives a result which is finite for $\Lambda \to \Lambda_0$. As a result, bare quantities are regularization-scheme-dependent, but the relations between physical quantities are regularization-scheme-independent and finite.

A convenient way to perform the renormalization is to introduce *renormalized parameters* by a suitable reparametrization

$$g_0 = g_0(g, \Lambda) = Z_g(g, \Lambda)\, g, \qquad \psi_0 = Z_\psi^{1/2}(g, \Lambda)\, \psi. \tag{2.5.1}$$

The renormalization of the fields allows to introduce appropriately normalized fields and Green functions. The *renormalization transformation* (2.5.1) applied to $\mathcal{L}$ gives

$$\mathcal{L}(\psi_0, g_0) = \mathcal{L}(Z_\psi^{1/2} Z_\psi^{-1/2} \psi_0, Z_g Z_g^{-1} g_0) = \mathcal{L}(Z_\psi^{1/2} \psi, Z_g g). \tag{2.5.2}$$

In perturbation theory we write

$$Z_\psi(g, \Lambda) = 1 + \delta Z_\psi(g, \Lambda), \qquad Z_g(g, \Lambda) = 1 + \delta Z_g(g, \Lambda), \tag{2.5.3}$$

yielding

$$\begin{aligned} \mathcal{L}(\psi_0, g_0) &= \mathcal{L}\Big((1 + \delta Z_\psi)^{1/2} \psi, (1 + \delta Z_g) g\Big) \\ &= \mathcal{L}(\psi, g) + \mathcal{L}_{\text{ct}}(\psi, g; \delta Z_\psi, \delta Z_g). \end{aligned} \tag{2.5.4}$$

The functional dependence of $\mathcal{L}(\psi, g)$ on ψ and g is the same as that of $\mathcal{L}(\psi_0, g_0)$ on ψ_0, g_0, and $\mathcal{L}(\psi, g)$ yields the same Feynman rules for the renormalized field and parameter as $\mathcal{L}(\psi_0, g_0)$ does for the bare ones. The *counterterm Lagrangian* $\mathcal{L}_{\text{ct}}$ summarizes all terms containing the renormalization constants δZ and generates the counterterm Feynman rules. These allow to calculate the Green functions and from them the S-matrix elements of the theory. For this purpose we define the *renormalized Green functions* G_{R}. They are obtained from the corresponding unrenormalized ones G by multiplying them with field renormalization constants

$$G_{\text{R}}(k_1, \dots, k_n) = Z_\psi^{-n/2} G(k_1, \dots, k_n). \tag{2.5.5}$$

They depend on the renormalized coupling constant g and the renormalization constant δZ_g.

The *renormalized truncated Green functions* $G_{\text{trunc,R}}$, are obtained by truncating all external legs with renormalized propagators,

$$G_{\text{trunc,R}}(k_1, \dots, k_n) = Z_\psi^{n/2} G_{\text{trunc}}(k_1, \dots, k_n), \tag{2.5.6}$$

and the S-matrix elements are according to (2.1.35) given by

$$\begin{aligned} \langle -p_{s+1}, \dots, -p_n | S | p_1, \dots, p_s \rangle &= R_{\text{R}}^{n/2} \tilde{G}_{\text{trunc,R}}(p_1, \dots, p_n)\Big|_{p_i^2 = M_i^2} \\ &= R_{\text{R}}^{n/2} Z_\psi^{n/2} \tilde{G}_{\text{trunc}}(p_1, \dots, p_n)\Big|_{p_i^2 = M_i^2}, \end{aligned} \tag{2.5.7}$$

where R_{R} is determined analogously to (2.1.34) from the renormalized two-point function. The product $R = R_{\text{R}} Z_\psi$, the unrenormalized LSZ factor, is independent of the renormalization scheme.

The *renormalized vertex functions* Γ_{R}, the one-particle irreducible components of G_{R}, are obtained from the renormalized Green functions by a Legendre transformation as in Sect. 2.2.2.2. They are related to the unrenormalized ones by

$$\Gamma_{\text{R}}(k_1, \dots, k_n) = Z_\psi^{n/2} \Gamma(k_1, \dots, k_n). \tag{2.5.8}$$

The requirement that the renormalization constants Z absorb the divergences fixes those only up to finite parts. The latter are fixed by *renormalization conditions* for the renormalized vertex functions. This allows to relate them directly to measurable quantities at suitable points in momentum space. The renormalization constants are fixed by prescribing values for certain renormalized vertex functions at chosen values of the momenta. The choice of the renormalization conditions fixes a *renormalization scheme.*

The renormalization procedure has to guarantee the absorption of divergences, the definition of the physical parameters, and a convenient normalization of the field ψ.

More complicated field theories involve more complicated renormalization procedures, but the basic ideas are the same as in the simple example discussed here.

2.5.1.2 The superficial degree of divergence

The renormalization of gauge theories has to handle the divergences. As shown in Sect. 2.1.4 all Feynman graphs can be composed by multiplication of one-particle-irreducible (1PI) graphs. Moreover, all ultraviolet (UV) divergences result from loop integrations in 1PI graphs or subgraphs. Therefore, it is sufficient to consider the UV divergences of 1PI graphs.

One of the simplest divergent Feynman diagrams is the one-loop self-energy diagram for a scalar field theory with ϕ^3 interaction,

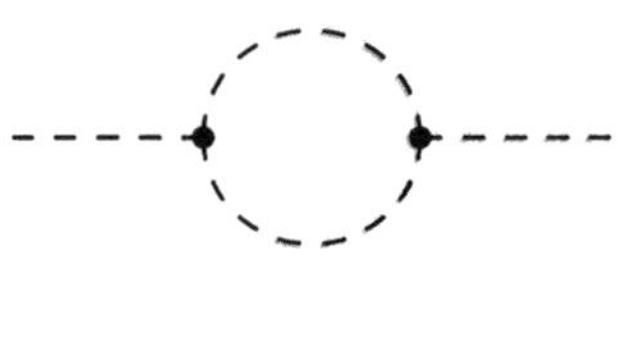

From the Feynman rules in App. A.1.3.3 we obtain for the self-energy

$$\Sigma(p) = \frac{(-\mathrm{i}g)^2}{2} \int \frac{\mathrm{d}^4 k}{(2\pi)^4} \frac{\mathrm{i}}{k^2 - m_1^2 + \mathrm{i}\varepsilon} \frac{\mathrm{i}}{(k-p)^2 - m_2^2 + \mathrm{i}\varepsilon}. \tag{2.5.9}$$

For large values of the integration momentum ($|k| \gg m_1, m_2, |p|$) the integrand behaves as $k^3 \mathrm{d}k/k^4$. The integral is logarithmically divergent. As this divergence originates from the large momentum region it is called an *UV divergence.*

It is interesting to observe that the integral is convergent for dimensions $D < 4$. This is the basis of the *dimensional regularization* scheme [tH72a,

Bo72, Co84] in which the Feynman integrals are calculated in $4-2\delta$ dimensions, where δ is the regularization parameter. This scheme is presented in Sect. 2.5.2.

In order to study the general structure of the UV divergences and to obtain a *criterion for renormalizability* we perform a systematic *power counting* of the Feynman integrals in D dimensions. Consider a graph G with E external lines with momenta $p_1, \dots, p_E$ and L internal loop momenta $k_1, \dots, k_L$ which must be integrated out. The integrand I_G is a product of vertices and propagators which depend on $p_1, \dots, p_E$ and $k_1, \dots, k_L$. The expression for the graph has the form

$$F_G(p_1, \dots, p_E) = \int^{\Lambda} I_G(p_1, \dots, p_E; k_1, \dots, k_L)\, \mathrm{d}^D k_1 \cdots \mathrm{d}^D k_L. \tag{2.5.10}$$

The possibly divergent integration is assumed to be cut off at momentum Λ. Simple counting of powers provides a definition of the *superficial degree of divergence* of sub-integrations. For this purpose consider the behaviour of I_G when a subset $S = \{k_{i'}\}$ of M loop momenta in a general configuration (i.e. no partial sum of the momenta $p_1, \dots, p_E; k_1, \dots, k_L$ must disappear) is scaled by a factor λ, $k_{i'} \to \lambda k_{i'}$. There is an upper bound for I_G as a function of λ

$$|I_G(\lambda)| \leq \lambda^{c(S)} \times \text{constant}, \tag{2.5.11}$$

[in our example (2.5.9), $M = L = 1, S = \{k\}$ and $c = -4$]. The integral over the momenta $k_{i'} \in S$ is bounded by

$$\left| \int^{\Lambda} I_G\, \mathrm{d}^D k_{i'_1} \cdots \mathrm{d}^D k_{i'_M} \right| < \Lambda^{\omega_D(S)} f(\Lambda), \tag{2.5.12}$$

where the function f grows slower than any power of Λ for $\Lambda \to \infty$, i.e. at most logarithmic (*Weinberg's theorem* [We60]). The quantity $\omega_D(S) = DM + c(S)$ is called the *superficial degree of divergence* [Dy49] of the integration with respect to the momenta $k_{i'} \in S$ in D space–time dimensions. The integration over the momenta in S is convergent if $\omega_D(S) < 0$. The superficial degree of divergence with respect to the set of all loop momenta of a graph G is called the superficial degree of divergence of the graph $\omega_D(G)$.

A graph G is called *primitively divergent* [Dy49] if only the final integration causes a divergence, i.e. if $\omega_D(S) < 0$ for all $m(S) < L$ but $\omega_D(G) \geq 0$. We can define a subgraph G_S of G by contracting all lines in G that do not depend on one of the momenta in S to a point. It can be shown that all divergences of an arbitrary divergent graph G originate from the primitive divergences associated with its subgraphs G_S [We60]. In particular, any graph G is convergent, if G and all its subgraphs are superficially convergent. On the other hand, a superficially divergent graph can nevertheless be convergent. For more details we refer to [Co84].

2.5.1.3 Criteria for renormalizability

Since renormalization proceeds order by order, and the inductive proof assumes that all subdivergences are already absorbed, it is sufficient to consider the primitive divergences of *one-particle irreducible* diagrams. The set of primitively divergent graphs can be deduced from dimensional analysis. Let us consider a general quantum field theory with scalars ϕ, spin-1/2 fermions ψ, and vector bosons A_μ in D dimensions. As the action $S = \int \mathrm{d}^D x\, \mathcal{L}$ is dimensionless in the natural system of units, $\hbar = c = 1$, the Lagrangian has mass dimension D, i.e. $\dim \mathcal{L} = [\mathcal{L}] = [M]^D$ (length and mass are measured in inverse units: $[x] = [M]^{-1}$). Using $[\partial] = [M]$, the canonical mass dimensions of the fields can be derived from the kinetic terms in the Lagrangian (cf. the upper part of Table 2.1).

In the same way the dimensions of the coupling constants follow from the requirement that the corresponding interaction terms in $\mathcal{L}$ have dimension D. For a general interaction term of the generic form

$$\mathcal{L}_v = g_v \partial^{d_v} \phi^{s_v} \psi^{f_v} A^{a_v}, \tag{2.5.13}$$

where d_v, s_v, f_v, and a_v denote the number of derivatives, scalar fields, fermion fields, and vector fields, respectively, the dimension of the coupling g_v is obtained as

$$[g_v] = D - d_v - \frac{D-2}{2} s_v - \frac{D-1}{2} f_v - \frac{D-2}{2} a_v. \tag{2.5.14}$$

Some examples are listed in the second part of Table 2.1.

Let us consider an arbitrary one-particle irreducible Feynman diagram G with E_ϕ, E_ψ, and E_A (I_ϕ, I_ψ, and I_A) external (internal) scalar, fermion,

Dirac field	$\bar{\psi}(\mathrm{i}\not{\partial} - M)\psi$	$\to [\psi] = [M]^{(D-1)/2} \to [M]^{3/2}$
scalar field	$(\partial_\mu \phi)^2$	$\to [\phi] = [M]^{(D-2)/2} \to [M]^1$
vector field	$(\partial_\nu A^\mu)^2$	$\to [A_\mu] = [M]^{(D-2)/2} \to [M]^1$
	$(F^{\mu\nu})^2$	$\to [F_{\mu\nu}] = [M]^{D/2} \to [M]^2$
minimal coupling	$g\bar{\psi}\gamma_\mu\psi A^\mu$	$\to [g] = [M]^{(4-D)/2} \to [M]^0$
three-gauge-boson coupl.	$gA^\mu A_\mu \partial_\nu A^\nu$	$\to [g] = [M]^{(4-D)/2} \to [M]^0$
four-gauge-boson coupl.	$g^2(A^\mu A_\mu)^2$	$\to [g] = [M]^{(4-D)/2} \to [M]^0$
four-fermion coupling	$G_\mu(\bar{\psi}\psi)^2$	$\to [G_\mu] = [M]^{2-2D} \to [M]^{-2}$
Yukawa coupling	$g\bar{\psi}\psi\phi$	$\to [g] = [M]^{(4-D)/2} \to [M]^0$
couplings of scalars	$gA^\mu A_\mu \phi$	$\to [g] = [M]^{(6-D)/2} \to [M]^1$
	$gA^\mu(\partial_\mu\phi)\phi$	$\to [g] = [M]^{(4-D)/2} \to [M]^0$
	$g^2\phi^2 A^\mu A_\mu$	$\to [g] = [M]^{(4-D)/2} \to [M]^0$

Tab. 2.1 Canonical mass dimensions of field operators and coupling constants in D and 4 dimensions

and vector-boson lines, respectively, V vertices, and L loops. For large momentum λk the *fermion propagators* (1.2.33) behave as λ^{-1} and the *scalar propagators* (1.2.31) as λ^{-2}. The behaviour of the *vector-boson propagators* is of the form $\lambda^{-\alpha}$ and depends on the theory and in gauge theories also on the gauge. In general $\alpha \leq 2$. Using the Feynman gauge propagator (1.2.34), we have $\alpha = 2$ in QED. Finally, each vertex v scales as $\lambda^{\hat{d}_v}$, where $\hat{d}_v$ denotes the number of derivatives associated with v and acting on internal lines. Collecting all contributions, we find for the *superficial degree of divergence* of G in D dimensions

$$\omega_D(G) = DL - 2I_\phi - I_\psi - \alpha I_A + \sum_v \hat{d}_v. \tag{2.5.15}$$

Using Euler's loop equation

$$L = I_\phi + I_\psi + I_A - V + 1 = I_\phi + I_\psi + I_A - \Big(\sum_v 1\Big) + 1, \tag{2.5.16}$$

the relations

$$\sum_v s_v = 2I_\phi + E_\phi, \qquad \sum_v f_v = 2I_\psi + E_\psi,$$
$$\sum_v a_v = 2I_A + E_A, \tag{2.5.17}$$

and (2.5.14), we obtain

$$\begin{aligned} \omega_D(G) &= D + (D-2)I_\phi + (D-1)I_\psi + (D-\alpha)I_A + \sum_v (\hat{d}_v - D) \\ &= D - \frac{D-2}{2}E_\phi - \frac{D-1}{2}E_\psi - \frac{D-2}{2}E_A \\ &\quad + (2-\alpha)I_A - \sum_v [g_v] - d, \end{aligned} \tag{2.5.18}$$

where $d = \sum_v (d_v - \hat{d}_v)$ is the total number of derivatives acting on external lines or—in momentum space—the total number of external momenta. In four dimensions this reduces to

$$\omega_4(G) = 4 - E_\phi - \frac{3}{2}E_\psi - E_A + (2-\alpha)I_A - \sum_v [g_v] - d. \tag{2.5.19}$$

From this result we can derive the set of primitively divergent Feynman graphs and vertex functions.

In general, a field theory is called *renormalizable* if it involves only a finite number of primitively divergent vertex functions. Inspecting (2.5.19) and (2.5.18), we see that this implies $[g_i] \geq 0$, $\alpha = 2$ ($\alpha > 2$ does not appear), and $D \geq 2$. In this case the superficial degree of divergence of any Feynman graph contributing to a given vertex function is bounded from above, and we can define the superficial degree of divergence for the vertex function by this upper bound. In four dimensions we thus find

$$\omega_4(\Gamma) = 4 - E_\phi - E_A - \frac{3}{2}E_\psi - d(\Gamma), \tag{2.5.20}$$

where $d(\Gamma)$ is the minimum number of derivatives acting on external lines in all graphs contributing to Γ. If, on the other hand, $[g_i] < 0$ for at least one coupling or $\alpha < 2$, any vertex function gets divergent contributions from diagrams with a large enough number of vector-boson lines or vertices associated with couplings of negative mass dimension, and the theory is non-renormalizable according to power counting.

These considerations can be summarized as follows: a *Lagrangian field theory is renormalizable* if all coupling constants have non-negative mass dimensions and if all boson propagators behave as k^{-2} and all fermion propagators as k^{-1} for large momenta k.

2.5.1.4 Renormalizability of specific theories

We start with the *QED* Lagrangian specified in (1.2.16). It involves only fermion fields, vector fields with propagators that behave as λ^{-2} because of gauge invariance, i.e. $\alpha = 2$, and one interaction vertex without derivative. The corresponding coupling constant has dimension $[e] = (4 - D)/2$, i.e. it is dimensionless in four dimensions. Thus, (2.5.19) and (2.5.20) reduce to

$$\begin{aligned} \omega_D(G) &= D - \frac{D-1}{2}E_\psi - \frac{D-2}{2}E_A - V\frac{4-D}{2}, \\ \omega_4(G) &= 4 - \frac{3}{2}E_\psi - E_A. \end{aligned} \tag{2.5.21}$$

For the physical space–time dimension $D = 4$, the superficial degree of divergence of any Feynman graph depends only on the number of external lines and thus is a property of the vertex functions $\omega_4(\Gamma) = \omega_4(G)$. The number of primitively divergent vertex functions of QED, i.e. those in which $\omega_4(\Gamma) \geq 0$, is limited to the following cases (because of fermion-number conservation E_ψ must be even)

- $E_\psi = 2$, $E_A = 0$: electron self-energy,
- $E_\psi = 2$, $E_A = 1$: electron–photon vertex,
- $E_\psi = 0$, $E_A = 1$: photon tadpole, vanishes because of Lorentz invariance,
- $E_\psi = 0$, $E_A = 2$: photon self-energy,
- $E_\psi = 0$, $E_A = 3$: triple photon vertex, vanishes in QED because of charge-conjugation invariance (Furry's theorem),
- $E_\psi = 0$, $E_A = 4$: photon–photon scattering, finite owing to gauge invariance.

Gauge invariance makes QED renormalizable and reduces the number of primitively divergent vertex functions to three.

In the *Fermi model* mentioned in Sect. 1.4.1, only fermions propagate and have a *four-fermion contact interaction* with a coupling constant G_μ of dimension $[G_\mu] = -2$ (cf. Table 2.1). This implies $\omega_4(G) = 4L - I_\psi = 4 - 3E_\psi/2 + 2V$ for the superficial degree of divergence. The number of primitively divergent vertex functions increases ad infinitum with increasing number of vertices V; therefore the Fermi model is not renormalizable.

The W and Z propagators in the *intermediate-vector-boson model* for the weak interaction have—as can be seen from (1.4.13)—an asymptotic power behaviour of zeroth order in the momentum k ($\alpha = 0$). Therefore, $\omega_4(G) = 4-3E_\psi/2-2(E_\mathrm{W}+E_\mathrm{Z})+2(I_\mathrm{W}+I_\mathrm{Z})$ represents the superficial degree of divergence of primitively divergent graphs. The number of primitively divergent vertex functions grows with the number of internal vector-boson propagators and the intermediate-vector-boson theory of the weak interaction is not renormalizable.

The difference between interacting, massive vector particles, as encountered in the previous example, and QED is gauge invariance. It allows to bring the photon propagator into the *Feynman-gauge* form $-g_{\mu\nu}/k^2$ in which QED is renormalizable. This suggests a formulation of the weak interaction theory as a gauge-field theory (cf. Chapt. 4), which is an extension of the intermediate-vector-boson model. The decisive step in this program is the Higgs-Kibble mechanism (Sect. 4.1). It allows to express massive vector fields as gauge fields. A suitable choice of gauge leads to propagators that behave as $1/k^2$ at large k^2 and thus—exactly as in QED—to a renormalizable theory.

Non-abelian gauge theories, e.g. QCD, can be evaluated in various gauges. Because physical quantities are gauge-invariant it is sufficient to prove renormalizability within one particular gauge. We consider the primitively divergent graphs in the Feynman gauge where the gauge-boson propagators behave as λ^{-2}. Let I_A, I_ψ, and I_G, (E_A, E_ψ, and E_G) be the numbers of internal (external) lines of gauge bosons, fermions, and Faddeev–Popov ghosts, respectively. With respect to power counting, the ghosts behave just as scalars. According to Table 2.1 all couplings are dimensionless and we obtain from (2.5.20)

$$\omega_4(\Gamma) = 4 - E_A - E_G - \frac{3}{2}E_\psi - d, \tag{2.5.22}$$

where d is the minimal number of external momenta present in all graphs contributing to the vertex function Γ. The number of primitively divergent vertex functions is finite in four dimensions. Since gauge-theories involve unphysical degrees of freedom one still has to show that these decouple from the S matrix. To this end, one must show that the renormalized theory is invariant under a renormalized gauge transformation. This is done in Sect. 2.5.4.1.

2.5.1.5 Structure of the counterterms

We add a few remarks on the *structure of the counterterms.* If a theory has been renormalized to $L-1$ loops, then at the L-loop level all subdivergences are already cancelled and only the primitive divergences are left. If these can be compensated by L-loop counterterms the theory can be renormalized at the L-loop level. Therefore, the counterterms are directly linked to the primitive divergences.

Differentiating a Feynman graph with respect to one of its external momenta reduces its superficial degree of divergence by one [Co84], and all superficial divergences disappear by differentiating sufficiently often. Consequently, the primitive divergence of a vertex function must be a polynomial in the external momenta with degree $\omega(\Gamma)$, and the counterterms, which have to cancel these divergences, must have the same structure [We60]. Such local counterterms can be generated from terms in the Lagrangian which are polynomial in space–time derivatives, ∂, with degree $\omega(\Gamma)$. Thus, the counterterms that compensate the primitive divergences of a vertex function Γ^{n_s,n_f,n_A} with n_s, n_f, and n_A external scalar, fermion, and vector-boson lines, respectively, have the general structure

$$\delta\mathcal{L} = c_k \partial^{\gamma_k} \phi^{n_s} \psi^{n_f} A^{n_A} \quad \text{with} \quad \gamma_k \le \omega(\Gamma^{n_s,n_f,n_A}). \tag{2.5.23}$$

In particular, we find for the dimensions of the coupling constants of the counterterms using (2.5.20)

$$\begin{aligned}[c_k] &= D - \gamma_k - \frac{D-2}{2} n_s - \frac{D-1}{2} n_f - \frac{D-2}{2} n_A \\ &= \omega(\Gamma^{n_s,n_f,n_A}) - \gamma_k \ge 0.\end{aligned} \tag{2.5.24}$$

Thus, the counterterms themselves are renormalizable.

2.5.1.6 Renormalization of QED

As an example, we perform the *renormalization of QED*. Since all types of possible counterterms are already contained in the unrenormalized Lagrangian, the subtraction of divergences can be carried out by *multiplicative renormalization*. The bare QED Lagrangian for the electron–photon system,

$$\begin{aligned}\mathcal{L}_{\mathrm{QED}} &= -\frac{1}{4}(\partial_\mu A_{0,\nu} - \partial_\nu A_{0,\mu})^2 - \frac{1}{2\xi_0}(\partial^\mu A_{0,\mu})^2 \\ &\quad + \bar{\psi}_0(\mathrm{i}\partial\!\!\!/ - m_0)\psi_0 + e_0 A_{0,\mu}\bar{\psi}_0\gamma^\mu\psi_0, \end{aligned} \tag{2.5.25}$$

where $A_{\mu,0}$, ψ_0, e_0, m_0, and ξ_0 are the unrenormalized fields, the bare charge, mass, and gauge-fixing parameter, respectively, is rewritten in terms of multiplicatively renormalized quantities by the substitutions

$$\begin{aligned} A_{\mu,0} &= Z_A^{1/2} A_\mu, \qquad \psi_0 = Z_\psi^{1/2}\psi, \\ e_0 &= Z_e e, \qquad m_0 = Z_m m, \qquad \xi_0 = Z_\xi \xi. \end{aligned} \tag{2.5.26}$$

This *renormalization transformation* yields

$$\begin{aligned}\mathcal{L}_{\mathrm{QED}} &= -\frac{1}{4}Z_A(\partial_\mu A_\nu - \partial_\nu A_\mu)^2 - \frac{1}{2\xi}Z_A Z_\xi^{-1}(\partial^\mu A_\mu)^2 \\ &\quad + Z_\psi\bar{\psi}\mathrm{i}\partial\!\!\!/\psi - Z_\psi Z_m\bar{\psi}m\psi + Z_e Z_A^{1/2} Z_\psi e A_\mu\bar{\psi}\gamma^\mu\psi. \end{aligned}$$

Writing $Z_i = 1 + \delta Z_i$, the Lagrangian is split into the original one with renormalized fields and parameters and into the counterterm Lagrangian

$$\begin{aligned}\mathcal{L}_{\mathrm{QED}} &= -\frac{1}{4}(\partial_\mu A_\nu - \partial_\nu A_\mu)^2 - \frac{1}{2\xi}(\partial^\mu A_\mu)^2 \\ &\quad + \bar{\psi}(\mathrm{i}\partial\!\!\!/ + m)\psi + eA_\mu\bar{\psi}\gamma^\mu\psi \\ &\quad - \frac{1}{4}\delta Z_A(\partial_\mu A_\nu - \partial_\nu A_\mu)^2 - \frac{1}{2\xi}(Z_A Z_\xi^{-1} - 1)(\partial^\mu A_\mu)^2 \\ &\quad + \delta Z_\psi\bar{\psi}\mathrm{i}\partial\!\!\!/\psi - (Z_\psi Z_m - 1)\bar{\psi}m\psi \\ &\quad + (Z_e Z_A^{1/2} Z_\psi - 1)eA_\mu\bar{\psi}\gamma^\mu\psi. \end{aligned} \tag{2.5.27}$$

The counterterm Lagrangian gives rise to counterterm vertices which have to be added to the ordinary Feynman rules and lead to counterterm graphs, which vanish for $Z_i \to 1$. Before we determine the QED renormalization constants Z_i, we first consider relations between them that follow from gauge invariance, i.e. from the Ward identities.

The *Ward identity for the renormalized photon propagator* [cf. (2.4.85)]

$$-\frac{1}{Z_\xi}\frac{1}{\xi}q^2 q^\mu Z_A G^{AA}_{\mathrm{R,c},\mu\nu}(q,-q) = \mathrm{i}q_\nu \tag{2.5.28}$$

implies that Z_A/Z_ξ must be finite. We can fix the finite part by choosing

$$Z_\xi = Z_A. \tag{2.5.29}$$

The *Ward identity for the renormalized electron–photon vertex function* $G^{A\psi\bar\psi}_{\mathrm{R,c},\mu}$ and the electron propagator $G^{\psi\bar\psi}_{\mathrm{R,c}}$ (2.4.88) reads with $Z_\xi = Z_A$

$$\begin{aligned} -\frac{\mathrm{i}}{Z_A\xi}q^2 q^\mu Z_\psi Z_A^{1/2} G^{A\psi\bar\psi}_{\mathrm{R,c},\mu}(q,p,p') &= \\ = Z_e e\, Z_\psi\Big(G^{\psi\bar\psi}_{\mathrm{R,c}}(-p',p') &- G^{\psi\bar\psi}_{\mathrm{R,c}}(p,-p)\Big). \end{aligned} \tag{2.5.30}$$

Therefore, $Z_e Z_A^{1/2}$ must be finite, and we can choose

$$Z_e = Z_A^{-1/2}. \tag{2.5.31}$$

Thus, charge renormalization can be related to the renormalization of the photon field.

The renormalization constants Z_A, Z_ψ, and Z_m are sufficient to cancel all UV divergences of QED (cf. Sect. 2.5.1.4). In the *on-shell renormalization scheme*, they are determined from the conditions that the photon propagator and the electron propagator have residue one, and that m is the physical electron mass.

The *photon-field renormalization* is fixed by requiring

$$\lim_{q^2\to 0}\left[\frac{\Gamma^{AA}_{\mathrm{R},\mu\nu}(q,-q)\varepsilon^\nu(q)}{q^2} = -\varepsilon_\mu(q)\right]. \tag{2.5.32}$$

Since $\Gamma^{AA}_{\mathrm{R},\mu\nu}(q,-q)$ gets contributions of ordinary and counterterm Feynman diagrams this condition fixes the renormalization constant Z_A. Equation

(2.5.32) enforces that the pole at $q^2 = 0$ of the renormalized photon propagator has residue 1. With this normalization the explicit wave-function renormalization constant R in the reduction formula (2.1.33) can be omitted. The polarization vectors $\varepsilon^\mu(q)$, $\varepsilon^\mu q_\mu = 0$, $\varepsilon^\mu \varepsilon^*_\mu = -1$, project in (2.5.32) onto the physical degrees of freedom of an on-shell photon.

In order to analyse the Ward identity (2.4.100), which is equivalent to (2.5.28), and the renormalization condition (2.5.32) in more detail we decompose $\Gamma^{AA}_{\mathrm{R},\mu\nu}$ into Lorentz tensors and scalar functions,

$$\Gamma^{AA}_{\mathrm{R},\mu\nu}(q,-q) = \left(g_{\mu\nu} - \frac{q_\mu q_\nu}{q^2}\right)\Gamma^{AA}_{\mathrm{R,T}}(q^2) + \frac{q_\mu q_\nu}{q^2}\Gamma^{AA}_{\mathrm{R,L}}(q^2). \tag{2.5.33}$$

The Ward identity (2.5.28) imposes

$$q^\mu \Gamma^{AA}_{\mathrm{R},\mu\nu}(q,-q) = q_\nu \Gamma^{AA}_{\mathrm{R,L}}(q^2) = -\frac{1}{\xi} q^2 q_\nu ,$$

or

$$\Gamma^{AA}_{\mathrm{R,L}}(q^2) = -\frac{1}{\xi} q^2 , \tag{2.5.34}$$

i.e. the longitudinal part gets no higher-order corrections. The Ward identity and the absence of poles in vertex functions, in particular in $\Gamma^{AA}_{\mathrm{R},\mu\nu}(q,-q)$, enforce that the transverse part of the photon vertex function vanishes at $q^2 = 0$,

$$\Gamma^{AA}_{\mathrm{R,T}}(0) = \Gamma^{AA}_{\mathrm{R,L}}(0) = 0, \tag{2.5.35}$$

i.e. no photon mass term is generated by higher-order corrections. Inserting (2.5.33) into (2.5.32) yields the renormalization condition

$$\lim_{q^2 \to 0}\left(\frac{\Gamma^{AA}_{\mathrm{R,T}}(q^2)}{q^2}\right) = \left.\frac{\partial \Gamma^{AA}_{\mathrm{R,T}}(q^2)}{\partial q^2}\right|_{q^2=0} = -1. \tag{2.5.36}$$

The constants Z_m and Z_ψ for the *renormalization of the electron mass and field* are fixed by

$$\lim_{p^2 \to m^2}\left[\frac{(\not p + m)\Gamma^{\psi\bar\psi}_{\mathrm{R}}(p,-p)u(p)}{p^2 - m^2} = u(p)\right]. \tag{2.5.37}$$

Here $u(p)$ denotes the electron spinor. Equation (2.5.37) implies in particular

$$\Gamma_{\mathrm{R}}^{\psi\bar{\psi}}(p,-p)u(p)|_{p^2=m^2} = 0, \tag{2.5.38}$$

i.e. m is the physical electron mass. Moreover, it guarantees the usual normalization of the electron propagator, i.e. the renormalized electron propagator has residue one.

Using again Lorentz covariance,

$$\begin{aligned}\Gamma_{\mathrm{R}}^{\psi\bar{\psi}}(p,-p) &= \not{p}\Gamma_{\mathrm{R,V}}^{\psi\bar{\psi}}(p^2) + m\Gamma_{\mathrm{R,S}}^{\psi\bar{\psi}}(p^2) \\ &= (\not{p}-m)\Gamma_{\mathrm{R,V}}^{\psi\bar{\psi}}(p^2) + m\Big(\Gamma_{\mathrm{R,V}}^{\psi\bar{\psi}}(p^2) + \Gamma_{\mathrm{R,S}}^{\psi\bar{\psi}}(p^2)\Big),\end{aligned} \tag{2.5.39}$$

expanding around $p^2 = m^2$,

$$\Gamma_{\mathrm{R,\{S,V\}}}^{\psi\bar{\psi}}(p^2) = \Gamma_{\mathrm{R,\{S,V\}}}^{\psi\bar{\psi}}(m^2) + (p^2-m^2)\Gamma_{\mathrm{R,\{S,V\}}}^{\prime\psi\bar{\psi}}(m^2) + \dots\,,$$

and inserting into (2.5.37) yields the two conditions

$$\Gamma_{\mathrm{R,V}}^{\psi\bar{\psi}}(m^2) + \Gamma_{\mathrm{R,S}}^{\psi\bar{\psi}}(m^2) = 0, \tag{2.5.40}$$

$$\Gamma_{\mathrm{R,V}}^{\psi\bar{\psi}}(m^2) + 2m^2\Big(\Gamma_{\mathrm{R,V}}^{\prime\psi\bar{\psi}}(m^2) + \Gamma_{\mathrm{R,S}}^{\prime\psi\bar{\psi}}(m^2)\Big) = 1. \tag{2.5.41}$$

The mass renormalization constant Z_m is determined from the first, the field renormalization constant Z_ψ from the second of these equations.

We have fixed the *charge renormalization* constant by putting $Z_e = Z_A^{-1/2}$. This is equivalent to the following condition

$$\bar{u}(p)\Gamma_{\mathrm{R},\mu}^{A\psi\bar{\psi}}(0,p,-p)u(p) = e\bar{u}(p)\gamma_\mu u(p) \quad \text{for} \quad p^2 = m^2. \tag{2.5.42}$$

This can be seen by using the Ward identity (2.4.104), the decomposition (2.5.39), the condition (2.5.41), the identity $2p_\mu = \{\not{p}, \gamma_\mu\}$, and the Dirac equation. Thus, the theory is fitted to physical parameters in such a way that a photon scattered with momentum transfer zero on a real, physical electron ($p^2 = m^2$) feels a coupling strength e, ($e^2/4\pi = 1/137.036\ldots$). Then, according to *Thirring's theorem* [Th53] all radiative corrections to Compton scattering vanish in the low-energy limit and the *Thomson cross section* acquires its usual form.

For a fermion with relative charge Q_f, one obtains in the same way

$$\bar{u}(p)\Gamma^{A\psi_f\bar{\psi}_f}_{\mathrm{R},\mu}(0,p,-p)u(p) = -eQ_f\bar{u}(p)\gamma_\mu u(p) \quad \text{for} \quad p^2 = m^2. \tag{2.5.43}$$

Equation (2.5.43) expresses the *charge universality* in QED: the renormalized charge, i.e. the on shell coupling to the photon, is independent of the fermion species.

Since Γ^{AA}_{R}, $\Gamma^{\psi\bar{\psi}}_{\mathrm{R}}$, and $\Gamma^{A\psi\bar{\psi}}_{\mathrm{R}}$ are the only primitively divergent vertex functions of QED, and since the renormalization conditions guarantee that the divergences can be compensated by the counterterms, the multiplicative renormalization constants Z_i of fields and parameters, which become infinite in the limit $\Lambda \to \Lambda_0$ or within dimensional regularization for $2\delta = 4 - D \to 0$, are sufficient to absorb all divergences. Therefore, all renormalized Green functions and vertex functions are UV-finite.

The one-loop results for the renormalization constants are calculated in Sect. 2.5.3.2.

2.5.1.7 Renormalization of a non-abelian gauge theory

The *renormalization of a non-abelian gauge theory* with fermions proceeds along the same lines as that of QED. The Lagrangian

$$\begin{aligned}\mathcal{L} = &-\frac{1}{4}(\partial_\mu A^a_{0,\nu} - \partial_\nu A^a_{0,\mu} + g_0 f^{abc} A^b_{0,\mu} A^c_{0,\nu})^2 - \frac{1}{2\xi_0}(\partial^\mu A^a_{0,\mu})^2 \\ &+ \bar{\psi}_0(\mathrm{i}\partial\!\!\!/ - m_0)\psi_0 + g_0 A^a_{0,\mu} T^a \bar{\psi}_0 \gamma^\mu \psi_0 \\ &- \bar{u}^a_0(\delta^{ab}\Box - g_0 f^{abc}\partial^\mu A^c_{0,\mu})u^b_0\end{aligned} \tag{2.5.44}$$

can be multiplicatively renormalized (cf. Sect. 2.5.4.2). Again, no gauge-fixing counterterm is needed, and we choose $Z_\xi = Z_A$. Because of the gauge symmetry it is possible to have only one renormalized coupling constant. Therefore, the *renormalization transformation* reads [cf. (2.5.149) and (2.5.150)]

$$\begin{aligned}&A^a_{0,\mu} = Z_A^{1/2} A^a_\mu, \qquad \psi_0 = Z_\psi^{1/2}\psi, \qquad u^a_0 = Z_u u^a, \qquad \bar{u}^a_0 = \bar{u}^a, \\ &g_0 = Z_g g, \qquad m_0 = Z_m m, \qquad \xi_0 = Z_A \xi,\end{aligned} \tag{2.5.45}$$

and the Lagrangian becomes

$$\begin{aligned}
\mathcal{L} = &-\frac{1}{4}(\partial_\mu A^a_\nu - \partial_\nu A^a_\mu + g f^{abc} A^b_\mu A^c_\nu)^2 - \frac{1}{2\xi}(\partial^\mu A^a_\mu)^2 \\
&+ \overline{\psi}(\mathrm{i}\partial\!\!\!/ + m)\psi + g A^a_\mu \overline{\psi}\gamma^\mu T^a \psi \\
&- \bar{u}^a \partial^\mu (\delta^{ab}\partial_\mu - g f^{abc} A^c_\mu) u^b \\
&+ \delta Z_\psi \overline{\psi}\mathrm{i}\partial\!\!\!/\psi - (Z_\psi Z_m - 1)\overline{\psi} m \psi \\
&+ (Z_g Z_A^{1/2} Z_\psi - 1) g A^a_\mu \overline{\psi}\gamma^\mu T^a \psi \\
&- \frac{1}{4}\delta Z_A (\partial_\mu A^a_\nu - \partial_\nu A^a_\mu)^2 - (Z_g Z_A^{3/2} - 1) g \partial_\mu A^a_\nu f^{abc} A^{b,\mu} A^{c,\nu} \\
&- \frac{1}{4}(Z_g^2 Z_A^2 - 1) g^2 f^{abc} A^b_\mu A^c_\nu f^{ade} A^{d,\mu} A^{e,\nu} \\
&- \delta Z_u \bar{u}^a \Box u^a + (Z_g Z_A^{1/2} Z_u - 1) g \bar{u}^a f^{abc} \partial^\mu A^c_\mu u^b. \qquad (2.5.46)
\end{aligned}$$

From this Lagrangian the complete Feynman rules including the counterterms follow. The renormalization constants are determined from the primitively divergent vertex functions. Because of the Ward identities it is sufficient to consider the gauge-boson propagator, the Faddeev–Popov ghost propagator, the fermion propagator, and the fermion–gauge-boson vertex.

A particularly important non-abelian gauge theory is QCD. Since the fundamental degrees of freedom of QCD, quarks and gluons, are confined in the hadrons and do not exist as free particles, an on-shell renormalization does not make sense. Commonly used renormalization schemes are the *minimal-subtraction scheme* (MS scheme), where just the divergences are subtracted, or the *modified minimal-subtraction scheme* ($\overline{\mathrm{MS}}$ scheme), where in addition remnants of the regularization procedure are subtracted. Both are defined after (2.5.63) and in Sects. 2.5.3.3 and 2.6.3. Here we present the so-called *momentum-subtraction scheme* (MOM scheme). In this scheme, the renormalized vertex functions are fixed at the renormalization point $k^2 = p^2 = -\mathcal{M}^2$, which is situated at space-like momenta $\mathcal{M}^2 > 0$ in order to avoid possible infrared divergences,

$$\begin{aligned}
\Gamma^{AA,ab}_{\mathrm{R},\mu\nu}(k,-k)\Big|_{k^2=-\mathcal{M}^2} &= (-g_{\mu\nu}k^2 + k_\mu k_\nu(1-\xi))\delta^{ab}, \\
\Gamma^{u\bar{u},ab}_{\mathrm{R}}(k,-k)\Big|_{k^2=-\mathcal{M}^2} &= k^2\delta^{ab}, \\
\Gamma^{q\bar{q}}_{\mathrm{R}}(p,-p)\Big|_{p^2=-\mathcal{M}^2} &= (p\!\!\!/ - m),
\end{aligned}$$

$$\Gamma^{Aq\bar{q},a}_{\mathrm{R},\mu}(p_1,p_2,p_3)\Big|_{p_i^2=-\mathcal{M}^2} = g\frac{\lambda^a}{2}\gamma_\mu. \tag{2.5.47}$$

These conditions determine the vector-boson-field renormalization constant Z_A, the Faddeev–Popov-ghost-field renormalization constant Z_u, the fermion mass and field renormalization constants Z_m and Z_ψ, and the renormalization Z_g of the gauge coupling constant. The one-loop results for the renormalization constants and the renormalized self-energies are derived in the following sections. Other renormalization schemes are discussed in Sect. 2.5.3.3.

2.5.1.8 Renormalization of a non-abelian gauge theory in the background-field method

We consider the renormalization of the non-abelian gauge theory with fermions discussed in the previous section within the background-field method introduced in Sect. 2.4.4. The corresponding Lagrangian is given in (2.4.108)–(2.4.113). It involves fermion fields ψ, background gauge fields $\hat{A}_\mu$, quantum gauge fields A_μ, and Faddeev–Popov ghost fields u. In order to render all Green functions, including those involving quantum fields, finite all these fields have to be renormalized.

We introduce the following renormalization constants

$$\begin{aligned}
\hat{A}^a_{0,\mu} &= Z_{\hat{A}}^{1/2}\hat{A}^a_\mu, \qquad \psi_0 = Z_\psi^{1/2}\psi, \\
A^a_{0,\mu} &= Z_A^{1/2}A^a_\mu, \qquad u^a_0 = Z_u u^a, \qquad \bar{u}^a_0 = \bar{u}^a, \\
g_0 &= Z_g g, \qquad m_0 = Z_m m, \qquad \xi_0 = Z_\xi \xi.
\end{aligned} \tag{2.5.48}$$

We note that owing to the non-renormalization of the gauge-fixing term the renormalization of the gauge-parameter is actually related to the normalization of the quantum gauge fields ($Z_\xi = Z_A$).

For the gauge-invariant effective action only background-field Green functions are needed. As far as these Green functions are concerned the renormalization of the quantum gauge fields and Faddeev–Popov fields is irrelevant and need not to be considered, i.e. we can choose $Z_A = Z_u = 1$. Because these fields appear in background-field Green functions only inside loops their renormalization constants cancel exactly between propagators and vertices. The renormalization of the gauge parameter ξ does not cancel in Green functions, but enters only beyond the one-loop level.

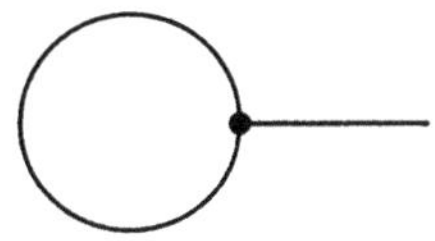

Fig. 2.6 The tadpole graph

Because explicit gauge invariance is retained in the background-field method, the renormalization constants $Z_{\hat{A}}$ and Z_g are related. The infinities appearing in the gauge-invariant effective action $\Gamma\{\hat{\mathbf{A}}, \psi, \overline{\psi}\}$ must be proportional to the gauge-invariant operators $F^a_{\mu\nu} F^{a,\mu\nu}$ and $\overline{\psi}(x)(\mathrm{i}\not{D} - m)\psi(x)$. This requires that the renormalization constants of the background-gauge field and the coupling constant are related by

$$Z_g = Z_{\hat{A}}^{-1/2}. \tag{2.5.49}$$

This relation is analogous to (2.5.31) in QED and can also be derived from the Ward identities (2.4.121).

2.5.2 Calculation of one-loop corrections

Higher-order corrections to Green functions or S-matrix elements are obtained from Feynman diagrams with loops. Starting from the Feynman rules, all diagrams of a given order in perturbation theory, together with the corresponding counterterm diagrams, have to be generated and translated into analytical expressions. This can be done by hand or—especially for more complicated problems—using computer-algebraic programs (for a review see [Ha99] and references therein). The result contains integrals over products of propagators and coupling terms. In this section we describe the systematic *evaluation of one-loop diagrams.*

2.5.2.1 The tadpole integral in dimensional regularization

Let us start with the treatment of the simplest one-loop diagram, the scalar *tadpole diagram* (Fig. 2.6). It already shows typical features of one-loop integrals, but is free of the technical complications that occur for diagrams with more external legs. The tadpole diagram involves the following integral

$$A_0(m) = \frac{1}{\mathrm{i}\pi^2} \int \mathrm{d}^4 k \, \frac{1}{k^2 - m^2 + \mathrm{i}\varepsilon}. \tag{2.5.50}$$

The factor $(\mathrm{i}\pi^2)^{-1}$ is added for convenience. According to power counting, this integral is quadratically divergent in four dimensions. Therefore, it has to be regularized.

We use the method of *dimensional regularization* [tH72a, Bo72] which has the advantage of respecting Lorentz and gauge invariance. In this regularization scheme calculations are performed in D instead of four dimensions. Since the loop integrals converge for small enough D ($D < 2$ for A_0) the usual calculational rules for integrals, such as linearity, translational and rotational invariance, and the usual scaling can be used. The analytic structure of these integrals allows for an analytic continuation to arbitrary complex values of D. The UV divergences manifest themselves as poles at integer values of D. Changing the dimension of the integral also changes the dimension of $A_0(m)$. This effect gets compensated by multiplying the integral with the factor $(2\pi\mu)^{4-D}$, where μ has the dimension of a mass.

A precise definition of dimensional regularization can be found in Ref. [Co84]. Especially, it implies that integrals that do not depend on any scale vanish in dimensional regularization,

$$\int \mathrm{d}^D k\, (k^2)^\alpha = 0, \tag{2.5.51}$$

although being formally infinite for all α and integer D. For later applications we consider A_0 with an arbitrary power of the propagator and define

$$A_0^{(r)}(m) = \frac{\mu^{4-D}}{\mathrm{i}\pi^2} \int \frac{\mathrm{d}^D k}{(2\pi)^{D-4}} \frac{1}{(k^2 - m^2 + \mathrm{i}\varepsilon)^r}, \qquad r \geq 1. \tag{2.5.52}$$

The momentum-space integration over $k^\mu = (k^0, k^1, \dots, k^{D-1})$, $k^2 = (k^0)^2 - \mathbf{k}^2$, contains a k^0 integration along the real axis from $-\infty$ to $+\infty$. The integrand has poles in k^0 at

$$k^0 = \pm\sqrt{\mathbf{k}^2 + m^2 - \mathrm{i}\varepsilon} = \pm\left(\sqrt{\mathbf{k}^2 + m^2} - \frac{\mathrm{i}\varepsilon}{2\sqrt{\mathbf{k}^2 + m^2}} + \mathcal{O}(\varepsilon^2)\right).$$

These singularities lie in the second and fourth quadrant of the complex k^0 plane (cf. the crosses in Fig. 2.7). Therefore, the integral over the closed contour in Fig. 2.7 vanishes owing to Cauchy's theorem. For $D < 2r$, where

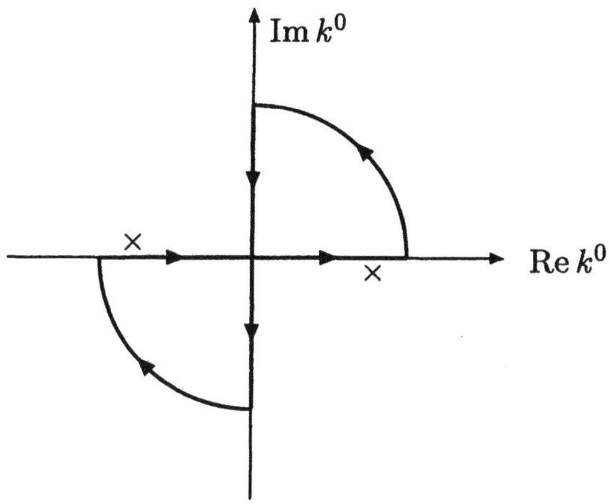

Fig. 2.7 Singularities and integration contour for the tadpole integral in the k^0 plane.

$A_0^{(r)}$ is convergent, the integrand vanishes fast enough so that the contribution of the arcs drops out for $|k^0| \to \infty$. Therefore, we can replace the integral over the real axis by one over the imaginary axis

$$\begin{aligned} A_0^{(r)}(m) &= \frac{(2\pi\mu)^{4-D}}{\mathrm{i}\pi^2} \int_{-\infty}^{\infty} \mathrm{d}k^0 \int \mathrm{d}^{D-1}\mathbf{k} \, \frac{1}{[(k^0)^2 - \mathbf{k}^2 - m^2 + \mathrm{i}\varepsilon]^r} \\ &= \frac{(2\pi\mu)^{4-D}}{\mathrm{i}\pi^2} \int_{-\mathrm{i}\infty}^{\mathrm{i}\infty} \mathrm{d}k^0 \int \mathrm{d}^{D-1}\mathbf{k} \, \frac{1}{[(k^0)^2 - \mathbf{k}^2 - m^2 + \mathrm{i}\varepsilon]^r} \end{aligned}$$

Substituting $k^0 = \mathrm{i}k_{\mathrm{E}}^0$ and $\mathbf{k} = \mathbf{k}_{\mathrm{E}}$ gives

$$\begin{aligned} A_0^{(r)}(m) &= \frac{(2\pi\mu)^{4-D}}{\mathrm{i}\pi^2} \mathrm{i} \int_{-\infty}^{\infty} \mathrm{d}k_{\mathrm{E}}^0 \int \frac{\mathrm{d}^{D-1}\mathbf{k}_{\mathrm{E}}}{[-(k_{\mathrm{E}}^0)^2 - \mathbf{k}_{\mathrm{E}}^2 - m^2 + \mathrm{i}\varepsilon]^r} \\ &= \frac{(2\pi\mu)^{4-D}}{\pi^2} \int \mathrm{d}^D k_{\mathrm{E}} \, \frac{(-1)^r}{[k_{\mathrm{E}}^2 + m^2 - \mathrm{i}\varepsilon]^r}, \end{aligned} \tag{2.5.53}$$

where $k_{\mathrm{E}} = (k_{\mathrm{E}}^0, \mathbf{k}_{\mathrm{E}})$ is a euclidean vector, and $\mathrm{d}^D k_{\mathrm{E}}$ the euclidean volume element. This procedure, the *Wick rotation*, transforms integrals over Minkowski space to integrals over euclidean space.

It is possible to use the rotational symmetry of the euclidean integral to decompose it into a radial and an angular part by introducing spherical coordinates

$$\mathrm{d}^D k_{\mathrm{E}} = k_{\mathrm{E}}^{D-1} \mathrm{d}k_{\mathrm{E}} \, \mathrm{d}\Omega_D = \frac{1}{2} (k_{\mathrm{E}}^2)^{(D-2)/2} \, \mathrm{d}k_{\mathrm{E}}^2 \, \mathrm{d}\Omega_D. \tag{2.5.54}$$

The integral over the surface of the D-dimensional unit sphere yields

$$\int \mathrm{d}\Omega_D = \Omega_D = \frac{2\pi^{D/2}}{\Gamma(D/2)}. \tag{2.5.55}$$

Here Γ is *Euler's Gamma function*

$$\Gamma(z) = \int_0^\infty t^{z-1}\mathrm{e}^{-t}\mathrm{d}t,$$

which has the basic properties

$$\Gamma(z+1) = z\Gamma(z), \qquad \Gamma(n+1) = n!\,, \qquad \Gamma\left(\tfrac{1}{2}\right) = \sqrt{\pi}, \tag{2.5.56}$$

and first-order poles for $z \in \{0, -1, -2, \ldots\}$.

Putting everything together, we have

$$\begin{aligned} A_0^{(r)}(m) &= \frac{(2\pi\mu)^{4-D}}{\pi^2}\frac{2\pi^{D/2}}{\Gamma(D/2)}\frac{1}{2}\int_0^\infty \mathrm{d}k_\mathrm{E}^2\,\frac{(-1)^r (k_\mathrm{E}^2)^{(D-2)/2}}{(k_\mathrm{E}^2 + m^2 - i\varepsilon)^r} \\ &= (4\pi\mu^2)^{(4-D)/2}\frac{\Gamma(r - D/2)}{\Gamma(r)}(-1)^r (m^2 - \mathrm{i}\varepsilon)^{(D/2-r)} \end{aligned} \tag{2.5.57}$$

for $D < 2r$, where we used the fact that the integral in the first line is a representation of Euler's Beta function $B(x, y) = \Gamma(x)\Gamma(y)/\Gamma(x+y)$. The Gamma function $\Gamma(r - D/2)$ in (2.5.57) contains poles for $r - D/2 = 0, -1, -2, \ldots$, which can be identified with the UV singularities. The analyticity of A_0 as a function of D allows to calculate the integral in a region of the complex D plane where it is convergent and to define an analytic continuation to other dimensions. All UV singularities show up as poles in D. In order to work this out in more detail for the physical dimension four, we write $D = 4 - 2\delta$ and obtain

$$A_0^{(r)}(m) = (4\pi\mu^2)^\delta\,\frac{\Gamma(r - 2 + \delta)}{\Gamma(r)}(-1)^r \left(m^2 - \mathrm{i}\varepsilon\right)^{(2-r-\delta)}. \tag{2.5.58}$$

The original tadpole integral is obtained by putting $r = 1$,

$$A_0(m) = -\left(\frac{4\pi\mu^2}{m^2}\right)^\delta \Gamma(-1+\delta)\,m^2. \tag{2.5.59}$$

Because δ is small in the vicinity of $D = 4$, we expand

$$z^{\delta} = \exp(\delta \log z) = 1 + \delta \log z + \mathcal{O}(\delta^2),$$
$$\Gamma(\delta) = \frac{1}{\delta} - \gamma + \mathcal{O}(\delta), \quad \text{(Euler constant } \gamma = 0.577\ldots\text{)}, \tag{2.5.60}$$

and obtain

$$A_0(m) = m^2 \left(\Delta - \log \frac{m^2}{\mu^2} + 1 \right) + \mathcal{O}(\delta) \tag{2.5.61}$$

with

$$\Delta = \frac{1}{\delta} - \gamma + \log 4\pi. \tag{2.5.62}$$

The term $1/\delta - \gamma + \log 4\pi$ is typical for dimensional regularization. It clearly shows the singularity for $\delta \to 0$, i.e. $D \to 4$. As a consequence we have

$$(D - 4) A_0(m) = -2m^2 + \mathcal{O}(D - 4). \tag{2.5.63}$$

In the *minimal-subtraction scheme* (MS) the terms with $1/\delta$, in the $\overline{\text{MS}}$ *scheme* the terms with $1/\delta - \gamma + \log 4\pi$ are subtracted leaving a finite expression for the renormalized quantities.

2.5.2.2 Scalar Feynman integrals

The tadpole integral or scalar one-point integral is the simplest of the *scalar n-point integrals*. These correspond to scalar one-loop diagrams with n external legs joined to the loop via three-vertices, schematically depicted in Fig. 2.8. These integrals are Lorentz-invariant and depend on the scalar products of the external momenta and on the masses of the inner propagators.

The two-point function, the *scalar self-energy integral*, is defined as

$$B_0(p^2; m_1, m_2) = \frac{(2\pi\mu)^{4-D}}{\mathrm{i}\pi^2} \int \frac{\mathrm{d}^D k}{[k^2 - m_1^2][(k+p)^2 - m_2^2]}, \tag{2.5.64}$$

where we have suppressed the $\mathrm{i}\varepsilon$ terms. As discussed in Sect. 2.5.1.2, B_0 is logarithmically divergent in four dimensions.

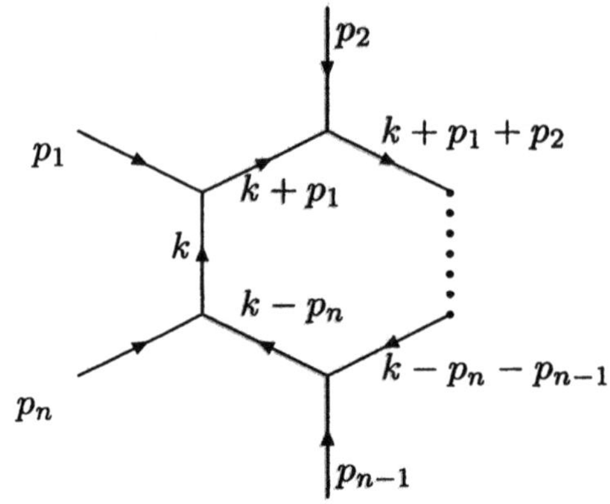

Fig. 2.8 Conventions for the n-point integral

The higher scalar n-point functions are finite in four dimensions. The *triangle integral* is defined as

$$C_0(p_1^2, p_2^2, p_3^2; m_1, m_2, m_3) = \frac{1}{\mathrm{i}\pi^2} \int \mathrm{d}^4 k \, \frac{1}{[k^2 - m_1^2][(k+p_1)^2 - m_2^2][(k+p_1+p_2)^2 - m_3^2]}, \tag{2.5.65}$$

where the incoming momenta are p_1, p_2 and $p_3 = -p_1 - p_2$.

The *box integral* reads

$$D_0(p_1^2, p_2^2, p_3^2, p_4^2; (p_2+p_3)^2, (p_1+p_2)^2; m_1, m_2, m_3, m_4) = \frac{1}{\mathrm{i}\pi^2} \int \mathrm{d}^4 k \, \frac{1}{[k^2 - m_1^2][(k+p_1)^2 - m_2^2][(k+p_1+p_2)^2 - m_3^2]} \times \frac{1}{[(k+p_1+p_2+p_3)^2 - m_4^2]}. \tag{2.5.66}$$

The pentagon, hexagon, ... integrals (with five, six, ... propagators) are defined in a similar way. In four dimensions, they can be algebraically reduced to linear combinations of box integrals D_0 by using the linear dependence of more than four momenta [Me65, vN84]. Therefore, only A_0, B_0, C_0, and D_0 have to be calculated directly.

The momentum-space integrals for B_0, C_0, and D_0 can be reduced to those of the tadpole integral (2.5.52) using the *Feynman parametrization* of the integrand

$$\frac{1}{D_1^{a_1} D_2^{a_2} \cdots D_n^{a_n}} = \frac{\Gamma(a_1 + a_2 + \ldots + a_n)}{\Gamma(a_1)\Gamma(a_2)\cdots\Gamma(a_n)} \times \int_0^1 \mathrm{d}x_1 \cdots \int_0^1 \mathrm{d}x_n \frac{\delta(1 - x_1 - \ldots - x_n) x_1^{a_1-1} \cdots x_n^{a_n-1}}{[x_1 D_1 + \ldots + x_n D_n]^{a_1+\ldots+a_n}}. \tag{2.5.67}$$

In the case of the B_0 function we apply this formula to the product of two propagators,

$$\frac{1}{D_1 D_2} = \int_0^1 \frac{\mathrm{d}x}{[(1-x)D_1 + xD_2]^2}, \tag{2.5.68}$$

to obtain

$$B_0 = \frac{(2\pi\mu)^{4-D}}{\mathrm{i}\pi^2} \int_0^1 \mathrm{d}x \int \mathrm{d}^D k \times \left[k^2 + 2xkp + x(p^2 - m_2^2) - (1-x)m_1^2\right]^{-2}. \tag{2.5.69}$$

Using translational invariance of the momentum-space integral ($k \to k - xp$) gives a tadpole integral (2.5.52) with $r = 2$ and thus using (2.5.58)

$$\begin{aligned} B_0 &= \frac{(2\pi\mu)^{4-D}}{\mathrm{i}\pi^2} \int_0^1 \mathrm{d}x \int \frac{\mathrm{d}^D k}{[k^2 - x^2p^2 + x(p^2 - m_2^2) - (1-x)m_1^2]^2} \\ &= (4\pi\mu^2)^\delta\, \Gamma(\delta) \int_0^1 \mathrm{d}x \left[x^2p^2 - x(p^2 - m_2^2 + m_1^2) + m_1^2\right]^{-\delta}. \end{aligned} \tag{2.5.70}$$

Expanding in $\delta \ll 1$ and keeping the singular and constant terms, we obtain

$$B_0(p^2; m_1, m_2) = \Delta - \int_0^1 \mathrm{d}x \log \frac{x^2p^2 - x(p^2 - m_2^2 + m_1^2) + m_1^2 - \mathrm{i}\varepsilon}{\mu^2} + \mathcal{O}(\delta), \tag{2.5.71}$$

where we have restored the iε term via $m_i^2 \to m_i^2 - \mathrm{i}\varepsilon$ and used Δ from (2.5.62). The terms in the first line reflect the logarithmic singularity of the momentum-space integral and lead to

$$(D-4)B_0 = -2 + \mathcal{O}(D-4). \tag{2.5.72}$$

The parameter integral is elementary and yields

$$\begin{aligned} B_0 &= \Delta + 2 - \log\frac{m_1 m_2}{\mu^2} \\ &\quad + \frac{m_1^2 - m_2^2}{p^2}\log\frac{m_2}{m_1} - \frac{m_1 m_2}{p^2}\left(\frac{1}{r} - r\right)\log r, \end{aligned} \tag{2.5.73}$$

where r and $1/r$ are the solutions of the quadratic equation

$$r^2 + \frac{p^2 - m_1^2 - m_2^2 + \mathrm{i}\varepsilon}{m_1 m_2} r + 1 = 0, \tag{2.5.74}$$

namely

$$r = \frac{-p^2 + m_1^2 + m_2^2 - \mathrm{i}\varepsilon \pm \sqrt{(p^2 - m_1^2 - m_2^2 + \mathrm{i}\varepsilon)^2 - 4m_1^2 m_2^2}}{2m_1 m_2}. \tag{2.5.75}$$

For real, negative r, the iε prescription yields $\operatorname{Im} r = \varepsilon \operatorname{sgn}(r - 1/r)$. The two-point function B_0 is an analytic function of p^2 with a cut from the physical threshold $p^2 = (m_1 + m_2)^2$ to ∞. Above threshold it gets (for real p^2) an imaginary part which is related to the decay of a particle with mass $\sqrt{p^2}$ into two particles with masses m_1 and m_2.

The triangle and box integrals can be calculated in a similar way. For illustration we present the Feynman parameter integral for C_0

$$\begin{aligned} &C_0(p_1^2, p_2^2, p_3^2; m_1, m_2, m_3) = \\ &\quad -\int_0^1 \mathrm{d}x \int_0^x \mathrm{d}y \Big[p_2^2 x^2 + p_1^2 y^2 + (p_3^2 - p_1^2 - p_2^2)xy \\ &\qquad + (m_2^2 - m_3^2 - p_2^2)x + (m_1^2 - m_2^2 + p_2^2 - p_3^2)y + m_3^2 - \mathrm{i}\varepsilon\Big]^{-1}. \end{aligned} \tag{2.5.76}$$

Twofold integration yields an expression which contains Euler's *dilogarithm* $\mathrm{Li}_2(z)$ [Le58] (also called *Spence function*)

$$\mathrm{Li}_2(z) = -\int_0^1 \mathrm{d}t \, \frac{\log(1-zt)}{t}, \qquad |\arg(1-z)| < \pi. \tag{2.5.77}$$

For the explicit expressions of C_0 and D_0 we refer to the literature [tH79, De91, De93].

2.5.2.3 Dirac algebra and tensor integrals

The consideration of scalar integrals is sufficient for calculations in scalar field theories. Gauge theories contain also gauge bosons and fermions. Therefore, techniques for the treatment of the Dirac algebra and tensor integrals in D dimensions are needed when dimensional regularization is used (cf. Ref. [Co84]).

For positive integer D dimensions the *metric tensor* has the following form

$$g^{\mu\nu} = \mathrm{diag}(1, \underbrace{-1, \ldots, -1}_{D-1}) \tag{2.5.78}$$

and obeys for arbitrary D

$$g^{\mu}_{\mu} = D. \tag{2.5.79}$$

The *Dirac algebra* can be extended to D dimensions. The spinor representation of the pseudo-orthogonal group has anticommutation relations which are formally the same as in four dimensions

$$\{\gamma^\mu, \gamma^\nu\} = 2g^{\mu\nu}\mathbf{1}, \qquad \mu = \{0, 1, \ldots, D-1\}. \tag{2.5.80}$$

Since μ runs from 0 to $D-1$ we have

$$\gamma^\mu\gamma_\mu = D\mathbf{1}, \qquad \gamma^\mu\gamma^\alpha\gamma_\mu = (2-D)\gamma^\alpha, \qquad \text{etc.} \tag{2.5.81}$$

The trace of the unit matrix can be chosen as in the four-dimensional case, $\mathrm{Tr}\,\mathbf{1} = 4$.

Problems arise with the definition of the matrix γ_5 in D dimensions. Since this is related to chiral anomalies, and since γ_5 is only needed in theories with parity breaking, we postpone this topic to Sect. 2.7.2.2.

The propagators for gauge bosons and fermions as well as the couplings that contain derivatives generate tensors of the integration momenta in the numerators of the Feynman integrals. A systematic method to reduce these *tensor integrals* to the set of basic scalar integrals A_0, B_0, C_0, and D_0 was worked out in Refs. [Pa79, Ol90, De93]. We illustrate this technique here for the case of the self-energy tensor integrals. The vector two-point integral reads

$$B^\mu(p; m_1, m_2) = \frac{(2\pi\mu)^{4-D}}{\mathrm{i}\pi^2} \int \frac{k^\mu\, \mathrm{d}^D k}{[k^2 - m_1^2][(k+p)^2 - m_2^2]}, \tag{2.5.82}$$

and the tensor integral

$$B^{\mu\nu}(p; m_1, m_2) = \frac{(2\pi\mu)^{4-D}}{\mathrm{i}\pi^2} \int \frac{k^\mu k^\nu\, \mathrm{d}^D k}{[k^2 - m_1^2][(k+p)^2 - m_2^2]}. \tag{2.5.83}$$

These integrals are Lorentz-covariant. Therefore, they can be built up from a complete set of Lorentz tensors and invariant scalar coefficient functions. The Lorentz tensors are constructed from the external momenta and the D-dimensional metric tensor $g^{\mu\nu}$,

$$\begin{aligned} B^\mu &= p^\mu B_1, \\ B^{\mu\nu} &= g^{\mu\nu} B_{00} + p^\mu p^\nu B_{11}. \end{aligned} \tag{2.5.84}$$

These linear equations can be inverted

$$\begin{aligned} B_1 &= \frac{1}{p^2} p_\mu B^\mu, \\ B_{00} &= \frac{1}{D-1} \left(g_{\mu\nu} - \frac{p_\mu p_\nu}{p^2} \right) B^{\mu\nu}, \\ B_{11} &= \frac{1}{D-1} \frac{1}{p^2} \left(-g_{\mu\nu} + D \frac{p_\mu p_\nu}{p^2} \right) B^{\mu\nu}, \end{aligned} \tag{2.5.85}$$

and the expressions on the right-hand side directly reduced to scalar integrals as follows. Writing

$$2kp = [(k+p)^2 - m_2^2] - [k^2 - m_1^2] - p^2 + m_2^2 - m_1^2$$

yields

$$B_1 = \frac{(2\pi\mu)^{4-D}}{i\pi^2}\frac{1}{2p^2}\int \frac{2pk\, d^D k}{[k^2-m_1^2][(k+p)^2-m_2^2]}$$
$$= \frac{1}{2p^2}\Big(A_0(m_1)-A_0(m_2)-(p^2+m_1^2-m_2^2)B_0(p^2;m_1,m_2)\Big), \tag{2.5.86}$$

and in a similar way

$$\begin{aligned} B_{00} &= \frac{1}{2(D-1)}\Big(A_0(m_2)+2m_1^2 B_0(p^2;m_1,m_2) \\ &\quad + (p^2+m_1^2-m_2^2)B_1(p^2;m_1,m_2)\Big), \\ B_{11} &= \frac{1}{2(D-1)p^2}\Big((D-2)A_0(m_2)-2m_1^2 B_0(p^2;m_1,m_2) \\ &\quad - D(p^2+m_1^2-m_2^2)B_1(p^2;m_1,m_2)\Big). \end{aligned} \tag{2.5.87}$$

These equations, inserted in (2.5.84), give finally

$$\begin{aligned} B^\mu &= \frac{p^\mu}{2p^2}\Big(A_0(m_1)-A_0(m_2)-(p^2+m_1^2-m_2^2)B_0(p^2;m_1,m_2)\Big), \\ B^{\mu\nu} &= \frac{1}{2}\Big[\frac{1}{D-1}\left(g_{\mu\nu}-\frac{p_\mu p_\nu}{p^2}\right)\Big(A_0(m_2)+2m_1^2B_0(p^2;m_1,m_2) \\ &\quad + (p^2+m_1^2-m_2^2)B_1(p^2;m_1,m_2)\Big) \\ &\quad + \frac{p_\mu p_\nu}{p^2}\left(A_0(m_2)-(p^2+m_1^2-m_2^2)B_1(p^2;m_1,m_2)\right)\Big]. \end{aligned} \tag{2.5.88}$$

The treatment of the triangle and box diagrams is basically the same. The corresponding results can be found in [Pa79, De93].

In this way every one-loop diagram is expressible by the scalar integrals A_0, B_0, C_0, and D_0 and thus finally by elementary functions and dilogarithms. From the inversion of the linear equations negative powers of determinants of the scalar products of the external momenta (Gram determinants) occur. For the two-point functions considered above these are the factors $1/p^2$. For those values of the variables where these determinants vanish, i.e. $p^2 = 0$ for the two-point case, the linear equations cannot be inverted. However, in this case the linear equations yield relations between the coefficient functions corresponding to tensor integrals with fewer indices. These can be solved for these tensor integrals [St88]. For instance for $p^2 = 0$, the second equation (2.5.87) determines B_1 in terms of B_0 and A_0. Thus, for vanishing determinants the reduction to scalar integrals can be performed by investigating the tensor integrals with one additional Lorentz index, i.e. $B_{\mu\nu}$ for the reduction of B_μ. In this case the final results become actually simpler.

The evaluation of the Dirac and tensor structures of the integrals and the reduction to scalar basic integrals follows clear algebraic rules. They can be implemented in algebraic computer programs [Ha99] which then perform these calculations automatically.

2.5.2.4 Remarks on the calculation of higher-order corrections

The preceding pages show that the calculation of one-loop Feynman diagrams is based on a well-worked-out procedure. This includes the generation of the diagrams, their reduction to standard integrals, and their numerical evaluation. In contrast to this, the situation in the two-loop case is much more involved. Complications arise because of the large number of diagrams which makes a treatment by hand almost impossible. Moreover, with the exception of self-energy diagrams, no general reduction procedure to scalar two-loop integrals with arbitrary masses is available, and the scalar integrals are in general not expressible in terms of well-known functions of mathematical physics. Some results on the reduction of two-loop tensor integrals can be found in Ref. [We94, Ta96, Gh98]. Numerical methods for the evaluation of massive two-loop two-point, three-point, and four-point integrals have been given in Refs. [Ba95, Ba95a, Ba95b], [Cz95a, Fr97], and [Us93], respectively.

Two-loop calculations exist for many problems in massless theories like QCD. However, in the electroweak case we only have results for special problems such as the calculation of the anomalous magnetic moments of leptons, contributions to μ decay, or certain limits (small momenta, large momenta, etc.). In view of this, we restrict ourselves to some remarks on the calculation of higher-order contributions and their renormalization.

The leading parts of the higher-order corrections can in many cases be obtained from the renormalization-group equation (cf. Sect. 2.6).

The many Feynman diagrams can be generated automatically using algebraic computer programs including the counterterm diagrams. Renormalization has to be performed in an iterative way. In the L-loop case it is assumed that the theory is renormalized up to $L-1$ loops. The counterterm diagrams up to $L-1$ loops are sufficient to cancel all non-local divergent terms. This allows to compensate the remaining local ones in a way similar to the one-loop case, i.e. by the terms that have been generated by the ordinary renormalization transformation.

2.5.3 One-loop renormalization of gauge theories

As an application and illustration of the renormalization program discussed in Sect. 2.5.1 we perform the explicit renormalization of an abelian gauge theory, QED, and of a non-abelian gauge theory with the gauge group SU(N) at the one-loop level. Because of the Ward identities, in QED all renormalization constants can be determined from the photon and electron propagators (cf. Sect. 2.5.1.6). Consequently, we calculate the one-loop corrections to these propagators for QED, using the methods of Sect. 2.5.2, and perform their on-shell renormalization. The renormalization of the non-abelian gauge theory, which for instance applies to QCD, proceeds along similar lines. It is presented in Sect. 2.5.3.3.

2.5.3.1 QED self-energies

We decompose the photon–photon vertex function $\Gamma^{AA}_{\mu\nu}$ into its free part and the *photon self-energy* $\Sigma^{A}_{\mu\nu}$,

$$\Gamma^{AA}_{\mu\nu} = -\Big(g_{\mu\nu}q^2 - q_\mu q_\nu(1-\xi)\Big) - \Sigma^{A}_{\mu\nu}. \tag{2.5.89}$$

The photon self-energy is calculated from the one-loop diagram

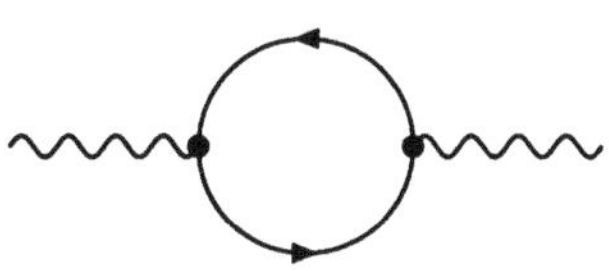

.

Using the Feynman rules of QED this translates to

$$\Sigma^{A}_{\mu\nu}(q) = -\mathrm{i}e^2\mu^{4-D}\int \frac{\mathrm{d}^D k}{(2\pi)^D} \frac{\mathrm{Tr}\Big[\gamma_\mu(\not k + m)\gamma_\nu(\not k + \not q + m)\Big]}{[k^2-m^2][(k+q)^2-m^2]}.$$

Using (A.1.30) the Dirac trace gives

$$\begin{aligned}\mathrm{Tr}\Big[\gamma_\mu(\not{k}+m)\gamma_\nu(\not{k}+\not{q}+m)\Big] &= \\ &= 4\Big[2k_\mu k_\nu + k_\mu q_\nu + q_\mu k_\nu - g_{\mu\nu}(k^2+qk-m^2)\Big].\end{aligned}$$

Writing $2(k^2+qk-m^2) = [k^2-m^2] + [(k+q)^2-m^2] - q^2$ yields

$$\begin{aligned}\Sigma^A_{\mu\nu}(q) &= -4\mathrm{i}e^2\mu^{4-D}\int\frac{\mathrm{d}^Dk}{(2\pi)^D} \\ &\quad\times\frac{2k_\mu k_\nu+k_\mu q_\nu+q_\mu k_\nu-\frac{1}{2}g_{\mu\nu}\Big((k^2-m^2)+((k+q)^2-m^2)-q^2\Big)}{[k^2-m^2][(k+q)^2-m^2]} \\ &= \frac{e^2}{4\pi^2}\Big(2B_{\mu\nu}+q_\mu B_\nu+B_\mu q_\nu-\frac{g_{\mu\nu}}{2}\Big(A_0(m)+A_0(m)-q^2B_0\Big)\Big),\end{aligned}$$

where the arguments of the B functions are $(q^2;m,m)$. Using the reduction formulas for the tensor integrals, (2.5.84), we find

$$\begin{aligned}\Sigma^A_{\mu\nu}(q) = \frac{e^2}{4\pi^2}\Big\{&\Big(g_{\mu\nu}-\frac{q_\mu q_\nu}{q^2}\Big)\Big(2B_{00}-A_0+\frac{q^2}{2}B_0\Big) \\ &+\frac{q_\mu q_\nu}{q^2}\Big(2B_{00}+2q^2(B_{11}+B_1)+\frac{q^2}{2}B_0-A_0\Big)\Big\}.\end{aligned} \tag{2.5.90}$$

With help of (2.5.88) this can be expressed in terms of the basic scalar integrals A_0 and B_0,

$$\Sigma^A_{\mu\nu}(q) = \frac{e^2}{4\pi^2}\Big(g_{\mu\nu}-\frac{q_\mu q_\nu}{q^2}\Big)\frac{(2-D)(2A_0-q^2B_0)+4m^2B_0}{2(D-1)}. \tag{2.5.91}$$

In agreement with (2.4.100) the one-loop photon self-energy is purely transverse.

Introducing the *fine-structure constant* $\alpha = e^2/4\pi$, putting $D = 4-2\delta$, and using the relation

$$B_0(0;m,m) = \frac{\partial A_0(m)}{\partial m^2} = (1-\delta)\frac{A_0(m)}{m^2}, \tag{2.5.92}$$

which can be read off from (2.5.64), (2.5.52), and (2.5.59), we obtain

$$\begin{aligned}
\Sigma_{\mathrm{T}}^{A}(q^2) &= \frac{\alpha}{\pi}\frac{1}{3-2\delta}\Big\{(1-\delta)q^2 B_0(q^2;m,m) + \\
&\quad + 2m^2\Big[B_0(q^2;m,m) - B_0(0;m,m)\Big]\Big\} \\
&= \frac{\alpha}{3\pi}\Big\{q^2 B_0(q^2;m,m) + 2m^2\Big(B_0(q^2;m,m) - B_0(0;m,m)\Big) \\
&\quad - \frac{q^2}{3}\Big\} + \mathcal{O}(\delta), \\
\Sigma_{\mathrm{L}}^{A}(q^2) &= 0
\end{aligned} \tag{2.5.93}$$

for the transverse and longitudinal self-energies [defined as in (2.5.33)]. This shows that indeed $\Sigma_{\mathrm{T}}^{A}(q^2) \sim q^2$ for small q^2 in accordance with (2.5.35). As a consequence of gauge invariance no photon mass renormalization is needed.

The *electron self-energy* $\Sigma^{\psi}(p)$ is the higher-order contribution to the corresponding vertex function and is decomposed into a vector and a scalar part,

$$\begin{aligned}
\Gamma^{\psi\bar{\psi}}(p,-p) &= (\not{p} - m) + \Sigma^{\psi}(p) \\
&= (\not{p} - m) + \Big(\not{p}\,\Sigma_{\mathrm{V}}^{\psi}(p^2) + m\Sigma_{\mathrm{S}}^{\psi}(p^2)\Big).
\end{aligned} \tag{2.5.94}$$

At the one-loop order it results from the Feynman diagram

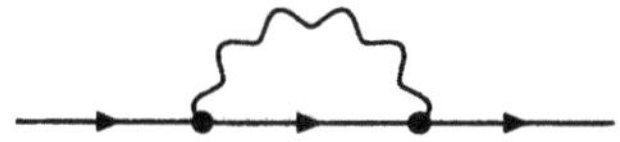

.

Introducing a small photon mass λ in order to regularize possible infrared (IR) divergences and taking the photon propagator in the Feynman gauge, gives[10]

$$\Sigma^{\psi}(p) = \mathrm{i}e^2\mu^{4-D}\int \frac{\mathrm{d}^D k}{(2\pi)^D}\frac{\gamma_\mu(\not{k}+m)\gamma^\mu}{[k^2-m^2][(k-p)^2-\lambda^2]}. \tag{2.5.95}$$

[10]The introduction of the photon mass breaks gauge invariance. However, the gauge-invariance-breaking terms are proportional to λ and thus vanish for physical finite quantities in the limit $\lambda \to 0$.

Working out the algebra and the integration with the methods of Sect. 2.5.2 and neglecting terms proportional to λ^2 (which vanish for $\lambda \to 0$, since the IR divergence is only logarithmic) yields

$$\Sigma^{\psi}(p) = -\frac{\alpha}{4\pi}\bigg\{\not{p}(1-\delta)\frac{A_0(m)-(p^2+m^2)B_0(p^2;m,\lambda)}{p^2} + m(4-2\delta)B_0(p^2;m,\lambda)\bigg\}, \tag{2.5.96}$$

and for the scalar and vector part of the self-energy

$$\begin{aligned}\Sigma^{\psi}_{\mathrm{V}}(p^2) &= -\frac{\alpha}{4\pi}(1-\delta)\frac{A_0(m)-(p^2+m^2)B_0(p^2;m,\lambda)}{p^2},\\ \Sigma^{\psi}_{\mathrm{S}}(p^2) &= -\frac{\alpha}{4\pi}(4-2\delta)B_0(p^2;m,\lambda).\end{aligned} \tag{2.5.97}$$

Next, we verify the *Ward identity* (2.4.102) between the photon–electron vertex

$$\Gamma^{A\psi\bar{\psi}}_{\mu}(q,p,\bar{p}) = e\gamma_\mu + e\Lambda_\mu(q,p,\bar{p}) \tag{2.5.98}$$

and the fermion self-energy by explicit calculation. The *vertex correction* is given by the Feynman diagram

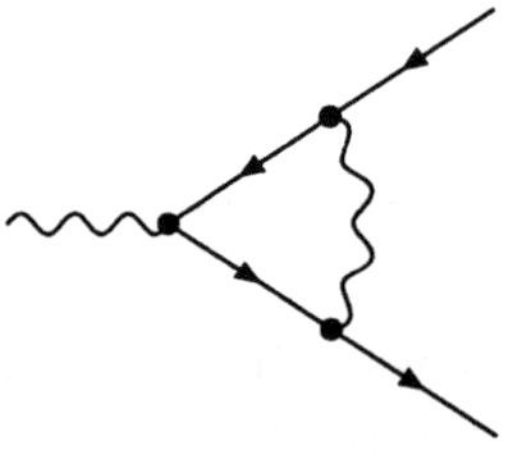

and reads

$$\Lambda_\mu(q,p,\bar{p}) = -\mathrm{i}e^2\mu^{4-D}\int\frac{\mathrm{d}^Dk}{(2\pi)^D} \gamma_\rho\frac{1}{-\not{\bar{p}}+\not{k}-m}\gamma_\mu\frac{1}{\not{p}+\not{k}-m}\gamma^\rho\frac{1}{k^2-\lambda^2}. \tag{2.5.99}$$

Multiplying by $-\bar{p}^\mu - p^\mu$ and writing $-\not{\bar{p}} - \not{p} = -\not{\bar{p}} + \not{k} - m - \not{p} - \not{k} + m$ indeed yields

$$(-\bar{p}^\mu - p^\mu)\Lambda_\mu(q,p,\bar{p}) = -\mathrm{i}e^2\mu^{4-D}\int \frac{\mathrm{d}^D k}{(2\pi)^D}$$
$$\left(\gamma_\rho \frac{1}{\not{p}+\not{k}-m}\gamma^\rho \frac{1}{k^2-\lambda^2} - \gamma_\rho \frac{1}{-\not{\bar{p}}+\not{k}-m}\gamma^\rho \frac{1}{k^2-\lambda^2}\right)$$
$$= \Sigma^\psi(-\bar{p}) - \Sigma^\psi(p), \tag{2.5.100}$$

i.e. the one-loop contribution to (2.4.102).

2.5.3.2 Renormalization constants of QED

The QED Lagrangian (2.5.27) contains the counterterms which are needed for the renormalization procedure. In one-loop order, we obtain for the renormalized photon and electron vertex functions ($Z_\xi = Z_A$)

$$\Gamma^{AA}_{\mathrm{R},\mu\nu}(q,-q) = \Gamma^{AA}_{\mu\nu}(q,-q) - \delta Z_A\Big(g_{\mu\nu}q^2 - q_\mu q_\nu\Big),$$
$$\Gamma^{\psi\bar{\psi}}_{\mathrm{R}}(p,-p) = \Gamma^{\psi\bar{\psi}}(p,-p) - \delta Z_m m + \delta Z_\psi(\not{p} - m). \tag{2.5.101}$$

Using the decompositions (2.5.89) and (2.5.94) together with (2.5.33) and (2.5.39) this can be written as

$$\Gamma^{A}_{\mathrm{R,T}}(q^2) = -\Big(q^2 + \Sigma^A_{\mathrm{T}}(q^2) + q^2\delta Z_A\Big), \tag{2.5.102}$$
$$\Gamma^{\psi}_{\mathrm{R,V}}(q^2) = 1 + \Sigma^\psi_{\mathrm{V}}(q^2) + \delta Z_\psi,$$
$$\Gamma^{\psi}_{\mathrm{R,S}}(q^2) = -1 + \Sigma^\psi_{\mathrm{S}}(q^2) - \delta Z_m - \delta Z_\psi. \tag{2.5.103}$$

The *photon-field renormalization constant* is obtained from (2.5.36). Inserting (2.5.102) gives

$$\frac{\partial}{\partial q^2}\Big[-\Big(q^2 + \Sigma^A_{\mathrm{T}}(q^2) + q^2\delta Z_A\Big)\Big]\bigg|_{q^2=0} = -1$$

or

$$\delta Z_A = -\frac{\partial \Sigma^A_{\mathrm{T}}(q^2)}{\partial q^2}\bigg|_{q^2=0}. \tag{2.5.104}$$

Taking the result (2.5.93) for the photon self-energy yields

$$\delta Z_A = -\frac{\alpha}{\pi}\frac{(1-\delta)B_0(0;m,m) + 2m^2 B_0'(0;m,m)}{3-2\delta}, \tag{2.5.105}$$

where $B_0'(q^2;m,m) = \partial/\partial q^2 B_0(q^2;m,m)$. Using (2.5.70) we obtain

$$B_0(0;m,m) = \left(\frac{4\pi\mu^2}{m^2}\right)^\delta \Gamma(\delta),$$
$$B_0'(0;m,m) = \left(\frac{4\pi\mu^2}{m^2}\right)^\delta \frac{\delta\,\Gamma(\delta)}{6m^2}, \tag{2.5.106}$$

and thus for the photon-field renormalization constant

$$\delta Z_A = -\frac{\alpha}{3\pi}\left(\frac{4\pi\mu^2}{m^2}\right)^\delta \Gamma(\delta) = -\frac{\alpha}{3\pi}\left(\Delta + \log\frac{\mu^2}{m^2}\right) + \mathcal{O}(\delta). \tag{2.5.107}$$

The renormalized photon self-energy is finite for $\delta \to 0$,

$$\Sigma^A_{\mathrm{R,T}}(q^2) = \frac{\alpha}{3\pi}\left[(q^2+2m^2)(B_0(q^2)-B_0(0))-2q^2m^2B_0'(0)\right], \tag{2.5.108}$$

and behaves like $(q^2)^2$ for small q^2.

The conditions (2.5.40) and (2.5.41) fix the *electron mass and field renormalization* in the on-shell scheme. Inserting (2.5.103) gives

$$\delta Z_m = \frac{\delta m}{m} = \Sigma^\psi_{\mathrm{S}}(m^2) + \Sigma^\psi_{\mathrm{V}}(m^2), \tag{2.5.109}$$
$$\delta Z_\psi = -\Sigma^\psi_{\mathrm{V}}(m^2) - 2m^2\left(\Sigma'^\psi_{\mathrm{S}}(m^2) + \Sigma'^\psi_{\mathrm{V}}(m^2)\right). \tag{2.5.110}$$

The explicit result for the *electron mass renormalization constant* is obtained by inserting (2.5.97) in (2.5.109) and using (2.5.59), (2.5.70), and (2.5.62),

$$\begin{aligned}\frac{\delta m}{m} &= -\frac{\alpha}{4\pi}\left((1-\delta)\frac{A_0(m)}{m^2} + 2B_0(m^2;m,0)\right)\\ &= -\frac{\alpha}{4\pi}\left(\frac{4\pi\mu^2}{m^2}\right)^\delta \Gamma(\delta)\frac{3-2\delta}{1-2\delta}\\ &= -\frac{3\alpha}{4\pi}\left(\Delta + \log\frac{\mu^2}{m^2} + \frac{4}{3}\right) + \mathcal{O}(\delta).\end{aligned} \tag{2.5.111}$$

The *electron-field renormalization constant* δZ_ψ is both UV- and IR-divergent in the on-shell scheme. Equation (2.5.110) gives

$$\delta Z_\psi = \frac{\alpha}{4\pi}\left((1-\delta)\left[\frac{A_0(m)}{m^2} - 2B_0\right] + 4m^2 B_0'(m^2;m,\lambda)\right). \quad (2.5.112)$$

The UV-singular term simplifies to

$$\frac{\alpha}{4\pi}(1-\delta)\left[\frac{A_0(m)}{m^2} - 2B_0\right] = -\frac{\alpha}{4\pi}\left(\frac{4\pi\mu^2}{m^2}\right)^\delta \Gamma(\delta)\frac{1}{1-2\delta}.$$

The IR-singular part

$$\frac{\alpha}{\pi}m^2 B_0' = \frac{\alpha}{\pi}\left(\frac{4\pi\mu^2}{m^2}\right)^\delta \Gamma(1+\delta)\int_0^1 \mathrm{d}x\, \frac{x-x^2}{\left[x^2+\frac{\lambda^2}{m^2}(1-x)\right]^{1+\delta}}$$

can be evaluated in two ways:

i) keep the small photon mass λ, $(0 < \lambda^2 \ll m^2)$ and let $\delta \to 0$:

$$\begin{aligned}\frac{\alpha}{\pi}m^2 B_0' &= \frac{\alpha}{\pi}\int_0^1 \mathrm{d}x\, \frac{x-x^2}{x^2+\frac{\lambda^2}{m^2}(1-x)} \\ &= -\frac{\alpha}{2\pi}\left(\log\frac{\lambda^2}{m^2}+2\right) + \mathcal{O}(\lambda^2), \end{aligned} \quad (2.5.113)$$

ii) regularize the IR-divergent integral dimensionally ($\lambda = 0$, $\delta = \delta_{\mathrm{IR}} < 0$):

$$\begin{aligned}\frac{\alpha}{\pi}m^2 B_0' &= \frac{\alpha}{\pi}\left(\frac{4\pi\mu^2}{m^2}\right)^{\delta_{\mathrm{IR}}} \Gamma(1+\delta_{\mathrm{IR}})\int_0^1 \mathrm{d}x\, \frac{x-x^2}{(x^2)^{1+\delta_{\mathrm{IR}}}} \\ &= -\frac{\alpha}{2\pi}\left(\frac{1}{\delta_{\mathrm{IR}}} - \gamma + \log\frac{4\pi\mu^2}{m^2} + 2\right) + \mathcal{O}(\delta_{\mathrm{IR}}). \end{aligned} \quad (2.5.114)$$

This allows to compare the two schemes

$$\log\lambda^2 \leftrightarrow \frac{1}{\delta_{\mathrm{IR}}} - \gamma + \log 4\pi\mu^2.$$

Using the photon-mass regularization we obtain

$$\delta Z_\psi = -\frac{\alpha}{4\pi}\left(\Delta + \log\frac{\mu^2}{m^2} + 2\left(\log\frac{\lambda^2}{m^2}+2\right)\right) + \mathcal{O}(\delta,\lambda^2) \quad (2.5.115)$$

with Δ from (2.5.62).

This completes the determination of the one-loop renormalization constants Z_ψ, Z_m, and $Z_A = Z_e^{-2}$ of QED in the on-shell scheme. The IR singularities are treated in Sect. 2.8.2.

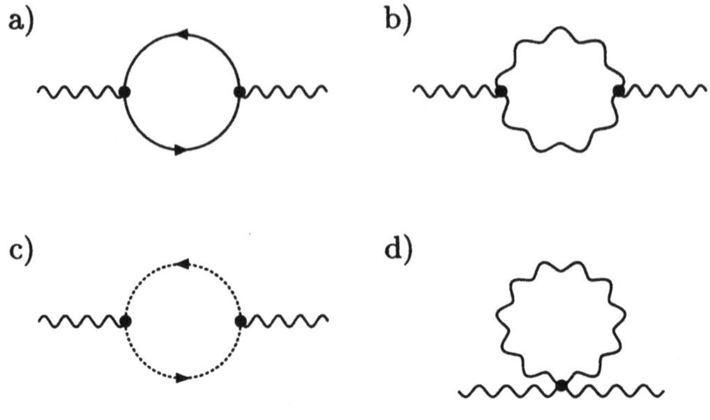

Fig. 2.9 Graphs contributing to the gauge-boson self-energy in one-loop approximation

2.5.3.3 One-loop renormalization of a non-abelian gauge theory

We consider the one-loop *renormalization of a non-abelian gauge theory* with gauge group SU(N) and N_F multiplets of fermions in the fundamental representation. The masses of the fermions of multiplet f are denoted by m_f. For $N = 3$ this corresponds to QCD with N_F quark flavours. For the renormalization we need the one-loop contributions to the gauge-boson and fermion propagators and to the gauge-boson–fermion vertex. We do not discuss the field renormalization of the Faddeev–Popov ghost, which can be determined from the ghost–gauge-boson vertex, because this is not relevant for the calculation of physical quantities at the one-loop level. We present the calculations for the *gauge-boson self-energy* in some detail, for the other quantities we list only the results. We use the Feynman gauge in this section.

The graphs contributing to the gauge-boson self-energy in one-loop approximation are given in Fig. 2.9.

We start by calculating the *vacuum polarization* induced by fermion pairs $(f\bar{f})$ [graph (a) of Fig. 2.9]. This differs from the QED photon self-energy only by a group-theoretical factor (the colour factor in QCD) of the fermions T_F [$= 1/2$ for SU(N)] [cf. (1.2.52)] and the sum over the N_F fermion multiplets. The results are

$$\Sigma^{A,ab}_{T,(f\bar{f})}(q^2) = \frac{\alpha_s}{\pi}\frac{N_F T_F \delta^{ab}}{(3-2\delta)}\sum_f \Big\{(1-\delta)q^2 B_0(q^2; m_f, m_f) +$$

$$+ 2m^2\Big[B_0(q^2; m_f, m_f) - B_0(0; m_f, m_f)\Big]\Big\},$$

$$\Sigma^{A,ab}_{\mathrm{L},(f\bar{f})}(q^2) = 0. \tag{2.5.116}$$

This contribution to the gauge-boson self-energy is again transverse. It is diagonal in the group indices and gets equivalent contributions from all fermion multiplets. We have introduced the shorthand $\alpha_s = g^2/4\pi$, which is the *strong fine-structure constant* in the case of QCD.

In many cases it is sufficient to treat the fermions as massless. Using (2.5.70), we find

$$B_0(q^2; 0, 0) = \left(\frac{4\pi\mu^2}{-q^2 - \mathrm{i}\varepsilon}\right)^{\delta} \Gamma(\delta)\frac{\Gamma^2(1-\delta)}{\Gamma(2-2\delta)},$$

and (2.5.116) simplifies to

$$\begin{aligned}\Sigma^{A,ab}_{\mathrm{T},(f\bar{f})}(q^2) &= \frac{\alpha_s}{\pi}\delta^{ab} N_{\mathrm{F}} T_{\mathrm{F}} q^2 \left(\frac{4\pi\mu^2}{-q^2 - \mathrm{i}\varepsilon}\right)^{\delta} \Gamma(\delta)\frac{\Gamma^2(1-\delta)}{\Gamma(2-2\delta)}\frac{1-\delta}{3-2\delta} \\ &= \frac{\alpha_s}{\pi}\delta^{ab} N_{\mathrm{F}} T_{\mathrm{F}} \frac{q^2}{3}\left(\Delta + \log\frac{\mu^2}{-q^2 - \mathrm{i}\varepsilon} + \frac{5}{3}\right) + \mathcal{O}(\delta).\end{aligned} \tag{2.5.117}$$

The other graphs must be calculated in a similar manner. The vacuum polarization by gauge-boson-pair creation (AA) (Fig. 2.9b) is particularly interesting. It results from the self-interaction of coloured gauge bosons and is not present in an abelian gauge theory with neutral gauge bosons such as QED. In Feynman gauge, the contribution of this graph to the self-energy is

$$\begin{aligned}\Sigma^{A,ab}_{\mathrm{T},(AA)}(q^2) &= \frac{\alpha_s}{4\pi}\delta^{ab}\left(\frac{-C_{\mathrm{A}}}{4}\right) q^2 \left(\frac{4\pi\mu^2}{-q^2 - \mathrm{i}\varepsilon}\right)^{\delta} \\ &\quad \times\Gamma(\delta)\frac{\Gamma^2(1-\delta)}{\Gamma(2-2\delta)}\frac{19-12\delta}{3-2\delta}, \\ \Sigma^{A,ab}_{\mathrm{L},(AA)}(q^2) &= \frac{\alpha_s}{4\pi}\delta^{ab}\left(\frac{C_{\mathrm{A}}}{4}\right) q^2 \left(\frac{4\pi\mu^2}{-q^2 - \mathrm{i}\varepsilon}\right)^{\delta} \Gamma(\delta)\frac{\Gamma^2(1-\delta)}{\Gamma(2-2\delta)}.\end{aligned} \tag{2.5.118}$$

This result was obtained with the help of (1.2.56), $\sum f^{cda} f^{cdb} = C_{\mathrm{A}}\delta^{ab}$, with $C_{\mathrm{A}} = 3$ for SU(3).

The gauge-boson contribution (2.5.118) is not transverse. The contribution containing the quartic coupling of the gauge bosons (Fig. 2.9d) gives zero in dimensional regularization and therefore does not compensate the longitudinal part. The transversality of the gauge-boson self-energy which is guaranteed by the Ward identity (2.4.68) is restored by the Faddeev–Popov-ghost contribution $(u\bar{u})$ (Fig. 2.9c) which leads to

$$\begin{aligned}
\Sigma^{A,ab}_{\mathrm{T},(u\bar{u})}(q^2) &= \frac{\alpha_s}{4\pi}\delta^{ab}\left(\frac{-C_\mathrm{A}}{4}\right)q^2\left(\frac{4\pi\mu^2}{-q^2-\mathrm{i}\varepsilon}\right)^\delta \\
&\quad\times\Gamma(\delta)\frac{\Gamma^2(1-\delta)}{\Gamma(2-2\delta)}\frac{1}{3-2\delta}, \qquad (2.5.119)\\
\Sigma^{A,ab}_{\mathrm{L},(u\bar{u})}(q^2) &= \frac{\alpha_s}{4\pi}\delta^{ab}\left(\frac{-C_\mathrm{A}}{4}\right)q^2\left(\frac{4\pi\mu^2}{-q^2-\mathrm{i}\varepsilon}\right)^\delta\Gamma(\delta)\frac{\Gamma^2(1-\delta)}{\Gamma(2-2\delta)}.
\end{aligned}$$

Adding (2.5.118) and (2.5.119) indeed gives a vanishing longitudinal gauge-boson self-energy,

$$\begin{aligned}
\Sigma^{A,ab}_{\mathrm{T},(AA+u\bar{u})}(q^2) &= \frac{\alpha_s}{4\pi}\delta^{ab}\left(-C_\mathrm{A}\right)q^2\left(\frac{4\pi\mu^2}{-q^2-\mathrm{i}\varepsilon}\right)^\delta \\
&\quad\times\Gamma(\delta)\frac{\Gamma^2(1-\delta)}{\Gamma(2-2\delta)}\frac{5-3\delta}{3-2\delta}, \\
\Sigma^{A,aab}_{\mathrm{L},(AA+u\bar{u})}(q^2) &= 0. \qquad (2.5.120)
\end{aligned}$$

Collecting all contributions, we finally find

$$\begin{aligned}
\Sigma^{A,ab}_{\mathrm{T}}(q^2) &= \frac{\alpha_s}{4\pi}\delta^{ab}q^2\left(\frac{4\pi\mu^2}{-q^2-\mathrm{i}\varepsilon}\right)^\delta\Gamma(\delta)\frac{\Gamma^2(1-\delta)}{\Gamma(2-2\delta)} \\
&\quad\times\left[\frac{4(1-\delta)}{3-2\delta}N_\mathrm{F}T_\mathrm{F}-\frac{(5-3\delta)}{3-2\delta}C_\mathrm{A}\right], \\
\Sigma^{A,ab}_{\mathrm{L}}(q^2) &= 0 \qquad (2.5.121)
\end{aligned}$$

for the *self-energy of a non-abelian gauge boson.*

The renormalized gauge-boson self-energy is obtained by adding the counterterm to Σ^A_T. The gauge-boson-field renormalization constant has to be fixed by a suitable renormalization condition. In contrast to the photon, which exists as a real physical particle, the non-abelian gauge bosons of QCD, the gluons, are screened inside hadrons. This renders an on-shell renormalization of QCD impossible. Instead, the following schemes are used.

In the *minimal-subtraction scheme*, MS, the renormalization constant consists only of the UV-singular term,

$$\delta Z_A^{\mathrm{MS}} = -\frac{\alpha_s}{4\pi}\frac{1}{\delta}\left[\frac{4}{3}T_{\mathrm{F}}N_{\mathrm{F}} - \frac{5}{3}C_{\mathrm{A}}\right]. \qquad (2.5.122)$$

The modified minimal-subtraction ($\overline{\mathrm{MS}}$) scheme differs from this by replacing

$$\frac{1}{\delta} \to \Delta = \frac{1}{\delta} - \gamma + \log 4\pi.$$

A third possibility is the *momentum subtraction* (MOM) *scheme* where the renormalization is performed for space-like momenta $q^2 = -\mathcal{M}^2 < 0$. The corresponding result for the gauge-boson renormalization constant reads

$$\begin{aligned}\delta Z_A^{\mathrm{MOM}} &= -\frac{\alpha_s}{4\pi}\left(\frac{4\pi\mu^2}{\mathcal{M}^2}\right)^{\delta}\Gamma(\delta)\frac{\Gamma^2(1-\delta)}{\Gamma(2-2\delta)} \\ &\times\left[\frac{4(1-\delta)}{3-2\delta}N_{\mathrm{F}}T_{\mathrm{F}} - \frac{(5-3\delta)}{3-2\delta}C_{\mathrm{A}}\right]. \qquad (2.5.123)\end{aligned}$$

The *fermion self-energy* is calculated as in QED, the only difference being the group-theoretical factor

$$\sum_{a,j} T^a_{ij}T^a_{jk} = C_{\mathrm{F}}\delta_{ik} \quad \text{with} \quad C_{\mathrm{F}} = \frac{4}{3} \quad \text{for} \quad \mathrm{SU}(3). \qquad (2.5.124)$$

For massless fermions in the Feynman gauge, the result reads

$$\begin{aligned}\Sigma_{\mathrm{V}}^{qq}(q^2) &= \frac{\alpha_s}{4\pi}C_{\mathrm{F}}\left(\frac{4\pi\mu^2}{-q^2-\mathrm{i}\varepsilon}\right)^{\delta}\Gamma(\delta)\frac{\Gamma^2(1-\delta)}{\Gamma(2-2\delta)}(1-\delta) \\ &= \frac{\alpha_s}{4\pi}C_{\mathrm{F}}(1-\delta)B_0(q^2,0,0), \\ \Sigma_{\mathrm{S}}^{qq}(q^2) &= 0. \qquad (2.5.125)\end{aligned}$$

For the fermion-field renormalization constant this implies

$$\begin{aligned}\delta Z_\psi^{\mathrm{MS}} &= -\frac{\alpha_s}{4\pi}C_{\mathrm{F}}\frac{1}{\delta}, \qquad \delta Z_\psi^{\mathrm{MSbar}} = -\frac{\alpha_s}{4\pi}C_{\mathrm{F}}\Delta, \\ \delta Z_\psi^{\mathrm{MOM}} &= -\frac{\alpha_s}{4\pi}C_{\mathrm{F}}\left(\frac{4\pi\mu^2}{\mathcal{M}^2}\right)^{\delta}\Gamma(\delta)\frac{\Gamma^2(1-\delta)}{\Gamma(2-2\delta)}(1-\delta) \qquad (2.5.126)\end{aligned}$$

in the different schemes. Finally, we need the renormalization of the coupling constant g. It is obtained from the vertex corrections, e.g. the fermion–gauge-boson vertex. There are two diagrams, the first one is similar to that of QED, the second one contains the non-abelian triple gauge-boson couplings:

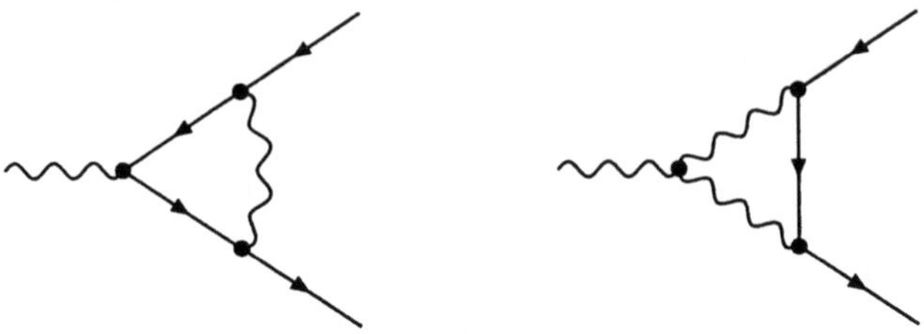

For massless on-shell fermions ($p_1^2 = p_2^2 = 0$) and within dimensional regularization the following expressions for the renormalization of the fermion–gauge-boson vertex are obtained,

$$\begin{aligned}
\left(\delta Z_g + \frac{1}{2}\delta Z_A + \delta Z_\psi\right)^{\mathrm{MS}} &= -\frac{\alpha_s}{4\pi}(C_{\mathrm{A}} + C_{\mathrm{F}})\frac{1}{\delta}, \\
\left(\delta Z_g + \frac{1}{2}\delta Z_A + \delta Z_\psi\right)^{\mathrm{MOM}} &= -\frac{\alpha_s}{4\pi}(C_{\mathrm{A}} + C_{\mathrm{F}})\left(\frac{4\pi\mu^2}{\mathcal{M}^2}\right)^\delta \\
&\quad \times\Gamma(\delta)\frac{\Gamma^2(1-\delta)}{\Gamma(2-2\delta)} + \text{finite terms},
\end{aligned} \tag{2.5.127}$$

where we have not explicitly given the finite terms in the second expression. From this and the results (2.5.122), (2.5.123), and (2.5.126) one finds

$$\begin{aligned}
\delta Z_g^{\mathrm{MS}} &= \frac{\alpha_s}{4\pi}\frac{1}{\delta}\left[\frac{2}{3}N_{\mathrm{F}}T_{\mathrm{F}} - \frac{11}{6}C_{\mathrm{A}}\right], \\
\delta Z_g^{\mathrm{MOM}} &= \frac{\alpha_s}{4\pi}\left(\frac{4\pi\mu^2}{\mathcal{M}^2}\right)^\delta \Gamma(\delta)\frac{\Gamma^2(1-\delta)}{\Gamma(2-2\delta)}\left[\frac{2}{3}N_{\mathrm{F}}T_{\mathrm{F}} - \frac{11}{6}C_{\mathrm{A}}\right] \\
&\quad + \text{finite terms}
\end{aligned} \tag{2.5.128}$$

for the renormalization of the coupling constant. Note that the finite terms are relevant when calculating physical observables.

This completes the determination of the renormalization constants in the physical sector of a non-abelian gauge theory. They are sufficient to obtain

UV-finite results in one-loop order. The calculation of one-loop corrections to observables in QCD is discussed in Sect. 3.3.

The arbitrariness of the renormalization prescription does not lead to physically different results. This can be formulated using the renormalization-group equation, which is investigated in Sect. 2.6.

2.5.4 Sketch of the proof of renormalizability of gauge theories

In the previous sections we used QED and QCD in one-loop approximation as examples to show the essential steps of the perturbative evaluation of a quantized gauge-field theory:

1. gauge-invariant, e.g. dimensional, regularization;
2. subtraction of divergences by introducing counterterms into the Lagrangian (equivalent to multiplicative renormalization of fields and parameters in the examples in Sect. 2.5.3);
3. adjustment of the renormalized (finite) parameters of the theory to their measured values (renormalization conditions).

In this section, we discuss the general proof of the renormalizability of gauge theories. We do not give a complete treatment but rather illustrate the main ingredients and, in particular, the rôle of gauge invariance. Since in non-abelian gauge theories with massless gauge bosons the S matrix is not well-defined owing to severe IR divergences, our considerations are formal. They can, however, be directly transferred to spontaneously broken gauge theories where, if all gauge bosons become massive, this problem is absent, and our arguments become justified.

By introducing gauge-fixing terms and associated ghost terms, gauge theories are defined in such a way that they can be evaluated in perturbation theory. As the gauge couplings are dimensionless, gauge theories are renormalizable according to power counting (cf. Sect. 2.5.1.3) in suitably chosen (renormalizable) gauges. The generating functional Γ of vertex functions is well-defined in $D = 4 - 2\delta$ dimensions, and the addition of a finite number of counterterms to the Lagrangian results in a renormalized vertex functional Γ which remains finite for $\delta \to 0$.

However, gauge theories involve unphysical (negative-norm) states and arbitrary gauge parameters. One has to show that the unphysical states do not contribute to the physical S matrix, and that the latter is independent of the gauge parameters, i.e. one must prove that the physical S matrix is unitary and gauge-independent. To this end, one needs the gauge invariance, or more precisely the BRS invariance, of the theory and the associated Ward identities. Therefore, one has to show that the gauge theory can be renormalized in such a way that a (renormalized) BRS invariance still holds.

This is proven by induction in the number of loops L in the perturbative expansion. The expansion in powers of the loop parameter, i.e. the Planck constant $\hbar$, has the advantage that it preserves any symmetry of the Lagrangian order by order, because the loop parameter multiplies the whole Lagrangian [cf. (2.2.78)]. On the other hand, splitting up the Lagrangian into a free (g-independent) part and an interaction part is not gauge-invariant as the non-abelian gauge transformations explicitly contain the coupling constant [cf. (2.3.32)].

2.5.4.1 Recursive construction of gauge-invariant counterterms

We consider a pure gauge theory (without matter fields) and assume that it is regularized in such a way that the BRS invariance is not broken. A practical regularization scheme with these properties is dimensional regularization. Then, the theory is well-defined in D dimensions, and the Lee identities (2.4.77) and (2.4.78) hold in the regularized theory. According to power counting (cf. Sect. 2.5.1.3) the theory is renormalizable, and the counterterms are a local polynomial of degree four in the fields (Sect. 2.5.1.5). We determine the counterterms in the MS scheme, i.e. the counterterms are chosen in such a way that they cancel exactly the divergences of the loop integrals but contain no finite contributions, apart from those required by gauge invariance. This is sufficient to prove renormalizability. Other renormalization schemes can be obtained by adding appropriate finite counterterms that respect the symmetry of the renormalized action. We restrict ourselves to linear gauges of the form (2.4.54).

We have to show that by adding a well-defined set of counterterms to the Lagrangian, the resulting generating functional is finite for $D \to 4$ and satisfies the *Lee identities*.

To this end, we follow Ref. [Zi75] and consider the loop expansion of the *bare vertex functional* [without gauge fixing terms (2.4.76)]

$$\tilde{\Gamma}_0 = \sum_{L=0}^{\infty} \tilde{\Gamma}_0^{(L)}, \tag{2.5.129}$$

where L is the number of loops. The lowest-order contribution consists of the classical action, the ghost contribution, and the terms involving the sources $\mathbf{K}$ and $\mathbf{K}_u$ of the BRS-transformed gauge and ghost fields, respectively [cf. (2.4.17), (2.4.53), (2.4.55) and (2.4.76)]

$$\tilde{\Gamma}_0^{(0)} = \tilde{S}_0 = \int \mathrm{d}^4x \, [\mathcal{L}_{\mathrm{gauge}}(\mathbf{A}) + (K^{a,\mu} - \bar{u}^b \Phi^{ba,\mu}) s A_\mu^a - K_u^a s u^a]. \tag{2.5.130}$$

The linear relation (2.4.78) implies that also in higher orders the vertex functional depends on $\bar{u}$ and K only through the combination $K^{a,\mu} - \bar{u}^b \Phi^{ba,\mu}$. Taking this into account, it suffices to consider the non-linear relation (2.4.77). If we introduce the compact notation

$$\tilde{\Gamma}_1 * \tilde{\Gamma}_2 = \int \mathrm{d}^4x \left[\frac{\delta \tilde{\Gamma}_1}{\delta A_\mu^a} \frac{\delta \tilde{\Gamma}_2}{\delta K^{a,\mu}} + \frac{\delta \tilde{\Gamma}_1}{\delta u^a} \frac{\delta \tilde{\Gamma}_2}{\delta K_u^a} \right] \tag{2.5.131}$$

for arbitrary functionals Γ_1 and Γ_2, the non-linear Lee identities (2.4.77) can be written as

$$\tilde{\Gamma}_0 * \tilde{\Gamma}_0 = 0. \tag{2.5.132}$$

and the contribution of order L reads

$$\sum_{p+q=L} \tilde{\Gamma}_0^{(p)} * \tilde{\Gamma}_0^{(q)} = 0. \tag{2.5.133}$$

We seek counterterms so that the renormalized[11] vertex functional $\tilde{\Gamma} = \sum_{L=0}^{\infty} \tilde{\Gamma}^{(L)}$ satisfies

$$\sum_{p+q=L} \tilde{\Gamma}^{(p)} * \tilde{\Gamma}^{(q)} = 0. \tag{2.5.134}$$

[11] We do not use an index R to denote the renormalized functional in this section in order to avoid too many indices.

Because the tree-level action is BRS-invariant and does not need to be renormalized ($\tilde{\Gamma}_0^{(0)} = \tilde{\Gamma}^{(0)}$), both relations are fulfilled in the tree-graph approximation

$$\tilde{S}_0 * \tilde{S}_0 = \tilde{\Gamma}_0^{(0)} * \tilde{\Gamma}_0^{(0)} = \tilde{\Gamma}^{(0)} * \tilde{\Gamma}^{(0)} = 0. \tag{2.5.135}$$

Let us first consider the one-loop level. We split $\tilde{\Gamma}_0^{(1)}$ into a finite part and a (for $D \to 4$) divergent part, $\tilde{\Gamma}_0^{(1)} = \tilde{\Gamma}_{0,\mathrm{fin}}^{(1)} + \tilde{\Gamma}_{0,\mathrm{div}}^{(1)}$, which satisfy the Lee identities in the one-loop approximation separately. Substituting (2.5.129) and $\tilde{\Gamma}_0^{(0)} = \tilde{S}_0$ into the Lee identities (2.5.133) we find that the one-loop divergent part satisfies

$$\tilde{S}_0 * \tilde{\Gamma}_{0,\mathrm{div}}^{(1)} + \tilde{\Gamma}_{0,\mathrm{div}}^{(1)} * \tilde{S}_0 = 0. \tag{2.5.136}$$

In order to obtain a finite renormalized one-loop vertex functional we add counterterms to the action that cancel the one-loop divergences. This results in the one-loop renormalized action

$$\tilde{S}_1' = \tilde{S}_0 - \tilde{\Gamma}_{0,\mathrm{div}}^{(1)}. \tag{2.5.137}$$

Equation (2.5.136) guarantees that $\tilde{S}_1'$ satisfies the Lee identity to one-loop order, i.e. $\tilde{S}_1' * \tilde{S}_1' = \mathcal{O}(\hbar^2)$. However, the recursive construction works only if the one-loop renormalized action satisfies the Lee identity $\tilde{S}_1 * \tilde{S}_1 = 0$ exactly. This can be achieved by adding appropriate higher-order counterterms Δ_1, which do not modify the functional $\tilde{S}_1'$ at one-loop order, i.e. we define

$$\tilde{S}_1 = \tilde{S}_0 - \tilde{\Gamma}_{0,\mathrm{div}}^{(1)} + \Delta_1. \tag{2.5.138}$$

Here Δ_1 is a local polynomial in the fields of degree four of order $\hbar^2$ and determined such that $\tilde{S}_1 * \tilde{S}_1 = 0$ exactly. The explicit construction of $\tilde{S}_1$ is performed in Sect. 2.5.4.2. Then, $\tilde{S}_1$ yields a finite one-loop vertex functional $\tilde{\Gamma}_1$ that satisfies the Lee identities.

We now proceed by induction. We assume that we have constructed a renormalized action $\tilde{S}_n$ that satisfies the Lee identities and generates a finite perturbation series up to order n. Then, the corresponding vertex functional $\tilde{\Gamma}_n$ satisfies the Lee identities (2.5.132) as well and has a loop expansion similar to (2.5.129). The contributions $\tilde{\Gamma}_n^{(L)}$ in this expansion are finite for $L \leq n$ and the divergences of $\tilde{\Gamma}_n^{(n+1)}$ originate only from overall divergences. The divergent part of $\tilde{\Gamma}_n^{(n+1)}$ is therefore an integral over a local polynomial of degree four in A, u, $\bar{u}$, K, and K_u that satisfies the Lee identities. At order $n+1$ these identities yield [cf. (2.5.133)]

$$\tilde{S}_0 * \tilde{\Gamma}_{n,\text{div}}^{(n+1)} + \tilde{\Gamma}_{n,\text{div}}^{(n+1)} * \tilde{S}_0 = 0 \tag{2.5.139}$$

for the divergent part, since all other contributions are finite. Now we can continue as in the one-loop case. The divergences of $\tilde{\Gamma}_n^{(n+1)}$ can be removed by adding a counterterm to $\tilde{S}_n$ so that the renormalized action at order $n+1$, $\tilde{S}_{n+1}$, still satisfies the *Lee identities* This concludes the inductive proof.

2.5.4.2 Structure of the renormalized action

According to the preceding section, the renormalized action in the MS scheme is obtained from the bare action by adding appropriate local polynomials of degree four so that the Lee identities hold in each order of perturbation theory. Consequently, the renormalized action $\tilde{S}_\text{R}$ is the most general local polynomial of degree four that satisfies the identities

$$0 = \tilde{S}_0 * \tilde{S}_\text{R} + \tilde{S}_\text{R} * \tilde{S}_0 \equiv \mathcal{B}_{\tilde{S}_0} \tilde{S}_\text{R}. \tag{2.5.140}$$

Here we introduced the operator $\mathcal{B}_{\tilde{S}_0}$, which is defined as

$$\mathcal{B}_{\tilde{S}_0} = \int \mathrm{d}^4x \left[\frac{\delta \tilde{S}_0}{\delta A_\mu^a} \frac{\delta}{\delta K^{a,\mu}} + \frac{\delta \tilde{S}_0}{\delta K^{a,\mu}} \frac{\delta}{\delta A_\mu^a} + \frac{\delta \tilde{S}_0}{\delta u^a} \frac{\delta}{\delta K_u^a} + \frac{\delta \tilde{S}_0}{\delta K_u^a} \frac{\delta}{\delta u^a} \right]. \tag{2.5.141}$$

Using the Lee identity for the tree-level action, $\mathcal{B}_{\tilde{S}_0} \tilde{S}_0 = 0$, and the anticommutativity of the Grassmann variables it can be easily shown that $\mathcal{B}_{\tilde{S}_0}$ is nilpotent [It80]

$$\mathcal{B}_{\tilde{S}_0} \mathcal{B}_{\tilde{S}_0} = 0. \tag{2.5.142}$$

Any gauge-invariant functional $G\{\mathbf{A}\}$ of A_μ^a is BRS-invariant and thus satisfies

$$\mathcal{B}_{\tilde{S}_0} G\{\mathbf{A}\} = \int \mathrm{d}^4x \frac{\delta \tilde{G}}{\delta A_\mu^a} \frac{\delta \tilde{S}_0}{\delta K^{a,\mu}} = \int \mathrm{d}^4x \frac{\delta \tilde{G}}{\delta A_\mu^a} s A_\mu^a = 0. \tag{2.5.143}$$

Therefore, any expression of the form

$$\tilde{S}_{\mathrm{R}} = G\{\mathbf{A}\} + \mathcal{B}_{\tilde{S}_0} F\{\mathbf{A}, \mathbf{u}, \bar{\mathbf{u}}, \mathbf{K}, \mathbf{K}_u\} \tag{2.5.144}$$

with an arbitrary functional F is a solution of (2.5.140). In fact this is the general solution as has been first proven in Ref. [Jo76].

Since the only local gauge-invariant functional of dimension 4 is proportional to the action,[12] we have $G\{\mathbf{A}\} = \alpha \int \mathrm{d}^4x\, \mathcal{L}_{\mathrm{gauge}}(\mathbf{A})$. Because $\tilde{S}_{\mathrm{R}}$ is a local functional of dimension 4 and ghost number 0, and because $\mathcal{B}_{\tilde{S}_0}$ has dimension 1 and ghost number 1, F must be a local functional of dimension 3 and ghost number -1. Requiring that the global invariance is not broken by the choice of gauge ($\mathcal{B}_{\tilde{S}_0}$ is globally invariant) and using (2.4.78), the most general form is

$$\begin{aligned}
\tilde{S}_{\mathrm{R}} &= \alpha \int \mathrm{d}^4x\, \mathcal{L}_{\mathrm{gauge}}(\mathbf{A}) + \mathcal{B}_{\tilde{S}_0} \int \mathrm{d}^4x \Big[\beta(K^a_\mu - \bar{u}^a \Phi^{ba}_\mu) A^{a,\mu} - \gamma K^a_u u^a\Big] \\
&= \int \mathrm{d}^4x \left[\alpha \mathcal{L}_{\mathrm{gauge}}(\mathbf{A}) + \beta\left(A^a_\mu \frac{\delta \tilde{S}_0}{\delta A^a_\mu} - K^a_\mu \frac{\delta \tilde{S}_0}{\delta K^a_\mu} - \bar{u}^a \frac{\delta \tilde{S}_0}{\delta \bar{u}^a}\right)\right. \\
&\qquad \left. + \gamma\left(u^a \frac{\delta \tilde{S}_0}{\delta u^a} - K^a_u \frac{\delta \tilde{S}_0}{\delta K^a_u}\right)\right],
\end{aligned} \tag{2.5.145}$$

where α, β, and γ are arbitrary dimensionless constants that can be used to absorb the divergences that appear when the regularization is removed.

Using the relations

$$\begin{aligned}
2 \int \mathrm{d}^4x\, \mathcal{L}_{\mathrm{gauge}}(\mathbf{A}) &= A^a_\mu \frac{\delta \tilde{S}_0}{\delta A^a_\mu} + K^a_u \frac{\delta \tilde{S}_0}{\delta K^a_u} - g \frac{\partial \tilde{S}_0}{\partial g}, \\
u^a \frac{\delta \tilde{S}_0}{\delta u^a} &= 2K^a_u \frac{\delta \tilde{S}_0}{\delta K^a_u} + K^a_\mu \frac{\delta \tilde{S}_0}{\delta K^a_\mu} + \bar{u}^a \frac{\delta \tilde{S}_0}{\delta \bar{u}^a},
\end{aligned} \tag{2.5.146}$$

which can be easily verified from the explicit form of the lowest-order action (2.5.130), we find the general solution to (2.5.140)

$$\begin{aligned}
\tilde{S}_{\mathrm{R}} = \Bigg\{ &\left(\frac{\gamma}{2} - \frac{\beta}{2}\right)\left(u^a \frac{\delta}{\delta u^a} + \bar{u}^a \frac{\delta}{\delta \bar{u}^a} + K^a_\mu \frac{\delta}{\delta K^a_\mu}\right) \\
&+ \left(\frac{\alpha}{2} + \beta\right)\left(A^a_\mu \frac{\delta}{\delta A^a_\mu} + K^a_u \frac{\delta}{\delta K^a_u}\right) - \frac{\alpha}{2} g \frac{\partial}{\partial g} \Bigg\} \\
&\times \tilde{S}_0\{\mathbf{A}, \mathbf{u}, \bar{\mathbf{u}}, \mathbf{K}, \mathbf{K}_u; g\}.
\end{aligned}$$

[12]There is actually a further gauge-invariant functional, $\varepsilon^{\mu\nu\rho\sigma} \int \mathrm{d}^4x\, F_{\mu\nu} F_{\rho\sigma}$. Because this term is a topological invariant and irrelevant in perturbation theory, it is dropped.

All counterterms arise from renormalization of the parameters, the sources, and the fields of the effective action. Moreover, A^a_μ and K^a_u as well as u^a, $\bar{u}^a$, and K^a_μ are renormalized in the same way.

If we identify

$$Z_A^{1/2} = \frac{\alpha}{2} + \beta, \qquad Z_u^{1/2} = \frac{\gamma - \beta}{2}, \qquad Z_g^{1/2} = -\frac{\alpha}{2}, \tag{2.5.147}$$

this implies

$$\begin{aligned} &\tilde{S}_{\mathrm{R}}\{\mathbf{A}, \mathbf{u}, \bar{\mathbf{u}}, \mathbf{K}, \mathbf{K}_u; g\} = \\ &\quad = \tilde{S}_0\{Z_A^{1/2}\mathbf{A}, Z_u^{1/2}\mathbf{u}, Z_u^{1/2}\bar{\mathbf{u}}, Z_u^{1/2}\mathbf{K}, Z_A^{1/2}\mathbf{K}_u; Z_g g\} \\ &\quad = \tilde{S}_0\{\mathbf{A}_0, \mathbf{u}_0, \bar{\mathbf{u}}_0, \mathbf{K}_0, \mathbf{K}_{u,0}; g_0\}. \end{aligned} \tag{2.5.148}$$

By adding the lowest-order gauge-fixing term we finally obtain the complete *renormalized action* S. The non-renormalization of the gauge-fixing term in linear gauges can be accounted for by choosing the renormalization constant of the gauge parameter ξ (2.5.45) as $Z_\xi = Z_A$.

Thus, the counterterms can be generated by the following *multiplicative renormalization transformations* of the fields and parameters:

$$\begin{aligned} A^a_{0,\mu} &= Z_A^{1/2} A^a_\mu, \qquad K^a_{0,u} = Z_A^{1/2} K^a_u, \qquad \xi_0 = Z_A \xi, \\ u^a_0 &= Z_u^{1/2} u^a, \qquad \bar{u}^a_0 = Z_u^{1/2} \bar{u}^a, \qquad K^a_{0,\mu} = Z_u^{1/2} K^a_\mu, \\ g_0 &= Z_g g. \end{aligned} \tag{2.5.149}$$

Because of ghost-number conservation, which is expressed by the second line of (2.5.146), there is a freedom in the renormalization of the fields with non-vanishing ghost number. This can be used in such a way that the antighost field is not renormalized and (2.5.149) is changed to

$$u^a_0 = Z_u u^a, \qquad \bar{u}^a_0 = \bar{u}^a, \qquad K^a_{0,\mu} = K^a_\mu, \qquad K^a_{0,u} = \frac{Z_A^{1/2}}{Z_u} K^a_u. \tag{2.5.150}$$

The renormalized action and the renormalized vertex functional are obtained from the bare quantities as

$$\begin{aligned} S_{\mathrm{R}}\{\mathbf{A}, \mathbf{u}, \bar{\mathbf{u}}, \mathbf{K}, \mathbf{K}_u; g, \xi\} &= S_0\{\mathbf{A}_0, \mathbf{u}_0, \bar{\mathbf{u}}_0, \mathbf{K}_0, \mathbf{K}_{u,0}; g_0, \xi_0\}, \\ \Gamma_{\mathrm{R}}\{\mathbf{A}, \mathbf{u}, \bar{\mathbf{u}}, \mathbf{K}, \mathbf{K}_u; g, \xi\} &= \Gamma_0\{\mathbf{A}_0, \mathbf{u}_0, \bar{\mathbf{u}}_0, \mathbf{K}_0, \mathbf{K}_{u,0}; g_0, \xi_0\}. \end{aligned} \tag{2.5.151}$$

These relations imply that the renormalized action is invariant under renormalized BRS transformations. Since $\mathbf{A}\,\mathbf{K}$, $\mathbf{u}\,\mathbf{K}_u$, and $\bar{\mathbf{u}}\,\mathbf{K}$ are renormalized in the same way, the renormalized Lee identities are obtained by multiplying the bare ones by the corresponding renormalization constants.

The renormalization constants have the loop expansion

$$Z_A = 1 + \sum_{L=1}^{\infty} \delta Z_A^{(L)}, \quad Z_u = 1 + \sum_{L=1}^{\infty} \delta Z_u^{(L)}, \quad Z_g = 1 + \sum_{L=1}^{\infty} \delta Z_g^{(L)}, \tag{2.5.152}$$

and can be determined order by order in perturbation theory.

Note that the action, renormalized to order L, S_L, fulfils the Lee identities to all orders as required in the inductive proof in Sect. 2.5.4.1. The required terms of order $\hbar^i$, $i > L$ are automatically generated by the non-linear appearance of the fields in S.

After having performed the proof we can set $K_\mu^a = K_u^a = 0$.

The proof can be extended to gauge theories involving matter fields with gauge-invariant couplings of dimension less than or equal to 4. The relations between the renormalization constants of the various fields stay intact and, in particular, the renormalization of the gauge coupling remains universal.

As the proof required the existence of a gauge-invariant regularization scheme, it breaks down for chiral symmetries, in particular for gauge theories involving the matrix γ_5. But also for chiral gauge theories, such as the Electroweak Standard Model, the existence of a gauge-invariant renormalized action can be proven as long as no anomalies occur [Be74, Be76b, Pi81, Kr98]. Anomalies are discussed in Sect. 2.7.

2.5.4.3 Physical states and physical sources

The covariant quantization of a gauge-field theory in a renormalizable gauge, as described in Sect. 2.4.1, requires the introduction of unphysical degrees of freedom:

1. (zero-mass) gauge fields give rise to negative-norm states (2.1.26), and
2. fictitious ghost degrees of freedom have to be introduced to provide the formal gauge covariance of the theory (Sects. 2.4.1.3 and 2.4.3.1).

The physical *S matrix* transforms physical initial states into physical final states. Thus, a definition of ***physical states*** must be made before any further investigation of the S matrix is carried out [tH72, Be74, Co75, Ku78]. Quite apart from the question as to whether such states can actually be dynamically realized, the definition of physical states is phrased as follows:

Physical degrees of freedom comprise all degrees of freedom invariant under the renormalized gauge group.

The definition of physical states is most conveniently done in the Hamiltonian picture after canonical quantization of the effective Lagrangian (2.4.24) [Ku78].

In gauges that do not involve ghosts, e.g. in Coulomb-like gauges, this amounts to the requirement that physical states are invariant under (renormalized) gauge transformations

$$U|\psi\rangle_{\text{phys}} = 0. \tag{2.5.153}$$

In general renormalizable gauges this condition is generalized to the requirement that physical states are BRS-invariant [Ku78], i.e. annihilated by the *BRS charge* Q_{BRS} (2.4.52),

$$Q_{\text{BRS}}|\psi\rangle_{\text{phys}} = 0, \tag{2.5.154}$$

which acts on anticommuting/commuting fields as

$$[Q_{\text{BRS}}, \psi]_{\pm} = s\psi. \tag{2.5.155}$$

The BRS operator s was defined in (2.4.33). We implicitly assume here the formalism with the auxiliary field B^a (cf. end of Sect. 2.4.3.1). Owing to the nilpotency of the BRS transformation, the BRS charge is nilpotent and, as a conserved charge, commutes with the effective Hamiltonian H and thus with the S matrix

$$(Q_{\text{BRS}})^2 = 0, \qquad [Q_{\text{BRS}}, H] = 0, \qquad [Q_{\text{BRS}}, S] = 0. \tag{2.5.156}$$

Moreover, it is hermitian provided that the hermiticity properties of the Faddeev–Popov ghosts are chosen appropriately, i.e.

$$(u^a)^\dagger = u^a, \qquad (\bar{u}^a)^\dagger = -\bar{u}^a. \tag{2.5.157}$$

The minus sign appearing in the hermitian conjugation of the antighost is required because of the anticommuting nature of the ghost fields [Ku78].

The BRS charge can be used to divide the eigenstates of H into three subspaces:

1. the subspace of states that are not annihilated by Q_{BRS}:
   $\mathcal{V}_2 = \{|\psi_a\rangle \mid Q_{\text{BRS}}|\psi_a\rangle \neq 0\}$,
2. the subspace of states that are obtained by applying Q_{BRS} to $\mathcal{V}_2$ and thus are annihilated by Q_{BRS}:
   $\mathcal{V}_0 = \{|\psi_a\rangle \mid |\psi_a\rangle = Q_{\text{BRS}}|\psi'_a\rangle, |\psi'_a\rangle \in \mathcal{V}_2\}$,
3. the subspace of states that are annihilated by Q_{BRS} but are not in $\mathcal{V}_0$:
   $\mathcal{V}_1 = \{|\psi_a\rangle \mid Q_{\text{BRS}}|\psi_a\rangle = 0,\ |\psi_a\rangle \notin \mathcal{V}_0\}$.

In the following we denote the states in the subspaces $\mathcal{V}_0$, $\mathcal{V}_1$, and $\mathcal{V}_2$ by $|\psi_a^{(0)}\rangle$, $|\psi_a^{(1)}\rangle$, and $|\psi_a^{(2)}\rangle$, respectively.

The states in $\mathcal{V}_2$ are characterized as unphysical by (2.5.154). The other two subspaces form the space of physical states $\mathcal{V}_{\text{phys}} = \mathcal{V}_0 \cup \mathcal{V}_1$. Owing to (2.5.154) and (2.5.156), $\mathcal{V}_{\text{phys}}$ is invariant under the S matrix, $S\mathcal{V}_{\text{phys}} \subseteq \mathcal{V}_{\text{phys}}$. For gauge theories the inner product in this space is positive semi-definite [Ku78].

It is easily seen that all states in $\mathcal{V}_0$ have zero norm and are orthogonal to all states in $\mathcal{V}_1$, and thus also in $\mathcal{V}_{\text{phys}}$

$$\begin{aligned} \langle\psi_a^{(0)}|\psi_b^{(0)}\rangle &= \langle\psi_a^{(2)}|Q_{\text{BRS}}|\psi_b^{(0)}\rangle = 0, \\ \langle\psi_a^{(0)}|\psi_b^{(1)}\rangle &= \langle\psi_a^{(2)}|Q_{\text{BRS}}|\psi_b^{(1)}\rangle = 0. \end{aligned} \tag{2.5.158}$$

As a consequence these states decouple, i.e. they disappear from any physical matrix element, and physical states that differ only by zero-norm states are physically equivalent. The physical Hilbert space $\mathcal{H}_{\text{phys}}$ can be defined as the quotient space of $\mathcal{V}_{\text{phys}}$ with respect to the subspace $\mathcal{V}_0$

$$\mathcal{H}_{\text{phys}} \equiv \mathcal{V}_{\text{phys}}/\mathcal{V}_0, \tag{2.5.159}$$

i.e. the distinguishable physical states are defined as equivalence classes of states with strictly positive norm. Besides, physical states are required to have vanishing ghost number. This immediately characterizes all Faddeev–Popov ghosts as unphysical.

In quantum field theory, within perturbation theory, the space of states is the Fock space of asymptotic fields. This Fock space is generated by the creation operators appearing in the Fourier representation of the asymptotic field operators (1.2.10). According to our definition of physical states (2.5.154) and the relation between states and asymptotic fields (cf. Sect. 1.2.1), the *physical components of asymptotic fields* must have vanishing BRS variation

$$s\Psi^{\mathrm{as,phys}} = 0. \tag{2.5.160}$$

As discussed in the following, the BRS variation of asymptotic fields is essentially obtained by discarding all terms that involve composite operators (products of fields) from the BRS variation of the fields (2.4.33) and (2.4.35), i.e. the BRS variation of the asymptotic fields is linear. This implies, in particular, that fields transforming inhomogeneously under gauge transformations, and therefore have contributions involving only a ghost field in their BRS transforms, e.g. gauge fields or Higgs fields (cf. Sect. 4.1.2), involve unphysical components.

In order to determine the physical degrees of freedom of the fields we have to derive the BRS transformations of the asymptotic fields [Ku78]. The asymptotic fields are related to the discrete poles of Green functions via the reduction formula (2.2.46). Consequently, the properties of the asymptotic fields can be determined by investigating the Green functions. To this end, we consider an arbitrary Green function with one of the external fields Ψ_I (the index I includes the space–time coordinates x_I, generically summarizes all field and Lorentz indices, and distinguishes different fields) replaced by its BRS variation $\langle T(s\Psi_I)\Psi_{I_1}\ldots\Psi_{I_n}\rangle$. We put the corresponding external line on-shell and extract the pole terms. Since the presence of a pole requires one-particle intermediate states, we find (underlining indicates truncation)

$$\begin{aligned}&\langle T(s\Psi_I)\Psi_{I_1}\ldots\Psi_{I_n}\rangle = \\ &\qquad \sum_K \langle T(s\Psi_I)\underline{\Psi_K}\rangle\langle T\Psi_K\Psi_{I_1}\ldots\Psi_{I_n}\rangle + \text{non-pole terms},\end{aligned} \tag{2.5.161}$$

or in diagrams

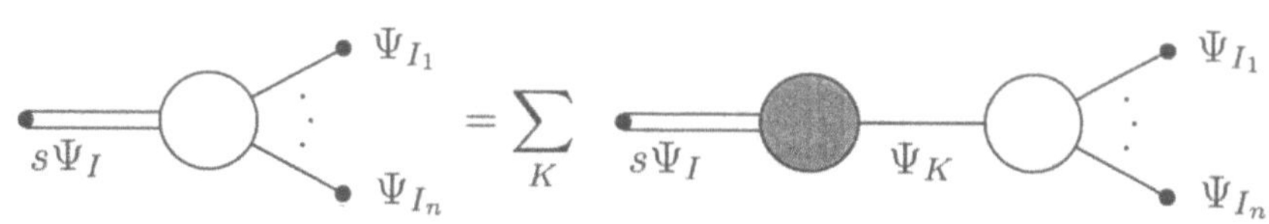

$$+ \text{non-pole terms}, \tag{2.5.162}$$

where we indicated the BRS-transformed fields by a double line. Note that $\sum_K$ in the preceding equations includes also the integration over the corresponding space–time coordinates x_K. Equation (2.5.161) suggests to define the BRS transformation of the asymptotic fields as

$$s\Psi_I \underset{\text{on-shell}}{\longrightarrow} s\Psi_I^{\text{as}} = C_{IK}\Psi_K^{\text{as}} \tag{2.5.163}$$

with

$$C_{IK} = \langle T(s\Psi_I)\underline{\Psi_K}\rangle\Big|_{\text{on-shell}} = \frac{\mathrm{i}\delta\Gamma}{\delta J_{s\Psi_I}\delta\Psi_K}\Bigg|_{\text{on-shell}}$$

$$= \tag{2.5.164}$$

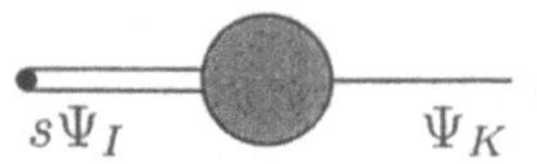

and $J_{s\Psi_I}$ is a source coupled to $s\Psi_I$. By construction the BRS transformation of asymptotic fields is linear, and a field Ψ_K can only contribute if its mass is equal to the one of Ψ_I. In practice the field Ψ_K is an antighost field since $s\Psi_I$ involves a ghost field, and the ghost number is conserved.

We define *physical sources* J_I^{phys} as those sources that couple only to physical fields. Because of (2.5.160) we immediately find that these satisfy

$$J_I^{\text{phys}} s\Psi_I^{\text{as,phys}} = J_I^{\text{phys}} \frac{\mathrm{i}\delta\Gamma}{\delta J_{s\Psi_I}\delta\Psi_K}\Bigg|_{\text{on-shell}} \Psi_K^{\text{as}} = 0, \tag{2.5.165}$$

which is equivalent to [tH72]

$$J_I^{\text{phys}}\langle 0|s\Psi_I|\Psi_K\rangle = 0, \tag{2.5.166}$$

where $|\Psi_K\rangle$ is the state belonging to the asymptotic field Ψ_K^{as}.

Let us consider gauge fields. Owing to ghost-number conservation the (linear) BRS transformation of the asymptotic gauge fields must have the form[13]

$$sA_\mu^{\text{as},a}(x) = \int \mathrm{d}^4y\, R_\mu^{ab}(x,y)u^{\text{as},b}(y), \tag{2.5.167}$$

[13]Note our conventions in App. A.1.3.2.

where

$$R^{ab}_{\mu}(x,y) = \left\langle T\left(sA^{\mathrm{as},a}_{\mu}(x)\right)\underline{\bar{u}^b(y)}\right\rangle\Big|_{\text{on-shell}} = \frac{\mathrm{i}\delta\Gamma}{\delta K^{a,\mu}(x)\delta u^b(y)}\Big|_{\text{on-shell}} \tag{2.5.168}$$

because of (2.5.163) and (2.5.164). The fact that the ghost masses coincide with those of the gauge-boson fields follows from BRS invariance. Equations (2.4.62), (2.4.33), (2.4.35) and (2.4.54) imply

$$\begin{aligned}\langle T(D_\mu u)^a \bar{u}^b\rangle &= \langle T(sA^a_\mu)\bar{u}^b\rangle = -\langle TA^a_\mu s\bar{u}^b\rangle \\ &= \langle TA^a_\mu A^c_\nu\rangle \Phi^{bc,\nu}\frac{1}{\xi} = G^{ac}_{\mu\nu}\Phi^{bc,\nu}\frac{1}{\xi}.\end{aligned} \tag{2.5.169}$$

The poles in the expression on the left-hand side are determined by the ghost propagator, those on the right-hand side by the gauge-boson propagator. Equality of poles on both sides yields equality of masses for those components of the gauge-boson fields that are not orthogonal to $\Phi^{bc,\nu}$.

Considering diagrams, (2.5.167) implies

$\langle T(sA^a_\mu)\Psi_{I_1}\cdots\Psi_{I_n}\rangle =$ 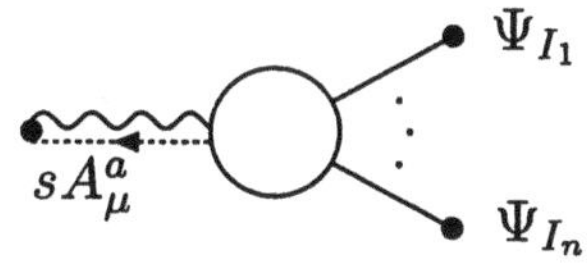

$= \partial_\mu$ 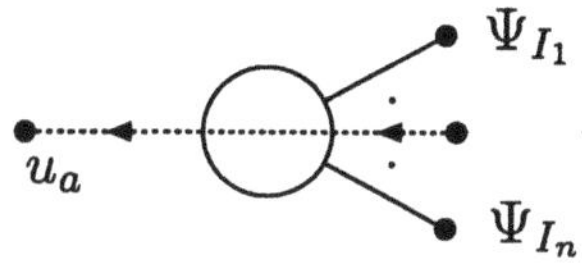

$- gf^{acd}$ 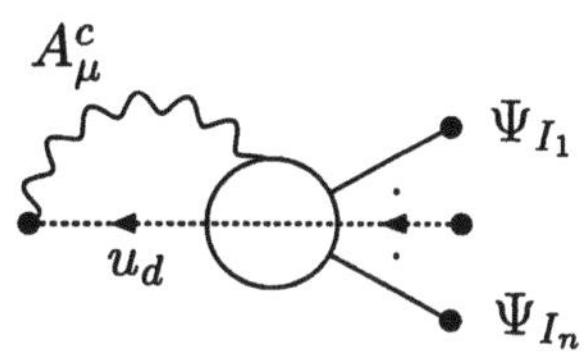

$= \partial_\mu$ 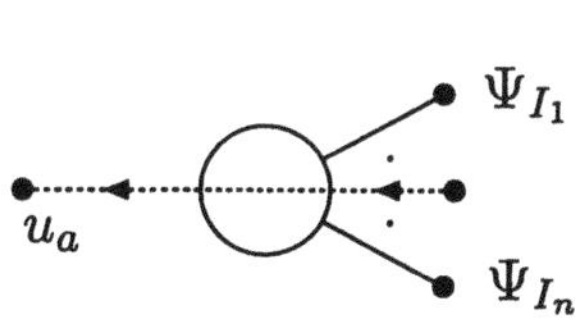

$- gf^{acd}$ 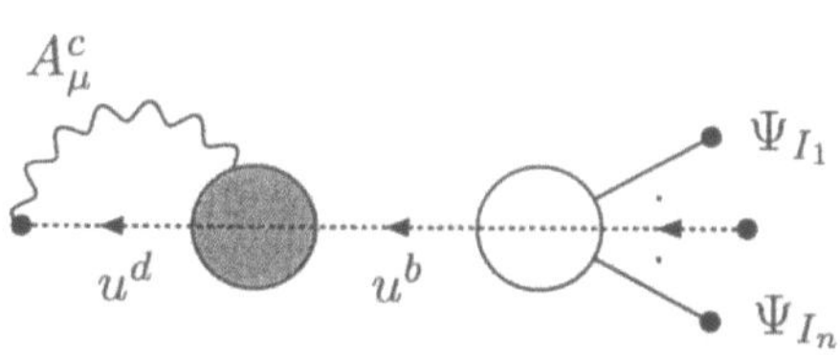

$$+ \text{non-pole terms}$$

$$= R^{ab}_{\mu} \langle T u^b \Psi_{I_1} \cdots \Psi_{I_n} \rangle + \text{non-pole terms} \tag{2.5.170}$$

with

$$R^{ab}_{\mu}(x, y) = \delta^{ab} \partial^{\mu} \delta^{(4)}(x - y) - g f^{acd}$$

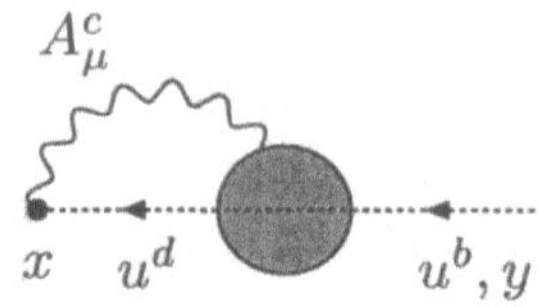

(2.5.171)

Lorentz invariance and global gauge invariance restrict R^{ab}_{μ} to the form

$$R^{ab}_{\mu}(x, y) = \delta^{ab} \partial_{\mu} R(x, y), \tag{2.5.172}$$

because δ^{ab} is the only gauge-invariant tensor with two group indices. The linear Lee identity (2.4.75) implies

$$\Phi^{ab}_{\mu} \frac{\mathrm{i}\delta\Gamma}{\delta K^b_{\mu}(x) \delta u^c(y)} = \frac{\mathrm{i}\delta\Gamma}{\delta \bar{u}^a(x) \delta u^c(y)}, \tag{2.5.173}$$

and the renormalization condition for the (renormalized) ghost two-point function can be fixed in such a way that this two-point function is on shell equal to its tree-level contribution

$$\left. \frac{\mathrm{i}\delta\Gamma}{\delta \bar{u}^a(x) \delta u^c(y)} \right|_{\text{on-shell}} = \Phi^{ac}_{\mu} \partial^{\mu} \delta^{(4)}(x - y). \tag{2.5.174}$$

Putting all this together yields

$$\Phi^{ab}_{\mu} R^{bc,\mu}(x, y) = \Phi^{ac}_{\mu} \partial^{\mu} R(x, y) = \Phi^{ac}_{\mu} \partial^{\mu} \delta^{(4)}(x - y), \tag{2.5.175}$$

or

$$R(x, y) = \delta^{(4)}(x - y), \tag{2.5.176}$$

and thus

$$s A^{\mathrm{as},a}_{\mu}(x) = \partial_{\mu} u^{\mathrm{as},a}(x). \tag{2.5.177}$$

The BRS transformation of the asymptotic ghost field vanishes because there is no single particle state with ghost number two, and the BRS transformation of the asymptotic antighost field follows directly from the linear BRS transformation of the antighost field, so that we find

$$\begin{aligned} sA_\mu^{\mathrm{as},a}(x) &= \partial_\mu u^{\mathrm{as},a}(x), \\ su^{\mathrm{as},a}(x) &= 0, \\ s\bar{u}^{\mathrm{as},a}(x) &= B^{\mathrm{as},a}(x). \end{aligned} \tag{2.5.178}$$

Note that these have exactly the form of the BRS transformations of the free fields.

The preceding relations allow to read off the *physical fields of a pure gauge theory*. The antighost $\bar{u}^{\mathrm{as},a}$ and the longitudinal component of the asymptotic gauge-boson field ($\varepsilon_\mu = k_\mu$) are not annihilated[14] by the BRS operator and are thus unphysical fields ($\in \mathcal{V}_2$). The ghost field $u^{\mathrm{as},a}$ and the auxiliary field $B^{\mathrm{as},a}$ are BRS transformations and consequently zero-norm states ($\in \mathcal{V}_0$). As the equation of motion

$$B^{\mathrm{as},a}(x) = -\frac{1}{\xi}\Phi_\mu^{ab} A^{\mathrm{as},b,\mu}(x) \tag{2.5.179}$$

identifies the auxiliary field with the component $\Phi_\mu^{ab} A^{\mathrm{as},b,\mu}$ of the gauge-boson field, also this component is a zero-norm state. Consequently, the only physical states ($\in \mathcal{V}_1$) are the two remaining (transverse) components of the gauge field. The corresponding polarization vectors are orthogonal to the momentum and to the gauge-fixing function.

There are exactly four unphysical modes (including zero-norm modes), the antighost, the longitudinal gauge-boson mode, the ghost and the B mode. This is an example of the so-called *quartet mechanism* [Ku78, Ku97]. According to this, unphysical states appear always as quartets, and only the combinations of the quartet states with zero norm can appear in the physical space. The quartet mechanism can be derived from the BRS algebra.

Let us specialize to the *Lorentz gauge*, $\Phi_\mu^{ab} = \delta^{ab}\partial_\mu$. The unphysical longitudinal *polarization vector* reads

$$\varepsilon_\mu(k, \mathrm{L}) = k_\mu = (|\mathbf{k}|, \mathbf{k}), \tag{2.5.180}$$

[14] This can be easily seen by transforming (2.5.178) to momentum space.

the zero-norm (scalar) polarization vector ($\varepsilon^\mu k_\mu \neq 0$) reads

$$\varepsilon_\mu(k,0) = (|\mathbf{k}|, -\mathbf{k}), \tag{2.5.181}$$

and the physical polarization vectors are given by

$$\begin{aligned} &\varepsilon^\mu_{\text{phys}}(k,1) = (0, \hat{\mathbf{k}}^{(1)}_\perp), \qquad \varepsilon^\mu_{\text{phys}}(k,2) = (0, \hat{\mathbf{k}}^{(2)}_\perp), \\ &\hat{\mathbf{k}}^{(i)}_\perp \mathbf{k} = 0, \qquad \hat{\mathbf{k}}^{(i)}_\perp \hat{\mathbf{k}}^{(j)}_\perp = \delta^{ij}, \qquad i,j = 1,2. \end{aligned} \tag{2.5.182}$$

All physical polarization vectors are transverse and have non-vanishing norm. All vectors $\varepsilon^\mu(k,i) = \varepsilon^\mu_{\text{phys}}(k,i) + c\varepsilon^\mu(k,0)$, $i = 1,2$, with arbitrary c describe the same physical state and are members of the equivalence class characterized by $\varepsilon^\mu_{\text{phys}}(k,i)$. While the space $\mathcal{V}_1$ spanned by $\varepsilon_{\text{phys}}(k,i)$, $i = 1,2$, is not Lorentz-invariant, the space of physical states $\mathcal{V}_{\text{phys}}$ spanned by $\varepsilon_{\text{phys}}(k,i)$, $i = 0,1,2$, and consequently the physical Hilbert space $\mathcal{H}_{\text{phys}}$ is. Note that the time-like excitation $\varepsilon_\mu = (1,0,0,0)$, which introduces negative norm-states (2.1.26) is a linear combination of the unphysical and the zero-norm polarization and thus does not contribute to physical matrix elements.

In the case of fermion fields, the BRS transformation (2.4.33) in combination with ghost-number and fermion-number conservation exclude one-particle intermediate states in $\langle (s\psi)\Phi_{I_1} \ldots \Phi_{I_n}\rangle$. Consequently, the BRS transformation of the asymptotic fermion fields vanishes and the fermion fields involve only physical degrees of freedom.

In gauge theories with spontaneous symmetry breaking, scalar fields occur that transform inhomogeneously under gauge transformations. The corresponding BRS transformation involves a linear (ghost-field) term. Owing to the mass degeneracy of the scalar field and the ghost field this term survives in the BRS transformation of the asymptotic field. Consequently, these scalar fields are unphysical degrees of freedom (cf. Sect. 4.4.2).

2.5.4.4 Gauge independence of the physical S matrix

The introduction of gauge-fixing terms along with the corresponding ghost terms introduces a gauge dependence in the theory. Therefore, the gauge invariance of physical results has to be demonstrated by proving the gauge independence of the S matrix. This can be done with help of the Ward identities that follow from BRS invariance. It is essential that the renormalization has been carried out in such a way that the renormalized theory is invariant under (renormalized) BRS transformations.

Consider the gauge theory defined by (2.4.20) and (2.4.21). Since we are only interested in the S matrix and thus in Green functions involving only physical fields, we put the sources $\mathbf{J}_u$, $\mathbf{J}_{\bar{u}}$, $\mathbf{K}$, $\mathbf{K}_u$, $\mathbf{K}_\psi$, and $\mathbf{K}_{\bar{\psi}}$ to zero. We want to study the gauge dependence of the resulting generating functional $Z\{\mathbf{J},\mathbf{J}_\psi,\mathbf{J}_{\bar{\psi}}\}$. We assume that the theory is renormalized and invariant under BRS transformations.

Let us perform an infinitesimal local change of the gauge-fixing functional $C^b \to C^b + \Delta C^b$. To first order in ΔC, we have

$$\begin{aligned} Z_{C+\Delta C}\{\mathbf{J},\mathbf{J}_\psi,\mathbf{J}_{\bar{\psi}}\} - Z_C\{\mathbf{J},\mathbf{J}_\psi,\mathbf{J}_{\bar{\psi}}\} &= \int \mathcal{D}[\mathbf{A}]\mathcal{D}[\mathbf{u}]\mathcal{D}[\bar{\mathbf{u}}]\mathcal{D}[\psi]\mathcal{D}[\bar{\psi}] \\ \left[-\frac{1}{\xi}C^b\{\mathbf{A};y\}\Delta C^b\{\mathbf{A};y\} - \bar{u}^b(y)\int \mathrm{d}^4z\,\frac{\delta\Delta C^b\{\mathbf{A};y\}}{\delta A^c_\mu(z)}D^{cd}_\mu u^d(z)\right]& \\ \times\exp\left(\mathrm{i}\int \mathrm{d}^4x\left[\mathcal{L}_{\text{eff}} + J^{\mu,a}A^a_\mu + J^i_\psi\psi_i - \bar{\psi}_i J^i_{\bar{\psi}}\right]\right). \end{aligned} \tag{2.5.183}$$

In order to simplify this expression we use the Slavnov–Taylor identities (2.4.56). Differentiating (2.4.56) with respect to $J^b_{\bar{u}}(y)$ and putting $\mathbf{J}_u = \mathbf{J}_{\bar{u}} = 0$ afterwards yields

$$\begin{aligned} &\int \mathcal{D}[\mathbf{A}]\mathcal{D}[\mathbf{u}]\mathcal{D}[\bar{\mathbf{u}}]\mathcal{D}[\psi]\mathcal{D}[\bar{\psi}] \\ &\quad\times\exp\left(\mathrm{i}\int \mathrm{d}^4x\left[\mathcal{L}_{\text{eff}} + J^{a,\mu}A^a_\mu + J^i_\psi\psi_i - \bar{\psi}_i J^i_{\bar{\psi}}\right]\right) \\ &\quad\times\left\{-s\bar{u}^b(y) + \mathrm{i}\bar{u}^b(y)\int \mathrm{d}^4z\left[J^{a,\mu}sA^a_\mu - J^i_\psi s\psi_i - (s\bar{\psi})_i J^i_{\bar{\psi}}\right]\right\} = 0. \end{aligned} \tag{2.5.184}$$

Operating on this equation with $\Delta C^b\{\delta/\mathrm{i}\delta\mathbf{J};y\}$, using

$$\Delta C^b\left\{\frac{\delta}{\mathrm{i}\delta\mathbf{J}},y\right\}\mathrm{i}J^{a,\mu}(z) = \frac{\delta\Delta C^b}{\delta A^a_\mu(z)}\left\{\frac{\delta}{\mathrm{i}\delta\mathbf{J}},y\right\} + \mathrm{i}J^{a,\mu}(z)\Delta C^b\left\{\frac{\delta}{\mathrm{i}\delta\mathbf{J}},y\right\}$$

and $s\bar{u}^b(y) = -C^b\{\mathbf{A};y\}/\xi$, leads to

$$\begin{aligned}&\int \mathcal{D}[\mathbf{A}]\mathcal{D}[\mathbf{u}]\mathcal{D}[\bar{\mathbf{u}}]\mathcal{D}[\psi]\mathcal{D}[\overline{\psi}] \\ &\quad\times \exp\left(\mathrm{i}\int \mathrm{d}^4x\left[\mathcal{L}_{\text{eff}} + J^a_\mu A^a_\mu + J^i_\psi\psi_i - \overline{\psi}_i J^i_{\overline{\psi}}\right]\right) \qquad (2.5.185)\\ &\quad\times\left\{\frac{1}{\xi}C^b\{\mathbf{A};y\}\Delta C^b\{\mathbf{A};y\} + \bar{u}^b(y)\int \mathrm{d}^4z\,\frac{\delta\Delta C^b\{\mathbf{A};y\}}{\delta A^c_\mu(z)}sA^c_\mu(z)\right.\\ &\quad\left. + \mathrm{i}\Delta C^b\{\mathbf{A};y\}\bar{u}^b(y)\int \mathrm{d}^4z\left[J^{a,\mu}sA^a_\mu - J^i_\psi s\psi_i - (s\overline{\psi})_i J^i_{\overline{\psi}}\right]\right\} = 0.\end{aligned}$$

Adding (2.5.183) results in

$$\begin{aligned}&Z_{C+\Delta C}\{\mathbf{J},\mathbf{J}_\psi,\mathbf{J}_{\overline{\psi}}\} - Z_C\{\mathbf{J},\mathbf{J}_\psi,\mathbf{J}_{\overline{\psi}}\} = \int \mathcal{D}[\mathbf{A}]\mathcal{D}[\mathbf{u}]\mathcal{D}[\bar{\mathbf{u}}]\mathcal{D}[\psi]\mathcal{D}[\overline{\psi}]\\ &\quad\exp\left(\mathrm{i}\int \mathrm{d}^4x\left[\mathcal{L}_{\text{eff}} + J^{\mu,a}A^a_\mu + J^i_\psi\psi_i - \overline{\psi}_i J^i_{\overline{\psi}}\right]\right) \qquad (2.5.186)\\ &\quad\times\left\{\mathrm{i}\Delta C^b\{\mathbf{A};y\}\bar{u}^b(y)\int \mathrm{d}^4z\left[J^{a,\mu}sA^a_\mu - J^i_\psi s\psi_i - (s\overline{\psi})_i J^i_{\overline{\psi}}\right]\right\}\end{aligned}$$

or to first order in ΔC

$$\begin{aligned}&Z_{C+\Delta C}\{\mathbf{J},\mathbf{J}_\psi,\mathbf{J}_{\overline{\psi}}\} = \int \mathcal{D}[\mathbf{A}]\mathcal{D}[\mathbf{u}]\mathcal{D}[\bar{\mathbf{u}}]\mathcal{D}[\psi]\mathcal{D}[\overline{\psi}]\\ &\quad\times\exp\left\{\mathrm{i}\int \mathrm{d}^4x\left[\mathcal{L}_{\text{eff}} + J^{\mu,a}\left(A^a_\mu - sA^a_\mu\int \mathrm{d}^4y\,\bar{u}^b(y)\Delta C^b\{\mathbf{A};y\}\right)\right.\right.\\ &\quad + J^i_\psi\left(\psi_i + s\psi_i\int \mathrm{d}^4y\,\bar{u}^b(y)\Delta C^a\{\mathbf{A};y\}\right)\\ &\quad\left.\left. - \left(\overline{\psi}_i + s\overline{\psi}_i\int \mathrm{d}^4y\,\bar{u}^b(y)\Delta C^a\{\mathbf{A};y\}\right)J^i_{\overline{\psi}}\right]\right\}. \qquad (2.5.187)\end{aligned}$$

As a result, the generating functional $Z_{C+\Delta C}$ differs from Z_C only in the source terms. An infinitesimal change in the gauge condition corresponds to a change in the source terms by an infinitesimal amount.

To first order in ΔC, the difference of Green functions calculated with the two different gauge-fixing terms reads

$$\left\langle T\prod_l \Psi_{I_l}\right\rangle_{C+\Delta C} - \left\langle T\prod_l \Psi_{I_l}\right\rangle_C = \tag{2.5.188}$$
$$-\sum_k \left\langle T\left(\prod_{l<k}\Psi_{I_l}\right)(s\Psi_{I_k})\int \mathrm{d}^4y\,\bar{u}^b(y)\Delta C^b\{\mathbf{A};y\}\prod_{m>k}\Psi_{I_m}\right\rangle,$$

$$= -\sum_k \tag{2.5.189}$$

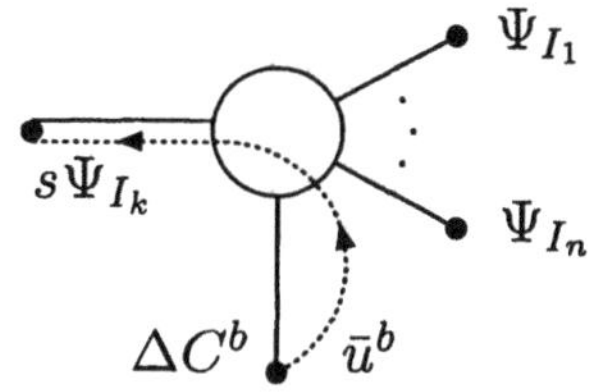

where Ψ_I represents anyone of the fields A^a_μ, ψ_i, or $\bar{\psi}_i$, and the multi-index I_l includes also the space–time coordinate x_l.

According to Sect. 2.1.3, the S-matrix elements are obtained from the Green functions by truncating the external legs and multiplying with the appropriate physical polarization vectors and wave-function renormalization constants. Thus, only terms that involve poles in each of the external lines contribute. These terms correspond to diagrams that are one-particle reducible in each of the external lines, i.e. in which each external line is connected to the rest of the diagram via a single line. This must hold, in particular, for the line with the BRS variation at the end. All diagrams in which the single line is located between the BRS insertion $(s\Psi_{i_k})$ and the composite operator $\int \mathrm{d}^4y\,\bar{u}^b(y)\Delta C^a\{\mathbf{A};y\}$ do not contribute to the S matrix because such contributions are only non-zero for unphysical field components [cf. (2.5.165)]

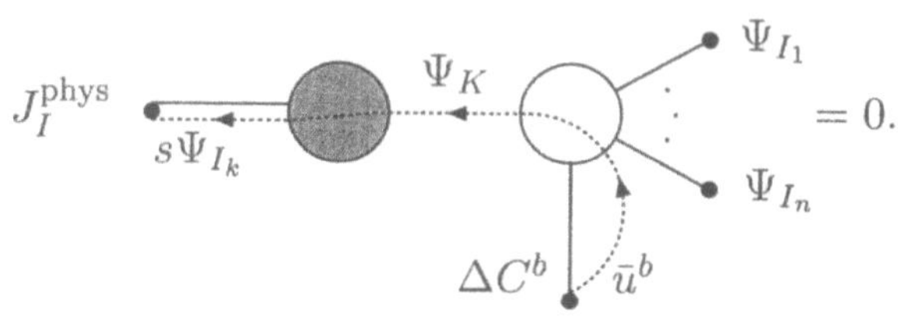

(2.5.190)

Therefore, the only contributions to the S matrix arise from diagrams in which the product of operators $(s\Psi_{I_k})\int \mathrm{d}^4y\,\bar{u}^b(y)\Delta C^a\{\mathbf{A};y\}$ is separated from the rest of the diagram by a single line, i.e. from diagrams of the structure

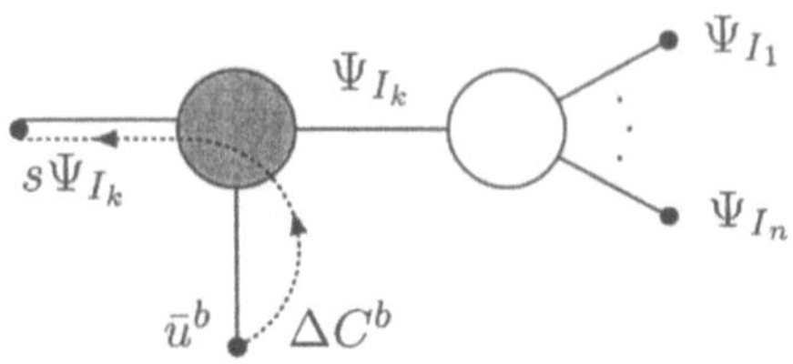

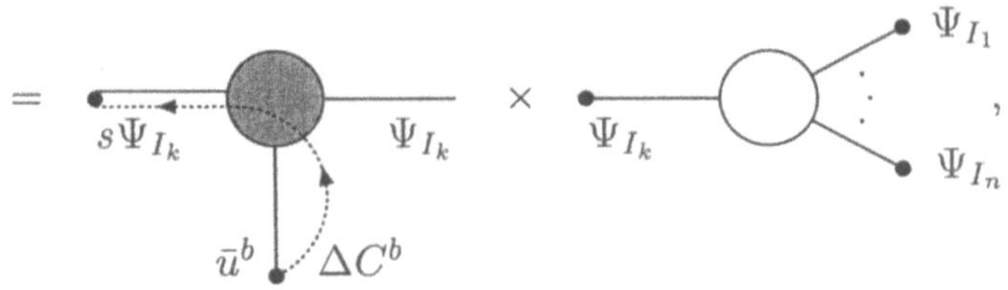

(2.5.191)

which factorize into the original Green function and a factor

$r_{I_k} =$

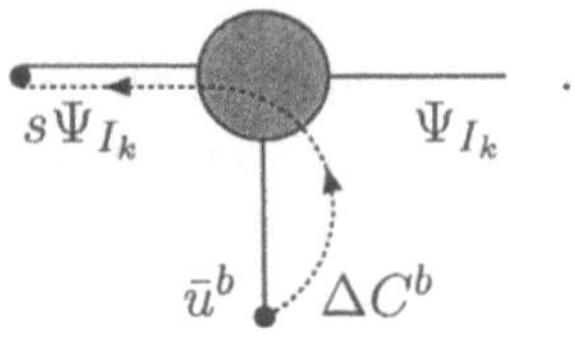

(2.5.192)

We have assumed that the mass of the field Ψ_{I_k} does not coincide with any other physical mass so that only ψ_{I_k} contributes. On shell, the factor r_{I_k} is a constant. Thus, we arrive at

$$\left\langle T \prod_l \Psi_{I_l} \right\rangle_{C+\Delta C} = \left(\prod_l (1 + r_{I_l}) \right) \left\langle T \prod_l \Psi_{I_l} \right\rangle_C$$
$$+ \text{ on-shell-irrelevant terms.} \tag{2.5.193}$$

The wave-function renormalization constants are determined from the two-point functions according to (2.1.34) and are related by

$$R_I^{C+\Delta C} = (1 + r_I)^2 R_I^C . \tag{2.5.194}$$

Thus, when going to the S matrix [cf. (2.1.33)] the extra (gauge-dependent) constants r_I cancel between the Green functions and the wave-function renormalization constants and the S matrix is gauge-independent [Le76].

2.5.4.5 Unitarity of the S matrix

Formally the unitarity of the S matrix can be proven very easily once the physical states have been defined [Ku78]. If the hermiticity assignment of the Faddeev–Popov ghosts is done appropriately (2.5.157), the effective Lagrangian is hermitian and the full S matrix is unitary, i.e. $SS^\dagger = 1$. The physical S matrix is defined[15] as the full S matrix projected onto physical states, represented by states $|\psi_a^{(1)}\rangle \in \mathcal{V}_1$. Since the *BRS charge* $Q_{\rm BRS}$ commutes with S, we have

$$Q_{\rm BRS} S|\psi_a^{(1)}\rangle = 0, \qquad \text{i.e.} \quad S|\psi_a^{(1)}\rangle \in \mathcal{V}_{\rm phys}. \tag{2.5.195}$$

Thus, we have, owing to (2.5.158),

$$\langle\psi_b^{(1)}|\psi_a^{(1)}\rangle = \langle\psi_b^{(1)}|S^\dagger S|\psi_a^{(1)}\rangle = \sum_c \langle\psi_b^{(1)}|S^\dagger|\psi_c^{(1)}\rangle\langle\psi_c^{(1)}|S|\psi_a^{(1)}\rangle. \tag{2.5.196}$$

The states $\psi^{(0)}$ do not contribute in (2.5.196) owing to (2.5.195) and (2.5.158). Therefore, (2.5.196) expresses the unitarity of the S matrix projected to the subspace of physical states, i.e. of the physical S matrix.

Unitarity can also be proven directly by exploiting the Ward identities [tH71, Ta71, Co75]. Having established the gauge independence of the S matrix, this can be done by considering a convenient gauge. Of course, also in this approach the definition of physical states or sources is inevitable.

For gauge theories in which all gauge fields acquire a mass by spontaneous breaking of the gauge symmetry (cf. Sect. 4.1), a simple proof of unitarity is possible [tH72]. In this case, a gauge exists in which the masses of all corresponding unphysical degrees of freedom reside at infinity. Therefore, they do not contribute to the S matrix and the latter is unitary. This so-called unitary gauge is non-renormalizable according to power counting because

[15] More rigorously, the physical S matrix $S_{\rm phys}$ is defined as the quotient mapping of the full S matrix into the physical Hilbert space (2.5.159) [Ku78].

the corresponding propagators (1.4.13) do not behave as $1/k^2$ for $k \to \infty$. However, this special gauge can be continuously transformed into a renormalizable gauge in such a way that in all intermediate gauges the theory is well-defined. Then, the gauge independence of the S matrix ensures that the theory is unitary also in renormalizable gauges.

2.6 Renormalization group

Relativistic quantum field theories require, as we have seen, regularization and renormalization. Renormalization is a reparametrization of the theory which at least has to compensate possible divergences. This reparametrization is not unique. It can be fixed by renormalization conditions in different ways. Physical results are independent of the choice of renormalized parameters, i.e. a reparametrization of the coupling constant g and mass m, $g \to g'$, $m^2 \to m'^2$, only changes the explicit expressions for physical quantities but not the relations between them. This fact is the basis of the *renormalization group* (RG) which quite generally describes the composition law of the renormalization transformations. For simple transformations like changes in the *renormalization point*, $\mathcal{M} \to \mathcal{M}+\mathrm{d}\mathcal{M}$, and corresponding changes of the parameters, $g \to g + \mathrm{d}g$, $m^2 \to m^2 + \mathrm{d}m^2$, etc., one can derive a differential equation, the *renormalization-group equation* [St53, Ge54] which describes the effects of this change on Green functions or vertex functions.

2.6.1 Renormalization-group equation

In this section we derive and discuss the equation which describes the behaviour of vertex functions under a shift $\mathcal{M} \to \mathcal{M} + \mathrm{d}\mathcal{M}$ of the renormalization point. First we consider a massive scalar field theory. Starting point is the independence of physical quantities under reparametrizations of the theory (cf. Sect. 2.5.1). This condition reads for S-matrix elements

$$\mathcal{M}\frac{\mathrm{d}}{\mathrm{d}\mathcal{M}}\langle \mathrm{out}|S|\mathrm{in}\rangle = 0. \tag{2.6.1}$$

2.6.1.1 Derivation of the renormalization-group equation

The connected parts of the S-matrix elements are obtained from the corresponding truncated renormalized Green functions $\tilde{G}^{(n)}_{\mathrm{R,trunc}}(p_i; g, m; \mathcal{M})$ multiplied with the wave-function renormalization constants $R^{1/2}$ according to (2.1.35). The truncated Green functions are products of vertex functions and propagators. As a consequence, the vertex functions behave exactly as the truncated Green functions under infinitesimal changes of the renormalization point $\mathcal{M}$. The *invariance of the S-matrix elements* (2.6.1) implies

$$\begin{aligned} R^{n/2}(\mathcal{M})\Gamma^n_{\mathrm{R}}\Big(p_k; g(\mathcal{M}), m(\mathcal{M}); \mathcal{M}\Big) &= R^{n/2}(\mathcal{M}+\mathrm{d}\mathcal{M}) \\ \times\Gamma^n_{\mathrm{R}}\Big(p_k; g(\mathcal{M}+\mathrm{d}\mathcal{M}), m(\mathcal{M}+\mathrm{d}\mathcal{M}); \mathcal{M}+\mathrm{d}\mathcal{M}\Big), \end{aligned}$$

where

$$g = g\Big(g_0\mu^\delta, m_0, \mu, \delta; \mathcal{M}\Big), \qquad m = m\Big(g_0\mu^\delta, m_0, \mu, \delta; \mathcal{M}\Big) \tag{2.6.2}$$

with $\delta = 2 - D/2$. The bare parameters g_0, m_0 and the parameters for dimensional regularization μ, δ are kept fixed.[16] Expanding (2.6.2) gives

$$\mathcal{M}\left(\frac{\partial}{\partial\mathcal{M}} + \frac{\partial g}{\partial\mathcal{M}}\frac{\partial}{\partial g} + \frac{\partial m}{\partial\mathcal{M}}\frac{\partial}{\partial m} + \frac{n}{2}\frac{1}{R}\frac{\partial R}{\partial\mathcal{M}}\right)\ \Gamma^{(n)}_{\mathrm{R}}\Big(p_k; g, m; \mathcal{M}\Big) = 0. \tag{2.6.3}$$

Defining the *β-function*, also called *renormalization-group function*,

$$\beta(g, m; \mathcal{M}) = \lim_{\delta\to 0}\mathcal{M}\frac{\partial}{\partial\mathcal{M}}g\Big(g_0\mu^\delta, m_0, \mu, \delta; \mathcal{M}\Big), \tag{2.6.4}$$

the *renormalization-group coefficient for the mass* term

$$\gamma_m(g, m; \mathcal{M}) = -\lim_{\delta\to 0}\frac{1}{m}\mathcal{M}\frac{\partial}{\partial\mathcal{M}}m\Big(g_0\mu^\delta, m_0, \mu, \delta; \mathcal{M}\Big) \tag{2.6.5}$$

and the *anomalous dimension*

$$\gamma(g, m; \mathcal{M}) = -\lim_{\delta\to 0}\frac{1}{2R}\mathcal{M}\frac{\partial}{\partial\mathcal{M}}R\Big(g_0\mu^\delta, m_0, \mu, \delta; \mathcal{M}\Big), \tag{2.6.6}$$

[16] In other regularization schemes they have to be replaced by the corresponding cut-off quantities.

we obtain the *renormalization group equation*

$$\left(\mathcal{M}\frac{\partial}{\partial\mathcal{M}}+\beta\frac{\partial}{\partial g}-\gamma_m m\frac{\partial}{\partial m}-n\gamma\right)\Gamma^n_{\mathrm{R}}\left(p_k;g,m;\mathcal{M}\right)=0. \tag{2.6.7}$$

The functions $\beta(g,m;\mathcal{M})$, $\gamma_m(g,m;\mathcal{M})$, and $\gamma(g,m;\mathcal{M})$ are the same for all vertex functions and thus a characteristic of the theory.

The anomalous dimension $\gamma(g,m;\mathcal{M})$ reflects the anomaly connected with the dilatation transformation of massless theories (cf. Sect. 2.7). This anomalous breaking is controlled by the Callan–Symanzik equation (2.6.13).

The renormalization-group (RG) equation can alternatively be obtained from the fact that the unrenormalized vertex functions

$$\begin{aligned}\Gamma^n&\left(p_k;g_0\mu^\delta,m_0,\mu,\delta\right)=\\&Z^{-n/2}\left(g(g_0\mu^\delta,m_0,\mu,\delta;\mathcal{M}),m(g_0\mu^\delta,m_0,\mu,\delta;\mathcal{M});\mathcal{M}\right)\\&\times\Gamma^n_{\mathrm{R}}\left(p_k;g(g_0\mu^\delta,m_0,\mu,\delta;\mathcal{M}),m(g_0\mu^\delta,m_0,\mu,\delta;\mathcal{M});\mathcal{M}\right)\end{aligned} \tag{2.6.8}$$

are independent of the renormalization point $\mathcal{M}$. Applying $\mathcal{M}\partial/\partial\mathcal{M}$ yields (2.6.7) with (2.6.4), (2.6.5), and

$$\gamma(g,m;\mathcal{M})=\lim_{\delta\to 0}\frac{1}{2}Z^{-1}\mathcal{M}\frac{\partial}{\partial\mathcal{M}}Z\left(g_0\mu^\delta,m_0,\mu,\delta;\mathcal{M}\right). \tag{2.6.9}$$

Using the definition of R, (2.1.34), and the relation between the renormalized and unrenormalized propagator, we find

$$\begin{aligned}R&=-\mathrm{i}(p^2-m^2)G_{\mathrm{R}}(p,-p)|_{p^2=m^2}\\&=-\mathrm{i}(p^2-m^2)Z^{-1}G(p,-p)|_{p^2=m^2}.\end{aligned} \tag{2.6.10}$$

The independence of the bare propagator $G(p,-p)$ on $\mathcal{M}$ implies the RG invariance of $R\times Z$ and thus the equality of (2.6.6) and (2.6.9). This shows that the RG invariance of the S matrix and of the bare Green functions are equivalent.

We have derived the RG equation for the case of a scalar theory with one coupling and one mass parameter. Generalization to several couplings g_i, masses m_α, n_B bosonic fields, and n_f fermionic fields is straightforward and yields

$$\left(\mathcal{M}\frac{\partial}{\partial\mathcal{M}} + \beta_j\frac{\partial}{\partial g_j} - \gamma_{m,\beta}m_\beta\frac{\partial}{\partial m_\beta} - n_B\gamma_B - n_f\gamma_f\right) \\ \times\Gamma_{\mathrm{R}}^{n_B,n_f}\left(p_k; g_i, m_\alpha; \mathcal{M}\right) = 0. \tag{2.6.11}$$

In the case of gauge theories the vertex functions also depend on the gauge-fixing parameter ξ, and an additional term $\gamma_\xi\xi\partial/\partial\xi$ has to be incorporated in (2.6.11) [Gr73a].

An elementary dimensional consideration can be used to reshape (2.6.11). The usual mass dimensions for fermion fields, $\dim\psi = M^{3/2}$, and boson fields, $\dim B = M^1$, given in Table 2.1, imply that $\dim\Gamma^{n_B,n_f} = M^{4-n_B-3/2n_f}$ (the δ-function for energy-momentum conservation is split off). Euler's differential equation for homogeneous functions gives

$$\left(\lambda\frac{\partial}{\partial\lambda} + m_\beta\frac{\partial}{\partial m_\beta} + \mathcal{M}\frac{\partial}{\partial\mathcal{M}} + n_B + \frac{3}{2}n_f - 4\right) \\ \times\Gamma_{\mathrm{R}}^{n_B,n_f}(\lambda p_k,; g_i, m_\alpha; \mathcal{M}) = 0. \tag{2.6.12}$$

Thus, if $\mathcal{M}\partial/\partial\mathcal{M}$ in (2.6.11) is replaced by $\lambda\partial/\partial\lambda$ using (2.6.12), we obtain the *Callan–Symanzik equation* [Ca70, Sy70]

$$\left(\lambda\frac{\partial}{\partial\lambda} - \beta_j\frac{\partial}{\partial g_j} + (\gamma_{m,\beta}+1)m_\beta\frac{\partial}{\partial m_\beta} + n_B(1+\gamma_B) \right. \\ \left. + n_f\left(\frac{3}{2}+\gamma_f\right) - 4\right)\Gamma_{\mathrm{R}}^{n_B,n_f}(\lambda p_k; g_i, m_\alpha; \mathcal{M}) = 0. \tag{2.6.13}$$

Here, the renormalization point $\mathcal{M}$ is kept fixed. The Callan–Symanzik equation (2.6.13) describes the behaviour of the vertex functions under scale transformations of the four-momenta. The breaking of the naïve scaling behaviour is contained in the anomalous dimensions γ_B and γ_f. This becomes clear when (2.6.13) is solved.

2.6.1.2 Solution of the RG equation

The RG equation (2.6.13) is a linear, partial differential equation of first order which can be solved by the method of *characteristic curves* [Co68]. The characteristic curves $\bar{g}_i(t)$ and $\overline{m}_\alpha(t)$, where $t = \log\lambda$, are determined from the system of ordinary differential equations

$$\frac{\mathrm{d}\bar{g}_i(t)}{\mathrm{d}t} = \beta_i(\bar{g}_j, \overline{m}_\beta, t), \quad \frac{\mathrm{d}\log\overline{m}_\alpha}{\mathrm{d}t} = -\gamma_{m,\alpha}(\bar{g}_j, \overline{m}_\beta, t) - 1, \quad (2.6.14)$$

together with the initial conditions $\bar{g}_i(0) = g_i$, $\overline{m}_\alpha(0) = m_\alpha$.

The explicit solution of (2.6.13) reads with $t = \log\lambda$

$$\begin{aligned}\Gamma_{\mathrm{R}}^{n_B,n_f}(\lambda p_k; g_i, m_\alpha; \mathcal{M}) &= \Gamma_{\mathrm{R}}^{n_B,n_f}(p_k; \bar{g}_i(t), \overline{m}_\alpha(t); \mathcal{M})\, \lambda^{4-n_B-3n_f/2} \\ &\times \exp\Big(-\int_0^t \mathrm{d}\tau \Big[n_B\gamma_B\big(\bar{g}_j(\tau), \overline{m}_\beta(\tau), \tau\big) \\ &\quad + n_f\gamma_f\big(\bar{g}_j(\tau), \overline{m}_\beta(\tau), \tau\big)\Big]\Big). \qquad (2.6.15)\end{aligned}$$

The solution of the RG equation shows that a scale change in the momenta by a factor λ can be expressed by a change in the coupling constant and in the mass, and by a multiplicative factor. This factor differs from that of homogeneous functions with canonical dimension, i.e. $\lambda^{4-n_B-3n_f/2}$. The contribution coming from the dynamical dimensions γ_B and γ_f *violate naïve scaling*.

The RG equation is very similar to the continuity equation between a density $\rho(\mathbf{x}, t)$ and a current density $\mathbf{j}(\mathbf{x}, t) = \rho(\mathbf{x}, t)\mathbf{v}(\mathbf{x}, t)$,

$$\frac{\partial\rho}{\partial t} + \nabla(\rho\mathbf{v}) = \frac{\partial\rho}{\partial t} + v_i\frac{\partial\rho}{\partial x_i} + \rho\nabla\mathbf{v} = 0. \qquad (2.6.16)$$

In this case the characteristic curves are the flow lines.

As an example we solve the RG equation for the simple case of one field with one coupling constant g and one mass m, where the β- and γ-functions depend only on g, $\beta = \beta(g)$, $\gamma_m = \gamma_m(g)$, and $\gamma = \gamma(g)$. The differential equations for the characteristic curves

$$\frac{\mathrm{d}\bar{g}}{\mathrm{d}t} = \beta(\bar{g}), \qquad \frac{\mathrm{d}\log\overline{m}}{\mathrm{d}t} = -\gamma_m(\bar{g}) - 1$$

decouple in this case and can easily be solved

$$t = \int_g^{\bar{g}(t)} \frac{dg'}{\beta(g')}, \tag{2.6.17}$$

$$\log\left(\frac{\overline{m}(t)}{m}\right) = -\int_g^{\bar{g}(t)} \frac{dg'}{\beta(g')}\Big(\gamma_m(g')+1\Big), \tag{2.6.18}$$

implicitly yielding the *running coupling* $\bar{g}(t)$ and *running mass* $\overline{m}(t)$. Inserting this in (2.6.15) we finally obtain

$$\Gamma^n_{\mathrm{R}}(\lambda p_k; g, m) = \Gamma^n_{\mathrm{R}}(p_k; \bar{g}(t), \overline{m}(t))\lambda^{4-n} \exp\left[-n\int_g^{\bar{g}(t)} \frac{dg'}{\beta(g')}\gamma(g')\right]. \tag{2.6.19}$$

The typical behaviour of the RG functions in one-loop approximation is

$$\beta(g) = -\beta_0 \frac{g^3}{16\pi^2}, \qquad \gamma_m(g) = \gamma_{m,0}\frac{g^2}{16\pi^2}, \qquad \gamma(g) = \gamma_0 \frac{g^2}{16\pi^2}, \tag{2.6.20}$$

leading to

$$t = \frac{8\pi^2}{\beta_0}\left(\frac{1}{\bar{g}(t)^2} - \frac{1}{g^2}\right) \tag{2.6.21}$$

or

$$\bar{g}^2(t) = \frac{g^2}{1+\frac{g^2}{16\pi^2}2\beta_0 t}, \tag{2.6.22}$$

and

$$\overline{m}(t) = m\left(\frac{\bar{g}^2(t)}{g^2}\right)^{\gamma_{m,0}/2\beta_0} \mathrm{e}^{-t}. \tag{2.6.23}$$

The vertex functions have the scaling behaviour

$$\Gamma^n_{\mathrm{R}}(\lambda p_k; g, m) = \Gamma^n_{\mathrm{R}}(p_k; \bar{g}(t), \overline{m}(t))\lambda^{4-n}\left(\frac{\bar{g}^2(t)}{g^2}\right)^{n\gamma_0/2\beta_0}. \tag{2.6.24}$$

The scaling behaviour (2.6.24) holds also for truncated Green functions. Note that on the right-hand side of (2.6.24) the possibly large terms involving $\log\lambda$ are re-summed [cf. (2.6.22)]. Because of this *re-summation*, the RG can be used to improve finite-order predictions.

2.6.1.3 Fixed points

The structure of quantum field theories for large or small momenta is given by the limits of the running coupling constant $\bar{g}(\lambda, g)$ for large or small λ. Concerning large momenta, it is important to note that mass terms are insignificant for the behaviour of the vertex functions for large euclidean values of the momenta, $p^2 \to -\infty$. This has been confirmed in perturbation theory [We60, Sy70]. Therefore, the behaviour for large momenta can be studied neglecting the mass terms in the Lagrangian and in the RG equation. Thus, for large λ we can put $\gamma_m(g) = 0$. Then the β-function and anomalous dimension of the corresponding massless theory, $\beta(g)$ and $\gamma(g)$, provide the leading terms for the massive theory [We73a, Sy70]. The running coupling constant $\bar{g}$ and the anomalous dimension $\gamma(g)$ thus determine the behaviour of the vertex functions for large values of λ. The corresponding values of the running coupling constant are determined by the fixed points of the characteristic equation (2.6.14), i.e. by those values of g for which $\beta(g) = 0$.

In order to determine the behaviour of $\bar{g}$ around a fixed point g^*, $\beta(g)$ is expanded around g^*

$$\beta(g) = (g - g^*)^r \beta', \tag{2.6.25}$$

where in perturbation theory $r \geq 1$. Equation (2.6.17) gives

$$\begin{aligned} \bar{g} - g^* &= (g - g^*)\, \lambda^{\beta'}, \quad r = 1; \\ (\bar{g} - g^*)^{r-1} &= \frac{(g - g^*)^{r-1}}{1 - \beta'(r-1)(g - g^*)^{r-1} \log \lambda}, \quad r = 2, 3, \ldots \end{aligned} \tag{2.6.26}$$

for the behaviour of the running coupling constant around the fixed point. Depending on the sign of β' the fixed point g^* is approached in the following way

$$\bar{g}(\lambda, g) \to g^* \quad \text{for} \quad \begin{cases} \lambda \to \infty, & \text{if} \quad \beta' < 0, \\ \lambda \to 0, & \text{if} \quad \beta' > 0. \end{cases} \tag{2.6.27}$$

In the first case, $\beta' < 0$, g^* is called an *ultraviolet-stable fixed point*, for $\beta' > 0$ it is an *infrared-stable fixed point.*

The vertex functions are homogeneous functions of the order $4 - (3/2 + \gamma_f(g^*))n_f - (1 + \gamma_B(g^*))n_B$ at the stable fixed point, in agreement with the scale-change equation (2.6.15). Thus, the shape of the RG function $\beta(g)$,

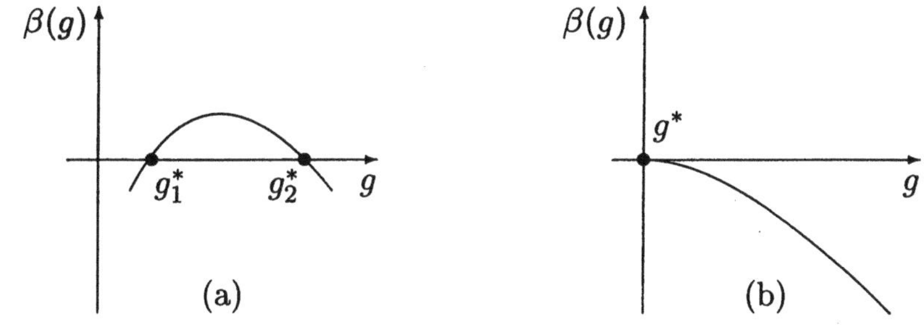

Fig. 2.10 Examples for fixed points of the RG equation

illustrated in Fig. 2.10, the *renormalization group flow*, determines the general properties of a field theory.

Theory (a) has an UV-stable fixed point g_2^* and an IR-stable one g_1^* with $r = 1$. If the renormalized coupling constant g lies between g_1^* and g_2^*, then the theory approaches the corresponding limits with homogeneous vertex functions for large and small distances, respectively.

Theory (b) —for which as we shall see QCD is an example—has an UV-stable fixed point at $g^* = 0$ with $\beta' < 0$. This type of theory is called *asymptotically free* since for large momenta, i.e. small distances, its Green functions have the scaling behaviour of a free theory, which is modified only by logarithmic terms for $r > 1$ as (2.6.26) shows (in perturbation theory $r = 3$). Equation (2.6.27) implies that $\bar{g}(\lambda, g) \to \infty$ for $\lambda \to 0$. Theory (b) is IR-unstable.

The important special cases of QED and QCD are treated in Sects. 2.6.2.1 and 2.6.2.2.

If $\beta(g)$ and $\gamma(g)$ are calculated in one-loop approximation, the RG equation provides the asymptotic *summation of the leading logarithms* to all orders of g and thus improves the perturbation series in the asymptotic region. For small momenta, mass terms can have a considerable effect on the behaviour of Green functions.

2.6.1.4 Renormalization invariance of physical quantities

The starting point of our considerations was the fact that S-matrix elements are invariant under RG transformations (2.6.1). This holds for all physical quantities. So, if a measurable physical quantity O, e.g. a mass, is calculated in gauge theories, it depends on g, m, and $\mathcal{M}$, i.e. $O = O(g, m, \mathcal{M})$, but only in a combination that is invariant under renormalization [Gr74, Ca78]

$$\left(\mathcal{M}\frac{\partial}{\partial \mathcal{M}} + \beta(g)\frac{\partial}{\partial g} - \gamma_m m\frac{\partial}{\partial m}\right) O(g, m, \mathcal{M}) = 0, \tag{2.6.28}$$

as a change in $\mathcal{M}$ must be compensated by a change in the meaning of g and m.

In a massless theory like QCD, any mass M must be proportional to $\mathcal{M}$ for dimensional reasons. Therefore, it must have the *renormalization–invariant form*

$$M = \mathcal{M} \exp\left(-\int^g \frac{dg'}{\beta(g')}\right). \tag{2.6.29}$$

The specific value of M is determined by the integration constant.

2.6.2 Renormalization-group function and anomalous dimensions of massless gauge theories

As an example we consider *massless gauge theories* with fermions. In order to determine the RG function $\beta(g)$ and the anomalous dimensions γ^f and γ^A of fermion and gauge-boson fields in one-loop approximation, we need the *renormalization constants* Z_g, Z_ψ, and Z_A. For QCD in Feynman gauge these were calculated in Sect. 2.5.3.3. All three renormalization constants depend on the renormalization point $\mathcal{M}$ in the form $f(\delta)(4\pi\mu^2/\mathcal{M}^2)^\delta$. We need the logarithmic derivative of this quantity

$$\mathcal{M}\frac{\partial}{\partial \mathcal{M}}\left[f(\delta)\left(\frac{4\pi\mu^2}{\mathcal{M}^2}\right)^\delta\right] = -2\delta f(\delta)\left(\frac{4\pi\mu^2}{\mathcal{M}^2}\right)^\delta \tag{2.6.30}$$

for $\delta \to 0$, and consequently only the $1/\delta$ term of $f(\delta)$. In a general massless gauge theory with fermions in a representation R the relevant terms read

$$\begin{aligned}
Z_g &= 1 - \frac{g^2}{16\pi^2}\left(\frac{4\pi\mu^2}{\mathcal{M}^2}\right)^\delta \left(\frac{11}{6}C_A - \frac{2}{3}T_R\right)\frac{1}{\delta}, \\
Z_A &= 1 - \frac{g^2}{16\pi^2}\left(\frac{4\pi\mu^2}{\mathcal{M}^2}\right)^\delta \left\{\frac{4}{3}T_R - C_A\left[\frac{5}{3} + \frac{1}{2}(1-\xi)\right]\right\}\frac{1}{\delta}, \\
Z_\psi &= 1 - \frac{g^2}{16\pi^2}\left(\frac{4\pi\mu^2}{\mathcal{M}^2}\right)^\delta C_R\xi\frac{1}{\delta}
\end{aligned} \tag{2.6.31}$$

for arbitrary values of the gauge parameter ξ.[17] The Casimir operators C_{A}, C_R, and the Dynkin index T_R are defined in Sect. 1.2.2.2. The renormalization of the coupling constant g is, as it should be, independent of the gauge parameter ξ.

Using (2.6.31) and the independence of the bare coupling $g_0 = Z_g g$ of $\mathcal{M}$, we obtain

$$\begin{aligned}\beta(g) &= \lim_{\delta\to 0} \mathcal{M}\frac{\partial}{\partial\mathcal{M}} g(g_0, \mu^\delta, \mu, \delta; \mathcal{M}) = -g\mathcal{M}\left(\frac{\partial}{\partial\mathcal{M}}\log Z_g\right) \\ &= -\frac{g^3}{16\pi^2}\left(\frac{11}{3}C_{\mathrm{A}} - \frac{4}{3}T_{\mathrm{R}}\right) \qquad (2.6.32)\end{aligned}$$

for the *β-function of a massless gauge theory*, and in a similar way

$$\begin{aligned}\gamma_A(g) &= -\frac{g^2}{16\pi^2}\left(\left[\frac{5}{3} + \frac{1}{2}(1-\xi)\right]C_{\mathrm{A}} - \frac{4}{3}T_{\mathrm{R}}\right), \\ \gamma_f(g) &= \frac{g^2}{16\pi^2}C_{\mathrm{R}}\xi \qquad (2.6.33)\end{aligned}$$

for the *anomalous dimensions* of the gauge-boson and fermion fields.

According to the criterion for the β-function formulated in Sect. 2.6.1.3, $g = 0$ is a fixed point. Non-abelian, massless gauge theories are *asymptotically free*, provided the number of fermions is not too large. The gauge-boson self-coupling, which gives the term proportional to C_{A} and thus the negative sign to the RG function, is responsible for the asymptotic freedom of non-abelian gauge theories.

The β-function of non-abelian, massless gauge theories has been calculated up to the four-loop approximation. The corresponding expansions of the RG function $\beta(g)$

$$\begin{aligned}\beta(g) &= -\beta_0\frac{g^3}{16\pi^2} - \beta_1\frac{g^5}{(16\pi^2)^2} - \beta_2\frac{g^7}{(16\pi^2)^3} + \dots \\ &= -g\left(\beta_0\frac{\alpha_s}{4\pi} + \beta_1\left(\frac{\alpha_s}{4\pi}\right)^2 + \beta_2\left(\frac{\alpha_s}{4\pi}\right)^3 + \dots\right) \qquad (2.6.34)\end{aligned}$$

[17] $\xi = 1$ and $\xi = 0$ correspond to the *Feynman and Landau gauge*, respectively.

and of the anomalous dimensions

$$\gamma(g) = \gamma_0 \frac{g^2}{16\pi^2} + \gamma_1 \frac{g^4}{(16\pi^2)^2} + \dots \tag{2.6.35}$$

defines the coefficients β_0, β_1, $\beta_2, \dots$ and $\gamma_0, \gamma_1, \dots$. The one-loop coefficients β_0 and γ_0 read

$$\begin{aligned} \beta_0 &= \frac{11}{3} C_{\mathrm{A}} - \frac{4}{3} T_R, \\ \gamma_{A,0} &= -\frac{5}{3} C_{\mathrm{A}} - \frac{1-\xi}{2} C_{\mathrm{A}} + \frac{4}{3} T_R, \\ \gamma_{f,0} &= C_{\mathrm{R}} \xi. \end{aligned} \tag{2.6.36}$$

The two-loop coefficient β_1 is given by [Ca74, Jo74]

$$\beta_1 = \frac{34}{3} (C_{\mathrm{A}})^2 - \frac{20}{3} C_{\mathrm{A}} T_R - 4 C_R T_R. \tag{2.6.37}$$

The three-loop coefficient β_2 is scheme-dependent. Its value in the MS-scheme is [Ta80]

$$\begin{aligned} \beta_2 = {} & \frac{2857}{54} (C_{\mathrm{A}})^3 - \frac{1415}{27} (C_{\mathrm{A}})^2 T_R + \frac{158}{27} C_{\mathrm{A}} (T_R)^2 \\ & - \frac{205}{9} C_{\mathrm{A}} C_R T_R + \frac{44}{9} C_R (T_R)^2 + 2 (C_R)^2 T_R. \end{aligned} \tag{2.6.38}$$

The four-loop coefficient β_3 has been calculated in Ref. [Ri97].

2.6.2.1 QED

Since QED is abelian we have $C_{\mathrm{A}} = 0$, and the β-function has a positive sign. This means that QED is not asymptotically free. The effective coupling constant

$$\bar{\alpha}(Q^2) = \frac{\alpha}{1 - \frac{\alpha}{3\pi} \log\left(\frac{Q^2}{m_{\mathrm{e}}^2}\right)} \tag{2.6.39}$$

increases as distance becomes smaller, i.e. as momenta become larger. This leads to a screening of the unrenormalized (bare) charge [La55, Be76, No78]. Vacuum polarization by virtual electron–positron pairs screens the bare charge as the latter attracts particles with opposite charge and repels those with equal charge. So, from a given distance, the bare charge is seen surrounded by a cloud of charges with opposite sign. The physical electrical charge measured in a sphere with this radius is smaller than the bare point charge. The one-loop result for the running coupling constant of QED diverges for $Q^2 \approx m_e^2 \exp(3\pi/\alpha)$. This Landau pole is an indication of the breakdown of perturbation theory of pure QED at very high energies.

On the other hand QED gets free for large distances, i.e. small energies. This and *Thirring's theorem* [Th53] guarantee that the value of the electric charge e can be determined from the *Thomson cross section* without any corrections.

2.6.2.2 QCD

The gauge group of QCD is $SU(3)_C$. The coloured quark fields occur in N_f fundamental representations of $SU(3)_C$, i.e. in N_f flavours. This means $C_A = 3$, $C_R = C_F = 4/3$, and $T_R = N_f T_F = N_f/2$. Thus, we obtain for the one-loop RG coefficients

$$\begin{aligned} \beta_0 &= 11 - \frac{2}{3} N_f, \\ \gamma_{A,0} &= -\left(\frac{13}{2} - \frac{3}{2}\xi - \frac{2}{3} N_f\right), \\ \gamma_{q,0} &= \frac{4}{3}\xi, \end{aligned} \tag{2.6.40}$$

and the coefficients β_1, β_2

$$\beta_1 = 102 - \frac{38}{3} N_f, \qquad \beta_2 = \frac{2857}{2} - \frac{5033}{18} N_f + \frac{325}{54} (N_f)^2 . \tag{2.6.41}$$

The sign of the one-loop β-function of QCD changes for $N_f = 33/2$. If the number of quarks is below 33/2, then QCD is asymptotically free at one-loop order. Provided that quarks are not too numerous, $\beta(g)$ remains negative in higher orders.

The negative sign of the β-function of QCD may be explained by the following physical mechanism. Also in non-abelian gauge theories vacuum polarization by the Coulomb interaction of fermions and charged, transverse, i.e.

physically polarized, gauge-boson pairs causes colour-charge screening. However, the self-interaction of the gauge bosons generates an additional dipole term between a transverse and a longitudinal gauge-boson. This dipole is oriented so that it makes an antiscreening contribution which is larger by a factor of 12 [Ap77].

2.6.2.3 Theories with scalars

All known renormalizable theories which are asymptotically free are non-abelian gauge theories. Gauge theories for the electroweak interaction and the unified strong and electroweak interaction contain scalar fields in addition to fermion fields. Their gauge-invariant coupling with the gauge bosons makes a positive contribution to the RG function which is similar in structure to that of the fermions and which destroys asymptotic freedom when the number of fermions and scalars is too high. Moreover, the electroweak theories require scalar-field self-interactions. This implies that they are no longer asymptotically free [Co73].

2.6.2.4 Calculation of the β-function in the background-field method

The β-functions of gauge couplings can be calculated very efficiently within the background-field method [Ab81]. As discussed in Sect. 2.5.1.8, in the background-field method the renormalization constant of the gauge coupling is related to the renormalization constant of the background gauge field according to

$$Z_g = Z_{\hat{A}}^{-1/2}. \tag{2.6.42}$$

As a consequence, the β-function is directly linked to the *anomalous dimension* of the background gauge field

$$\begin{aligned}\beta(g) &= -g\mathcal{M}\left(\frac{\partial}{\partial\mathcal{M}}\log Z_g\right)\\ &= \frac{g}{2}\mathcal{M}\left(\frac{\partial}{\partial\mathcal{M}}\log Z_{\hat{A}}\right) = g\gamma_{\hat{A}}(g).\end{aligned} \tag{2.6.43}$$

The renormalization constant for the background field $\gamma_{\hat{A}}(g)$ can be calculated from the background-gauge-field two-point function similarly as in the conventional formalism (cf. Sect. 2.5.3.3). The non-vanishing Feynman graphs are given by Fig. 2.9a), b), and c) on page 226 with external background fields. While graph (a) gives the same result as in the conventional formalism, graphs (b) and (c) yield different results owing to the modified Feynman rules. It turns out that each of the graphs (b) and (c) yields a transverse contribution. We finally obtain for the background-field renormalization constant

$$Z_{\hat{A}} = 1 + \frac{g^2}{32\pi^2}\left(\frac{4\pi\mu^2}{\mathcal{M}^2}\right)^{\delta}\left(\frac{11}{6}C_{\mathrm{A}} - \frac{2}{3}T_R\right)\frac{1}{\delta}, \tag{2.6.44}$$

and for the corresponding anomalous dimension

$$\gamma_{\hat{A}}(g) = -\frac{g^2}{16\pi^2}\left(\frac{11}{3}C_{\mathrm{A}} - \frac{4}{3}T_R\right). \tag{2.6.45}$$

From this the β-function (2.6.32) is directly obtained using (2.6.43).

Note that only the background-field two-point function had to be computed. This is in contrast to the conventional approach where the gauge-field two-point function, the fermion two-point function and the fermion–gauge-boson vertex function (or the ghost two-point function and the ghost–gauge-boson vertex function) have to be calculated. Using the background-field method also the two-loop β-function has been calculated [Ab81, Ab83a]. In this calculation the technical advantages of the background-field method become even more obvious.

2.6.3 Relation between different renormalization schemes

The various definitions of the running coupling constant in asymptotically free theories often lead to different second-order formulas. The importance of controlling this problem in the framework of QCD is discussed in Sect. 3.1.2. Therefore, a systematic review of the transformation of coupling constants is necessary [Sy73, Gr73a]. If the Lagrangian contains a set of independent, dimensionless running coupling constants $g_i(\mathcal{M})$, these satisfy differential equations of the form

$$\mathcal{M}\frac{\mathrm{d}g_i}{\mathrm{d}\mathcal{M}} = \beta_i(g_j), \tag{2.6.46}$$

If a different renormalization procedure leads to a different set of coupling constants $\tilde{g}_i(\tilde{\mathcal{M}})$ which satisfy (2.6.46) with $\tilde{\beta}_i(\tilde{g}_j)$, then within the theory a transformation formula between these coupling constants exists which for dimensional reasons must be of the form

$$g_i(\mathcal{M}) = F_i\left(\tilde{g}_j(\tilde{\mathcal{M}}), \frac{\tilde{\mathcal{M}}}{\mathcal{M}}\right). \tag{2.6.47}$$

This yields the transformation formula for the RG functions β_i

$$\beta_i(g_k) = \sum_j \frac{\partial g_i}{\partial \tilde{g}_j} \tilde{\beta}_j(\tilde{g}_k). \tag{2.6.48}$$

Consequently, the transformation (2.6.47) maps fixed points into fixed points.

As a result of (2.6.48), in perturbative transformations of coupling constants in non-abelian gauge-theories,

$$\begin{aligned} g &= \tilde{g} + \rho\frac{\tilde{g}^3}{4\pi} + \dots, \\ \frac{\mathrm{d}g}{\mathrm{d}\tilde{g}} &= 1 + 3\rho\frac{\tilde{g}^2}{4\pi} + \dots, \end{aligned} \tag{2.6.49}$$

the first two coefficients β_0 and β_1 of $\beta(g)$ in (2.6.34) remain invariant [Sy73],

$$\beta_0 = \tilde{\beta}_0, \qquad \beta_1 = \tilde{\beta}_1. \tag{2.6.50}$$

These transformations are needed when results from different schemes are compared. So far, we have used the MS and $\overline{\text{MS}}$ schemes and the MOM scheme defined by the renormalization conditions (2.5.47). For both practical and historical reasons, these and other schemes are used in the literature. Let us summarize these schemes here for convenience.

In the *minimal-subtraction scheme* (MS) of 't Hooft [tH73a], only the singular terms, e.g. the terms $\sim 1/\delta$, are subtracted. This simplifies higher-order calculations. However, as the transition to a dimensionless coupling constant in the form $g^2 \to g^2\mu^\delta$ is not unique, the MS method can be generalized [Co74]. The substitution $g^2 \to g^2\mu^\delta f(\delta, g^2)$ with $f(\delta, g^2) = 1 - \delta(\gamma - \log 4\pi)$,

(γ: Euler's constant) yields the *modified minimal-subtraction scheme* ($\overline{\text{MS}}$) where to one-loop order

$$\frac{1}{\delta} \to \frac{1}{\delta} - \gamma + \log 4\pi. \tag{2.6.51}$$

The determination of finite renormalization terms in the *momentum-subtraction procedure* (MOM) can of course also be done by fixing the three-gauge-boson or four-gauge-boson vertex normalization instead of that of the quark–gauge-boson vertex at symmetric momenta. All these various ways of renormalization constitute different renormalization schemes but give the same relations between physical quantities. In perturbation theory, the relevant renormalized coupling constants are linked by transformations as in (2.6.48). These leave the β-functions invariant up to the second order.

2.6.4 Running unrenormalized coupling constant

The RG equations allow for the renormalization procedure of a quantum field theory to be viewed from a different angle [Gl76]. We assume that the singular expressions for Green functions can be made finite by any regularization procedure with a cut-off parameter ε. Physical results then follow after subtraction, i.e. renormalization, in the limit $\varepsilon \to 0$. This is of importance for the lattice-approximation of gauge theories (Sect. 2.10), where the cut-off parameter ε is given by the lattice constant.

Then, instead of the running renormalized coupling constant $\bar{g}$ an unrenormalized coupling constant $g_0(\varepsilon)$ and its dependence on the cut-off parameter ε is considered [Wi75, Sy77, Sy80]

$$\begin{aligned} \frac{1}{\varepsilon}\frac{\mathrm{d}}{\mathrm{d}(1/\varepsilon)} g_0(\varepsilon) &= -\varepsilon \frac{\mathrm{d}}{\mathrm{d}\varepsilon} g_0(\varepsilon) = \tilde{\beta}(g_0(\varepsilon)) \\ &= -\beta_0' \frac{g_0^3}{(4\pi)^2} - \beta_1' \frac{g_0^5}{(4\pi)^4} + \dots . \end{aligned} \tag{2.6.52}$$

For finite $1/\varepsilon$, the renormalization procedure relates renormalized and unrenormalized coupling constants by a perturbative expansion. Thus, (2.6.50) implies that the lowest-order coefficients of the $\tilde{\beta}$-function agree with those given in (2.6.36) and (2.6.37) for non-abelian gauge theories.

2.7 Anomalies

We have seen in Sects. 2.2.6 and 2.4.3 that a symmetry manifests itself in relations between the Green functions of the theory, the Ward identities. We derived these relations in a purely formal way not caring about the possibility that they could get destroyed or modified by the UV divergences which occur in higher-order calculations.

If a regularization scheme exists that is compatible with the symmetry relations, then the Ward identities are valid also in the quantized theory. For example, we have seen in (2.5.100) that the Ward identities for the vector currents of QED are exactly fulfilled in dimensional regularization in one-loop order, so far justifying the applied renormalization procedure. If a symmetry is broken by the regularization scheme, two cases are possible. In the first case, the symmetry is restored after renormalization and switching off the cut-off. An example is lattice regularization of pure gauge theories. Poincaré invariance is broken to a discrete symmetry but restored in the continuum limit.

If the symmetry remains broken after renormalization, i.e. if the quantum fluctuations violate the classical symmetry relations, this is called an *anomaly* (for an exhausting review see [Be96]). This happens, for instance, for axial or chiral currents of global symmetries. The corresponding anomalies may have physical meaning. For instance, they determine the matrix element for the decay of the neutral π meson: $\pi^0 \to \gamma\gamma$. A further example is dilatation symmetry ($x^\mu \to \lambda x^\mu$) which is broken by all regularization schemes since they introduce a dimensionful parameter (e.g. a momentum cut-off Λ). The corresponding anomalies lead to scaling violations. These can be formulated in renormalization-group equations, the Callan–Symanzik equations, and lead to important effects in hard-scattering processes.

If anomalies occur in the Noether currents of local gauge symmetries, i.e. *internal currents*, they destroy the validity of the classical Ward identities. The Ward identities have to be modified by including extra anomalous, non-renormalizable terms. In gauge theories, Ward identities are crucial to prove renormalizability. Consequently in the presence of anomalies, unitarity and gauge invariance of the S matrix are no more guaranteed. Such anomalous theories are physically not acceptable. The following result of Ref. [Ba69] is important in this context. Theories that are invariant under space reflection or charge conjugation are free of anomalies of the Ward identities related to gauge invariance. This class of gauge theories includes pure Yang–Mills theories, QED, and QCD, but in general no theories in which the gauge bosons are coupled to axial or chiral fermionic currents.

2.7.1 The triangle-graph anomaly

2.7.1.1 Current conservation in non-abelian gauge theories

Noether's theorem for current conservation in classical field theory was derived in Sect. 1.2.3.1. As preparation for the study of anomalies, we investigate the conservation of the currents of a multiplet of charged fermion fields $\psi(x) = (\psi_i(x))$ which transforms under gauge transformations as $\delta\psi_i = \mathrm{i}(T^a\psi)_i\delta\theta^a$, $\delta\overline{\psi}_i = -\mathrm{i}(\overline{\psi}T^a)_i\delta\theta^a$ and has vector couplings to non-abelian gauge fields. The fermionic part of the Lagrangian,

$$\mathcal{L} = \overline{\psi}(x)(\mathrm{i}\not{D} - m)\psi(x), \qquad D_\mu = \partial_\mu - \mathrm{i}gT^aA^a_\mu(x), \tag{2.7.1}$$

yields the Dirac equation

$$(\mathrm{i}\not{D} - m)\psi(x) = (\mathrm{i}\not{\partial} + gT^a\not{A}^a(x) - m)\psi(x) = 0 \tag{2.7.2}$$

and its adjoint

$$\overline{\psi}(x)(\mathrm{i}\overleftarrow{\not{\partial}} - gT^a\not{A}^a + m) = 0. \tag{2.7.3}$$

We define the *vector and axial-vector non-singlet currents*

$$j^a_\mu(x) = \overline{\psi}(x)\gamma_\mu T^a\psi(x), \qquad j^a_{5,\mu}(x) = \overline{\psi}(x)\gamma_\mu\gamma_5 T^a\psi(x), \tag{2.7.4}$$

and the *singlet currents*

$$j^0_\mu(x) = \overline{\psi}(x)\gamma_\mu\psi(x), \qquad j^0_{5,\mu}(x) = \overline{\psi}(x)\gamma_\mu\gamma_5\psi(x). \tag{2.7.5}$$

Using the Dirac equation (2.7.2) and its adjoint (2.7.3), we calculate the divergence of the non-singlet vector current

$$\begin{aligned}
\partial^\mu j^a_\mu &= \partial^\mu(\overline{\psi}\gamma_\mu T^a\psi) = \overline{\psi}\overleftarrow{\not{\partial}}T^a\psi + \overline{\psi}T^a\not{\partial}\psi \\
&= -\mathrm{i}\overline{\psi}(gT^b\not{A}^b - m)T^a\psi - \mathrm{i}\overline{\psi}T^a(-gT^b\not{A}^b + m)\psi \\
&= \mathrm{i}g\overline{\psi}[T^a, T^b]\not{A}^b\psi = -gf^{abc}A^b_\mu\overline{\psi}\gamma^\mu T^c\psi = -gf^{abc}A^{\mu,b}j^c_\mu.
\end{aligned} \tag{2.7.6}$$

The non-singlet fermion current is not conserved. This is not surprising since Noether's theorem guarantees only the conservation of the total current, the sum of the fermionic and the gauge-boson current. The latter does not vanish in the non-abelian case because the gauge bosons are charged. Since the structure constants f^{abc} vanish in the abelian case, the vector current is conserved in QED and similar models.

However, the non-abelian fermion current is *conserved covariantly.* This can be seen by moving the term on the right-hand side of (2.7.6) to the left and identifying the resulting expression as the covariant derivative,

$$(D^{\mu} j_{\mu})^{a}(x) = 0. \tag{2.7.7}$$

A similar calculation yields

$$(D^{\mu} j_{5,\mu})^{a}(x) = 2im\bar{\psi}(x)\gamma_{5}T^{a}\psi(x) \equiv 2imp^{a}(x) \tag{2.7.8}$$

for the divergence of the non-singlet axial-vector current with the pseudoscalar density $p^{a}(x)$.

The singlet vector current is conserved,

$$\partial^{\mu} j^{0}_{\mu}(x) = 0. \tag{2.7.9}$$

The divergence of the singlet axial-vector current is given by the pseudoscalar density,

$$\partial^{\mu} j^{0}_{,5\mu}(x) = 2im\bar{\psi}(x)\gamma_{5}\psi(x) = 2imp^{0}(x), \tag{2.7.10}$$

and vanishes in a massless theory.

For theories with parity violation like the Electroweak Standard Model, we have to consider chiral couplings between the fermions and the non-abelian gauge bosons. For this purpose we use *left- and right-handed fields and currents*,

$$\begin{aligned} \psi_{\mathrm{L}} &= \frac{1-\gamma_5}{2}\psi, & \psi_{\mathrm{R}} &= \frac{1+\gamma_5}{2}\psi, \\ j^{a}_{\mathrm{L},\mu} &= \frac{1}{2}(j^{a}_{\mu} - j^{a}_{5,\mu}), & j^{a}_{\mathrm{R},\mu} &= \frac{1}{2}(j^{a}_{\mu} + j^{a}_{5,\mu}). \end{aligned} \tag{2.7.11}$$

In the massless case, the chiral fermion Lagrangian

$$\mathcal{L} = \overline{\psi}_{\mathrm{L}} \mathrm{i} \not{D}_{\mathrm{L}} \psi_{\mathrm{L}} + \overline{\psi}_{\mathrm{R}} \mathrm{i} \not{D}_{\mathrm{R}} \psi_{\mathrm{R}} \tag{2.7.12}$$

implies the covariant conservation of the non-singlet currents,

$$(D^{\mu}_{\mathrm{L}} j_{\mathrm{L},\mu})^a(x) = 0, \qquad (D^{\mu}_{\mathrm{R}} j_{\mathrm{R},\mu})^a(x) = 0, \qquad \text{for } m = 0, \tag{2.7.13}$$

and the ordinary conservation of the singlet currents,

$$\partial^{\mu} j^0_{\mathrm{L},\mu}(x) = 0, \qquad \partial^{\mu} j^0_{\mathrm{R},\mu}(x) = 0, \qquad \text{for } m = 0. \tag{2.7.14}$$

2.7.1.2 Ward identities of current Green functions

Anomalies were found first in the Green functions of axial-vector currents in abelian theories, e.g. QED, [St49, Sc51a, Su66, Ad69, Be69]. Let us consider Green functions of abelian local current operators, i.e. vacuum expectation values of time-ordered products of the currents. From (2.7.7) and (2.7.8), Ward identities for the current Green functions follow just like those discussed in Sect. 2.2.6.4.

The simplest Green function in which anomalies occur is the three-point function of two vector and one axial-vector currents,[18]

$$T^{\mathrm{VVA}}_{\mu\nu\rho}(x_1, x_2, x_3) = \langle 0 | T j_{\mu}(x_1) j_{\nu}(x_2) j_{5,\rho}(x_3) | 0 \rangle. \tag{2.7.15}$$

This Green function is symmetric under the exchange of the two vector currents. Its naïve Ward identities read

$$\begin{aligned} \partial^{\mu}_{x_1} T^{\mathrm{VVA}}_{\mu\nu\rho}(x_1, x_2, x_3) &= \partial^{\nu}_{x_2} T^{\mathrm{VVA}}_{\mu\nu\rho}(x_1, x_2, x_3) = 0, \\ \partial^{\rho}_{x_3} T^{\mathrm{VVA}}_{\mu\nu\rho}(x_1, x_2, x_3) &= 2m\mathrm{i}\, \langle 0 | T j_{\mu}(x_1) j_{\nu}(x_2) p(x_3) | 0 \rangle \\ &= 2m\mathrm{i}\, T^{\mathrm{VVP}}_{\mu\nu}(x_1, x_2, x_3). \end{aligned} \tag{2.7.16}$$

The equivalent relations in momentum space are

$$\begin{aligned} p_1^{\mu} T^{\mathrm{VVA}}_{\mu\nu\rho}(p_1, p_2, p_3) &= p_2^{\nu} T^{\mathrm{VVA}}_{\mu\nu\rho}(p_1, p_2, p_3) = 0, \\ p_3^{\rho} T^{\mathrm{VVA}}_{\mu\nu\rho}(p_1, p_2, p_3) &= 2m\, T^{\mathrm{VVP}}_{\mu\nu}(p_1, p_2, p_3) \end{aligned} \tag{2.7.17}$$

with $p_1 + p_2 + p_3 = 0$. Similar naïve Ward identities may be derived for other Green functions.

[18] As will be seen in the following, the chiral anomalies involve the tensor $\varepsilon_{\mu\nu\rho\sigma}$. Consequently, Green functions of two currents are anomaly-free since they have only two Lorentz indices and depend only on one momentum.

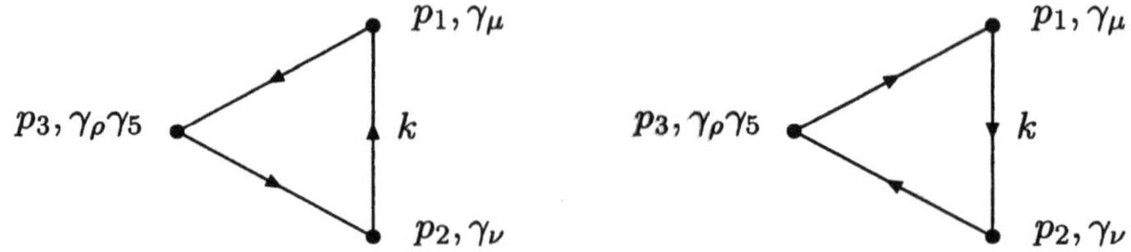

Fig. 2.11 Feynman diagrams for the triangle anomaly

2.7.1.3 Calculation of the triangle-graph anomaly

The first-order quantum corrections to these relations are obtained from the one-loop diagrams of Fig. 2.11.

The two diagrams for $T^{\mathrm{VVA}}_{\mu\nu\rho}(p_1, p_2, p_3)$ are linearly UV-divergent. Consequently, they must be regularized. We use the *Pauli–Villars regularization*,[19] i.e. we subtract the same diagrams with mass $M \gg m$. From the Feynman rules we get (each current contributes a factor γ_μ, γ_ν, or $\gamma_\rho\gamma_5$)

$$\begin{aligned} T^{\mathrm{VVA}}_{\mu\nu\rho}(p_1,p_2,p_3) &= -\mathrm{i}^3 \int \frac{\mathrm{d}^4k}{(2\pi)^4} \\ &\Big\{\mathrm{Tr}\Big[\gamma_\mu(\not{k}-m)^{-1}\gamma_\nu(\not{k}-\not{p}_2-m)^{-1}\gamma_\rho\gamma_5(\not{k}+\not{p}_1-m)^{-1}\Big] \\ &+\mathrm{Tr}\Big[\gamma_\nu(\not{k}-m)^{-1}\gamma_\mu(\not{k}-\not{p}_1-m)^{-1}\gamma_\rho\gamma_5(\not{k}+\not{p}_2-m)^{-1}\Big] \\ &-\ (m\to M)\Big\}. \end{aligned} \tag{2.7.18}$$

After subtraction, the integral is finite and can be treated in the usual way. In order to test the vector Ward identities we multiply with p_1^μ, decompose

$$\not{p}_1 = (\not{k}+\not{p}_1-m) - (\not{k}-m) = -(\not{k}-\not{p}_1-m) + (\not{k}-m),$$

and obtain

$$\begin{aligned} p_1^\mu T^{\mathrm{VVA}}_{\mu\nu\rho} = -\mathrm{i}\int \frac{\mathrm{d}^4k}{(2\pi)^4}\Big\{&\mathrm{Tr}\Big[-(\not{k}-m)^{-1}\gamma_\nu(\not{k}-\not{p}_2-m)^{-1}\gamma_\rho\gamma_5 \\ &+(\not{k}+\not{p}_1-m)^{-1}\gamma_\nu(\not{k}-\not{p}_2-m)^{-1}\gamma_\rho\gamma_5\Big] \end{aligned}$$

[19] For dimensional regularization see Sect. 2.7.1.5.

$$+\operatorname{Tr}\Big[(\not{k}+\not{p}_2-m)^{-1}\gamma_\nu(\not{k}-m)^{-1}\gamma_\rho\gamma_5 \tag{2.7.19}$$
$$-(\not{k}+\not{p}_2-m)^{-1}\gamma_\nu(\not{k}-\not{p}_1-m)^{-1}\gamma_\rho\gamma_5\Big]-(m\to M)\Big\}.$$

Since the integral is finite we are allowed to perform the shifts $k \to k + p_2$, $k \to k + p_2 - p_1$ and find a vanishing integrand. Consequently, the vector Ward identity is fulfilled. This is the case for all correlation functions involving only vector currents.

For the study of the axial-vector case, we use

$$\begin{aligned}\not{p}_3\gamma_5 = -(\not{p}_1+\not{p}_2)\gamma_5 &= (\not{k}-\not{p}_2-m)\gamma_5+\gamma_5(\not{k}+\not{p}_1-m)+2m\gamma_5\\ &= (\not{k}-\not{p}_1-m)\gamma_5+\gamma_5(\not{k}+\not{p}_2-m)+2m\gamma_5,\end{aligned}$$

which follows from the anticommutativity of γ_5 in four dimensions. With help of translation invariance and Dirac algebra we find

$$p_3^\rho T_{\mu\nu\rho}^{\mathrm{VVA}} = 2\mathrm{i}\int\frac{\mathrm{d}^4k}{(2\pi)^4} \tag{2.7.20}$$
$$\Big\{m\operatorname{Tr}\Big[\gamma_\mu(\not{k}-m)^{-1}\gamma_\nu(\not{k}-\not{p}_2-m)^{-1}\gamma_5(\not{k}+\not{p}_1-m)^{-1}\Big]$$
$$+m\operatorname{Tr}\Big[\gamma_\nu(\not{k}-m)^{-1}\gamma_\mu(\not{k}-\not{p}_1-m)^{-1}\gamma_5(\not{k}+\not{p}_2-m)^{-1}\Big]$$
$$-M\operatorname{Tr}\Big[\gamma_\mu(\not{k}-M)^{-1}\gamma_\nu(\not{k}-\not{p}_2-M)^{-1}\gamma_5(\not{k}+\not{p}_1-M)^{-1}\Big]$$
$$-M\operatorname{Tr}\Big[\gamma_\nu(\not{k}-M)^{-1}\gamma_\mu(\not{k}-\not{p}_1-M)^{-1}\gamma_5(\not{k}+\not{p}_2-M)^{-1}\Big]\Big\}.$$

This has to be compared with $T_{\mu\nu}^{\mathrm{VVP}}$. The corresponding diagrams are those of Fig. 2.11 with $\gamma_\rho\gamma_5$ replaced by γ_5. They give an UV-finite expression which does not require regularization,

$$T_{\mu\nu}^{\mathrm{VVP}} = \mathrm{i}\int\frac{\mathrm{d}^4k}{(2\pi)^4} \tag{2.7.21}$$
$$\Big\{\operatorname{Tr}\Big[\gamma_\mu(\not{k}-m)^{-1}\gamma_\nu(\not{k}-\not{p}_2-m)^{-1}\gamma_5(\not{k}+\not{p}_1-m)^{-1}\Big]$$
$$+\operatorname{Tr}\Big[\gamma_\nu(\not{k}-m)^{-1}\gamma_\mu(\not{k}-\not{p}_1-m)^{-1}\gamma_5(\not{k}+\not{p}_2-m)^{-1}\Big]\Big\},$$

and is identical to the first two lines of (2.7.20). Thus, we arrive at

$$\mathrm{i}p_3^\rho T_{\mu\nu\rho}^{\mathrm{VVA}} = 2m\mathrm{i}\, T_{\mu\nu}^{\mathrm{VVP}} + A_{\mu\nu}. \tag{2.7.22}$$

The integral

$$\begin{aligned} A_{\mu\nu} = \lim_{M\to\infty} 2M \int \frac{\mathrm{d}^4k}{(2\pi)^4} \\ \times \Bigg\{ \frac{\mathrm{Tr}\left[\gamma_\mu(\not k + M)\gamma_\nu(\not k - \not p_2 + M)\gamma_5(\not k + \not p_1 + M)\right]}{(k^2 - M^2)[(k-p_2)^2 - M^2][(k+p_1)^2 - M^2]} \\ + (\mu, 1) \leftrightarrow (\nu, 2) \Bigg\} \end{aligned} \tag{2.7.23}$$

involves the regulator terms. It is UV-finite but non-vanishing. We have an anomaly. The Dirac trace yields $4\mathrm{i}M\varepsilon_{\mu\nu\rho\sigma}p_1^\rho p_2^\sigma$, independent of k. The second term gives the same contribution. In the remaining scalar integral, p_1 and p_2 can be neglected relative to M leaving a special case of the tadpole integral (2.5.57), namely $A_0^{(3)}(M) = -1/(2M^2)$. Thus, we find

$$A_{\mu\nu} = \lim_{M\to\infty} 4M\, 4\mathrm{i}M\varepsilon_{\mu\nu\rho\sigma}p_1^\rho p_2^\sigma \frac{\mathrm{i}}{16\pi^2}\frac{-1}{2M^2} = \frac{1}{2\pi^2}\varepsilon_{\mu\nu\rho\sigma}p_1^\rho p_2^\sigma. \tag{2.7.24}$$

This anomaly is finite and independent of the masses in the triangle diagram. A similar anomaly occurs in the triangle diagram with three axial-vector currents.

The modified *(anomalous) Ward identities* for the regularized one-loop vertex functions read

$$\begin{aligned} p_1^\mu T_{\mu\nu\rho}^{\mathrm{VVA}} &= p_2^\nu T_{\mu\nu\rho}^{\mathrm{VVA}} = 0, \\ \mathrm{i}p_3^\rho T_{\mu\nu\rho}^{\mathrm{VVA}} &= 2m\mathrm{i}\, T_{\mu\nu}^{\mathrm{VVP}} + \frac{1}{2\pi^2}\varepsilon_{\mu\nu\rho\sigma}p_1^\rho p_2^\sigma. \end{aligned} \tag{2.7.25}$$

As they stand, they may be interpreted in the following way: the vector currents are anomaly-free whereas the axial-vector current acquires an anomaly from the quantum fluctuations.

In spite of the fact that the anomaly is finite one might wonder whether it can be compensated by suitable renormalization counterterms for the composite operator $T_{\mu\nu\rho}^{\mathrm{VVA}}$. These counterterms must have the appropriate Lorentz and symmetry structure and must be renormalizable (cf. Sect. 2.5.1.5). The only choice is

$$T^{\mathrm{VVA}}_{\mu\nu\rho,\mathrm{ct}} = C\varepsilon_{\mu\nu\rho\sigma}(p_1 - p_2)^\sigma . \tag{2.7.26}$$

Since this term contributes to all divergences $p_1^\mu T^{\mathrm{VVA}}_{\mu\nu\rho}$, $p_2^\mu T^{\mathrm{VVA}}_{\mu\nu\rho}$, and $p_3^\mu T^{\mathrm{VVA}}_{\mu\nu\rho}$, also the vector Ward identities get anomalies for $C \neq 0$, but for a specific choice of C the Ward identity for the axial-vector current can be made anomaly-free. The counterterm redistributes the anomaly but cannot compensate it. For chiral currents, the renormalization can be used to distribute the anomalies equally over all currents. There is no renormalization scheme that is anomaly-free for this simple model of a charged spinor field and its currents.

In the case of non-abelian currents, the coupling matrices T^a enter. The general expression for the triangle-graph anomaly of the *non-singlet axial-vector current* reads

$$A^{abc}_{\mu\nu} = \frac{\mathrm{Tr}(\{T^a, T^b\}T^c)}{2}\frac{1}{2\pi^2}\varepsilon_{\mu\nu\rho\sigma}p_1^\rho p_2^\sigma . \tag{2.7.27}$$

Also in the general case, the anomaly appears as an extra term in the Ward identities of Sect. 2.7.1.2.

2.7.1.4 An application: $\pi^0 \to \gamma\gamma$

As an example for the physical relevance of the anomaly of the axial-vector current we calculate the decay width of the neutral π meson. (A discussion of chiral symmetry in QCD can be found in Sect. 3.5.1.) The invariant S-matrix element is obtained from the corresponding Green function using the LSZ formalism (cf. Sect. 2.1.3),

$$\begin{aligned}\langle \gamma, p_1; \gamma, p_2 | S | \pi^0, p\rangle &= \mathrm{i} p_1^2 p_2^2 (p^2 - M_\pi^2)\varepsilon^\mu(p_1)\varepsilon^\nu(p_2) \\ &\times \int \mathrm{d}^4x_1\,\mathrm{d}^4x_2\,\mathrm{d}^4y\,\mathrm{e}^{-\mathrm{i}(py - p_1x_1 - p_2x_2)}\langle T A_\mu(x_1)A_\nu(x_2)\pi^0(y)\rangle .\end{aligned} \tag{2.7.28}$$

This Green function can be written as a three-current Green function using the Maxwell equation $\Box A_\mu = e j^\gamma_\mu$ and the *PCAC (partially conserved axial-vector current) hypothesis.* This relates the divergence of the SU(2)-flavour-isotriplet axial-vector current, i.e. a pseudoscalar density, to the corresponding π-meson field operator [Na60],

$$\partial^\mu j^a_{5,\mu}(x) = \frac{f_\pi}{\sqrt{2}} M^2_\pi \pi^a(x), \qquad j^a_{5,\mu} = \bar{\psi}\gamma_\mu\gamma_5 \frac{\tau^a}{2}\psi \tag{2.7.29}$$

with the pion decay constant $f_\pi = 130\,\mathrm{MeV}$ and the Pauli matrices τ^a (A.1.11). We obtain

$$\begin{aligned}\langle\gamma,p_1;\gamma,p_2|S|\pi^0,p\rangle &= \mathrm{i}e^2 \frac{\sqrt{2}}{f_\pi M^2_\pi}(p^2 - M^2_\pi)\varepsilon^\mu(p_1)\varepsilon^\nu(p_2) \\ &\quad\times \int \mathrm{d}^4x_1\,\mathrm{d}^4x_2\,\mathrm{d}^4y\,\mathrm{e}^{-\mathrm{i}(py-p_1x_1-p_2x_2)}\langle T j^\gamma_\mu(x_1) j^\gamma_\nu(x_2)\partial^\rho j^0_{\rho,5}(y)\rangle. \\ &= \mathrm{i}e^2 \frac{\sqrt{2}}{f_\pi M^2_\pi}(p^2 - M^2_\pi)\varepsilon^\mu(p_1)\varepsilon^\nu(p_2) \\ &\quad\times \mathrm{i}p^\rho T^{\mathrm{V},\gamma;\mathrm{V},\gamma;\mathrm{A},0}_{\mu\nu\rho}(-p_1,-p_2,p)(2\pi)^4\delta^{(4)}(p-p_1-p_2),\end{aligned} \tag{2.7.30}$$

where we have used the fact that the commutator terms vanish when extracting the derivative of the axial-vector current from the time ordering. Neglecting the anomaly, this expression can be evaluated in the soft-pion approximation, i.e. for $p^2 = 0$, as follows: we write the most general decomposition of the tensor $T^{\mathrm{V},\gamma;\mathrm{V},\gamma;\mathrm{A},0}_{\mu\nu\rho}(-p_1,-p_2,p)$ taking into account Bose symmetry, electromagnetic current conservation [the first two identities of (2.7.25)], and the negative parity of the axial-vector current. Contracting this expression with p^ρ yields

$$p^\rho T^{\mathrm{V},\gamma;\mathrm{V},\gamma;\mathrm{A},0}_{\mu\nu\rho}(-p_1,-p_2,p) = \varepsilon_{\mu\nu\rho\sigma}p_1^\mu p_2^\nu p^2 f(p^2) \tag{2.7.31}$$

with some coefficient function $f(p^2)$. Since no strongly interacting zero-mass particle is around ($m^2_\pi \neq 0$), $T^{\mathrm{V},\gamma;\mathrm{V},\gamma;\mathrm{A},0}_{\mu\nu\rho}$ and thus $f(p^2)$ does not contain any poles for $p^2 \to 0$. Therefore, $p^\rho T^{\mathrm{V},\gamma;\mathrm{V},\gamma;\mathrm{A},0}_{\mu\nu\rho}$ has to vanish for $p^2 \to 0$ and the pion decay would be strongly suppressed. This is the Sutherland–Veltman theorem [Su67, Ve67].

However, in the presence of gauge fields, the PCAC relation has to be modified by the anomaly (cf. Sect. 2.7.2.1),

$$\partial^\mu j^a_{5,\mu}(x) = \frac{f_\pi}{\sqrt{2}} M^2_\pi \pi^a(x) - \frac{e^2}{16\pi^2}\varepsilon^{\mu\nu\rho\sigma}F_{\mu\nu}(x)F_{\rho\sigma}(x)\,\mathrm{Tr}\left(Q^2\frac{\tau^a}{2}\right), \tag{2.7.32}$$

where $F_{\mu\nu}$ is the electromagnetic field strength. The trace corresponds to the one in (2.7.27) and results from the charges in the electromagnetic currents and the SU(2) matrix in the isospin current. As a consequence, we obtain a contribution of the anomaly in (2.7.30) which gives a non-vanishing result in the soft-pion limit. Thus, the amplitude is given in this limit by the anomaly, and we find for the invariant decay matrix element,

$$\mathcal{M}(\pi^0 \to \gamma\gamma) = \frac{2\alpha\sqrt{2}}{\pi f_\pi} \operatorname{Tr}\left(Q^2 \frac{\tau^3}{2}\right) \varepsilon_{\mu\nu\rho\sigma} \varepsilon^\mu(p_1) \varepsilon^\nu(p_2) p_1^\rho p_2^\sigma . \quad (2.7.33)$$

The trace gets contributions from the constituents of the pion, the up and down quarks with charge Q_q, third component of isospin $I^3 = \tau^3/2$, and N_C colours,

$$\operatorname{Tr}(Q^2 I^3) = \sum_q (Q_q^2 I_q^3) = N_C \left(\frac{4}{9}\frac{1}{2} - \frac{1}{9}\frac{1}{2}\right) = \frac{N_C}{6}.$$

The total decay width is given by [cf. (1.3.12)]

$$\Gamma(\pi^0 \to \gamma\gamma) = \frac{1}{2M_\pi}\frac{1}{32\pi^2}\frac{1}{2}\int \mathrm{d}\Omega \sum_{\text{pols.}} |\mathcal{M}|^2 \quad (2.7.34)$$

(The factor 1/2 accounts for the two identical particles in the final state). The polarization sum is performed with $\sum_\lambda \varepsilon^\mu(\lambda)\varepsilon^\nu(\lambda) = -g^{\mu\nu}$, leading to

$$\sum_{\lambda_1\lambda_2} |\varepsilon_{\mu\nu\rho\sigma} \varepsilon^\mu(p_1,\lambda_1) \varepsilon^\nu(p_2,\lambda_2) p_1^\rho p_2^\sigma|^2 = \frac{M_\pi^4}{2}.$$

The resulting distribution is isotropic, and the angular integral gives 4π. Putting everything together, we finally obtain

$$\Gamma(\pi^0 \to \gamma\gamma) = \frac{\alpha^2}{32\pi^3}\frac{M_\pi^3}{f_\pi^2}\frac{N_C^2}{9}. \quad (2.7.35)$$

Inserting $f_\pi = 130\,\text{MeV}$ and $m_\pi = 135\,\text{MeV}$, this yields $\Gamma(\pi^0 \to \gamma\gamma) = 7.81\,\text{eV}$ in good agreement with the experimental result [PDG00] $\Gamma_{\text{exp}}(\pi^0 \to \gamma\gamma) = 8.3 \pm 0.6\,\text{eV}$. Depending on the attitude, this agreement is used as an argument for three colours, the PCAC hypothesis, or the importance of axial anomalies.

2.7.1.5 γ_5 in dimensional regularization

The existence of anomalies of axial or chiral currents shows that no chirally invariant regularization scheme exists. Chirally invariant theories are massless. The Pauli–Villars scheme breaks chiral symmetry by introducing heavy fermions. In the dimensional regularization scheme, the anomaly of the axial-vector current is related to the properties of the matrix γ_5. This is defined in four dimensions as the completely antisymmetric product of all four Dirac matrices,

$$\gamma_5^{(4)} = \frac{-\mathrm{i}}{4!}\varepsilon_{\mu\nu\rho\sigma}\gamma^\mu\gamma^\nu\gamma^\rho\gamma^\sigma , \tag{2.7.36}$$

where the totally antisymmetric tensor $\varepsilon_{\mu\nu\rho\sigma}$ is given in (A.1.9). This definition implies the properties

$$\begin{aligned} \left(\gamma_5^{(4)}\right)^2 &= 1, \qquad \{\gamma_5^{(4)}, \gamma^\mu\} = 0, \\ \mathrm{Tr}\left(\gamma_5^{(4)}\right) &= 0, \qquad \mathrm{Tr}\left(\gamma_5^{(4)}\gamma^\mu\gamma^\nu\right) = 0, \\ \mathrm{Tr}\left(\gamma_5^{(4)}\gamma^\mu\gamma^\nu\gamma^\rho\gamma^\sigma\right) &= -4\mathrm{i}\varepsilon^{\mu\nu\rho\sigma} , \end{aligned} \tag{2.7.37}$$

which we have used in the calculation of the triangle-graph anomaly. If the D-dimensional $\gamma_5^{(D)}$ matrix is defined in such a way that it anticommutes with γ_μ,

$$\{\gamma_5^{(D)}, \gamma^\mu\} = 0, \tag{2.7.38}$$

the following trace relations can be derived (see e.g. Ref. [Co84])

$$\begin{aligned} D\,\mathrm{Tr}\left(\gamma_5^{(D)}\right) &= 0, \qquad (D-2)\,\mathrm{Tr}\left(\gamma_5^{(D)}\gamma^\mu\gamma^\nu\right) = 0, \\ (D-4)\,\mathrm{Tr}\left(\gamma_5^{(D)}\gamma^\mu\gamma^\nu\gamma^\rho\gamma^\sigma\right) &= 0, \quad \text{etc.} \end{aligned} \tag{2.7.39}$$

The last equation shows that this $\gamma_5^{(D)}$ is not a suitable analytic continuation of the four-dimensional object since it enforces a vanishing trace for the product of four Dirac matrices in clear contradiction to the four-dimensional case (2.7.37).

Several ways out of this dilemma have been proposed. We mention the following ones:

- The four-dimensional $\gamma_5^{(4)}$ is used also in D dimensions. This 't Hooft–Veltman or Breitenlohner–Maison scheme can be handled by splitting D-dimensional space into ordinary four-dimensional physical space plus an extra $(D-4)$-dimensional space [tH72a, Br77, Co84].
- The cyclicity of the trace of Dirac matrices is given up [Ko92].
- In one-loop calculations in the (anomaly-free) Electroweak Standard Model it is possible to use—with certain care—the anticommuting $\gamma_5^{(D)}$ in D-dimensional calculations together with the last relation in (2.7.37) [Je00], despite the fact that this scheme is in general inconsistent.

The existence of axial anomalies can easily be seen in the 't Hooft–Veltman dimensional regularization scheme. The divergence of the axial-vector current contains a term of the form $\bar{\psi}(x)\{\gamma_\mu, \gamma_5^{(4)}\}\partial^\mu\psi(x)$. The anticommutator is effectively proportional to $D-4$. Therefore, it does not contribute in lowest order but gives a finite result if it gets multiplied with UV singular ($\propto (D-4)^{-1}$) expressions. In general, operators that vanish in 4 dimensions but are non-vanishing in $D > 4$ dimensions are called *evanescent operators.*

Let us finish this section with a general remark. The non-existence of a chirally invariant regularization scheme is the formal origin of the axial or chiral anomalies. In the path-integral formulation, the anomalies result from the non-invariance of the integration measure under chiral transformations [Fu80]. The anomalies can be obtained directly from the path integral and coincide with those resulting from one-loop calculations. As an important by-product of this calculation one obtains the relation of the anomaly to topological charges and to the number of zero modes of the covariant Dirac operator $\not{D}$. More details are discussed in Sect. 2.9.3.8.

2.7.2 Anomalies in gauge theories

In the last section we have calculated the triangle-graph anomalies of the Green functions of fermionic currents. In gauge theories, the anomalies of those fermionic currents which are the Noether currents of the local gauge group can be expressed in terms of the gauge fields. This can be seen most clearly in the path-integral formalism [Fu80] or within the point-splitting technique [Sc62].

2.7.2.1 Anomalies and gauge fields

In the *non-abelian case*, anomalies do not only arise in the triangle diagrams with an odd number of axial-vector currents. They occur also in box and pentagon one-loop diagrams with an odd number of of axial couplings. Explicit calculations [Ba69, We71] show that the total anomaly can be expressed in terms of an operator composed of gauge fields. For the *singlet axial-vector current* in massless gauge theories, where the fermions are coupled by their vector currents to the gauge fields, it reads

$$\partial^\mu j^0_{5,\mu}(x) = -\frac{g^2}{16\pi^2}\varepsilon^{\mu\nu\rho\sigma} F^a_{\mu\nu}(x) F^b_{\rho\sigma}(x)\,\mathrm{Tr}(T^a T^b) \tag{2.7.40}$$

with the non-abelian field strengths

$$F^a_{\mu\nu}(x) = \partial_\mu A^a_\nu(x) - \partial_\nu A^a_\mu(x) + g f^{abc} A^b_\mu(x) A^c_\nu(x). \tag{2.7.41}$$

Using $\mathcal{F}_{\mu\nu}(x) = gT^a F^a_{\mu\nu}(x)$, this can be written in the form

$$\partial^\mu j^0_{5,\mu}(x) = -\frac{1}{16\pi^2}\varepsilon^{\mu\nu\rho\sigma}\,\mathrm{Tr}(\mathcal{F}_{\mu\nu}(x)\mathcal{F}_{\rho\sigma}(x)). \tag{2.7.42}$$

This expression is proportional to the *Pontryagin density* indicating a close relation between anomalies and topological properties of gauge theories. This is considered explicitly in Sect. 2.9.3.

The result (2.7.42) can be rewritten in terms of the potentials $\mathcal{A}_\mu(x) = gT^a A^a_\mu(x)$,

$$\partial^\mu j^0_{5,\mu} = -\frac{1}{4\pi^2}\varepsilon^{\mu\nu\rho\sigma}\partial_\mu\,\mathrm{Tr}\left(\mathcal{A}_\nu\partial_\rho\mathcal{A}_\sigma - \frac{2}{3}\mathrm{i}\mathcal{A}_\nu\mathcal{A}_\rho\mathcal{A}_\sigma\right). \tag{2.7.43}$$

For the *non-singlet current*, assuming covariant conservation of the vector currents, Bardeen [Ba69] found the result

$$(D^\mu j_{5,\mu})^a = -\frac{1}{16\pi^2}\varepsilon^{\mu\nu\rho\sigma}\,\mathrm{Tr}\left(T^a\mathcal{F}_{\mu\nu}(x)\mathcal{F}_{\rho\sigma}(x)\right). \tag{2.7.44}$$

For *chiral theories*, assuming that the classical conservation laws are broken in a symmetric way, Bardeen [Ba69] obtained for the non-singlet currents:

$$(D^\mu j_{\mathrm{L},\mu})^a = \tag{2.7.45}$$
$$-\frac{1}{24\pi^2}\varepsilon^{\mu\nu\rho\sigma}\partial_\mu\,\mathrm{Tr}\left[T^a\left(\mathcal{A}_{\mathrm{L},\nu}\partial_\rho\mathcal{A}_{\mathrm{L},\sigma} - \frac{\mathrm{i}}{2}\mathcal{A}_{\mathrm{L},\nu}\mathcal{A}_{\mathrm{L},\rho}\mathcal{A}_{\mathrm{L},\sigma}\right)\right].$$

For right-handed currents, L has to be replaced by R and the total expression multiplied by -1.

In the abelian case, (2.7.40) remains valid but the field strength is linear in $A_\mu(x)$. Expressed in terms of the gauge field, the anomaly reads

$$\partial^\mu j^0_{5,\mu}(x) = -\frac{g^2}{4\pi^2}\varepsilon^{\mu\nu\rho\sigma}(\partial_\mu A_\nu(x))\partial_\rho A_\sigma(x). \tag{2.7.46}$$

The Feynman rules for this local operator give

$$A_{\mu\nu} = \frac{g^2}{2\pi^2}\varepsilon_{\mu\nu\rho\sigma}p_1^\rho p_2^\sigma, \tag{2.7.47}$$

i.e. the same result (2.7.24) as for the one-loop triangle diagram.

2.7.2.2 Cancellation of anomalies

If anomalies occur in the internal currents of gauge theories, i.e. in those currents that constitute the fundamental interaction between fermions and gauge bosons, they enter the basic Ward identities (2.4.75) and destroy the renormalizability of the theory. Therefore, it is important to find criteria for the absence of these anomalies in gauge theories. For this purpose it is sufficient to consider the algebraic prefactors of the anomalies (2.7.27) of the left- and right-handed internal currents of the Dirac fields. These contain the hermitian representation matrices T^a_{L} and T^a_{R} of the generators of the gauge group, i.e. of the couplings of the fermions to the gauge bosons, in the form

$$A^{abc} \propto \mathrm{Tr}\left(T^a_{\mathrm{L}}\left\{T^b_{\mathrm{L}}, T^c_{\mathrm{L}}\right\}\right) - \mathrm{Tr}\left(T^a_{\mathrm{R}}\left\{T^b_{\mathrm{R}}, T^c_{\mathrm{R}}\right\}\right), \tag{2.7.48}$$

where the trace extends over all fermionic degrees of freedom.

The anomaly vanishes if the trace vanishes,

$$\mathrm{Tr}\left(T^a_{\mathrm{L}}\left\{T^b_{\mathrm{L}}, T^c_{\mathrm{L}}\right\}\right) - \mathrm{Tr}\left(T^a_{\mathrm{R}}\left\{T^b_{\mathrm{R}}, T^c_{\mathrm{R}}\right\}\right) = 0. \tag{2.7.49}$$

The algebraic anomaly factor is zero if these traces vanish separately, or if the contributions of the left-handed and the right-handed couplings compensate each other. In the first case the representations are called *safe*. The safe representations include, in particular, all real representations. A representation is called real if $(T^a)^* = -UT^aU$ with a unitary matrix U, i.e. if the negative of the conjugate-complex representation is unitary equivalent to the representation matrix.[20] Hermiticity implies that $T^a = (T^a)^\dagger = (T^a)^{*\mathrm{T}}$ and consequently for real representations

$$\begin{aligned}\mathrm{Tr}\Big(T^a\left\{T^b,T^c\right\}\Big) &= \mathrm{Tr}\Big((T^a)^{*\mathrm{T}}\left\{(T^b)^{*\mathrm{T}},(T^c)^{*\mathrm{T}}\right\}\Big)\\ &= \mathrm{Tr}\Big((T^a)^{*}\left\{(T^b)^{*},(T^c)^{*}\right\}\Big) = -\,\mathrm{Tr}\Big(T^a\left\{T^b,T^c\right\}\Big) = 0.\end{aligned}$$

Algebras that have only safe representation are called *safe algebras*. These include, in particular, those that have only real representations. The simple Lie algebras have been classified by Cartan [GT]. The following ones are safe [Ge72a]:

$$\begin{array}{ll} \mathrm{SU}(2) \simeq \mathrm{SO}(3), & \mathrm{G}(2), \\ \mathrm{SO}(N),\ N \geq 4,\ N \neq 6, & \mathrm{F}(4), \\ \mathrm{Sp}(2N), & \mathrm{E}(6),\ \mathrm{E}(7),\ \mathrm{E}(8). \end{array} \tag{2.7.50}$$

Direct sums of these algebras are also safe.

Consequently, anomalies can only occur for gauge algebras that involve abelian factors U(1) (our example with one fermion field was an U(1) model) or unitary algebras SU(N), $N \geq 3$ [SO(6) $\simeq$ SU(4)]. The symmetry groups of the Standard Model are the non-safe SU(2) $\times$ U(1) and SU(3). Therefore, cancellation of anomalies must be obtained in another way.

The second possibility, the cancellation of anomalies between left-handed and right-handed contributions, is guaranteed if the left- and right-handed couplings are unitary equivalent, $T^a_{\mathrm{L}} = VT^a_{\mathrm{R}}V^{-1}$. Theories of these type are called *vector-like*. The internal currents that are coupled to the gluons in QCD are covariantly conserved vector currents. Therefore, the Lee identities of QCD are anomaly-free. On the other hand, the gluons can be coupled to *external* axial-vector currents. In this case, the external, singlet axial-vector current $j^0_{5,\mu}$ has an anomaly owing to $\mathrm{Tr}(T^aT^a) \neq 0$. But the renormalizability of QCD is not destroyed by this external current. Physical effects of this anomaly are discussed in Sect. 3.5.

The *Electroweak Standard Model* is built on the non-safe group SU(2) $\times$ U(1) and contains left- and right-handed currents which are coupled to the

[20] The notion comes from mathematics where $\mathrm{i}T^a$ is used instead of T^a.

gauge bosons. Compensation of anomalies can be realized only with special representations for the leptons and quarks. Since these are identical for the different families it is sufficient to discuss cancellation in one family. The generators of the group are weak isospin I_{w}^{a}, hypercharge Y_{w}, and the electric charge $Q = I_{\mathrm{w}}^{3} + Y_{\mathrm{w}}/2$. While the electric charge Q is the same for left-handed and right-handed fermions, i.e. vector-like, the isospin I^{a} is only non-vanishing for left-handed fermions. The following combinations occur in the triangle diagrams:

$$\mathrm{Tr}(I_{\mathrm{w}}^{a}\{I_{\mathrm{w}}^{b}, I_{\mathrm{w}}^{c}\}), \quad \mathrm{Tr}(Q\{I_{\mathrm{w}}^{a}, I_{\mathrm{w}}^{b}\}), \quad \mathrm{Tr}(I_{\mathrm{w}}^{a}Q^{2}), \quad \mathrm{Tr}(Q^{3}).$$

Since the group SU(2) is safe, the first term vanishes. The last term is purely vectorial and therefore gives a vanishing contribution. The second term gives

$$\mathrm{Tr}\Big(Q\{I_{\mathrm{w}}^{a}, I_{\mathrm{w}}^{b}\}\Big) = \frac{1}{2}\delta^{ab}\,\mathrm{Tr}\Big(Q\Big) = \frac{1}{2}\delta^{ab}\sum_{f} Q_{f}. \tag{2.7.51}$$

Because the charges of the members of all isospin doublets differ by one, the third term gives no independent constraint. Altogether, the *condition for cancellation of the anomalies* in the Electroweak Standard Model reads

$$\sum_{f} Q_{f} = 0. \tag{2.7.52}$$

This requires for one family, e.g. $(\nu_{\mathrm{e}}, \mathrm{e}, \mathrm{u}, \mathrm{d})$,

$$(0-1) + N_{\mathrm{C}}\left(\frac{2}{3} - \frac{1}{3}\right) = -1 + \frac{N_{\mathrm{C}}}{3} = 0.$$

Thus, three quark colours are needed for the compensation of anomalies.

In this context also the *non-renormalization theorem* [Ad69a, Ba69] is important. If the one-loop diagrams are anomaly-free, all higher-order diagrams are anomaly-free, i.e. the chiral anomalies are determined by the one-loop diagrams. This is a consequence of the fact that a chirally invariant regularization (method of higher covariant derivatives) exists for all diagrams beyond one loop.

2.7.2.3 Structure of anomalies in non-abelian gauge theories

Our discussion of anomalies was based on the evaluation of one-loop diagrams. The direct treatment of the regularized path-integral formula gives the same results. An even more general approach considers the Lee identity (2.4.77) for the generating functional of vertex functions $\tilde{\Gamma}$. This can be written in the form [cf. (2.5.140)]

$$\mathcal{B}_{\tilde{\Gamma}}\tilde{\Gamma} = 0 \tag{2.7.53}$$

with $\mathcal{B}_{\tilde{\Gamma}}$ defined as in (2.5.141). In the presence of anomalies the right-hand side does not vanish,

$$\mathcal{B}_{\tilde{\Gamma}}\tilde{\Gamma} = \Delta. \tag{2.7.54}$$

Applying $\mathcal{B}_{\tilde{\Gamma}}$ and using $\mathcal{B}_{\tilde{\Gamma}}\mathcal{B}_{\tilde{\Gamma}} = 0$ (2.5.142), we find that the anomalous term Δ must obey

$$\mathcal{B}_{\tilde{\Gamma}}\Delta = 0. \tag{2.7.55}$$

Power counting implies that Δ consists of polynomials in the fields with dimension 5 and ghost number +1. An analysis similar to the one in Sect. 2.5.4.2 of all possible polynomials reveals that Δ must contain the ghost field $u^a(x)$ and be of the form [Be76a]

$$\Delta = \int \mathrm{d}^4x\, p_A^a(x) u^a(x), \tag{2.7.56}$$

where $p_A^a(x)$ depends only on the gauge fields $A_\mu^a(x)$. The infinitesimal generator of gauge transformations of the gauge field

$$w^a(x) = D_\mu^{ca}\frac{\delta}{\delta A_\mu^c} = \partial_\mu\frac{\delta}{\delta A_\mu^a} - gf^{cab}A_\mu^b\frac{\delta}{\delta A_\mu^c} \tag{2.7.57}$$

has the property

$$w^a(x)w^b(y) - w^b(x)w^a(y) = gf^{abc}w^c(x)\delta^{(4)}(x-y). \tag{2.7.58}$$

Inserting this into (2.7.55) and using the definition of $\mathcal{B}_{\tilde{\Gamma}}$ as in (2.5.141) and (2.7.56), yields the *Wess–Zumino consistency condition* [We71] for $p_A^a(x)$,

$$w^a(x)p_A^b(y) - w^b(x)p_A^a(y) = gf^{abc}p_A^c(x)\delta^{(4)}(x-y). \tag{2.7.59}$$

This condition fixes the anomaly up to normalization. The result for the anomaly is identical with the symmetric anomaly (2.7.45) obtained by the evaluation of Feynman diagrams.

More results on the relation between anomalies and the topological properties of gauge theories using methods of differential forms or fibre bundles are presented in Sect. 2.9.3. An almost complete treatment is contained in [Be96].

2.8 Infrared and collinear singularities

Besides UV singularities, which result from infinite loop momenta and are removed by renormalization, quantum field theories contain singularities that are connected with finite momenta. The singularities of the Feynman integrals are called *Landau singularities* [La59] and occur in general only for special values of the external momenta. Some of these singularities occur, however, independently of the external momenta. Since they are always related to vanishing masses they are generically called *mass singularities.* There are two types of mass singularities. The *infrared (IR) or soft singularities* are connected with vanishing momenta and are thus long-distance effects. The *collinear singularities,* on the other hand, are related to collinear light-like momenta.

Soft and collinear singularities result not only from virtual soft or collinear massless particles in the loop integrals. The real emission of soft or collinear massless particles gives rise to similar singularities when integrating over the phase space of the emitted particles. The singularities resulting from real and virtual particles cancel each other in suitably defined observables.

Starting from the general singularity structure of Feynman integrals we discuss the origin of these singularities in loop integrals and their form in Sect. 2.8.1. In Sect. 2.8.2 we show that the infrared singularities cancel between virtual and real contributions and that the remaining terms exponentiate. The treatment of collinear singularities in QED is discussed in Sect. 2.8.3.

2.8.1 The origin of mass singularities

Mass singularities of Feynman amplitudes have been investigated extensively in Ref. [Ki62]. In order to classify these singularities it is useful to consider the general singularities of an arbitrary Feynman graph G. The corresponding amplitude has the generic structure

$$F_G(p_j, m_n) = \int \left(\prod_{l=1}^{L} \mathrm{d}^4 k_l \right) N(p_j, k_l, m_n) \prod_{i=1}^{I} \frac{\mathrm{i}}{q_i^2 - m_i^2 + \mathrm{i}\varepsilon}. \tag{2.8.1}$$

Here p_j denote the E external momenta, k_l the L loop momenta, and m_i the internal masses. The momenta q_i that flow through the I internal propagators are linear combinations of p_j and k_l. The numerator $N(p_j, k_l, m_n)$, which results from vertices and fermion propagators (in gauge theories we work in the Feynman gauge), is assumed to be a regular function of the momenta and masses.

2.8.1.1 Singularities of complex integrals

The possible singularities of Feynman amplitudes have been discussed in Ref. [Ed66]. Leaving aside the UV singularities at infinity, i.e. $k_l \to \infty$, the function F_G can only become singular if the integrand has a singularity on the integration path. This requires that one or more propagators develop a pole. Since the integrals in (2.8.1) are actually contour integrals in the complex plane, these poles can in general be avoided by deforming the integration contour. For any set of external momenta and masses for which this is the case, the function F_G is analytic in all these variables.

There are two situations where the poles cannot be circumvented by contour deformation. This can be best illustrated by a simple example. Consider a general function $I(z)$ of a complex variable z defined by an integral along a contour C in the complex w plane with fixed end points w_a and w_b

$$I(z) = \int_{C(w_a, w_b)} \mathrm{d}w \, f(w, z). \tag{2.8.2}$$

The integrand $f(w, z)$ is assumed to be analytic in the variables w and z, except for singularities located at $w = w_s(z)$. Evidently, $I(z)$ is analytic for any z for which the integration contour encounters no singularities. Even if one of the singularities $w_s(z)$ happens to lie on C, $I(z)$ remains analytic as long as the contour can be deformed to avoid it. Therefore, singularities of $I(z)$ can only occur in two cases.

1. One of the singularities $w_s(z)$ coincides with one of the endpoints of the contour C, $w_s(z) \to w_a$ or w_b for $z \to z_0$. Then, z_0 is called an *end-point singularity.*

2. The contour is trapped between at least two singularities w_1 and w_2, i.e. as $z \to z_0$, w_1 and w_2 approach the contour from opposite sides and coincide; in this case z_0 is called a *pinch singularity.*

The singularities produced by the above mechanisms are usually branch points.

Let us now turn to a function of several complex variables $\mathbf{w} = (w_i)$ and $\mathbf{z} = (z_j)$, $I(\mathbf{z}) = \int_H (\prod_i \mathrm{d}w_i) f(\mathbf{w}, \mathbf{z})$ [Ed66]. The boundary of the integration domain, the hyper-contour H, is specified by a set of analytic functions $\tilde{\mathcal{S}}_r(\mathbf{w}, \mathbf{z}) = 0$, and the singularities of the integrand take place on analytic manifolds $\mathcal{S}_s(\mathbf{w}, \mathbf{z}) = 0$. The integral becomes singular if the hyper-contour H is pinched between two or more surfaces of singularities or when a singularity surface meets a boundary surface. More precisely, it can be shown that a necessary condition for a singularity is the existence of a set of complex parameters λ_s and $\tilde{\lambda}_r$ which are not all equal to zero so that simultaneously

$$
\begin{aligned}
&\lambda_s \mathcal{S}_s(\mathbf{w}_0, \mathbf{z}_0) = 0 \quad \text{for all } s, \qquad \tilde{\lambda}_r \tilde{\mathcal{S}}_r(\mathbf{w}_0, \mathbf{z}_0) = 0 \quad \text{for all } r, \\
&\frac{\partial}{\partial w_i} \left[\sum_s \lambda_s \mathcal{S}_s(\mathbf{w}, \mathbf{z}) + \sum_r \tilde{\lambda}_r \tilde{\mathcal{S}}_r(\mathbf{w}, \mathbf{z}) \right] \Bigg|_{\mathbf{w}=\mathbf{w}_0, \mathbf{z}=\mathbf{z}_0} = 0 \quad \text{for all } i.
\end{aligned}
\tag{2.8.3}
$$

The last condition requires that at the pinching point $(\mathbf{w}_0, \mathbf{z}_0)$, the normals to the hyper-surfaces are linearly-dependent, i.e. tangent for two hyper-surfaces. Whether there is actually a singularity at this point requires a detailed study.

2.8.1.2 The Landau equations

Let us now apply these results to the Feynman integral (2.8.1). The boundaries of the integrand are at infinity. In the following, we disregard the endpoint singularities, which are actually UV singularities, and introduce no term $\tilde{\mathcal{S}}_r(\mathbf{w}, \mathbf{z})$. The singularities of the integrand are given by the equations $\mathcal{S}_i = q_i^2 - m_i^2 = 0$, i.e. the mass-shell conditions, and the necessary conditions (2.8.3) read

$$\lambda_i(q_i^2 - m_i^2) = 0 \qquad \text{for all } i = 1, \ldots, I, \tag{2.8.4}$$

and

$$\sum_i \lambda_i q_i \frac{\partial q_i}{\partial k_l} = 0 \qquad l = 1, \ldots, L. \tag{2.8.5a}$$

These are the *Landau equations* [La59].

Using the fact that the internal momenta are linear combinations of the external and loop momenta with coefficients ± 1 and that those internal momenta that depend on the loop momentum k_l lie on a single loop $\mathcal{L}_l$ the second equation can be written as

$$\sum_{i \in \mathcal{L}_l} \pm \lambda_i q_i = 0. \tag{2.8.5b}$$

Equation (2.8.4) implies that each internal line is either on its mass shell or the corresponding λ_i vanishes. In the latter case this line does not appear in the other Landau equations. Thus, each singularity can be associated with a *reduced diagram* where a subset of lines has been contracted to points and all remaining ones are on their mass shell.

Landau equations can also be derived for other representations of the Feynman integrals, their solutions being of course unique. The solution corresponding to $\lambda_i \neq 0$ for all i is called the leading singularity, while all others are non-leading. The non-leading ones are leading in the corresponding reduced diagrams.

The Landau equations determine the possible locations of all singularities of Feynman integrals, the *Landau singularities*, as a function of the external momenta p_i. In general, the singularities are branch points, such as threshold singularities, but do not necessarily give rise to divergences in transition amplitudes.

2.8.1.3 Singularities in physical regions

We now restrict ourselves to singularities that occur in the physical regions. For these singularities, it can be shown that the loop momenta k_l must be real and the variables λ_i real and positive at the solutions of the Landau equations. Since the external momenta p_j are real in the physical regions and each q_i is a linear combination of the k_l and p_j, also the q_i are real.

For the leading Landau singularity, as a consequence of (2.8.4) all the q_i are on the mass shell and can therefore be regarded as momenta of physical particles. Moreover, at the leading singularity, the diagram may be considered as a real physical process, i.e. the vertices may be regarded as real space–time points and the internal lines as trajectories of real on-shell particles [Co65]. To see this we must identify $\lambda_i m_i$ ($\lambda_i > 0$) with the proper time of flight of particle i between the two vertices at the endpoints of line i. Then, the space–time displacement between these two vertices equals $\Delta x = \lambda_i q_i$, and (2.8.5) guarantees that the total space–time displacement along a closed loop is zero. For non-leading singularities, the corresponding reduced diagrams provide physical pictures in the same way.

2.8.1.4 Mass singularities

A special kind of Landau singularities are the *mass singularities*. Firstly, they give rise to divergences and not only to branch points. Secondly, they appear for arbitrary directions of the external momenta, and their appearance depends only on the masses of the external and internal particles. An extensive discussion of mass singularities of Feynman amplitudes has been given in Ref. [Ki62]. We sketch the main results in the following. We assume that the amplitude contains no identical propagators.[21]

Let us consider the case of one-loop diagrams, where the integrals have the general structure (cf. Fig. 2.8)

$$F_G(p_j, m_n) = \int \mathrm{d}^4 k \, N(p_j, k, m_n) \prod_{i=1}^{I} \frac{\mathrm{i}}{q_i^2 - m_i^2 + \mathrm{i}\varepsilon} \tag{2.8.6}$$

[21] Amplitudes with identical propagators can be obtained from the considered ones by differentiation with respect to appropriate masses.

with $q_i = k+\sum_{n=1}^{i-1} p_n = k+P_i$. If the singularities have to be independent of the directions of the external momenta, the corresponding *Landau equations*

$$\lambda_i(q_i^2 - m_i^2) = 0 \qquad \text{for all } i = 1, \dots, I, \tag{2.8.7a}$$

$$\sum_{i=1}^{I} \lambda_i q_i = 0, \tag{2.8.7b}$$

must not involve scalar products of different external momenta. Equation (2.8.7a) implies that each q_i for which $\lambda_i \neq 0$ must have a norm independent of the orientation of the external momenta, i.e. it must be either light-like (for $m_i = 0$) or proportional to one of the external on-shell momenta. Since $q_i - q_j = \sum_{n=j}^{i-1} p_n$, this can be fulfilled for at most three neighbouring propagators, provided the loop momentum is chosen appropriately. Since (2.8.7b) relates the three corresponding momenta q_i, the squares of only two neighbouring momenta can in fact be independent of the directions of the external momenta.

The Landau conditions are necessary for a singularity. However, the actual appearance of a singularity depends on the multiplicity of the zeros in the numerator and denominator. The degree of the singularity is enhanced if some of the propagators for which $\lambda_i = 0$ become singular in addition. Since these poles should also be independent of the directions of the external momenta, at most three neighbouring propagators (including those with $\lambda_i = 0$) can contribute to the singularity.

We now consider the possible cases: if the Landau singularity results from only *one propagator*, $\lambda_r \neq 0$ but $\lambda_i = 0$ for $i \neq r$, the Landau equations imply $q_r = 0$ and $m_r = 0$, which is evidently independent of the external momenta. Choosing $q_r = k$, simple power counting shows that the integral behaves as $\int \mathrm{d}^4k/k^2$ for $q_r = k \to 0$ and does not diverge unless the possible singularity is enhanced by additional propagators. Each of the propagators $i \neq r$ can enhance the degree of the singularity by one if $(P_i - P_r)^2 - m_i^2 = 0$, since then $q_i^2 - m_i^2 = k^2 + 2k(P_i - P_r)$. However, these conditions depend on the orientation of the external momenta except for the two propagators $r+1$ and $r-1$ that are neighbours of the propagator r. If both contribute to the enhancement, the integral becomes logarithmically divergent. The singularity results from subdiagrams as the one shown in Fig. 2.12a and is

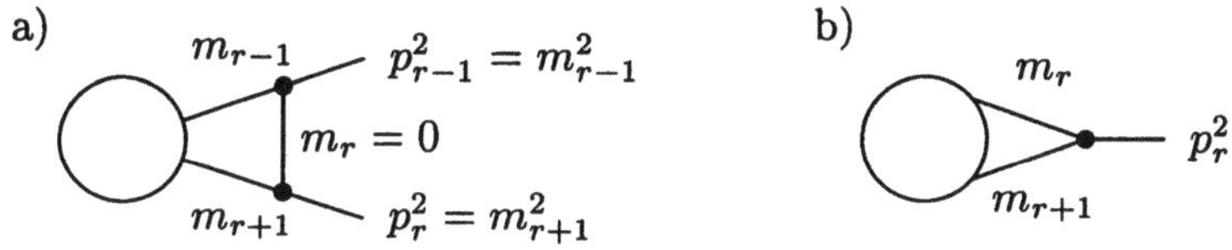

Fig. 2.12 Diagrams leading to IR (left) and collinear (right) singularities

called *infrared singularity* and defined by the conditions

$$m_r = 0, \qquad \begin{aligned} p_{r-1}^2 &= (P_r - P_{r-1})^2 = m_{r-1}^2, \\ p_r^2 &= (P_r - P_{r+1})^2 = m_{r+1}^2. \end{aligned} \tag{2.8.8}$$

Typically, this singularity occurs for photon or gluon exchange between external particles. It gives rise to divergences proportional to $\log m_r$ for $m_r \to 0$.

The IR singularity is absent, if the pole is cancelled by a zero in the numerator N, i.e. by a momentum k in the numerator. This may typically arise from the numerator of the propagator r or the two vertices at its endpoints. Therefore, in particular, massless neutrinos do not lead to IR singularities.

If the Landau singularity results from *two propagators*, i.e. $\lambda_r \neq 0$ and $\lambda_s \neq 0$ but $\lambda_i = 0$ for $i \neq r, s$, the Landau equations imply

$$q_r^2 - m_r^2 = 0, \qquad q_s^2 - m_s^2 = 0, \qquad \lambda_r q_r + \lambda_s q_s = 0. \tag{2.8.9}$$

The corresponding subdiagram is shown in Fig. 2.12b. The momenta q_r and q_s have to be proportional to each other, i.e. collinear, and

$$q_r = \pm\frac{m_r}{m_s} q_s, \qquad q_r^2 = m_r^2, \qquad (q_r - q_s)^2 = (m_r \pm m_s)^2. \tag{2.8.10}$$

The last condition is only angular-independent if r and s are neighbouring propagators, i.e. $s = r \pm 1$. Then $(q_r - q_s)^2$ is the square of the external momentum p_{r-1} or p_r, i.e. an external mass squared. In general, these singularities are the well-known branch points resulting from two-particle thresholds.

In order to determine whether a divergence actually occurs, we set $s = r+1$, parametrize the loop integral as $q_r = k$ and $q_{r+1} = k + p_r$, and use the Sudakov decomposition for the integration momentum [Su56],

$$k = yp_r + zn + k_\perp,$$
$$n^2 = 0, \quad p_r n \neq 0, \quad p_r k_\perp = 0 = nk_\perp, \quad k_\perp = (0, \mathbf{k}_\perp), \tag{2.8.11}$$

where n is an appropriate vector. Using

$$\int \mathrm{d}^4 k = |p_r n| \int \mathrm{d}^2\mathbf{k}_\perp \mathrm{d}y\, \mathrm{d}z = \pi |p_r n| \int \mathrm{d}\mathbf{k}_\perp^2 \mathrm{d}y\, \mathrm{d}z, \tag{2.8.12}$$

the relevant part of the integral reads

$$\begin{aligned} &\int \mathrm{d}^4 k \frac{1}{(k^2 - m_r^2)[(k+p_r)^2 - m_{r+1}^2]} = \\ &\pi|p_r n| \int \mathrm{d}^2\mathbf{k}_\perp \mathrm{d}y\, \mathrm{d}z \frac{1}{[p_r^2 y^2 + 2p_r n yz - \mathbf{k}_\perp^2 - m_r^2]} \\ &\qquad \times \frac{1}{[p_r^2(1+y)^2 + 2p_r n(1+y)z - \mathbf{k}_\perp^2 - m_{r+1}^2]}. \end{aligned} \tag{2.8.13}$$

For

$$p_r^2 = (q_{r+1} - q_r)^2 = 0, \qquad m_r^2 = 0, \qquad m_{r+1}^2 = 0, \tag{2.8.14}$$

this integral diverges logarithmically. The divergence originates from the part of the integration region where $\mathbf{k}_\perp^2 \to 0$ and $z \to 0$, i.e. $k \to yp_r$. Since this singularity is associated with collinear momenta, $k \propto p_r$, and massless particles $m_r = m_{r+1} = 0$, it is called *collinear singularity*. It yields singular terms of the form $\log m = \{\log m_r, \log m_s, \log \sqrt{|p_r^2|}\ \}$, depending on the variable that is sent to zero at the end. Condition (2.8.14) implies that all three lines meeting at a three vertex must be massless to produce a mass singularity. The mass singularity cannot be enhanced by other propagators, since the corresponding requirements $q_i^2 = m_i^2$ depend on the external momenta.

In practice, IR and collinear singularities can *overlap*, leading to divergences of the form $(c \log \lambda + \log m) \log m$.[22] The mass singularities of any one-loop

[22] Here λ regularizes the IR singularities [as m_r in (2.8.8)] and m the collinear singularities.

diagram can be written as a sum of terms of this kind and single logarithms $\log \lambda$ and $\log m$.

This analysis can be extended to the case of more loops [Ki62]. The requirement, that the singularity should be independent of the relative orientation of the external momenta implies immediately that only one vertex with an external line may exist between any pair of internal propagators that define the Landau singularity. Thus, the singularity is localized to a neighbourhood of one external vertex. Without enhancement, only the IR and collinear singularities discussed above are found in multi-loop diagrams. However, if there is strong enough enhancement, new types of mass singularities may occur.

At a mass singularity, a Feynman amplitude can be reduced to a product of irreducible components corresponding to reduced diagrams. For example, the reduced diagrams of Fig. 2.12 are those where the bubbles are contracted to a vertex. The strongest mass singularities are found in those reduced diagrams that involve only a single loop. Loops with two lines contribute primarily a $\log m$ factor, loops with three lines may contribute a factor $(c \log \lambda + \log m) \log m$. The parts of Feynman diagrams that give rise to these singularities are associated with one or two external lines corresponding to collinear or IR singularities. The relevant diagrams are shown in Fig. 2.12 and become singular if the conditions (2.8.8) or (2.8.14) are fulfilled.

Mass singularities do not only result from loop integrals but also from the integration of the phase space of physical particles created in some reaction. The mass singularities in the phase-space integrals can be related to the absorptive (imaginary) parts of loop integrals via the optical theorem. It can be shown, that the mass singularity of the absorptive part may not be as strong as that of the dispersive (real) part. Furthermore, the relation between mass singularities in virtual and real diagrams leads to cancellations when these contributions are added (cf. Sects. 2.8.2.3 and 2.8.3.9).

The physical origin of mass singularities is the presence of degeneracies in initial and/or final states. In the case of IR singularities, the states of a charged particle with an arbitrary number of soft photons are nearly degenerate. They are indistinguishable for a detector with finite energy resolution. In the case of a collinear singularity, the state of a light-like particle with momentum p is degenerate with the states of an arbitrary number of light-like particles with the same overall quantum numbers with momenta $\alpha_i p_i$

if $p = \sum_i \alpha_i p_i$. These states are indistinguishable for a detector with finite angular resolution. In physical observables all these states contribute simultaneously and the singularities cancel.

A general theorem concerning mass singularities is the *Kinoshita–Lee–Nauenberg (KLN) theorem* [Ki62, Le64]. It states that, as a consequence of unitarity, transition probabilities are finite when summed over all degenerate states in initial and final states. This is true order by order in perturbation theory for unrenormalized quantities or in renormalization schemes that do not introduce mass singularities via the renormalization constants. For the cancellation of IR singularities in QED with finite fermion masses it is sufficient to sum over degenerate final states, i.e. all states with arbitrary numbers of soft photons in the final state. This is the *Bloch–Nordsieck theorem* [Bl37].

In general one has to include contributions involving real radiation of soft and collinear particles in order to obtain observables that are free of mass singularities. The mass singularities typically cancel between real and virtual contributions. The cancellation of infrared and collinear singularities is discussed in the Sects. 2.8.2.3 and 2.8.3.9.

2.8.2 Infrared singularities

In the previous section we have seen [cf. Fig. 2.12a] that IR singularities in loop diagrams result from the exchange of a massless particle between two on-shell external legs, provided that the emission of this massless particle does not change the masses of the emitting particles. In physically relevant theories these massless particles are gluons or photons. In this section, we study the IR singularities in QED resulting from both virtual and real corrections. We give their general form and show how they are re-summed to all orders. We demonstrate that in QED the IR singularities resulting from virtual and real corrections cancel in physical observables. The complete treatment of IR singularities has been given by Yennie, Frautschi and Suura [Ye61]. We follow the simplified analysis of Weinberg [We65].

2.8.2.1 Amplitudes for soft-photon emission

Consider a process $i \to f$ involving any number of charged (and neutral) particles of any types. The corresponding amplitude is denoted by $\mathcal{M}^{i\to f}(p_j)$,

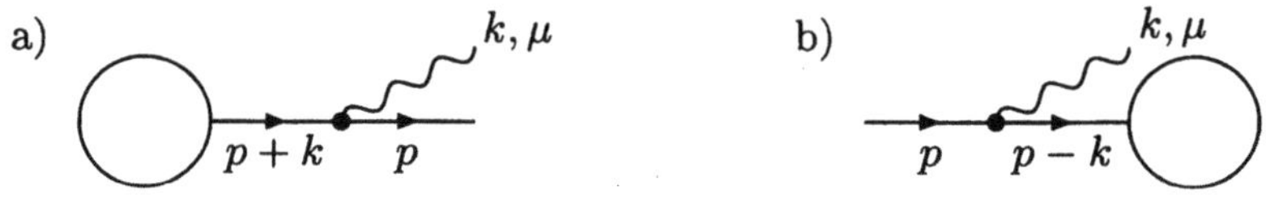

Fig. 2.13 Graphs for the emission of a photon (wavy line) from an outgoing (left) or incoming (right) charged particle

where the p_j denote the momenta of the external particles. We want to derive a universal formula that gives the amplitude for the processes $i \to f+n\gamma$ where n is any number of soft photons, i.e. photons with energies small compared to all energy scales in the hard process. Since we do not assume that the photon lines are on-shell and contracted with polarization vectors we can use the resulting formula to obtain directly the virtual IR singularities in Sect. 2.8.2.2 and the real singularities in Sect. 2.8.2.3.

Let us start with the emission of a single soft photon with momentum k and Lorentz index μ from an outgoing charged particle with momentum p (Fig. 2.13a). The corresponding amplitude is obtained from the one of the hard process $i \to f$ by inserting an additional charged particle propagator with momentum $k+p$ and an additional charged-particle–photon vertex. For an outgoing fermion with mass m and charge $-Qe$, the amplitude of the hard process can be written as (we single out the momentum p from the p_j for the time being)

$$\mathcal{M}^{i\to f}(p,p_j) = \bar{u}(p)A(p,p_j), \tag{2.8.15}$$

where $\bar{u}(p)$ is the spinor of the fermion and A the remaining part of the amplitude. The amplitude with photon emission reads

$$\mathcal{M}_\mu^{i\to f\gamma}(p,p_j,k) = \bar{u}(p)[-\mathrm{i}eQ\gamma_\mu]\left[\frac{\mathrm{i}(\not{p}+\not{k}+m)}{(p+k)^2-m^2+\mathrm{i}\varepsilon}\right]A(p+k,p_j). \tag{2.8.16}$$

If we assume that $A(p,p_j)$ is sufficiently smooth[23] in p, this becomes in the limit $k\to 0$, i.e. in the *soft-photon approximation*,

$$\mathcal{M}_\mu^{i\to f\gamma}(p,p_j,k) \underset{k\to 0}{\sim} \bar{u}(p)\frac{2eQp_\mu}{2pk+\mathrm{i}\varepsilon}A(p,p_j) = \frac{2eQp_\mu}{2pk+\mathrm{i}\varepsilon}\mathcal{M}^{i\to f}(p,p_j). \tag{2.8.17}$$

[23]This result has, for instance, to be modified in the presence of a resonance unless the energy of the soft photon is small compared to the width of the resonance.

The vanishing denominator leads to IR singularities for $k \to 0$. For a photon emitted from an outgoing scalar particle, we find

$$\begin{aligned}
&\mathcal{M}_{\mu}^{i\to f\gamma}(p,p_j,k) = \\
&\quad = \left[-\mathrm{i}eQ(2p+k)_{\mu}\right]\left[\frac{\mathrm{i}}{(p+k)^2-m^2+\mathrm{i}\varepsilon}\right]\mathcal{M}^{i\to f}(p+k,p_j) \\
&\quad \underset{k\to 0}{\sim} \frac{2eQp_{\mu}}{2pk+\mathrm{i}\varepsilon}\mathcal{M}^{i\to f}(p,p_j),
\end{aligned} \tag{2.8.18}$$

i.e. the same result. In fact, independently of the spin of the charged particle the emission of a soft photon from an outgoing particle yields a factor

$$\frac{2eQp_{\mu}}{2pk+\mathrm{i}\varepsilon}, \tag{2.8.19}$$

in the soft-photon approximation. Any spin-dependent terms are IR-regular.

The emission of a photon from an incoming charged particle can be treated in the same way. Since after the emission of the photon the charged particle has momentum $p-k$ we find a factor

$$\frac{2eQp_{\mu}}{-2pk+\mathrm{i}\varepsilon}. \tag{2.8.20}$$

instead of (2.8.19). If the soft photon is emitted from an (off-shell) internal line, no IR-singular factor results, since no IR-singular propagator occurs.

Summing over all external particles, we obtain the amplitude for the emission of a single soft photon from the considered process

$$\mathcal{M}_{\mu}^{i\to f\gamma}(p_j,k) \underset{k\to 0}{\to} \mathcal{M}^{i\to f}(p_j)\sum_{l}\frac{2eQ_l p_{l,\mu}}{2\eta_l p_l k+\mathrm{i}\varepsilon}, \tag{2.8.21}$$

where Q_l and p_l are the relative charges and momenta of the lth external particle and η_l is a sign factor that equals $+1$ for outgoing and -1 for incoming particles. Note that the charges of antiparticles have to be taken opposite to the charges of particles in (2.8.21).

Let us now consider the emission of two photons with momenta k_1 and k_2 and Lorentz indices μ_1 and μ_2 (Fig. 2.14). If the two photons are emitted from different external lines, each of them gives a factor like in (2.8.19) or (2.8.20) to the matrix element of the hard process $i \to f$. If the two

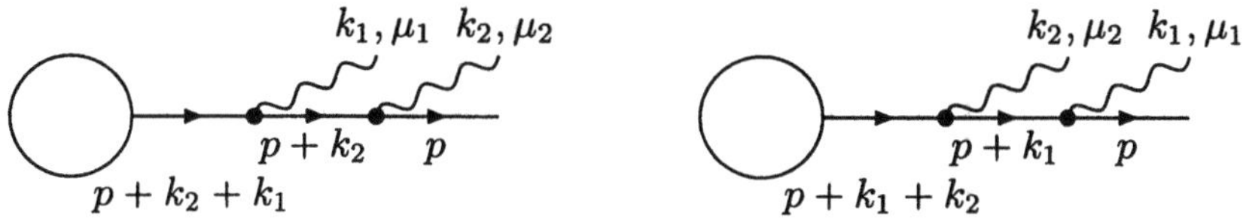

Fig. 2.14 Graphs for the emission of two photons from the same outgoing charged particle

photons are emitted from the same external line we have two contributions, one where photon 1 is emitted first and one where photon 2 is emitted first. Adding the factors from these two contributions yields in the limit $k_i \to 0$

$$\begin{aligned} &\left(\frac{2eQp_{\mu_1}}{2\eta pk_1 + \mathrm{i}\varepsilon}\right)\left(\frac{2eQp_{\mu_2}}{2\eta p(k_1+k_2) + \mathrm{i}\varepsilon}\right) \\ &\quad + \left(\frac{2eQp_{\mu_2}}{2\eta pk_2 + \mathrm{i}\varepsilon}\right)\left(\frac{2eQp_{\mu_1}}{2\eta p(k_1+k_2) + \mathrm{i}\varepsilon}\right) \\ &= \left(\frac{2eQp_{\mu_2}}{2\eta pk_2 + \mathrm{i}\varepsilon}\right)\left(\frac{2eQp_{\mu_1}}{2\eta pk_1 + \mathrm{i}\varepsilon}\right), \end{aligned} \tag{2.8.22}$$

i.e. the product of two factors (2.8.19). The corresponding result for the emission of an arbitrary number of photons can be derived by induction. As a consequence, soft photons are emitted independently.

Thus, we find the amplitude for the emission of n soft photons with momenta $k_1, \ldots, k_n$ and Lorentz indices $\mu_1, \ldots, \mu_n$ in the process $i \to f$ in the limit $k_i \to 0$ by multiplying the matrix element of the hard process by a product of factors like the one in (2.8.21),

$$\mathcal{M}^{i\to fn\gamma}_{\mu_1\ldots\mu_n}(p_j, k_i) \underset{k_i\to 0}{\sim} \prod_{m=1}^{n}\left(\sum_l \frac{2Q_l e p_{l,\mu}}{2\eta_l p_l k_m + \mathrm{i}\varepsilon}\right)\mathcal{M}^{i\to f}(p_j). \tag{2.8.23}$$

2.8.2.2 Virtual soft-photon amplitudes

Let us use the results of the previous section to calculate the virtual radiative corrections originating from soft photons for the process $i \to f$. To this end, we have to consider the amplitude (2.8.23) for an even number of soft photons, join pairs of soft-photon lines to form virtual photons, and sum over all different pairings. For each virtual photon we must multiply with a propagator factor $-\mathrm{i}g^{\mu\nu}/(k^2+\mathrm{i}\varepsilon)$ and integrate over the photon momentum. In this way we get the *virtual corrections* owing to n soft photons to the process $i \to f$,

$$\mathcal{M}^{i\to f}_{n\gamma,\mathrm{soft}}(p_j) = \frac{1}{n!}\left[\frac{1}{2}\sum_{l,m} e^2 Q_l Q_m J_{lm}\right]^n \mathcal{M}^{i\to f}(p_j) \qquad (2.8.24)$$

with

$$J_{lm} = -\mathrm{i}\int \frac{\mathrm{d}^4k}{(2\pi)^4}\frac{4p_l p_m}{(k^2+\mathrm{i}\varepsilon)(2\eta_l p_l k+\mathrm{i}\varepsilon)(-2\eta_m p_m k+\mathrm{i}\varepsilon)}. \qquad (2.8.25)$$

The factor $1/n!$ cancels the $n!$ equivalent permutations of one of the ends of the photon lines and the factor $1/2$ in the bracket the 2 equivalent contributions resulting from the interchanges of the two ends of each of the photon lines. The sign of $p_m k$ has been changed in one of the denominators of (2.8.25) because the photon momentum must be incoming in one of the photon lines of each pair. The sum in (2.8.24) involves also terms where both ends of the photon lines are attached to the same external leg ($l = m$). It has been shown in Ref. [Ye61] that (2.8.25) is also valid for this case. These terms originate from the wave-function renormalization constants that appear in the LSZ formula relating truncated Green functions and S-matrix elements and/or in the field renormalization constants of the external fields [cf. Sect. 2.1.3 and (2.5.7)]. The factor $1/2$ in (2.8.24) results in this case from the fact that these renormalization constants are determined from the square-roots of the residues of the propagators (cf. Sect. 2.1.3).

Summing over the number of virtual photons n in (2.8.24), we find for the radiative corrections owing to any number of virtual photons

$$\mathcal{M}^{i\to f}_{\mathrm{virt,soft}}(p_j, k_i) = \exp\left[\frac{1}{2}\sum_{l,m} e^2 Q_l Q_m J_{lm}\right]\mathcal{M}^{i\to f}(p_j). \qquad (2.8.26)$$

The integral (2.8.25) is both UV- and IR-divergent. The UV divergence can be eliminated by restoring the k^2 terms, which are irrelevant in the IR limit, in the denominator, and the IR divergences can be regularized by an infinitesimal photon mass λ. Then, the integral (2.8.25) becomes

$$J_{lm} = -\mathrm{i} \int \frac{\mathrm{d}^4 k}{(2\pi)^4} \frac{1}{(k^2 - \lambda^2 + \mathrm{i}\varepsilon)} \times \frac{4 p_l p_m}{(k^2 + 2\eta_l p_l k + \mathrm{i}\varepsilon)(k^2 - 2\eta_m p_m k + \mathrm{i}\varepsilon)}. \tag{2.8.27}$$

This is just an ordinary scalar three-point function (2.5.65). Using standard techniques it can be evaluated to

$$\begin{aligned} J_{lm} &= \frac{\eta_l \eta_m}{8\pi^2} \frac{1}{\beta_{lm}} \log \frac{\lambda^2}{m_l m_m} \log\left(\frac{\eta_l \eta_m p_l p_m (\beta_{lm} - 1)}{m_l m_m} + \mathrm{i}\varepsilon \right) \\ &\quad + \text{IR-finite terms} \\ &= \frac{\eta_l \eta_m}{16\pi^2} \frac{1}{\beta_{lm}} \log \frac{\lambda^2}{m_l m_m} \left(\log \frac{1 - \beta_{lm}}{1 + \beta_{lm}} + 2\mathrm{i}\pi\theta(\eta_l \eta_m) \right) \\ &\quad + \text{IR-finite terms} \end{aligned} \tag{2.8.28}$$

for $l \neq m$, where

$$\beta_{lm} = \sqrt{1 - \frac{m_l^2 m_m^2}{(p_l p_m)^2}} \tag{2.8.29}$$

is the relative velocity of the particles l and m in the rest frame of either one, and we used the fact that the scalar product of two physical momenta is positive, $p_l p_m > 0$.

The imaginary part in (2.8.28) leads to an IR-divergent phase factor in (2.8.26), which drops out when taking the absolute square of the matrix element to calculate the cross section for the process $i \to f$. This phase factor is the relativistic counterpart of the well-known *Coulomb phase.*

The integral J_{ll} can be obtained from (2.8.28) by putting $\eta_l \eta_m = 1$ and taking the limit $\beta_{lm} \to 0$. We thus find for the real part,

$$\operatorname{Re} J_{ll} = -\frac{1}{8\pi^2} \log \frac{\lambda^2}{m_l^2} + \text{IR-finite terms}, \tag{2.8.30}$$

while the irrelevant imaginary part diverges for $\beta_{lm} \to 0$.

For the study of the IR-singular terms it is more convenient to regularize the UV and IR divergences by introducing an upper and lower cut-off on the momenta of the virtual photons, i.e. $k_{\min} < |\mathbf{k}| < k_{\max}$. The integral over k_0 in (2.8.25) can be done by the method of residues. The integrand is analytic in k_0 except for the poles at

$$\begin{aligned} k^0 &= |\mathbf{k}| - \mathrm{i}\varepsilon, & k^0 &= -|\mathbf{k}| + \mathrm{i}\varepsilon, \\ k^0 &= \frac{2\mathbf{p}_l\mathbf{k} - \eta_l \mathrm{i}\varepsilon}{2p_{l,0}}, & k^0 &= \frac{2\mathbf{p}_m\mathbf{k} + \eta_m \mathrm{i}\varepsilon}{2p_{k,0}}. \end{aligned} \tag{2.8.31}$$

The first two poles are called photon poles and the last two particle poles. The integral can be evaluated by closing the integration contour at infinity. If particle m is outgoing ($\eta_m = +1$) and particle l is incoming ($\eta_l = -1$) or vice versa the particle poles can be avoided by closing the contour in the upper or lower half plane, respectively. Consequently, the integral is determined by one of the photon poles only and turns out to be purely real. If particles l and m are both incoming or both outgoing, the particle poles lie on opposite sides of the real k_0 axis and one of them contributes to the integral. Thus, we find

$$\begin{aligned} J_{lm} &= -\mathrm{i}\int_{k_{\min}<|\mathbf{k}|<k_{\max}} \frac{\mathrm{d}^4k}{(2\pi)^4} \frac{4p_lp_m}{(k^2+\mathrm{i}\varepsilon)(2\eta_lp_lk+\mathrm{i}\varepsilon)(-2\eta_mp_mk+\mathrm{i}\varepsilon)} \\ &= \int_{k_{\min}<|\mathbf{k}|<k_{\max}} \frac{\mathrm{d}^3k}{(2\pi)^3}\left[\frac{p_lp_m\eta_l\eta_m}{2|\mathbf{k}|(p_{l,0}|\mathbf{k}| - \mathbf{p}_l\mathbf{k})(p_{m,0}|\mathbf{k}| - \mathbf{p}_m\mathbf{k})}\right. \\ &\quad \left. - \theta(\eta_l\eta_m)\frac{p_{l,0}^2}{[p_{l,0}^2|\mathbf{k}|^2 - (\mathbf{p}_l\mathbf{k})^2]}\frac{p_lp_m}{[p_{m,0}(\mathbf{p}_l\mathbf{k}) - p_{l,0}(\mathbf{p}_m\mathbf{k}) - \mathrm{i}\varepsilon]}\right]. \end{aligned} \tag{2.8.32}$$

Integration over $|\mathbf{k}|$ makes the IR singularity explicit,

$$\begin{aligned} J_{lm} &= \frac{1}{(2\pi)^3}\log\frac{k_{\max}}{k_{\min}}\int \mathrm{d}\Omega_{\mathbf{k}}\left[\frac{p_lp_m\eta_l\eta_m}{2(p_{l,0} - \mathbf{p}_l\hat{\mathbf{k}})(p_{m,0} - \mathbf{p}_m\hat{\mathbf{k}})}\right. \\ &\quad \left. - \theta(\eta_l\eta_m)\frac{p_{l,0}^2}{[p_{l,0}^2 - (\mathbf{p}_l\hat{\mathbf{k}})^2]}\frac{p_lp_m}{[p_{m,0}(\mathbf{p}_l\hat{\mathbf{k}}) - p_{l,0}(\mathbf{p}_m\hat{\mathbf{k}}) - \mathrm{i}\varepsilon]}\right]. \end{aligned} \tag{2.8.33}$$

The angular integration finally leads to

$$J_{lm} = \frac{1}{8\pi^2}\frac{1}{\beta_{lm}}\left[\eta_l\eta_m\log\frac{1+\beta_{lm}}{1-\beta_{lm}} - 2\pi\mathrm{i}\theta(\eta_l\eta_m)\right]\log\frac{k_{\max}}{k_{\min}} \tag{2.8.34}$$

for $l \neq m$, and the real part of J_{ll} can easily be found from the limit $\beta_{lm} \to 0$. For high energies, $E \gg m$, (2.8.34) contains a product of two large logarithms, called Sudakov double logarithms. Note that the terms proportional to $\log k_{\min}$ in (2.8.34) are equal to the terms proportional to $\log \lambda$ in (2.8.28). In fact, the effect of virtual soft photons on the absolute square of the matrix element is obtained from (2.8.26) and (2.8.34) as

$$|\mathcal{M}^{i\to f}_{\text{virt,soft}}|^2 = \left(\frac{k_{\min}}{k_{\max}}\right)^{\alpha B^{i\to f}} |\mathcal{M}^{i\to f}|^2 \tag{2.8.35}$$

with the exponent

$$\begin{aligned} B^{i\to f} &= -\frac{1}{4\pi^2}\sum_{l,m} Q_l Q_m \int \mathrm{d}\Omega_{\mathbf{k}} \frac{p_l p_m \eta_l \eta_m}{(p_{l,0} - \mathbf{p}_l\hat{\mathbf{k}})(p_{m,0} - \mathbf{p}_m\hat{\mathbf{k}})} \\ &= -\frac{1}{2\pi}\sum_{l,m} Q_l Q_m \eta_l \eta_m \frac{1}{\beta_{lm}} \log\frac{1+\beta_{lm}}{1-\beta_{lm}}, \end{aligned} \tag{2.8.36}$$

where the contributions for $l = m$ have to be obtained from the limit $\beta_{lm} \to 0$.

For instance, for a process with a single charged particle with charge Q in the initial and final state we find

$$B = \frac{Q^2}{2\pi}\left[\frac{2}{\beta}\log\frac{1+\beta}{1-\beta} - 4\right] \tag{2.8.37}$$

from (2.8.36). The logarithmic term originates from the virtual photon exchange between the initial-state and final-state charged particle ($\beta_{lm} = \beta$ and $\eta_l\eta_m = -1$) and the constant from the virtual photon attached to the final-state particle and the initial-state particle only ($\beta_{lm} = 0$ and $\eta_l\eta_m = 1$). The exponent is positive for all $0 < \beta < 1$. For $\beta = 0$, B vanishes, and for $\beta \to 1$, it is given by

$$B = \frac{2Q^2}{\pi}\left[\log\frac{(p-p')^2}{m^2} - 1\right], \tag{2.8.38}$$

where p and p' are the momenta of the incoming and outgoing charged particles.

Also in the general case B is always positive. As a consequence, the effect of the IR divergences summed to all orders makes the rate for any charged particle process $i \to f$ vanish in the limit $\lambda \to 0$.

2.8.2.3 Real soft-photon contributions and cancellation of IR divergences

The IR divergences lead to divergent decay rates and cross sections for processes involving charged particles and definite numbers of soft photons in finite orders of perturbation theory. When summed to all orders, the results vanish. However, processes with a fixed number of photons are unobservable, since they cannot be distinguished experimentally from those involving additional soft external photons. Because photons are massless, their energies can be arbitrarily small and thus less than the resolution of any detector. Therefore, the only processes that can be observed are those where in addition to a definite number of charged particles and observed photons with energies greater than some minimal photon energy ΔE, an arbitrary number of soft photons with energies less than ΔE, and some small total energy E_{tot} are produced.

The S-matrix element for the emission of n real soft photons in a process $i \to f$ is obtained by contracting each of the n Lorentz indices $\mu_1, \dots, \mu_n$ of the amplitude (2.8.23) with a photon polarization vector $\varepsilon^*_\mu(k_i, \lambda_i)$, where λ_i denotes the helicity of the corresponding photon,

$$\mathcal{M}^{i\to f n\gamma}_{\text{soft}}(p_j, k_i) = \mathcal{M}^{i\to f}(p_j) \prod_{m=1}^{n} \left(\sum_l \frac{2eQ_l\, p_l \varepsilon^*(k_m, \lambda_m)}{2\eta_l p_l k_m + \mathrm{i}\varepsilon} \right). \quad (2.8.39)$$

Since $p_l^2 \neq 0$, the denominators are always non-zero for kinematical reasons, and $\mathrm{i}\varepsilon$ can be omitted. The factor

$$j^\mu(k) = \sum_l \frac{eQ_l p_l^\mu}{\eta_l p_l k} \quad (2.8.40)$$

appearing in this matrix element is sometimes called the *eikonal* or *convection current*. The corresponding term for a single charged particle in the initial and final state,

$$j^\mu(k) = eQ \left(\frac{p'^\mu}{p'k} - \frac{p^\mu}{pk} \right), \quad (2.8.41)$$

equals the Fourier transform of the classical current resulting from a classical charge that is instantaneously accelerated from momentum p to momentum

p'. Because of the factorization of the soft-photon contributions discussed in Sect. 2.8.2.1, (2.8.39) holds also in the presence of virtual soft photons.

The cross section is obtained by squaring the matrix element [cf. (1.2.24)]. Since the soft photons are not detected, we must sum over their polarizations and integrate their momenta over the available phase space. Owing to charge conservation, $\sum_l \eta_l Q_l = 0$, the terms in the polarization sum (A.1.46) involving the photon momenta do not contribute, and the polarization sum reduces to

$$\sum_\lambda \varepsilon^*_\mu(k,\lambda)\varepsilon_\nu(k,\lambda) = -g_{\mu\nu}. \tag{2.8.42}$$

Putting everything together, we arrive at the differential cross section for the process $i \to f$ with n undetected real soft photons,

$$\begin{aligned} \mathrm{d}\sigma^{i\to fn\gamma}_{\text{soft}}(p_j) &= \mathrm{d}\sigma^{i\to f}(p_j)\frac{1}{n!}\prod_{i=1}^{n}\int\frac{\mathrm{d}^3\mathbf{k}_i}{(2\pi)^3 2E_i}\sum_{l,m}\frac{-e^2 Q_l Q_m\, p_l p_m}{\eta_l\eta_m\, p_l k_i\, p_m k_i} \\ &= \mathrm{d}\sigma^{i\to f}(p_j)\frac{1}{n!}\left(\sum_{l,m} e^2 Q_l Q_m I_{lm}\right)^n, \end{aligned} \tag{2.8.43}$$

where E_i is the energy of photon i,

$$I_{lm} = -\int\frac{\mathrm{d}^3\mathbf{k}}{(2\pi)^3 2E}\frac{\eta_l\eta_m p_l p_m}{(p_l k)(p_m k)}, \tag{2.8.44}$$

and the factor $1/n!$ originates from the n identical photons in the final state. Note that in the soft-photon approximation the photon momenta are neglected in the overall momentum-conservation δ-function. For simplicity we disregard the fact that the sum of the energies of the soft photons is bounded by E_{tot}. The general case can be found in [Ye61, We65]. Summing over n we obtain the differential cross section for the process $i \to f$ with an arbitrary number of undetected real soft photons

$$\mathrm{d}\sigma^{i\to f}_{\text{real,soft}}(p_j) = \mathrm{d}\sigma^{i\to f}(p_j)\exp\left[\sum_{l,m} e^2 Q_l Q_m I_{lm}\right]. \tag{2.8.45}$$

Because the photon energy is restricted by $E < \Delta E$ in a certain reference frame, the integral (2.8.44) is not Lorentz-invariant. Moreover, it is

IR-divergent. For a finite photon mass as IR regulator it has been worked out, for instance, by 't Hooft and Veltman [tH79],

$$I_{lm} = \frac{\eta_l \eta_m}{8\pi^2} \log \frac{2\Delta E}{\lambda} \frac{1}{\beta_{lm}} \log \frac{1-\beta_{lm}}{1+\beta_{lm}} + \text{IR-finite terms.} \tag{2.8.46}$$

Note that the IR singularity contained in this integral is just the negative of the IR singularity contained in (2.8.28).

If we regularize the IR divergence instead with a photon mass with a lower cut-off on the photon energy, $E > k_{\text{min}}$, we find [cf. (2.8.33) and (2.8.34)]

$$\begin{aligned} I_{lm} &= -\int_{k_{\text{min}}<|\mathbf{k}|<\Delta E} \frac{\mathrm{d}^3 k}{(2\pi)^3} \frac{\eta_l \eta_m p_l p_m}{2|\mathbf{k}|(p_{l,0}|\mathbf{k}| - \mathbf{p}_l \mathbf{k})(p_{m,0}|\mathbf{k}| - \mathbf{p}_m \mathbf{k})} \\ &= -\frac{1}{(2\pi)^3} \log \frac{\Delta E}{k_{\text{min}}} \int \mathrm{d}\Omega_{\mathbf{k}} \frac{\eta_l \eta_m p_l p_m}{2(p_{l,0} - \mathbf{p}_l \hat{\mathbf{k}})(p_{m,0} - \mathbf{p}_m \hat{\mathbf{k}})} \\ &= -\frac{1}{8\pi^2} \frac{1}{\beta_{lm}} \eta_l \eta_m \log \frac{1+\beta_{lm}}{1-\beta_{lm}} \log \frac{\Delta E}{k_{\text{min}}} \end{aligned} \tag{2.8.47}$$

with the same angular integral as in (2.8.33). Evidently, I_{lm} is just the negative of the photon-pole contribution to J_{lm} if both the maximal and the minimal photon energies are chosen equal. In any case, the IR singularity in I_{lm} equals the one in J_{lm} up to a minus sign if both integrals are regularized in the same way.[24]

The contribution of real soft photons to a hard process can be written as

$$\mathrm{d}\sigma^{i\to f}_{\text{real,soft}}(p_j) = \mathrm{d}\sigma^{i\to f}(p_j) \left(\frac{\Delta E}{k_{\text{min}}}\right)^{\alpha B^{i\to f}} \tag{2.8.48}$$

with $B^{i\to f}$ from (2.8.36).

Because of the factorization discussed at the end of Sect. 2.8.2.1, real and virtual soft-photon corrections yield independent factors to the hard-scattering matrix element. Consequently, the differential cross section for the process

[24] Note that k_{min} has to be defined in the same reference frame for both virtual and real photons.

$i \to f$ with an arbitrary number of virtual and undetected real soft photons reads

$$\begin{aligned} d\sigma^{i\to f}_{\text{soft}}(p_j) &= d\sigma^{i\to f}(p_j) \exp\left[\sum_{l,m} e^2 Q_l Q_m (\operatorname{Re} J_{lm} + I_{lm})\right] \\ &= d\sigma^{i\to f}(p_j) \left(\frac{\Delta E}{k_{\min}}\right)^{\alpha B^{i\to f}} \left(\frac{k_{\min}}{k_{\max}}\right)^{\alpha B^{i\to f}} \\ &= d\sigma^{i\to f}(p_j) \left(\frac{\Delta E}{k_{\max}}\right)^{\alpha B^{i\to f}} . \end{aligned} \tag{2.8.49}$$

Thus, the correction factor is independent of the IR cut-off $k_{\min}$ and in general independent of the way the IR singularities are regularized. Since $k_{\max}$ can be chosen arbitrarily large, the correction factor is always between 0 and 1 and decreases with the soft-photon cut-off ΔE. If photons with energies larger than $k_{\max}$ are taken into account in the hard cross section $d\sigma^{i\to f}(p_j)$ the dependence on $k_{\max}$ cancels exactly. The fact that IR singularities cancel between virtual and real soft-photonic corrections in QED is known as the *Bloch–Nordsieck theorem* [Bl37].

2.8.3 Collinear singularities in QED

Radiative corrections by soft real and virtual photons were discussed in lowest order and higher orders in the last section. A further source of mass singularities is the emission of collinear but not soft photons. These singularities are regularized by the mass m of the emitting particles which is non-zero in QED. However, for scattering processes at high energies, $E \gg m$, the corresponding contributions are important since they are of the order of $\alpha \log(E^2/m^2)$. They are usually denoted as *leading-logarithmic* (LL) *contributions*. The logarithmic enhancement may require the summation of the multiple radiation of collinear photons.

2.8.3.1 Radiation of a collinear photon

We start by considering the emission of a photon from an incoming electron in a general scattering process, i.e. $e + i \to \gamma + f$, where i and f denote sets of other incoming and outgoing particles, respectively (cf. Fig. 2.13b). We want to extract the LL contribution that results from the emission of collinear photons, i.e. photons that are radiated almost parallel to the high-energetic electron and remain undetected. To this end we have to integrate over the phase space of these collinear photons.

For the invariant matrix element we use a notation similar to (2.8.15) and (2.8.16)

$$\begin{aligned}\mathcal{M}^{ei\to\gamma f}_{\kappa\lambda}(p,k,p_j) &=\\ &= \bar{A}^{ei\to f}(p-k,p_j)\frac{\mathrm{i}(\not{p}-\not{k}+m)}{(p-k)^2-m^2+\mathrm{i}\varepsilon}[\mathrm{i}e\not{\varepsilon}^*(k,\lambda)]u(p,\kappa),\end{aligned} \tag{2.8.50}$$

where p, k, and p_j denote the momenta of the electron, the photon, and the other particles, respectively, and $\kappa = \pm$ and $\lambda = \pm$ the helicities of the incoming electron and outgoing photon, respectively. When extracting the LL contributions at high energies, $E \gg m$, we can neglect the mass of the incoming electron in the matrix element,[25] $p^2 = m^2 = 0$. The mass singularity then appears as a logarithmically divergent integral and the electron mass has to be reintroduced in the phase-space integral as lower cut-off. Then, the four-momenta of the incoming electron, the photon, and the internal electron can be parametrized as

$$\begin{aligned}p^\mu &= (E,0,0,E),\\ k^\mu &= \left((1-x)E+\frac{\mathbf{k}_\perp^2}{4(1-x)E},0,0,(1-x)E-\frac{\mathbf{k}_\perp^2}{4(1-x)E}\right)+k_\perp^\mu,\\ p'^\mu &= p^\mu-k^\mu = \left(xE-\frac{\mathbf{k}_\perp^2}{4(1-x)E},0,0,xE+\frac{\mathbf{k}_\perp^2}{4(1-x)E}\right)-k_\perp^\mu,\\ k_\perp^\mu &= (0,\mathbf{k}_\perp) = (0,|\mathbf{k}_\perp|\cos\phi,|\mathbf{k}_\perp|\sin\phi,0),\end{aligned} \tag{2.8.51}$$

and obey

$$p^2 = 0, \qquad k^2 = 0, \qquad p'^2 = -\frac{\mathbf{k}_\perp^2}{1-x}. \tag{2.8.52}$$

The variable x determines the energy of the emitted photon $E_k \approx (1-x)E$ and of the intermediate fermion $E' \approx xE$, i.e. the energy for the scattering

[25] Terms involving the electron mass lead to integrals of the form $\int \mathrm{d}|\mathbf{k}_\perp|^2 m^i/|\mathbf{k}_\perp|^{i+2}$, $i > 0$ and thus do not give rise to large logarithms. They are, however, relevant for the non-logarithmic finite terms.

process $\mathrm{e}+i \to f$. The soft-photon limit corresponds to $x \to 1$ and $\mathbf{k}_\perp \to 0$. The transverse momentum $\mathbf{k}_\perp$ is a measure of the acollinearity of the photon and determines the off-shellness of the intermediate fermion $p'^2 = -\mathbf{k}_\perp^2/(1-x)$ (instead of the on-shell value $p'^2 = 0$). The LL contributions come from small values of the denominator $p'^2 \sim \mathbf{k}_\perp^2 \sim 0$, i.e. from the emission of collinear photons, $|\mathbf{k}_\perp| \ll (1-x)E$.

Up to terms of the order of $p'^2 = \mathcal{O}(\mathbf{k}_\perp^2)$, the numerator of the fermion propagator can be replaced by the polarization sum of on-shell fermions with momentum $p' = p - k$,

$$\not{p}' = \sum_\kappa u(p',\kappa)\bar{u}(p',\kappa), \tag{2.8.53}$$

resulting in

$$\begin{aligned}\mathcal{M}^{ei\to\gamma f}_{\kappa\lambda}(p,k,p_j) &= -\bar{A}^{ei\to f}(p-k,p_j)\\ &\quad\times\sum_{\kappa'} u(p',\kappa')\frac{1}{(p-k)^2}e\bar{u}(p',\kappa')\not{\varepsilon}^*(k,\lambda)u(p,\kappa) + \mathcal{O}(|\mathbf{k}_\perp|^0)\\ &= \sum_{\kappa'} \mathcal{M}^{ei\to f}_{\kappa'}(xp,p_j)\frac{1-x}{\mathbf{k}_\perp^2}V_{\mathrm{em},\kappa',\kappa,\lambda} + \mathcal{O}(|\mathbf{k}_\perp|^0).\end{aligned} \tag{2.8.54}$$

This matrix element consists of the matrix element $\mathcal{M}^{ei\to f}$ with reduced momentum xp, the scalar propagator, and the electromagnetic vertex

$$V_{\mathrm{em},\kappa',\kappa,\lambda} = e\bar{u}(p',\kappa')\not{\varepsilon}^*(k,\lambda)u(p,\kappa) \tag{2.8.55}$$

with helicity spinors $u(p,\kappa)$ and $u(p',\kappa')$.

Neglecting terms of the order of $|\mathbf{k}_\perp|^3$, the electromagnetic vertex can be calculated, for instance, using the explicit expressions for the circular polarization vectors,

$$\begin{aligned}\varepsilon^{*\mu}_\pm(k) &= \frac{1}{\sqrt{2}}\left(\frac{|\mathbf{k}_\perp|\mathrm{e}^{\mp\mathrm{i}\phi}}{2(1-x)E}, 1, \mp\mathrm{i}, -\frac{|\mathbf{k}_\perp|\mathrm{e}^{\mp\mathrm{i}\phi}}{2(1-x)E}\right),\\ \varepsilon_\pm\varepsilon^*_\pm &= -1, \qquad \varepsilon^*_\pm k = 0,\end{aligned} \tag{2.8.56}$$

and the helicity *Dirac spinors* [in the Weyl representation (A.1.25)],

$$u(p',-) = \left(0, 0, \frac{|\mathbf{k}_\perp| \mathrm{e}^{-\mathrm{i}\phi}}{\sqrt{2xE}}, \sqrt{2xE}\right)^{\mathrm{T}}, \quad u(p,-) = \left(0,0,0,\sqrt{2E}\right)^{\mathrm{T}},$$
$$u(p',+) = \left(\sqrt{2xE}, \frac{-|\mathbf{k}_\perp| \mathrm{e}^{\mathrm{i}\phi}}{\sqrt{2xE}}, 0, 0\right)^{\mathrm{T}}, \quad u(p,+) = \left(\sqrt{2E}, 0,0,0\right)^{\mathrm{T}}. \tag{2.8.57}$$

This yields

$$V_{\mathrm{em},\kappa',\kappa,\lambda} = e\frac{\sqrt{2x}|\mathbf{k}_\perp|}{x(1-x)}[\delta_{\kappa\lambda} + x\delta_{-\kappa\lambda}]\delta_{\kappa'\kappa}\mathrm{e}^{-\mathrm{i}\lambda\phi} + \mathcal{O}(|\mathbf{k}_\perp|^3), \tag{2.8.58}$$

and thus

$$\sum_{\lambda=\pm} V_{\mathrm{em},\kappa',\kappa,\lambda} V^*_{\mathrm{em},\kappa'',\kappa,\lambda} = 2e^2 \frac{\mathbf{k}_\perp^2}{x(1-x)} \frac{1+x^2}{1-x} \delta_{\kappa'\kappa}\delta_{\kappa''\kappa} + \mathcal{O}(|\mathbf{k}_\perp|^4), \tag{2.8.59}$$

independent of ϕ.

Inserting this into (2.8.54), the absolute square of the matrix element summed over photon polarizations is obtained as

$$\sum_{\lambda=\pm} |\mathcal{M}^{\mathrm{e}i\to\gamma f}_{\kappa\lambda}|^2 = \sum_{\lambda,\kappa',\kappa''} V_{\mathrm{em}\kappa',\kappa,\lambda} V^*_{\mathrm{em},\kappa'',\kappa,\lambda} \mathcal{M}^{\mathrm{e}i\to f}_{\kappa'} (\mathcal{M}^{\mathrm{e}i\to f}_{\kappa''})^* \frac{(1-x)^2}{(\mathbf{k}_\perp^2)^2}$$
$$= 2e^2 \frac{(1-x)}{x\mathbf{k}_\perp^2} \frac{1+x^2}{1-x} |\mathcal{M}^{\mathrm{e}i\to f}_{\kappa}|^2 + \mathcal{O}(|\mathbf{k}_\perp|^{-1}), \tag{2.8.60}$$

i.e. it is proportional to the matrix element squared for the subprocess $\mathrm{e}i \to f$.

As a consequence, after integrating over the phase space of the invisible photon and summing over its polarizations, the cross section resulting from (2.8.54) is composed of the cross section for $\mathrm{e} + i \to f$ at momentum xp,

$$\mathrm{d}\sigma^{\mathrm{e}i\to\gamma f}_{\kappa}(p, p_j) = 2e^2 \int \frac{\mathrm{d}^3 k}{(2\pi)^3 2E_k} \frac{(1-x)}{x\mathbf{k}_\perp^2} \frac{1+x^2}{1-x}$$
$$\times \sum_{\lambda} \left[x\,\mathrm{d}\sigma^{\mathrm{e}i\to f}_{\kappa'}(xp, p_j) + \mathcal{O}(|\mathbf{k}_\perp|)\right]. \tag{2.8.61}$$

Here several comments are in order: the cross sections $\mathrm{d}\sigma^{ei\to\gamma f}$ and $\mathrm{d}\sigma^{ei\to f}$ are to be taken in the same reference frame, i.e. if $\mathrm{d}\sigma^{ei\to\gamma f}(p,p_j)$ is in the CM frame, $\mathrm{d}\sigma^{ei\to f}(xp,p_j)$ is not in its CM frame for $x\neq 1$. The factor x in front of $\mathrm{d}\sigma^{ei\to f}$ corrects the flux factor in the definition of the cross section [cf. (1.2.24)] since the incoming electron in the non-radiative process has momentum xp instead of p. When integrating over the photon momentum not all momenta p, p_j can be fixed. Since we are only interested in the collinear singularity, which comes from the region $k\approx(1-x)p$, we can neglect $k_\perp$ in $\mathrm{d}\sigma^{ei\to f}$ and replace $p-k$ by xp. Although this is incorrect for large $k_\perp$, it still yields the correct coefficient of the collinear singularity. We can either integrate over the complete photon phase space or over the part of this phase space that contains the singular region.

The volume element of the photon phase space can be decomposed as

$$\frac{\mathrm{d}^3 k}{(2\pi)^3 2E_k}=\frac{\mathrm{d}x\,\mathrm{d}\phi\,\mathrm{d}\mathbf{k}_\perp^2}{4(2\pi)^3(1-x)}+\mathcal{O}(\mathbf{k}_\perp^2). \tag{2.8.62}$$

Since $\mathrm{d}\sigma_{\kappa'}^{ei\to f}(xp,p_j)$ is independent of ϕ, the integration over ϕ yields simply a factor 2π and the cross section is given by

$$\begin{aligned}
&\mathrm{d}\sigma_\kappa^{ei\to\gamma f}(p,p_j)=\\
&=\frac{\alpha}{2\pi}\int\frac{\mathrm{d}\mathbf{k}_\perp^2}{\mathbf{k}_\perp^2}\int_0^1\mathrm{d}x\left[\frac{1+x^2}{1-x}\,\mathrm{d}\sigma_\kappa^{ei\to f}(xp,p_j)+\mathcal{O}(|\mathbf{k}_\perp|)\right]\\
&=\frac{\alpha}{2\pi}\log\left(\frac{\mathbf{k}_{\perp,\max}^2}{\mathbf{k}_{\perp,\min}^2}\right)\int_0^1\mathrm{d}x\,\frac{1+x^2}{1-x}\,\mathrm{d}\sigma_\kappa^{ei\to f}(xp,p_j)\\
&\qquad+\text{non-logarithmic terms.}
\end{aligned} \tag{2.8.63}$$

The upper limit of the transverse-momentum integral is determined by the phase space boundaries (experimental cuts) and given by some scale $Q=|\mathbf{k}_{\perp,\max}|$. If the cuts are not too strict, Q is of the order of the energy E of the process. Since we are only interested in the LL contributions, the precise definition of the scale Q is not relevant. For massless electrons, the lower limit is zero, reflecting the collinear singularity. When calculating with finite electron mass, this regularizes the singularity, and the effective lower limit is actually of the order of the electron mass and in LL approximation can

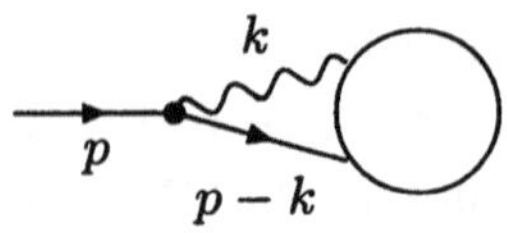

Fig. 2.15 Graph with virtual collinear photon

replaced by the electron mass, $|\mathbf{k}_{\perp,\mathrm{min}}| = m$. Thus, we find

$$\begin{aligned} &\mathrm{d}\sigma^{\mathrm{e}i\to\gamma f}_{\mathrm{LL},\kappa}(p,p_j) = \\ &= \frac{\alpha}{2\pi}\log\left(\frac{Q^2}{m^2}\right)\int_0^1 \mathrm{d}x\,\frac{1+x^2}{1-x}\,\mathrm{d}\sigma^{\mathrm{e}i\to f}_{\kappa}(xp,p_j) \end{aligned} \tag{2.8.64}$$

The collinear singularity manifests itself in the logarithm $\log Q^2/m^2 \sim \log E^2/m^2$. Summing over κ yields the same relation for the unpolarized cross section.

The soft-photon singularity at $x \to 1$ in (2.8.64) is of the type that was discussed in Sect. 2.8.2. As was shown there, this singularity is compensated by the virtual radiative corrections.

2.8.3.2 Virtual collinear photons

Leading-logarithmic corrections from virtual collinear photons result from field and/or wave-function renormalization constants as well as diagrams of the form shown in Fig. 2.15. The corresponding amplitude reads

$$\begin{aligned} \mathcal{M}^{\mathrm{e}i\to f}_{\mathrm{virt},\kappa}(p,p_j) &= \int \frac{\mathrm{d}^4k}{(2\pi)^4}\bar{\mathcal{A}}^{\mathrm{e}\gamma i\to f}_{\mu}(p-k,k,p_j) \\ &\times \frac{\mathrm{i}(\not{p}-\not{k}+m)}{(p-k)^2-m^2+\mathrm{i}\varepsilon}\,\mathrm{i}e\gamma_\nu\,u(p,\kappa)\frac{-\mathrm{i}N^{\mu\nu}}{k^2+\mathrm{i}\varepsilon}, \end{aligned} \tag{2.8.65}$$

where $N^{\mu\nu}$ denotes the numerator of the photon propagator, e.g. $g^{\mu\nu}$ in the Feynman gauge, and $\bar{\mathcal{A}}^{\mathrm{e}\gamma i\to f}$ represents the amplitude with the on-shell electron replaced by an off-shell electron and photon. Because the matrix element $\mathcal{M}^{\mathrm{e}i\to f}$ is fully truncated, $\bar{\mathcal{A}}^{\mathrm{e}\gamma i\to f}$ does not contain contributions which are one-particle reducible between the electron–photon pair and the rest of the diagram, i.e. it is given by the diagrams in the brackets on the left-hand side of (2.4.98).

Since in this section we are only interested in mass singularities, we calculate all loop integrals in four dimensions and regularize the UV singularities by a simple cut-off Λ if necessary. As the UV singularities are cancelled by other UV-divergent but non-mass-singular contributions, we can set the cut-off equal to the scale Q of order of the energy E, $\Lambda = Q \sim E$. Like for the real collinear singularities, we can put the electron mass to zero for the extraction of the virtual collinear singularities. The collinear singularity can be extracted by using the Sudakov variables (2.8.11). In the collinear-singular part, we can use $k = yp$ in the numerator and obtain

$$\begin{aligned} \mathcal{M}^{ei\to f}_{\mathrm{virt},\kappa}(p,p_j) &= \mathrm{i}e\frac{1}{(2\pi)^4}|pn| \int \mathrm{d}^2\mathbf{k}_\perp \mathrm{d}y\,\mathrm{d}z \\ &\times \bar{A}^{e\gamma i\to f}_\mu((1-y)p, yp, p_j)\frac{\not{p}(1-y)\gamma_\nu u(p,\kappa)N^{\mu\nu}}{(k^2+\mathrm{i}\varepsilon)[(p-k)^2+\mathrm{i}\varepsilon]} \\ &= 2\mathrm{i}e\frac{1}{(2\pi)^4}|pn| \int \mathrm{d}^2\mathbf{k}_\perp \mathrm{d}y\,\mathrm{d}z\,(1-y) \\ &\times \bar{A}^{e\gamma i\to f}_\mu((1-y)p, yp, p_j)u(p,\kappa)\frac{N^{\mu\nu}p_\nu}{(k^2+\mathrm{i}\varepsilon)[(p-k)^2+\mathrm{i}\varepsilon]}, \end{aligned} \tag{2.8.66}$$

where we applied the massless Dirac equation in the last line. If we choose the gauge in such a way that the numerator of the photon propagator is transverse, i.e. $N^{\mu\nu}k_\nu = 0$, this vanishes owing to $k = yp$. In such a gauge the only collinear-singular virtual contributions originate from the wave-function renormalization constants and/or field renormalization constants of the external fields [cf. (2.5.7)], where the above does not apply. In the following we assume the on-shell scheme of Sect. 2.5.1.6, where the field renormalization is chosen in such a way that no wave-function renormalization constants are needed.

An appropriate[26] transverse gauge is the *axial gauge*, where

$$N^{\mu\nu} = g^{\mu\nu} - \frac{n^\mu k^\nu + k^\mu n^\nu}{nk} + \frac{n^\mu n^\nu}{(nk)^2}k^2, \qquad n^2 = 0. \tag{2.8.67}$$

In order to calculate the fermion-field renormalization constant for massless fermions in this gauge we consider the *electron self-energy* [cf. (2.5.95) on page 221]

$$\Sigma^{\mathrm{e}}(p) = \mathrm{i}e^2 \int \frac{\mathrm{d}^4k}{(2\pi)^4}\frac{\gamma_\mu(\not{p}-\not{k})\gamma_\nu N^{\mu\nu}}{(k^2+\mathrm{i}\varepsilon)[(p-k)^2+\mathrm{i}\varepsilon]} \tag{2.8.68}$$

[26]The Landau gauge, $N^{\mu\nu} = g^{\mu\nu} - k^\mu k^\nu/k^2$, introduces further singularities owing to the $1/k^2$ term, so that the approximation $k = xp$ used in (2.8.66) is not valid.

with the covariant decomposition

$$\Sigma^{\psi}(p) = \not{p}\,\Sigma^{\psi}_{\mathrm{V}}(p^2) + \frac{p^2}{2pn}\not{n}\,\Sigma^{\psi}_{\mathrm{n}}(p^2). \tag{2.8.69}$$

Then, the *electron-field renormalization* constant is obtained using the condition (2.5.37) as

$$\begin{aligned}\delta Z_{\psi} &= -\Sigma^{\psi}_{\mathrm{V}}(0) - \Sigma^{\psi}_{\mathrm{n}}(0)\\ &= \lim_{p^2\to 0}\left(\frac{1}{4pn}\,\mathrm{Tr}(\Sigma^{\psi}\not{n}) - \frac{1}{2p^2}\,\mathrm{Tr}(\Sigma^{\psi}\not{p})\right).\end{aligned} \tag{2.8.70}$$

Inserting (2.8.68) with (2.8.67) and calculating the traces yields

$$\begin{aligned}\delta Z_{\psi} = \lim_{p^2\to 0} 4\mathrm{i}e^2 \int \frac{\mathrm{d}^4k}{(2\pi)^4}&\frac{1}{(k^2+\mathrm{i}\varepsilon)[(p-k)^2+\mathrm{i}\varepsilon]}\\ \times &\left\{\frac{1}{2pn}(kn-pn) - \frac{1}{p^2}\left[pk\left(1-2\frac{np}{nk}\right) + k^2\frac{(np)^2}{(nk)^2}\right]\right\}.\end{aligned} \tag{2.8.71}$$

Terms that involve $(p-k)^2$ or k^2 in the numerator cancel one of the zeros of the denominator and therefore give not rise to logarithmic singularities. After substituting $pk = (p^2+k^2-(p-k)^2)/2 \to p^2/2$ in the numerator, we find

$$\begin{aligned}\delta Z_{\psi} = 4\mathrm{i}e^2 \int \frac{\mathrm{d}^4k}{(2\pi)^4}&\frac{1}{(k^2+\mathrm{i}\varepsilon)[(p-k)^2+\mathrm{i}\varepsilon]}\left[\frac{kn}{2pn} - 1 + \frac{np}{nk}\right]\Bigg|_{p^2=0}\\ &+ \text{non-logarithmic terms.}\end{aligned} \tag{2.8.72}$$

We evaluate this integral using Sudakov variables (2.8.11), where $kn/pn = y$. After integrating over z by closing the integration contour in the upper half plane, we obtain

$$\begin{aligned}\delta Z_{\psi} = -\frac{e^2}{(2\pi)^2}\int\frac{\mathrm{d}\mathbf{k}_{\perp}^2}{\mathbf{k}_{\perp}^2}&\int_0^1 \mathrm{d}y\left\{\frac{y}{2} - 1 + \frac{1}{y}\right\}\\ &+ \text{non-logarithmic terms}\end{aligned} \tag{2.8.73}$$

As in (2.8.63), the lower limit of the transverse-momentum integral is given by the electron mass. At the upper limit this integral diverges owing to the

UV singularity. Since we are only interested in the mass-singular contributions, we set the upper limit equal to the scale Q, as explained above, and arrive at

$$\delta Z_{\psi} = -\frac{\alpha}{2\pi}\log\left(\frac{Q^2}{m^2}\right)\int_0^1 \mathrm{d}x\,\frac{1+x^2}{1-x} + \text{non-logarithmic terms}, \tag{2.8.74}$$

where we substituted $y \to (1-x)$. The pole at $x = 1$ results again from the IR singularity.

Finally, we find

$$\begin{aligned} \mathrm{d}\sigma^{ei\to f}_{\mathrm{virt},\kappa}(p,p_j) &= \delta Z_{\psi}\mathrm{d}\sigma^{ei\to f}_{\kappa}(p,p_j) \\ &= -\frac{\alpha}{2\pi}\log\left(\frac{Q^2}{m^2}\right)\mathrm{d}\sigma^{ei\to f}_{\kappa}(p,p_j)\int_0^1 \mathrm{d}x\,\frac{1+x^2}{1-x} \\ &\quad + \text{non-logarithmic terms} \end{aligned} \tag{2.8.75}$$

for the contributions of one-loop collinear singularities to the cross section in the axial gauge.

For comparison we calculate these contributions in the Feynman gauge. Then, the *electron-field renormalization* constant is obtained as

$$\delta Z_{\psi} = -\frac{\alpha}{2\pi}\log\left(\frac{Q^2}{m^2}\right)\int_0^1 \mathrm{d}x\,(1-x) + \text{non-logarithmic terms}. \tag{2.8.76}$$

On the other hand, in this gauge the diagrams in Fig. 2.15 yield a non-vanishing contribution given by (2.8.66) with $N^{\mu\nu}p_{\nu} = p^{\mu}$. After setting $p = k/y$, which holds in the collinear limit, we can use the Ward identity (2.4.97) on page 176, which translates to

$$k^{\mu}\bar{A}^{e\gamma i\to f}_{\mu}(p-k,k,p_j) = e\bar{A}^{ei\to f}(p,p_j) \tag{2.8.77}$$

because the second term in (2.4.97) subtracts the one-particle reducible terms that are not present in $\mathcal{M}^{ei\to f}_{\mathrm{virt}}$. Thus, (2.8.66) turns into

$$\begin{aligned} &\mathcal{M}^{ei\to f}_{\mathrm{virt},\kappa}(p,p_j) = \\ &= 2\mathrm{i}e^2\frac{1}{(2\pi)^4}|pn|\int \mathrm{d}^2\mathbf{k}_{\perp}\mathrm{d}y\,\mathrm{d}z\,\frac{1-y}{y}\,\frac{\bar{A}^{ei\to f}(p,p_j)u(p,\kappa)}{(k^2+\mathrm{i}\varepsilon)[(p-k)^2+\mathrm{i}\varepsilon]} \end{aligned}$$

$$
\begin{aligned}
&= -\frac{\alpha}{2\pi}\int\frac{\mathrm{d}\mathbf{k}_\perp^2}{\mathbf{k}_\perp^2}\int_0^1 \mathrm{d}y\,\frac{1-y}{y}\,\mathcal{M}_\kappa^{\mathrm{e}i\to f}(p,p_j)\\
&= -\frac{\alpha}{2\pi}\log\left(\frac{Q^2}{m^2}\right)\int_0^1 \mathrm{d}x\,\frac{x}{1-x}\,\mathcal{M}_\kappa^{\mathrm{e}i\to f}(p,p_j) \qquad (2.8.78)
\end{aligned}
$$

up to non-logarithmic terms. Adding the contribution of the field renormalization constant, $\mathcal{M}_\kappa^{\mathrm{e}i\to f}\delta Z_\psi/2$, and calculating the cross section yields the same result as in the axial gauge (2.8.75).

2.8.3.3 Electron–electron splitting function

Summing the contributions of real and virtual collinear photons, (2.8.63) and (2.8.75), the IR singularities at $x=1$ cancel, and we find the one-loop LL corrections originating from an incoming electron,

$$
\begin{aligned}
\mathrm{d}\sigma_{\kappa,\mathrm{LL}}^{\mathrm{e}i\to\gamma f}(p,p_j) &= \frac{\alpha}{2\pi}\log\left(\frac{Q^2}{m^2}\right)\left[\int_0^1 \mathrm{d}x\,\frac{1+x^2}{1-x}\,\mathrm{d}\sigma_\kappa^{\mathrm{e}i\to f}(xp,p_j)\right.\\
&\qquad\left.-\int_0^1 \mathrm{d}x\,\frac{1+x^2}{1-x}\,\mathrm{d}\sigma_\kappa^{\mathrm{e}i\to f}(p,p_j)\right]\\
&= \frac{\alpha}{2\pi}\log\left(\frac{Q^2}{m^2}\right)\int_0^1 \mathrm{d}x\left(\frac{1+x^2}{1-x}\right)_+\mathrm{d}\sigma_\kappa^{\mathrm{e}i\to f}(xp,p_j). \qquad (2.8.79)
\end{aligned}
$$

Here we have introduced the (+)-*distribution* which is defined by

$$
\int_0^1 \mathrm{d}x\, f(x)(h(x))_+ = \int_0^1 \mathrm{d}x\,\Big(f(x)-f(1)\Big)h(x), \qquad (2.8.80)
$$

or

$$
(h(x))_+ = \lim_{\beta\to 0}\left[\theta(1-x-\beta)h(x) - \delta(1-x-\beta)\int_0^{1-\beta} h(y)\,\mathrm{d}y\right] \qquad (2.8.81)
$$

and IR-finite by construction. Defining the lowest order *electron–electron splitting function*

$$
P_{\mathrm{ee}}^{(0)}(x) = \left(\frac{1+x^2}{1-x}\right)_+ = \frac{1+x^2}{(1-x)_+} - \frac{3}{2}\delta(1-x), \qquad (2.8.82)
$$

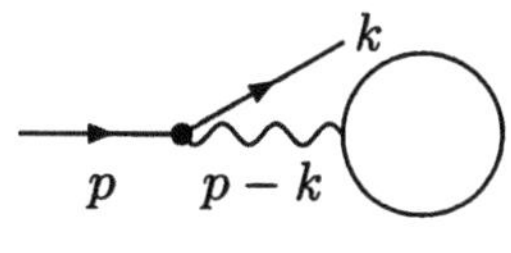

Fig. 2.16 Graph for collinear electron emission

we finally obtain the *leading-logarithmic approximation* (LL approximation) of the radiative cross section

$$\mathrm{d}\sigma_{\kappa,\mathrm{LL}}^{ei\to\gamma f}(p,p_j) = \frac{\alpha}{2\pi}\log\left(\frac{Q^2}{m^2}\right)\int_0^1 \mathrm{d}x\, P_{\mathrm{ee}}^{(0)}(x)\,\mathrm{d}\sigma_\kappa^{\mathrm{e}i\to f}(xp,p_j). \tag{2.8.83}$$

In the collinear approximation the radiative cross section factorizes into the logarithm of the collinear singularity and the convolution of the lowest-order cross section at momentum xp with the splitting function $P_{\mathrm{ee}}^{(0)}(x)$. The splitting function is related to the probability of finding a fermion with momentum xp after the emission of a collinear photon with momentum $(1-x)p$ from the incoming fermion with momentum p.

2.8.3.4 Collinear electron emission

A second possible reaction mechanism is the one where the collinear photon undergoes a hard scattering $\gamma + i \to f$ and the collinear electron remains undetected (Fig. 2.16). This is part of the process $\mathrm{e} + i \to \mathrm{e} + f$. Using the conventions for the momenta as defined in Fig. 2.16 and (2.8.51) and denoting the helicity of the outgoing electron by κ', the corresponding invariant matrix element reads

$$\mathcal{M}_{\kappa\kappa'}^{\mathrm{e}i\to\mathrm{e}f}(p,k,p_j) = -e\bar{u}(k,\kappa')\gamma_\mu u(p,\kappa)\frac{-g^{\mu\nu}}{(p-k)^2}A_\nu^{\gamma i\to f}(p-k,p_j) \tag{2.8.84}$$

in the Feynman gauge. As in the previous case we can neglect the electron mass and terms of order $\mathbf{k}_\perp^2$ when extracting the LL contributions. In particular, in the numerator only the leading terms in $|\mathbf{k}_\perp|$ are needed, and the metric tensor in the numerator of the photon propagator can be expressed as

$$-g^{\mu\nu} = \sum_{\lambda=\pm} \varepsilon^{*\mu}(p',\lambda)\varepsilon^{\nu}(p',\lambda) - \frac{p'^{\mu}\bar{p}'^{\nu} + p'^{\nu}\bar{p}'^{\mu}}{p'\bar{p}'}, \tag{2.8.85}$$

where $\bar{p}' = (|\mathbf{p}'|, -\mathbf{p}')$ for $p' = (|\mathbf{p}'|, \mathbf{p}')$. The contraction of p'^{ν} with the amplitude A_ν vanishes owing to a Ward identity (2.4.93), and the contraction of p'^{μ} with the electromagnetic vertex gives zero. In other gauges, the numerator of the photon propagator contains additional terms involving p'^{μ} or p'^{ν} which also drop out. Therefore, independently of the gauge, the numerator of the photon propagator can be replaced by

$$-g^{\mu\nu} \to \sum_{\lambda=\pm} \varepsilon^{*\mu}(p',\lambda)\varepsilon^{\nu}(p',\lambda), \tag{2.8.86}$$

i.e. the polarization sum for the physical transverse polarization vectors only, and the matrix element turns into

$$\begin{aligned} \mathcal{M}_{\kappa\kappa'}^{ei\to ef}(p,k,p_j) &= -e\sum_{\lambda} \bar{u}(k,\kappa')\not{\varepsilon}^{*}(p',\lambda)u(p,\kappa) \\ &\quad \times \varepsilon^{\nu}(p',\lambda)A_{\nu}^{\gamma i\to f}(p-k,p_j)\frac{1}{(p-k)^2} + \mathcal{O}(|\mathbf{k}_\perp|^0) \\ &= -\sum_{\lambda} V_{\mathrm{em},\kappa',\kappa,\lambda}\mathcal{M}_{\lambda}^{\gamma i\to f}(p-k,p_j)\frac{1}{(p-k)^2} + \mathcal{O}(|\mathbf{k}_\perp|^0). \end{aligned} \tag{2.8.87}$$

The corresponding electromagnetic vertex

$$V_{\mathrm{em},\kappa',\kappa,\lambda} = e\bar{u}(k,\kappa')\not{\varepsilon}^{*}(p',\lambda)u(p,\kappa) \tag{2.8.88}$$

differs from (2.8.55) only by the interchange $p' \leftrightarrow k$, i.e. by the substitutions $x \leftrightarrow (1-x)$ and $k_\perp \to -k_\perp$ [cf. (2.8.51)]. Therefore, from (2.8.58) we immediately find

$$\begin{aligned} V_{\mathrm{em},\kappa,\kappa',\lambda} &= -e\frac{\sqrt{2(1-x)}|\mathbf{k}_\perp|}{x(1-x)}[\delta_{\kappa\lambda} + (1-x)\delta_{-\kappa\lambda}]\delta_{\kappa\kappa'}\mathrm{e}^{-\mathrm{i}\lambda\phi} \\ &\quad + \mathcal{O}(|\mathbf{k}_\perp|^3), \end{aligned} \tag{2.8.89}$$

and thus

$$\sum_{\kappa'=\pm} V_{\mathrm{em},\kappa',\kappa,\lambda} V^*_{\mathrm{em},\kappa',\kappa,\lambda'} = 2e^2 \frac{\mathbf{k}_\perp^2}{x(1-x)} \left[\frac{\delta_{\kappa\lambda} + (1-x)^2 \delta_{-\kappa\lambda}}{x} \delta_{\lambda\lambda'} \right.$$
$$\left. + \frac{1-x}{x} \delta_{-\lambda'\lambda} \mathrm{e}^{-2\mathrm{i}\lambda\phi} \right] + \mathcal{O}(|\mathbf{k}_\perp|^4). \tag{2.8.90}$$

When integrating over ϕ, the second term averages to zero[27] and the non-vanishing term is proportional to $\delta_{\lambda\lambda'}$. Therefore, we effectively obtain for the absolute square of the matrix element

$$\sum_{\kappa'=\pm} |\mathcal{M}^{\mathrm{e}i\to\mathrm{e}f}_{\kappa\kappa'}|^2 = \sum_{\kappa',\lambda,\lambda'} V_{\mathrm{em},\kappa',\kappa,\lambda} V^*_{\mathrm{em},\kappa',\kappa,\lambda'} \mathcal{M}^{\gamma i\to f}_{\lambda} (\mathcal{M}^{\gamma i\to f}_{\lambda'})^* \frac{1}{(p-k)^4}$$
$$= 2e^2 \sum_{\lambda} \frac{(1-x)}{x\mathbf{k}_\perp^2} \frac{\delta_{\kappa\lambda} + (1-x)^2 \delta_{-\kappa\lambda}}{x} |\mathcal{M}^{\gamma i\to f}_{\lambda}|^2 + \mathcal{O}(|\mathbf{k}_\perp|^{-1}), \tag{2.8.91}$$

i.e. a contribution proportional to the absolute squares of the photonic matrix element. Performing the same steps as in Sect. 2.8.3.1, we find the cross section in the LL approximation

$$\mathrm{d}\sigma^{\mathrm{e}i\to\mathrm{e}f}_{\mathrm{LL},\kappa}(p,p_j) =$$
$$= \frac{\alpha}{2\pi} \log\left(\frac{Q^2}{m^2}\right) \int_0^1 \mathrm{d}x \sum_{\lambda} P^{(0)}_{\gamma_\lambda \mathrm{e}_\kappa}(x) \mathrm{d}\sigma^{\gamma i\to f}_{\lambda}(xp,p_j) \tag{2.8.92}$$

with the polarized *electron–photon splitting function*

$$P^{(0)}_{\gamma_\lambda \mathrm{e}_\kappa}(x) = \frac{\delta_{\kappa\lambda} + (1-x)^2 \delta_{-\kappa\lambda}}{x}. \tag{2.8.93}$$

The $1/x$ behaviour of this splitting function does not lead to a singularity since the cross section $\mathrm{d}\sigma^{\gamma i\to f}$ vanishes for $x \to 0$, i.e. for zero photon energy.

[27] We have chosen the phases of the spinors and polarization vectors in such a way that the matrix element $\mathcal{M}^{\gamma i\to f}_{\lambda}(p-k,p_j)$ is independent of ϕ in the collinear limit $k = xp$.

Fig. 2.17 Graphs for collinear fermion emission

The case of photon splitting from an incoming positron gives the same result. Summing over the electron or positron helicities yields the corresponding results for unpolarized cross sections with the unpolarized *splitting function*

$$P^{(0)}_{\gamma \mathrm{e}} = \sum_{\kappa} P^{(0)}_{\gamma_\lambda \mathrm{e}_\kappa}(x) = \frac{1+(1-x)^2}{x}. \tag{2.8.94}$$

The cross section (2.8.92) is the product of the logarithmic collinear singularity and the convolution of the splitting function $P^{(0)}_{\gamma \mathrm{e}}(x)$ with the cross section of a real, transversely polarized photon of momentum xp. Since $\mathrm{d}\sigma^{\mathrm{e}i\to \mathrm{e}f}$ is calculated from the photonic cross section $\mathrm{d}\sigma^{\gamma i\to f}$, the result (2.8.92) is called the *equivalent-photon approximation* or *Weizsäcker–Williams approximation.*

2.8.3.5 Photon splitting

Scattering processes $\gamma + i \to \mathrm{e} + f$ with incoming high-energetic photons involve collinear-singular contributions originating from the splitting of the photon into fermion–antifermion pairs (Fig. 2.17). One of the collinear fermions is taken inclusively, i.e. its phase space is integrated out, the other one takes part in a hard-scattering process. For the case of an inclusive collinear electron (Fig. 2.17a), the matrix element (2.8.16) reads

$$\begin{aligned}\mathcal{M}^{\gamma i\to \mathrm{e}f}_{\lambda\kappa}(p,k,p_j) &=\\ &= e\bar{u}(k,\kappa)\not{\varepsilon}(p,\lambda)\frac{(\not{p}-\not{k}-m)}{(p-k)^2-m^2+\mathrm{i}\varepsilon}A^{\mathrm{e}i\to f}(p-k,p_j),\end{aligned} \tag{2.8.95}$$

where the momenta are defined in Fig. 2.17 and (2.8.51). As in Sect. 2.8.3.1, we neglect the electron mass and terms of order $\mathbf{k}_\perp^2$ and write the numerator of the fermion propagator in terms of fermion spinors [cf. (2.8.53)]. The resulting electromagnetic vertex involves an incoming photon and takes the form

$$V_{\mathrm{em},\kappa,\kappa',\lambda} = e\bar{u}(k,\kappa)\not{\varepsilon}(p,\lambda)u(p',\kappa'). \tag{2.8.96}$$

Using the explicit polarization vectors

$$\varepsilon^{\mu}_{\pm}(k) = \frac{1}{\sqrt{2}}\,(0,1,\pm\mathrm{i},0)\,, \tag{2.8.97}$$

and the helicity Dirac spinors

$$\begin{aligned} u(k,-) &= \left(0,0,\frac{-|\mathbf{k}_{\perp}|\mathrm{e}^{-\mathrm{i}\phi}}{\sqrt{2(1-x)E}},\sqrt{2(1-x)E}\right)^{\mathrm{T}}, \\ u(k,+) &= \left(\sqrt{2(1-x)E},\frac{|\mathbf{k}_{\perp}|\mathrm{e}^{\mathrm{i}\phi}}{\sqrt{2(1-x)E}},0,0\right)^{\mathrm{T}}, \end{aligned} \tag{2.8.98}$$

as well as $u(p',\pm)$ from (2.8.57), we find

$$V_{\mathrm{em},\kappa',\kappa,\lambda} = e\frac{\sqrt{2}|\mathbf{k}_{\perp}|}{\sqrt{x(1-x)}}[(1-x)\delta_{\kappa\lambda} - x\delta_{-\kappa\lambda}]\delta_{\kappa\kappa'}\mathrm{e}^{\mathrm{i}\lambda\phi} + \mathcal{O}(|\mathbf{k}_{\perp}|^3), \tag{2.8.99}$$

and

$$\sum_{\kappa'=\pm}\left|V_{\mathrm{em},\kappa,\kappa',\lambda}\right|^2 = 2e^2\frac{\mathbf{k}_{\perp}^2}{x(1-x)}[\delta_{\kappa\lambda}(1-x)^2 + \delta_{-\kappa\lambda}x^2] + \mathcal{O}(|\mathbf{k}_{\perp}|^4). \tag{2.8.100}$$

The calculation of the cross section in the collinear approximation proceeds as in Sect. 2.8.3.1 and leads to

$$\mathrm{d}\sigma_{\lambda}^{\gamma i\to \mathrm{e}f}(p,p_j) = \frac{\alpha}{2\pi}\log\left(\frac{Q^2}{m^2}\right)\int_0^1 \mathrm{d}x\sum_{\kappa}P^{(0)}_{\mathrm{e}_{\kappa}\gamma_{\lambda}}(x)\mathrm{d}\sigma_{\kappa}^{\mathrm{e}i\to f}(xp,p_j) \tag{2.8.101}$$

with the *photon–electron splitting function*

$$P^{(0)}_{\mathrm{e}_{\kappa}\gamma_{\lambda}}(x) = \delta_{\kappa\lambda}(1-x)^2 + \delta_{-\kappa\lambda}x^2, \tag{2.8.102}$$

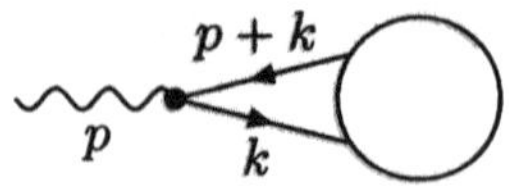

Fig. 2.18 Graph with virtual collinear electron–positron pair

or in the unpolarized case

$$P_{e\gamma}^{(0)}(x) = (1 - x)^2 + x^2. \tag{2.8.103}$$

For the case where the positron escapes (Fig. 2.17b) we find the same result with the same splitting function.

2.8.3.6 Virtual collinear electron–positron pairs

Unlike in the non-abelian case, where the splitting of a gauge-boson in two gauge-bosons is an important reaction mechanism, the splitting of a photon in two photons is not possible as an elementary process in abelian gauge theories. Therefore in QED only virtual diagrams like those in Fig. 2.18 contribute to the splitting function $P_{\gamma\gamma}^{(0)}$. By evaluating the diagram Fig. 2.18, one finds that all collinear-singular terms are either proportional to k^2 or $\varepsilon(k)k$ and therefore drop out for physical transverse photons. Thus, the only collinear-singular terms result from the photon-field renormalization. According to (2.5.104) and (2.5.93), this is given by

$$\delta Z_A = -\frac{\alpha}{3\pi}\left(B_0(p^2, 0, 0) - \frac{1}{3}\right). \tag{2.8.104}$$

Evaluating the B_0 function as in (2.8.72)–(2.8.74), yields

$$\delta Z_A = -\frac{\alpha}{3\pi}\log\left(\frac{Q^2}{m^2}\right) + \text{non-logarithmic terms}, \tag{2.8.105}$$

and the corresponding *photon–photon splitting function* reads

$$P_{\gamma\gamma}^{(0)}(x) = -\frac{2}{3}\delta(1 - x). \tag{2.8.106}$$

2.8.3.7 Probabilistic interpretation and distribution functions

The results for the QED cross sections in the LL approximation (2.8.83), (2.8.92), (2.8.101), and the one corresponding to (2.8.106) have the following physical interpretation [A177] which is closely related to the parton picture discussed in Sect. 1.5. The cross section factorizes into the probability $f_{j,i}(x, Q^2)$ of finding a parton j (an electron, positron, or photon) with longitudinal momentum fraction x in the incoming particle i, taking into account the radiation of collinear partons with transverse momentum $k_\perp^2 < Q^2$. The parton j enters the hard-scattering process. Up to order α, the probabilities are given by

$$
\begin{aligned}
f_{\mathrm{e,e}}(x,Q^2) &= f_{\bar{\mathrm{e}},\bar{\mathrm{e}}}(x,Q^2) = \delta(1-x) + \frac{\alpha}{2\pi}\log\left(\frac{Q^2}{m^2}\right) P^{(0)}_{\mathrm{ee}}(x),\\
f_{\bar{\mathrm{e}},\mathrm{e}}(x,Q^2) &= f_{\mathrm{e},\bar{\mathrm{e}}}(x,Q^2) = 0,\\
f_{\gamma,\mathrm{e}}(x,Q^2) &= f_{\gamma,\bar{\mathrm{e}}}(x,Q^2) = \frac{\alpha}{2\pi}\log\left(\frac{Q^2}{m^2}\right) P^{(0)}_{\gamma\mathrm{e}}(x),\\
f_{\gamma,\gamma}(x,Q^2) &= \delta(1-x) + \frac{\alpha}{2\pi}\log\left(\frac{Q^2}{m^2}\right) P^{(0)}_{\gamma\gamma}(x),\\
f_{\mathrm{e},\gamma}(x,Q^2) &= f_{\bar{\mathrm{e}},\gamma}(x,Q^2) = \frac{\alpha}{2\pi}\log\left(\frac{Q^2}{m^2}\right) P^{(0)}_{\mathrm{e}\gamma}(x).
\end{aligned}
\tag{2.8.107}
$$

The integration constant in the logarithm was chosen in such a way that for $Q^2 = m^2$ no splitting occurs. A simple integration, using (2.8.80), shows that the probabilities $f_{j,i}(x, Q^2)$, which are also called *distribution functions* respect the conservation of electron number and of the longitudinal momentum,

$$
\begin{aligned}
&\int_0^1 \mathrm{d}x\,[f_{\mathrm{e,e}}(x,Q^2) - f_{\bar{\mathrm{e}},\mathrm{e}}(x,Q^2)] = 1,\\
&\int_0^1 \mathrm{d}x\,[f_{\mathrm{e},\gamma}(x,Q^2) - f_{\bar{\mathrm{e}},\gamma}(x,Q^2)] = 0,\\
&\qquad \sum_{j=\mathrm{e},\bar{\mathrm{e}},\gamma} \int_0^1 \mathrm{d}x\, x f_{j,i}(x,Q^2) = 1.
\end{aligned}
\tag{2.8.108}
$$

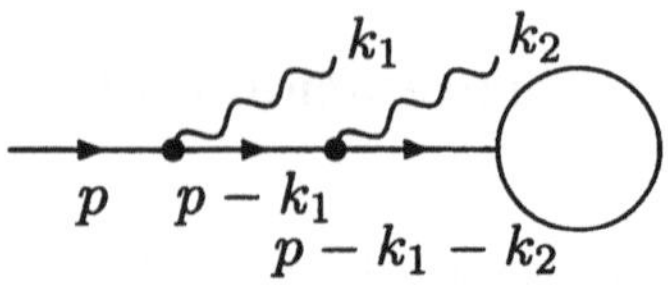

Fig. 2.19 Graphs for the emission of a two collinear photons from an incoming electron

Finally, we write down the expression for the cross section in a very condensed form,

$$\mathrm{d}\sigma^i = \sum_j \int_0^1 \mathrm{d}x\, f_{j,i}(x, Q^2)\mathrm{d}\sigma^j(x). \tag{2.8.109}$$

This equation holds also in higher orders and summarizes the factorization of mass singularities in QED. In this context, Q^2 is called factorization scale.

The generalization of (2.8.109) to non-abelian gauge theories, e.g. QCD, and more detailed calculations including the non-leading contributions is performed in Sect. 3.3.

2.8.3.8 Multiple splittings and evolution equations

The subprocesses with soft and collinear photons give important contributions to high-energy cross sections in QED since they are enhanced by large logarithms. Therefore, the investigation of higher-order soft and collinear processes is also of practical importance. The general results for the soft-photon case were derived and discussed in Sect. 2.8.2. Now we consider the higher-order collinear singularities.

The double-photon emission from an initial charged particle (Fig. 2.19) leads to large contributions if it causes two singular propagators. If the first emitted photon has a small acollinearity $|\mathbf{k}_{\perp,1}| \ll |\mathbf{k}_{\perp,2}|$, then the second one still can give an enhanced contribution. In this case, the first virtual electron is very close to its mass shell compared to $\mathbf{k}^2_{\perp,2}$ and we can ignore its virtuality in computing the emission of the second photon. The appearance of the leading logarithms is illustrated by the following simple calculation

$$\left(\frac{\alpha}{2\pi}\right)^2 \int_{m^2}^{Q^2} \frac{\mathrm{d}\mathbf{k}^2_{\perp,2}}{\mathbf{k}^2_{\perp,2}} \int_{m^2}^{|\mathbf{k}_{\perp,2}|^2} \frac{\mathrm{d}\mathbf{k}^2_{\perp,1}}{\mathbf{k}^2_{\perp,1}} = \frac{1}{2}\left(\frac{\alpha}{2\pi}\log\frac{Q^2}{m^2}\right)^2. \tag{2.8.110}$$

In the opposite limit, $|\mathbf{k}_{\perp,2}| \ll |\mathbf{k}_{\perp,1}|$, there is no denominator of order $\mathbf{k}^2_{\perp,2}$, and thus no double logarithm. A more detailed calculation supporting this result is performed in Sect. 3.3. The generalization to the LL contribution of the emission of n photons reads

$$\frac{1}{n!}\left(\frac{\alpha}{2\pi}\log\frac{Q^2}{m^2}\right)^n, \tag{2.8.111}$$

resulting from integration over the photon phase space corresponding to the strong ordering

$$|\mathbf{k}_{\perp,1}| \ll |\mathbf{k}_{\perp,2}| \ll |\mathbf{k}_{\perp,3}| \ll \ldots \ll |\mathbf{k}_{\perp,n}|. \tag{2.8.112}$$

The summation of these important multi-photon contributions leads to an integral equation which was first obtained by Gribov and Lipatov [Gr72]. We derive this equation using the probabilistic interpretation. Consider the *distribution function* $f(x, Q^2)$ which describes the probability of finding a certain particle in a given particle taking into account collinear photon emissions with transverse momenta $\mathbf{k}^2_\perp < Q^2$. Then, the distribution function $f(x, Q^2 + \mathrm{d}Q^2)$ at momentum transfer squared $Q^2 + \mathrm{d}Q^2$ must take into account the possibility of an additional emission of a parton with transverse momentum $Q^2 < \mathbf{k}^2_\perp < Q^2 + \mathrm{d}Q^2$. The differential probability to split off a parton that carries away the part $1 - x$ of the energy is

$$\frac{\alpha}{2\pi}\frac{\mathrm{d}\mathbf{k}^2_\perp}{\mathbf{k}^2_\perp}P(x), \tag{2.8.113}$$

and we can write

$$\begin{aligned} f\left(x, Q^2 + \mathrm{d}Q^2\right) &= f(x, Q^2) + \int_0^1 \mathrm{d}y \int_0^1 \mathrm{d}z\, \delta(x - yz) \\ &\quad \times \frac{\alpha}{2\pi}\frac{\mathrm{d}Q^2}{Q^2}P(y) f(z, Q^2) \\ &= f(x, Q^2) + \frac{\mathrm{d}Q^2}{Q^2}\frac{\alpha}{2\pi}\int_x^1 \frac{\mathrm{d}z}{z}P^{(0)}\left(\frac{x}{z}\right) f(z, Q^2). \end{aligned}$$

This leads to the integro-differential equation

$$\frac{\partial}{\partial(\log Q^2)} f(x, Q^2) = \frac{\alpha}{2\pi}\int_x^1 \frac{\mathrm{d}z}{z}P\left(\frac{x}{z}\right) f(z, Q^2). \tag{2.8.114}$$

It describes the Q^2 dependence of the distribution function if its initial value, e.g. $f(x, m^2) = \delta(1-x)$, is given. The *evolution equation* (2.8.114) contains on the right-hand side a *multiplicative convolution* of the evolution kernel and the distribution function. If we write for the multiplicative convolution of two functions $f(x)$ and $g(x)$

$$[f \otimes g](x) = \int_0^1 \mathrm{d}y \int_0^1 \mathrm{d}z\, \delta(x - yz) f(y) g(z) = \int_x^1 \frac{\mathrm{d}z}{z} f\left(\frac{x}{z}\right) g(z), \tag{2.8.115}$$

(2.8.114) reads

$$\frac{\partial}{\partial(\log Q^2)} f(x, Q^2) = \frac{\alpha}{2\pi} [P \otimes f](x, Q^2). \tag{2.8.116}$$

For the electron–positron–photon system, i.e. the simplest case of QED, the evolution equations are a system of three coupled equations for the electron, photon, and positron distribution functions $f_{\mathrm{e}}(x, Q^2)$, $f_{\bar{\mathrm{e}}}(x, Q^2)$, and $f_\gamma(x, Q^2)$. The kernels are the *splitting functions*

$$\begin{aligned}
P^{(0)}_{\mathrm{ee}}(x) &= P^{(0)}_{\bar{\mathrm{e}}\bar{\mathrm{e}}}(x) = \frac{1+x^2}{(1-x)_+} - \frac{3}{2}\delta(1-x) = \left(\frac{1+x^2}{1-x}\right)_+, \\
P^{(0)}_{\bar{\mathrm{e}}\mathrm{e}}(x) &= P^{(0)}_{\mathrm{e}\bar{\mathrm{e}}}(x) = 0, \\
P^{(0)}_{\gamma\mathrm{e}}(x) &= P^{(0)}_{\gamma\bar{\mathrm{e}}}(x) = \frac{1+(1-x)^2}{x}, \\
P^{(0)}_{\mathrm{e}\gamma}(x) &= P^{(0)}_{\bar{\mathrm{e}}\gamma}(x) = x^2 + (1-x)^2, \\
P^{(0)}_{\gamma\gamma}(x) &= -\frac{2}{3}\delta(1-x).
\end{aligned} \tag{2.8.117}$$

Using the symmetry properties of the splitting functions, the *evolution equations* read

$$\begin{aligned}
\frac{\partial}{\partial(\log Q^2)} f_{\mathrm{e}}(x, Q^2) &= \frac{\alpha}{2\pi}\left[P^{(0)}_{\mathrm{ee}} \otimes f_{\mathrm{e}} + P^{(0)}_{\mathrm{e}\gamma} \otimes f_\gamma\right](x, Q^2), \\
\frac{\partial}{\partial(\log Q^2)} f_{\bar{\mathrm{e}}}(x, Q^2) &= \frac{\alpha}{2\pi}\left[P^{(0)}_{\mathrm{ee}} \otimes f_{\bar{\mathrm{e}}} + P^{(0)}_{\mathrm{e}\gamma} \otimes f_\gamma\right](x, Q^2), \\
\frac{\partial}{\partial(\log Q^2)} f_\gamma(x, Q^2) &= \frac{\alpha}{2\pi}\left[P^{(0)}_{\gamma\mathrm{e}} \otimes (f_{\mathrm{e}} + f_{\bar{\mathrm{e}}}) + P^{(0)}_{\gamma\gamma} \otimes f_\gamma\right](x, Q^2).
\end{aligned} \tag{2.8.118}$$

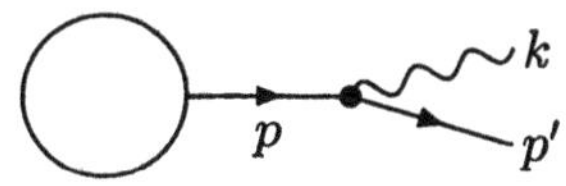

Fig. 2.20 Graph for collinear photon emission in the final state

These *evolution equations of QED* summarize the multi-photon-emission effects in the LL approximation. For given initial conditions, i.e.

$$f_{\mathrm{e}}(x, m^2) = \delta(1-x), \qquad f_{\bar{\mathrm{e}}}(x,, m^2) = f_\gamma(x, m^2) = 0, \qquad (2.8.119)$$

they yield the electron, positron, and photon distribution functions at arbitrary Q^2.

The evolution equations of QED are an important tool for the calculation of electromagnetic radiative corrections in high-energy scattering processes. Similar equations are valid in non-abelian gauge theories, e.g. QCD, where they play a basic rôle in determining the higher-order corrections to the parton model (cf. Sect. 3.2.3).

2.8.3.9 Collinear particles in the final state

We now turn to the discussion of collinear particles in the final state. The virtual collinear corrections are given by the same factors as in the initial state, e.g. (2.8.75) for electrons in the final state. For the emission of real collinear particles the situation is different. For initial-state particles, the incoming momentum is kept fixed and the momentum entering the hard-scattering process varies owing to the momentum carried away by the collinear emission. In the case of collinear emission from a final-state particle, the detector usually cannot distinguish the two collinear particles, and these are detected as one. Therefore, the momentum participating in the hard scattering, i.e. the sum of the momenta of the two collinear particles, is measured and has to be kept fixed when integrating over the momentum of the collinearly emitted particle.

For definiteness we consider an outgoing electron emitting a collinear photon, as depicted in Fig. 2.20, where also the momenta are defined. Using momentum conservation, the phase-space volume element of the two massless, collinear final-state particles can be rewritten as

$$
\begin{aligned}
\frac{1}{(2\pi)^6}\int\frac{\mathrm{d}^3\mathbf{k}}{2|\mathbf{k}|}\int\frac{\mathrm{d}^3\mathbf{p}'}{2|\mathbf{p}'|} &= \frac{1}{(2\pi)^6}\int\frac{\mathrm{d}^3\mathbf{k}}{2|\mathbf{k}|}\int\frac{\mathrm{d}^3\mathbf{p}'}{2|\mathbf{p}'|}\int \mathrm{d}^4p\,\delta^{(4)}(p-p'-k)\\
&= \frac{1}{(2\pi)^6}\int\frac{\mathrm{d}^3\mathbf{k}}{2|\mathbf{k}|}\int\frac{\mathrm{d}^3\mathbf{p}}{2|\mathbf{p}|}\frac{|\mathbf{p}|}{|\mathbf{p}-\mathbf{k}|}\bigg|_{p^0=|\mathbf{p}-\mathbf{k}|+|\mathbf{k}|}\,. \qquad (2.8.120)
\end{aligned}
$$

In the collinear limit, $k=(1-x)p$, the extra factor becomes $|\mathbf{p}|/|\mathbf{p}-\mathbf{k}| = 1/x$.

The diagram of Fig. 2.20 gives rise to the matrix element

$$
\begin{aligned}
&\mathcal{M}^{i\to e\gamma f}_{\kappa'\lambda}(p',k,p_j) =\\
&= -e\bar{u}(p',\kappa')\not{\epsilon}^*(k,\lambda)\frac{(\not{p}'+\not{k}+m)}{(p'+k)^2-m^2+\mathrm{i}\varepsilon}A^{i\to ef}(p,p_j). \qquad (2.8.121)
\end{aligned}
$$

As above we omit electron-mass terms. For the momenta we use the parametrization [neglecting terms of order $(\mathbf{k}_\perp^2)^2$]

$$
\begin{aligned}
p^\mu &= \left(E+\frac{\mathbf{k}_\perp^2}{2x(1-x)E},0,0,E\right),\\
k^\mu &= \left((1-x)E+\frac{\mathbf{k}_\perp^2}{4x(1-x)E},0,0,(1-x)E+\frac{\mathbf{k}_\perp^2(1-2x)}{4x(1-x)E}\right)+k_\perp^\mu,\\
p'^\mu &= p^\mu-k^\mu = \left(xE+\frac{\mathbf{k}_\perp^2}{4x(1-x)E},0,0,xE-\frac{\mathbf{k}_\perp^2(1-2x)}{4x(1-x)E}\right)-k_\perp^\mu,\\
p^2 &= \frac{\mathbf{k}_\perp^2}{x(1-x)},\qquad k^2=0,\qquad p'^2=0. \qquad (2.8.122)
\end{aligned}
$$

This parametrization differs from (2.8.51) because now p' is on-shell and p is off-shell. Replacing the numerator of the fermion propagator by the polarization sum [cf. (2.8.53)], results in

$$
\begin{aligned}
\mathcal{M}^{i\to e\gamma f}_{\kappa'\lambda}(p',k,p_j) &= -e\sum_\kappa \bar{u}(p',\kappa')\not{\epsilon}^*(k,\lambda)u(p,\kappa)\bar{u}(p,\kappa)\\
&\times\frac{x(1-x)}{\mathbf{k}_\perp^2}A^{i\to ef}(p,p_j)+\mathcal{O}(\mathbf{k}_\perp^2). \qquad (2.8.123)
\end{aligned}
$$

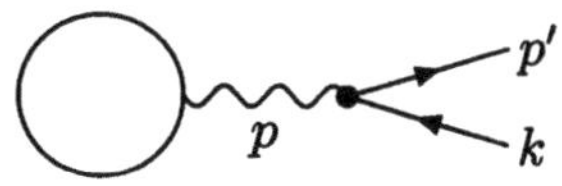

Fig. 2.21 Graph for collinear electron emission in the final state

Up to terms of order $\mathbf{k}_\perp^2$ the electromagnetic vertex is the same as in (2.8.55). The resulting cross section then reads

$$\mathrm{d}\sigma_{\kappa'}^{i\to e\gamma f}(p', p_j) = 2e^2 \int \frac{\mathrm{d}^3 k}{(2\pi)^3 2E_k} \frac{x(1-x)}{\mathbf{k}_\perp^2} \frac{1+x^2}{1-x} \times \sum_\lambda \left[\frac{1}{x} \mathrm{d}\sigma_{\kappa'}^{i\to ef}(xp, p_j) + \mathcal{O}(|\mathbf{k}_\perp|) \right], \tag{2.8.124}$$

where the factor $1/x$ results from the transformation of the phase space (2.8.120). Since we fix p but not p' when integrating over k, we can pull the lowest-order cross section out of the integration and obtain instead of (2.8.63) the result

$$\mathrm{d}\sigma_{\kappa}^{i\to e\gamma f}(p', p_j) = \frac{\alpha}{2\pi} \log\left(\frac{Q^2}{m^2}\right) \mathrm{d}\sigma_{\kappa}^{ei\to f}(p, p_j) \int_0^1 \mathrm{d}x \, \frac{1+x^2}{1-x} + \text{non-logarithmic terms.} \tag{2.8.125}$$

The logarithmic terms cancel exactly against those of the virtual contribution (2.8.75).

For final-state photons we similarly find from Fig. 2.21 a contribution proportional to the integral over the splitting function

$$\int_0^1 \mathrm{d}x \, P_{e\gamma}^{(0)}(x) = \frac{2}{3}, \tag{2.8.126}$$

which cancels exactly against the virtual contributions from (2.8.106). Since the final-state collinear fermions are not detected individually, we have to sum over their polarizations, and the unpolarized splitting function shows up.

The exact cancellation of the final-state collinear singularities in QED is a special case of the *Kinoshita–Lee–Nauenberg (KLN) theorem* (cf. page 294 in

Sect. 2.8.1.4). According to this theorem, the mass singularities associated with the final state must cancel, when we integrate over all degenerate final states.

2.9 Non-perturbative aspects of gauge theories

We mentioned at different places that there are important properties of gauge theories which cannot be treated by conventional perturbation theory. In this section we discuss some of these features.

- Classical gauge-field configurations decompose into different topological classes. These features become most transparent if we consider them on closed manifolds in the euclidean region. These different *Chern classes* are distinguished by *topological quantum numbers*. Perturbation theory considers only the trivial class.
- There is a particular relationship between modes with zero eigenvalues, the *zero modes* of the Dirac operator in a background gauge field, and the topological quantum numbers. This is expressed by the *index theorem*.
- It is not yet finally clarified how far the different topological aspects survive the quantization procedure. In the *semi-classical approximation* of the path-integral formulation, the decomposition of gauge-field configurations into Chern classes introduces an additional summation over the topological quantum numbers. The appearance of the zero modes requires a modification of the path integral.

The first item is well established in mathematics [Bo65, At68, Gi95] and theoretical physics [Be75a, Go87]. We treat some examples in Sects. 2.9.1 and 2.9.2.

In Sect. 2.9.3 we discuss non-perturbative methods of quantum gauge theories in semi-classical approximation. As an example, we consider 't Hooft's one-instanton calculation [tH76] in Sect. 2.9.3.2. Renormalization is treated analogously to perturbation theory in the background-field method. The range of validity of the semi-classical approximation within such a treatment of ultraviolet quantum fluctuations is not well-known. Maybe, this approach to non-perturbative problems of quantum gauge theories gives only qualitative insight.

However, the semi-classical approximation sheds some light on very important questions. For instance, the cluster properties of the vacuum lead to the introduction of the θ-vacuum (Sect. 2.9.3.4) and the strong CP problem (Sect. 2.9.3.9). New processes like *vacuum tunnelling* (Sect. 2.9.3.5) cannot be described by perturbation theory. It leads to new types of Green functions describing vacuum condensates and topological excitations (Sect. 2.9.3.7). These help to understand spontaneous breaking of chiral symmetry and the U(1) problem (Sect. 2.9.3.10), i.e. the approximately zero mass of the pion and the high mass of the η' (cf. Sect. 3.5.1). Index theorems exhibit the geometrical origin of the anomalies (Sect. 2.9.3.8).

It is of great importance that the *lattice approximation* allows another approach to the non-perturbative problems of gauge theories [Wi74]. Basically, it starts from a more systematic approach yielding a mathematically better defined path-integral formula. We treat this method in Sect. 2.10. However, compared with lattice approximation, in many cases the semi-classical approximation gives an intuitive picture of the physical problem. Thus, we consider lattice approximation supplemented with semi-classical approximation as the appropriate method for non-perturbative problems. Needless to say, this method has not yet reached the maturity of perturbation theory.

Non-perturbative physics is mainly discussed for problems of QCD. We treat the problem of confinement in Sect. 3.7 in the spirit just explained. There we use SU(2) symmetry as a simple example. Concerning the electroweak interaction, the perturbative calculations agree with experimental data at the per-mill level. There are presently no indications for non-perturbative effects.

2.9.1 Topological quantum numbers

2.9.1.1 Flux quantization on the torus

We begin with one of the simplest examples of a topological quantum number. To this end, we consider an abelian gauge field on a two-dimensional torus $\mathcal{T}$ of lengths L_1 and L_2. On the torus we introduce cartesian coordinates (x_1, x_2). The two sides a_1: $x_1 = 0$ and $x_1 = L_1$ as well as the two sides a_2: $x_2 = 0$ and $x_2 = L_2$ must be identified, and therefore the points B (cf. Fig. 2.22). The U(1) gauge potential $A_\mu(x)$, $\mu = 1, 2$, and the field

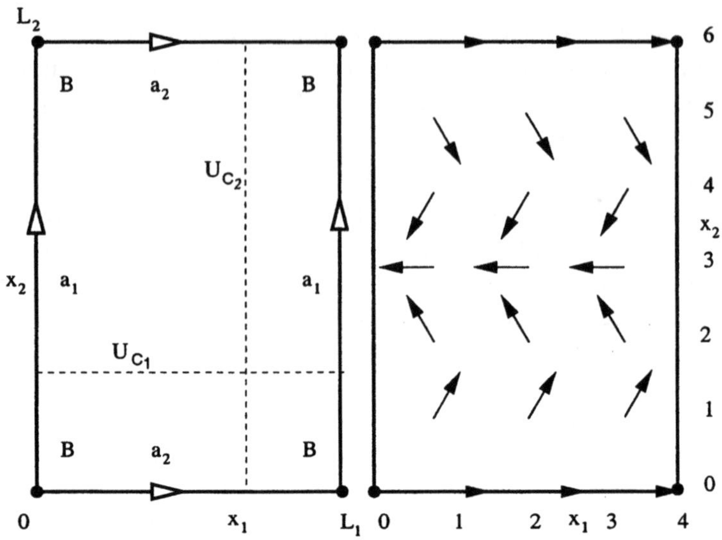

Fig. 2.22 Topology of the torus as explained in the text with $L_1 = 4$ units and $L_2 = 6$ units (left) and the phase of $\exp(-\mathrm{i}eL_1A(x))$ (right)

strength $F_{\mu\nu}(x)$ are related by

$$F_{\mu\nu}(x) = \partial_\mu A_\nu(x) - \partial_\nu A_\mu(x). \tag{2.9.1}$$

The torus topology, the identification of sides and points, requires *boundary conditions.* In Sect. 2.3.2 we introduced $A_\mu(x)$ as the components of an infinitesimal parallel transport. This implies that $A_\mu(x)$ should be double-periodic up to a gauge transformation,

$$A_\mu(x + \hat{L}_i) = A_\mu(x) - \frac{\mathrm{i}}{e} g_i^{-1}(x)\partial_\mu g_i(x), \qquad i = 1, 2. \tag{2.9.2}$$

Here $\hat{L}_i$ denotes the vectors $\hat{L}_1 = (L_1, 0)$ and $\hat{L}_2 = (0, L_2)$ and $g_i(x)$ continuous gauge transformations describing a parallel transport U_C along a path C traversing the torus parallel to a side a_i (cf. Fig. 2.22). Continuity around the point B requires

$$g_1(x)g_2(x + \hat{L}_1) = g_2(x)g_1(x + \hat{L}_2). \tag{2.9.3}$$

To be specific, we take the gauge

$$g_1(x) = 1, \qquad g_2(x) = \exp\left(-2\pi \mathrm{i} k \frac{x_1}{L_1}\right). \tag{2.9.4}$$

Then, periodicity (2.9.3) requires $\exp(-2\pi\mathrm{i}\, k) = 1$, and therefore $k =$ integer. The number k is called the *topological quantum number* of the gauge field on the torus.

We may express this quantum number by an integral over the field strength. Stokes' theorem transforms the integral over the torus $\mathcal{T}$ into one over the lines of the boundary. Using the boundary conditions (2.9.2)–(2.9.4) results in

$$\begin{aligned}\frac{e}{2\pi}\int_{\mathcal{T}} F_{12}\,\mathrm{d}x_1\mathrm{d}x_2 &= \frac{e}{2\pi}\int_{\partial\mathcal{T}} A_\mu\,\mathrm{d}s^\mu \\ &= \frac{\mathrm{i}}{2\pi}\int_0^{L_1} \mathrm{d}x\, g_2^{-1}(x, L_2)\partial_x g_2(x, L_2) = k.\end{aligned} \tag{2.9.5}$$

We note that k does not depend on a specific gauge potential.

There is a special solution with constant field strength and topological quantum number k, the *instanton solution* of the two-dimensional torus, which satisfies the Maxwell equation, ($\varepsilon_{12} = 1$, $\varepsilon_{\mu\nu} = -\varepsilon_{\nu\mu}$)

$$A_\mu^{(k)}(x) = -\frac{2\pi k}{eL_1L_2}x_2\delta_{\mu 1}, \qquad F_{\mu\nu}^{(k)} = \frac{2\pi k}{eL_1L_2}\varepsilon_{\mu\nu}. \tag{2.9.6}$$

This example describes the well-known flux quantization of a magnetic monopole enclosed by a torus.

2.9.1.2 The SU(2) instantons

As an example of a field configuration with non-trivial topological quantum number in four euclidean dimensions we discuss the *instanton solutions* of an SU(2) gauge theory on a sphere [Be75a]. The SU(2) gauge potential is described with help of the Pauli matrices τ^a, $\mathcal{A}_\mu(x) = gA_\mu^a\tau^a/2$, $\mu = 1,2,3,4$. The definitions of the field strength, the action, and the field equations follow from the scheme in Sects. 2.3.2 and 2.3.3,

$$\begin{aligned}\mathcal{F}_{\mu\nu} &= \partial_\mu\mathcal{A}_\nu - \partial_\nu\mathcal{A}_\mu - \mathrm{i}[\mathcal{A}_\mu, \mathcal{A}_\nu], \\ S_{\mathrm{E}}\{\mathcal{A}\} &= \frac{1}{2g^2}\int \mathrm{d}^4x\,\mathrm{Tr}(\mathcal{F}_{\mu\nu}\mathcal{F}^{\mu\nu}) = \frac{1}{2g^2}\frac{1}{2}(\mathcal{F}, \mathcal{F}),\end{aligned} \tag{2.9.7}$$

with the field equation

$$D^\mu \mathcal{F}_{\mu\nu}(x) = \partial^\mu \mathcal{F}_{\mu\nu} - \mathrm{i}[\mathcal{A}^\mu, \mathcal{F}_{\mu\nu}] = 0 \tag{2.9.8}$$

and the shorthand

$$(\mathcal{F}_1, \mathcal{F}_2) = 2 \int \mathrm{d}^4x \, \mathrm{Tr}(\mathcal{F}_{1,\mu\nu} \mathcal{F}_2^{\mu\nu}). \tag{2.9.9}$$

We are interested in solutions of the classical field equations with finite action $S_\mathrm{E}\{\mathcal{A}\} < \infty$. This means that $\mathcal{F}_{\mu\nu}(x) \to 0$ for $x^2 \to \infty$, or that $\mathcal{A}_\mu$ approaches a pure gauge,

$$\mathcal{A}_\mu = \mathrm{i}g(x)\partial_\mu g^{-1}(x) \quad \text{for} \quad x^2 \to \infty. \tag{2.9.10}$$

We consider the *dual field strength* $\breve{\mathcal{F}}_{\mu\nu}$

$$\breve{\mathcal{F}}_{\mu\nu}(x) = \frac{1}{2}\varepsilon_{\mathrm{E};\mu\nu\rho\sigma}\mathcal{F}^{\rho\sigma}(x), \qquad \mathcal{F}_{\mu\nu}(x) = \frac{1}{2}\varepsilon_{\mathrm{E};\mu\nu\rho\sigma}\breve{\mathcal{F}}^{\rho\sigma}(x), \tag{2.9.11}$$

with $\varepsilon_{\mathrm{E};\mu\nu\rho\sigma}$ the euclidean totally antisymmetric Levi–Civita tensor[28] defined in (A.1.48). The Bianchi identity [cf. (2.3.30)]

$$D_\rho \mathcal{F}_{\mu\nu}(x) + D_\mu \mathcal{F}_{\nu\rho}(x) + D_\nu \mathcal{F}_{\rho\mu}(x) = 0, \tag{2.9.12}$$

or

$$\frac{1}{2}\varepsilon_\alpha{}^{\rho\mu\nu} D_\rho \mathcal{F}_{\mu\nu}(x) = D^\rho \breve{\mathcal{F}}_{\alpha\rho}(x) = 0, \quad \alpha = 1, \ldots, 4 \tag{2.9.13}$$

implies that *self-dual (anti-self-dual) fields*

$$\breve{\mathcal{F}}_{\mu\nu}(x) = \pm \mathcal{F}_{\mu\nu}(x) \tag{2.9.14}$$

are solutions of the field equations

$$D^\mu \breve{\mathcal{F}}_{\mu\nu}(x) = 0 = \pm D^\mu \mathcal{F}_{\mu\nu}(x). \tag{2.9.15}$$

[28] If it is evident that we are in the euclidean region, we omit the index "E".

In order to solve the self-duality equation (2.9.14) we take as the asymptotic gauge in (2.9.10)

$$g(x) = \frac{x^\mu \tau_\mu}{|x|} = \frac{\mathrm{i}\boldsymbol{x\tau} + x_4}{|x|}, \tag{2.9.16}$$

where

$$(\tau_\mu) = (\mathrm{i}\boldsymbol{\tau}, 1), \qquad \bar{\tau}_\mu = (-\mathrm{i}\boldsymbol{\tau}, 1) \tag{2.9.17}$$

with

$$\tau_\mu \bar{\tau}_\nu + \tau_\nu \bar{\tau}_\mu = 2\delta_{\mu\nu}, \qquad \bar{\tau}_\mu \tau_\nu + \bar{\tau}_\nu \tau_\mu = 2\delta_{\mu\nu}. \tag{2.9.18}$$

We make the ansatz

$$\mathcal{A}_\mu(x) = \mathrm{i} f(x^2)\, g(x) \partial_\mu g^{-1}(x) = 2 f(x^2)\, \tau_{\mu\nu} \frac{x^\nu}{x^2} \tag{2.9.19}$$

with the self-dual (anti-self-dual) matrix tensor $\tau_{\mu\nu}$ ($\bar{\tau}_{\mu\nu}$)

$$\tau_{\mu\nu} = \frac{1}{4\mathrm{i}}(\tau_\mu \bar{\tau}_\nu - \tau_\nu \bar{\tau}_\mu), \qquad \bar{\tau}_{\mu\nu} = \frac{1}{4\mathrm{i}}(\bar{\tau}_\mu \tau_\nu - \bar{\tau}_\nu \tau_\mu). \tag{2.9.20}$$

Then, the corresponding field strength

$$\begin{aligned} \mathcal{F}_{\mu\nu}(x) &= \partial_\mu \mathcal{A}_\nu - \partial_\nu \mathcal{A}_\mu - \mathrm{i}[\mathcal{A}_\mu, \mathcal{A}_\nu] = -4 \frac{f(x^2)(1 - f(x^2))}{x^2} \tau_{\mu\nu} \\ &+ 4[f'(x^2) x^2 - f(x^2)(1 - f(x^2))] \frac{\tau_{\nu\rho} x_\mu x^\rho - \tau_{\mu\rho} x_\nu x^\rho}{x^4} \end{aligned} \tag{2.9.21}$$

is self-dual if the term in the second line disappears,

$$f'(x^2) x^2 - f(x^2)(1 - f(x^2)) = 0. \tag{2.9.22}$$

The solution of this equation is $f(x^2) = x^2/(x^2 + \lambda^2)$, which leads to

$$\mathcal{A}_\mu^{\mathrm{I},(1)}(x) = \frac{2x^2}{x^2 + \lambda^2} \tau_{\mu\nu} \frac{x^\nu}{x^2}, \qquad \mathcal{F}_{\mu\nu}^{\mathrm{I},(1)}(x) = -\frac{4\lambda^2}{(x^2 + \lambda^2)^2} \tau_{\mu\nu}. \tag{2.9.23}$$

The *instanton potential* $\mathcal{A}^{\mathrm{I},(1)}_{\mu,1}(x)$ is regular in the whole euclidean space but vanishes only as $|x|^{-1}$ for $x \to \infty$. Using the gauge transformation (2.9.16), we can change this feature. The resulting potential and field-strength,

$$\mathcal{A}^{\mathrm{I},(2)}_{\mu}(x) = g^{-1}(x)\mathcal{A}^{\mathrm{I},(1)}_{\mu}(x)g(x) + \mathrm{i}g^{-1}(x)\partial_\mu g(x) = \frac{2\lambda^2}{x^2+\lambda^2}\bar{\tau}_{\mu\nu}\frac{x^\nu}{x^2},$$
$$\mathcal{F}^{\mathrm{I},(2)}_{\mu\nu}(x) = -\frac{4\lambda^2}{(x^2+\lambda^2)^2}\left[\bar{\tau}_{\mu\nu} - 2\left(\bar{\tau}_{\mu\rho}\frac{x_\nu x^\rho}{x^2} - \bar{\tau}_{\nu\rho}\frac{x_\mu x^\rho}{x^2}\right)\right], \qquad (2.9.24)$$

are singular at $x = 0$, but vanish as $|x|^{-3}$ and $|x|^{-4}$, respectively, at infinity. In the language of *fibre bundles* (Sect. 2.3.2.7), we can express this fact in the following form: we have covered the euclidean space $\mathbb{R}^4$ by two coordinate systems. In both systems, the gauge potential as connection can be separated into the dependence on the base space $x \in \mathbb{R}^4 - \{\infty\} = \mathbb{R}^{(1)}$ or $x \in \mathbb{R}^4 - \{0\} = \mathbb{R}^{(2)}$ and the SU(2) adjoint representation space. The $\mathcal{A}^{\mathrm{I},(i)}_{\mu}(x)$ describe trivial bundles on these sub-manifolds. In the overlapping space the connections are related by a gauge transformation. Thus, on the whole $\mathbb{R}^4$ (or compactified to a sphere by a stereo-graphic transformation), our solutions describe a non-trivial bundle.

We characterize this non-trivial bundle by a topological quantum number, the *Pontryagin number*. It is obtained from the *topological charge density* or *Pontryagin density* (cf. Sect. 2.7.2.1)

$$\mathcal{P}_{\mathrm{E}}(x) = \frac{1}{16\pi^2}\,\mathrm{Tr}\,\mathcal{F}_{\mu\nu}(x)\breve{\mathcal{F}}^{\mu\nu}(x) = \frac{1}{64\pi^2}F^a_{\mu\nu}F^a_{\rho\sigma}\varepsilon^{\mu\nu\rho\sigma} \qquad (2.9.25)$$

as

$$k = \int \mathrm{d}^4x\, \mathcal{P}_{\mathrm{E}}(x). \qquad (2.9.26)$$

The Pontryagin density can be written as a divergence (cf. Sect. 2.7.2.1),

$$\mathcal{P}_{\mathrm{E}}(x) = \partial_\mu G^\mu(x),$$
$$G_\mu(x) = -\frac{\mathrm{i}}{8\pi^2}\varepsilon_{\mu\nu\rho\sigma}\,\mathrm{Tr}\left(\frac{2}{3}\mathcal{A}^\nu\mathcal{A}^\rho\mathcal{A}^\sigma + \mathrm{i}\mathcal{A}^\nu\partial^\rho\mathcal{A}^\sigma\right). \qquad (2.9.27)$$

This follows by direct calculation from the definition of the field strength and its dual.

For illustration, we calculate the topological quantum number using Gauss's theorem. We take the regular solution $\mathcal{A}^{\mathrm{I},(1)}(x)$ in $S^{(1)} = \{x|x^2 \leq 1\}$ and $\mathcal{A}^{\mathrm{I},(2)}(x)$ in $S^{(2)} = \{x|x^2 \geq 1\}$. The integral (2.9.26) then turns into one over the unit sphere,

$$\begin{aligned}
k &= \int \mathrm{d}^4x\, \mathcal{P}_{\mathrm{E}}(x) = \int_{S^{(1)}} \mathrm{d}^4x\, \mathcal{P}_{\mathrm{E}}(x) + \int_{S^{(2)}} \mathrm{d}^4x\, \mathcal{P}_{\mathrm{E}}(x) \\
&= -\frac{\mathrm{i}}{8\pi^2} \int_{S_3} \mathrm{d}S_\mu\, \varepsilon^{\mu\nu\rho\sigma} \left[\mathrm{Tr}\left(\frac{2}{3} \mathcal{A}^{\mathrm{I},(1)}_\nu \mathcal{A}^{\mathrm{I},(1)}_\rho \mathcal{A}^{\mathrm{I},(1)}_\sigma + \mathrm{i}\mathcal{A}^{\mathrm{I},(1)}_\nu \partial_\rho \mathcal{A}^{\mathrm{I},(1)}_\sigma \right) \right. \\
&\qquad \left. - \mathrm{Tr}\left(\frac{2}{3} \mathcal{A}^{\mathrm{I},(2)}_\nu \mathcal{A}^{\mathrm{I},(2)}_\rho \mathcal{A}^{\mathrm{I},(2)}_\sigma + \mathrm{i}\mathcal{A}^{\mathrm{I},(2)}_\nu \partial_\rho \mathcal{A}^{\mathrm{I},(2)}_\sigma \right) \right] \\
&= -\int_{S_3} \mathrm{d}S_\mu\, \frac{\varepsilon^{\mu\nu\rho\sigma}}{24\pi^2}\, \mathrm{Tr}\Big\{ [g^{-1}(x)\partial_\nu g(x)][g^{-1}(x)\partial_\rho g(x)][g^{-1}(x)\partial_\sigma g(x)] \Big\} \\
&= +1.
\end{aligned} \tag{2.9.28}$$

The final result was obtained using (2.9.16), (2.9.24), $g^{-1}(x)\partial_\mu g(x) = -(\partial_\mu g^{-1}(x))g(x)$, and the symmetries of $\varepsilon^{\mu\nu\rho\sigma}$ and of the trace.

This derivation contains some remarkable features. In the last integral of (2.9.28) only the gauge transformation $g(x) = (x_4 + \mathrm{i}\boldsymbol{x\tau})/|x|$ that connects the two regular solutions appears. Further, $g(x)$ describes a mapping $\mathbf{M}$ of the 3-sphere S_3 on the group SU(2). In order to see this, we introduce four-dimensional polar coordinates ρ, χ, ϑ, and ϕ,

$$\begin{aligned}
x^1 &= \rho \sin\chi \sin\vartheta \cos\phi, \qquad x^2 = \rho \sin\chi \sin\vartheta \sin\phi \\
x^3 &= \rho \sin\chi \cos\vartheta, \qquad x^4 = \rho \cos\chi, \\
0 &\leq \chi < \pi, \qquad 0 \leq \vartheta < \pi, \qquad 0 \leq \phi \leq 2\pi, \qquad 0 \leq \rho < \infty, \\
\mathrm{d}^4x &= \rho^3 \mathrm{d}\Omega_3 = \rho^3 \sin^2\chi \sin\vartheta\, \mathrm{d}\chi\, \mathrm{d}\vartheta\, \mathrm{d}\phi\, \mathrm{d}\rho.
\end{aligned} \tag{2.9.29}$$

Considering $g(x)$ in these coordinates and comparing it with the standard parametrization of $g(\theta', \vartheta', \phi') \in \mathrm{SU}(2)$ (1.2.63), we see that the mapping $\mathbf{M}$ can be described by

$$\mathbf{M}\colon \frac{1}{2}\theta' = \chi, \qquad \vartheta' = \vartheta, \qquad \phi' = \phi. \tag{2.9.30}$$

The SU(2) Haar measure (1.2.64) reads in the polar coordinates $1/(2\pi^2) \times \sin^2\chi \sin\vartheta \mathrm{d}\chi \mathrm{d}\vartheta \mathrm{d}\phi$, i.e. it is normalized to the volume of S_3. Applying the coordinate transformation (2.9.29) to the integral (2.9.28), it becomes transformed into the Haar measure of SU(2).

The expression (2.9.28) counts how often the image of S_3 under $g(x)$ covers SU(2), it is the *winding number* of $g(x)$ [Be96, Go87]. The concept of winding number is more familiar from the mapping of a circle $S_1 : \{\phi | 0 \leq \phi \leq 2\pi\}$ on the group U(1) with parameters $\{\exp(\mathrm{i}\phi') | 0 \leq \phi' \leq 2\pi\}$. The map $\mathbf{M}$: $\phi \to \phi' = n\phi$ covers U(1) n times, its winding number is $w(\mathbf{M}) = n$. Our description of $\mathbf{M}$, (2.9.30), is not complete. The spherical coordinates do not cover the "south pole" $x^4 = -\rho$, the SU(2) coordinates do not reach $g = -\mathbf{1}$. So we should describe the tangent spaces of S_3, or the infinitesimal transformations of SU(2) by overlapping coordinate systems, i.e. by *vector bundles*. We have described this procedure by the discussion of the instanton potential (2.9.24). A complete definition of the winding number refers to the map of the corresponding vector bundles. From these considerations we get a geometric interpretation of the topological quantum number $k = 1$ of the instanton: it is the winding number of a map of the sphere S_3 into the gauge group SU(2).

Other similar types of solutions, the so-called *antiinstanton solutions*, are obtained from the condition for anti-self-dual fields $\check{\mathcal{F}}_{\mu\nu} = -\mathcal{F}_{\mu\nu}$. The simplest antiinstanton solution with Pontryagin number -1 can be written in the regular gauge as

$$\mathcal{A}_\mu^{\bar{\mathrm{I}},(1)}(x) = \frac{2x^2}{x^2 + \lambda^2} \bar{\tau}_{\mu\nu} \frac{x_\nu}{x^2}, \qquad \mathcal{F}_{\mu\nu}^{\bar{\mathrm{I}},(1)}(x) = -\frac{4\lambda^2}{(x^2 + \lambda^2)^2} \bar{\tau}_{\mu\nu}, \tag{2.9.31}$$

and similarly to (2.9.24) in the singular gauge.

The definition of duality is translation-invariant. Therefore we may substitute x by $x - z$ in the expressions (2.9.23), (2.9.24), and (2.9.31) for getting other solutions of the field equations with $k = \pm 1$.

Self-dual solutions for higher Pontryagin numbers $k \geq 1$ can be obtained from the ansatz [Ja77]

$$\mathcal{A}_\mu^{(k)}(x) = -\bar{\tau}_{\mu\nu} \partial^\nu \log \rho(x) \tag{2.9.32}$$

in the singular gauge as a solution of the self-duality equation $\partial_\mu \partial^\mu \rho(x) = 0$ in the form

$$\rho(x) = 1 + \sum_{i=1}^{k} \frac{\lambda_i^2}{(x - z_i)^2} \tag{2.9.33}$$

with $5k$ free parameters λ_i and z_i. They can be interpreted as k instantons which are located at positions z_i and have sizes λ_i. The most general self-dual solution with topological charge k is characterized by $8k-3$ parameters α_i^k. No simple explicit representations have been found for these solutions. However, there are algebraic methods which allow for the construction of all solutions of the self-duality equation. This also applies for self-dual SU(N) gauge fields [At78, Ch78, Co78].

For self-dual fields $\mathcal{F}_{\mu\nu} = \check{\mathcal{F}}_{\mu\nu}$, the action agrees with the topological quantum number up to a factor

$$S_{\mathrm{E}} = \frac{1}{4g^2}(\mathcal{F},\mathcal{F}) = \frac{1}{4g^2}(\mathcal{F},\tilde{\mathcal{F}}) = \frac{8\pi^2}{g^2}k. \tag{2.9.34}$$

Moreover, Schwarz's inequality, $(\mathcal{F},\tilde{\mathcal{F}}) \leq \sqrt{(\mathcal{F},\mathcal{F})(\tilde{\mathcal{F}},\tilde{\mathcal{F}})}$, and the relation $(\mathcal{F},\mathcal{F}) = (\tilde{\mathcal{F}},\tilde{\mathcal{F}})$ imply $(\mathcal{F},\mathcal{F}) \geq (\mathcal{F},\tilde{\mathcal{F}})$ for arbitrary fields. Consequently, within the class of fields with topological quantum numbers k, *self-dual fields* give the minimal value of S_{E}. Thus, in agreement with Hamilton's principle of minimal action, self-dual fields are in fact solutions of the field equations.

2.9.1.3 The Chern classification of gauge fields

The Chern classification of classical gauge fields by topological quantum numbers is most straightforward for compact space–time manifolds. (In certain cases these can be replaced by boundary conditions at infinity [cf. (2.9.10) and (2.9.23)]). Such manifolds $\mathcal{M}$ require a covering by different local coordinate systems $\mathcal{U}^i$. A well-known example is the sphere where we have for instance two coordinate systems, one continuous at the north pole and one continuous at the south pole, respectively. If we consider the gauge fields as infinitesimal connections, then the representations of the gauge fields in the different coordinate systems are related by gauge transformations, (2.3.21),

$$\mathcal{A}_\mu^{(1)} = g^{12}\mathcal{A}_\mu^{(2)}(g^{12})^{-1} - \mathrm{i}(\partial_\mu g^{12})(g^{12})^{-1} \tag{2.9.35}$$

in the overlapping region. The $\mathcal{A}_\mu^{(i)}$ are the connections in the coordinate systems $\mathcal{U}^i$. In order to be consistent in the case of three overlappings $\mathcal{U}^i$, the *transition transformation* $g^{ik} \in G$ must satisfy the *cocycle conditions*

$$g^{ik} = g^{il}\, g^{lk}. \tag{2.9.36}$$

The systems of coordinates in $\mathcal{U}^i$, the gauge fields $\mathcal{A}_\mu^{(i)}$, and the transition transformation g^{ik} satisfying the cocycle conditions are called a *principle fibre bundle of the gauge group $\mathcal{G}$*. It is a concretization of classical gauge fields that includes the possibly twisted boundary conditions [Ch82, Na90].

In Sects. 2.9.1.1 and 2.9.1.2 we gave some illustrative examples of such fibre bundles. Our discussion contained also the main elements of their classification by topological quantum numbers. In the following we give a definition of *topological quantum numbers* for general gauge fields in arbitrary even euclidean dimensions $D = 2j$ with complex gauge groups. Such considerations play a rôle in theories beyond the standard model.

We consider the unitary group SU(N) and denote the hermitian generators by $T^{i_1 i_2}$, $i_\nu = 1, \ldots, N$. Gauge potentials $A_\mu^{i_1 i_2}(x)$ and the field strength $F_{\mu\nu}^{i_1 i_2}(x)$ are constructed according to the general scheme explained in Sect. 2.3. The generalization of the Pontryagin density for such cases in D dimensions reads [Gi95, Go87]

$$c_j(F^j) = \frac{1}{(4\pi)^j\, j!} \sum_{i_1,\ldots,i_j=1}^{m} \sum_{\pi} \operatorname{sgn}(\pi) \\ \times F_{\mu_1\mu_2}^{i_1 i_{\pi(1)}}(x) \cdots F_{\mu_{D-1}\mu_D}^{i_j i_{\pi(j)}}(x) \varepsilon^{\mu_1 \ldots \mu_D} . \tag{2.9.37}$$

This formula (2.9.37) contains a sum over the permutations π of the indices $i_1, i_2, \ldots, i_j$ with signature $\operatorname{sgn}(\pi) = \pm 1$. The topological quantum number $k(F^j)$ is defined as [cf. (2.9.26)]

$$k(F^j) = \int_{\mathcal{M}} c_j(F^j(x))\, \mathrm{d}^D x . \tag{2.9.38}$$

Without proof we add the following remarks on (2.9.37) and (2.9.38).

1. For general manifolds $\mathcal{M}$, $\varepsilon^{\mu_1 \cdots \mu_D}$ must be interpreted as the covariant volume tensor. Only in flat space, $\varepsilon^{\mu_1 \cdots \mu_D}$ is the constant antisymmetric tensor.

2. The coefficients in the definition of c_j are chosen such that the topological quantum numbers become integer in a compact manifold. Similar to (2.9.27) and (2.9.28), the proof of this fact follows from showing that $c_j(F^j)$ is a generalized divergence of some functional $G_\mu^j\{\mathcal{A}\}$.

3. The topological quantum number $k(F^j)$ is invariant under a continuous transformation of the gauge potential. All gauge potentials with the same topological quantum number belong by definition to the same Chern class.

Finally we want to show that (2.9.37) and (2.9.38) really cover our two examples. Restricting (2.9.37) to the two-dimensional SU(1) case (Sect. 2.9.1.1) gives directly $c_1(F^1) = 1/(4\pi)F_{\mu\nu}\varepsilon^{\mu\nu}$, i.e. (2.9.5). For the SU(2) case in 4 dimensions, we put $F^{il}_{\mu\nu} = \sum_a F^a_{\mu\nu}\tau^a_{il}/2$. The evaluation of the sums in (2.9.37) gives

$$\sum_{i_1,i_2=1}^{2}\sum_{\pi}\operatorname{sgn}(\pi)\frac{\tau^a_{i_1\pi(i_1)}}{2}\frac{\tau^b_{i_2\pi(i_2)}}{2} = \frac{1}{4}(\operatorname{Tr}\tau^a\operatorname{Tr}\tau^b - \operatorname{Tr}(\tau^a\tau^b)) = -\frac{\delta^{ab}}{2},$$

and therefore $c_2(F^2) = -1/(64\pi^2)F^a_{\mu\nu}F^a_{\rho\sigma}\varepsilon^{\mu\nu\rho\sigma}$ corresponding to (2.9.25). The instanton has[29] $k(F^2) = -1$.

These two examples are good illustrations of the general theory of topological quantum numbers.

2.9.2 The index theorem

There is a characteristic relationship between the numbers n^+ and n^- of right- and left-handed solutions of the massless Dirac equation in a background gauge field and the topological quantum number of this gauge field. Index theorems express this fact [At68, Eg80, Gi95]. Dirac equations on general manifolds have their own involved features. Therefore, we restrict ourselves to demonstrate the index theorem for the two- and four-dimensional cases where $n^+ - n^-$ is equal to the topological quantum number k as defined in (2.9.5) and (2.9.26), i.e.

$$n^+ - n^- = k. \tag{2.9.39}$$

We give some explicit examples and add some hints on the proof of the index theorem.

[29] In this section we use the mathematical notation where $k(F^2) = -1$ for a physical instanton with $k = 1$.

2.9.2.1 The zero modes of the Dirac equation on the torus

We consider again the solution of the two-dimensional euclidean Dirac equation on the torus $\mathcal{T}$ [Jo90, Jo94],

$$D\!\!\!/\psi(x) \equiv \gamma^\mu(\partial_\mu - \mathrm{i}e\,A_\mu)\psi(x) = 0 \tag{2.9.40}$$

with $\gamma^1 = \tau_1$, $\gamma^2 = \tau_2$, $\gamma^5 = -\mathrm{i}\gamma^1\gamma^2 = \tau_3$. The *normalizable* solutions of (2.9.40) are called zero modes, because they belong to the eigenvalue zero of the Dirac operator $D\!\!\!/$. In order to have an example of the index theorem, we consider the potential $A_\mu(x)$ (2.9.6) with a constant field strength $F_{\mu\nu}(x)$,

$$A_\mu = \left(-\frac{2\pi}{eL_1L_2}x_2,\ 0\right), \qquad F_{\mu\nu} = \varepsilon_{\mu\nu}\frac{2\pi}{eL_1L_2}, \tag{2.9.41}$$

as a physical background field. The flux is quantized and belongs according to (2.9.5) to the topological quantum number $k = 1$. The solutions must be continuous on the U(1) fibre bundle on the torus, i.e. they must be consistent with the periodicity conditions (2.9.2) and (2.9.4). They must satisfy

$$\psi(x + \hat{L}_i) = g_i(x)\psi(x) \tag{2.9.42}$$

with $g_1(x) = 1$ and $g_2(x) = \exp(-2\pi\mathrm{i}x_1/L_1)$. The ansatz

$$\psi(x) = \exp\left(-\mathrm{i}\pi\frac{x_1x_2}{L_1L_2}\right)\exp\left(-\frac{\pi}{2L_1L_2}(x_1^2 + x_2^2)\gamma^5\right)\psi^{(0)}(x), \tag{2.9.43}$$

where $\psi^{(0)}(x)$ is a solution of the free Dirac equation

$$\begin{pmatrix} 0 & \partial_1 - \mathrm{i}\partial_2 \\ \partial_1 + \mathrm{i}\partial_2 & 0 \end{pmatrix}\begin{pmatrix} \psi_1^{(0)}(x) \\ \psi_2^{(0)}(x) \end{pmatrix} = 0, \tag{2.9.44}$$

solves the Dirac equation locally. The solution of the free Dirac equation has to be chosen in such a way that the boundary conditions (2.9.42) are satisfied. This amounts to

$$\begin{aligned}
\psi_1^{(0)}(x+\hat{L}_1) &= \exp\left[\frac{\pi}{|\tau|}\left(z+\frac{1}{2}\right)\right]\psi_1^{(0)}(x), \\
\psi_1^{(0)}(x+\hat{L}_2) &= \exp\left[-\mathrm{i}\pi\left(z+\frac{\mathrm{i}|\tau|}{2}\right)\right]\psi_1^{(0)}(x), \\
\psi_2^{(0)}(x+\hat{L}_1) &= \exp\left[-\frac{\pi}{|\tau|}\left(\bar{z}+\frac{1}{2}\right)\right]\psi_2^{(0)}(x), \\
\psi_2^{(0)}(x+\hat{L}_2) &= \exp\left[-\mathrm{i}\pi\left(\bar{z}-\frac{\mathrm{i}|\tau|}{2}\right)\right]\psi_2^{(0)}(x)
\end{aligned} \tag{2.9.45}$$

with

$$z = \frac{x_1+\mathrm{i}x_2}{L_1}, \qquad \bar{z} = \frac{x_1-\mathrm{i}x_2}{L_1}, \qquad |\tau| = \frac{L_2}{L_1}. \tag{2.9.46}$$

Since $\gamma^5\psi_2(x) = -\psi_2(x)$, the solution with $\psi_2^{(0)}(x) \neq 0$ is exponentially growing and has to be discarded. Thus, the only consistent solution obeys $\psi_2^{(0)}(x) = 0$ or $\gamma^5\psi(x) = \psi(x)$.

A more explicit argument follows from the theory of *elliptic functions*. The function $\psi_1^{(0)}(x)$ is analytic in z and periodic up to an exponential factor in the real and imaginary direction. This leads to elliptic functions. Therefore, it is no surprise that the solutions of (2.9.40) can be expressed by Jacobi's θ-function $\theta_3(z,\tau) = \sum_n \exp(2\pi\mathrm{i}nz - |\tau|\pi n^2)$, namely

$$\begin{aligned}
\psi(x) &= \frac{\left|\frac{\tau}{2}\right|^{-1/4}}{L_1}\exp\left[-\mathrm{i}\pi\frac{x_1x_2}{L_1L_2}\right]\exp\left[\frac{\pi}{2|\tau|}(z^2-|z|^2)\right]\theta_3(z,\tau)\begin{pmatrix}1\\0\end{pmatrix} \\
&= \frac{\left|\frac{\tau}{2}\right|^{-1/4}}{L_1}\sum_{n=-\infty}^{n=+\infty}\exp\left[2\pi\mathrm{i}n\frac{x_1}{L_1} - \pi|\tau|\left(\frac{x_2}{L_2}+n\right)^2\right]\begin{pmatrix}1\\0\end{pmatrix}.
\end{aligned} \tag{2.9.47}$$

This is an example of the index theorem with $n^+ = 1$ and $n^- = 0$. In Fig. 2.23 we give the absolute value ρ and the phase of the zero mode. The size of the torus is that of Fig. 2.22. The physical interpretation of this zero mode is the following [Va84]: the four-dimensional Dirac equation with a constant magnetic field M_z with energy $E = 0$ and zero momentum in z direction can essentially be reduced to the two-dimensional equation (2.9.40). In this sense the wave functions (2.9.47) describe the lowest Landau

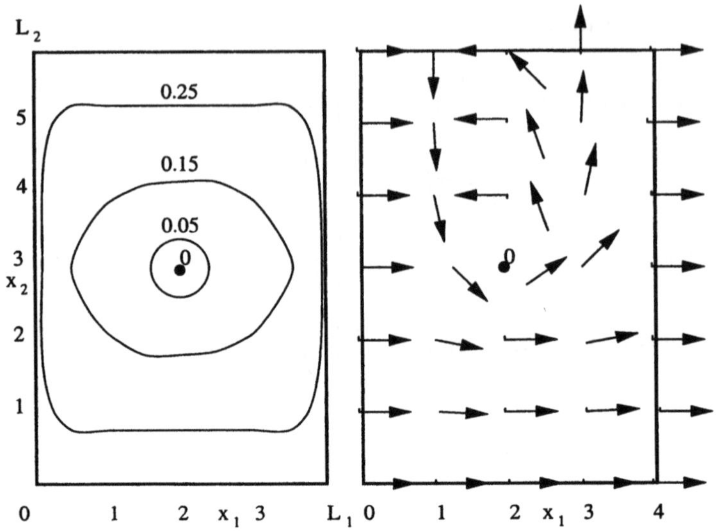

Fig. 2.23 Zero mode of the Dirac equation on the torus in the instanton background field: lines of constant absolute value ρ (left) and phase (right)

level of a charged Dirac particle in a constant magnetic field with aligned spin. The exact compensation of the orbital zero-point energy by the spin-alignment energy renders this a dynamically preferred mode. This effect is not described by perturbation theory.

We discuss the significance of (2.9.47) and of Fig. 2.23 for the path integral in Sect. 2.9.3.6.

2.9.2.2 A Dirac zero mode in the instanton background field

We consider the solution of the four-dimensional euclidean Dirac equation in the instanton background field [tH76]

$$\not{D}\psi^{(i)}(x) = \gamma^\mu \left(\partial_\mu - \mathrm{i}\mathcal{A}_\mu^{\mathrm{I},(i)}(x)\right) \psi^{(i)}(x) = 0. \tag{2.9.48}$$

Here $\mathcal{A}_\mu^{\mathrm{I},(i)}(x)$ are the instanton fields (2.9.23) and (2.9.24), which are regular in $\mathbb{R}^{(1)}$ and $\mathbb{R}^{(2)}$, respectively [cf. discussion after (2.9.24)]. In order that $\psi(x)$ is continuous in the SU(2) instanton bundle, the two solutions have to be related by the gauge transformation $g(x)$ in the intermediate region $0 < |x| < \infty$,

$$\psi^{(1)}(x) = g(x)\psi^{(2)}(x), \qquad g(x) = \frac{x_4 + \mathrm{i}\boldsymbol{x\tau}}{|x|}. \tag{2.9.49}$$

Because of the topological quantum number $k = 1$ of the instanton, the index theorem states that there is at least one normalizable solution of the massless Dirac equation with $\gamma^5\psi = \psi$. (There is only one such solution because in our case there is no solution for $\gamma^5\psi = -\psi$).

In order to describe explicitly the solution we should have in mind that $\psi^{(i)}_{\alpha,m}(x)$ has two types of indices, the spinor indices $\alpha = (\alpha_1, \alpha_2)$ and the SU(2) index $m = (m_1, m_2)$. These indices get mixed in the solution of (2.9.48). Our conventions for the euclidean Dirac matrices are given in (A.1.50). The gauge field $\mathcal{A}^{\mathrm{I},(i)}_\mu(x)$ is expressed with help of $(\tau^\mu)_{mn} = (\mathrm{i}\boldsymbol{\tau}, 1)_{mn}$. In the euclidean Weyl representation [cf. (A.1.50)], we can write the two solutions as

$$\psi^{(i),m}(x) = \begin{pmatrix} \kappa^{(i),m}_\alpha \\ 0 \end{pmatrix},$$

$$\kappa^{(1),m}_\alpha = \frac{\lambda^{3/2}}{\pi(x^2+\lambda^2)^{3/2}}\,\varepsilon_\alpha{}^m, \tag{2.9.50}$$

$$\kappa^{(2),m}_\alpha = \frac{\lambda^{3/2}}{\pi(x^2+\lambda^2)^{3/2}}\,\frac{(x_4 - \mathrm{i}\boldsymbol{x\tau})^m{}_{m'}}{|x|}\,\varepsilon_\alpha{}^{m'}. \tag{2.9.51}$$

In order to prove (2.9.48) for (2.9.50) by direct calculations it is useful to rewrite (2.9.23) as

$$\mathcal{A}^{\mathrm{I},(1)}_\mu(x) = \frac{1}{\mathrm{i}}\frac{x^2}{x^2+\lambda^2}\,\frac{\tau_\mu\bar{\tau}_\nu x^\nu - x_\mu}{x^2} \tag{2.9.52}$$

and make use of (2.9.18) and of the completeness relation of the Pauli matrices

$$\bar{\sigma}^\mu_{\alpha\beta}\tau^{mn}_\mu = \sigma^\mu_{\alpha\beta}\bar{\tau}^{mn}_\mu = 2\delta^n_\alpha\delta^m_\beta. \tag{2.9.53}$$

Equation (2.9.51) follows from (2.9.50) by using the gauge transformation (2.9.49). Thus, the validity of the Dirac equation (2.9.48) results for $i = 2$

directly from (2.9.49) and from the transformation property of the covariant derivative (2.3.23).

For the antiinstantons we get similarly a solution

$$\psi^{(\bar{i})}(x) = \begin{pmatrix} 0 \\ \kappa_\beta^{(i),m} \end{pmatrix} . \tag{2.9.54}$$

2.9.2.3 Remark on the proof of the index theorem

We derive the index theorem with help of the heat-kernel equation [Gi95]. The *heat kernel* G_Δ of a hermitian operator Δ with non-negative eigenvalues $\lambda_m^2 \geq 0$ and corresponding eigenfunctions $\phi_m(x,\alpha)$, where α denotes discrete indices, $\Delta\phi_m(x,\alpha) = \lambda_m^2\phi_m(x,\alpha)$, satisfies the *heat-kernel equation*

$$\begin{aligned} \left(\frac{\partial}{\partial\tau} + \Delta_x\right) G_\Delta(x,\alpha,y,\beta;\tau) &= 0, \\ G_\Delta(x,\alpha,y,\beta;0) &= \delta(x-y)\delta_{\alpha\beta} \end{aligned} \tag{2.9.55}$$

with the solution

$$\begin{aligned} G_\Delta(x,\alpha,y,\beta;\tau) &= \langle y,\beta|\mathrm{e}^{-\Delta\tau}|x,\alpha\rangle \\ &= \sum_m \mathrm{e}^{-\lambda_m^2\tau}\phi_m^*(y,\beta)\phi_m(x,\alpha), \qquad \text{for } \tau > 0. \end{aligned} \tag{2.9.56}$$

The orthonormality of the eigenfunctions implies

$$\sum_\alpha \int \mathrm{d}^D x\, G_\Delta(x,\alpha,x,\alpha;\tau) = \sum_m \mathrm{e}^{-\lambda_m^2\tau} = \mathrm{Tr}\left(\mathrm{e}^{-\Delta\tau}\right). \tag{2.9.57}$$

We consider now a differential Δ operator which acts on a vector field. The components of the vector are labelled by α. Here α contains besides the spinor index also the SU(N) index. In our application, Δ is the product of the *antihermitian euclidean Dirac operator* $\not{D} = \gamma^\mu(\partial_\mu - \mathrm{i}\mathcal{A}_\mu)$ [cf. (2.9.40) or (2.9.48)] with its adjoint

$$\begin{aligned} \Delta = \not{D}\not{D}^\dagger &= -\not{D}\not{D} = -D_\mu D^\mu + \frac{\mathrm{i}}{4}[\gamma^\mu,\gamma^\nu]\mathcal{F}_{\mu\nu} \\ &= -(\partial_\mu - \mathrm{i}\mathcal{A}_\mu)(\partial^\mu - \mathrm{i}\mathcal{A}^\mu) + \frac{\mathrm{i}}{4}[\gamma^\mu,\gamma^\nu]\mathcal{F}_{\mu\nu}. \end{aligned} \tag{2.9.58}$$

There is a well-known asymptotic expansion in τ for $G_\Delta(x,y;\tau)$. If the leading differential operator of Δ is the Laplacian $-\partial^\mu\partial_\mu$ in D dimensions, like in (2.9.58), this heat-kernel can be expanded in the asymptotic series

$$G_\Delta(x,y;\tau) \underset{\tau\to 0}{\sim} \frac{1}{(4\pi\tau)^{D/2}} \mathrm{e}^{-\frac{(x-y)^2}{4\tau}} \sum_{j=0}^{\infty} a_j(x,y)\tau^j \tag{2.9.59}$$

with $a_0(x,y) = 1$ owing to the initial condition (2.9.55) and $a_j(x,y) = a_j^\dagger(y,x)$ from hermiticity. For $\Delta_0 = -\partial^\mu\partial_\mu$ already the first term, $j=0$, is a solution of (2.9.55).

In the following, the chiralities of the eigenfunctions ψ_m of the Dirac operator,

$$\mathrm{i}\not{D}\psi_m = \mathrm{i}\gamma^\mu(\partial_\mu - \mathrm{i}\mathcal{A}_\mu)\psi_m = \lambda_m\psi_m, \tag{2.9.60}$$

play an important rôle. The relation $\{\gamma^5,\gamma^\mu\} = 0$ implies

$$\mathrm{i}\not{D}\gamma^5\psi_m = -\mathrm{i}\gamma^5\not{D}\psi_m = -\lambda_m\gamma^5\psi_m. \tag{2.9.61}$$

Hence, for $\lambda_m \neq 0$, λ_m and $-\lambda_m$ are both eigenvalues which belong to independent orthonormal eigenfunctions.

For $\lambda_m = 0$, we can distinguish between *left-handed and right-handed solutions* $\psi_{m,\pm}^{(0)} = \frac{1}{2}(1\pm\gamma^5)\psi_m^{(0)}$,

$$\not{D}\frac{1\pm\gamma^5}{2}\psi_m^{(0)} = \frac{1\mp\gamma^5}{2}\not{D}\psi_m^{(0)} = 0. \tag{2.9.62}$$

The *index* of $\not{D}$

$$\mathrm{ind}\,\not{D} = n_+ - n_- \tag{2.9.63}$$

counts the difference of the multiplicity of the left- and right-handed eigenvalues $\lambda_m = 0$ of $\not{D}$. It appears on the left-hand side of the index theorem (2.9.39).

With this information on the chirality dependence of the eigenfunctions ψ_m of $\not{D}$, and therefore of $\Delta = \not{D}\not{D}^\dagger$, we calculate

$$\begin{aligned}\int \mathrm{d}^D x\, \mathrm{Tr}(\gamma^5 G_\Delta(x,x;\tau)) &= \int \mathrm{d}^D x \sum_m \psi_m^*(x)\gamma^5\psi_m(x)\mathrm{e}^{-\lambda_m^2\tau}\\ &= \int \mathrm{d}^D x \sum_{m,\lambda_m=0} \psi_m^{(0)*}(x)\gamma^5\psi_m^{(0)}(x) = \mathrm{ind}\,\not{D}\\ &\underset{\tau\to 0}{\sim} \frac{1}{(4\pi\tau)^{D/2}}\sum_{j=1}\int \mathrm{d}^D x\, \tau^j\, \mathrm{Tr}[\gamma^5 a_j(x,x)].\end{aligned} \tag{2.9.64}$$

The trace refers to the summation over the spin and $\mathrm{SU}(N)$ indices of the matrices G_Δ and $a_j(x,x)$. In the transition from the first line to the second line we used the fact that for $\lambda_m^2 > 0$, ψ_m and $\gamma^5\psi_m$ are orthogonal degenerate eigenfunctions of Δ. Alternatively, if one uses the basis with eigenfunctions of γ^5, $(1\pm\gamma^5)\psi_m$, these contribute to the sum with opposite signs, and their contribution cancels. The remaining contribution with $\lambda_m^2 = 0$ is independent of τ. Therefore, we can apply (2.9.59) in the last line for arbitrary small τ. The proof of the index theorem is now reduced to the evaluation of

$$e_j = \int \mathrm{d}^D x\, \mathrm{Tr}[\gamma^5 a_j(x,x)]. \tag{2.9.65}$$

One can prove $e_j = 0$ for $j < D/2$.

For $a_j(x,x)$ we have the recursion relations

$$-\Delta a_{j-1}(x,x) = j a_j(x,x), \qquad a_0(x,x) = 1. \tag{2.9.66}$$

This follows from (2.9.55) and (2.9.59). It is immediately evident from the fact that the gaussian in (2.9.59) approaches $\delta(x-y)$ for $\tau\to 0$. For Δ we use the expression (2.9.58). Because of the trace formulas of γ^5 with the Dirac matrices only the commutator term in (2.9.58) contributes, and we obtain

$$e_1 = \int \mathrm{d}^2 x\, \varepsilon^{\mu\nu} F_{\mu\nu}(x) \tag{2.9.67}$$

for the two-dimensional case, and

$$e_2 = \frac{1}{4}\int \mathrm{d}^4 x\, \varepsilon^{\mu\nu\rho\sigma} F^a_{\mu\nu} F^a_{\rho\sigma} \tag{2.9.68}$$

for SU(2) in four dimensions.

We add some remarks to this short sketch of the proof of the index theorem. Equation (2.9.59) is mathematically proven only for some bounded region in x, y. But for all the assertions made here, there are precise mathematical proofs. The heat-kernel expansion is used here as a regularization scheme for not well-defined sums like $\sum_m \psi_m^*(x)\gamma^5\psi_m(x)$. This regularization is used for many other applications in perturbative and non-perturbative gauge theories (cf. Sect. 2.9.3.2 and Sect. 2.9.3.8).

2.9.3 Path integral and topology

In this section, we investigate the consequences of these topological properties by evaluating the euclidean path integral for the quantum-mechanical expectation value of an operator Ω,

$$\langle \Omega\{\mathbf{A}, \bar{\psi}, \psi\}\rangle = \frac{1}{Z}\int \mathcal{D}[\mathbf{A}, \psi, \bar{\psi}]\, \Omega\{\mathbf{A}, \bar{\psi}, \psi\} \mathrm{e}^{-S_{\mathrm{E}}\{\mathbf{A},\psi,\bar{\psi}\}},$$
$$Z = \int \mathcal{D}[\mathbf{A}, \psi, \bar{\psi}]\, \mathrm{e}^{-S_{\mathrm{E}}\{\mathbf{A},\psi,\bar{\psi}\}}. \tag{2.9.69}$$

The significance of the classical solutions, like the instantons, for the quantized theory might be inferred from the semi-classical evaluation of the path integral. In this approximation, the integral $\int \mathcal{D}[\mathbf{A}]$ over the gauge fields must be supplemented by a sum over all Chern classes. Zero modes for fermions appear as a consequence of the index theorems. These afford a more sophisticated consideration of the Grassmann integrals of the fermion integration. We first treat these two problems independently. In this section we discuss several physical consequences of this procedure. These are not yet firmly established since most of these considerations depend on the validity of the semi-classical approximation.

2.9.3.1 Semi-classical approximation

The semi-classical approximation is based on the saddle-point method. This method approximates integrals of the form

$$I(g) = \int_{-\infty}^{\infty} \mathrm{d}x\, \Phi(x) \exp\left(-\frac{f(x)}{g}\right) \tag{2.9.70}$$

for small g by gaussian integrals. Here $\Phi(x)$ and $f(x)$ are real functions of x. In general, the most important contribution to such an integral comes from the region in which $f(x)$ assumes its minimal value (a more precise statement must be based on specific assumptions). So if $f(x)$ is expanded around its minimum,

$$f(x) = f(x_0) + f''(x_0)\frac{(x-x_0)^2}{2}, \qquad f'(x_0) = 0, \qquad f''(x_0) > 0, \tag{2.9.71}$$

this leads to

$$\begin{aligned} I(g) &\approx \Phi(x_0) \int_{-\infty}^{+\infty} \mathrm{d}x \, \exp\left(-\frac{1}{g}\left[f(x_0) + f''(x_0)\frac{(x-x_0)^2}{2}\right]\right) \\ &= \sqrt{\frac{2\pi g}{f''(x_0)}}\, \Phi(x_0) \exp\left(-\frac{f(x_0)}{g}\right). \end{aligned} \tag{2.9.72}$$

If the function $f(x)$ has several minima, the contributions of all these minima have to be summed.

This method can be generalized to the evaluation of the path-integral formula (2.9.69). To this end one has to find the minima of the euclidean action $S_\mathrm{E} \geq 0$. According to Hamilton's principle, field configurations which minimize S_E are solutions of the classical field equations. This is, in particular, true for the self-dual or anti-self-dual solutions. Moreover, these yield the absolute minimum of S_E for field configurations with a specific topological quantum number k, in agreement with the discussion after (2.9.34).

Similarly to (2.9.71), we expand field configurations around classical solutions $\mathbf{A}_\mu^{\mathrm{cl}}$, here self-dual solutions with winding number k,

$$\begin{aligned} \mathbf{A}_\mu &= \mathbf{A}_\mu^{\mathrm{cl}} + \mathbf{A}_\mu^{\mathrm{qu}}, \\ S_\mathrm{E}\{\mathbf{A}\} &= S_\mathrm{E}\{\mathbf{A}^{\mathrm{cl}}, \mathbf{A}^{\mathrm{qu}}\} = S_\mathrm{E}\{\mathbf{A}^{\mathrm{cl}}\} + \mathbf{A}^{\mathrm{qu}}\Delta\mathbf{A}^{\mathrm{qu}} + \mathcal{O}(\{\mathbf{A}^{\mathrm{qu}}\}^3), \\ S_\mathrm{E}\{\mathbf{A}^{\mathrm{cl}}\} &= \frac{8\pi^2 k}{g^2}. \end{aligned} \tag{2.9.73}$$

The fluctuation operator Δ is defined by the second-order expansion of $S_\mathrm{E}(\mathbf{A})$ in $\mathbf{A}^{\mathrm{qu}}$. The partition sum Z and the expectation values are sums of gaussian integrals,

$$Z = \sum_k \exp\left(-\frac{8\pi^2 k}{g^2}\right) \int_{M_k} \mathrm{d}\mu(\alpha^k) \frac{1}{\sqrt{\det \Delta}}, \qquad (2.9.74)$$

$$\langle \Omega\{\mathbf{A}\}\rangle = \frac{1}{Z} \sum_k \exp\left(-\frac{8\pi^2 k}{g^2}\right) \int_{M_k} \mathrm{d}\mu(\alpha^k) \frac{\Omega\{\mathbf{A}^{\mathrm{cl}}(x, \alpha_i^k)\}}{\sqrt{\det \Delta}}. \qquad (2.9.75)$$

The integration extends over a volume M_k in the $(8k-3)$-dimensional space of parameters α_i^k of the solutions belonging to the minimal action [cf. remarks following (2.9.33)]. In general, the fluctuation operator Δ depends on these $(8k-3)$ instanton parameters. An eigenvalue zero of Δ is linked with each instanton parameter and must be split off when $\det \Delta$ is evaluated. This can be done by introducing collective instanton parameters [cf. Sect. 2.9.3.2,4)].

2.9.3.2 The one-instanton calculation

As an illustration of the rather general procedure of Sect. 2.9.3.1, we calculate the one-loop *contribution of one instanton to the vacuum-to-vacuum transition.* To this end we use the *background-field method* of Sect. 2.4.4 with the classical instanton field $\mathbf{A}^{\mathrm{I}}$ as background field $\hat{\mathbf{A}}$. In this approach, the renormalized semi-classical approximation can be considered as an approximation in the background-field method where the background field has the physical meaning of a classical instanton field.

The background-field method is presented in Sect. 2.4.4.1. Here, however, all quantities refer to the euclidean metric. The vacuum-to-vacuum amplitude about a single instanton can be represented by a path integral,

$$\frac{{}^{\mathrm{I}}\langle 0|0\rangle^{\mathrm{I}}}{\langle 0|0\rangle} = T\{\mathbf{A}^{\mathrm{I}}\} = \frac{1}{Z} \int \mathcal{D}[\mathbf{A}, \mathbf{u}, \bar{\mathbf{u}}]\, \mathrm{e}^{-S_{\mathrm{E}}\{\mathbf{A}, \mathbf{A}^{\mathrm{I}}\}},$$
$$Z = \int \mathcal{D}[\mathbf{A}, \mathbf{u}, \bar{\mathbf{u}}]\, \mathrm{e}^{-S_{\mathrm{E}}\{\mathbf{A}, 0\}}, \qquad (2.9.76)$$

where $S_{\mathrm{E}}(\mathbf{A}, \mathbf{A}^{\mathrm{I}})$ is the euclidean version of the action corresponding to the Lagrangian (2.4.113). We want to calculate

$$W^{(1)} = \left.\frac{{}^{\mathrm{I}}\langle 0|0\rangle^{\mathrm{I}}}{\langle 0|0\rangle}\right|_{\text{one-loop}} = T\{\mathbf{A}^{\mathrm{I}}\}\big|_{\text{one-loop}}. \qquad (2.9.77)$$

We follow 't Hooft's paper [tH76] but cannot treat all the details. However, we sketch the basic ideas, i.e.

1) the definition of the fluctuation matrix,
2) the calculation of the determinant,
3) the regularization of the determinant,
4) the treatment of the zero modes, and
5) the results.

1) For the discussion of the fluctuation matrix we take as classical solution the instanton potential $\mathbf{A}^{(1)}$ (2.9.23) and its field strength,

$$A_\mu^a(x) = A_\mu^{\mathrm{I},a}(x-z) + A_\mu^{\mathrm{qu},a}(x) = \frac{1}{g}\frac{2\tau_{\mu\nu}^a (x-z)^\nu}{(x-z)^2+\lambda^2} + A_\mu^{\mathrm{qu},a}(x),$$
$$F_{\mu\nu}^{\mathrm{I},a}(x) = \frac{1}{g}\frac{-4\lambda^2}{(x-z)^2+\lambda^2}\tau_{\mu\nu}^a, \tag{2.9.78}$$

where $\tau_{\mu\nu}$ of (2.9.20) is decomposed as $\tau_{\mu\nu} = \tau_{\mu\nu}^a \tau^a/2$. The five free parameters z^μ and λ are associated with translation and scale invariance. The splitting of $A_\mu^a = A_\mu^{\mathrm{I},a} + A_\mu^{\mathrm{qu},a}$ into its classical and its quantum part is formally equivalent to (2.4.107) in the background-field method (cf. Sect. 2.4.4) with the background field equal to the classical instanton field $\hat{A}_\mu^a \equiv A_\mu^{\mathrm{I},a}$.

The path integral (2.9.76) requires gauge fixing. For this the background-field gauge fixing (2.4.111) is used,

$$C^a\{A^{\mathrm{qu}};x\} = \hat{D}^{ac,\mu} A_\mu^{c,\mathrm{qu}}, \qquad \hat{D}_\mu^{ac} = \partial_\mu \delta^{ac} + g\varepsilon^{abc} A_\mu^{b,\mathrm{I}}. \tag{2.9.79}$$

According to (2.9.73), the Lagrangian including gauge-fixing (with $\xi = 1$) and the ghost term is expanded up to second order. The result reads

$$\begin{aligned}\mathcal{L} &= \mathcal{L}^{\mathrm{cl}} + \frac{1}{2}(\hat{D}_\mu^{ab} A_\nu^{\mathrm{qu},b})(\hat{D}^{ac,\mu} A^{\mathrm{qu},c,\nu}) + g A_\nu^{\mathrm{qu},a} \varepsilon^{abc} F^{\mathrm{I},b,\mu\nu} A_\mu^{\mathrm{qu},c} \\ &\quad - \bar{u}^a \hat{D}^{ab} \hat{D}^{bc} u^c + \bar{\psi}\gamma^\mu \hat{D}_\mu \psi + \mathcal{O}(\{A,u,\bar{u}\}^3) \\ &= \mathcal{L}^{\mathrm{cl}} + \frac{1}{2} A^{\mathrm{qu}} \Delta_A A^{\mathrm{qu}} + \bar{u}\Delta_{\mathrm{gh}} u + \bar{\psi}\Delta_\psi \psi + \mathcal{O}(\{A,u,\bar{u}\}^3).\end{aligned} \tag{2.9.80}$$

The terms linear in the quantum field $\mathbf{A}^{\text{qu}}$ are proportional to the equations of motion of the classical fields $\delta S/\delta A^{\text{I}}$ and vanish because $\mathbf{A}^{\text{I}}$ is a classical solution. From (2.9.80) we can read off the fluctuation matrices,

$$\Delta_A A_\mu^{\text{qu},a} = -\hat{D}^2 A_\mu^{\text{qu},a} + 2g\varepsilon^{abc} F_{\mu\nu}^{\text{I},b} A^{\text{qu},c,\nu}, \tag{2.9.81}$$

$$\Delta_{\text{gh}} u = -\hat{D}^2 u, \tag{2.9.82}$$

$$\Delta_\psi^2 = \hat{D}^2 \psi - \frac{\text{i}}{4} \tau^a F_{\mu\nu}^{\text{I},a} \gamma^\mu \gamma^\nu \psi. \tag{2.9.83}$$

We have chosen the instanton solution $\mathbf{A}^{(1)}$, which is regular in $\mathbb{R}_+$. For a complete discussion we should study $\mathbf{A}^{(1)}$ and $\mathbf{A}^{(2)}$ and their connection (2.9.24) including the transformations for the ghost fields. This is beyond the scope of this book.

2) In this paragraph, we discuss the one-loop vacuum-to-vacuum amplitude $W^{(1)}$ using the Lagrangian (2.9.80). We restrict ourselves in the following to a pure gauge theory.[30] Using (2.2.11) and (2.2.27), we get formally

$$W^{(1)} = \left(\frac{\det \Delta_A}{\det \Delta_{0,A}} \right)^{-1/2} \left(\frac{\det \Delta_{\text{gh}}}{\det \Delta_{0,\text{gh}}} \right), \tag{2.9.84}$$

where $\det \Delta_A$ and $\det \Delta_{\text{gh}}$ are the determinants for the instanton background and $\det \Delta_{0,A}$ and $\det \Delta_{0,\text{gh}}$ those for $\mathbf{A}^{\text{I}} = 0$.

The determinants of the fluctuation operators Δ can be calculated as a product of their eigenvalues,

$$\Delta \Phi_n = \lambda_n^2 \Phi_n, \qquad \det \Delta = \prod_n \lambda_n^2. \tag{2.9.85}$$

Zero modes in Δ_A lead to infinite fluctuations. In the path integral their (non-gaussian) integration must be treated separately. Therefore we start without the zero modes and consider

$$\det \Delta_A' = \prod_{\lambda_n^2 > 0} \lambda_n^2, \qquad \log \det \Delta_A' = \sum_{\lambda_n^2 > 0} \log \lambda_n^2 \tag{2.9.86}$$

[30] The contributions of fermions are considered in Sect. 2.9.3.3.

and

$$W'^{(1)} = \left(\frac{\det \Delta'_A}{\det \Delta'_{0,A}}\right)^{-1/2} \left(\frac{\det \Delta'_{\text{gh}}}{\det \Delta'_{0,\text{gh}}}\right) . \tag{2.9.87}$$

G. 't Hooft determined the non-vanishing eigenvalues, $\lambda_n^2 \neq 0$, by considering (2.9.85) as a type of Schrödinger equation in four dimensions including spin and isospin. Later it was discovered [Or77] that Δ can be treated by group-theoretical methods if one transforms the euclidean space $\mathbb{R}_4$ by a stereographic transformation to S_4, the unit sphere in $\mathbb{R}_5$. In this compact space Δ_A has discrete eigenvalues. Without going into further details, we give the results [Or77, Ch77] for the eigenvalues λ_n^2 for Δ_A and Δ_{gh} together with their multiplicities d_n,

$$\begin{aligned}
\Delta_A :\ & \lambda_n^2 = n(n+3) - 4, && d_n = \frac{1}{6}(2n+3)(n+1)(n+2), && n \geq 1, \\
& \lambda_n^2 = n(n+3) - 2, && d_n = \frac{1}{2}(2n+3)n(n+3), && n \geq 2, \\
& \lambda_n^2 = n(n+3) + 2, && d_n = \frac{5}{6}(2n+3)(n-1)(n+4), && n \geq 1, \\
\Delta_{0,A} :\ & \lambda_n^2 = n(n+3) + 2, && d_n = \frac{3}{2}(2n+3)n(n+3), && n \geq 1, \\
\Delta_{\text{gh}} :\ & \lambda_n^2 = n(n+3) - 2, && d_n = \frac{1}{2}(2n+3)n(n+3), && n \geq 1, \\
\Delta_{0,\text{gh}} :\ & \lambda_n^2 = n(n+3), && d_n = \frac{1}{2}(2n+3)(n+1)(n+2), && n \geq 0.
\end{aligned} \tag{2.9.88}$$

The determinant $\Delta_{0,A}$ contains five *zero modes* and the determinant $\Delta_{0,\text{gh}}$ contains three.

The eigenfunctions are composed of spherical harmonics, vector spherical harmonics, and tensor spherical harmonics of S_4. Since $d_n \sim n^3$ and $\log \lambda_n^2 \sim \log n$, the product of these eigenvalues (2.9.86), even when divided by $\det \Delta_0$, badly diverges, so that we must regularize.

3) Instead of dimensional regularization, it is more convenient to regularize the determinants by the Pauli–Villars method.[31] In this scheme (2.9.86) is substituted by

$$(\det \Delta')_{\text{reg}} = \det \Delta' \prod_{i=1}^{R} \det(\Delta' + M_i^2)^{e_i} \tag{2.9.89}$$

with sufficiently large R. The Pauli–Villars regulators with masses M_i are of alternating bosonic and fermionic character $e_i = \pm 1$. They play the rôle of effective cut-off parameters (cf. Sect. 2.5.1.1). The product (2.9.89) converges rapidly if the regulator conditions

$$\begin{gathered}\sum_{i=0}^{R} e_i = 0, \qquad \sum_{i=0}^{R} e_i M_i^2 = 0, \qquad \sum_{i=0}^{R} e_i M_i^4 = 0, \quad \ldots, \\ \sum_{i=1}^{R} e_i \log M_i^2 \equiv -\log M^2 = \text{finite}\end{gathered} \tag{2.9.90}$$

are satisfied.[32]

For the evaluation of (2.9.89) it is helpful to introduce the *zeta function* $\zeta_\Delta(s)$ of a differential operator Δ with positive eigenvalues $\lambda_n^2 > 0$ [Lu82a],

$$\zeta_\Delta(s) = \sum_{\lambda_n^2 > 0} \frac{1}{(\lambda_n^2)^s}. \tag{2.9.91}$$

The sum includes all eigenvalues with their degeneracies. As a consequence of an estimate on the growth of λ_n^2 of differential operators, this series is convergent for large enough $\text{Re}(s)$. Using

$$\frac{1}{(\lambda_n^2)^s} = \frac{1}{\Gamma(s)} \int_0^\infty \mathrm{d}t \, t^{s-1} \mathrm{e}^{-\lambda_n^2 t}, \tag{2.9.92}$$

[31] The rather involved dimensional regularization was discussed by M. Lüscher [Lu82, Lu82a].

[32] We have included the original term in the product by defining $e_0 = 1$ and $M_0 = 0$.

where $\Gamma(s)$ is Euler's Γ-function, $\zeta_\Delta(s)$ can be related to the *heat kernel* $G(x,y;\tau)$ [cf. (2.9.56)],

$$\begin{aligned}\zeta_\Delta(s) &= \frac{1}{\Gamma(s)}\int_0^\infty dt\, t^{s-1}\,\mathrm{Tr}\left(\mathrm{e}^{-\Delta t}\right) \\ &= \frac{1}{\Gamma(s)}\int_0^\infty dt\, t^{s-1}\sum_\alpha \int \mathrm{d}^D x\, G_\Delta(x,\alpha,x,\alpha;t). \end{aligned} \tag{2.9.93}$$

If Δ is a second-order differential operator in D-dimensional euclidean space–time the following results can be proven [Gi95]:

- $\zeta_\Delta(s)$ can be analytically continued into the entire complex plane with the exception of possible simple poles at $s = D/2 - j$, $j = 0, 1, \ldots$.
- The residues of $\zeta_\Delta(s)$ at these poles are

$$\mathrm{Res}\,\zeta_\Delta(s)\big|_{s=\frac{D}{2}-j} = \frac{A_j}{\Gamma\left(\frac{D}{2}-j\right)}. \tag{2.9.94}$$

  Here A_j is obtained from the heat-kernel coefficient defined in (2.9.59),

$$A_j = \frac{1}{(4\pi)^{D/2}}\int \mathrm{Tr}(a_j(x,x))\,\mathrm{d}^D x. \tag{2.9.95}$$

- If s is a non-positive integer, i.e. $s = 0, -1, \ldots$, then $\zeta_\Delta(s)$ is regular at these points with the value $A_j/\mathrm{Res}\,\Gamma(s)$.

Thus, the singularities of the ζ-functions are closely related to the heat-kernel expansion.

We can use the regular ζ-function in order to define the finite part of $\log\det\Delta'$ [Ha77],

$$\log\det\Delta' = \log\prod_{\lambda_n^2>0}\lambda_n^2 = -\sum_{\lambda_n^2>0}\frac{\mathrm{d}}{\mathrm{d}s}(\lambda_n^2)^{-s}\bigg|_{s=0} = -\frac{\mathrm{d}}{\mathrm{d}s}\zeta_\Delta(s)\bigg|_{s=0}. \tag{2.9.96}$$

In order to evaluate the regularized expression (2.9.89) for large M_i we write

$$\begin{aligned}\log(\det\Delta')_{\mathrm{reg}} &= \mathrm{Tr}\left(\sum_{i=0}^R e_i\log(\Delta' + M_i^2)\right) \\ &= -\int_0^\infty \frac{\mathrm{d}t}{t}\left(\sum_{i=0}^R e_i \mathrm{e}^{-M_i^2 t}\,\mathrm{Tr}\left(\mathrm{e}^{-\Delta' t}\right)\right), \end{aligned} \tag{2.9.97}$$

where we used $\int_x^\infty (\mathrm{d}t/t)\exp(-t) \underset{x\to 0}{\sim} -\gamma - \log x$ and the regularization condition $\sum_{i=0}^R e_i = 0$.

Equation (2.9.93) gives $\Gamma(s)\zeta_{\Delta'}(s)$ as *Mellin transformation* of $\mathrm{Tr}\exp(-\Delta' t)$. The *inverse Mellin transformation* [Er53a] reads

$$\mathrm{Tr}\left(\mathrm{e}^{-\Delta' t}\right) = \frac{1}{2\pi\mathrm{i}}\int_{c-\mathrm{i}\infty}^{c+\mathrm{i}\infty} \Gamma(s)\zeta_{\Delta'}(s)t^{-s}\mathrm{d}s, \tag{2.9.98}$$

where c has to be chosen in such a way that $\Gamma(s)\zeta_{\Delta'}(s)$ is regular on the right-hand side of the straight line from $c - \mathrm{i}\infty$ to $c + \mathrm{i}\infty$. Since $\Gamma(s)$ has poles at $s = 0, -1, \ldots$, and $\zeta_{\Delta'}(s)$ at $s = 2, 1$ for $D = 4$, we must choose $c > 2$. Now we put (2.9.98) into (2.9.97),

$$\begin{aligned}\log(\det\Delta')_{\mathrm{reg}} &= \\ &= -\int_0^\infty \frac{\mathrm{d}t}{t}\left(\sum_{i=0}^R e_i \mathrm{e}^{-M_i^2 t}\frac{1}{2\pi\mathrm{i}}\int_{c-\mathrm{i}\infty}^{c+\mathrm{i}\infty}\Gamma(s)\zeta_{\Delta'}(s)t^{-s}\mathrm{d}s\right),\end{aligned} \tag{2.9.99}$$

and shift the path to $c = \varepsilon > 0$, ε infinitesimally small, by taking into account the contributions of the poles at $s = 2, 1$. Next we shift the path for the part of the integral that comes from the regularization terms ($i > 0$) to $-1 < c < 0$ by taking into account the contributions of the pole at $s = 0$. At $c < 0$ the integral becomes of order $\mathcal{O}(M_i^{-2}\log M_i)$ for large M_i^2. The result of these operations is

$$\begin{aligned}\log(\det\Delta')_{\mathrm{reg}} &= \frac{A_0}{2}\sum_{i=1}^R e_i M_i^4 \log M_i^2 - A_1 \sum_{i=1}^R e_i M_i^2 \log M_i^2 \\ &- \zeta_{\Delta'}(0)\log M^2 - \frac{\mathrm{d}}{\mathrm{d}s}\zeta_{\Delta'}(s)|_{s=0} + \sum_{i\geq 1}\mathcal{O}(M_i^{-2}\log M_i).\end{aligned} \tag{2.9.100}$$

Equation (2.9.100) expresses the logarithm of the regularized determinant in terms of contributions of the Pauli–Villars cut-off parameters and a finite part.

We want to add some more details on the derivation of (2.9.100). The first two terms come from the residue (2.9.94) of the ζ-function, and the third term from the residue of $\Gamma(s)$ at $s = 0$ for the regularization fields $i = 1, \ldots, R$. Since the contribution of $i = 0$ is not included in the residue at $s = 0$, a term $-\gamma\zeta_{\Delta'}(0)$ remains. We used for the t integration the formula $\int_\epsilon^\infty t^\nu \exp(-t)\mathrm{d}t = \Gamma(\nu+1,\epsilon)$ [Er53]. The remaining integral for $i = 0$ can be evaluated as follows. First we transform the inversion of the Mellin transformation into one of the Fourier transformation by the substitutions $t = \exp(-\tau)$ and $s = \mathrm{i}x$. This is allowed by the regularization of the ultraviolet divergences at $t = 0$. Because of the resulting δ-function at $x = 0$ we may expand $\Gamma(\mathrm{i}x)\zeta_{\Delta'}(\mathrm{i}x)$ around zero. Thus, we find

$$\begin{aligned}
&-\int_0^\infty \frac{\mathrm{d}t}{t}\frac{1}{2\pi\mathrm{i}}\int_{\varepsilon-\mathrm{i}\infty}^{\varepsilon+\mathrm{i}\infty} \Gamma(s)\zeta_{\Delta'}(s)t^{-s}\mathrm{d}s = \\
&= -\frac{1}{2\pi}\int_{-\infty}^\infty \mathrm{d}\tau \int_{-\infty}^\infty \mathrm{d}x\, \mathrm{e}^{\mathrm{i}x\tau}\Gamma(\mathrm{i}x)\zeta_{\Delta'}(\mathrm{i}x) \\
&= -\int_{-\infty}^\infty \mathrm{d}x\, \delta(x)\Gamma(\mathrm{i}x)\zeta_{\Delta'}(\mathrm{i}x) \\
&= -\int_{-\infty}^\infty \mathrm{d}x\, \delta(x)\left(\frac{1}{\mathrm{i}x} - \gamma\right)[\zeta_{\Delta'}(0) + \mathrm{i}x\zeta'_{\Delta'}(0)] \\
&= \gamma\zeta_{\Delta'}(0) - \zeta'_{\Delta'}(0). \qquad (2.9.101)
\end{aligned}$$

For the Fourier inversion we have to use the principle value. The term $\gamma\zeta_{\Delta'}(0)$ gets compensated by a corresponding term of the regulator fields mentioned above. So we finally find the fourth term of (2.9.100).

The first term in (2.9.100) describes a divergent contribution of 4th order. In the difference $\log(\det\Delta')_{\mathrm{reg}} - \log(\det\Delta'_0)_{\mathrm{reg}}$ this term disappears. In a renormalizable theory, the second term in this difference drops out, too [tH76, Be77]. Thus, the final result for the regularized determinant of a renormalizable theory without zero modes reads [Ch77]

$$\begin{aligned}
\log\frac{(\det\Delta')_{\mathrm{reg}}}{(\det\Delta'_0)_{\mathrm{reg}}} &= -[\zeta_{\Delta'}(0) - \zeta_{\Delta'_0}(0)]\log M^2 \\
&\quad - \frac{\mathrm{d}}{\mathrm{d}s}(\zeta_{\Delta'}(s) - \zeta_{\Delta'_0}(s))]|_{s=0}. \qquad (2.9.102)
\end{aligned}$$

The $\log M^2$ term describes the cut-off dependence in the Pauli–Villars regulator scheme with respect to the eigenvalues of the fluctuation matrix. 't Hooft relates M to a fixed-mass regulator μ_0 [tH76],

$$\log\mu_0 = \log M + \log 2 - \frac{5}{6}. \qquad (2.9.103)$$

The formula (2.9.102) is the starting point for the calculation of the determinant of the fluctuation matrix in an instanton background field. The ζ-functions $\zeta(0)$ and $\zeta'(0)$ for eigenvalues of the type (2.9.88) were discussed by [Ch77, Or77a]. Putting everything together, one obtains 't Hooft's final result for the determinant without zero modes

$$W'^{(1)} = \mathrm{e}^{-\frac{2}{3}\log(\mu_0\lambda)-\alpha(1)} \tag{2.9.104}$$

with

$$\alpha(1) = -8\zeta_{\mathrm{R}}'(-1) + \frac{1}{3}\log 2 - \frac{10}{9} = 0.443307\ldots\,, \tag{2.9.105}$$

and the derivative of the Riemann ζ-function

$$\zeta_{\mathrm{R}}'(-1) = -0.165421\ldots\,. \tag{2.9.106}$$

The λ dependence follows from straightforward dimensional analysis.

4) There are 8 linearly independent *zero modes* $\mathbf{A}_0$ of the operator Δ_A. These are related to the symmetry transformations of the instanton parameters: four translations, one dilatation, and three global gauge rotations. The zero modes do not satisfy the necessary condition for gaussian integration, $\mathbf{A}^{\mathrm{qu}}\Delta\mathbf{A}^{\mathrm{qu}} > 0$. The functional integration $\mathcal{D}[\mathbf{A}]$ into the directions of the zero mode must be performed explicitly. Following a general procedure, we transform these integrations into integrals over the collective instanton parameters. This is achieved by adopting the Faddeev–Popov procedure (Sect. 2.4.1.2) for the finite symmetry groups.

i) Translations: the gauge fields

$$\begin{aligned} A^a_{0,\mu}(x-z;\nu) &= -\frac{4}{g}\frac{\tau^a_{\mu\nu}}{[(x-z)^2+\lambda^2]^2}\,, \qquad \nu = 1,2,3,4,\\ \|A^a_{0,\mu}(x;\nu)\|^2 &= \int \mathrm{d}^4x\, A^a_{0,\mu}(x;\nu)A^{a,\mu}_0(x;\nu) = \frac{8\pi^2}{g^2} \end{aligned} \tag{2.9.107}$$

are zero modes of Δ_A. Now $A^a_{0,\mu}(x-z;\nu)$ is up to a gauge transformation,

$$\Lambda(x;\nu) = -\frac{2}{g}\frac{\tau^a_{\nu\lambda}x^\lambda}{x^2+\lambda^2}\frac{\tau^a}{2}\,,$$

equal to the spatial derivative of the instanton field $\mathbf{A}_\mu^{(1)}(x)$,

$$A^a_{0,\mu}(x-z;\nu) = -\frac{\partial}{\partial z^\nu}\left[\frac{2}{g}\frac{\tau^a_{\mu\rho}(x-z)^\rho}{(x-z)^2+\lambda^2}\right] + (\hat{D}_\mu \Lambda)^a(x-z;\nu). \tag{2.9.108}$$

The gauge transformation appears because the gauge-fixing term depends on the background field, and the derivative of the classical field with respect to the collective coordinates, in general, is not in the background gauge [Be79].

After separating off the zero modes, the quantum-mechanical fluctuations should be orthogonal to the manifold of gauge fields connected by symmetry transformations. The separation can be done by the *Faddeev–Popov procedure* discussed in Sect. 2.4.1.2 (cf. Ref. [Mo85]). Like in (2.4.5), we insert the identity

$$\begin{aligned} 1 = {} & \Delta\{\mathbf{A}\} \\ & \times \int \mathrm{d}^4 z \prod_{\nu=1}^{4} \delta\left(\int \mathrm{d}^4x\, A_0^{a,\mu}(x-z;\nu)\left[A^a_\mu(x) - A^{\mathrm{I},a}_\mu(x-z)\right]\right) \end{aligned} \tag{2.9.109}$$

into the path integral. Using the well-known formula $\delta(f(x)) = \delta(x)/|f'|$, the Jacobian can be evaluated as

$$\begin{aligned} \Delta\{\mathbf{A}\} = {} & \prod_{\nu=1}^{4} \frac{\partial}{\partial z^\nu}\int \mathrm{d}^4x\, A^a_{0,\mu}(x-z;\nu) \\ & \times \left[A^{a,\mu}(x) - A^{\mathrm{I},a,\mu}(x-z)\right]\Big|_{\mathbf{A}(x)=\mathbf{A}^{\mathrm{cl}}(x-z;\nu)} \end{aligned} \tag{2.9.110}$$

After a change of the order of integration, we can apply the semi-classical approximation (2.9.75) for the evaluation of $W^{(1)}$, (2.9.87),

$$\begin{aligned} W^{(1)} = {} & \frac{1}{Z}\int \mathrm{d}^4 z \int \mathcal{D}[\mathbf{A}]\Delta\{\mathbf{A}\}\mathrm{e}^{-\int \mathrm{d}^4x\,\mathcal{L}} \\ & \times \prod_{\nu=1}^{4}\delta\Big(\int \mathrm{d}^4x\, A^a_{0,\mu}(x-z;\nu)\left[A^{a,\mu}(x) - A^{\mathrm{I},a,\mu}(x-z)\right]\Big) \\ \approx {} & \frac{\mathrm{e}^{-\frac{8\pi^2}{g_0^2}}}{Z}\int \mathrm{d}^4z\,\Delta\{\mathbf{A}^{\mathrm{I}}\}\int \mathcal{D}[\mathbf{A}^{\mathrm{qu}}] \\ & \times \prod_{\nu=1}^{4}\delta\Big(\int \mathrm{d}^4x\, A^a_{0,\mu}(x-z;\nu)A^{a,\mu,\mathrm{qu}}(x)\Big)\mathrm{e}^{-\frac{1}{2}\mathbf{A}^{\mathrm{qu}}\Delta_A \mathbf{A}^{\mathrm{qu}}}. \end{aligned} \tag{2.9.111}$$

We used $\mathbf{A}_\mu(x) = \mathbf{A}^{\mathrm{I}}_\mu(x-z;\nu) + \mathbf{A}^{\mathrm{qu}}_\mu(x)$ according to (2.9.78). For zero modes, the argument of the δ-function becomes proportional to their norm and does not vanish. Therefore, the δ-function in (2.9.111) ensures that $\mathbf{A}^{\mathrm{qu}}$ is orthogonal to the zero modes. The remaining functional integration over $\mathcal{D}[\mathbf{A}] = \mathcal{D}[\mathbf{A}^{\mathrm{qu}}]$ can be evaluated with the method explained above in 3), i.e. by expanding $\mathbf{A}^{\mathrm{qu}}$ in eigenfunctions and considering only the non-vanishing eigenvalues.

The evaluation of the Jacobian $\Delta\{\mathbf{A}^{\mathrm{I}}\}$ results in

$$\begin{aligned}\Delta\{\mathbf{A}^{\mathrm{I}}\} &= \prod_{\nu=1}^{4} \frac{\partial}{\partial y^\nu} \int \mathrm{d}^4x\, A^a_{0,\mu}(x-y;\nu) A^{\mathrm{cl},a,\mu}(x-z)\Big|_{y=z} \\ &= \prod_{\nu=1}^{4} \int \mathrm{d}^4x\, A^a_{0,\mu}(x-z;\nu) A^{a,\mu}_0(x-z;\nu) \\ &= \prod_{\nu=1}^{4} \|A^a_{0,\mu}(x;\nu)\|^2 .\end{aligned} \tag{2.9.112}$$

The expansion that leads to $W'^{(1)}$ is based on normalized eigenfunctions. Therefore, we have to add a factor $(\sqrt{2\pi}\|A^a_{0,\mu}(x;\nu)\|)^{-1}$ for each zero mode in order to be consistent with the evaluation of the factor $1/Z$. The factor $\sqrt{2\pi}$ results from the gaussian integration of the zero mode of Δ in Z. Moreover, we have disregarded the fact that we should use the regularized determinant (2.9.89) in the evaluation of (2.9.111). In the regularization factors $\det(\Delta_A + M_i^2)$, zero modes are not omitted. A detailed analysis shows that these contribute a factor μ_0 for each zero mode. Hence, the translational zero modes contribute a factor

$$I_{\mathrm{Trans}} = \prod_{\nu=1}^{4} \frac{\mu_0}{\sqrt{2\pi}} \|A^a_{0,\mu}(x;\nu)\| = \frac{16\pi^2\mu_0^4}{g^4} \tag{2.9.113}$$

to $W^{(1)}$.

The other zero modes have to be treated in the same manner.

ii) Dilatations: the configuration

$$A_{0,\mu}(x;\lambda) = \frac{4}{g}\frac{\lambda \tau^a_{\mu\nu} x^\nu}{[(x-z)^2+\lambda^2]^2}, \qquad \|A_{0,\mu}(x;\lambda)\| = \frac{4\pi}{g}, \tag{2.9.114}$$

is a zero mode of Δ_A. It is a pure infinitesimal dilatation of the classical solution

$$A^a_{0,\mu}(x;\lambda) = -\frac{\partial}{\partial\lambda}\left(\frac{2}{g}\frac{\tau^a_{\mu\nu}(x-z)^\nu}{(x-z)^2+\lambda^2}\right) \tag{2.9.115}$$

and yields the factor

$$I_{\rm Dil} = \frac{\mu_0}{\sqrt{2\pi}}\|A^a_{0,\mu}(x;\lambda)\| = \frac{2^{3/2}\pi^{1/2}\mu_0}{g} \tag{2.9.116}$$

to $W^{(1)}$.

iii) Global gauge rotations: the fields

$$A^a_{0,\mu}(x;b) = -2\tau^a_{\mu\rho}\bar{\tau}^{\rho\nu,b}\frac{x_\nu\lambda}{[(x-z)^2+\lambda^2]^2}, \qquad a=1,2,3, \tag{2.9.117}$$

are zero modes of Δ_A related to global SU(2) rotations. As 't Hooft showed, the relation of these zero modes to the global SU(2) rotations is somewhat hidden by the complexity of the gauge-fixing procedure. It was shown by Bernard [Be79] that these considerations could be simplified if one considers the zero modes in the singular gauge (2.9.24) with $\|A^a_{0,\mu}(x;b)\| = 4\pi\lambda/g$. These zero modes contribute a factor

$$I_{\rm Ga} = \prod_{b=1}^{3}\frac{\mu_0}{\sqrt{2\pi}}\|A^a_{0,\mu}(x;b)\| = \frac{2^{9/2}\pi^{3/2}\lambda^3\mu_0^3}{g^3} \tag{2.9.118}$$

to $W^{(1)}$ and a factor π^2 from the integration over the gauge group SU(2). Combining all contributions we find

$$W^{(1)} = \int {\rm d}^4 z \int \frac{{\rm d}\lambda}{\lambda^5}\,\pi^2 I_{\rm Zero}\,{\rm e}^{-\frac{8\pi^2}{g^2(\mu_0)}+\frac{22}{3}\log\mu_0\lambda-\alpha(1)} \tag{2.9.119}$$

with $I_{\text{Zero}} = I_{\text{Trans}} I_{\text{Dil}} I_{\text{Ga}} = 2^{10}\pi^4 g^{-8}$. We associated g^2 with $g^2(\mu_0)$. Note that the coefficient 22/3 multiplying $\log \mu_0 \lambda$ coincides with the first-order coefficient of the β-function (2.6.36) of the renormalization-group equation. Thus, (2.9.119) becomes independent of the subtraction point if we choose $g^2(\mu_0)$ to obey the renormalization-group equation [tH76]. We remark that $\beta_0 = 22/3$ results from μ_0^8 of the zero modes and from $\mu_0^{-2/3}$ of the determinant without zeros (2.9.104).

Finally, we translate the result into the $\overline{\text{MS}}$ *scheme*. To this end we have to perform the replacements [Ha81, tH86]

$$\frac{22}{3}\log\mu_0 \to \frac{22}{3}\log\mu + \frac{1}{3}, \qquad \alpha_s(\mu_0) \to \alpha_s(\mu), \tag{2.9.120}$$

where μ is the mass scale of dimensional regularization.

5) The final result for SU(2) in the $\overline{\text{MS}}$ scheme has the form

$$\begin{aligned} W^{(1)} &= \int_{\mathbb{R}^4} \mathrm{d}^4 z \int_0^\infty \frac{\mathrm{d}\lambda}{\lambda^5} (\mu_0\lambda)^c d \left(\frac{\pi}{\alpha(\mu_0)}\right)^{2N} \exp\left(-\frac{2\pi}{\alpha(\mu_0)}\right) \\ &\equiv \int_{\mathbb{R}^4} \mathrm{d}^4 z \int_0^\infty \frac{\mathrm{d}\lambda}{\lambda^5} D_{\mathrm{I}}(z,\lambda,\alpha(\mu_0)) \end{aligned} \tag{2.9.121}$$

with

$$\begin{aligned} c &= \beta_0 = \frac{22}{3}, \qquad N = 2, \\ d &= \frac{4}{\pi^2}\mathrm{e}^{-\alpha(1)+\frac{1}{3}} = \frac{2}{\pi^2}\mathrm{e}^{\frac{5}{6}}\mathrm{e}^{-4\alpha(\frac{1}{2})+\frac{1}{3}}, \\ \alpha\left(\frac{1}{2}\right) &= \frac{\alpha(1) - \log 2 + \frac{5}{6}}{4} = -2\zeta_{\mathrm{R}}'(-1) - \frac{1}{6}\log 2 - \frac{5}{72} \\ &= 0.145873\ldots . \end{aligned} \tag{2.9.122}$$

A direct calculation in the $\overline{\text{MS}}$ scheme was performed in Ref. [Lu82a] and confirmed this result. Comparing (2.9.121) with the semi-classical approximation (2.9.74) with $S_E = 8\pi^2/g^2$, we see that $g^2 = 4\pi\alpha$ is replaced by the running coupling constant [cf. (3.1.3) and (3.1.16)].

According to 't Hooft, the interpretation of (2.9.121) is best given in the language of path integrals: if $|0\rangle$ is the vacuum, and $|\tilde{0}\rangle$ is the gauge-rotated vacuum, then the density $D_{\mathrm{I}}(z, \lambda, \alpha(\mu_0))$ in (2.9.121) is the total contribution to $\langle 0|\tilde{0}\rangle$ from all fields in euclidean space that have an instanton located at z within $\mathrm{d}^4 z$ and a scale within λ and $\lambda + \mathrm{d}\lambda$. We give more details of this interpretation in Sect. 2.9.3.5.

The result (2.9.121) is also true for general SU(N), $N \geq 2$ [Be79] and N_{F} fermion multiplets in the fundamental representation [tH76]. In this case the constants c and d are in one-loop accuracy given by

$$\begin{aligned} c &= \beta_0 = \frac{11}{3}N - \frac{2}{3}N_{\mathrm{F}}, \\ d &= \frac{2\mathrm{e}^{5/6}}{\pi^2 (N-1)!(N-2)!} \mathrm{e}^{-2\alpha(\frac{1}{2})(N-N_{\mathrm{F}})+\frac{N}{6}}. \end{aligned} \tag{2.9.123}$$

The constant d is scheme-dependent. The above result holds in the $\overline{\mathrm{MS}}$ scheme. In the Pauli–Villars scheme, the term $N/6$ in the exponent is missing.

The two-loop contributions are also known [Mo85]. Including these via

$$c = \beta_0 + (\beta_1 - 4\beta_0 N)\frac{\alpha(\mu_0)}{4\pi}, \tag{2.9.124}$$

where the coefficient β_1 and the running coupling $\alpha_{\mathrm{s}}(\mu_0)$ were defined in (3.1.13) and (3.1.15), respectively, renders the instanton density two-loop RG-invariant, i.e. $D_{\mathrm{I}}^{-1}\mathrm{d}D_{\mathrm{I}}/\mathrm{d}\log\mu_0 = \mathcal{O}(\alpha^2)$.

The instanton contribution becomes exponentially small for $\alpha_{\mathrm{s}}(\mu_0) \ll 1$. However, the instanton density behaves as $D_{\mathrm{I}}(\lambda, \alpha(\mu_0)) \sim \lambda^{7/3}$ for SU(2), or $\sim \lambda^6$ for SU(3), and thus diverges for large λ. The running coupling constant $\alpha(\mu_0 \sim 1/\lambda)$ gets large. The semi-classical approximation is no longer applicable. There are some more intuitive models which try to describe the instanton sector in this region. As an example we mention the *instanton-liquid model* [Sc98]. This model assumes that the large-size instantons and antiinstantons interact with each other.

We postpone the question of the relevance of 't Hooft's calculation for physics to Sect. 3.7.4.2. There, we combine the semi-classical approximation with results gained from lattice approximation.

2.9.3.3 Fermionic zero modes

Zero modes for massless fermions exist in the topological non-trivial gauge sectors as a consequence of the index theorems. The Fubini formula (2.2.23) in the euclidean version allows to perform the *fermion integration* in the path integral for the partition sum

$$Z_\psi = \int \mathcal{D}[\bar{\psi}, \psi] \exp\left(- \int \mathrm{d}^4x \, \bar{\psi}(\not{D}+m)\psi \right) = \det[\not{D} + m]. \quad (2.9.125)$$

According to (2.9.86), a zero mode leads for $m = 0$ to a vanishing Z_ψ. Thus, only the topologically trivial sector, which is free of zero modes contributes to the partition sum.

However, we get a non-vanishing result for the Green functions, i.e. for the path integral including fermion sources. This follows from the integration rules for Grassmann variables, (2.2.20) and (2.2.21), as the following simple example shows:

$$\begin{aligned} \int \mathrm{d}\bar{a} \, \mathrm{d}a \, \mathrm{e}^{(-\varepsilon \bar{a} a + \bar{a}\eta - a\bar{\eta})} &= \int \mathrm{d}\bar{a} \, \mathrm{d}a \, (1 - \varepsilon \bar{a} a + \bar{a}\eta - a\bar{\eta} - \bar{a}\eta\bar{\eta}a) \\ &= \varepsilon + \eta\bar{\eta} \to \eta\bar{\eta} \neq 0 \quad \text{for} \quad \varepsilon = 0. \quad (2.9.126) \end{aligned}$$

The fermionic Green functions in an external gauge field A with k zero modes $\chi_i(x)$ and $\bar{\chi}_j(y)$ are obtained from [cf. (2.2.65)]

$$\begin{aligned} &\langle \psi(x_1)\bar{\psi}(y_1) \cdots \psi(x_n)\bar{\psi}(y_n) \rangle_A = \quad (2.9.127) \\ &= \frac{1}{Z} \int \mathcal{D}[\bar{\psi}, \psi] \, \psi(x_1)\bar{\psi}(y_1) \cdots \psi(x_n)\bar{\psi}(y_n) \exp\left(- \int \mathrm{d}^4x \, \bar{\psi}\not{D}\psi \right) \\ &= \frac{1}{Z'} \sum_{\text{perm}} \Big\{ (\pm 1)\chi_1(x_{i_1}) \cdots \chi_l(x_{i_l})\bar{\chi}_1(y_{j_1}) \cdots \bar{\chi}_l(y_{j_l}) \\ &\quad \times \mathcal{S}(x_{i_{l+1}}, y_{j_{l+1}}; A) \cdots \mathcal{S}(x_{i_n}, y_{j_n}; A) \Big\} \exp\left(\frac{-\Gamma\{\mathbf{A}\}}{2} \right) \end{aligned}$$

for $n > k$ and are zero for $n < k$. The signs (± 1) result from anticommutations of the fermionic operators. The ingredients of this formula are the zero-mode wave functions $\chi_i(x)$ and $\bar{\chi}_j(y)$, the effective action $\exp(-\Gamma\{\mathbf{A}\}/2) = \det'[\not{D}]$, and the Green function $\mathcal{S}(x, y; A)$ of the Dirac operator in the space orthogonal to the zero modes. This formula can be derived in a way similar to (2.9.126) by expanding $\psi(x)$ and $\bar{\psi}(y)$ in an orthogonal system which contains the zero-mode wave functions as base vectors. Equation (2.9.127) is a euclidean generalization of (2.2.65) with interaction that takes into account the existence of zero modes.

Finally, the calculation of the fermionic Green functions requires the averaging over the gauge fields,

$$\begin{aligned}
&\langle\psi(x_1)\bar{\psi}(y_1)\cdots\psi(x_n)\bar{\psi}(y_n)\rangle = \qquad (2.9.128)\\
&\quad \frac{1}{Z_0}\int \mathcal{D}[A,u,\bar{u}]\langle\psi(x_1)\bar{\psi}(y_1)\cdots\psi(x_n)\bar{\psi}(y_n)\rangle_A\, \mathrm{e}^{-\int \mathrm{d}^4x\,\mathcal{L}_{\mathrm{eff}}}
\end{aligned}$$

with $\mathcal{L}_{\mathrm{eff}} = \mathcal{L}_{\mathrm{gauge}} + \mathcal{L}_{\mathrm{fix}} + \mathcal{L}_{\mathrm{ghost}}$ as in (2.4.17).

With the general results (2.9.127) and (2.9.128) 't Hooft's calculation of the one-instanton contribution can be completed. We calculated the right-handed fermionic zero modes in an SU(2) background field in Sect. 2.9.2.2,

$$\psi_{\mathrm{I},\alpha}^{(1),m}(x) = \begin{pmatrix} \kappa_\alpha^{(1),m} \\ 0 \end{pmatrix}, \qquad \kappa_\alpha^{(1),m} = \frac{\lambda^{3/2}}{\pi(x^2+\lambda^2)^{3/2}}\,\varepsilon_\alpha{}^m . \qquad (2.9.129)$$

This zero mode is N_{F}-fold degenerate for fermions that carry a flavour index $f = 1,\ldots,N_{\mathrm{F}}$.

Therefore, we get

$$\begin{aligned}
&\langle\psi(x_1)\bar{\psi}(y_1)\cdots\psi(x_n)\bar{\psi}(y_n)\rangle_{\mathrm{Zero}} = \int_{\mathbb{R}_4}\mathrm{d}^4z\int_0^\infty\frac{\mathrm{d}\lambda}{\lambda^5} \qquad (2.9.130)\\
&\quad\times D_{\mathrm{I}}(z,\lambda,\alpha(\mu))\sum_{\mathrm{perm}}(\pm 1)\prod_{i=1}^{N_{\mathrm{F}}}\prod_{j=1}^{N_{\mathrm{F}}}\psi_{\mathrm{I}}^{(1)}(x_i-z)\bar{\psi}_{\mathrm{I}}^{(1)}(y_j-z)
\end{aligned}$$

for the zero-mode contribution of the fermionic Green functions and a corresponding expression for the antiinstanton sector. For $N_{\mathrm{F}} = 1$, (2.9.130) becomes

$$\begin{aligned}
&\langle\psi_{\mathrm{R}}^m(x)\bar{\psi}_{\mathrm{R}}^{m'}(y)\rangle = \qquad (2.9.131)\\
&\quad\frac{1}{2\pi^2}\int_{\mathbb{R}_4}\mathrm{d}^4z\int_0^\infty\frac{\mathrm{d}\lambda}{\lambda^5}\,\frac{\lambda^3(1+\gamma^5)\delta^{mm'}}{[(x-z)^2+\lambda^2]^{3/2}[(y-z)^2+\lambda^2]^{3/2}} .
\end{aligned}$$

This expression is formally similar to the introduction of a chiral-symmetry-violating one-point vertex with $\mathcal{L}_{\mathrm{I}} \sim C\lambda(1+\gamma^5)$ and a fermion propagator

$\sim [(x-z)^2+\lambda^2]^{-3/2}$ in configuration space (cf. Sect. 2.2.5.1). The Feynman rules of instanton perturbation theory [tH76, Mo85, Mo97] are based on this interpretation of (2.9.131).

2.9.3.4 Clustering and Θ-vacuum

In this section we investigate the qualitative implications of the topological properties of classical gauge theories for the evaluation of the quantum-mechanical path integral. The result of the semi-classical approximation (2.9.74) and (2.9.75) violates the *cluster-decomposition condition* (2.1.40), an important property of Green functions. The cluster decomposition of Green functions requires the uniqueness of the vacuum [Ha92]. Therefore, the semi-classical approximation must be reconciled with the uniqueness of the vacuum. This leads to the introduction of the Θ-vacuum.

The fact that the semi-classical approximation (2.9.73), (2.9.74), and (2.9.75) leads to a violation of the cluster property may be seen by the following argument [Lu80]. The instanton contributions $\langle \breve{F}_{\mu\nu}(0)F^{\mu\nu}(0)\rangle^{\mathrm{I}} = \rho \neq 0$ violate parity. Parity may be restored by adding the antiinstanton contribution $\langle \breve{F}_{\mu\nu}(0)F^{\mu\nu}(0)\rangle^{\bar{\mathrm{I}}}$. If the cluster property were valid, one would expect

$$\lim_{|x-y|\to\infty} \langle \breve{F}_{\mu\nu}(x)F^{\mu\nu}(x)\breve{F}_{\rho\sigma}(y)F^{\rho\sigma}(y)\rangle^{\mathrm{I}+\bar{\mathrm{I}}} = \rho^2 \neq 0,$$

where $\langle\ldots\rangle^{\mathrm{I}+\bar{\mathrm{I}}} = (\langle\ldots\rangle^{\mathrm{I}} + \langle\ldots\rangle^{\bar{\mathrm{I}}})/2$. However, because of $\breve{F}^{I}_{\mu\nu} = F^{I}_{\mu\nu}$ and $\breve{F}^{\bar{\mathrm{I}}}_{\mu\nu} = -F^{\bar{\mathrm{I}}}_{\mu\nu}$ [cf. (2.9.14)], we get

$$\langle \breve{F}_{\mu\nu}(0)F^{\mu\nu}(0)\rangle^{\mathrm{I}+\bar{\mathrm{I}}} = 0 \neq \rho. \tag{2.9.132}$$

This means a violation of the cluster property. Apparently, in the definition of the vacuum $|0\rangle^{\mathrm{I}+\bar{\mathrm{I}}}$ mixed instanton–antiinstanton states are missing. It is not yet known how to incorporate these. The *instanton-liquid model* mentioned above [Sc98] is an attempt in this direction.

Formally, gauge theories in temporal gauge allow for *canonical quantization* with canonical commutation relations, Hamilton function, and wave functions. Let us reconsider the problem of clustering in the semi-classical approximation in this more general quantum-mechanical context. To this end we consider the gauge field $\mathcal{A}_\mu$ enclosed in a large box with volume V. We consider the corresponding quantum-mechanical wave function as a functional $\Psi[\mathcal{A}_\mu]$ of the gauge potential. In order to make the quantum-mechanical scheme gauge-invariant, we proceed by the following steps. First we consider $\Psi[\mathcal{A}_\mu]$ for a *temporal gauge* $\mathcal{A}_4 = 0$ [cf. (2.3.38)]. Then, we apply subsidiary conditions which render $\Psi[\mathcal{A}_\mu]$ invariant against infinitesimal, time-independent, local gauge transformations

$$\delta A_\mu^a(x) = \frac{1}{g}\partial_\mu \delta\theta^a(x) + f^{abc} A_\mu^b(x)\delta\theta^c(x), \quad \partial_{x_4}\delta\theta(x) = 0. \qquad (2.9.133)$$

Accordingly, these gauge transformations respect the condition $\mathcal{A}_4 = 0$.

However, there are gauge transformations that cannot be generated infinitesimally. Such gauge transformations have a *winding number*. We have discussed this fact for the example $g(x)$ in (2.9.16) in Sect. 2.9.1.2. The winding number of $g(x)$ is $w(g) = k = 1$ [cf. (2.9.28)]. Let $v(x)$ be a gauge transformation[33] consistent with the temporal gauge of winding number $w(v) = 1$. The gauge transformation $v(x)$ acts on the wave function $\Psi[\mathcal{A}_\mu]$ as

$$U(v)\Psi[\mathcal{A}_\mu] = \Psi[v(x)(\mathcal{A}_\mu(x) + \mathrm{i}\partial_\mu)v^{-1}(x)]. \qquad (2.9.134)$$

The gauge transformation $[v(x)]^n \equiv v^n$ has winding number $w(v^n) = n$. Gauge transformations like v^n, $n = 1, 2, \ldots$, are called *large gauge transformations*, in contrast to the infinitesimally generated *small gauge transformations*. The space of wave functions $\{\Psi[\mathcal{A}_\mu]\}$ invariant against small gauge transformations decomposes into sections $\{\Psi[\mathcal{A}_\mu, n]\}$ labelled by the winding number n. Large gauge transformations $U^n \equiv U^n(v) = U(v^n)$ connect the different sections,

$$\Psi[\mathcal{A}_\mu, n + n'] = U^n(v)\Psi[\mathcal{A}_\mu, n']. \qquad (2.9.135)$$

Several of these intuitive arguments need more mathematical justification. These are based on the homotopy theory of the gauge group [Co79, Be96].

The postulate of invariance under large gauge transformations leads to some new features of the quantum theory of gauge fields. In each section, there is a state $|n\rangle$, the *n-vacuum*, with minimal classical energy E_n. The Hamiltonian commutes with large gauge transformations. Therefore, E_n is independent of n. On the other hand, the cluster condition requires the uniqueness of the quantum-mechanical vacuum $|\Omega\rangle$ [Ha92]. Therefore $|\Omega\rangle$ must be quite distinct from the n-vacua.

[33] We give an explicit example in Sect. 2.9.3.5.

In the following discussion of clustering we follow an argument by S. Coleman [Co78]. Let us consider a system in a box with volume V. The transition matrix element from some initial state with winding number 0 to some final state with winding number n is described by the euclidean path integral

$$\langle n|\mathrm{e}^{-HT}|0\rangle = \langle n|\frac{1}{Z_0}\int \mathcal{D}[\mathbf{A},n]\,\mathrm{e}^{-S_\mathrm{E}\{\mathbf{A}\}}|0\rangle = f(V,T,n),$$
$$S_\mathrm{E}\{\mathbf{A}\} = \int_0^T \mathrm{d}t \int_V \mathrm{d}^3x\,\mathcal{L}(\mathbf{A}), \tag{2.9.136}$$

where in semi-classical approximation only fields contribute that interpolate between a field configuration with winding number 0 and one with winding number n. This is indicated by the index n in $\mathcal{D}[\mathbf{A},n]$. It is not important that we take the n-vacuum for the transition matrix element. It is, however, important that the equation

$$f(V,T_1+T_2,n) = \sum_{n_1+n_2=n} f(V,T_1,n_1)f(V,T_2,n_2) \tag{2.9.137}$$

holds for large times T_1 and T_2. This follows from the fact that the winding number n is related via Stokes' theorem to the integral of a local quantity. We have shown this for the classical instanton solution on a sphere in (2.9.26) and (2.9.28). In the example above, it tells us that the way to put total winding number n in a large box is to put winding number n_1 in one part of the box and winding number $n_2 = n - n_1$ in the remainder of the box.[34] In this interpretation we may consider (2.9.137) as a special form of the cluster property.

Formula (2.9.137) is not what we expect from the transition matrix element of a single non-degenerate energy eigenstate like a vacuum. Such an object would be a simple exponential and would for large times obey a multiplicative composition law rather than the convolutive composition law (2.9.137). However, by forming a Fourier series

$$f(V,T,\Theta) = \sum_n \mathrm{e}^{in\Theta} f(V,T,n), \tag{2.9.138}$$

[34] Such counting misses field configurations with significant action density on the boundary between the two sub-boxes. For our rather qualitative argument, we assume this surface effect to be small for large boxes.

(2.9.137) turns into

$$f(V,T_1+T_2,\Theta)=f(V,T_1,\Theta)f(V,T_2,\Theta). \tag{2.9.139}$$

This is the desired composition law. Thus, we identify $f(V,T,\Theta)$ as being (up to a normalization constant) the expectation value of $\exp(-HT)$ in a vacuum state, in which the cluster property is valid, the Θ-*vacuum*

$$|\Theta\rangle=\sum_{n=-\infty}^{\infty}\mathrm{e}^{\mathrm{i}n\Theta}|n\rangle. \tag{2.9.140}$$

Following (2.9.136) and (2.9.140), we write formally

$$\langle\Theta|\mathrm{e}^{-HT}|\Theta\rangle=\frac{1}{Z'}\sum_n\int\mathcal{D}[\mathbf{A},n]\,\mathrm{e}^{-S_{\mathrm{E}}\{\mathbf{A}\}}\,\mathrm{e}^{\mathrm{i}n\Theta}, \tag{2.9.141}$$

where Z' is a new normalization factor. Replacing the topological quantum number n by (2.9.26), $n=1/(16\pi^2)\int\mathrm{d}^4x\,\mathrm{Tr}\,\mathcal{F}_{\mu\nu}\check{\mathcal{F}}^{\mu\nu}$, we get

$$\begin{aligned}\langle\Theta|\mathrm{e}^{-HT}|\Theta\rangle=&\frac{1}{Z'}\sum_n\int\mathcal{D}[\mathbf{A},n]\\&\times\exp\left(\int\mathrm{d}^4x\left(-\frac{1}{2g^2}\,\mathrm{Tr}[\mathcal{F}_{\mu\nu}\mathcal{F}^{\mu\nu}]+\mathrm{i}\frac{\Theta}{16\pi^2}\,\mathrm{Tr}[\mathcal{F}_{\mu\nu}\check{\mathcal{F}}^{\mu\nu}]\right)\right).\end{aligned} \tag{2.9.142}$$

Some far-reaching conclusions [Ca76, Ca78] follow from this treatment of the topological properties of a pure gauge theory:

- The physical Θ-vacuum is invariant up to a phase under a large gauge transformation U^m

$$U^m|\Theta\rangle=\sum_n\mathrm{e}^{\mathrm{i}n\Theta}U^m|n\rangle=\sum_n\mathrm{e}^{\mathrm{i}n\Theta}|n+m\rangle=\mathrm{e}^{-\mathrm{i}m\Theta}|\Theta\rangle. \tag{2.9.143}$$

- In a pure gauge theory the path integral[35] (2.4.16) or (2.4.19), has to be extended by summation over topological sectors (*Chern classes*)

$$Z_{\mathrm{E}}\{J\} = \sum_n \int \mathcal{D}[\mathbf{A}, n]\, \mathrm{e}^{\mathrm{i}n\Theta}\, \mathrm{e}^{-S_{\mathrm{E}}\{\mathbf{A}\}+\int \mathrm{d}^4x A_\mu^a J^{\mu,a}}. \qquad (2.9.144)$$

  Equation (2.9.144) is based on the fact that the Hamiltonian is invariant under small gauge transformations. Therefore, all other energy eigenstates can be chosen so that they change at most by this constant phase.

- The Lagrangian in the path integral becomes

$$\mathcal{L}(\mathcal{A}; \Theta) = \frac{1}{2g^2} \operatorname{Tr}[\mathcal{F}_{\mu\nu}\mathcal{F}^{\mu\nu}] - \mathrm{i}\frac{\Theta}{16\pi^2} \operatorname{Tr}[\mathcal{F}_{\mu\nu}\breve{\mathcal{F}}^{\mu\nu}]. \qquad (2.9.145)$$

- The inclusion of the topological properties (e.g. winding numbers) in the semi-classical approximation causes super-selection rules for gauge-invariant quantities, labelled by a parameter Θ. Only non-gauge-invariant operators transform states of different sectors into each other.

- In a complete theory one has to distinguish between a perturbative vacuum $|0\rangle$, which disregards topological excitations, and the non-perturbative, physical vacuum $|\Theta\rangle$.

2.9.3.5 Vacuum tunnelling

Large gauge transformations connect fields belonging to sectors with different winding numbers. In order to see this, we transform the instanton solution (2.9.23) into the *temporal gauge* $\mathcal{A}_0 = 0$ [Ca78, Hu92], using

$$g(\mathbf{x}, x_4) = \exp\left[\frac{\mathrm{i}\pi \mathbf{x}\boldsymbol{\tau}}{\sqrt{\mathbf{x}^2+\lambda^2}}\,\Phi(\mathbf{x}, x_4)\right],$$
$$\Phi(\mathbf{x}, x_4) = \frac{1}{2} + \frac{1}{\pi}\arctan\frac{x_4}{\sqrt{\mathbf{x}^2+\lambda^2}}. \qquad (2.9.146)$$

Note that

$$\Phi(\mathbf{x}, x_4) \to \begin{cases} 1 & \text{for } \; x_4 \to \infty, \\ 0 & \text{for } \; x_4 \to -\infty, \end{cases} \qquad (2.9.147)$$

[35] We omit gauge-fixing and ghost terms like those in (2.4.17).

and therefore

$$g(\mathbf{x}, x_4) \to \begin{cases} v(\mathbf{x}) = \exp\left(\dfrac{\mathrm{i}\pi \mathbf{x}\boldsymbol{\tau}}{\sqrt{\mathbf{x}^2+\lambda^2}}\right) & \text{for } x_4 \to \infty, \\ 1 & \text{for } x_4 \to -\infty. \end{cases} \tag{2.9.148}$$

Like $g(x)$ in (2.9.16), $v(\mathbf{x})$ has winding number $w(v) = 1$. This can be seen by straightforward calculation. Alternatively, one can express x in $g(x)$ in polar coordinates (2.9.29) and finds that $\chi = \pi|\mathbf{x}|/(\mathbf{x}^2+\lambda^2)^{1/2}$ is a one-to-one map between $0 \le \chi < \pi$ and $0 \le |\mathbf{x}| < \infty$. Hence, $v(\mathbf{x})$ can be smoothly deformed into $g(x)$.

Now we apply the gauge transformation $g(\mathbf{x}, x_4)$ to the instanton solution,

$$\mathcal{A}'_\mu(\mathbf{x}, x_4) = g(\mathbf{x}, x_4)\frac{2\tau_{\mu\nu}x^\nu}{x^2+\lambda^2}g^{-1}(\mathbf{x}, x_4) + \mathrm{i}g(\mathbf{x}, x_4)\partial_\mu g^{-1}(\mathbf{x}, x_4). \tag{2.9.149}$$

By direct calculation we get $\mathcal{A}'_4 = 0$. For $\mathcal{A}'_i(\mathbf{x}, x_4)$, $i = 1, 2, 3$, the first term vanishes for $|x_4| \to \infty$, which implies

$$\mathcal{A}_i(\mathbf{x}, x_4) \longrightarrow \begin{cases} \mathrm{i}v(\mathbf{x})\nabla_i v^{-1}(\mathbf{x}) & \text{for} \quad x_4 \to \infty, \\ 0 & \text{for} \quad x_4 \to -\infty. \end{cases} \tag{2.9.150}$$

As in (2.9.10), $\mathcal{A}_\mu(x)$ approaches a pure gauge. Thus, this calculation shows that the instanton solutions interpolate between gauge field configurations with different winding numbers ($k = 0$ for $x_4 \to -\infty$ and $k = 1$ for $x_4 \to \infty$ in this case).

The transition amplitude between neighbouring n-vacua is in semi-classical approximation [cf. (2.9.73), (2.9.74) and (2.9.136)] given by

$$\langle n+1|\exp(-HT)|n\rangle = \langle 1|\exp(-HT)|0\rangle \sim \frac{\mathrm{e}^{-\frac{8\pi^2}{g^2}}}{\sqrt{\det\Delta}}, \tag{2.9.151}$$

or less formal by 't Hooft's result (2.9.121). The formula (2.9.151) reminds us of the energy splitting between the lowest energy states of a double-well potential $V(x) = g^2(x^2 - a^2)^2/2$. It is known from elementary quantum mechanics that the tunnelling between the classical ground state at $x = -a$

to the classical ground state at $x = +a$ leads to a splitting between the symmetric and antisymmetric ground-state solutions with [Sc98]

$$E_{\pm} = E_0 \pm \Delta E = ag \pm 4ag\sqrt{\frac{3S_0}{2\pi}}\mathrm{e}^{-S_0}, \qquad S_0 = \frac{4}{3}a^3 g. \tag{2.9.152}$$

The difference of the energies is proportional to the barrier penetration factor $\exp(-S_0)$. In this sense instantons describe tunnelling between the classical n-vacua.

Now we assume that the density $N(z,\lambda)$ of one instanton or antiinstanton in a four-dimensional box with volume $V^{(4)} = VT$ is so small that only one instanton or antiinstanton contributes to the transitions between the n-vacua. Then, we can calculate the expectation values for the Θ-vacuum in the *dilute-gas approximation* of the instanton–antiinstanton gas. Denoting the numbers of instantons and antiinstantons by N_+ and N_-, respectively, the partition function has to be averaged over [cf. (2.9.121)]

$$\frac{1}{N_+!N_-!}\prod_{j=1}^{N_++N_-}\left\{\int \mathrm{d}^4 z_j\,\frac{\mathrm{d}\lambda_j}{\lambda_j}\,N(z_j,\lambda_j)\right\}. \tag{2.9.153}$$

In the simple approximation of (2.9.151), this leads to the Θ-vacuum expectation value

$$\begin{aligned}\langle\Theta|\mathrm{e}^{-HT}|\Theta\rangle \propto &\sum_{N_+,N_-=0}^{\infty}\frac{1}{N_+!N_-!}(KVT)^{N_++N_-}\\ &\times\exp\left[-\frac{8\pi^2}{g^2}(N_++N_-)+\mathrm{i}\Theta(N_+-N_-)\right]\\ &=\exp\left[2KVT\cos\Theta\,\mathrm{e}^{-\frac{8\pi^2}{g^2}}\right].\end{aligned} \tag{2.9.154}$$

The factor K represents $(\det\Delta)^{-1/2}$ in (2.9.151). For our qualitative discussion we let K undetermined. From (2.9.154) we can get the energy density

$$\frac{E(\Theta)}{V} = -2K\cos\Theta\,\mathrm{e}^{-\frac{8\pi^2}{g^2}}. \tag{2.9.155}$$

As in the double-well case (2.9.152), $E(\Theta)$ is the energy shift of the quantum-mechanical Θ-vacuum. It has its origin in the tunnelling between the perturbative n-vacua. This energy shift can be regarded as an effect of the Θ-term in the Lagrangian (2.9.145).

2.9.3.6 Fermions in the vacuum tunnelling picture

The picture of vacuum tunnelling in a pure gauge theory changes if one includes massless fermions [Ca78]. The zero modes of the Dirac equation in an (anti)instanton background field with topological quantum numbers $k = \pm 1$ (2.9.63) render the tunnelling amplitude zero in the absence of external fermion sources. Therefore, Green functions which contain at least one pair of Dirac fields $\psi(x)\overline{\psi}(y)$ have to be considered. Moreover, we have to include in the n-vacuum the occupation number of fermions and antifermions.

In Dirac's hole theory, these numbers count the occupied or empty fermion states, respectively. A perturbation-theoretical n-state is thus one for which the gauge field belongs to a winding number n, and all negative energy fermion levels are occupied. One can think of tunnelling as being an adiabatic process as far as the fermions are concerned. The time-independent Dirac equation in an instanton field has an eigenvalue $\epsilon_i(t)$ and eigenfunctions $\psi_i(t, \mathbf{x})$ for each value of the euclidean time t. Since the net effect of tunnelling is just a gauge transformation, the original eigenvalues of $\epsilon_i(0)$ must be in one-to-one correspondence with the final eigenvalues.

For the massless Dirac equation in an instanton background field this mapping $\epsilon_i(0) \to \epsilon_i(T)$ is non-trivial $\epsilon_i(0) \neq \epsilon_i(T)$. In particular, the highest right-handed negative-energy eigenstate crosses zero and becomes the lowest right-handed positive-energy eigenstate while the lowest left-handed positive-energy state becomes the highest left-handed negative-energy state. This is a special consequence of the *spectral-flow theorem* [Va84] which can be derived from the index theorem. Thus, if in the initial configuration all negative-energy eigenstates are occupied and all positive-energy eigenstates empty, then the final configuration contains one right-handed positive-energy eigenstate that is occupied and one left-handed negative-energy eigenstate that is empty. In the instanton case, the tunnelling process is $n \to n + 1 + \bar{q}_{\mathrm{L}} q_{\mathrm{R}}$ rather than $n \to n + 1$ as in (2.9.151). Strictly speaking this is only a virtual vacuum tunnelling $n \to n + 1 + \bar{q}_{\mathrm{L}} q_{\mathrm{R}} \to n$. This has important consequences for the strong CP problem. As shown in Sect. 2.9.3.9, the strong CP problem is absent for massless fermions.

It is beyond the scope of this book to go into the details of the spectrum-flow theorem. However, Fig. 2.23 on page 342 shows as an example of this theorem the zero mode (2.9.48) of the two-dimensional Dirac equation in a two-dimensional instanton background field (cf. Sect. 2.9.2.2). The U(1) phase is constant on the line $x_2 = 0$. At this line we have an n-vacuum with

$n = 0$. At the line $x_2 = L_2$, the phase is $\exp(-2\pi x_1/L_1)$ and $n = 1$. The zero mode interpolates between these two states.

2.9.3.7 Vacuum condensates

The considerations above show that non-perturbative effects change the vacuum structure of gauge theories considerably. As a consequence, new types of Green functions must be discussed. We consider the two most important examples.

The *condensate* $\langle g^2 FF \rangle \equiv \langle g^2 F^a_{\mu\nu}(0) F^{a,\mu\nu}(0) \rangle$ has its origin in the non-perturbative part of the physical vacuum.[36] Since it is most important for QCD, it is called generally *gluon condensate.* The formal definition of $\langle g^2 FF \rangle$ contains the product of field operators at the same space–time point, which needs regularization. Only the product $g^2 FF$ is renormalization-group-invariant, i.e. independent of the subtraction point [Va82].

The condensate $\langle g^2 FF \rangle$ plays an important rôle in the *sum-rule technique* introduced by Shifman, Vainshtein and Zhakharov [Sh79, Na89, Sh92]. Its physical results depend on some skillful combination of phenomenological data with perturbative and non-perturbative properties of two-point functions. We do not treat this technique here. A theoretically more satisfying calculation of $\langle g^2 FF \rangle$ is based on the lattice approximation of gauge theories. Within large errors the results of the sum-rule technique [Na96] and of the lattice approximation [DE97] agree:

$$\begin{aligned} &\langle g^2 FF \rangle \approx (0.89 \pm 0.11\,\mathrm{GeV})^4 && \text{from sum rules,} \\ &\langle g^2 FF \rangle \sim (1\,\mathrm{GeV})^4 && \text{from lattice.} \end{aligned} \tag{2.9.156}$$

One of the problems in the determination of $\langle g^2 FF \rangle$ is the not well-defined separation of perturbative and non-perturbative contributions. In the definition of $\langle g^2 FF \rangle$ we put $\Theta = 0$. However, the dependence on Θ might be helpful for such a separation [Ha98].

In gauge theories with massless fermions the *chiral condensate* $\langle \bar{\psi}(0)\psi(0) \rangle$ is the order parameter of chiral symmetry breaking (cf. Sect. 3.5.1.1). This is a non-perturbative effect. The connection with the vacuum picture is seen by the following qualitative argument.

[36] In perturbation theory we have $\langle g^2{:}FF{:} \rangle = 0$.

We consider $\langle\bar{\psi}(0)\psi(0)\rangle$ for a finite euclidean volume V and a finite fermion mass m. Then we have

$$\langle\bar{\psi}\psi\rangle = \frac{1}{V}\,\mathrm{Tr}\left[\frac{1}{m+\not{D}}\right]. \tag{2.9.157}$$

The density of the eigenvalues λ of the Dirac operator $\not{D}\psi = \mathrm{i}\lambda_n\psi$ is $V\rho(\lambda)$ with $\rho(\lambda) = \rho(-\lambda) = \langle\sum_n \delta(\lambda-\lambda_n)\rangle$. It becomes continuous for $V \to \infty$ and leads to

$$\langle\bar{\psi}\psi\rangle = \int_{-\infty}^{\infty} \mathrm{d}\lambda\,\frac{\rho(\lambda)}{m+\mathrm{i}\lambda} = 2\int_0^{\infty} \mathrm{d}\lambda\,\frac{m\rho(\lambda)}{m^2+\lambda^2}. \tag{2.9.158}$$

Using

$$\lim_{m\to 0}\frac{\mathrm{i}}{m+\mathrm{i}\lambda} = \mathrm{i}\pi\delta(\lambda) + \mathrm{PP}\left(\frac{1}{\lambda}\right), \tag{2.9.159}$$

we get in the limit $m \to 0$

$$\langle\bar{\psi}\psi\rangle = \lim_{m\to 0}\int_{-\infty}^{\infty} \mathrm{d}\lambda\,\frac{\rho(\lambda)}{m+\mathrm{i}\lambda} = \pi\int_{-\infty}^{\infty} \mathrm{d}\lambda\,\rho(\lambda)\delta(\lambda) = \pi\rho(0). \tag{2.9.160}$$

This relation [Ba80] states that for chiral symmetry breaking one must have an accumulation of eigenvalues at $\lambda = 0$. Now we know from the index theorem (Sect. 2.9.2) that we have such zero modes in an instanton background field. Thus, under the assumption of the dilute-gas approximation (2.9.153), the separated instantons can be a source of eigenfunctions at $\lambda = 0$.

Numerical values of $\langle\bar{\psi}\psi\rangle$ come from the sum-rule technique and from lattice approximation. We quote

$$\langle\bar{\psi}\psi\rangle \approx -(230\,\mathrm{MeV})^3 \tag{2.9.161}$$

from sum rules [Do98], and

$$\langle\bar{\psi}\psi\rangle \approx -(245\,\mathrm{MeV})^3 \tag{2.9.162}$$

from quenched lattice approximation [Gi99].

2.9.3.8 Anomaly and the index theorem

In Sect. 2.7.1 we found that anomalies of axial-vector currents are related to the lack of a consistent regularization scheme for operators containing γ^5 matrices. However, it is not only the high-energy behaviour of the triangle graph (Fig. 2.11 on page 272) which is responsible for the anomalous non-conservation of axial-vector currents in quantum field theory. It is a problem of the evaluation of the path integral which is related to the topological zero modes of the Dirac equation. We can see this from the equation for the singlet anomaly (2.7.40) in Minkowski space,

$$\partial^\mu j^0_{5,\mu}(x) = -\frac{g^2}{16\pi^2} F^{a,\mu\nu}(x)\breve{F}^a_{\mu\nu}(x). \tag{2.9.163}$$

By integrating over $\mathbb{R}^3$ we get

$$\frac{\mathrm{d}Q^5}{\mathrm{d}t} = \frac{\mathrm{d}}{\mathrm{d}t}\int_{\mathbb{R}^3} \mathrm{d}^3\mathbf{x}\, j^5_0(\mathbf{x},t) = -\frac{g^2}{16\pi^2}\int_{\mathbb{R}^3} \mathrm{d}^3\mathbf{x}\, F^{a,\mu\nu}(\mathbf{x},t)\breve{F}^a_{\mu\nu}(\mathbf{x},t), \tag{2.9.164}$$

and by integration over $\mathbb{R}^4$,

$$\int_{\mathbb{R}^4} \mathrm{d}^4x\, \partial^\mu j^5_\mu(x) = -\frac{g^2}{16\pi^2}\int_{\mathbb{R}^4} \mathrm{d}^4x\, F^{a,\mu\nu}(x)\breve{F}^a_{\mu\nu}(x). \tag{2.9.165}$$

Comparing (2.9.165) with (2.9.63), (2.9.64), (2.9.65), and (2.9.68), the right hand side of (2.9.163) might be considered as an *index density* [Ja76, Ni77], or the anomaly equation as a local form of the index theorem. The following description makes this considerations more precise.

According to Fujikawa [Fu80], the anomalous non-conservation of the axial-vector current can be traced back to the non-trivial transformation of the euclidean fermionic path-integral measure under chiral transformations. This measure was introduced in Sect. 2.2.1.4 as a generalization of the integration over a finite number of Grassmann variables (2.2.20)–(2.2.23) to an integration over functionals (2.2.27). In this spirit we expand the fermion fields $\psi(x)$ and $\bar{\psi}(x) = \psi^\dagger\gamma^4$ in orthonormalized eigenfunctions of the *euclidean Dirac operator*,

$$\psi(x) = \sum_m a_m\psi_m(x), \qquad \mathrm{i}\not{D}\psi_m(x) = \lambda_m\psi_m(x), \tag{2.9.166}$$

$$\bar{\psi}(x) = \sum_m \psi^\dagger_m(x)\bar{b}_m, \qquad \mathrm{i}\psi^\dagger_m(x)\not{D} = \lambda_m\psi^\dagger_m(x), \tag{2.9.167}$$

and define the measure by

$$\mathcal{D}[\psi(x)]\mathcal{D}[\bar{\psi}(x)] = \prod_m \mathrm{d}a_m \mathrm{d}\bar{b}_m, \tag{2.9.168}$$

where $\mathrm{d}a_m$ and $\mathrm{d}\bar{b}_m$ satisfy (2.2.20) and (2.2.21). The fields $\psi(x)$ and $\bar{\psi}(x)$ transform under local chiral transformations as

$$\psi'(x) = \mathrm{e}^{\mathrm{i}\beta(x)\gamma^5}\psi(x), \qquad \bar{\psi}'(x) = \bar{\psi}(x)\mathrm{e}^{\mathrm{i}\beta(x)\gamma^5}. \tag{2.9.169}$$

This corresponds to a linear transformation $C(\beta) = (c_{mn}(\beta))$ for the expansion coefficients a_m and $\bar{b}_m$,

$$a'_m = \sum_n c_{mn} a_n, \qquad \bar{b}'_m = \sum_n \bar{b}_n c_{nm}, \tag{2.9.170}$$

or for an infinitesimal transformation $C = 1 + \delta C$,

$$\delta c_{nm}(\beta) = \mathrm{i}\int \mathrm{d}^4x\, \beta(x)\psi_n^\dagger(x)\gamma^5\psi_m(x). \tag{2.9.171}$$

For the measure (2.9.168) this implies [cf. (2.2.22)]

$$\prod_m \mathrm{d}a'_m \mathrm{d}\bar{b}'_m = (\det C)^{-2} \prod_m \mathrm{d}a_m \mathrm{d}\bar{b}_m, \tag{2.9.172}$$

or formally

$$\mathcal{D}[\psi'(x)]\mathcal{D}[\bar{\psi}'(x)] = (\det C)^{-2}\mathcal{D}[\psi(x)]\mathcal{D}[\bar{\psi}(x)]. \tag{2.9.173}$$

We use $\det C = \exp(\mathrm{Tr}\log C) = \exp(\mathrm{Tr}\,\delta C)$ to get the Jacobian $J(\beta)$ of the chiral transformation (2.9.171),

$$J(\beta) = (\det C)^{-2} = \exp\left(-2\mathrm{i}\int \mathrm{d}^4x\, \beta(x) \sum_m \psi_m^\dagger(x)\gamma^5\psi_m(x)\right). \tag{2.9.174}$$

Of course this formal expression needs regularization. Fujikawa proposed the *heat-kernel regularization* (cf. Sect. 2.9.2.3),

$$\left(\sum_m \psi_m^\dagger(x)\gamma^5\psi_m(x)\right)_{\text{reg}} = \lim_{\tau\to 0}\left(\sum_m \psi_m^\dagger(x)\gamma^5 \mathrm{e}^{-\lambda_m^2\tau}\psi_m(x)\right)$$
$$= \lim_{\tau\to 0}\operatorname{Tr}\left[\gamma^5 \mathrm{e}^{-\not{D}^\dagger\not{D}\tau}\right] = \frac{\varepsilon^{\mu\nu\alpha\beta}}{32\pi^2}\operatorname{Tr}[\mathcal{F}_{\mu\nu}\mathcal{F}_{\alpha\beta}]. \tag{2.9.175}$$

The last equality results from (2.9.64)–(2.9.68), the proof of the index theorem. The result is therefore

$$J(\beta) = \exp\left(-\frac{\mathrm{i}}{16\pi^2}\int \mathrm{d}^4x\, \beta(x)\epsilon^{\mu\nu\alpha\beta}\operatorname{Tr}[\mathcal{F}_{\mu\nu}(x)\mathcal{F}_{\alpha\beta}(x)]\right). \tag{2.9.176}$$

The non-invariance of the Jacobian leads according to (2.2.90) to extra contributions in the Ward identities, resulting in anomalous Ward identities.

2.9.3.9 On the problem of strong CP violation

The inclusion of non-trivial topological sectors in the path integral requires the Θ-term of (2.9.145) in the euclidean Lagrangian

$$\mathcal{L}(\mathcal{A};\Theta) = \frac{1}{2g^2}\operatorname{Tr}[\mathcal{F}_{\mu\nu}\mathcal{F}^{\mu\nu}] - \mathrm{i}\frac{\Theta}{16\pi^2}\operatorname{Tr}[\mathcal{F}_{\mu\nu}\breve{\mathcal{F}}^{\mu\nu}] = \mathcal{L}+\mathcal{L}_\Theta. \tag{2.9.177}$$

This term $\mathcal{L}_\Theta$ is proportional to the Pontryagin density and does as a pure divergence [cf. (2.9.27) and Sect. 2.7.2.1] not affect the classical field equation. However, it contributes to quantum corrections. With respect to parity and charge-conjugation parity, $F^a_{\mu\nu}F^{a,\mu\nu}$ differs from $F^a_{\mu\nu}\breve{F}^{a,\mu\nu}$. Therefore, the Lagrangian (2.9.177) violates CP invariance. This produces the *CP problem of QCD*.

There are stringent limits on CP violation. For instance, the experimental value of the neutron dipole moment, $d_\mathrm{n} < 6.3\times 10^{-26}e\,\mathrm{cm}$ [PDG00], together with a calculation of the contribution of (2.9.177) to d_n, $d_\mathrm{n} = 4\times 10^{-16}|\Theta|\,\mathrm{cm}$ [Cr79a], yields $|\Theta| < 10^{-9}$. There are several attempts to explain this new coupling constant and its smallness in QCD [Pe77, El79].

For massless fermions, the chiral invariance of a theory implies that the physics is the same in all Θ-super-selection sectors. In the path integral we can always transform Θ to zero by a global chiral transformation (2.9.169) with $\beta = \Theta/2$. This follows from a comparison of the Jacobian for an infinitesimal chiral transformation (2.9.176) with an infinitesimal Θ in $\mathcal{L}_\Theta$ in (2.9.177). Therefore, the CP problem is absent for massless fermions. Maybe, the CP problem can be solved by a more fundamental theory which starts from massless fermions.

2.9.3.10 The U(1) problem

Because of the anomaly, the charge Q^5 of the singlet $\mathrm{U}(1)_\mathrm{A}$ axial-vector current is not conserved. However, the two-point function of the divergence of the corresponding current [cf. (2.7.42)],

$$\partial_\mu(\bar{\psi}(x)\gamma^\mu\gamma^5\psi(x)) \equiv p(x) = -\frac{1}{8\pi^2}\,\mathrm{Tr}[\mathcal{F}^{\mu\nu}(x)\breve{\mathcal{F}}_{\mu\nu}(x)], \qquad (2.9.178)$$

obeys an interesting relation, the *Witten–Veneziano relation* [Wi79, Ve79], originally derived in the $1/N$ approximation, where N is the number of quark flavours. It states that the *topological susceptibility* χ_top, which measures the fluctuation of the topological charge density $p(x)$, is proportional to the square of the pseudo-scalar flavour-singlet mass m_0 and its decay constant f_0,

$$\chi^{(0)}_\mathrm{top} \equiv \int \mathrm{d}^4x\,\langle p(x)p(0)\rangle = 2f_0^2 m_0^2. \qquad (2.9.179)$$

Here, the left-hand side refers to $\chi^{(0)}_\mathrm{top}$ in a pure gauge theory.[37] A sensible determination of m_0^2 requires a theory including fermions. Thus, the Witten–Veneziano relation compares quantities of different theories.

Equation (2.9.179) has important consequences for the pseudo-scalar meson spectrum (cf. Sect. 3.5.1.2). The main point is that the mass of the η meson in the theory with two flavours (the mass of the η' meson for 3 flavours) is large compared to the pion mass (m_π and m_K). This U(1) problem is solved by the instantons [tH76], i.e. by the Witten–Veneziano relation. If we take a value for $\chi_\mathrm{top} = (180\,\mathrm{MeV})^4$, coming from lattice calculation [DE97], we get a reasonable agreement with experiment.

[37] In a gauge theory with N_f massless fermions, we have $\chi^{(N_\mathrm{f})}_\mathrm{top} = 0$.

2.10 Lattice approximation of gauge theories

In order to treat reliably the non-perturbative aspects of gauge theories, it is necessary to find an approach for the evaluation of the basic gauge-invariant path-integral formula that is different from perturbation theory (Sect. 2.4.1.3) [Wi74]. Lattice approximation of gauge theory is such an attempt [Se82]. In this scheme, the $(D-1)$-dimensional configuration space or the D-dimensional euclidean space–time is approximated by discrete points of a lattice Γ. The approximation of a finite euclidean space–time by a finite lattice appears most promising. There are quite a few limiting processes necessary in order to get from such a field theory on Γ to the Green functions in the four-dimensional continuum. These limiting processes define the gauge-invariant path-integral formula.

In this section, we sketch the main features of the lattice approximation of gauge theories [LAT]. Based on the geometry of local gauge symmetry, we describe the gauge-invariant field theory on Γ in Sect. 2.10.1. The transition to the continuum is treated in Sect. 2.10.4. The infinite-volume limit is considered in Sect. 2.10.5. Because fermions pose problems which have not yet been solved completely, we postpone a short overview of these questions to Sect. 2.10.6. The computational procedures are strong-coupling approximation (Sect. 2.10.2), and above all, numerical methods (Sect. 2.10.3). Results for QCD are given in Sect. 3.6.

2.10.1 Basics of lattice approximation

2.10.1.1 Geometrical elements of pure gauge theories

The geometry of local gauge symmetry described in Sect. 2.3.2 can be transferred point by point to the lattice [Ma81]. First, we describe pure D-dimensional gauge theory. The description of Dirac particles has some particular difficulties about which we report on at the end of this section.

For the sake of simplicity, a cubic lattice with lattice constant a and N^D lattice points, where N is finite or infinite, is used. The various lattice elements are

points x with coordinates $x = a(n^1, n^2, \ldots, n^D)$, $\quad n^i = 1, 2, \ldots, N$,

links $b = [x;\mu]$ connect neighbouring points with coordinates $(x_1, x) = (x + e_\mu, x)$, where $e_\mu = a\hat{\mu}$ and $\hat{\mu}$ is oriented from x to x_1,

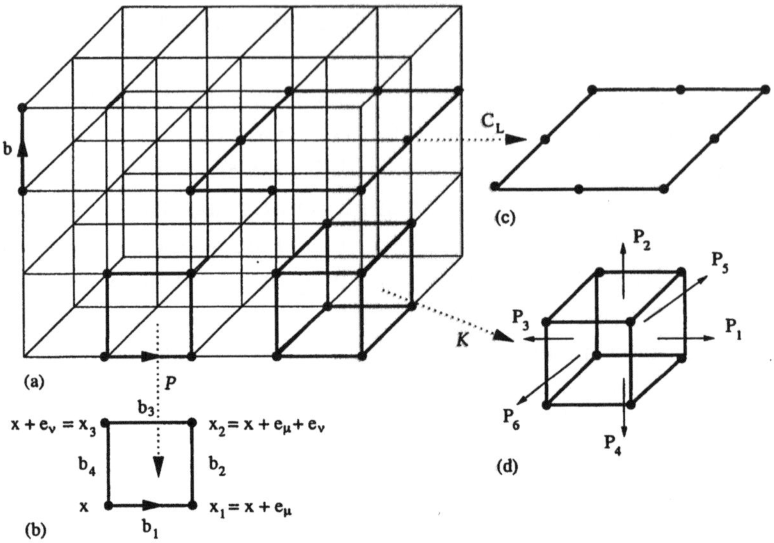

Fig. 2.24 Objects on the lattice

plaquettes $\mathcal{P} = [x;\mu,\nu]$ are the smallest closed loops with coordinates $(x_3, x_2, x_1, x) = (x+e_\nu, x+e_\mu+e_\nu, x+e_\mu, x)$, $(\mu \neq \nu)$, oriented by this sense of rotation,

cubes $\mathcal{K}$ with corresponding orientation, and

super-cubes with corresponding orientation.

A path $\mathcal{C}_\mathrm{L}$ on the lattice is a sum of connected links $\mathcal{C}_\mathrm{L} = \sum b_i$, a surface is a sum of plaquettes $\mathcal{B} = \sum \mathcal{P}_i$. The boundary of a path is denoted by $\partial\mathcal{C}_\mathrm{L}$, and that of a surface by $\partial\mathcal{B}$. Figure 2.24 illustrates these concepts.

The basic geometrical concepts of lattice gauge theory can be summarized as follows:

- Gauge fields describe the infinitesimal parallel displacement of charged fields [cf. (2.3.17) and (2.3.18)]. Hence, parallel displacements $U(b)$ (2.3.36) along links $b \in \Gamma$ correspond to *gauge fields on the lattice*,

$$\mathrm{P}\exp\left(\mathrm{i}\int_b \mathrm{d}t^\mu \mathcal{A}_\mu(t)\right) \mapsto U(b) = U([x;\mu]) \equiv e^{\mathrm{i}\mathcal{A}([x;\mu])}. \quad (2.10.1)$$

Lattice gauge fields $U(b) = U([x;\mu])$ are therefore functions that are defined on the links of the lattice and their values are group elements of G. They possess the property

$$U(-b) = U^{-1}(b) \qquad \text{with} \ -(x_1, x_2) = (x_2, x_1). \tag{2.10.2}$$

- *Parallel displacement* along a curve $\mathcal{C}$ [cf. (2.3.36)] reads in lattice approximation

$$U(\mathcal{C}) = \mathrm{P}\exp\left(\mathrm{i}\int_{\mathcal{C}} \mathrm{d}t^\mu \mathcal{A}_\mu(t)\right) \mapsto U(\mathcal{C}_\mathrm{L}) \equiv \prod_{b_i \in \mathcal{C}_\mathrm{L}} U(b_i), \tag{2.10.3}$$

where $\mathcal{C}_\mathrm{L} = \sum_i b_i \subset \Gamma$ is a path that approximates the continuous curve $\mathcal{C}$.

- In agreement with (2.3.25), the field strength describes the parallel displacement around a plaquette. Parallel displacement around the boundary $\partial\mathcal{P}_x$ of a plaquette with the links b_4, b_3, b_2, b_1 corresponds to

$$U(\partial\mathcal{P}_x) = \mathrm{P}\prod_{b_i \in \partial\mathcal{P}} U(b_i) = U(b_4)U(b_3)U(b_2)U(b_1) = U([x;\mu,\nu]). \tag{2.10.4}$$

As a result of path ordering, the initial point x is distinguished on the boundary $\partial\mathcal{P}$. The notions used in Fig. 2.24b are also used here.

The Baker–Hausdorff formula for the multiplication of exponential functions of non-commuting factors,

$$\mathrm{e}^{\mathrm{i}A}\mathrm{e}^{\mathrm{i}B} = \mathrm{e}^{\mathrm{i}A+\mathrm{i}B-\frac{1}{2}[A,B]+\dots}, \tag{2.10.5}$$

implies

$$U(\partial\mathcal{P}_x) = \mathrm{e}^{\left(\mathrm{i}\sum_i \mathcal{A}(b_i) - \frac{1}{2}\sum_{i>k}[\mathcal{A}(b_i),\mathcal{A}(b_k)]+\dots\right)} \equiv e^{\mathrm{i}\mathcal{F}(\mathcal{P})}. \tag{2.10.6}$$

The lattice analogue of the *field strength* is

$$\mathcal{F}_{\mu\nu}(x) \mapsto \mathcal{F}(\mathcal{P}_x) = \mathcal{F}([x;\mu,\nu]). \tag{2.10.7}$$

- Local gauge transformations $g(x) \in G$ as in (2.3.35) are defined at lattice points $x \in \Gamma$. The geometric quantities then transform in a way similar to the continuum

$$\begin{aligned} U(b) &\to g(x_2)U(b)g^{-1}(x_1) \qquad \text{with} \qquad b = (x_2, x_1), \\ U(\mathcal{C}_\mathrm{L}) &\to g(x')U(\mathcal{C}_\mathrm{L})g^{-1}(x), \end{aligned} \tag{2.10.8}$$

where C is a curve from x to x', and

$$\begin{aligned} U(\partial\mathcal{P}_x) &\to g(x)U(\partial\mathcal{P}_x)g^{-1}(x), \\ \mathcal{F}(\mathcal{P}_x) &\to g(x)\mathcal{F}(\mathcal{P}_x)g^{-1}(x). \end{aligned} \tag{2.10.9}$$

2.10.1.2 The Wilson action

The dynamics of a lattice gauge theory is determined by the definition of an action. According to K. Wilson [Wi74], the following expression is considered for a pure euclidean gauge theory with gauge group SU(N):

$$S_{\mathrm{W}} = -\beta \sum_{\mathcal{P}} \frac{1}{2N} \operatorname{Tr}[U(\partial\mathcal{P}) + U^{-1}(\partial\mathcal{P}) - 2]. \tag{2.10.10}$$

The sum extends over all non-oriented plaquettes of the euclidean space–time lattice, i.e. each plaquette appears in the sum only with one orientation,

$$\sum_{\mathcal{P}} = \sum_{x} \sum_{1\le\mu<\nu\le 4} = \frac{1}{2}\sum_{x}\sum_{\mu\nu}.$$

The trace $\operatorname{Tr}[U(\partial\mathcal{P}_x)]$ permits a cyclic permutation of the factors in (2.10.10) and is therefore independent of x. The gauge invariance of S_{W} follows from (2.10.9) and the trace property.

In order to investigate the correspondence with the continuum case we use (2.10.1)–(2.10.9), especially

$$\mathcal{A}([x,\mu]) \sim a g_0\, A^b_\mu(x)\frac{\lambda^b}{2}, \quad \mathcal{F}([x;\mu,\nu]) \sim a^2 g_0\, F^b_{\mu\nu}(x)\frac{\lambda^b}{2}, \tag{2.10.11}$$

where g_0 is the bare gauge coupling. Together with (2.10.6), (1.2.52), and $\sum_x \sim \int \mathrm{d}^4x/a^4$ one can show that S_{W} is an approximation of the action $S = \int \mathrm{d}^4x\, \mathcal{L}_{\mathrm{gauge}}$ (2.3.42) for small lattice constants,

$$\begin{aligned} S_{\mathrm{W}} &= -\beta \sum_{\mathcal{P}} \frac{1}{2N} \operatorname{Tr}[U(\partial\mathcal{P}) + U^{-1}(\partial\mathcal{P}) - 2] \\ &\sim \beta \sum_{x,\mu\nu} \frac{a^4}{4N} \operatorname{Tr}[\mathcal{F}([x;\mu,\nu])\mathcal{F}([x;\mu,\nu])] \\ &\sim \frac{\beta g_0^2}{4N}\frac{1}{2} \int \mathrm{d}^4x\, F^b_{\mu\nu}(x)F^{b,\mu\nu}(x) \\ &= \frac{1}{4}\int \mathrm{d}^4x\, F^b_{\mu\nu}(x)F^{b,\mu\nu}(x), \qquad \text{for} \quad \beta = \frac{2N}{g_0^2}, \end{aligned} \tag{2.10.12}$$

i.e. the continuum action. For abelian gauge groups $N = 1$ we have $\beta = 1/g_0^2$.

2.10.1.3 Path integral on the lattice

The quantum dynamics on a lattice can be derived from the path integral. It can be formulated with help of the action defined above. If $\Omega[U]$ is an observable which is given as a gauge-invariant functional of the lattice gauge fields, then its vacuum expectation value can be calculated from the *general euclidean path-integral formula* (cf. Sect. 2.2.4.4),

$$\langle\Omega[U]\rangle = \frac{1}{Z}\int \mathcal{D}[U]\,\Omega[U]\,\mathrm{e}^{-S_{\mathrm{W}}[U]}, \quad Z = \int \mathcal{D}[U]\,\mathrm{e}^{-S_{\mathrm{W}}[U]}. \tag{2.10.13}$$

The measure $\mathcal{D}[U]$ must be defined in the space of gauge fields. It can be given as

$$\mathcal{D}[U] = \prod_{b\in\Gamma} \mathrm{d}\mu(U(b)), \tag{2.10.14}$$

in which $\mathrm{d}\mu(g)$ is the Haar measure (1.2.64)–(1.2.65) for the symmetry group $G = \mathrm{SU}(N)$. The invariance of the Haar measure implies that a gauge-invariant quantum theory is described by the basic equation (2.10.13). The gauge invariance of the expectation values follows from the gauge invariance of $\Omega[U]$. Gauge-fixing terms as in (2.4.17) are not necessary for the integration with the finite Haar measure.

2.10.1.4 Lattice analogue of physical quantities

In Sects. 2.1 and 2.2 we have discussed in general how to derive in the continuum case dynamically relevant quantities like masses and cross sections from the vacuum expectation values of the corresponding operators. This can be translated in an appropriate way to the lattice. As examples, we indicate this for the determination of the Wilson loop and the lattice masses. The relation to the continuum quantities is discussed in Sect. 2.10.4. We report about the applications in QCD in Sect. 3.6.

In Sect. 3.7.1 we show that a gauge-invariant definition of a static potential can be derived from the *Wilson loop* [Wi74, We71]

$$W(C) = \langle \mathrm{Tr}\,[U(C)] \rangle = \frac{1}{Z} \int \mathcal{D}[\mathbf{A}]\, \mathrm{Tr}[U(C)] \exp(-S_{\mathrm{W}}\{\mathbf{A}\})$$

$$U(C) = \mathrm{P} \exp\left(\mathrm{i}g \int_C \mathrm{d}s^\mu A^a_\mu(s) \frac{\lambda^a}{2} \right). \qquad (2.10.15)$$

This expression can be treated in lattice approximation. To this end, we choose a rectangular path on the lattice which can be approximated by a sequence of lattice links

$$U[C] \mapsto U[C_{\mathrm{L}}] = \prod_{b_i \in C_{\mathrm{L}}} U(b_i). \qquad (2.10.16)$$

Then, the lattice approximation for the Wilson loop $W(C)$ reads

$$W(C) \mapsto W(C_{\mathrm{L}}) = \frac{1}{Z} \int \mathcal{D}[U]\ \mathrm{Tr}(U[C_{\mathrm{L}}])\, \mathrm{e}^{-S_{\mathrm{W}}[U]},$$

$$Z = \int \mathcal{D}[U]\, \mathrm{e}^{-S_{\mathrm{W}}[U]}. \qquad (2.10.17)$$

Poles of Green functions in an invariant mass variable P^2 at $P^2 = m^2$ point to the existence of free particles with mass m (Sect. 2.1.5). In coordinate space, these poles correspond to an exponential fall-off in space-like directions, and also in the euclidean time-like direction [cf. (2.1.16)]. This sets the frame for the calculations of masses in lattice approximations. Let us take the simplest case of a two-point function $\langle F(x)F(y) \rangle$ of a gauge-invariant field $F(x)$ exciting the vacuum to a one particle state. We assume $\langle F(x) \rangle = 0$ and $\langle 0|F(x)|M, p, j, j_3 \rangle \neq 0$. We consider the lattice approximation $F(x) \mapsto \Omega(\mathbf{x}, t)$. For a large lattice extent in the euclidean time direction, we expect the behaviour

$$\langle \Omega(\mathbf{0}, 0)\Omega(\mathbf{0}, t) \rangle \underset{t\to\infty}{\sim} A\mathrm{e}^{-ma\,|t|} + \dots, \qquad (2.10.18)$$

where the euclidean time t is measured in lattice units. Because of the finite-size effects, t should be large but much smaller than N, the length of the periodic lattice. The quantity $\xi = ma$ is called the *correlation length*. Measuring the correlation length of the two-point function determines the mass in lattice units. It is the main problem of the transition to the continuum to determine the lattice constant and the limit $a \to 0$.

This is only a first example of the translation of the structure of general field theory (Sect. 2.1) into the lattice approximation and the extraction of physical quantities.

2.10.2 Strong-coupling approximation

Next we discuss the strong-coupling approximation $g_0^2 \to \infty$, i.e. the limit $\beta \ll 1$ of the Wilson action (2.10.10) and the vacuum expectation value (2.10.13). According to the notion of Sect. 2.2.4.4 this may be called a high temperature expansion. This method played an important rôle for getting insight into the essential features of lattice field theory. However, as the continuum limit requires $\beta \to \infty$ (cf. Sect. 2.10.4.3), high temperature expansion is of limited value for physical problems.

2.10.2.1 A simple model

In lattice approximation it is possible to have gauge theories with finite gauge groups. We consider the Z_2 gauge theory on a two-dimensional lattice with N^2 points. The gauge group is the finite group with two elements $Z_2 = \{\sigma | \sigma = \pm 1\}$, $\sigma^2 = 1$. Lattice gauge fields $U(b) \equiv \sigma(b)$ are functions on links $b \in \Gamma^2$, which have values ± 1. In view of the commutativity of group multiplication, the field strength (2.10.4), $\sigma(\partial\mathcal{P}) = \sigma(b_4)\sigma(b_3)\sigma(b_2)\sigma(b_1)$, is a pure plaquette function. The following definitions correspond to (2.10.10), (2.10.13), (2.10.17), and (2.10.14):

- Action: $S_{\mathrm{W}} = \beta \sum_{\mathcal{P}} (1 - \sigma(\partial\mathcal{P}))$
- Partition sum: $Z = \int \mathcal{D}[\sigma] \exp(-S_{\mathrm{W}}[\sigma])$
- Wilson loop: $W(\mathcal{C}) = \frac{1}{Z} \int \mathcal{D}[\sigma]\, \sigma(\mathcal{C}) \exp(-S_{\mathrm{W}}[\sigma])$
- Haar measure: $\int \mathrm{d}\mu(\sigma) f(\sigma) = \frac{1}{2} \sum_{\sigma=\pm 1} f(\sigma)$
  with $\int \mathrm{d}\mu(\sigma) = 1$, $\int \mathrm{d}\mu(\sigma)\sigma = 0$.

The integrals for the partition function Z and the Wilson loop $W(\mathcal{C})$ for small β, i.e. the high-temperature expansion, are best evaluated by the following steps. First, the Boltzmann factor for an individual plaquette is calculated,

$$e^{-\beta(1-\sigma)} = e^{-\beta}\cosh\beta + \sigma e^{-\beta}\sinh\beta = C(1+\sigma v) \qquad (2.10.19)$$

with $C = \exp(-\beta)\cosh\beta$ and $v = \tanh\beta$. This result is an example of a Fourier representation of the Boltzmann factor. The first term C is the contribution of the identity representation of the Z_2 group, the second $\sigma C v$ that of the defining representation.

Since $v \approx \beta$ for $\beta \ll 1$, a power series in v can be considered instead of one in β. The expansion of the Boltzmann factor reads

$$\begin{aligned}\exp\Big(-\beta\sum_{\mathcal{P}\in\Gamma^2}\Big[1-\sigma(\partial\mathcal{P})\Big]\Big) &= \prod_{\mathcal{P}\in\Gamma^2} C\Big[1+v\sigma(\partial\mathcal{P})\Big] \qquad (2.10.20)\\ &= C^{N^2}\prod_{\mathcal{P}\in\Gamma^2}\Big[1+v\sigma(\partial\mathcal{P})\Big] = C^{N^2}\sum_{n=0}^{N^2} v^n\Big(\sum_{G^{(n)}}\prod_{\mathcal{P}\in G^{(n)}}\sigma(\partial\mathcal{P})\Big).\end{aligned}$$

After evaluation of the product $\prod_{\mathcal{P}\in G^{(n)}}$, the coefficient of v^n emerges as a sum of the contributions of all possible lattice graphs $G^{(n)}$. A *lattice graph* $G^{(n)}$ is a set of n non-oriented plaquettes. Its contribution to the Boltzmann factor is the product of all $\sigma(\partial\mathcal{P})$ of these plaquettes. The high-temperature (strong-coupling) expansion for the partition sum is obtained by inserting (2.10.20) in its expression above,

$$Z = C^{N^2}\sum_n v^n\left(\sum_{G^{(n)}} Z(G^{(n)})\right), \qquad (2.10.21)$$

in which $Z(G^{(n)})$ is the contribution of a lattice graph to the partition sum,

$$Z(G^{(n)}) = \int \mathcal{D}[\sigma]\prod_{\mathcal{P}\in G^{(n)}}\sigma(\partial\mathcal{P}), \qquad \mathcal{D}[\sigma] = \prod_{\mathcal{P}\in\Gamma^2} d\mu(b). \qquad (2.10.22)$$

The properties of the Haar measure imply that $Z(G^{(n)})$ is only different from zero if $\sigma(b)$ in $\prod_{\mathcal{P}\in G^{(n)}}\sigma(\partial\mathcal{P})$ appears in even powers. In view of $\sigma(\partial\mathcal{P}) =$

$\prod_{b\in\partial\mathcal{P}}\sigma(b)$, this means that each link must be the boundary of at least two plaquettes of $G^{(n)}$, i.e. $Z(G^{(n)}) \neq 0$ only when the boundary $\partial G^{(n)}$ of $G^{(n)}$ vanishes. In a two-dimensional finite lattice this only applies for $G^{(0)}$, i.e. the empty lattice graph, and hence (2.10.21) implies $Z = C^{N^2}$. Taking a periodic lattice in which opposite links of Γ^2 are identified, then $\partial\Gamma$ is also equal to 0, and one gets the result

$$Z = C^{N^2}(1+v^{N^2}) \approx C^{N^2} \qquad \text{for} \quad N^2 \gg 1, \quad v < 1. \tag{2.10.23}$$

This shows that in this limit there is no difference between a lattice with no boundary condition and one with periodic boundary condition. The influence of the boundary disappears for large N.

The corresponding strong-coupling approximation for the Wilson integral is obtained by substituting the Boltzmann factor (2.10.20) into the expression of the Wilson loop,

$$W(\mathcal{C}) = \sum_n v^n \left(\sum_{G^{(n)}} W(\mathcal{C}; G^{(n)}) \right), \qquad N \gg 1, \tag{2.10.24}$$

with

$$W(\mathcal{C}; G^{(n)}) = \int \mathcal{D}[\sigma]\, \sigma(\mathcal{C}) \prod_{\mathcal{P}\in G^{(n)}} \sigma(\partial\mathcal{P}). \tag{2.10.25}$$

The conclusion is similar to the one from the discussion of the contributions to the partition sum, i.e. only those $G^{(n)}$ contribute in which $\partial G^{(n)} = \mathcal{C}$ applies. In two dimensions and with no periodic boundary conditions, this is a single graph consisting of $\mathrm{Ar}[\mathcal{C}]$ plaquettes, where $\mathrm{Ar}[\mathcal{C}]$ is the number of plaquettes which are enclosed by $\mathcal{C}$, i.e. the area enclosed by $\mathcal{C}$. The result obtained is

$$W(\mathcal{C}) = v^{\mathrm{Ar}[\mathcal{C}]} = e^{\mathrm{Ar}[\mathcal{C}]\log v}. \tag{2.10.26}$$

Thus, the area law of the Wilson confinement condition (cf. Sect. 3.7.1.1) is satisfied with a string constant $\kappa = -\log v$ in lattice-constant units for a two-dimensional Z_2 lattice gauge theory.

2.10.2.2 Generalities of the strong-coupling approximation of SU(N)

This example of a Z_2 gauge theory shows the important steps for the general strong-coupling approximation:

- group-theoretical expansion of the Boltzmann factor for the individual plaquette, i.e. its *character expansion* (2.10.20);
- expansion of the total Boltzmann factor into lattice-graph contributions (2.10.21);
- determination and calculation of graphs which give a non-vanishing contribution;
- graph summation.

The expansion of the Boltzmann factor (2.10.19) is a special case of an expansion in group characters $\chi^{(k)}(g) = \operatorname{Tr} D^{(k)}(g)$ of the irreducible, unitary representations (k) of dimension $\dim(k)$ of the gauge group G. For this expansion the orthogonality and completeness relations of the group characters are important. We write the action as a sum over the contributions of the plaquettes $S[U] = \beta \sum_{\mathcal{P}} S_{\mathcal{P}}[U]$. Then, the Boltzmann factor of a single plaquette is represented by the sum

$$\mathrm{e}^{-\beta S_{\mathcal{P}}[U]} = \sum_{(k)} \dim(k) \lambda_k(\beta) \chi^{(k)}(U) \tag{2.10.27}$$

with

$$\lambda_k(\beta) = \frac{1}{\dim(k)} \int \mathcal{D}[U]\, \mathrm{e}^{-\beta S_{\mathcal{P}}[U]} \chi^{(k)*}(U). \tag{2.10.28}$$

The expansion of the contributions of all plaquettes in effective couplings $v(k)$ leads to lattice graphs as in the simple example of the Z_2 model.

Since in the modern treatment of lattice gauge theories the strong-coupling approximation no longer plays a dominant rôle, we only give some short hints for QCD. For details we refer to the literature [Ba75, Dr78].

For an SU(3) group, the effective coupling constant $v(3)$ associated with the fundamental three-dimensional representation (3) is approximately $v(3) \sim$

β. Therefore, it is often used as a general expansion parameter for the strong-coupling series. As in (2.10.20), the expansion in lattice graphs is carried out with the Boltzmann factor $\exp(-S_W)$ written as a product. The fact that the strong-coupling approximation of the SU(3) gauge theory on a four-dimensional lattice is more complex than the above example is due to the greater number of SU(3) representations with associated coupling constants $v(k)$, and to the geometric diversity of simple lattice graphs. It follows from the orthogonality relations that $Z\{G^{(n)}; k(\mathcal{P}_i)\} \neq 0$ only if the boundary $\partial G^{(n)}$ vanishes, i.e. $G^{(n)}$ is closed. Two plaquettes of $G^{(n)}$ are called connected if they have a common link. A general graph can be decomposed into connected components. Connected graphs $G^{(n)}$ without boundary form closed surfaces which may have branch lines.

The simplest example is a graph in the form of an elementary cube $\mathcal{K}$ in a four dimensional lattice with the fundamental representation (3). It has 12 links and 6 plaquettes. Its contribution $Z\{\mathcal{K};(3)\}$ to the partition sum can be evaluated according to the general method described above. The corresponding geometrical notations are given in Fig. 2.24d. The result is

$$\begin{aligned} Z\{\mathcal{K};(3)\} &= v^6[\dim(3)]^6 \int \mathrm{d}\mu(b_1)\cdots \mathrm{d}\mu(b_{12})\, \chi^{(3)}(\partial\mathcal{P}_1)\cdots\chi^{(3)}(\partial\mathcal{P}_6) \\ &= v^6[\dim(3)]^2 = 9v^6. \end{aligned} \tag{2.10.29}$$

If the contribution of all graphs of a certain order v^n is considered in a finite, four-dimensional lattice Γ_N^4 with periodic boundary conditions, the summation must be taken with respect to graphs in the various positions which follow from each other through N^4 translations. For example, Z is given in lowest order by the cubic graph described by (2.10.29). Multiplicity by translation and four-fold orientation of the three-dimensional cube in four-dimensional space results in

$$Z = (1 + 36N^4v^6 + \ldots)c. \tag{2.10.30}$$

Here c is an irrelevant constant, corresponding to C^{N^2} in (2.10.21), which drops out in the calculation of expectation values.

Already the N^4 dependence of the lowest-order term shows that the convergence of the series for Z, i.e. (2.10.30), and the transition to the infinite lattice $N \to \infty$, the *thermodynamic limit*, causes non-trivial problems.

These can be solved by considering the strong-coupling expansion for the free energy

$$F = -\frac{1}{\beta N^4} \log Z \tag{2.10.31}$$

instead of the one for the partition sum Z. In this case, the thermodynamic limit $N \to \infty$ exists in any order, and the resulting power series in β converges for sufficiently small β. The transition from Z to F leads to a representation of the strong-coupling approximation for F by means of *graph clusters*. If the local variables depend only on gauge fields $U(b)$ on a finite number of links, a theorem guarantees the existence of the thermodynamic limit and the convergence of the strong-coupling approximation for the expectation values of local observables $\langle \Omega[U] \rangle$. Roughly speaking, the N dependence in the numerator and the denominator in the path-integral representation of $\langle \Omega[U] \rangle$ is compensated in the thermodynamic limit.

2.10.3 Numerical methods

The calculation of the euclidean expectation value of $\Omega[U]$,

$$\langle \Omega[U] \rangle = \frac{1}{Z} \int \mathcal{D}[U]\, \Omega[U]\, \mathrm{e}^{-S_{\mathrm{E}}[U]}, \qquad Z = \int \mathcal{D}[U]\, \mathrm{e}^{-S_{\mathrm{E}}[U]}, \tag{2.10.32}$$

on a finite lattice Γ permits the application of numerical methods developed in the context of statistical mechanics. By now, these methods have reached a degree of refinement that is far beyond the scope of this book. We give only a short introduction into the main concepts. For more details we have to refer to the literature [Cr83, Mo94, Fr00]. Some of our description we owe to the book by I. Montvay and G. Münster [Mo94].

2.10.3.1 Monte Carlo integration

In order to work out the similarities of (2.10.32) with statistical mechanics, we write $S_{\mathrm{E}}[U] = \beta E[U] > 0$. Then, $\beta = 2N/g_0^2$ corresponds to the inverse temperature $1/kT$, $E[U]$ is the energy, and (2.10.32) describes the expectation value of an observable in the canonical ensemble. The integration with respect to the measure $\mathcal{D}[U]$ involves a large number of variables. For instance, on a 10^4 lattice with the gauge group SU(3) we have typically

$8\times4\times10^4$-dimensional integrations. In this case, the only general integration procedure is a *Monte Carlo integration.*

Monte Carlo integration in the simplest case consists in calculating the statistical average of randomly generated gauge-field configurations $[U]$. However, this is very inefficient, because the Boltzmann factor $\exp(-S_{\mathrm{E}}[U])$ is sharply peaked at some specific configurations. Therefore an *importance sampling* with the weight $W[U] \propto \exp(-S_{\mathrm{E}}[U])$ of the configurations $[U]$ is needed. We consider an ensemble, i.e. an infinite number of field configurations $[U]$ with density $W[U]$, defined with respect to the measure $\mathcal{D}[U]$. For our purpose $W[U]$ should be proportional to the density of the canonical ensemble,

$$W[U] \propto W_{\mathrm{c}}[U] = \frac{\mathrm{e}^{-S_{\mathrm{E}}[U]}}{Z}. \qquad (2.10.33)$$

2.10.3.2 Numerical simulation and Markov processes

The task in numerical simulation is therefore to generate samples consisting of a large number N of configurations $[U_n]$, $n = 1, \ldots, N$, in such a way that its distribution approximates the desired distribution of the *canonical ensemble* W_{c}. Then, the sample average of a quantity $\Omega[U]$,

$$\bar{\Omega}_N[U] \equiv \frac{1}{N}\sum_{n=1}^{N}\Omega[U_n], \qquad (2.10.34)$$

approximates the canonical ensemble average which according to (2.10.32) is equal to the euclidean expectation value $\langle\Omega[U]\rangle$,

$$\lim_{N\to\infty}\bar{\Omega}_N[U] = \langle\Omega[U]\rangle. \qquad (2.10.35)$$

The difference between $\bar{\Omega}_N[U]$ and $\langle\Omega[U]\rangle$ is given by an error estimate (cf. Sect. 2.10.3.3).

The configuration samples are generated in numerical simulations as a sequence $[U_n]$, $n = 1, \ldots, N$. This sequence is prepared by *updating*, i.e. repeated application of an algorithm which creates from a configuration the next one,

$$[U_n] \to [U_{n+1}]. \qquad (2.10.36)$$

Updating might be considered as a *stochastic process* $[U] \to [U']$ with transition probability

$$P([U'] \leftarrow [U]), \qquad \int \mathcal{D}[U'] \, P([U'] \leftarrow [U]) = 1. \tag{2.10.37}$$

An *updating step* changes the ensemble densities $W[U]$ as

$$W'[U'] = \int \mathcal{D}[U] \, P([U'] \leftarrow [U]) \, W[U], \tag{2.10.38}$$

or in short matrix notation $W' = PW$. The label n of the sequence of updatings, i.e. of the configurations, is sometimes called (computer) time.

The following conditions on P and W are requested:

- Every configuration can be reached with finite probability from any other, i.e. strong *ergodicity* holds,

$$P([U'] \leftarrow [U]) > 0, \qquad \text{for all } U \text{ and } U', \tag{2.10.39}$$

- The ensemble density $W[U]$ is normalized,

$$\int \mathcal{D}[U] \, W[U] = 1. \tag{2.10.40}$$

Equations (2.10.37), (2.10.39), and (2.10.40) define a *Markov process*, the sequence of configurations is a *Markov chain.*

An important condition for the updating process is that for any initial distribution W_0 (having an overlap with W_c) the repeated application of updating steps P leads to the canonical ensemble,

$$\lim_{k \to \infty} P^k W_0 = W_c. \tag{2.10.41}$$

This implies that W_c is an eigenvector of P with eigenvalue 1,

$$P W_c = W_c, \tag{2.10.42}$$

i.e. a fixed point. We require that this fixed point is unique.

A sufficient condition for (2.10.42) is *detailed balance*,

$$P([U'] \leftarrow [U]) \, W_c[U] = P([U] \leftarrow [U']) \, W_c[U']. \tag{2.10.43}$$

Together with (2.10.37) this implies

$$\int \mathcal{D}[U]\, P([U'] \leftarrow [U])\, W_{\rm c}[U] = \int \mathcal{D}[U]\, P([U] \leftarrow [U'])\, W_{\rm c}[U'] = W_{\rm c}[U'], \qquad (2.10.44)$$

i.e. (2.10.42). According to theorems on positive matrices [Bh88], the eigenvalue 1 of the matrix P is non-degenerate and all other eigenvalues are smaller than 1. Therefore, the repeated updating process damps all other components and converges to the canonical ensemble, and (2.10.41) holds.

Examples of updating procedures satisfying detailed balance are the *Metropolis algorithm* and the *heat-bath algorithm* discussed below.

2.10.3.3 Error estimates

The sample average $\bar{\Omega}_N[U]$ converges for large samples ($N \to \infty$) to the expectation value $\langle \Omega[U] \rangle$. A reliable result requests that N is large compared to the degrees of freedom of the system. In order that a finite sample average $\bar{\Omega}_N[U]$ with not unreasonably large N is a good estimator of $\langle \Omega[U] \rangle$ we must request smoothness properties[38] of $\Omega[U]$. For reasonable $\Omega[U]$ one can estimate the error. In the ideal case, when the configurations U_n contained in a sample are all statistically independent, the sample average is normally distributed around the mean value. We get in the standard manner the error estimate by the *variance* σ_Ω,

$$\langle \Omega[U] \rangle = \bar{\Omega}_N \pm \sigma_{\Omega,N}, \qquad \sigma^2_{\Omega,N} = \frac{\overline{\Omega^2}_N - (\bar{\Omega}_N)^2}{N-1}. \qquad (2.10.45)$$

However, the assumption that the configurations $[U_n]$ generated by the usual updating process are independent is unrealistic. Quite often one step in the updating consists only of a single local change of the configuration or of a sweep of local changes over the whole lattice. There are *autocorrelations.*

For a (computer-)time-independent updating procedure and infinitely large samples, the autocorrelations depend only on the time difference τ. In this case, the true variance of $\bar{\Omega}_N$ is for large N given by

$$\sigma^2_{\Omega,N} \approx \left(\overline{\Omega^2}_N - (\bar{\Omega}_N)^2 \right) \frac{2\tau_{{\rm int},\Omega}}{N} \qquad (2.10.46)$$

[38] An example for a non-smooth $\Omega[U]$ is one where $\Omega[U] \neq 0$ only for a single plaquette.

with the effective *autocorrelation time*

$$\tau_{\text{int},\Omega} = \frac{1}{2} \sum_{\tau=-\infty}^{+\infty} \frac{\langle (\Omega_n - \bar{\Omega}_N)(\Omega_{n+\tau} - \bar{\Omega}_N) \rangle}{\overline{\Omega^2}_N - (\bar{\Omega}_N)^2}, \tag{2.10.47}$$

where $\Omega_n = \Omega[U_n]$. Comparing (2.10.45) with (2.10.46) one can see that owing to autocorrelations the effective number of independent measurements is $N/(2\tau_{\text{int},\Omega})$.

One can get a numerical estimate of $\tau_{\text{int},\Omega}$ by *binning*. This means one divides large ensembles in sub-ensembles which can be considered statistically independent and evaluates the results in the sub-ensembles.

2.10.3.4 Metropolis algorithm

We end this short review of numerical methods by describing the main features of two of the most popular updating procedures, i.e. the Metropolis algorithm and the heat-bath algorithm. The transition probability matrix for the *Metropolis algorithm* for a system with $\mathcal{N}$ discrete possible configurations is defined for $[U] \neq [U']$ by

$$\begin{aligned} P([U'] \leftarrow [U]) &= \frac{1}{\mathcal{N}} \min\left\{1, \frac{W_\mathrm{c}[U']}{W_\mathrm{c}[U]}\right\} \\ &= \frac{1}{\mathcal{N}} \min\left\{1, \exp(S_\mathrm{E}[U] - S_\mathrm{E}[U'])\right\}. \end{aligned} \tag{2.10.48}$$

This transition matrix is realized by the following iterative procedure:

1. choose a trial configuration from the $\mathcal{N}$ configurations;
2. accept it as the next configuration if the action decreases;
3. if the action increases, then accept the change with a probability of the ratio of the Boltzmann factors.

The *accept–reject step* can be implemented in a computer program by comparing the ratio of the Boltzmann factors to a pseudo-random number between 0 and 1. The detailed balance condition (2.10.43) can be verified immediately. In the form above, strong ergodicity (2.10.39) is also fulfilled.

In practice it is not convenient to consider all possible changes of a configuration simultaneously. Often one considers only local changes of a single link variable. In order to have ergodicity one has to *sweep* through all links of the lattice. The transition probability P can be generated as a two-step process $P = P_A P_C$ consisting of a proposed change and an accept–reject step, where

- $P_C([U'] \leftarrow [U])$ is an arbitrary probability distribution for the proposed configuration change $[U] \to [U']$, and
- the acceptance probability $P_A([U'] \leftarrow [U])$ is defined as a generalization of (2.10.48) in such a way that the detailed balance condition (2.10.43) holds for $P = P_A P_C$, i.e.

$$P_A([U'] \leftarrow [U]) \propto \min\left\{1, \frac{P_C([U] \leftarrow [U'])\, W_{\mathrm{c}}[U']}{P_C([U'] \leftarrow [U])\, W_{\mathrm{c}}[U]}\right\}. \quad (2.10.49)$$

2.10.3.5 Heat-bath algorithm

The aim of an updating procedure is to create a canonical (equilibrium) distribution. In the heat-bath algorithm this goal is approached by local steps. In each step one link variable $U([x,\mu])$ is distributed canonically, while all other variables are kept fixed. In thermodynamical language, $U([x,\mu])$ is in contact with a heat bath provided by all the other variables. This elementary step has to sweep consecutively over all link variables of the lattice, and the whole process has to be repeated many times.

We want to put the heat-bath procedure into the general form of an updating process (Sect. 2.10.3.2). Let $U_x = U([x,\mu])$ be the link variable to be updated and $\check{U}_x$ all the other variables forming the heat bath. By $W_{\mathrm{c}}[U_x;\check{U}_x]$ we denote the conditional probability distribution of U_x in the canonical ensemble for fixed $\check{U}_x$. The whole canonical distribution can be written as

$$W_{\mathrm{c}}[U] = W_{\mathrm{c}}[U_x;\check{U}_x]\, \check{W}_{\mathrm{c}}[\check{U}_x]. \quad (2.10.50)$$

Similarly, the conditional transition probability for the updated variables is denoted by $P_x(U'_x \leftarrow U_x;\check{U}_x)$. The transition probability matrix for the whole system then is

$$P_x([U'] \leftarrow [U]) = P_x(U'_x \leftarrow U_x;\check{U}_x)\, \delta(\check{U}'_x - \check{U}_x). \quad (2.10.51)$$

It is assumed that $P_x([U'] \leftarrow [U])$ satisfies the conditions (2.10.37) and (2.10.42). A sufficient condition for (2.10.42) is *local detailed balance*,

$$P_x(U'_x \leftarrow U_x; \check{U}_x)\, W_{\mathrm{c}}[U_x; \check{U}_x] = P_x(U_x \leftarrow U'_x; \check{U}_x)\, W_{\mathrm{c}}[U'_x; \check{U}_x]. \tag{2.10.52}$$

The local updating step described by $P_x([U'] \leftarrow [U])$ is of course not ergodic because it acts only on U_x. One does require *local ergodicity*,

$$P_x(U'_x \leftarrow U_x; \check{U}_x) > 0 \quad \text{for all } U_x \text{ and } U'_x. \tag{2.10.53}$$

In order to achieve ergodicity for the whole configuration one can perform a cycle of local updates, called *sweep*, over all links consecutively,

$$P([U'] \leftarrow [U]) = \prod_x P_x([U'] \leftarrow [U]). \tag{2.10.54}$$

The local heat-bath algorithm corresponds to the conditional probability matrix $P_x(U'_x \leftarrow U_x; \check{U}_x) = W_{\mathrm{c}}(U_x; \check{U}_x)$.

The main task in a heat-bath updating is to generate the distribution $W_{\mathrm{c}}(U_x; \check{U}_x)$. For the SU(2) gauge group the simple properties of SU(2) matrices help. Here $U_x \in \mathrm{SU}(2)$ may be represented by the Pauli matrices (A.1.11), $U_x = a_{x0} + \mathrm{i}\boldsymbol{\tau}\mathbf{a}_x$, $U_x^{-1} = a_{x0} - \mathrm{i}\boldsymbol{\tau}\mathbf{a}_x$ with $a_{x0}^2 + \mathbf{a}_x^2 = 1$ and $\mathbf{a}_x = (a_{xi})$. A sum of SU(2) elements U_x is up to a normalization factor k proportional to an SU(2) element S_x. This allows us to write the Wilson action (2.10.10) as

$$S_{\mathrm{W}}[U] = -\frac{\beta}{2} \operatorname{Re} \operatorname{Tr}[U_x S_x] + \check{S}_{\mathrm{W}}[\check{U}], \tag{2.10.55}$$

where S_x represents the sum of the products of link variables other than U_x which are contained in the boundaries of all the plaquettes that have the link $[x, \mu]$ in common. Then, the conditional probability distribution of the link variable U_x can be written as

$$\begin{aligned} W_{\mathrm{c}}(U_x; \check{U}_x)\, \mathrm{d}U_x &\propto \exp\left[\frac{\beta}{2} \operatorname{Re} \operatorname{Tr}(U_x S_x)\right] \mathrm{d}U_x \\ &= \exp\left[\beta \sum_{i=0}^{3} a_{xi} s_{xi}\right] \mathrm{d}U_x, \end{aligned} \tag{2.10.56}$$

where s_{xi} are the components of $S_x = s_{x0} - \mathrm{i}\boldsymbol{\tau}\mathbf{s}_x$.

The heat-bath algorithms based on (2.10.56) may be further simplified by using the invariance of the Haar measure. Here we present only the principle of this algorithm, for details we have to refer to the literature [Cr79, Cr80]. By using sufficient SU(2) subgroups of SU(N), one can generalize the SU(2) heat-bath formalism to a heat-bath algorithm for SU(N) [Ca82].

The canonical distribution hides often conjectured long-range structures like instantons (Sect. 2.9.1.2) under short-range fluctuations. It is a widespread practice to suppress the short-range fluctuations by *cooling*. This means one puts between the different updating sweeps cooling sweeps. These are heat-bath sweeps with very high β corresponding to low temperature [Be81, Ho87].

2.10.4 Transition to the continuum

The lattice approximation of a field theory corresponds to a regularization by a momentum cut-off, as described in the discussions on divergent Feynman graphs in Sect. 2.5.1.1. Before discussing the *continuum limit* within the renormalization program, we illustrate this aspect of lattice approximation using the Hamiltonian of a free, scalar field on a infinite lattice as an example.

2.10.4.1 Lattice approximation of the Hamiltonian of a free field

The Hamilton operator of a free scalar field $A(x)$ is

$$H = \frac{1}{2}\int \mathrm{d}^3x \left[\pi^2(\mathbf{x},t) + (\nabla A(\mathbf{x},t))^2 + m^2 A^2(\mathbf{x},t)\right]. \tag{2.10.57}$$

The commutation relations for the canonically conjugate quantities $A(\mathbf{x},t)$ and $\pi(\mathbf{x},t) = \partial A(\mathbf{x},t)/\partial t$ read

$$\begin{aligned}\left[\pi(\mathbf{x},t), A(\mathbf{x}',t)\right] &= -\mathrm{i}\delta(\mathbf{x}-\mathbf{x}'),\\ \left[A(\mathbf{x},t), A(\mathbf{x}',t)\right] &= \left[\pi(\mathbf{x},t), \pi(\mathbf{x}',t)\right] = 0.\end{aligned} \tag{2.10.58}$$

We approximate these equations on a three-dimensional lattice, putting $\mathbf{x} = \varepsilon\mathbf{n}$, $\mathbf{n} = (n_1, n_2, n_3)$ with lattice constant ε. Using

$$\begin{aligned} A(\mathbf{n}) &= \varepsilon A(\mathbf{x}, t)|_{\mathbf{x}=\varepsilon\mathbf{n}}, \qquad \pi(\mathbf{n}) = \varepsilon^2 \pi(\mathbf{x}, t)|_{\mathbf{x}=\varepsilon\mathbf{n}}, \\ (\nabla_{\mathrm{L}}^1 A)(\mathbf{n}) &= A(n_1+1, n_2, n_3) - A(n_1, n_2, n_3), \qquad \text{etc.}, \end{aligned} \tag{2.10.59}$$

the equations (2.10.57) and (2.10.58) can be translated to

$$H = \frac{1}{2\varepsilon} \sum_{\mathbf{n}} \left[\pi^2(\mathbf{n}) + (\nabla_{\mathrm{L}} A(\mathbf{n}))^2 + \mu^2 A^2(\mathbf{n})\right], \qquad \mu = \varepsilon m,$$

$$\left[\pi(\mathbf{n}), A(\mathbf{n}')\right] = -\mathrm{i}\delta_{\mathbf{n}\mathbf{n}'}. \tag{2.10.60}$$

As in the continuum, the free Hamilton operator on the lattice is diagonalized using the *lattice Fourier transformation*

$$\begin{aligned} A(\mathbf{n}) &= (2\pi)^{-3/2} \int_{-\pi}^{\pi}\int_{-\pi}^{\pi}\int_{-\pi}^{\pi} \tilde{A}(\boldsymbol{\beta})\, \mathrm{e}^{+\mathrm{i}\mathbf{n}\boldsymbol{\beta}}\, \mathrm{d}^3\beta, \\ \pi(\mathbf{n}) &= (2\pi)^{-3/2} \int_{-\pi}^{\pi}\int_{-\pi}^{\pi}\int_{-\pi}^{\pi} \tilde{\pi}(\boldsymbol{\beta})\, \mathrm{e}^{-\mathrm{i}\mathbf{n}\boldsymbol{\beta}}\, \mathrm{d}^3\beta, \\ \tilde{A}^\dagger(\boldsymbol{\beta}) &= \tilde{A}(-\boldsymbol{\beta}), \qquad \tilde{\pi}^\dagger(\boldsymbol{\beta}) = \tilde{\pi}(-\boldsymbol{\beta}), \end{aligned} \tag{2.10.61}$$

and introducing *creation and annihilation operators*

$$\begin{aligned} a(\boldsymbol{\beta}) &= \sqrt{\frac{\omega(\boldsymbol{\beta})}{2}} \left(\tilde{A}(\boldsymbol{\beta}) + \frac{\mathrm{i}}{\omega(\boldsymbol{\beta})}\tilde{\pi}(-\boldsymbol{\beta})\right), \\ a^\dagger(\boldsymbol{\beta}) &= \sqrt{\frac{\omega(\boldsymbol{\beta})}{2}} \left(\tilde{A}(-\boldsymbol{\beta}) - \frac{\mathrm{i}}{\omega(\boldsymbol{\beta})}\tilde{\pi}(\boldsymbol{\beta})\right). \end{aligned} \tag{2.10.62}$$

A direct calculation, in which the essential steps are based on the orthogonality and completeness relations

$$\begin{aligned} &\frac{1}{2\pi}\int_{-\pi}^{\pi} \mathrm{d}\beta\, \mathrm{e}^{\mathrm{i}(n-m)\beta} = \delta_{nm}, \qquad n, m = 0, \pm 1, \pm 2, \dots, \\ &\frac{1}{2\pi}\sum_{n=-\infty}^{+\infty} \mathrm{e}^{\mathrm{i}n(\beta-\beta')} = \delta(\beta-\beta'), \qquad -\pi < \beta \le +\pi, \end{aligned} \tag{2.10.63}$$

shows that the canonical commutation relations (2.10.58) are equivalent to the commutation relations for the creation and annihilation operators,

$$[a(\boldsymbol{\beta}), a^\dagger(\boldsymbol{\beta}')] = \delta(\boldsymbol{\beta} - \boldsymbol{\beta}'), \qquad [a(\boldsymbol{\beta}), a(\boldsymbol{\beta}')] = [a^\dagger(\boldsymbol{\beta}), a^\dagger(\boldsymbol{\beta}')] = 0. \tag{2.10.64}$$

Moreover, using

$$\nabla_{\mathrm{L}}^k \, \mathrm{e}^{\mathrm{i}\mathbf{n}\boldsymbol{\beta}} = \mathrm{e}^{\mathrm{i}(n_k+1)\beta_k} - \mathrm{e}^{\mathrm{i}n_k\beta_k} = 2\mathrm{i}\, \mathrm{e}^{\mathrm{i}\mathbf{n}\boldsymbol{\beta}} \sin\left(\frac{\beta_k}{2}\right) \mathrm{e}^{\mathrm{i}\beta_k/2} \tag{2.10.65}$$

the form

$$H = \frac{1}{\varepsilon} \int_{|\beta_i| \le \pi} \mathrm{d}^3\beta \, \omega(\boldsymbol{\beta}) a^\dagger(\boldsymbol{\beta}) a(\boldsymbol{\beta}) + \text{constant} \tag{2.10.66}$$

with

$$\omega(\boldsymbol{\beta}) = \left[\mu^2 + 4\sum_{k=1}^{4} \sin^2\left(\frac{\beta_k}{2}\right)\right]^{1/2} \tag{2.10.67}$$

follows for the lattice Hamilton operator. Writing $\exp(\mathrm{i}\mathbf{n}\boldsymbol{\beta}) = \exp(\mathrm{i}\mathbf{p}\mathbf{x})$ in physical quantities $\mathbf{x} = \varepsilon\mathbf{n}$ and $\mathbf{p} = \boldsymbol{\beta}/\varepsilon$, the periodicity in $\boldsymbol{\beta}$, i.e. $|\beta_i| \le \pi$, corresponds to a maximal physical momentum on the lattice $|p_i| \le \pi/\varepsilon$. Using the physical quantities which approximate the continuum quantities on the lattice,

$$a(\mathbf{p}) = \varepsilon^{3/2} a(\boldsymbol{\beta}), \qquad \omega(\mathbf{p}) = \omega(\boldsymbol{\beta})/\varepsilon, \tag{2.10.68}$$

the Hamiltonian operator becomes

$$H = \int_{|p_i| \le \pi/\varepsilon} \mathrm{d}^3p \left[m^2 + \frac{4}{\varepsilon^2} \sum_{k=1}^{4} \sin^2\left(\frac{p_k\varepsilon}{2}\right)\right]^{1/2} a^\dagger(\mathbf{p}) a(\mathbf{p}). \tag{2.10.69}$$

This shows that also the energy is bounded. The lattice acts as cut-off to all quantities derived from creation and annihilation operators and the Hamiltonian operator, e.g. to the Feynman propagator as described in Sect. 2.1.2.1.

There is also a *lattice perturbation theory* [Re88] which is, however, more tedious than the perturbation theory in the continuum. Finally, (2.10.69) shows how the Hamilton operator of the continuum is approximated formally for decreasing lattice constant $\varepsilon \to 0$ and fixed m^2.

2.10.4.2 Critical points in statistical mechanics

For the general discussion of the continuum limit, the analogy between the lattice approximation of euclidean quantum field theory and statistical mechanics has turned out to be very fruitful. In Sect. 2.10.1.4 we have indicated that the correlation length ξ, which governs the asymptotic large-distance behaviour of the lattice approximation of the two-point function, is related to the lowest particle mass by $\xi = 1/ma$. Therefore, in order to extract a finite (renormalized) mass from lattice approximation, the correlation length must become infinite. In statistical mechanics, a point in the space of parameters where the correlation length diverges is called a *critical point* corresponding to a *second-order phase transition.* Thus, a continuum quantum field theory is obtained from lattice approximation at the critical point of the lattice coupling constants and lattice masses. Near critical points the lattice spacing can be neglected, and one expects a restoration of euclidean symmetry.

In statistical mechanics, various methods exist to find critical points and to discuss the corresponding physics. A general approach is the *Kadanoff–Wilson renormalization-group theory* [Ka65, Ka77, Wi71, Wi74a]. This theory embeds the cut-off dependence of the action in an infinite parameter space of actions. It explains why in the neighbourhood of critical points different statistical systems behave similarly. They fall into a relatively small number of universality classes. The renormalization-group approach which we discussed in field theory (Sect. 2.6) may be considered as a reduction of the infinite space of actions to a small number of relevant parameters, like the renormalized coupling constants and masses. We restrict our discussion of the continuum limit of lattice gauge theory to this more special frame.

2.10.4.3 Renormalization group on the lattice

In contrast to Sect. 2.6.1 we start with the renormalization-group equation for the bare vertex functions. We explain the main features by considering a scalar theory with a renormalized mass m_{R} and a renormalized coupling constant g_{R}, defined by the renormalized two-point vertex function, $m_{\mathrm{R}}^2 = -\Gamma_{\mathrm{R}}^2(p^2 = 0)$, and the renormalized four-point vertex function, $g_{\mathrm{R}} = -\Gamma_{\mathrm{R}}^4(0,0,0,0)$, respectively. The renormalized mass m_{R} just fixes the scale of all dimensionful quantities. In particular, the cut-off is characterized by $1/m_{\mathrm{R}}a$, a being the lattice constant. Fixed $g_{\mathrm{R}} = g_{\mathrm{R}}(g_0, m_{\mathrm{R}}a)$ describes a curve in the plane of the bare lattice coupling constant g_0 and the cut-off a, the curve of constant physics. One can show in k-loop order lattice perturbation theory that the renormalized vertex function obeys

$$\Gamma^n_{\mathrm{R}}(p_i; g_{\mathrm{R}}, m_{\mathrm{R}}, m_{\mathrm{R}}a) = \Gamma^n_{\mathrm{R}}(p_i; g_{\mathrm{R}}, m_{\mathrm{R}}, 0) + \mathcal{O}\left(a^2(\log a)^k\right) \quad (2.10.70)$$

or

$$a\frac{\mathrm{d}}{\mathrm{d}a}\Gamma^n_{\mathrm{R}}(p_i; g_{\mathrm{R}}, m_{\mathrm{R}}, m_{\mathrm{R}}a) = \mathcal{O}\left(a^2(\log a)^k\right). \quad (2.10.71)$$

Neglecting the scaling violation term $\mathcal{O}\left(a^2(\log a)^k\right)$, using the relation between the renormalized and unrenormalized vertex function,

$$\Gamma^n_{\mathrm{R}} = Z^{n/2}(g_0, m_{\mathrm{R}}a)\Gamma^n_0(g_0, m_{\mathrm{R}}, m_{\mathrm{R}}a), \quad (2.10.72)$$

we derive the renormalization-group equation for the bare coupling constant similar to Sect. 2.6.1.1 with the help of the chain rule of differentiation. The independence of Γ^n_{R} on a implies

$$\left(a\frac{\partial}{\partial a} - \beta_{\mathrm{L}}\frac{\partial}{\partial g_0} + n\gamma_{\mathrm{L}}\right)\Gamma^n_0 = 0, \quad (2.10.73)$$

where the lattice β-function β_{L} and the anomalous dimension γ_{L} are given by

$$\begin{aligned} \beta_{\mathrm{L}}(g_0, m_{\mathrm{R}}a) &= -\, a\frac{\partial g_0}{\partial a}\bigg|_{g_{\mathrm{R}}}, \\ \gamma_{\mathrm{L}}(g_0, m_{\mathrm{R}}a) &= +\,\frac{1}{2}a\frac{\partial \log Z}{\partial a}\bigg|_{g_{\mathrm{R}}}. \end{aligned} \quad (2.10.74)$$

The derivatives have to be taken for fixed g_{R}. Equation (2.10.73) describes the change of the bare vertex function under the change of g_0 while g_{R} is kept fixed. The equation is similar to the renormalization-group equation of Sect. 2.6.1. The discussion of the fixed points follows that of Sect. 2.6.1.3. Neglecting the scaling-violation terms, the functions β_{L} and γ_{L} depend only on g_0. The function $\beta_{\mathrm{L}}(g_0)$ determines the change of g_0 with the cut-off, i.e. the lattice constant a is related to a change of g_0 in order to stay on the curve of constant physics.

We discuss the *transition to the continuum* with help of the renormalization group for the bare coupling constant. More refined improvement methods are mentioned in Sect. 2.10.6.6. The derivation of the continuum limit of a pure SU(N) gauge theory follows similar lines. The renormalized coupling constant in the momentum subtraction scheme g_{MOM} is defined by the three-gluon vertex. Lattice perturbation theory in one-loop approximation gives the result

$$g_{\mathrm{MOM}}^2 = g_0^2 \left(1 - \frac{g_0^2}{16\pi^2}\left(\frac{11N}{3}\log(a^2\mathcal{M}^2) + C\right)\right)(1 + \mathcal{O}(a^2\mathcal{M}^2)) \tag{2.10.75}$$

with $C = 6\pi^2/(11N^2) - 9.44598$. The renormalization point $\mathcal{M}$ plays the rôle of m_{R} in the scalar theory. Equation (2.10.75) gives the first term of the expansion of β_{L} for $a \approx 0$,

$$\begin{aligned} \beta_{\mathrm{L}} &= -\beta_0 \frac{g_0^3}{16\pi^2} - \beta_1 \frac{g_0^5}{(16\pi^2)^2} + \ldots, \\ \beta_0 &= \frac{11}{3}N, \qquad \beta_1 = \frac{34}{3}N^2. \end{aligned} \tag{2.10.76}$$

These coefficients agree with the expansion of the β-function in the continuum (2.6.34)–(2.6.37) since $N = C_{\mathrm{A}}$ for SU(N). In Sect. 2.6.3 we have given arguments for the regularization-procedure independence of the first two coefficients of the β-function in asymptotically free theories.

Integrating the differential equation (2.10.74), results in

$$\begin{aligned} a &= \exp\left(-\int^{g_0} \frac{\mathrm{d}g'}{\beta_{\mathrm{L}}(g')}\right) \\ &= \Lambda_{\mathrm{L}}^{-1} \exp\left(-\frac{8\pi^2}{\beta_0 g_0^2}\right)\left(\frac{\beta_0 g_0^2}{16\pi^2}\right)^{-\beta_1/(2\beta_0^2)} (1 + \mathcal{O}(g_0^2)) \end{aligned} \tag{2.10.77}$$

or

$$\frac{16\pi^2}{g_0^2} = \beta_0 \log\frac{1}{a^2\Lambda_{\mathrm{L}}^2} + \frac{\beta_1}{\beta_0}\log\log\frac{1}{a^2\Lambda_{\mathrm{L}}^2} + \mathcal{O}\left(\frac{1}{\log a^2\Lambda_{\mathrm{L}}^2}\right). \tag{2.10.78}$$

This equation, which is equivalent to (3.1.16), describes how the lattice constant a has to be changed as a function of the lattice coupling constant

g_0 in order to stay on the curve of constant physics. The integration constant Λ_L with the dimension of a mass,

$$\Lambda_L = \lim_{g_0 \to 0} \frac{1}{a} \exp\left(-\frac{8\pi^2}{\beta_0 g_0^2}\right) \left(\frac{\beta_0 g_0^2}{16\pi^2}\right)^{-\beta_1/(2\beta_0^2)}, \tag{2.10.79}$$

can be related to similar constants of different regularization schemes [cf. (3.1.14)]. Comparing with the continuum definition of Λ_{MOM},

$$\Lambda_{\text{MOM}} = \lim_{g_{\text{MOM}} \to 0} \mathcal{M} \exp\left(-\frac{8\pi^2}{\beta_0 g_{\text{MOM}}^2}\right) \left(\frac{\beta_0 g_{\text{MOM}}^2}{16\pi^2}\right)^{-\beta_1/(2\beta_0^2)}, \tag{2.10.80}$$

we get, using (2.10.75),

$$\frac{\Lambda_{\text{MOM}}}{\Lambda_L} = \lim_{g_0 \to 0} a\mathcal{M} \exp\left\{\frac{8\pi^2}{\beta_0}\left(\frac{1}{g_0^2} - \frac{1}{g_{\text{MOM}}^2}\right)\right\} = \mathrm{e}^{-C/2}. \tag{2.10.81}$$

Thus, the numerical constant C in (2.10.75) determines the relation between the Λ parameters in one-loop approximation. For QCD, $N = 3$, the relation to $\Lambda_{\overline{\text{MS}}}$ is more interesting. In the case of zero flavours it is found as [Ha80, We81]

$$\Lambda_{\overline{\text{MS}}}/\Lambda_L = 28.809. \tag{2.10.82}$$

The Lagrangian of a pure gauge theory contains no parameter with dimension of a mass. Therefore, every physical quantity with the dimension of a mass must be proportional to Λ (*dimensional transmutation*): $m = C_m \Lambda_L$. In (2.10.18) we described how the mass of particles can be calculated in lattice units ma. If g_0 is sufficiently close to the continuum limit, i.e. g_0 is in the *asymptotic scaling* region, the leading term of (2.10.78) is a good approximation, and [cf. (2.10.79)]

$$ma = C_m \exp\left(-\frac{8\pi^2}{\beta_0 g_0^2}\right) \left(\frac{\beta_0 g_0^2}{16\pi^2}\right)^{-\beta_1/2\beta_0^2}. \tag{2.10.83}$$

Hence, the ratio of two different masses m_1 and m_2 is independent of g_0 in the asymptotic scaling region, and thus according to (2.10.77) independent of the lattice constant.

In order to calculate the continuum values of physical quantities from lattice quantities, it is necessary to reach the asymptotic scaling region for $g_0 \to 0$. On the other hand, the strong-coupling approximation is an expansion in $\beta = 2N/g_0^2$. So a high-order strong-coupling expansion allows for a reliable calculation of continuum quantities only for exceptional cases.

For numerical methods, the transition to the continuum causes some difficulties because of autocorrelations. The correlation length $\xi = 1/ma$ diverges near critical points. The autocorrelation time (2.10.47) depends on the only relevant length scale $\tau_{\text{int},\Omega} \propto \xi^{z(\Omega)}$. Here $z(\Omega)$ is the dynamical critical exponent, which is characteristic for the updating process and may also depend on the observed quantity Ω. In local updatings one has usually $z \simeq 2$. Obviously, for $z(\Omega) \neq 0$ we have a deterioration of the efficiency of the updating process near critical points. This is called *critical slowing down*.

Let us repeat the essential points of this subsection. Lattice approximation of field theory is a method to calculate physical continuum quantities. In the asymptotic scaling region, the ratio of mass values is independent of the lattice constant [up to terms of $\mathcal{O}\left((a\Lambda_{\text{L}})^{-2}\right)$]. More complicated Green functions with a possibly non-perturbative long-range behaviour may be calculated according to (2.10.72). In principle, the continuum limit always follows the line described here. In view of the various difficulties, large computers and refined methods, like improvement (cf. Sect. 2.10.6.6), are necessary for getting reliable results. We mention some results for QCD in Sect. 3.6.

2.10.5 Finite-size effects

Numerical simulations of quantum field theories on a lattice refer to the approximation of the fields in a finite volume. In order to get small deviations from the continuum limit, one has to choose a small lattice constant a, i.e. according to (2.10.78) a large $1/g_0^2$ and thus a large β. For a lattice with an extension of N_{s} points in space direction and N_{t} points in time direction, this gives a small physical three-space volume $V_{\text{s}} = L^3$ and a small time interval T,

$$V_{\text{s}} = a^3 N_{\text{s}}^3, \qquad T = aN_{\text{t}}. \tag{2.10.84}$$

However, generally one is interested in the large-volume limit $V_{\text{s}} \to \infty$, and in large times T [cf. (2.10.18)]. Since the computer power limits N_{s}

and N_t, one has to compromise between a reasonable approximation of the continuum and a sufficiently large space–time volume.

The optimal choice for N_s and N_t depends on the physical problem. The characteristic quantity is the correlation length ξ. We have seen how $\xi = ma$ determines the masses in lattice approximation. Similarly, it determines the spatial extension of a state. Good approximation of the continuum on the on hand and of the large-volume limit on the other hand requires $a < \xi < L$. We mention as an example the determination of a mass $m \approx 1\,\text{GeV}$, i.e. m_ρ or $m_\mathcal{N}$, in an $N_s^3 \times N_t = 48^3 \times 64$ lattice in Sect. 3.6.1.1. For much smaller masses, like the pion mass m_π, finite-size effects might become large.

M. Lüscher studied the behaviour of finite-size effects for asymptotically large L [Lu84]. For massive fields he finds an exponential fall off for the mass-correction term $\Delta m = m(L) - m$,

$$\Delta m \propto L^{-3/2} e^{-mL}, \tag{2.10.85}$$

where the details depend on the particular dynamics.

Spontaneous symmetry breaking and tunnelling near first-order phase transitions is another source of finite-size effects. The resulting energy splitting is due to tunnelling between vacuum states in finite volumes (Sect. 2.9.3.5). The analysis is based on a picture of domains in space and time as discussed in Sects. 2.9.3.4 and 2.9.3.5. From this picture a prediction for the energy splitting of the form

$$\Delta E \propto e^{-\sigma L^3} \tag{2.10.86}$$

is obtained, where σ is the interface tension associated with the domain walls [Fi69, Pr83, Mo94].

One can get a rough estimate on the errors caused by finite volumes by comparing the values of different numerical lattice approximations with the same lattice constant and different physical volumes. In this context we only mention that there is a method which uses controlled finite-size effects, *finite-size scaling*. It uses a finite-volume renormalization scheme for matching a sequence of lattices of different sizes. By this procedure one can study the evolution of physical quantities over large length scales [Lu99].

2.10.6 Lattice approximation of fermionic interactions

In order to treat QCD in lattice approximation one has to include quarks on the lattice. Surprisingly, the lattice approximation of fermions leads to conceptional and practical difficulties which get solved only slowly. We mention some of the problems in the following. This should explain why important problems of non-perturbative QCD, like the calculation of the hadron spectrum (Sect. 3.6.1), are still not completely solved.

2.10.6.1 Naïve lattice approximation of the Dirac equation

Let us first follow formally the scheme of lattice approximation for gauge fields explained above. We consider coloured Dirac fields on a four-dimensional lattice as spinors at a lattice point $n = (n_1, n_2, n_3, n_4)$ (cf. Sect. 2.10.1.1),

$$\psi_{\alpha,c}(n) = a^{3/2}\,\psi_{\alpha,c}(x)|_{x=an}, \qquad \bar{\psi}_{\alpha,c}(n) = a^{3/2}\,\bar{\psi}_{\alpha,c}(x)|_{x=an}, \tag{2.10.87}$$

where $\alpha = 1, \dots, 4$ is the spinor index, $c = \mathrm{r}, \mathrm{g}, \mathrm{b}$ the SU(3)-colour index, and we suppress the flavour index. The lattice spinors are made dimensionless (cf. Table 2.1 on page 195). According to the geometrical interpretation of gauge transformations (Fig. 2.4 on page 137), the forward and backward covariant derivatives of Dirac fields on the lattice (L) correspond to

$$\begin{aligned}(D^{\mathrm{fo}}_{\mathrm{L},\mu}\psi)_{\alpha,c} &= U^{-1}_{cc'}([n;\mu])\psi_{\alpha c'}(n+\hat{\mu}) - \psi_{\alpha,c}(n),\\ (D^{\mathrm{ba}}_{\mathrm{L},\mu}\psi)_{\alpha,c} &= \psi_{\alpha,c}(n) - U_{cc'}([n-\hat{\mu};\mu])\psi_{\alpha c'}(n-\hat{\mu})\end{aligned} \tag{2.10.88}$$

with $a\,\hat{\mu} = e_\mu$ (cf. Sect. 2.10.1.1). The coupling of the quarks to the gluon fields is described by the *naïve lattice Dirac equation*

$$\begin{aligned}(Q_{\mathrm{L}}\psi)_{\alpha,c} &\equiv \gamma^\mu_{\alpha\beta}(D_{\mathrm{L},\mu}\psi)_{\beta,c}(n) + \overline{m}\psi_{\alpha,c}(n) = 0,\\ D_{\mathrm{L},\mu} &= \frac{1}{2}(D^{\mathrm{fo}}_{\mathrm{L},\mu} + D^{\mathrm{ba}}_{\mathrm{L},\mu}),\end{aligned} \tag{2.10.89}$$

where $\overline{m} = am$ is the mass in lattice units. The corresponding part of the action reads

$$S_{\mathrm{q}}[U,\psi,\bar{\psi}] = \sum_{n\in\Gamma}\Bigg[\sum_{\mu=1}^{4}\frac{1}{2}\Big(\bar{\psi}(n)\gamma^\mu U^{-1}([n;\mu])\psi(n+\hat{\mu})$$

$$
\begin{aligned}
&\left. - \bar{\psi}(n)\gamma^\mu U([n-\hat{\mu}];\mu])\psi(n-\hat{\mu})\Big) + \bar{m}\bar{\psi}(n)\psi(n)\right] \\
&= \sum_{n\in\Gamma}\left[- \sum_{\mu=\pm 1}^{\pm 4} [\bar{\psi}(n+\hat{\mu})\gamma^\mu U([n;\mu])\psi(n)] + \bar{m}\bar{\psi}(n)\psi(n)\right] \\
&\equiv (\bar{\psi} Q_{\rm L} \psi). \qquad (2.10.90)
\end{aligned}
$$

For the transition from the first to the second line we used $U^{-1}([n;\mu]) = U([n+\hat{\mu};-\mu])$ [cf. (2.10.2)] and introduced the γ matrix in opposite direction $\gamma^{-\mu} = -\gamma^\mu$. We suppress the sum over the spin and SU(3) indices.

Now we can extend the path integral to include the quark fields,

$$
\begin{aligned}
\langle \Omega[U,\psi,\bar{\psi}]\rangle &= \frac{1}{Z}\int\int \mathcal{D}[U]\mathcal{D}[\psi,\bar{\psi}]\,\Omega[U,\psi,\bar{\psi}]\mathrm{e}^{-S_{\rm W}[U]-(\bar{\psi}Q_{\rm L}\psi)}, \\
Z &= \int\int \mathcal{D}[U]\mathcal{D}[\psi,\bar{\psi}]\,\mathrm{e}^{-S_{\rm W}[U]-(\bar{\psi}Q_{\rm L}\psi)}. \qquad (2.10.91)
\end{aligned}
$$

Like in the continuum, the gaussian integral over the quark fields has to be understood as integration over Grassmann variables (Sect. 2.2.1.4). Thus, *fermion integration* as in (2.2.65) leads to the final result for gauge-invariant Green functions of quark fields $\bar{\psi}(n_i)$ and $\psi(n_j)$ connected by a parallel displacement $U(C_{i,j})$,

$$
\begin{aligned}
&\left\langle \prod_{i,j}[\bar{\psi}(n_i)U(C_{i,j})\psi(n_j)]\right\rangle \\
&\quad = \frac{1}{Z}\int \mathcal{D}[U]\prod_{i,j}\mathcal{G}(n_i,n_j;U)U(C_{i,j})\mathrm{e}^{-\frac{1}{2}\Gamma[U]}\mathrm{e}^{-S_{\rm W}[U]}. \qquad (2.10.92)
\end{aligned}
$$

In order to evaluate the quark Green functions on the lattice, one has to calculate the propagator $\mathcal{G}(n_i,n_j;U)$ in the background field $U(n)$,

$$
\mathcal{G}(n_i,n_j;U) = \langle n_i,\alpha,c|Q_{\rm L}^{-1}|n'_j,\alpha',c'\rangle, \qquad (2.10.93)
$$

and the determinant

$$
\mathrm{e}^{-\frac{1}{2}\Gamma[U]} \equiv \det(\langle n_i,\alpha,c|Q_{\rm L}|n'_j,\alpha',c'\rangle). \qquad (2.10.94)
$$

of the massive Dirac operator $Q_{\rm L}$. Finally, one has to perform the gauge-field integration.

2.10.6.2 The spectrum-doubling problem

What are the specific problems of this scheme? There is the problem of *spectrum doubling*. In order to investigate this, we consider the free Dirac equation in momentum space,

$$\left(\mathrm{i}\sum_{\mu=1}^{4}\gamma^\mu \sin(p_\mu a) + \overline{m}\right)\tilde{\psi}(p) = 0, \tag{2.10.95}$$

where $\tilde{\psi}(p)$ is the Fourier transform of $\psi(n)$ on the lattice (2.10.61). We used

$$\partial_{\mathrm{L},\mu} e^{\mathrm{i}pna} = \frac{1}{2}(\partial^{\mathrm{fo}}_{\mathrm{L},\mu} + \partial^{\mathrm{ba}}_{\mathrm{L},\mu})e^{\mathrm{i}pna} = \mathrm{i}\sin(p_\mu a)e^{\mathrm{i}pna}, \tag{2.10.96}$$

with $\partial^{\mathrm{fo}}_{\mathrm{L},\mu}$ and $\partial^{\mathrm{ba}}_{\mathrm{L},\mu}$ given by (2.10.88) with $U = 1$. The propagator of S_q becomes

$$\tilde{\mathcal{G}}(p) \sim \left(\sum_\mu \gamma^\mu \frac{\mathrm{i}}{a}\sin p_\mu a + m\right)^{-1} = \frac{m - \sum_\mu \gamma^\mu \frac{\mathrm{i}}{a}\sin p_\mu a}{m^2 + \sum_\mu (\frac{1}{a}\sin p_\mu a)^2} \tag{2.10.97}$$

with p_μ varying in the first Brillouin zone $|p_\mu| \leq \pi/a$. For lattice spacing $a \to 0$, $\tilde{\mathcal{G}}(p)$ approaches the continuum expression $G_{\mathrm{cont}}(p) \approx (\mathrm{i}\not{p} + m)^{-1}$ not only for $ap_\mu \approx 0$ but for 16 different regions in the Brillouin zone where either $ap_\mu \approx 0$ or $ap_\mu \approx \pi$ for $\mu = 1,2,3,4$ independently. The lattice propagator $\tilde{\mathcal{G}}(p)$ has poles at $p^2 = m^2$ at these 16 different regions. Thus, the naïve action describes 16 Dirac particles instead of one. In the free case, this might be not so serious, however in an interacting theory, the additional fermions influence the physical content of the theory.

2.10.6.3 The Wilson fermions

There are two proposals to overcome this problem. K. Wilson suggested to add to the fermion action an additional term [Wi77],

$$\begin{aligned} S_{\mathrm{qW}} &= S_\mathrm{q} + r\sum_{n\in\Gamma}\left(-\sum_{\mu=\pm1}^{\pm4}[\overline{\psi}(n+\hat{\mu})U([n,\mu])\psi(n)] + 4\overline{\psi}(n)\psi(n)\right) \\ &= (\overline{\psi}Q_{\mathrm{LW}}\psi). \end{aligned} \tag{2.10.98}$$

In the free case, $U([n,\mu]) \equiv 1$, the propagator of (2.10.98) becomes

$$\begin{aligned}\tilde{\mathcal{G}}(p) &\sim \frac{m + \frac{2r}{a}\sum_\mu \sin^2\frac{p^\mu a}{2} - \sum_\mu \gamma^\mu \frac{\mathrm{i}}{a}\sin p_\mu a}{\left(m + \frac{2r}{a}\sum_\mu \sin^2\frac{p^\mu a}{2}\right)^2 + \sum_\mu \left(\frac{1}{a}\sin p_\mu a\right)^2} \\ &\to \frac{m - \mathrm{i}\not{p}}{m^2 + p^2} \quad \text{for} \quad p \to 0. \end{aligned} \tag{2.10.99}$$

For small p, this propagator has the form of the free continuum propagator, and the doubler poles at $p \neq 0$ disappear.

However, the Wilson expression Q_{LW} has some shortcomings. For $\overline{m} = 0$, Q_{LW} is no longer invariant under chiral transformations,

$$\gamma^5 Q_{\mathrm{LW}} + Q_{\mathrm{LW}}\gamma^5 \neq 0, \tag{2.10.100}$$

a symmetry which for example plays an important rôle in the understanding of the pion as a Goldstone particle (Sect. 3.5.1.1).

2.10.6.4 The staggered fermions

One may try to find another geometrical interpretation of the Dirac equation which is more appropriate for the lattice approximation. In Sect. 2.10.1 we introduced the geometrical elements $\mathcal{P}^{(n)}$ of a lattice: points, links, plaquettes, cubes, and super-cubes, for $n = 0, \ldots, 4$. The lattice approximations of the gauge potential and the field strength are defined as an average of the continuum quantities over such elements, i.e. (2.10.1) and (2.10.7), respectively. The lattice approximation of fermions described by the Dirac–Kähler equation follows this line of approach [Be82].

We use the formulation of differential calculus [Go87]. Antisymmetric covariant tensors are described by linear n-forms, $\Phi = \Phi_{\mu_1\ldots\mu_n}\mathrm{d}x^{\mu_1}\wedge\ldots\wedge\mathrm{d}x^{\mu_n}$, where the wedge product obeys $\mathrm{d}x^{\mu_1}\wedge\mathrm{d}x^{\mu_2} = -\mathrm{d}x^{\mu_2}\wedge\mathrm{d}x^{\mu_1}$. Therefore, $\Phi_{\mu_1\ldots\mu_n}$ is completely antisymmetric. The gauge potential $\mathcal{A}_\mu\mathrm{d}x^\mu$ and the field strength $(1/2)\mathcal{F}_{\mu\nu}\mathrm{d}x^\mu\wedge\mathrm{d}x^\nu$ are examples. In differential calculus the operator d of exterior differentiation and its adjoint δ are the natural generalization of the curl and the divergence to higher dimensions,

$$\begin{aligned}\mathrm{d}\Phi &= \frac{\partial}{\partial x^\mu}\Phi_{\mu_1\ldots\mu_n}\mathrm{d}x^\mu\wedge\mathrm{d}x^{\mu_1}\wedge\ldots\wedge\mathrm{d}x^{\mu_n}, \\ \delta\Phi &= -n\frac{\partial}{\partial x_\mu}\Phi_{\mu\mu_2\ldots\mu_n}\mathrm{d}x^{\mu_2}\wedge\ldots\wedge\mathrm{d}x^{\mu_n}. \end{aligned} \tag{2.10.101}$$

These definitions and antisymmetry imply

$$\mathrm{d}^2 = \delta^2 = 0, \qquad \mathrm{d}\delta + \delta\mathrm{d} = -\partial_\mu \partial^\mu = -\Box. \tag{2.10.102}$$

The Dirac–Kähler fields are described by inhomogeneous differential forms,

$$\Phi = \sum_n \frac{1}{n!} \Phi_{\mu_1 \ldots \mu_n}(x)\, \mathrm{d}x^{\mu_1} \wedge \ldots \wedge \mathrm{d}x^{\mu_n}. \tag{2.10.103}$$

For fermions we get the free *Dirac–Kähler equation* [Ka62, Be82],

$$(\mathrm{d} - \delta + m)\Phi = 0, \tag{2.10.104}$$

with $(\mathrm{d} - \delta + m)(\mathrm{d} - \delta - m)\Phi = (\Box - m^2)\Phi$. This means, the operator $(\mathrm{d} - \delta)$ shares the property $(d - \delta)^2 = \Box$ with the operator $\not\partial$. But this was the original motivation by Dirac to introduce the Dirac matrices. With the formal complex conjugate $\overline{\Phi}$, the Dirac–Kähler equation is related to the action

$$S_{\mathrm{qK}} = \int \mathrm{d}^4x \left\{ \sum_n \frac{1}{n!} \overline{\Phi}^{\mu_1,\ldots,\mu_n}(x) [(\mathrm{d} - \delta + m)\Phi]_{\mu_1,\ldots,\mu_n}(x) \right\}. \tag{2.10.105}$$

This action can be made gauge-covariant by replacing the partial derivative ∂_μ by the covariant derivative D_μ in the definition of d and δ, (2.10.101).

The advantage of differential forms is that they provide a straightforward lattice approximation which maps Φ onto lattice elements $\mathcal{P}^{(n)}$,

$$\chi(y) = \int_{\mathcal{P}^{(n)}(y)} \Phi_{\mu_1,\ldots,\mu_n}\, \mathrm{d}V^{(n)}, \qquad \mu_1 < \ldots < \mu_n. \tag{2.10.106}$$

The n-dimensional volume element is denoted by $\mathrm{d}V^{(n)}$. The central points of the lattice elements $\mathcal{P}^{(n)}$: $y = \{x + 1/2(e_{i_1} + \ldots + e_{i_k})\}$, (points, links, ... in the notation of Sect. 2.10.1) form a lattice with the lattice constant $a/2$. Thus, on this finer lattice the *staggered fermion fields* $\chi(y)$ represent the average of the Dirac–Kähler fields over $\mathcal{P}^{(n)}$. In (2.10.106) and in the following we set $a = 1$. The definition (2.10.106) leads to a lattice approximation of the gauge-invariant action (2.10.105) of the staggered fermion fields,

$$S_{\mathrm{qs}}[U, \chi, \bar{\chi}] = \overline{m} \sum_{y} \bar{\chi}(y)\chi(y) \tag{2.10.107}$$
$$+ \sum_{y,\mu} \frac{\rho_{x,\mu}}{2} \left[\bar{\chi}(y) U^{-1}([y;\mu]) \chi(y+e_\mu) - \bar{\chi}(y+e_\mu) U([y;\mu]) \chi(y) \right],$$

where the sign factor $\rho_{x,\mu} = (-1)^{x_1+\ldots+x_{\mu-1}}$ comes from the ordering in (2.10.106). If one does not use the language of algebraic topology, the derivation of this formula is straightforward but clumsy [Be82].

The propagator of the lattice approximation of the Dirac–Kähler action has no doublers. However, the free Dirac–Kähler equation represents a four-fold degeneracy of the Dirac field in the continuum and on the lattice. Furthermore, a reduction with respect to the flavour degeneracy of the naïve lattice approximation of the Dirac equation (Sect. 2.10.6.2) leads also to (2.10.107), [Su77, Sh81]. The action (2.10.107) has interesting symmetries which can be understood as restriction of the symmetries of the Dirac–Kähler equation [Jo87]. The natural four-fold flavour is sometimes used to study the flavour dependence of lattice approximation. A physical interpretation of this flavour degeneracy is not yet known.

2.10.6.5 Chiral symmetry and the Nielson–Ninomiya theorem

There is no simple and completely satisfactory lattice approximation to the Dirac equation. This is a consequence of the Nielson–Ninomiya theorem [Ni81], which states that the following properties cannot hold simultaneously for the Fourier transform of the free massless Dirac operator[39] on the lattice $\tilde{D}_{\mathrm{L}}(p)$ [Lu98]:

(a) $\tilde{D}_{\mathrm{L}}(p)$ is an analytic periodic function of the momentum p_μ with the period $2\pi/a$. This is necessary for $\tilde{D}_{\mathrm{L}}$ to be an essentially local operator.

(b) For momenta $p_\mu \to 0$ we have $\tilde{D}_{\mathrm{L}}(p) = \mathrm{i}\not{p}$ up to terms of order ap^2.

[39] It is customary to denote the Dirac operator Q_{L} by D_{L} for massless fermions.

(c) $\tilde{D}_{\mathrm{L}}(p)$ is invertible at all non-zero momenta in the first Brillouin zone. This ensures, together with (b), that the correct continuum limit is obtained.

(d) D_{L} anticommutes with γ^5: $\gamma^5 D_{\mathrm{L}} + D_{\mathrm{L}}\gamma^5 = 0$. This guarantees that the fermion action is invariant under continuous chiral transformations.

The Nielson–Ninomiya theorem was considered as prohibiting a lattice approximation of chiral fermions. However, an old suggestion by Ginsparg and Wilson [Gi82] to substitute (d) on the lattice by

$$\gamma^5 D_{\mathrm{L}} + D_{\mathrm{L}}\gamma^5 = aD_{\mathrm{L}}\gamma^5 D_{\mathrm{L}} \tag{2.10.108}$$

regained interest recently [Ha98a, Lu98, Ne98]. Examples of such a free Dirac operator can be constructed. The construction of gauge-invariant D_{L} might be complicated [Lu99a].

2.10.6.6 Remark on the improvement program

In order to approach the continuum limit more rapidly, one can add terms δS to the action that depend on the lattice constant a,

$$S_{\mathrm{imp}} = S_{\mathrm{W}}[U] + S_{\mathrm{q}}[U, \psi, \bar{\psi}] + \delta S. \tag{2.10.109}$$

Based on ideas for perturbation theory by K. Symanzik [Sy83], M. Lüscher started such an *improvement program* for the lattice approximation of QCD [Lu85]. To improve the action for the Wilson fermions S_{qW}, (2.10.98), one should add the Sheikholeslami–Wohlert term [Sh85]

$$\delta S = a^5 \sum_x c_{\mathrm{SW}} \bar{\psi}(x) \frac{\mathrm{i}}{4} \sigma_{\mu\nu} \hat{F}^{\mu\nu} \psi(x). \tag{2.10.110}$$

In (2.10.110), $\hat{F}_{\mu\nu}$ is a lattice approximation of the gluon field-strength tensor involving four plaquettes. The additional a-dependent term (2.10.110) compensates the linear term in a in the corresponding improved Green functions of field operators. The coupling $c_{\mathrm{SW}}(g_0)$ depends on the bare coupling constant g_0. A one-loop calculation gives $c_{\mathrm{SW}}(g_0) = 1 + 0.265 g_0^2 + \mathcal{O}(g_0^4)$ [Wo87]. There are schemes to determine it non-perturbatively.

2.10.6.7 Quenched and unquenched approximation

After choosing an appropriate lattice approximation of the Dirac operator Q_L, one follows the scheme described in (2.10.92), (2.10.93), and (2.10.94) for the calculation of quark Green functions.

In *quenched approximation*, the fermion determinant is set equal to one, $\det(Q_L) = 1$. This means that the effects of fermion loops are neglected. For some problems this might be a good approximation (cf. Sect. 3.6), however, the corrections are essentially unknown.

What remains is the inversion of the Dirac operator Q_L, i.e. the solution of the equation

$$Q_L\psi = \phi. \tag{2.10.111}$$

On a finite lattice this is a finite-dimensional linear equation. It is certainly beyond the scope of this book to describe the methods of numerical mathematics like *Jacobi iteration* or *conjugate-gradient method* which are used and refined for the purpose of lattice approximation [Mo94].

The numerical simulation of the lattice approximation of the complete *gauge theory with dynamical fermions* requests a procedure for the evaluation of the fermion determinant in equation (2.10.94). Many different procedures are developed for this difficult task. We sketch the main ideas of the *hybrid Monte Carlo procedure* which is frequently used.

The fermion determinant of two degenerate fermions can be described with help of a complex bosonic *pseudo-fermion field* $\Phi(x), \Phi^\dagger(x)$,

$$[\det(Q_L)]^2 = \int \mathcal{D}[\Phi, \Phi^\dagger] e^{-(\Phi^\dagger [Q_L^\dagger Q_L]^{-1} \Phi)}. \tag{2.10.112}$$

Here we used the different expressions for the gaussian integrals with bosonic variables (2.2.11) and fermionic variables (2.2.27). The lattice approximation of the bosonic gaussian integral over the pseudo-fermion fields can be evaluated by the Metropolis algorithm (Sect. 2.10.3.4) or the heat-bath method (Sect. 2.10.3.5). The transition to the *effective action* for the pseudo-fermion fields requires again the computer-time consuming inversion of $Q_L Q_L^\dagger$ like in (2.10.111). This favours the integration over the pseudo-fermions by the heat-bath method. The effective action of the pseudo-fermions is non-local.

This makes a straightforward Metropolis algorithm for the remaining integration over the gauge fields U very slow. One has to evaluate the fermion determinant many times. Therefore one introduces a further step. The update of the gauge fields U in the Metropolis step is produced by *molecular dynamics*. Roughly speaking this method has the advantage to propose comparatively uncorrelated configurations without a breakdown of the acceptance rate in the Metropolis decision.

This very short description of the hybrid Monte Carlo algorithm shows that a numerical treatment of the lattice approximation of gauge theories with fermions is possible. However, this method requires much more computer time. Thus, only relatively small lattices can be treated up to now. A non-trivial numerical simulation of QCD with three light quarks at the physical value of quark masses needs to perform at least 10^{20} floating-point operations. With the availability of teraflop computers in the coming years this seems to be a feasible task [Sa00].

References to Chapter 2

[Ab70] M. Abramowitz and I.A. Stegun, *Handbook of Mathematical Functions with Formulas, Graphs, and Mathematical Tables*, (Dover Publications, New York, 1970).

[Ab73] E. Abers and B.W. Lee, *Phys. Rept.* **9** (1973) 1.

[Ab81] L.F. Abbott, *Nucl. Phys.* **B185** (1981) 189 and *Acta Phys. Pol.* **B13** (1982) 33.

[Ab83a] L.F. Abbott, M.T. Grisaru and R.K. Schaefer, *Nucl. Phys.* **B229** (1983) 372.

[Ad69] S.L. Adler, *Phys. Rev.* **177** (1969) 2426.

[Ad69a] S.L. Adler and W.A. Bardeen, *Phys. Rev.* **182** (1969) 1517.

[Al77] G. Altarelli and G. Parisi, *Nucl. Phys.* **B126** (1977) 298.

[Ap77] T. Appelquist, M. Dine and I. Muzinich, *Phys. Lett.* **69B** (1977) 231.

[At68] M.F. Atiyah and I.M. Singer, *Ann. Math.* **87** (1968) 484, 531, and 546, **93** (1971) 119 and 139.

[At78] M.F. Atiyah, N.J. Hitchin, V.G. Drinfeld and Yu.I. Manin, *Phys. Lett.* **65A** (1978) 185.

[Ba69] W.A. Bardeen, *Phys. Rev.* **184** (1969) 1848.

[Ba75] R. Balian, J.M. Drouffe and C. Itzykson, *Phys. Rev.* **D11** (1975) 2104, E:**D19** (1979) 2514.

[Ba80] T. Banks and A. Casher, *Nucl. Phys.* **B169** (1980) 103.

[Ba95] S. Bauberger, F.A. Berends, M. Böhm and M. Buza, *Nucl. Phys.* **B434** (1995) 383.

[Ba95a] S. Bauberger, F.A. Berends, M. Böhm, M. Buza and G. Weiglein, *Nucl. Phys. B (Proc. Suppl.)* **37** (1994) 95.

[Ba95b] S. Bauberger and M. Böhm, *Nucl. Phys.* **B445** (1995) 25.

[Be51] H.A. Bethe and E.E. Salpeter, *Phys. Rev.* **84** (1951) 1232.

[Be66] F.A. Berezin, *The Method of Second Quantization*, (Academic Press, New York, 1966).

[Be69] J.S. Bell and R. Jackiw, *Nuovo Cimento* **60A** (1969) 47.

[Be74] C. Becchi, A. Rouet and R. Stora, *Phys. Lett.* **52B** (1974) 344 and *Commun. Math. Phys.* **42** (1975) 127.

[Be75a] A.A. Belavin, A.M. Polyakov, A.S. Schwartz and Yu.S. Tyupkin, *Phys. Lett.* **59B** (1975) 85.

[Be76] V.B. Berestetskiĭ, *Usp. Fiz. Nauk.* **120** (1976) 439, *Sov. Phys. Usp.* **19** (1976) 934.

[Be76a] C. Becchi, A. Rouet and R. Stora, *Renormalizable Theories with Symmetry Breaking* in *Field Theory, Quantization and Statistical Physics*, ed. E. Tirapegui (Dortrecht, 1976) p. 3.

[Be76b] C. Becchi, A. Rouet and R. Stora, *Ann. Phys.* **98** (1976) 287.

[Be77] A.A. Belavin and A.M. Polyakov, *Nucl. Phys.* **B123** (1977) 429.

[Be79] C. Bernard, *Phys. Rev.* **D19** (1979) 3013.

[Be81] B. Berg, *Phys. Lett.* **104B** (1981) 475.

[Be82] P. Becher and H. Joos, *Z. Phys.* **C15** (1982) 343.

[Be96] R.A. Bertlmann, *Anomalies in Quantum Field Theories*, (Clarendon Press, Oxford, 1996).

[Bh88] G. Bhanot, *Rept. Prog. Phys.* **51** (1988) 429.

[Bl37] F. Bloch and A. Nordsieck, *Phys. Rev.* **52** (1937) 54.

[Bo65] R. Bott and S.S. Chern, *Acta Math.* **114** (1965) 71.

[Bo72] C.G. Bollini and J.J. Giambiagi, *Phys. Lett.* **40B** (1972) 566.

[Bo73] M. Böhm, H. Joos and M. Krammer, *Nuovo Cimento* **7A** (1972) 21, *Nucl. Phys.* **B51** (1973) 397, *Dynamical Problems of the Relativistic Quark Model* in *Recent Developments in Mathematical Physics*, Acta Physica Austriaca, Suppl. XI, ed. P. Urban (Wien, 1973) p. 3.

[Bo81] D.G. Boulware, *Phys. Rev.* **D23** (1981) 389.

[Br77] P. Breitenlohner and D. Maison, *Commun. Math. Phys.* **52** (1977) 11, 39, and 55.

[Ca70] C.G. Callan, *Phys. Rev.* **D2** (1970) 1541.

[Ca74] W.E. Caswell, *Phys. Rev. Lett.* **33** (1974) 244.

[Ca76] C.G. Callan, Jr., R.F. Dashen and D.J. Gross, *Phys. Lett.* **63B** (1976) 334.

[Ca77] J. Calmet, S. Narison, M. Perrottet and E. de Rafael, *Rev. Mod. Phys.* **49** (1977) 21.

[Ca78] C.G. Callan, Jr., R.F. Dashen and D.J. Gross, *Phys. Rev.* **D17** (1978) 2717.

[Ca82] N. Cabibbo and E. Marinari, *Phys. Lett.* **119B** (1982) 387.

[Ch77] S. Chadha, A. D'Adda, P. Di Vecchia and F. Nicodemi, *Phys. Lett.* **72B** (1977) 103.

[Ch78] N. Christ, E. Weinberg and N. Stanton, *Phys. Rev.* **D18** (1978) 2013.

[Ch82] Y. Choquet-Bruhat and C. DeWitt-Morette, *Analysis, Manifolds and Physics*, (North Holland, Amsterdam, etc., 1982).

[Co65] S. Coleman and R.E. Norton, *Nuovo Cimento* **38** (1965) 438.

[Co68] R. Courant and D. Hilbert, *Mathematische Methoden der Physik I and II*, (Springer, Heidelberg, 1968).

[Co73] S. Coleman and E. Weinberg, *Phys. Rev.* **D7** (1973) 1888.

[Co74] J.C. Collins and A.J. MacFarlane, *Phys. Rev.* **D10** (1974) 1201.

[Co75] G. Costa and M. Tonin, *Revista del Nuovo Cim.* **5** (1975) 29.

[Co78] E.F. Corrigan, D.B. Fairlie, S. Templeton and P. Goddard, *Nucl. Phys.* **B140** (1978) 31.

[Co79] S. Coleman, *The Uses of Instantons* in *The Whys of Subnuclear Physics*, Proceedings of the International School of Subnuclear Physics, *Erice, 1977*, ed. A. Zichichi (New York, London, 1979) p. 805.

[Co84] J.C. Collins, *Renormalization*, (Cambridge University Press, Cambridge, 1984).

[Cr79] M. Creutz, L. Jacobs and C. Rebbi, *Phys. Rev. Lett.* **42** (1979) 1390.

[Cr79a] R.J. Crewther, P. Di Vecchia, G. Veneziano and E. Witten, *Phys. Lett.* **88B** (1979) 123; E:**91B** (1980) 487.

[Cr80] M. Creutz, *Phys. Rev.* **D21** (1980) 2308.

[Cr83] M. Creutz, *Quarks, Gluons, and Lattices*, (Cambridge University Press, Cambridge, 1983).

[Cz95a] A. Czarnecki, U. Kilian and D. Kreimer, *Nucl. Phys.* **B433** (1995) 259.

[De61] N.G. De Bruijn, *Asymptotic Methods in Analysis*, (North Holland, Amsterdam, 1961).

[De67] B.S. DeWitt, *Phys. Rev.* **162** (1967) 1195; *Dynamical Theory of Groups and Fields*, (Gordon and Breach, New York, 1965); *A Gauge-invariant Effective Action* in *Quantum Gravity 2*, eds. C.J. Isham, R. Penrose and D.W. Sciama (New York, 1981) p. 449.

[De91] A. Denner, U. Nierste and R. Scharf, *Nucl. Phys.* **B367** (1991) 637.

[De93] A. Denner, *Fortschr. Phys.* **41** (1993) 307.

[De96] A. Denner and S. Dittmaier, *Phys. Rev.* **D54** (1996) 4499.

[DE97] M. D'Elia, A. Di Giacomo and E. Meggiolaro, *Phys. Lett.* **B408** (1997) 315.

[Do98] H.G. Dosch and S. Narison, *Phys. Lett.* **B417** (1998) 173.

[Dr78] J.M. Drouffe and C. Itzykson, *Phys. Rept.* **38** (1978) 133.

[Dy49] F.J. Dyson, *Phys. Rev.* **75** (1949) 1736.

[Ed66] R.J. Eden, P.V. Landshoff, D.I. Olive and J.C. Polkinghorne, *The Analytic S Matrix*, (Cambrigde University Press, Cambrigde, 1966).

[Eg80] T. Eguchi, P.B. Gilkey and A.J. Hanson, *Phys. Rev.* **66** (1980) 213.

[El79] J. Ellis and M.K. Gaillard, *Nucl. Phys.* **B150** (1979) 141.

[Er53] F. Erdély (Bateman Manuscript Project) *Higher Transcendental Functions, Chapt. IX*, (Mac Graw-Hill, New York, 1953)

[Er53a] F. Erdély (Bateman Manuscript Project) *Tables of Integral Transforms, Chapt. VII*, (Mac Graw-Hill, New York, 1953)

[Fa67] L.D. Faddeev and V.N. Popov, *Phys. Lett.* **25B** (1967) 29.

[Fe48] R.P. Feynman, *Rev. Mod. Phys.* **20** (1948) 267.

[Fe51] R.P. Feynman, *Phys. Rev.* **84** (1951) 108.

[Fe65] R.P. Feynman and A.R. Hibbs, *Quantum Mechanics and Path Integrals*, (McGraw-Hill, New York, 1965).

[Fi69] M.E. Fischer, *Suppl. J. Phys. Soc. Japan* **26** (1969) 87.

[Fr72] H.M. Fried, *Functional Methods and Models in Quantum Field Theory*, (MIT, Cambridge, Massachusetts, 1972).

[Fr73] H. Fritzsch, M. Gell-Mann and H. Leutwyler, *Phys. Lett.* **47B** (1973) 365.

[Fr97] A. Frink, U. Kilian and D. Kreimer, *Nucl. Phys.* **B488** (1997) 426.

[Fr00] A. Frommer, T. Lippert, B. Medeke and K. Schilling (eds.), *Numerical Challenges in Lattice Quantum Chrodynamics*, (Springer, Berlin, Heidelberg, 2000).

[Fu80] K. Fujikawa, *Phys. Rev.* **D21** (1980) 2848, E:**D22** (1980) 1499.

[Ge51] M. Gell-Mann and F. Low, *Phys. Rev.* **84** (1951) 350.

[Ge54] M. Gell-Mann and F. Low, *Phys. Rev.* **95** (1954) 1300.

[Ge72a] H. Georgi and S.L. Glashow, *Phys. Rev.* **D6** (1972) 429.

[Gh98] A. Ghinculov and Y.-P. Yao, *Nucl. Phys.* **B516** (1998) 385 and hep-ph/0006314.

[Gi82] P.H. Ginsparg and K.G. Wilson, *Phys. Rev.* **D25** (1982) 2649.

[Gi95] P.B. Gilkey, *Invariance Theory, the Heat Equation, and the Atiyah–Singer Theorem*, (CRC Press, Boca Raton, etc., 1995).

[Gi99] L. Giusti, F. Rapuano, M. Talevi and A. Vladikas, *Nucl. Phys.* **B538** (1999) 249.

[Gl61] S.L. Glashow, *Nucl. Phys.* **22** (1961) 579.

[Gl70] S.L. Glashow, J. Iliopoulos and L. Maiani, *Phys. Rev.* **D2** (1970) 1285.

[Gl76] J. Glimm and A. Jaffe, *Commun. Math. Phys.* **51** (1976) 1.

[Go87] M. Göckeler and T. Schücker, *Differential Geometry, Gauge Theory, and Gravity*, (Cambridge University Press, Cambridge, 1987).

[Gr72] V.N. Gribov and L.N. Lipatov, *Sov. J. Nucl. Phys.* **15** (1972) 438 and 675.

[Gr73a] D.J. Gross and F. Wilczek, *Phys. Rev.* **D8** (1973) 3633 and **D9** (1974) 980.

[Gr73b] D.J. Gross and F. Wilczek, *Phys. Rev. Lett.* **30** (1973) 1343.

[Gr74] D.J. Gross and A. Neveu, *Phys. Rev.* **D10** (1974) 3235.

[Gr78] V.N. Gribov, *Nucl. Phys.* **B139** (1978) 1.

[Ha77] S.W. Hawking, *Comm. Math. Phys.* **55** (1977) 133.

[Ha80] A. Hasenfratz and P. Hasenfratz, *Phys. Lett.* **93B** (1980) 165.

[Ha81] A. Hasenfratz and P. Hasenfratz, *Nucl. Phys.* **B193** (1981) 210.

[Ha83] C.F. Hart, *Phys. Rev.* **D28** (1983) 1993.

[Ha92] R. Haag, *Local Quantum Physics: Fields, Particles, Algebras (Chapt. 3)*, (Springer, Berlin, etc., 1992).

[Ha98] I. Halperin and A. Zhitnitsky, *Mod. Phys. Lett.* **A13** (1998) 1955.

[Ha98a] P. Hasenfratz, *Nucl. Phys. B (Proc. Suppl.)* **63** (1998) 53.

[Ha99] R. Harlander and M. Steinhauser, *Prog. Part. Nucl. Phys.* **43** (1999) 167.

[Ho87] J. Hoek, M. Teper and J. Waterhous, *Nucl. Phys.* **B288** (1987) 589.

[Hu87] K. Huang, *Statistical Mechanics*, Vol. 1–3, (John Wiley & Sons, New York, 1987).

[Hu92] K. Huang, *Quarks, Leptons and Gauge fields*, (World Scientific, Singapore, 1992).

[It80] C. Itzykson and J.B. Zuber, *Quantum Field Theory*, (McGraw–Hill, New York, 1980).

[Ja76] R. Jackiw and C. Rebbi, *Phys. Rev.* **D14** (1976) 517 and **D16** (1977) 1052.

[Ja77] R. Jackiw, C. Nohl and C. Rebbi, *Phys. Rev.* **D15** (1977) 1642.

[Je00] F. Jegerlehner, hep-th/0005255, DESY 00-075.

[Jo65] R. Jost, *The General Theory of Quantized Fields*, (A.M.S., Providence, Rhode Island, 1965).

[Jo74] D.R.T. Jones, *Nucl. Phys.* **B75** (1974) 531.

[Jo76] S.D. Joglekar and B.W. Lee, *Ann. Phys.* **97** (1976) 160.

[Jo87] H. Joos and M. Schäfer, *Z. Phys.* **C34** (1987) 465.

[Jo90] H. Joos, *Helv. Phys. Acta* **63** (1990) 670.

[Jo94] H. Joos and S.I. Azakov, *Helv. Phys. Acta* **67** (1994) 723.

[Ka62] E. Kähler, *Ren. Mat. Ser. V* **21** (1962) 435.

[Ka65] L.P. Kadanoff, *Physica* **2** (1966) 263.

[Ka77] L.P. Kadanoff, *Rev. Mod. Phys.* **49** (1977) 267.

[Ki62] T. Kinoshita, *J. Math. Phys.* **3** (1962) 650.

[Kl75] H. Kluberg-Stern and J. Zuber, *Phys. Rev.* **D12** (1975) 482 and 3159.

[Ko92] J.G. Körner, D. Kreimer and K. Schilcher, *Z. Phys.* **C54** (1992) 503.

[Kr98] E. Kraus, *Ann. Phys.* **262** (1998) 155.

[Ku78] T. Kugo and I. Ojima, *Prog. Theor. Phys.* **60** (1978) 1869, **61** (1979) 294 and *Prog. Theor. Phys. Suppl.* **66** (1979) 1.

[Ku97] T. Kugo, *Eichtheorie*, (Springer, Berlin, etc., 1997).

[La55] L.D. Landau and I.Ya. Pomeranchuk, *Doklady Akad. Nauk. SSSR* **102** (1955) 489.

[La59] L.D. Landau, *Nucl. Phys.* **13** (1959) 181.

[La67] B.E. Lautrup, *Mat. Fys. Medd. Kon. Dan. Vid.-Sel. Medd.* **35** (1967) 29.

[Le55] H. Lehmann, K. Symanzik and W. Zimmermann, *Nuovo Cimento* **1** (1955) 1425.

[Le58] L. Lewin, *Dilogarithms and Associated Functions*, (Macdonald, London, 1958).

[Le64] T.D. Lee and M. Nauenberg, *Phys. Rev.* **133** (1964) 1549.

[Le72] B.W. Lee and J. Zinn-Justin, *Phys. Rev.* **D5** (1972) 3121, 3137, 3155 and **D7** (1973) 1049.

[Le76] B.W. Lee, *Gauge Theories* in *Methods in Field Theory*, Les Houches 1975, eds. R. Balian and J. Zinn-Justin (Amsterdam, 1976) p. 77.

[Lu80] M. Lüscher, *Exact Instanton Gases* in *Recent Developments in Gauge Theories, Proceedings of the 1979 Cargese Summer Institute*, Nato Advanced Study Institute Series: Series B, Vol. 59, eds. G. 't Hooft et al. (New York, London, 1980) p. 313.

[Lu82] M. Lüscher, *Ann. Phys.* **142** (1982) 359.

[Lu82a] M. Lüscher, *Nucl. Phys.* **B205** (1982) 483.

[Lu84] M. Lüscher, *On the Relation Between Finite Size Effects and Elastic Scattering Processes* in *Progress in Gauge Field Theory* (Cargese 1983), eds. G. 't Hooft et al. (New York, 1984) p. 451.

[Lu85] M. Lüscher, *Nucl. Phys.* **B254** (1985) 52.

[Lu98] M. Lüscher, *Phys. Lett.* **B428** (1998) 342.

[Lu99] M. Lüscher, *Advanced Lattice QCD* in *Probing the Standard Model of Particle Interactions*, Les Houches 1997, eds. R. Gupta et al. (Amsterdam, 1999) p. 229.

[Lu99a] M. Lüscher, *Nucl. Phys.* **B549** (1999) 295.

[Ma81] G. Mack, *Fortschr. Phys.* **29** (1981) 135.

[Me65] D.B. Melrose, *Nuovo Cimento* **XL A** (1965) 181.

[Mo85] T.R. Morris, D.A. Ross and C.T. Sachrajda, *Nucl. Phys.* **B255** (1985) 115.

[Mo94] I. Montvay and G. Münster, *Quantum Fields on a Lattice*, (Cambridge University Press, Cambridge, 1994).

[Mo97] S. Moch, A. Ringwald, and F. Schrempp, *Nucl. Phys.* **B507** (1997) 134.

[Na60] Y. Nambu, *Phys. Rev. Lett.* **4** (1960) 380.

[Na66] Y. Nambu, *Phys. Lett.* **26B** (1966) 626.

[Na66a] N. Nakanashi, *Prog. Theor. Phys.* **35** (1966) 1111.

[Na89] S. Narison, *QCD Spectral Sum Rules*, (World Scientific, Singapore, 1989).

[Na90] M. Nakahara, *Geometry, Topology and Physics*, Graduate Student Series in Physics, (Hilger, Bristol, etc., 1990).

[Na96] S. Narison, *Phys. Lett.* **B387** (1996) 162.

[Ne98] H. Neuberger, *Phys. Lett.* **B417** (1998) 141 and **B427** (1998) 353.

[Ni77] N.K. Nielsen and B. Schroer, *Nucl. Phys.* **B127** (1977) 493.

[Ni81] H.B. Nielsen and M. Ninomiya, *Nucl. Phys.* **B185** (1981) 20, E:**B195** (1982) 541.

[No78] V.A. Novikov et al., *Phys. Rept.* **41** (1978) 1.

[Ol90] G.J. van Oldenborgh and J.A.M. Vermaseren, *Z. Phys.* **C46** (1990) 425.

[Or77] F.R. Ore, Jr., *Phys. Rev.* **D15** (1977) 470.

[Or77a] F.R. Ore, Jr., *Phys. Rev.* **D16** (1977) 2577.

[Os75] K. Osterwalder and R. Schrader, *Commun. Math. Phys.* **42** (1975) 281.

[Pa49] W. Pauli and F. Villars, *Rev. Mod. Phys.* **21** (1949) 434.

[Pa79] G. Passarino and M. Veltman, *Nucl. Phys.* **B160** (1979) 151.

[PDG00] D.E. Groom et al. (Particle Data Group), *Eur. Phys. J.* **C15** (2000) 1.

[Pe77] R.D. Peccei and H.R. Quinn, *Phys. Rev.* **D16** (1977) 1791.

[Pi81] O. Piguet and A. Roulet, *Phys. Rept.* **76** (1981) 1.

[Pr83] V. Privman and M.E. Fischer, *J. Stat. Phys.* **33** (1983) 385.

[Re85] A. Rebhan and G. Wirthumer, *Z. Phys.* **C28** (1985) 269.

[Re88] T. Reisz, *Commun. Math. Phys.* **116** (1988) 81 and 573, **117** (1988) 79 and 639.

[Ri97] T. van Ritbergen, J.A.M Vermaseren and S.A. Larin, *Phys. Lett.* **B400** (1997) 379.

[Ro94] G. Roepstorff, *Path Integral Approach to Quantum Physics: an Introduction*, (Springer, Berlin, etc., 1994).

[Sa68] A. Salam, *Weak and Electromagnetic Interactions* in *Elementary Particle Theory*, ed. N. Svartholm (Stockholm, 1968) p. 367.

[Sa00] C.T. Sachrajda, *Nucl. Phys. B (Proc. Suppl.)* **83-84** (2000) 93.

[Sc51] J. Schwinger, *Proc. Nat. Acad. Sci. (U.S.A.)* **37** (1951) 452.

[Sc51a] J. Schwinger, *Phys. Rev.* **82** (1951) 914.

[Sc62] J. Schwinger, *Phys. Rev.* **128** (1962) 2425.

[Sc98] T. Schäfer and E.V. Shuryak, *Rev. Mod. Phys.* **70** (1998) 323.

[Se82] E. Seiler, *Gauge Theories as a Problem of Constructive Quantum Field Theory and Statistical Mechanics*, (Springer, Berlin, etc., 1982).

[Sh79] M.A. Shifman, A.I. Vainshtein and V.I. Zakharov, *Nucl. Phys.* **B147** (1979) 385.

[Sh81] H.S. Sharatchandra, H.J. Thun and P. Weisz, *Nucl. Phys.* **B192** (1981) 205.

[Sh85] B. Sheikholeslami and R. Wohlert, *Nucl. Phys.* **B259** (1985) 572.

[Sh92] M.A. Shifman, *Vacuum Structure and QCD Sum Rules*, (North Holland, Amsterdam, 1992).

[Sl72] A.A. Slavnov, *Theor. and Math. Phys.* **10** (1972) 99.

[St49] J. Steinberger, *Phys. Rev.* **76** (1949) 1180.

[St53] E.C.G. Stueckelberg and A. Petermann, *Helv. Phys. Acta* **26** (1953) 499.

[St88] R.G. Stuart, *Comp. Phys. Commun.* **48** (1988) 367 and **56** (1990) 337.

[Su56] V.V. Sudakov, *Sov. Phys. JETP* **3** (1956) 65.

[Su66] D.G. Sutherland, *Phys. Lett.* **23** (1966) 384.

[Su67] D.G. Sutherland, *Nucl. Phys.* **B2** (1967) 433.

[Su77] L. Susskind, *Phys. Rev.* **D16** (1977) 3031.

[Sy54] K. Symanzik, *Zeitschr. Naturforschung* **9a** (1954) 809.

[Sy60] K. Symanzik, *J. Math. Phys.* **1** (1960) 249.

[Sy66] K. Symanzik, *J. Math. Phys.* **7** (1966) 510.

[Sy67] K. Symanzik, *Many-Particle Structure of Green's Functions* in *Symposia on Theoretical Physics, Vol. 3*, ed. A. Ramakrishnan (New York, 1967) p. 121.

[Sy70] K. Symanzik, *Commun. Math. Phys.* **18** (1970) 227.

[Sy73] K. Symanzik, *Lett. Nuovo Cimento* **6** (1973) 77.

[Sy77] K. Symanzik, *Regularized Quantum Field Theory* in *New Developments in Quantum Field Theory and Statistical Mechanics*, eds. M. Lévy and P. Mitter (New York, 1977) p. 265.

[Sy80] K. Symanzik, *Cutoff Dependence in Lattice Φ_4^4 Theory* in *Recent Developments in Gauge Theories, Proceedings of the 1979 Cargese Summer Institute*, Nato Advanced Study Institute Series: Series B, Vol. 59, eds. G. 't Hooft et al. (New York, London, 1980) p. 313.

[Sy83] K. Symanzik, *Nucl. Phys.* **B226** (1983) 187 and 205.

[Ta57] Y. Takahashi, *Nuovo Cimento* **6** (1957) 370.

[Ta71] J.C. Taylor, *Nucl. Phys.* **B33** (1971) 436.

[Ta80] O.V. Tarasov, A.A. Vladimirov and A.Yu. Zharkov, *Phys. Lett.* **93B** (1980) 429.

[Ta96] O.V. Tarasov, *Phys. Rev.* **D54** (1996) 6479 and *Nucl. Phys.* **B502** (1997) 455.

[Th53] W. Thirring, *Helv. Phys. Acta* **26** (1953) 33.

[tH71] G. 't Hooft, *Nucl. Phys.* **B33** (1971) 173 and **B35** (1971) 167.

[tH72] G. 't Hooft and M. Veltman, *Nucl. Phys.* **B50** (1972) 318.

[tH72a] G. 't Hooft and M. Veltman, *Nucl. Phys.* **B44** (1972) 189.

[tH73] G. 't Hooft and M. Veltman, *DIAGRAMMAR* in *Particle Interactions at Very High Energies, Part B*, eds. D. Speiser et al. (New York, 1973) p. 177 and CERN-report 73-9 (1973).

[tH73a] G. 't Hooft, *Nucl. Phys.* **61** (1973) 455.

[tH76] G. 't Hooft, *Phys. Rev.* **D14** (1976) 3432, E:*Phys. Rev.* **D18** (1978) 2199, E:*Phys. Rept.* **142** (1986) 357, App. B.

[tH76a] G. 't Hooft, *Acta Universitatis Wratislavensis* **368** (1976) 345.

[tH79] G. 't Hooft and M. Veltman, *Nucl. Phys.* **B153** (1979) 365.

[tH86] G. 't Hooft, *Phys. Rept.* **142** (1986) 357.

[Us93] N.I. Usyukina and A.I. Davydychev, *Phys. Lett.* **B298** (1993) 363 and **B305** (1993) 136.

[Ut56] R. Utiyama, *Phys. Rev.* **101** (1956) 1597.

[Va82] A.I. Vainshtein, V.I. Zakharov, V.A. Nokvikov and M.A. Shifman, *Usp. Fiz. Nauk* **136** (1982) 553, *Sov. Phys. Usp.* **24** (1982) 195.

[Va84] C. Vafa and E. Witten, *Commun. Math. Phys.* **95** (1984) 257.

[Ve67] M. Veltman, *Proc. Roy. Soc.* **A301** (1967) 107.

[Ve79] G. Veneziano, *Nucl. Phys.* **B159** (1979) 213.

[vN84] W.L. van Neerven and J.A.M Vermaseren, *Phys. Lett.* **137B** (1984) 241.

[Wa50] J.C. Ward, *Phys. Rev.* **78** (1950) 1824.

[Wa53] I. Watanabe, *Prog. Theor. Phys.* **4** (1953) 371.

[We60] S. Weinberg, *Phys. Rev.* **118** (1960) 838.

[We65] S. Weinberg, *Phys. Rev.* **140** (1965) B516.

[We67] S. Weinberg, *Phys. Rev. Lett.* **19** (1967) 1264.

[We71] J. Wess and B. Zumino, *Phys. Lett.* **37B** (1971) 95.

[We73] S. Weinberg, *Phys. Rev. Lett.* **31** (1973) 494 and *Phys. Rev.* **D8** (1973) 4482.

[We73a] S. Weinberg, *Phys. Rev.* **D8** (1973) 3497.

[We81] P. Weisz, *Phys. Lett.* **100B** (1981) 331.

[We94] G. Weiglein, R. Scharf and M. Böhm, *Nucl. Phys.* **B416** (1994) 606.

[Wi56] A.S. Wightman, *Phys. Rev.* **101** (1956) 860.

[Wi71] K.G. Wilson, *Phys. Rev.* **B4** (1971) 3174 and 3184.

[Wi74] K.G. Wilson, *Phys. Rev.* **D10** (1974) 2445.

[Wi74a] K.G. Wilson and J. Kogut, *Phys. Rept.* **C12** (1974) 75.

[Wi75] K.G. Wilson, *Rev. Mod. Phys.* **47** (1975) 773.

[Wi77] K.G. Wilson, *Quark and Strings on a Lattice* in *New Phenomena in Subnuclear Physics*, ed. A. Zichichi (New York, 1977) p. 69.

[Wi79] E. Witten, *Nucl. Phys.* **B156** (1979) 269.

[Wo87] R. Wohlert, DESY preprint 87-069 (1987), unpublished.

[Ya54] C.N. Yang and R.L. Mills, *Phys. Rev.* **96** (1954) 191.

[Ye61] D.R. Yennie, S.C. Frautschi and H. Suura, *Ann. Phys.* **13** (1961) 379.

[Zi59] W. Zimmermann, *Nuovo Cimento* **13** (1959) 503.

[Zi75] J. Zinn-Justin, *Renormalization of Gauge Theories* in *Trends in Elementary Particle Theory,* Lecture Notes in Physics Vol. 37, eds. J. Ehlers et al. (Berlin, 1975) p. 2.

3 Quantum Chromodynamics

The theory of the strong interaction of elementary particles, Quantum Chromodynamics (QCD), is a non-abelian gauge theory with SU(3) as gauge group. The degrees of freedom corresponding to this SU(3) are called *colour*. The quarks have besides the flavour also colour degrees of freedom and transform according to the fundamental representation of colour SU(3). The eight coloured gauge bosons are called *gluons*. The colour degree of freedom was introduced in Refs. [Gr64, Ha65, Ge72]. Coloured gluons were proposed by H. Fritzsch, M. Gell-Mann, and H. Leutwyler [Fr73]. The colour degree of freedom solved the following three puzzles of the quark model: the problems with the lifetime of the neutral π meson, the total hadronic cross section in e^+e^- annihilation, and the "wrong" quark statistics in the baryon bound states.

QCD is renormalizable, as was shown in Sect. 2.5.1.4, and therefore treatable by perturbative methods. Perturbation theory is appropriate for QCD at high energy or high momentum transfers as was shown already in 1973 by D.J. Gross and F. Wilczek [Gr73a], H.D. Politzer [Po73], and G. 't Hooft (unpublished). They discovered[1] the asymptotic freedom of QCD (Sect. 3.1). This opened the way to the successful applications of perturbative QCD to deep-inelastic scattering (Sect. 3.2), hard hadronic scattering processes (Sect. 3.3), jet physics (Sect. 3.3.5), and many other problems in particle physics.

On the other hand, QCD as the theory of the strong interaction becomes strongly interacting at low energies. In this regime, the non-perturbative effects and the topological properties of a non-abelian gauge theory (Sect. 2.9) are essential. Non-perturbative techniques for the evaluation of QCD include the strong-coupling approximation (Sect. 2.10.2), the semi-classical approximation (Sect. 2.9.3.1), and, above all, lattice methods (Sect. 2.10). They enable the calculation of the masses and of the inner structure of the hadrons (Sect. 3.6) and lead to an understanding of confinement of quarks and gluons (Sect. 3.7).

[1]For the history of the discovery of asymptotic freedom see [tH85].

An important issue is the connection between low-energy physics and the region of asymptotic freedom. This gap is bridged by lattice gauge theory, (Sect. 3.6.3), factorization theorems (Sect. 3.3.2), and skillful application of the renormalization group. Indeed, the present status of QCD is characterized by the consistency and success of all these different methods and their results (Sect. 3.8).

Models based on QCD like the heavy-quark effective theory for quarkonia (Sect. 3.4), chiral perturbation theory for low-energy physics (Sect. 3.5), or sum rules are the basis for further applications in phenomenology.

3.1 Asymptotic freedom of QCD

In Sect. 2.6.2 we have seen that QCD as a non-abelian gauge theory is asymptotically free if the number of quarks is not too big. This allows the application of perturbation theory for hard-scattering processes. This section deals with physical consequences of the renormalization-group-improved scaling behaviour and introduces the concepts of the running coupling and the running quark masses of QCD.

3.1.1 The running coupling constant of QCD

The renormalization-group equation for the vertex functions in QCD can be solved with the help of (2.6.22) and (2.6.24). First of all, (2.6.22) and (2.6.36) yield the *one-loop approximation* of the scale dependence of the *running coupling constant* of massless QCD

$$\bar{g}_{\mathrm{s}}^2(\lambda, g_{\mathrm{s}}) = \frac{g_{\mathrm{s}}^2}{1 + \frac{g_{\mathrm{s}}^2}{16\pi^2}\beta_0 \log \lambda^2} \tag{3.1.1}$$

with [cf. (2.6.36) and (2.6.40)]

$$\beta_0 = \frac{11}{3}C_{\mathrm{F}} - \frac{2}{3}T_{\mathrm{F}}N_{\mathrm{f}} = \frac{11}{3}N - \frac{2}{3}N_{\mathrm{f}} \overset{N=3}{\longrightarrow} 11 - \frac{2}{3}N_{\mathrm{f}}, \tag{3.1.2}$$

where N denotes the number of colours [corresponding to the gauge group $SU(N)$] and N_{f} the number of effective massless quark flavours (in the fundamental representation). The running coupling constant $\bar{g}_{\mathrm{s}}$ vanishes for $\lambda \to \infty$. This is the *asymptotic freedom* of QCD.

The value of the strong fine-structure constant $\alpha_s = g_s^2/4\pi$ is linked to the renormalization mass parameter $\mathcal{M}$ by the renormalization condition (2.5.47). Using the fact that $\lambda^2 = Q^2/\mathcal{M}^2$ is the scale factor with which the four-momentum transfer at the vertex is multiplied, (3.1.1) yields

$$\bar{\alpha}_s(Q^2) = \frac{\bar{\alpha}_s(\mathcal{M}^2)}{1 + \frac{\bar{\alpha}_s(\mathcal{M}^2)}{4\pi}\beta_0 \log\frac{Q^2}{\mathcal{M}^2}} \tag{3.1.3}$$

with the initial condition $\alpha_s = \bar{\alpha}_s(\mathcal{M}^2)$. This allows to calculate in one-loop approximation the running coupling $\bar{\alpha}_s(Q^2)$ from the experimentally determined value of $\bar{\alpha}_s$ at $\mathcal{M}^2$. Defining

$$\Lambda^2 = \mathcal{M}^2 \exp\left[-\frac{4\pi}{\beta_0\bar{\alpha}_s(\mathcal{M}^2)}\right], \tag{3.1.4}$$

the momentum dependence of the running strong fine-structure constant $\bar{\alpha}_s(Q^2)$ can be brought into the simple form

$$\bar{\alpha}_s(Q^2) = \frac{4\pi}{\beta_0 \log\frac{Q^2}{\Lambda^2}}. \tag{3.1.5}$$

The *asymptotic scale parameter* Λ of QCD is a renormalization-group invariant quantity, as described in Sect. 2.6.1.4. The introduction of a dimensionful parameter into a dimensionless field theory like massless QCD by the renormalization procedure is known as *dimensional transmutation* [Co73]. The dimensional parameter Λ is closely related to the dimensionless coupling constant and determines like $\alpha_s(\mathcal{M}^2)$ the coupling strength of the strong interaction.

The determination of the running coupling constant from the first-order formula (3.1.3) from two different asymptotic scale parameters Λ and Λ' gives two coupling constants that differ by a second-order term,

$$\begin{aligned}\bar{\alpha}'_s(Q^2) &= \frac{4\pi}{\beta_0 \log Q^2/\Lambda'^2} = \frac{4\pi}{\beta_0(\log Q^2/\Lambda^2 + \log \Lambda^2/\Lambda'^2)} \\ &= \bar{\alpha}_s(Q^2) + \frac{\beta_0}{4\pi}\log\left(\frac{\Lambda'^2}{\Lambda^2}\right)\bar{\alpha}_s^2(Q^2) + \dots .\end{aligned} \tag{3.1.6}$$

Thus, different values of Λ lead to the same running coupling constant in first order. A second-order calculation is needed to fix Λ.

Upon replacing the renormalized coupling constant $\alpha_s(\mathcal{M}^2)$ by the running coupling constant $\bar{\alpha}_s(Q^2)$ in the quark–gluon vertex, contributions of arbitrary order in α_s are partially summed up, as it is obvious from (3.1.3). These $(\log Q^2/\Lambda^2)^n$ contributions come from higher-loop corrections to the elementary vertex. In this sense, application of the renormalization group (RG) leads beyond perturbation theory of finite order since it sums up the *leading-logarithmic* (LL) terms. Methods for partial summation of logarithmic contributions to the perturbation series are described in Sects. 2.8.3.8 and 3.3.3.2.

According to (2.6.34) and (2.6.35) the β- and γ-functions of QCD have the structure of (2.6.20) in one-loop order. Therefore, the explicit solution of the RG equation of a vertex function with n_g gluon and n_q quark fields is according to (2.6.24) given by

$$\begin{aligned}\Gamma_R^{n_g,n_q}(\lambda k_i, \lambda p_j; \alpha_s; \mathcal{M}) &= \Gamma_R^{n_g,n_q}(k_i, p_j; \bar{\alpha}_s(\lambda, \alpha_s); \mathcal{M}) \\ &\quad \times \lambda^{4-n_g-3n_q/2} \left(\frac{\bar{\alpha}_s}{\alpha_s}\right)^{(n_g\gamma_{g,0}+n_q\gamma_{q,0})/2\beta_0} \\ &= \Gamma_R^{n_g,n_q}(k_i, p_j; \bar{\alpha}_s(\lambda, \alpha_s); \mathcal{M}) \\ &\quad \times \lambda^{4-n_g-3n_q/2} \left(1 + \frac{\alpha_s}{4\pi}\beta_0 \log \lambda^2\right)^{-(n_g\gamma_{g,0}+n_q\gamma_{q,0})/2\beta_0}\end{aligned} \tag{3.1.7}$$

in one-loop approximation. The QCD vertex functions show an asymptotic scaling behaviour in which the dynamical dimensions occur in the form $(\log \lambda^2)^{-(n_g\gamma_{g,0}+n_q\gamma_{q,0})/2\beta_0}$.

We discuss as an example the two-point vertex function of massless quarks, the inverse of the *quark propagator*

$$\Gamma_R^{0,2}(p', -p'; \alpha_s; \mathcal{M}) = \Gamma_R^{0,2}(p, -p; \bar{\alpha}_s(\lambda, \alpha_s); \mathcal{M})\lambda \left(\frac{\bar{\alpha}_s(p'^2)}{\bar{\alpha}_s(p^2)}\right)^{\gamma_{q,0}/\beta_0}, \tag{3.1.8}$$

where $p' = \lambda p$, $\gamma_{q,0} = C_F\xi = 4\xi/3$, and $\beta_0 = 11 - 2N_f/3$ [cf. (2.6.36)]. Introducing the *anomalous dimension of the quark propagator*

$$d_{q,0} = \frac{\gamma_{q,0}}{\beta_0} = \frac{3\xi C_F}{11C_A - 4T_F N_f} = \frac{4\xi}{33 - 2N_f}, \tag{3.1.9}$$

which depends on the gauge parameter ξ (vanishing only in the Landau gauge), we obtain for the quark propagator

$$S_{\mathrm{F,R}}(p', \alpha_{\mathrm{s}}) = S_{\mathrm{F,R}}(p, \bar{\alpha}_{\mathrm{s}})\lambda^{-1}\left(\frac{\log\frac{p'^2}{\Lambda^2}}{\log\frac{p^2}{\Lambda^2}}\right)^{d_{\mathrm{q},0}} . \tag{3.1.10}$$

The asymptotic behaviour ($p^2 \gg \Lambda^2$) then is

$$S_{\mathrm{F,R}}(p) \propto \frac{\mathrm{i}}{\not{p}}\left(\log\frac{p^2}{\Lambda^2}\right)^{\frac{2\xi}{33-2N_{\mathrm{f}}}} . \tag{3.1.11}$$

3.1.2 Importance of higher-order corrections

In Sect. 3.1.1, we outlined asymptotic freedom under the most simple conditions, i.e. lowest order in perturbation theory and a mass-independent renormalization scheme like the MS or $\overline{\mathrm{MS}}$ scheme. There are many results available in the literature which go beyond this. At the two-loop level the running coupling is obtained from (2.6.17) and (2.6.34) as

$$\begin{aligned}\log\frac{Q^2}{\mathcal{M}^2} &= 2\int_{\bar{g}_{\mathrm{s}}(\mathcal{M}^2)}^{\bar{g}_{\mathrm{s}}(Q^2)}\frac{dg'}{\beta(g')} = -\frac{32\pi^2}{\beta_0}\int_{\bar{g}_{\mathrm{s}}(\mathcal{M}^2)}^{\bar{g}_{\mathrm{s}}(Q^2)}\frac{dg'}{g'^3+\frac{\beta_1}{16\pi^2\beta_0}g'^5}\\ &= \frac{16\pi^2}{\beta_0}\left(\frac{1}{\bar{g}_{\mathrm{s}}^2(Q^2)}-\frac{1}{\bar{g}_{\mathrm{s}}^2(\mathcal{M}^2)}\right)\\ &\quad+\frac{\beta_1}{\beta_0^2}\left(\log\frac{\bar{g}_{\mathrm{s}}^2(Q^2)}{1+\frac{\beta_1\bar{g}_{\mathrm{s}}^2(Q^2)}{16\pi^2\beta_0}}-\log\frac{\bar{g}_{\mathrm{s}}^2(\mathcal{M}^2)}{1+\frac{\beta_1\bar{g}_{\mathrm{s}}^2(\mathcal{M}^2)}{16\pi^2\beta_0}}\right)\end{aligned} \tag{3.1.12}$$

with [cf. (2.6.37) and (2.6.41)]

$$\begin{aligned}\beta_1 &= \frac{34}{3}(C_{\mathrm{A}})^2-\frac{20}{3}C_{\mathrm{A}}T_R-4C_RT_R\\ &= \frac{34}{3}N^2-\left(\frac{13}{3}N-\frac{1}{N}\right)N_{\mathrm{f}} \overset{N=3}{\longrightarrow} 102-\frac{38}{3}N_{\mathrm{f}} .\end{aligned} \tag{3.1.13}$$

Defining the *asymptotic scale parameter* Λ in this approximation as

$$\Lambda^2 = \mathcal{M}^2 \exp\left(-2\int^{g_s} \frac{dg'}{\beta(g')}\right)$$
$$= \mathcal{M}^2 \exp\left(-\frac{16\pi^2}{\beta_0 \bar{g}_s^2(\mathcal{M}^2)} - \frac{\beta_1}{\beta_0^2}\log\frac{\frac{\beta_1 \bar{g}_s^2(\mathcal{M}^2)}{16\pi^2\beta_0}}{1+\frac{\beta_1 \bar{g}_s^2(\mathcal{M}^2)}{16\pi^2\beta_0}}\right), \tag{3.1.14}$$

yields the *two-loop equation for the running coupling constant*

$$\frac{1}{\bar{\alpha}_s(Q^2)} + \frac{\beta_1}{4\pi\beta_0}\log\frac{\frac{\beta_1}{4\pi\beta_0}\bar{\alpha}_s(Q^2)}{1+\frac{\beta_1}{4\pi\beta_0}\bar{\alpha}_s(Q^2)} = \frac{\beta_0}{4\pi}\log\frac{Q^2}{\Lambda^2}. \tag{3.1.15}$$

An approximate solution of this transcendental equation can be obtained by expanding into powers of $(\log Q^2/\Lambda^2)^{-1}$,

$$\frac{\bar{\alpha}_s(Q^2)}{4\pi} = \frac{1}{\beta_0 \log Q^2/\Lambda^2} - \frac{\beta_1}{\beta_0^3}\frac{\log(\log Q^2/\Lambda^2)}{(\log Q^2/\Lambda^2)^2} + \frac{\beta_1}{\beta_0^3}\frac{\log\frac{\beta_1}{4\pi\beta_0^2}}{(\log Q^2/\Lambda^2)^2}$$
$$+\mathcal{O}\left(\frac{(\log\log Q^2/\Lambda^2)^2}{(\log Q^2/\Lambda^2)^3}\right). \tag{3.1.16}$$

Taking only the first two terms means neglecting $(\log Q^2/\Lambda^2)^{-2}$ terms, i.e. squares of the first-order result, a convention which was often used in the literature. The values of the asymptotic scale parameter determined from the full equation (3.1.15), Λ, and from (3.1.16) omitting the $(\log Q^2/\Lambda^2)^{-2}$ term, Λ', are related by

$$\Lambda = \left(\frac{\beta_0^2}{\beta_1}\right)^{\frac{\beta_1}{2\beta_0^2}} \Lambda'. \tag{3.1.17}$$

For five quark flavours this is a 10% difference.

The scale parameter Λ depends on the regularization (dimensional, Lattice, ...) and renormalization scheme (MS, $\overline{\text{MS}}$, MOM, ...) (cf. Sects. 2.5.3.3 and 2.10.4.3). Since the β-function is scheme-invariant up to the second order, the transition to another scheme can be compensated up to this order by rescaling Λ, e.g. $\Lambda_{\overline{\text{MS}}} = k\Lambda_{\text{MS}}$.

The asymptotic scale parameter Λ of QCD determines the strength of the strong interaction and defines the energy scale where an evaluation of QCD based on perturbative calculations is possible. Therefore, the higher-order coefficients of the loop expansion of the β-function of QCD are important quantities. They are known in the $\overline{\text{MS}}$ scheme analytically in the third [Ta81], (2.6.41), and the fourth order [Ri97].

Precise data come from experiments at the Z-boson resonance. Therefore, it has become customary to use the value of the running QCD coupling constant at $Q^2 = M_Z^2$ instead of Λ. The present value for $\bar{\alpha}_s(M_Z^2)$ is [PDG00]

$$\bar{\alpha}_s(M_Z^2) = 0.1181 \pm 0.0020. \tag{3.1.18}$$

This corresponds to

$$\Lambda^{(5)}_{\overline{\text{MS}}} = 208 \begin{array}{l} +25 \\ -23 \end{array} \text{MeV}, \tag{3.1.19}$$

in the $\overline{\text{MS}}$ scheme for five active flavours [PDG00].

In Fig. 3.1 we present $\beta(\alpha_s)$ (left figure) and the running $\bar{\alpha}_s(Q^2)$ of QCD for $\bar{\alpha}_s(M_Z^2) = 0.115$ (right figure) in one- to four-loop approximation taking into account the quark thresholds (variable N_f) [vN00]. The two-loop result (NLO) for $\beta(\alpha_s)$ still gives a sensible correction to the one-loop result (LO). The three-loop corrections (NNLO) are already small, and the curve which includes the four-loop corrections (N^{3}LO) is indistinguishable from the one including the three-loop contributions. The situation is similar for $\bar{\alpha}_s(Q^2)$.

3.1.3 The running quark masses of QCD

The renormalization-group equation of QCD determines besides the running coupling constant also the *running quark masses*. Their behaviour is governed according to (2.6.23) by the β-function and the renormalization-group coefficient γ_m of the mass term. The latter follows from the mass renormalization constant Z_m by

$$\gamma_m = \frac{\partial \log Z_m}{\partial \log \mathcal{M}}. \tag{3.1.20}$$

The mass renormalization constant Z_m is gauge-independent; the one-loop result was derived for QED in (2.5.111). In QCD it is given by

$$\delta Z_m = -\frac{g_s^2}{16\pi^2} 3C_F \left(\frac{1}{\delta} - \gamma + \log\frac{4\pi\mu^2}{\mathcal{M}^2}\right). \tag{3.1.21}$$

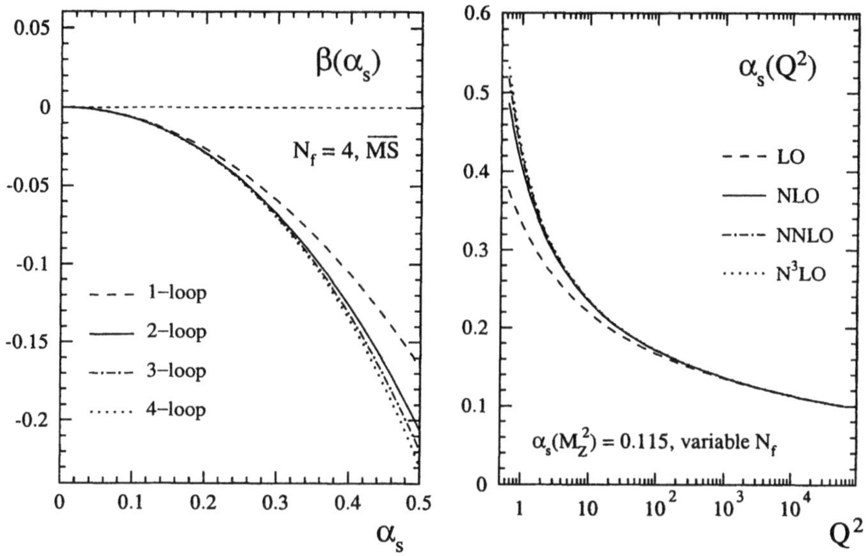

Fig. 3.1 β-function and running of $\alpha_s(Q^2)$ in different orders of perturbation theory (taking into account threshold effects) (from Ref. [vN00])

Writing

$$\gamma_m = \frac{g_s^2}{16\pi^2}\gamma_{m,0} + \left(\frac{g_s^2}{16\pi^2}\right)^2 \gamma_{m,1} + \dots\,, \tag{3.1.22}$$

(3.1.20) and (3.1.21) yield

$$\gamma_{m,0} = 6C_F \overset{SU(3)}{\longrightarrow} 8. \tag{3.1.23}$$

Inserting this in (2.6.23) finally gives for the running mass in one-loop approximation

$$\frac{\overline{m}(Q^2)}{m} = \left(\frac{1}{2}\log\frac{Q^2}{\Lambda^2}\right)^{-d_{m,0}} \left(\frac{4\pi}{\beta_0\alpha_s}\right)^{d_{m,0}} \frac{\mathcal{M}}{\sqrt{Q^2}} \tag{3.1.24}$$

with the *anomalous dimension of the mass*

$$d_{m,0} = \frac{\gamma_{m,0}}{2\beta_0} = \frac{9C_\mathrm{F}}{11C_\mathrm{A} - 4T_\mathrm{F}N_\mathrm{f}} \overset{\mathrm{SU(3)}}{\longrightarrow} \frac{12}{33 - 2N_\mathrm{f}}. \tag{3.1.25}$$

Introducing a renormalization-group-invariant integration constant $\hat{m}$ (similar to Λ), (3.1.24) can be written as

$$\frac{\overline{m}(Q^2)}{\hat{m}} = \left(\frac{1}{2}\log\frac{Q^2}{\Lambda^2}\right)^{-d_{m,0}}. \tag{3.1.26}$$

The anomalous dimension $d_{m,0}$ is flavour-independent in leading order [cf. (3.1.25)]. Consequently in this approximation, ratios of running quark masses are scale- and scheme-independent. The running mass gets smaller for $Q^2 \to \infty$.

The two-loop result for the anomalous mass dimension in the $\overline{\mathrm{MS}}$ scheme and the running mass is also known [Ta81, Na81]. As a typical result we give

$$\frac{\overline{m}(1\,\mathrm{GeV})}{\overline{m}(M_\mathrm{Z})} = 2.30 \pm 0.05. \tag{3.1.27}$$

The treatment of the quark thresholds and the corresponding matching conditions can be found in Refs. [Be83, Ma84]. The three-loop coefficient $\gamma_{m,2}$ has been calculated in Ref. [Ta82], and the four-loop coefficient $\gamma_{m,3}$ in Refs. [Ve97, Ch97].

The values of the quark masses are not only scale-dependent but, since quarks do not exist as free particles, also scheme-dependent. Following Ref. [PDG00] we give the $\overline{\mathrm{MS}}$ quark masses for the light quarks (u, d, s) at the scale 2 GeV and for the heavy quarks at the scale of the mass,

$$\begin{aligned} m_\mathrm{u} &= 1 \text{ to } 5\,\mathrm{MeV}, & m_\mathrm{d} &= 3 \text{ to } 9\,\mathrm{MeV}, \\ m_\mathrm{c} &= 1.15 \text{ to } 1.35\,\mathrm{GeV}, & m_\mathrm{s} &= 75 \text{ to } 170\,\mathrm{MeV}, \\ m_\mathrm{t} &= 174.3 \pm 5.1\,\mathrm{GeV}, & m_\mathrm{b} &= 4.0 \text{ to } 4.4\,\mathrm{GeV}. \end{aligned} \tag{3.1.28}$$

The masses of the u, d, s, c, and b quarks at other scales are obtained in one-loop order by using the scaling factor of (3.1.26). There is some experimental evidence for the running of the b-quark mass [Ab98, Br99].

The fermion propagator and other Green functions are, as we have seen, gauge-dependent. Therefore, it is necessary to introduce a *running gauge parameter* $\xi(Q^2)$.[2] The result is [Na82]

$$\xi(Q^2) = \hat{\xi}\frac{(\frac{1}{2}\log\frac{Q^2}{\Lambda^2})^{-d_{\xi,0}}}{1+\hat{\xi}\frac{3N}{13N-4N_\mathrm{f}}(\frac{1}{2}\log\frac{Q^2}{\Lambda^2})^{-d_{\xi,0}}} \tag{3.1.29}$$

where $\hat{\xi}$ is a RG-invariant parameter and the *anomalous gauge dimension* reads

$$d_{\xi,0} = \frac{13N-4N_\mathrm{f}}{2(11N-2N_\mathrm{f})} = \frac{39-4N_\mathrm{f}}{2(33-2N_\mathrm{f})}. \tag{3.1.30}$$

3.2 QCD in deep-inelastic scattering

The renormalizability of QCD allows to work out a consistent perturbation theory. Getting asymptotically free at high momentum transfers, QCD can be applied to hard-scattering processes. This is carried out for deep-inelastic lepton–nucleon scattering in this section.

The basic concept of the parton model (cf. Sect. 1.5) is visualized once again in Fig. 3.2a. Accordingly, the inclusive cross section is obtained by calculating the parton cross sections for the scattering of leptons off free, massless quarks and multiplying it with the probability of finding these quarks with momentum fraction x in the nucleon $\mathcal{N}$. In the framework of QCD this model has to be extended in order α_s by *gluon bremsstrahlung* (Fig. 3.2b), quark-pair creation (Fig. 3.2c), and internal radiative corrections (Fig. 3.2d). In higher orders more and more diagrams contribute.

In this section we first present the field-theoretical formulation of the parton model (Sect. 3.2.1). Then the operator-product expansion together with the renormalization-group is used for a partial summation of the perturbation series in Sect. 3.2.2. Finally, the evolution equations for the Q^2 dependence of the structure functions are discussed in Sect. 3.2.3 and applied to the experimental results in Sect. 3.2.4.

[2] See the remarks after (2.6.11).

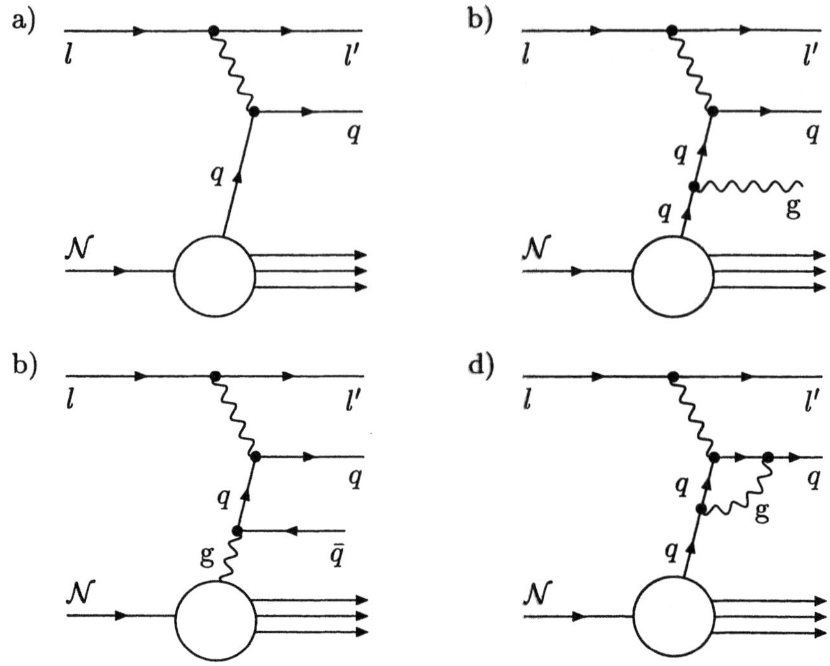

Fig. 3.2 Deep-inelastic lepton–nucleon scattering in the naïve parton model (a) and QCD corrections in order α_s (b,c,d)

3.2.1 The field-theoretical approach to the parton model

3.2.1.1 Properties of the hadronic tensor

As in Sect. 1.5.1, the starting point is the *spin-averaged hadronic tensor* $W_{\mu\nu}(p,q)$ expressed by the current matrix elements between the incoming hadron $|p,\lambda\rangle$, $p^2 = M$, and an arbitrary hadronic final state $|X,p_X\rangle$,

$$W_{\mu\nu}(p,q) = \frac{1}{4\pi}\sum_X (2\pi)^4 \delta^{(4)}(q+p-p_X) \times \frac{1}{2}\sum_\lambda \langle p,\lambda|h_\mu^\dagger(0)|X,p_X\rangle\langle X,p_X|h_\nu(0)|p,\lambda\rangle. \qquad (3.2.1)$$

Using the Fourier representation of the δ-function,

$$(2\pi)^4\delta^{(4)}(p-p_X+q) = \int \mathrm{d}^4y\, \mathrm{e}^{\mathrm{i}qy}\, \mathrm{e}^{\mathrm{i}(p-p_X)y}, \qquad (3.2.2)$$

translation invariance,

$$\langle p|h^\dagger_\mu(y)|p_X\rangle = \mathrm{e}^{\mathrm{i}y(p-p_X)}\langle p|h^\dagger_\mu(0)|p_X\rangle, \tag{3.2.3}$$

and the completeness of the states $|X, p_X\rangle$, this can be written as[3]

$$W_{\mu\nu}(p,q) = \int \frac{\mathrm{d}^4y}{4\pi}\,\mathrm{e}^{\mathrm{i}qy}\langle p|\,[h^\dagger_\mu(y), h_\nu(0)]\,|p\rangle. \tag{3.2.4}$$

The *current–current commutator* instead of the product is obtained by subtracting from (3.2.1) a similar expression where the δ-function (3.2.2) is replaced by $\delta^{(4)}(p - p_X - q)$, and the matrix element by $\langle p|h_\nu(0)h^\dagger_\mu(y)|p\rangle$. This contribution vanishes, since in the nucleon rest frame $q^0 = \nu > 0$ and $p^0 - p^0_X \leq 0$ imply $q^0 - p^0 + p^0_X \neq 0$. In the following we assume $p^2 = M^2 \ll pq$ and neglect p^2 where appropriate.

The next step is the decomposition of the hadronic tensor into Lorentz covariants and *structure functions* [cf. (1.5.12) and (1.5.13)]

$$\begin{aligned} W_{\mu\nu}(p,q) &= -g^{\mathrm{T}}_{\mu\nu}F_1 + \frac{p^{\mathrm{T}}_\mu p^{\mathrm{T}}_\nu}{pq}F_2 - \mathrm{i}\varepsilon_{\mu\nu\rho\sigma}\frac{p^\rho q^\sigma}{pq}F_3 \\ &= \left(-\frac{g^{\mathrm{T}}_{\mu\nu}}{2x} + \frac{p^{\mathrm{T}}_\mu p^{\mathrm{T}}_\nu}{pq}\right)F_2 + \frac{g^{\mathrm{T}}_{\mu\nu}}{2x}F_{\mathrm{L}} - \mathrm{i}\varepsilon_{\mu\nu\rho\sigma}\frac{p^\rho q^\sigma}{pq}F_3 \end{aligned} \tag{3.2.5}$$

with

$$g^{\mathrm{T}}_{\mu\nu} = g_{\mu\nu} - \frac{q_\mu q_\nu}{q^2}, \qquad p^{\mathrm{T}}_\mu = p_\mu - q_\mu\frac{pq}{q^2}, \tag{3.2.6}$$

and

$$x = \frac{-q^2}{2pq} = \frac{Q^2}{2pq}. \tag{3.2.7}$$

---

[3]The average over the hadron spin $1/2\sum_\lambda$ is not written explicitly in the following.

In D dimensions, the inversion of this decomposition reads

$$\begin{aligned}
F_1(x,Q^2) &= \frac{1}{D-2}\left(-g^{\mathrm{T}}_{\mu\nu} + 2x\frac{p^{\mathrm{T}}_{\mu}p^{\mathrm{T}}_{\nu}}{pq}\right) W^{\mu\nu}(p,q),\\
\frac{1}{2x}F_2(x,Q^2) &= \frac{1}{D-2}\left(-g^{\mathrm{T}}_{\mu\nu} + (D-1)2x\frac{p^{\mathrm{T}}_{\mu}p^{\mathrm{T}}_{\nu}}{pq}\right) W^{\mu\nu}(p,q),\\
F_{\mathrm{L}}(x,Q^2) &= -2xF_1(x,q^2) + F_2(x,q^2) = 4x^2\frac{p^{\mathrm{T}}_{\mu}p^{\mathrm{T}}_{\nu}}{pq} W^{\mu\nu}(p,q),\\
F_3(x,Q^2) &= \mathrm{i}\varepsilon_{\mu\nu\rho\sigma}\frac{p^{\rho}q^{\sigma}}{2pq} W^{\mu\nu}(p,q). \qquad (3.2.8)
\end{aligned}$$

3.2.1.2 Light-cone behaviour of the free current commutator

We are interested in the behaviour of $W_{\mu\nu}$, (3.2.4), in the *Bjorken limit*, i.e. for $Q^2 \to \infty$, $\nu = pq/M \to \infty$, $x = Q^2/2M\nu$ fixed. According to the method of stationary phase this comes from that part of the integration region where qy remains finite. In the laboratory frame, where $q^{\mu} = (\nu, 0, 0, \sqrt{\nu^2 + Q^2})$, we obtain

$$|qy| = \nu\left|y_0 - y_3\sqrt{1 + 2Mx/\nu}\right| \underset{\nu\to\infty}{\sim} \nu|y_0 - y_3|. \qquad (3.2.9)$$

In order to keep $|qy|$ finite for $\nu \to \infty$ this requires $|y_0 - y_3| = 0$. Hence the essential contribution to the current commutator in the Bjorken limit originates from the vicinity of the *light cone*.

Next we rederive the results of the quark-parton model for the hadronic tensor $W^{VV'N}_{\mu\nu}$ of deep-inelastic lepton–nucleon scattering (cf. Sect. 1.5.2) in a field-theoretical manner using the commutation relations for free, massless quark fields. Here we consider general vector and axial-vector quark currents

$$\begin{aligned}
h^a_{\mu}(y) &= \sum_{q_1,q_2,c} :\bar{\psi}_{q_1,c}(y)\gamma_{\mu}\lambda^a_{q_1q_2}\psi_{q_2,c}(y):,\\
h^a_{5,\mu}(y) &= \sum_{q_1,q_2,c} :\bar{\psi}_{q_1,c}(y)\gamma_{\mu}\gamma_5\lambda^a_{q_1q_2}\psi_{q_2,c}(y):, \qquad (3.2.10)
\end{aligned}$$

where $\lambda^a_{q_1 q_2}$ denotes the Gell-Mann matrix (cf. Sect. 1.2.2.2) in the flavour space of the quarks.[4] The *electromagnetic, charged, and neutral weak currents* (cf. Sect. 1.4.2.1) are linear combinations of these general flavour currents

$$\begin{aligned} h^\gamma_\mu(y) &= \sum_q \bar{\psi}_q(y)\gamma_\mu Q_q \psi_q(y) = \bar{\psi}(y)\gamma_\mu Q_{\mathrm{q}} \psi(y), \\ h^Z_\mu(y) &= \sum_q \bar{\psi}_q(y)\gamma_\mu (v_q - \gamma_5 a_q)\psi_q(y), \\ h^+_\mu(y) &= \sum_{ij} \bar{\psi}_{u_i}(y)\gamma_\mu \frac{1-\gamma_5}{2} \mathbf{V}_{ij}\psi_{d_j}(y). \end{aligned} \tag{3.2.11}$$

For illustration we give the relation between the Gell-Mann-matrix notation in (3.2.10) and the one in (3.2.11),

$$Q_{\mathrm{q}} = \mathrm{diag}\left(\frac{2,-1,-1}{3}\right) = \frac{\lambda^3 + \lambda^8/\sqrt{3}}{2}, \tag{3.2.12}$$

for the electromagnetic current in the case of three quark flavours.

The Bjorken limit of the structure functions is obtained from the commutator of the quark currents[5]

$$\begin{aligned} \left[h^{a\dagger}_\mu(x), h^b_\nu(y)\right] = \sum_{q_i} \Big(& \bar{\psi}_{q_1}(x)\gamma_\mu \lambda^a_{q_1 q_2}\{\psi_{q_2}(x), \bar{\psi}_{q_3}(y)\}\gamma_\nu \lambda^b_{q_3 q_4}\psi_{q_4}(y) \\ & - \bar{\psi}_{q_3}(y)\gamma_\nu \lambda^b_{q_3 q_4}\{\psi_{q_4}(y), \bar{\psi}_{q_1}(x)\}\gamma_\mu \lambda^a_{q_1 q_2}\psi_{q_2}(x)\Big). \end{aligned} \tag{3.2.13}$$

Inserting the anticommutator of free massless quark fields, (1.2.12),

$$\{\psi_{q_1}(x), \bar{\psi}_{q_2}(y)\} = \delta_{q_1 q_2}(\mathrm{i}\gamma_\lambda \partial^\lambda_x)\mathrm{i}\Delta(x-y), \tag{3.2.14}$$

and using the relation

$$\lambda^a \lambda^b = (\mathrm{i}f^{abc} + d^{abc})\lambda^c + \frac{2}{3}\delta^{ab}\mathbf{1} \tag{3.2.15}$$

[4]In the following we suppress the symbols : . . . : for normal ordering and the colour index c.

[5]We here restrict the treatment to vector currents.

for the product of Gell-Mann matrices yields

$$\begin{aligned}\left[h_\mu^{a\dagger}(x), h_\nu^b(y)\right] = {}& \\ & - f^{abc}\left(\bar\psi(x)\gamma_\mu\gamma_\lambda\gamma_\nu\lambda^c\psi(y)\, \mathrm{i}\partial_x^\lambda\Delta(x-y) + (x\leftrightarrow y, \mu\leftrightarrow\nu)\right) \\ & + \mathrm{i}d^{abc}\left(\bar\psi(x)\gamma_\mu\gamma_\lambda\gamma_\nu\lambda^c\psi(y)\, \mathrm{i}\partial_x^\lambda\Delta(x-y) - (x\leftrightarrow y, \mu\leftrightarrow\nu)\right) \\ & + \mathrm{i}\frac{2}{3}\delta^{ab}\left(\bar\psi(x)\gamma_\mu\gamma_\lambda\gamma_\nu\mathbf{1}\psi(y)\, \mathrm{i}\partial_x^\lambda\Delta(x-y) - (x\leftrightarrow y, \mu\leftrightarrow\nu)\right).\end{aligned} \tag{3.2.16}$$

Decomposing the product of Dirac matrices with help of the Chisholm identity

$$\begin{aligned}\gamma_\mu\gamma_\lambda\gamma_\nu &= (g_{\mu\lambda}g_{\nu\rho} + g_{\lambda\nu}g_{\mu\rho} - g_{\mu\nu}g_{\lambda\rho})\,\gamma^\rho - \mathrm{i}\varepsilon_{\mu\lambda\nu\rho}\gamma_5\gamma^\rho \\ &= \sigma_{\mu\lambda\nu\rho}\gamma^\rho - \mathrm{i}\varepsilon_{\mu\lambda\nu\rho}\gamma_5\gamma^\rho,\end{aligned} \tag{3.2.17}$$

finally gives for the current commutator

$$\begin{aligned}\left[h_\mu^{a\dagger}(x), h_\nu^b(y)\right] &= \partial_x^\lambda\Delta(x-y) \\ \times\Big\{ & -\mathrm{i}f^{abc}\Big[\sigma_{\mu\lambda\nu\rho}\left(\bar\psi(x)\gamma^\rho\lambda^c\psi(y) + \bar\psi(y)\gamma^\rho\lambda^c\psi(x)\right) \\ & \qquad - \mathrm{i}\varepsilon_{\mu\lambda\nu\rho}\left(\bar\psi(x)\gamma^\rho\gamma_5\lambda^c\psi(y) - \bar\psi(y)\gamma^\rho\gamma_5\lambda^c\psi(x)\right)\Big] \\ & - d^{abc}\Big[\sigma_{\mu\lambda\nu\rho}\left(\bar\psi(x)\gamma^\rho\lambda^c\psi(y) - \bar\psi(y)\gamma^\rho\lambda^c\psi(x)\right) \\ & \qquad - \mathrm{i}\varepsilon_{\mu\lambda\nu\rho}\left(\bar\psi(x)\gamma^\rho\gamma_5\lambda^c\psi(y) + \bar\psi(y)\gamma^\rho\gamma_5\lambda^c\psi(y)\right)\Big] \\ & - \frac{2}{3}\delta^{ab}\Big[\sigma_{\mu\lambda\nu\rho}\left(\bar\psi(x)\gamma^\rho\psi(y) - \bar\psi(y)\gamma^\rho\psi(x)\right) \\ & \qquad - \mathrm{i}\varepsilon_{\mu\lambda\nu\rho}\left(\bar\psi(x)\gamma^\rho\gamma_5\psi(y) + \bar\psi(y)\gamma^\rho\gamma_5\psi(x)\right)\Big]\Big\}.\end{aligned} \tag{3.2.18}$$

This provides an example, which already shows the typical features, of an *operator-product expansion* (OPE). The expansion consists of singular functions, in this case the derivative of the free scalar commutator function (1.2.9)

$$\begin{aligned}\Delta(x-y) &= -\frac{\mathrm{i}}{(2\pi)^3}\int \mathrm{d}^4q\ \mathrm{sgn}(q^0)\delta(q^2)\exp^{-\mathrm{i}q(x-y)} \\ &= -\frac{1}{2\pi}\,\mathrm{sgn}(x^0-y^0)\delta\left((x-y)^2\right),\end{aligned} \tag{3.2.19}$$

flavour tensors f^{abc}, d^{abc}, δ^{ab}, Lorentz tensors $\sigma_{\mu\lambda\nu\rho}$, $\varepsilon_{\mu\lambda\nu\rho}$, and regular bilocal operators.

The *bilocal operators* are of the vector- and axial-vector type, symmetrized and antisymmetrized,

$$\begin{aligned} \mathcal{O}_{\rho}^{qq',\pm}(x,y) &= \bar{\psi}_q(x)\gamma_\rho\psi_{q'}(y) \pm \bar{\psi}_q(y)\gamma_\rho\psi_{q'}(x), \\ \mathcal{O}_{3,\rho}^{qq',\pm}(x,y) &= \bar{\psi}_q(x)\gamma_\rho\gamma_5\psi_{q'}(y) \pm \bar{\psi}_q(y)\gamma_\rho\gamma_5\psi_{q'}(x), \end{aligned} \tag{3.2.20}$$

such that the total expression (3.2.18) is antisymmetric.

The electroweak currents transform according to the adjoint representation of the *flavour group* [cf. (3.2.10)]. Their commutators, i.e. the bilocal operators, lie in the adjoint representation (λ^c terms) with symmetric (d^{abc}) or antisymmetric (f^{abc}) couplings, or in the singlet representation (**1** terms) with coupling δ^{ab} and are organized in expressions with definite charge-conjugation parity. The corresponding contributions are denoted *flavour non-singlet* (NS)

$$\mathcal{O}_{\rho}^{\mathrm{NS},c,\pm} = \sum_{qq'} \lambda_{qq'}^{c} \mathcal{O}_{\rho}^{qq',\pm}, \qquad \mathcal{O}_{3,\rho}^{\mathrm{NS},c,\pm} = \sum_{qq'} \lambda_{qq'}^{c} \mathcal{O}_{3,\rho}^{qq',\pm}, \tag{3.2.21}$$

and *flavour singlet* (S),

$$\mathcal{O}_{\rho}^{\mathrm{S,q},\pm} = \sum_{q} \mathcal{O}_{\rho}^{qq,\pm}, \qquad \mathcal{O}_{3,\rho}^{\mathrm{S,q},\pm} = \sum_{q} \mathcal{O}_{3,\rho}^{qq,\pm}, \tag{3.2.22}$$

respectively, with *charge-conjugation parity* $\pm$. Inserting these expressions into (3.2.18) gives

$$\begin{aligned} \left[h_\mu^{a\dagger}(x), h_\nu^b(y)\right] &= \partial_x^\lambda \Delta(x-y) \\ &\times\Big\{-\mathrm{i}f^{abc}\Big[\sigma_{\mu\lambda\nu}{}^{\rho}\mathcal{O}_{\rho}^{\mathrm{NS},c,+}(x,y) - \mathrm{i}\varepsilon_{\mu\lambda\nu}{}^{\rho}\mathcal{O}_{3,\rho}^{\mathrm{NS},c,-}(x,y)\Big] \\ &- d^{abc}\Big[\sigma_{\mu\lambda\nu}{}^{\rho}\mathcal{O}_{\rho}^{\mathrm{NS},c,-}(x,y) - \mathrm{i}\varepsilon_{\mu\lambda\nu}{}^{\rho}\mathcal{O}_{3,\rho}^{\mathrm{NS},c,+}(x,y)\Big] \\ &- \frac{2}{3}\delta^{ab}\Big[\sigma_{\mu\lambda\nu}{}^{\rho}\mathcal{O}_{\rho}^{\mathrm{S,q},-}(x,y) - \mathrm{i}\varepsilon_{\mu\lambda\nu}{}^{\rho}\mathcal{O}_{3,\rho}^{\mathrm{S,q},+}(x,y)\Big]\Big\}. \end{aligned} \tag{3.2.23}$$

This decomposition into irreducible representations of the flavour group with definite charge-conjugation parity is important when higher-order QCD corrections are calculated (cf. Sect. 3.2.3).

Similar results are obtained when commutators with axial-vector currents or chiral currents are calculated. This and a more detailed discussion of (3.2.18) is postponed to Sect. 3.2.1.3 where also the generalization to the interacting case in QCD is considered.

We are interested in the contribution of the free current commutator to the nucleon structure functions. In order to work out the essential points we evaluate the contribution of the electromagnetic current $h_\mu^\gamma(x) = \sum_q \bar{\psi}_q(x)\gamma_\mu Q_q \psi_q(x)$. To extract the Bjorken limit of the unpolarized structure functions, the expression (3.2.18) has to be substituted into (3.2.4). For the case of the electromagnetic current, the terms with the structure constants f^{abc} and with γ_5 are absent in the unpolarized case, and using translational invariance ($x = 0$) the hadronic tensor reads

$$W_{\mu\nu}^{\gamma\gamma\mathcal{N}}(p,q) = \frac{1}{4\pi}\int \mathrm{d}^4y\, \mathrm{e}^{\mathrm{i}qy} \sigma_{\mu\lambda\nu}{}^{\rho} \left(\partial_y^\lambda \Delta(y)\right) \sum_q Q_q^2 \times \langle p|\mathcal{O}_\rho^{qq,-}(y,0)|p\rangle, \tag{3.2.24}$$

where $\Delta(y) = -\Delta(-y)$. Written in the group-theoretic notation of (3.2.18), (3.2.21), and (3.2.22), this reads

$$\begin{aligned} W_{\mu\nu}^{\gamma\gamma\mathcal{N}}(p,q) = \frac{1}{4\pi}\int \mathrm{d}^4y\, \mathrm{e}^{\mathrm{i}qy} \sigma_{\mu\lambda\nu}{}^{\rho} \left(\partial_y^\lambda \Delta(y)\right) \\ \times \left(\frac{1}{6}\langle p|\mathcal{O}_\rho^{\mathrm{NS},3,-}(y,0)|p\rangle + \frac{\sqrt{3}}{18}\langle p|\mathcal{O}_\rho^{\mathrm{NS},8,-}(y,0)|p\rangle \right. \\ \left. + \frac{2}{9}\langle p|\mathcal{O}_\rho^{\mathrm{S,q},-}(y,0))|p\rangle\right). \end{aligned} \tag{3.2.25}$$

This representation of $W_{\mu\nu}^{\gamma\gamma\mathcal{N}}$ shows that in the electromagnetic case both non-singlet and singlet operators contribute. The same occurs in neutral- and charged-current reactions.

The spin-averaged nucleon matrix element can be decomposed into Lorentz covariants p^ρ, y^ρ and invariant functions of py, p^2, and y^2,

$$\langle p|\mathcal{O}_\rho^{qq,-}(y,0)|p\rangle = 2p_\rho \tilde{g}_{q,\mathcal{N}}(py,y^2,p^2) + y_\rho M^2 \tilde{h}_{q,\mathcal{N}}(py,y^2,p^2). \tag{3.2.26}$$

Because of the $\delta(y^2)$-function in $\partial_y^\lambda \Delta(y)$ [cf. (3.2.19)] and $p^2 = M^2 \ll Q^2$, this effectively reduces to

$$\langle p|O_\rho^{qq,-}(y,0)|p\rangle = 2p_\rho \tilde{g}_{q,\mathcal{N}}(py), \tag{3.2.27}$$

since the term $\tilde{h}_{q,\mathcal{N}}$ gives a contribution which is suppressed by a factor M^2/pq in momentum space and therefore can be neglected in the Bjorken limit. The form (3.2.27) of the matrix element of the bilocal operator has the consequence that the longitudinal structure function [cf. (3.2.8)]

$$F_{\mathrm{L}} = 4x^2 \frac{p^\mu p^\nu}{pq} W_{\mu\nu} \tag{3.2.28}$$

is proportional to $p^2 = M^2 \ll Q^2$ and can be neglected in the Bjorken limit. The *Callan–Gross relation* (1.5.29) is valid.

Introducing the Fourier representation of $\tilde{g}$,

$$\tilde{g}_{q,\mathcal{N}}(py) = \int_{-\infty}^{+\infty} \mathrm{d}\xi\, g_{q,\mathcal{N}}(\xi) \mathrm{e}^{\mathrm{i}\xi(py)}, \tag{3.2.29}$$

yields

$$\begin{aligned} W_{\mu\nu}^{\gamma\gamma\mathcal{N}}(p,q) &= \\ &= \frac{1}{4\pi}\sum_q Q_q^2 \int \mathrm{d}^4y\, \mathrm{d}\xi\, \mathrm{e}^{\mathrm{i}(q+\xi p)y} g_{q,\mathcal{N}}(\xi) 2p_\rho \sigma_{\mu\lambda\nu}{}^\rho \left(\partial_y^\lambda \Delta(y)\right), \end{aligned} \tag{3.2.30}$$

where the sum over q runs over all quarks. Equation (3.2.8) then yields for the structure function

$$F_1^{\gamma\gamma\mathcal{N}}(pq,q^2) = \sum_q Q_q^2 \int \frac{\mathrm{d}^4 y}{2\pi} \int_{-\infty}^{+\infty} \mathrm{d}\xi\, e^{\mathrm{i}y(q+\xi p)} g_{q,\mathcal{N}}(\xi)\,(p\partial_y)\Delta(y), \tag{3.2.31}$$

and by using after partial integration again the integral (3.2.19),

$$F_1^{\gamma\gamma\mathcal{N}}(pq,q^2) = pq \sum_q Q_q^2 \int \mathrm{d}\xi\, \mathrm{sgn}(q^0 + \xi p^0)\, \delta\Big((q+\xi p)^2\Big) g_{q,\mathcal{N}}(\xi). \tag{3.2.32}$$

Neglecting the nucleon mass, we have in the Bjorken limit, since $q^0, p^0 > 0$,

$$\mathrm{sgn}(q^0 + \xi p^0)\delta\Big((q+\xi p)^2\Big) \approx \delta\Big(q^2 + 2\xi pq\Big) = \frac{1}{2pq}\delta(\xi - x) \quad (3.2.33)$$

with the scaling variable x. Kinematics implies $0 \le x \le 1$ [cf. (1.5.3)]. Thus, performing the ξ integration with help of the δ-function, the Bjorken limit of the structure functions $F_i^{\gamma\gamma\mathcal{N}}$ is obtained as

$$F_1^{\gamma\gamma\mathcal{N}}(x) = \frac{1}{2}\sum_q Q_q^2 g_{q,\mathcal{N}}(x), \qquad F_2^{\gamma\gamma\mathcal{N}}(x) = 2xF_1^{\gamma\gamma\mathcal{N}}(x). \quad (3.2.34)$$

The functions $g_{q,\mathcal{N}}(x)$ are the nucleon expectation values of the bilocal operators $\mathcal{O}_\rho^{qq,-}$ [cf. (3.2.27)]. Writing these operators in terms of quark creation and annihilation operators the physical meaning of $g_{q,\mathcal{N}}(x)$ becomes obvious: $g_{q,\mathcal{N}}(x)$ is the density of quarks and antiquarks with momentum fraction x in the nucleon and thus identical with the quark distribution functions, $g_{q,\mathcal{N}}(x) = f_{q,\mathcal{N}}(x) + f_{\bar{q},\mathcal{N}}(x)$. Therefore the final result is

$$F_1^{\gamma\gamma\mathcal{N}}(x) = \frac{1}{2}\sum_q Q_q^2(f_{q,\mathcal{N}}(x) + f_{\bar{q},\mathcal{N}}(x)),$$
$$F_2^{\gamma\gamma\mathcal{N}}(x) = x\sum_q Q_q^2\,(f_{q,\mathcal{N}}(x) + f_{\bar{q},\mathcal{N}}(x)). \quad (3.2.35)$$

Altogether we found: the light-cone singularity of a free quark field gives Bjorken scaling (and vice versa, see [Bj69]). The Callan–Gross relation (1.5.29) is found to be valid. In addition we obtained a *field-theoretical definition of the quark and antiquark distribution functions* $f_{q,\mathcal{N}}(x)$ and $f_{\bar{q},\mathcal{N}}(x)$. These are the Fourier transforms of the reduced hadron expectation values of the bilocal operators on the *light cone.* In a similar way the parton model results for neutral and charged weak current processes (1.5.28) can be obtained field theoretically.

3.2.1.3 Wilson expansion

Expanding the bilocal operators of (3.2.20), e.g.

$$\mathcal{O}_\rho^{qq,-}(y,0) = \bar{\psi}_q(y)\gamma_\rho\psi_q(0) - \bar{\psi}_q(0)\gamma_\rho\psi_q(y) \quad (3.2.36)$$

into a power series in y,

$$\mathcal{O}_\rho^{qq,-}(y,0) = \sum_{n=0}^{\infty} \frac{y_{\mu_1}\cdots y_{\mu_n}}{n!} \mathcal{O}_{\rho\mu_1\ldots\mu_n}^{qq,-}(0),$$

$$\mathcal{O}_{\rho\mu_1\ldots\mu_n}^{qq,-}(0) = \partial_{\mu_1}\cdots\partial_{\mu_n}\overline{\psi}_q(y)\gamma_\rho\psi_q(0)\Big|_{y=0}, \tag{3.2.37}$$

and substituting this into (3.2.18), one obtains the *Wilson expansion* for the free current commutator, i.e. its representation by singular functions, Lorentz tensors, and local operators, the *Wilson operators* [Wi69].

The mass dimension of the operators $\mathcal{O}_{\rho\mu_1\ldots\mu_n}(0)$ can be calculated as $\dim[\mathcal{O}] = n+3$ according to the above equation. On the other hand, the $\mathcal{O}_{\rho\mu_1\ldots\mu_n}$ are tensors of rank $(n+1)$ and thus possess the maximum angular momentum $l_{\max} = n+1$. The difference between dimension and angular momentum of an operator is called *twist*,

$$\tau = \dim[\mathcal{O}] - l. \tag{3.2.38}$$

Hence the operators $\mathcal{O}$ have minimal twist 2.

The spin-averaged nucleon matrix element of the Wilson operators can be decomposed into covariants and reduced matrix elements,

$$\langle p|\mathcal{O}_{\rho\mu_1\ldots\mu_n}^{qq,-}(0)|p\rangle = \mathrm{i}^n\Big(p_\rho p_{\mu_1}\cdots p_{\mu_n} 2A_{\mathcal{N},n+1}^q \tag{3.2.39}$$

$$+ M^2 p_\rho g_{\mu_1\mu_2} p_{\mu_3}\cdots p_{\mu_n} B_{\mathcal{N},n+1}^q + \text{terms with more } g_{\mu\nu}\text{'s}\Big).$$

The terms containing $g_{\mu\nu}$ have higher twist, $\tau \geq 4$. In the Bjorken limit their contribution to the structure functions is suppressed by the factor (M^2/Q^2). Therefore, their effect is small compared with the leading twist-2 terms.

Again we discuss the electromagnetic case in more detail. Comparing (3.2.27) with (3.2.37) and (3.2.39) shows that the coefficients $A_{\mathcal{N},n+1}^q$ are obtained from the invariant functions $\tilde{g}_{q,\mathcal{N}}(py)$ by power-series expansion,

$$A_{\mathcal{N},n+1}^q = \left(\frac{\mathrm{d}}{\mathrm{d}(\mathrm{i}py)}\right)^n \tilde{g}_{q,\mathcal{N}}(py)\Big|_{py=0}.$$

Using the Fourier transform of $\tilde{g}_{q,\mathcal{N}}(py)$ yields

$$\begin{aligned} A^q_{\mathcal{N},n} &= \left(\frac{\mathrm{d}}{\mathrm{d}(\mathrm{i}py)}\right)^{n-1} \int_0^1 \mathrm{d}x\, g_{q,\mathcal{N}}(x) \mathrm{e}^{\mathrm{i}(py)x} \Big|_{py=0} \\ &= \int_0^1 \mathrm{d}x\, x^{n-1} g_{q,\mathcal{N}}(x) = \int_0^1 \mathrm{d}x\, x^{n-1} (f_{q,\mathcal{N}}(x) + f_{\bar{q},\mathcal{N}}(x)). \end{aligned} \tag{3.2.40}$$

The x integral extends from 0 to 1 only, since $g_{q,\mathcal{N}}(x)$ vanishes for $x < 0$ and $x > 1$ for kinematical reasons [cf. (1.5.3)].

The reduced matrix elements $A^q_{\mathcal{N},n}$ of the Wilson operators $\mathcal{O}^{qq,-}_{\rho\mu_1\ldots\mu_n}$ are the sums of the *Mellin moments* $f_{q,\mathcal{N},n}$ and $f_{\bar{q},\mathcal{N},n}$ of the *quark and antiquark distribution functions.* They determine the *moments of the structure functions* in the Bjorken limit,

$$F^{\gamma\gamma\mathcal{N}}_{1,n} = \int_0^1 \mathrm{d}x\, x^{n-1} F^{\gamma\gamma\mathcal{N}}_1(x) = \frac{1}{2} \sum_q Q_q^2 (f_{q,\mathcal{N},n} + f_{\bar{q},\mathcal{N},n}). \tag{3.2.41}$$

The inversion of (3.2.41) is obtained by performing an inverse Mellin transformation.

Putting everything together, we can write

$$\begin{aligned} &\int_0^1 \mathrm{d}x\, x^{n-1} W^{\gamma\gamma\mathcal{N}}_{\mu\nu}(p,q) = \\ &= \left(-g^{\mathrm{T}}_{\mu\nu} + 2x \frac{p^{\mathrm{T}}_\mu p^{\mathrm{T}}_\nu}{pq}\right) \frac{1}{2} \sum_q Q_q^2 (f_{q,\mathcal{N},n} + f_{\bar{q},\mathcal{N},n}) \end{aligned} \tag{3.2.42}$$

for the electromagnetic structure function in the case of free massless quarks as partons. Similar expressions result for the case of neutral and charged weak-current deep-inelastic processes.

3.2.2 QCD corrections to the parton model

In Sect. 3.2.1, the parton-model results for the structure functions in deep-inelastic scattering were derived for free quarks using the Wilson expansion of the current commutators. In this section we generalize this procedure to the case of interacting quarks in the one-loop approximation. We consider the quark currents (3.2.10), which contain as special cases the electromagnetic as well as the charged and neutral weak currents (3.2.11). Then, all three structure functions $F_1^{VV'\mathcal{N}}$, $F_2^{VV'\mathcal{N}}$, and $F_3^{VV'\mathcal{N}}$ occur. Moreover, besides the quarks also the gluons have to be taken into account as partons. As a consequence, the Callan–Gross relation is violated. Therefore, we consider the longitudinal structure function $F_{\mathrm{L}} = F_2 - 2xF_1$ instead of $F_1^{VV'\mathcal{N}}$ and introduce the index $a = \{2, \mathrm{L}, 3\}$ corresponding to the projectors (3.2.8) on the structure functions F_2, F_{L}, and the parity-violating part F_3.

3.2.2.1 Operator-product expansion in higher orders

The operator-product expansion for the general case has been introduced in Ref. [Wi69, Ka69]. Using this and the facts discussed above, the moments of the hadronic tensor can be in general expressed as

$$\int_0^1 \mathrm{d}x\, x^{n-1} W_{\mu\nu}^{VV'\mathcal{N}}(Q^2, x) = \frac{1}{2}\sum_k \Bigg[\left(-g_{\mu\nu}^{\mathrm{T}} + 2x\frac{p_\mu^{\mathrm{T}} p_\nu^{\mathrm{T}}}{pq}\right) C_{2,n}^{VV'k} A_{2,\mathcal{N},n}^k + g_{\mu\nu}^{\mathrm{T}} C_{\mathrm{L},n}^{VV'k} A_{\mathrm{L},\mathcal{N},n}^k - \mathrm{i}\varepsilon_{\mu\nu\rho\sigma}\frac{p^\rho q^\sigma}{pq} C_{3,n}^{VV'k} A_{3,\mathcal{N},n}^k\Bigg]. \tag{3.2.43}$$

The *Wilson coefficients* $C_{a,n}^{VV'k}(Q^2, \alpha_{\mathrm{s}}, \mathcal{M})$ describe the hard-scattering process between the virtual electroweak gauge bosons and the partons, and can be calculated in perturbation theory [Gr73a]. They depend on the flavour and parity of the currents and on the parton species but not on the properties of the external hadrons. In the electromagnetic case they are given by the quark charges, e.g. $C_{2,n}^{\gamma\gamma q}(Q^2, \alpha_{\mathrm{s}} = 0, \mathcal{M}) = Q_q^2$, in lowest order.

The coefficients $A_{a,\mathcal{N},n}^k(\alpha_{\mathrm{s}}, \mathcal{M})$ are the reduced hadronic matrix elements of the Wilson operators. They are given by the moments of the parton distribution functions and cannot be calculated in perturbation theory.

In one-loop order of QCD it is necessary to use the decomposition into non-singlet (NS) and singlet (S) contributions (3.2.25) with definite symmetry under charge conjugation. The singlet quark contribution can mix with gluonic contributions with the same quantum numbers. The index $k = \{\mathrm{NS},c;\ \mathrm{S,q};\ \mathrm{S,g}\}$ discriminates between all these contributions: quark contributions are denoted in the group-theoretic notation by NS,c and S,q in the non-singlet and singlet case, respectively. The gluonic contributions, which occur in one-loop order, are labelled by S,g.

The following *Wilson operators of QCD* with twist two generalize the expressions (3.2.20), (3.2.21), (3.2.22), and (3.2.37) of the free quark case:

$$\begin{aligned}
\mathcal{O}^{\mathrm{NS},c}_{\rho\mu_1\ldots\mu_n}(0) &= \bar{\psi}_q(0)\mathcal{S}\{\gamma_\rho D_{\mu_1}\ldots D_{\mu_n}\}\lambda^c_{qq'}\psi_{q'}(0), \\
\mathcal{O}^{\mathrm{S,q}}_{\rho\mu_1\ldots\mu_n}(0) &= \sum_q \bar{\psi}_q(0)\mathcal{S}\{\gamma_\rho D_{\mu_1}\ldots D_{\mu_n}\}\psi_q(0), \\
\mathcal{O}^{\mathrm{NS},c}_{3,\rho\mu_1\ldots\mu_n}(0) &= \bar{\psi}_q(0)\mathcal{S}\{\gamma_\rho\gamma_5 D_{\mu_1}\ldots D_{\mu_n}\}\lambda^c_{qq'}\psi_{q'}(0), \\
\mathcal{O}^{\mathrm{S,q}}_{3,\rho\mu_1\ldots\mu_n}(0) &= \sum_q \bar{\psi}_q(0)\mathcal{S}\{\gamma_\rho\gamma_5 D_{\mu_1}\ldots D_{\mu_n}\}\psi_q(0).
\end{aligned} \tag{3.2.44}$$

Here $\mathcal{S}$ denotes symmetrization of the Lorentz indices and taking out traces in order to obtain, with respect to the Lorentz group, irreducible tensor operators. The covariant derivative with respect to the colour SU(3) group D_μ occurs because of gauge invariance. In QCD also the gluon fields contribute to the singlet Wilson operators with twist two,

$$\mathcal{O}^{\mathrm{S,g}}_{\rho\mu_1\ldots\mu_n}(0) = \mathcal{S}\{\mathrm{Tr}\, F_\rho{}^\alpha(0)D_{\mu_1}\ldots D_{\mu_{n-1}}F_{\alpha\mu_n}(0)\}, \tag{3.2.45}$$

where the trace refers to the colour indices. The operators $\mathcal{O}^{\mathrm{S,q}}$ and $\mathcal{O}^{\mathrm{S,g}}$ have identical quantum numbers and dimensions and therefore can mix. The operators $\mathcal{O}^{\mathrm{NS},c}_3$ and $\mathcal{O}^{\mathrm{S,q}}_3$ occur only in parity-violating processes. They contribute to the structure function F_3.

The spin-averaged hadronic matrix element of the Wilson operators can be represented by Lorentz covariants and the coefficients $A^k_{a,\mathcal{N},n}(\alpha_\mathrm{s},\mathcal{M})$ describing the nucleon structure [cf (3.2.39)],

$$\langle p|\mathcal{O}^k_{a,\rho\mu_1\ldots\mu_n}(0)|p\rangle = \mathrm{i}^n p_\rho p_{\mu_1}\ldots p_{\mu_n} 2A^k_{a,\mathcal{N},n+1}(\alpha_\mathrm{s},\mathcal{M}). \tag{3.2.46}$$

A comparison of (3.2.43) with the general decomposition of the hadronic tensor (3.2.5) provides relations between the moments of the structure functions and the coefficient functions,

$$\begin{aligned}
F_{2,n}^{VV'\mathcal{N}}(Q^2) &= \int_0^1 \mathrm{d}x\, x^{n-2} F_2^{VV'\mathcal{N}}(Q^2,x) \\
&= \sum_k C_{2,n}^{VV'k}(Q^2,\alpha_s,\mathcal{M}) A_{2,\mathcal{N},n}^k(\alpha_s,\mathcal{M}), \\
F_{\mathrm{L},n}^{VV'\mathcal{N}}(Q^2) &= \int_0^1 \mathrm{d}x\, x^{n-2} F_{\mathrm{L}}^{VV'\mathcal{N}}(Q^2,x) \\
&= \sum_k C_{\mathrm{L},n}^{VV'k}(Q^2,\alpha_s,\mathcal{M}) A_{\mathrm{L},\mathcal{N},n}^k(\alpha_s,\mathcal{M}), \\
F_{3,n}^{VV'\mathcal{N}}(Q^2) &= \int_0^1 \mathrm{d}x\, x^{n-1} F_3^{VV'\mathcal{N}}(Q^2,x) \\
&= \frac{1}{2}\sum_k C_{3,n}^{VV'k}(Q^2,\alpha_s,\mathcal{M}) A_{3,\mathcal{N},n}^k(\alpha_s,\mathcal{M}).
\end{aligned} \tag{3.2.47}$$

Thus, the *moments of the structure functions* are determined by the Fourier-transformed Wilson coefficients and the nucleon matrix elements of the Wilson operators.

In lowest order of the strong coupling constant α_s, the covariant derivative D_μ is replaced by the partial one ∂_μ, and there is no contribution of the flavour-neutral gluon fields. Consequently, we have only a spin-1/2 contribution, and F_{L} vanishes. The matrix elements of the Wilson operators $A_{2,\mathcal{N},n}^q$ and $A_{3,\mathcal{N},n}^q$ are the sums and differences of the moments $f_{q,\mathcal{N},n} \pm f_{\bar{q},\mathcal{N},n}$ of the quark and antiquark distribution functions [cf. (3.2.42)]. The Wilson coefficients reduce to the products of the quark couplings (1.5.11). In this way the results of the parton model are reproduced.

3.2.2.2 Renormalization-group equation

The quark currents $h_\mu(x)$ and the Wilson operators $\mathcal{O}_n^k(0) \equiv \mathcal{O}_{\rho\mu_1\cdots\mu_n}^k(0)$ are *local composite operators* built up from quark and gluon fields to which the multiplicative renormalization procedure of QCD (cf. Sect. 2.5.1.7) can be generalized [Zi73, Di74, Jo76, Co84]. However, mixing of operators of the same quantum numbers and the same or lower dimensions may occur. In our case the flavour-singlet Wilson operators $\mathcal{O}_n^{\mathrm{S,q}}$ and $\mathcal{O}_n^{\mathrm{S,g}}$ mix. Hence in general, a renormalization-constant matrix $(Z_{\mathcal{O}_n})^{kl}$ relates the renormalized Wilson operators $\mathcal{O}^k_{\mathrm{R},n}(\alpha_\mathrm{s},\mathcal{M})$ to the unrenormalized ones,

$$\begin{aligned} h^\mu_\mathrm{R}(x,\alpha_\mathrm{s},\mathcal{M}) &= Z_h^{-1}(\alpha_\mathrm{s},\mathcal{M})h^\mu(x), \\ \mathcal{O}^k_{\mathrm{R},n}(0,\alpha_\mathrm{s},\mathcal{M}) &= \sum_l [Z^{-1}_{\mathcal{O}_n}(\alpha_\mathrm{s},\mathcal{M})]_{kl}\mathcal{O}^l_n(0). \end{aligned} \tag{3.2.48}$$

From the corresponding Green functions,

$$\begin{aligned} G^{\psi\bar\psi h^\dagger h} &= \langle 0|\, T\psi(x_1)\, [h^\dagger_\mu(x), h_\nu(0)]\, \bar\psi(x_2)\, |0\rangle, \\ G^{\psi\bar\psi\mathcal{O}^k_n} &= \langle 0|\, T\psi(x_1)\, \mathcal{O}^k_n(0)\, \bar\psi(x_2)\, |0\rangle, \end{aligned} \tag{3.2.49}$$

the renormalized vertex functions $\Gamma_\mathrm{R}^{\psi\bar\psi h^\dagger h}$ and $\Gamma_\mathrm{R}^{\psi\bar\psi\mathcal{O}^k_n}$ are derived by truncation of the quark and antiquark legs. These vertex functions satisfy the renormalization-group (RG) equations[6]

$$\begin{aligned} \left[\mathcal{M}\frac{\mathrm{d}}{\mathrm{d}\mathcal{M}} - 2\gamma^\psi + 2\gamma^h\right]\Gamma_\mathrm{R}^{\psi\bar\psi h^\dagger h}(\alpha_\mathrm{s},\mathcal{M}) &= 0, \\ \sum_l \left[\left(\mathcal{M}\frac{\mathrm{d}}{\mathrm{d}\mathcal{M}} - 2\gamma^\psi\right)\delta^{kl} + \gamma^{kl}_n\right]\Gamma_\mathrm{R}^{\psi\bar\psi\mathcal{O}^l_n}(\alpha_\mathrm{s},\mathcal{M}) &= 0. \end{aligned} \tag{3.2.50}$$

The coefficients γ^h and γ^{kl}_n are the anomalous dimensions of the current operator and the Wilson operator, respectively, [cf. (2.6.9)]

$$\begin{aligned} \gamma^h(g_\mathrm{s}) &= Z_h^{-1}\mathcal{M}\frac{\partial}{\partial\mathcal{M}}Z_h, \\ \gamma^{kl}_n(g_\mathrm{s}) &= \sum_m (Z^{-1}_{\mathcal{O}_n})^{km}\mathcal{M}\frac{\partial}{\partial\mathcal{M}}(Z_{\mathcal{O}_n})^{ml}. \end{aligned} \tag{3.2.51}$$

The charges of the quark currents h^μ are the generators of the flavour symmetry group. Because h^μ is a conserved current, it does not need renormalization, i.e. $Z_h = 1$, and its anomalous dimension vanishes, i.e. $\gamma^h = 0$.

The current–current and Wilson-operator vertex functions are linked together by the *Wilson expansion*, written in a condensed form as

$$\Gamma_\mathrm{R}^{\psi\bar\psi h^\dagger h} \sim \sum_{k,n} C^k_n(Q^2,\alpha_\mathrm{s},\mathcal{M})\Gamma_\mathrm{R}^{\psi\bar\psi\mathcal{O}^k_n}(\alpha_\mathrm{s},\mathcal{M}). \tag{3.2.52}$$

[6]The different signs in front of γ^ψ and γ^{kl}_n result from the fact that the ψ legs are truncated.

This allows to derive a RG equation for the Wilson coefficients C_n^k by substituting (3.2.52) into (3.2.50),

$$\begin{aligned} 0 &= \left[\mathcal{M}\frac{\mathrm{d}}{\mathrm{d}\mathcal{M}} - 2\gamma^{\psi}\right] \sum_{k,n} C_n^k(Q^2, \alpha_s, \mathcal{M}) \Gamma_R^{\psi\bar{\psi}\mathcal{O}_n^k}(\alpha_s, \mathcal{M}) \\ &= \sum_{k,n} \left[\Gamma_R^{\psi\bar{\psi}\mathcal{O}_n^k} \mathcal{M}\frac{\mathrm{d}}{\mathrm{d}\mathcal{M}} C_n^k + C_n^k \left(\mathcal{M}\frac{\mathrm{d}}{\mathrm{d}\mathcal{M}} - 2\gamma^{\psi}\right) \Gamma_R^{\psi\bar{\psi}\mathcal{O}_n^k}\right] \\ &= \sum_{k,l,n} \Gamma_R^{\psi\bar{\psi}\mathcal{O}_n^l}(\alpha_s, \mathcal{M}) \left(\delta^{lk}\mathcal{M}\frac{\mathrm{d}}{\mathrm{d}\mathcal{M}} - \gamma_n^{lk}\right) C_n^k(Q^2, \alpha_s, \mathcal{M}). \quad (3.2.53) \end{aligned}$$

For irreducible Lorentz tensors this equation must be fulfilled for each n and l separately,

$$0 = \sum_k \left(\delta^{lk}\mathcal{M}\frac{\mathrm{d}}{\mathrm{d}\mathcal{M}} - \gamma_n^{lk}\right) C_n^k(Q^2, \alpha_s, \mathcal{M}). \quad (3.2.54)$$

As the Wilson coefficients are dimensionless, the associated scale-transformation equation reads (in the massless case)

$$\sum_k \left[\delta^{lk}\left(\lambda\frac{\partial}{\partial\lambda} - \beta(g_s)\frac{\partial}{\partial g_s}\right) + \gamma_n^{lk}(g_s)\right] C_n^k\left((\lambda Q)^2, \alpha_s\right) = 0. \quad (3.2.55)$$

This general RG equation is a coupled system of linear, partial differential equations. It allows for the application of perturbation theory to the calculation of the momentum dependence of the Wilson coefficients, because QCD is asymptotically free.

No mixing of operators occurs in the *flavour non-singlet case*, where $\mathcal{O}^{\mathrm{NS},c}$ and $\mathcal{O}_3^{\mathrm{NS},c}$ contribute. The corresponding one-loop anomalous dimensions are flavour- and parity-independent in massless QCD. Therefore, we denote them by $\gamma_n^{\mathrm{NS}}(g_s)$. The RG equation is a scalar equation and (suppressing flavour indices) solved by

$$C_n^{\mathrm{NS}}\left((\lambda Q)^2, \alpha_s\right) = C_n^{\mathrm{NS}}\left(Q^2, \bar{\alpha}_s((\lambda Q)^2)\right) \exp\left(-\int_{g_s}^{\bar{g}_s} \mathrm{d}g' \, \frac{\gamma_n^{\mathrm{NS}}(g')}{\beta(g')}\right), \quad (3.2.56)$$

where $\bar{\alpha}_s$ is the running coupling constant. With (3.2.47) this implies

$$F_n^{\rm NS}\left((\lambda Q)^2, \alpha_s\right) = F_n^{\rm NS}\left(Q^2, \bar{\alpha}_s((\lambda Q)^2)\right) \exp\left(-\int_{g_s}^{\bar{g}_s} \mathrm{d}g' \, \frac{\gamma_n^{\rm NS}(g')}{\beta(g')}\right). \tag{3.2.57}$$

for the Q^2 dependence of the *non-singlet moments*. The breaking of scaling in the *moments of the structure functions* is determined by the anomalous dimensions $\gamma_n^{\rm NS}$ of the Wilson operators and the β-function of QCD.

Now we treat the *flavour singlet case*, e.g. contributions to neutral current scattering, at one loop. Because of mixing among the operators $\mathcal{O}^{\rm S,q}$ and $\mathcal{O}^{\rm S,g}$ in the singlet case, $\gamma_n^{\rm S}(g_s)$ is a constant 2×2 matrix,

$$\gamma_n^{\rm S} = \begin{pmatrix} \gamma_n^{\rm qq} & 2N_{\rm f}\gamma_n^{\rm qg} \\ \gamma_n^{\rm gq} & \gamma_n^{\rm gg} \end{pmatrix}, \tag{3.2.58}$$

which can be diagonalized

$$U^{-1}\gamma_n^{\rm S} U = \begin{pmatrix} \gamma_n^- & 0 \\ 0 & \gamma_n^+ \end{pmatrix}, \qquad \begin{pmatrix} C_n^{\rm S,+} \\ C_n^{\rm S,-} \end{pmatrix} = U^{-1} \begin{pmatrix} C_n^{\rm S,q} \\ C_n^{\rm S,g} \end{pmatrix} \tag{3.2.59}$$

with

$$\gamma_n^{\pm} = \frac{1}{2}\left(\gamma_n^{\rm qq} + \gamma_n^{\rm gg} \pm \sqrt{(\gamma_n^{\rm qq} - \gamma_n^{\rm gg})^2 + 8N_{\rm f}\gamma_n^{\rm qg}\gamma_n^{\rm gq}}\right),$$
$$U = \begin{pmatrix} \gamma_n^{\rm qq} - \gamma_n^+ & \gamma_n^{\rm qq} - \gamma_n^- \\ \gamma_n^{\rm gq} & 2N_{\rm f}\gamma_n^{\rm qg} \end{pmatrix} \frac{1}{\gamma_n^- - \gamma_n^+}. \tag{3.2.60}$$

Here $N_{\rm f}$ denotes the number of flavours. This procedure decouples the two differential equations for $C_n^{\rm S,q}$ and $C_n^{\rm S,g}$. The linear combinations $C_n^{\rm S,+}$ and $C_n^{\rm S,-}$ then behave in the usual way,

$$C_n^{\rm S,\pm}\left((\lambda Q)^2, \alpha_s\right) = C_n^{\rm S,\pm}\left(Q^2, \bar{\alpha}_s((\lambda Q)^2)\right) \exp\left(-\int_{g_s}^{\bar{g}_s} \mathrm{d}g' \, \frac{\gamma_n^{\pm}(g')}{\beta(g')}\right). \tag{3.2.61}$$

This treatment of the singlet case can be generalized to higher orders if a decomposition into irreducible representations of the flavour and charge-conjugation symmetry is performed. This will be treated in Sect. 3.2.3.

3.2.2.3 Structure of QCD corrections

Evaluation of (3.2.56) and (3.2.61) by means of perturbation theory starts from a power-series expansion of the relevant quantities, explicitly

$$\beta(g_s) = -\beta_0 g_s \frac{g_s^2}{16\pi^2} - \beta_1 g_s \left(\frac{g_s^2}{16\pi^2}\right)^2 + \dots, \tag{3.2.62}$$

$$\gamma_n(g_s) = \gamma_{n,0} \frac{g_s^2}{16\pi^2} + \gamma_{n,1} \left(\frac{g_s^2}{16\pi^2}\right)^2 + \dots. \tag{3.2.63}$$

The expansion of the Wilson coefficients $C_{a,n}^{VV'k}(\bar{\alpha}_s)$ starts with the parton-model values

$$C_{a,n}^{VV'k}(\alpha_s) = C_{a,n,0}^{VV'k} + \frac{\alpha_s}{4\pi} C_{a,n,1}^{VV'k} + \dots \tag{3.2.64}$$

with

$$\begin{aligned} C_{2,n,0}^{VV'q} &= G_v^{q,VV'}, & C_{2,n,0}^{VV'g} &= 0, \\ C_{3,n,0}^{VV'q} &= G_a^{q,VV'}, & C_{3,n,0}^{VV'g} &= 0, \\ C_{L,n,0}^{VV'q} &= 0, & C_{L,n,0}^{VV'g} &= 0 \end{aligned} \tag{3.2.65}$$

in the product base and similar expressions in the decomposition into the adjoint and singlet representation of the flavour group. The coupling strengths G_v and G_a of the quarks are given in (1.5.11). Since $F_L = 0$ in the parton model, and since the gluons do not couple directly to the flavour currents, the expansion of the corresponding quantities starts with the α_s terms.

From (3.2.64)–(3.2.65), the calculation of the behaviour of the moments of the structure function proceeds. Using only the first terms of the expansions above to work out (3.2.57), one gets the leading-order results. There is a difference between the non-singlet case, where only one Wilson operator of order n exists, and the singlet case, where mixing of operators occurs. From the corresponding RG equations one obtains

$$\begin{aligned} F_{2,n}^{VV'\mathcal{N},\mathrm{NS}}(Q^2) &= \left(\frac{\bar{\alpha}_s(Q_0^2)}{\bar{\alpha}_s(Q^2)}\right)^{d_{n,0}^{\mathrm{NS}}} F_{2,n}^{VV'\mathcal{N},\mathrm{NS}}(Q_0^2), \\ F_{3,n}^{VV'\mathcal{N},\mathrm{NS}}(Q^2) &= \left(\frac{\bar{\alpha}_s(Q_0^2)}{\bar{\alpha}_s(Q^2)}\right)^{d_{n,0}^{\mathrm{NS}}} F_{3,n}^{VV'\mathcal{N},\mathrm{NS}}(Q_0^2) \end{aligned} \tag{3.2.66}$$

for the Q^2 dependence of the non-singlet moments $F_{2,n}^{\mathrm{NS}}$ and the parity-violating non-singlet moments $F_{3,n}^{\mathrm{NS}}$, where Q_0 is the reference scale and

$$d_{n,0}^{\mathrm{NS}} = -\frac{\gamma_{n,0}^{\mathrm{NS}}}{2\beta_0}. \tag{3.2.67}$$

The Q^2 dependence of the singlet moments $F_{2,n}^{\mathrm{S},\pm}$ is given by

$$F_{2,n}^{VV'\mathcal{N},\mathrm{S},\pm}(Q^2) = \left(\frac{\bar{\alpha}_{\mathrm{s}}(Q_0^2)}{\bar{\alpha}_{\mathrm{s}}(Q^2)}\right)^{d_{n,0}^{\mathrm{S},\pm}} F_{2,n}^{VV'\mathcal{N},\mathrm{S},\pm}(Q_0^2) \tag{3.2.68}$$

with

$$d_{n,0}^{\mathrm{S},\pm} = -\frac{\gamma_{n,0}^{\mathrm{S},\pm}}{2\beta_0}. \tag{3.2.69}$$

The *anomalous dimensions* $d_{n,0}$ are calculated in the following section.

3.2.2.4 Anomalous dimensions of Wilson operators in QCD

In order to calculate the anomalous dimensions of the Wilson operators $\mathcal{O}$ in order α_{s}, the QCD Feynman rules must be extended to the case of local composite operators. The following field points are assigned to the Wilson operators (3.2.44) and (3.2.45) (cf. [Gr73a] for further details):

$$\bar{\psi}_{q,c}(x)\mathcal{S}\left\{\gamma_{\mu_1}\lambda_{qq'}^f\partial_{\mu_2}\dots\partial_{\mu_n}\right\}\psi_{q',c}(x)$$
$$\to \mathcal{S}\left\{\gamma_{\mu_1}\lambda_{qq'}^f k_{\mu_2}\dots k_{\mu_n}\right\}\delta_{cc'} \to$$

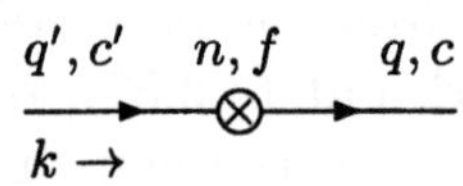

, (3.2.70)

$$\bar{\psi}_{q,c}(x)\mathcal{S}\left\{\gamma_{\mu_1}\lambda_{qq'}^f\partial_{\mu_2}\dots g_{\mathrm{s}}A_{\mu_i}^a(x)\frac{\lambda_{cc'}^a}{2}\dots\partial_{\mu_n}\right\}\psi_{q',c'}(x) \tag{3.2.71}$$
$$\to g_{\mathrm{s}}\frac{\lambda_{cc'}^a}{2}\lambda_{qq'}^f\mathcal{S}\left\{\gamma_{\mu_1}k_{\mu_2}\dots\varepsilon_{\mu_i}\dots k'_{\mu_n}\right\} \to$$

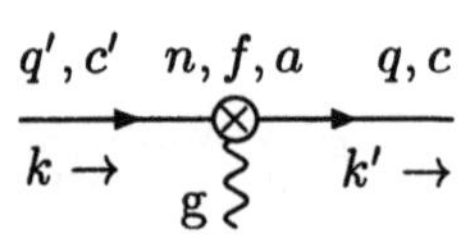

,

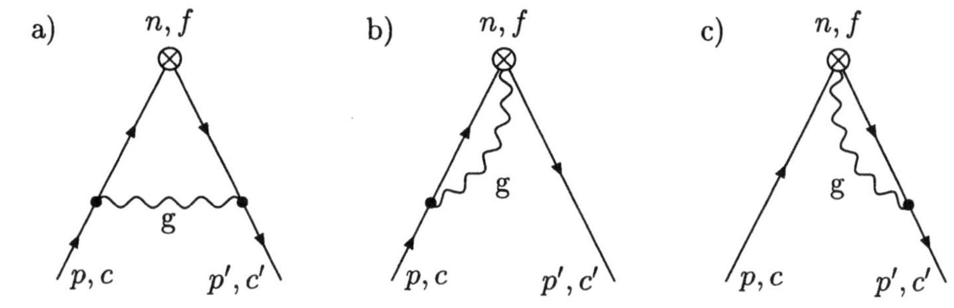

Fig. 3.3 One-loop diagrams for Wilson operators of QCD

$$
\begin{aligned}
&\mathcal{S}\left\{F_{\mu_1}{}^{a}(x)\partial_{\mu_2}\dots\partial_{\mu_{n-1}}F_{a\mu_n}(x)\right\} \\
&\quad\to \mathcal{S}\Big\{g_{\mu\nu}k_{\mu_1}\dots k_{\mu_n} + k^2 g_{\mu\mu_1}g_{\nu\mu_2}k_{\mu_3}\dots k_{\mu_n} \\
&\qquad - k_\mu g_{\nu\mu_1}k_{\mu_2}\dots k_{\mu_n} - k_\nu g_{\mu\mu_1}k_{\mu_2}\dots k_{\mu_n}\Big\}\delta^{ab}
\end{aligned}
$$

$\to$ 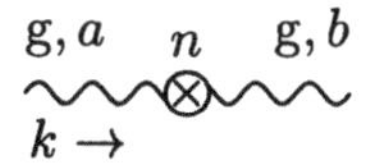

. (3.2.72)

Similar rules apply for axial operators. The diagrams representing the one-loop corrections to the operator (3.2.70) are shown in Fig. 3.3. The calculation is similar to that for the one-loop diagrams in Sect. 2.5.2. It was worked out initially by Gross and Wilczek [Gr73a] and Politzer [Po74]. Symmetrization and taking out traces can be most easily achieved by contracting the operator with the tensor $\Delta_{\mu_1}\cdots\Delta_{\mu_n}$ formed by light-like vector Δ_μ. In Feynman gauge, using dimensional regularization, the Feynman rules give

$$
\begin{aligned}
\mathcal{M}^{(a)} &= \mathrm{i}^5 g_\mathrm{s}^2 C_\mathrm{F}\lambda^f\mu^{4-D}\int\frac{\mathrm{d}^Dk}{(2\pi)^D}\frac{\gamma^\mu \not{k}\not{\Delta}(\Delta k)^{n-1}\not{k}\gamma^\nu}{k^2(p-k)^2k^2}(-g_{\mu\nu}) \\
&= -\mathrm{i}g_\mathrm{s}^2 C_\mathrm{F}\lambda^f(2-D)\mu^{4-D}\int\frac{\mathrm{d}^Dk}{(2\pi)^D}\frac{\not{k}\not{\Delta}(\Delta k)^{n-1}\not{k}}{k^2(p-k)^2k^2} \\
&= -\mathrm{i}g_\mathrm{s}^2 C_\mathrm{F}\lambda^f(2-D)\mu^{4-D}\int\frac{\mathrm{d}^Dk}{(2\pi)^D}\int_0^1\mathrm{d}z \\
&\quad\times 2(1-z)\frac{(\not{k}+z\not{p})\not{\Delta}(\not{k}+z\not{p})(\Delta k+z\Delta p)^{n-1}}{[k^2+z(1-z)p^2]^3}
\end{aligned}
\tag{3.2.73}
$$

for diagram (a). In the last step we used the Feynman parametrization of the integrand (2.5.67). For the anomalous dimension γ_n^{NS} of the Wilson operator only the residue of the pole at $D = 4$ is needed. This is determined by the $k_\mu k_\nu$ terms in the numerator.[7] These act in the integral like $g_{\mu\nu}k^2/D$. Since $\Delta^2 = 0$, there is only one contribution,

$$\begin{aligned}\mathcal{M}^{(a)} &= -\mathrm{i}g_s^2 C_F \lambda^f \not{\Delta}(\Delta p)^{n-1}\frac{(D-2)^2}{D}2\int_0^1 \mathrm{d}z\,(1-z)z^{n-1} \\ &\quad\times \mu^{4-D}\int\frac{\mathrm{d}^D k}{(2\pi)^D}\frac{1}{[k^2+z(1-z)p^2]^2} \\ &= \frac{\alpha_s}{4\pi}C_F\mathcal{M}_0\frac{(D-2)^2}{D} \\ &\quad\times 2\int_0^1 \mathrm{d}z\,(1-z)z^{n-1}A_0^{(2)}(z(1-z)p^2),\end{aligned} \tag{3.2.74}$$

where we have inserted the lowest-order expression

$$\mathcal{M}_0^{(a)} = \lambda^f \not{\Delta}(\Delta p)^{n-1}. \tag{3.2.75}$$

The tadpole integral $A_0^{(2)}$ was calculated in Sect. 2.5.2.1. Writing as usual $D = 4 - 2\delta$, it reads [cf. (2.5.58)]

$$A_0^{(2)}(z(1-z)p^2) = \left(\frac{4\pi\mu^2}{p^2}\right)^\delta \Gamma(\delta)(z(1-z))^{-\delta}. \tag{3.2.76}$$

The z integration gives Euler's β-function, i.e. a ratio of Γ functions. The result of diagram (a) finally is

$$\mathcal{M}^{(a)} = \frac{\alpha_s}{4\pi}C_F\mathcal{M}_0^{(a)}\,2\,\frac{(2-2\delta)^2}{4-2\delta}\frac{\Gamma(n-\delta)\Gamma(2-\delta)}{\Gamma(n+2-2\delta)}\left(\frac{4\pi\mu^2}{p^2}\right)^\delta\Gamma(\delta). \tag{3.2.77}$$

For the anomalous dimension we need only the singular part of this expression [cf. (2.6.30)],

$$\mathcal{M}_{\text{sing}}^{(a)} = \frac{\alpha_s}{4\pi}C_F\mathcal{M}_0^{(a)}\frac{2}{n(n+1)}\frac{1}{\delta}. \tag{3.2.78}$$

[7] The contributions of the terms with more k's in the numerator vanish owing to $\Delta^2 = 0$.

The second and the third graph of Fig. 3.3 yield

$$\mathcal{M}^{(b)}_{\text{sing}} = \mathcal{M}^{(c)}_{\text{sing}} = -\frac{\alpha_s}{4\pi} C_F \mathcal{M}^{(a)}_0 \left(\sum_{j=2}^{n} \frac{2}{j} \right) \frac{1}{\delta}. \tag{3.2.79}$$

Adding these contributions and the renormalization constants for the external fields (2.6.31) yields the renormalization constant of the non-singlet operators in the MS scheme,

$$\delta Z^{\text{NS}}_n = \frac{\alpha_s}{4\pi} C_F \left(\frac{2}{n(n+1)} - 4 \sum_{j=2}^{n} \frac{1}{j} - 1 \right) \frac{1}{\delta}, \tag{3.2.80}$$

and with (3.2.51) and (2.6.30) the one-loop coefficient of the anomalous dimension,

$$\gamma^{\text{NS}}_{n,0} = 2C_F \left(1 - \frac{2}{n(n+1)} + 4 \sum_{j=2}^{n} \frac{1}{j} \right). \tag{3.2.81}$$

Hence for the colour group SU(3) the exponent in (3.2.66) is given by

$$d^{\text{NS}}_{n,0} = \frac{4}{33 - 2N_f} \left(-1 + \frac{2}{n(n+1)} - 4 \sum_{j=2}^{n} \frac{1}{j} \right). \tag{3.2.82}$$

The elements of the 2×2 matrix for the singlet case have also been worked out in [Gr73a],

$$\begin{aligned}
\gamma^{\text{qq}}_{n,0} &= 2C_F \left(1 - \frac{2}{n(n+1)} + 4 \sum_{j=2}^{n} \frac{1}{j} \right) = \gamma^{\text{NS}}_{n,0}, \\
\gamma^{\text{qg}}_{n,0} &= -8T_F \frac{n^2+n+2}{n(n+1)(n+2)}, \\
\gamma^{\text{gq}}_{n,0} &= -4C_F \frac{n^2+n+2}{n(n^2-1)}, \\
\gamma^{\text{gg}}_{n,0} &= -4C_A \left(\frac{2}{n(n-1)} + \frac{2}{(n+1)(n+2)} - \frac{1}{6} - 2 \sum_{j=2}^{n} \frac{1}{j} \right) \\
&\quad - \frac{8}{3} T_F N_f,
\end{aligned} \tag{3.2.83}$$

where $C_F = 4/3$, $C_A = 3$ and $T_F = 1/2$ for SU(3).

The exponents of the logarithms that determine the scaling violations in asymptotically free gauge theories are given by (3.2.82) and (3.2.69) together with (3.2.60) and (3.2.83). For light quarks they are flavour-independent. Since the Q^2 dependence of the moments is only logarithmic, it is weak compared with that of non-gauge theories.

The anomalous dimensions of the Wilson operators at two loops have been calculated in Refs. [Fl77, Fl79, Go79, Ha92a]. For the three-loop case some specific moments are known [La97].

3.2.3 Evolution equations

Using the Wilson expansion and the renormalization group we have derived QCD predictions for the moments of structure functions. In particular, logarithmic corrections to the scaling behaviour of the parton model have been found. In the naïve parton model the structure functions are independent of the momentum transfer Q^2, i.e. the probability of finding a parton with momentum fraction x in the nucleon does not depend on the resolution the nucleon is probed with.[8]

The scaling violations in QCD emerge from the fact that the interaction at short distances may convert a quark into a quark and a gluon (cf. Fig. 3.2 on page 436), and a gluon into two gluons or into a quark–antiquark pair. The observation of these fluctuations depends on the resolution. Increasing the momentum transfer reveals the structure at shorter distances. More *virtual quarks and gluons* are found, each of them carrying a smaller fraction of the original *valence-quark* momentum. This is the parton picture of scaling violations.

The Q^2 dependence of the structure functions can be controlled by evolution equations derived by Dokshitzer, Gribov and Lipatov, and Altarelli and Parisi [Gr72, Li75, Al77, Do77]. We start from the solution (3.2.66) of the RG equation for the *flavour non-singlet moments* $F_{2,n}^{\mathrm{NS}}(Q^2)$. They satisfy the differential equation

$$\frac{\mathrm{d}}{\mathrm{d}t} F_{2,n}^{\mathrm{NS}}(t) = d_{n,0}^{\mathrm{NS}} \frac{F_{2,n}^{\mathrm{NS}}(t)}{t} \quad \text{with} \quad t = \log Q^2/\Lambda^2. \tag{3.2.84}$$

[8] The three-momentum of the virtual photon in the nucleon rest frame is $(Q^2+\nu^2)^{1/2}$. According to the uncertainty principle it probes the nucleon structure at the corresponding length scale.

Inserting $d_{n,0}^{\mathrm{NS}} = -\gamma_{n,0}^{\mathrm{NS}}/(2\beta_0)$ and $\alpha_s(t) = 4\pi/\beta_0 t$ yields

$$\frac{\mathrm{d}}{\mathrm{d}t} F_{2,n}^{\mathrm{NS}}(t) = -\frac{\alpha_{\mathrm{s}}(t)}{4\pi} \frac{\gamma_{n,0}^{\mathrm{NS}}}{2} F_{2,n}^{\mathrm{NS}}(t). \tag{3.2.85}$$

In the quark–parton model the structure functions $F_2^{VV'\mathcal{N}}(x,t)$ are products of the coupling strengths $G^{qVV'}$ and the quark distribution functions $f_{q,\mathcal{N}}(x,t)$ and $f_{\bar{q},\mathcal{N}}(x,t)$ [cf. (3.2.35)],

$$F_2^{VV'\mathcal{N},\mathrm{NS}}(x,t) = \sum_q x \left(G^{qVV'} f_{q,\mathcal{N}}(x,t) + G^{\bar{q}VV'} f_{\bar{q},\mathcal{N}}(x,t) \right). \tag{3.2.86}$$

This implies for the *moments of the quark distribution functions*

$$\frac{\mathrm{d}}{\mathrm{d}t} \int_0^1 \mathrm{d}x\, x^{n-1} f_{q,\mathcal{N}}(x,t) = -\frac{\alpha_{\mathrm{s}}(t)}{4\pi} \frac{\gamma_{n,0}^{\mathrm{NS}}}{2} \int_0^1 \mathrm{d}x\, x^{n-1} f_{q,\mathcal{N}}(x,t). \tag{3.2.87}$$

This set of equations is equivalent to an integro-differential equation, the *evolution equation* for the *quark distribution functions* $f_{q,\mathcal{N}}(x,t)$, also called *Dokshitzer–Gribov–Lipatov–Altarelli–Parisi (DGLAP) equation,*

$$\begin{aligned} \frac{\mathrm{d}f_{q,\mathcal{N}}(x,t)}{\mathrm{d}t} &= \frac{\alpha_{\mathrm{s}}(t)}{2\pi} \int_0^1 \mathrm{d}y \int_0^1 \mathrm{d}z\, \delta(x-yz) P_{\mathrm{qq}}^{(0)}(z) f_{q,\mathcal{N}}(y,t) \\ &\equiv \frac{\alpha_{\mathrm{s}}(t)}{2\pi} \left[P_{\mathrm{qq}}^{(0)} \otimes f_{q,\mathcal{N}} \right](x,t), \end{aligned} \tag{3.2.88}$$

provided

$$\gamma_{n,0}^{\mathrm{NS}} = -4 \int_0^1 \mathrm{d}z\, z^{n-1} P_{\mathrm{qq}}^{(0)}(z). \tag{3.2.89}$$

The structure of (3.2.88) is identical to the one of QED for the emission of hard, collinear photons (2.8.116). This is no surprise since both equations describe the same physical situation, *collinear emission of massless gauge bosons.*

The *evolution kernel* $P^{(0)}_{\mathrm{qq}}(x)$, also called *splitting function*, is calculated in one-loop approximation in Sect. 2.8.3.1 for QED and in Sect. 3.3 for QCD. The result for QCD is [cf. (3.3.1)]

$$P^{(0)}_{\mathrm{qq}}(z) = C_{\mathrm{F}}\left((1+z^2)\left(\frac{1}{1-z}\right)_+ + \frac{3}{2}\delta(1-z)\right) = C_{\mathrm{F}}\left(\frac{1+z^2}{1-z}\right)_+ , \tag{3.2.90}$$

with the (+)-*distribution* defined in (2.8.80) or (2.8.81).

Multiplicative convolutions [cf. (2.8.115)]

$$\begin{aligned} h(x) = [f \otimes g](x) &= \int_0^1 \mathrm{d}y \int_0^1 \mathrm{d}z\, \delta(x-yz) f(y) g(z) \\ &= \int_x^1 \frac{\mathrm{d}z}{z} f\left(\frac{x}{z}\right) g(z) \end{aligned} \tag{3.2.91}$$

can be factorized by taking *Mellin moments*

$$h_n = \int_0^1 \mathrm{d}x\, x^{n-1} h(x), \tag{3.2.92}$$

as the following calculation shows:

$$\begin{aligned} h_n &= \int_0^1 \mathrm{d}x\, x^{n-1} h(x) = \int_0^1 \mathrm{d}x \int_0^1 \mathrm{d}y \int_0^1 \mathrm{d}z\, \delta(x-yz) x^{n-1} f(y) g(z) \\ &= \int_0^1 \mathrm{d}y\, y^{n-1} f(y) \int_0^1 \mathrm{d}z\, z^{n-1} g(z) = f_n g_n . \end{aligned} \tag{3.2.93}$$

The relation (3.2.89) between the anomalous dimension (3.2.81) and the splitting function (3.2.90) can be verified as follows:

$$\int_0^1 \mathrm{d}z\, z^{n-1} P^{(0)}_{\mathrm{qq}}(z) = C_{\mathrm{F}} \int_0^1 \mathrm{d}z\, z^{n-1} \left(\frac{1+z^2}{(1-z)_+} + \frac{3}{2}\delta(1-z)\right). \tag{3.2.94}$$

The first integral is elementary,

$$\begin{aligned} \int_0^1 \mathrm{d}z\, \frac{z^{n-1}+z^{n+1}-2}{1-z} &= \int_0^1 \mathrm{d}z\, \frac{(z-1)(z^n+z^{n-1}+2\sum_{j=0}^{n-2} z^j)}{1-z} \\ &= -\left(\frac{1}{n+1}+\frac{1}{n}+2\sum_{j=1}^{n-1}\frac{1}{j}\right) = \frac{1}{n(n+1)} - 2 - 2\sum_{j=2}^{n}\frac{1}{j} . \end{aligned}$$

Adding the contribution of the δ-function term finally gives

$$\int_0^1 \mathrm{d}z\, z^{n-1} P^{(0)}_{\mathrm{qq}}(z) = C_{\mathrm{F}} \left(-\frac{1}{2} + \frac{1}{n(n+1)} - 2\sum_{j=2}^{n} \frac{1}{j} \right) = -\frac{\gamma^{\mathrm{NS}}_{n,0}}{4} \tag{3.2.95}$$

in agreement with (3.2.81).

In general, the *evolution equation* couples the *gluon distribution function* $f_{\mathrm{g},\mathcal{N}}(x,t)$ with the quark and antiquark distribution functions[9]

$$\begin{aligned}
\frac{\mathrm{d}f_{q_i,\mathcal{N}}}{\mathrm{d}t} &= \frac{\alpha_{\mathrm{s}}(t)}{2\pi} \left[\sum_j (P_{q_i q_j} \otimes f_{q_j,\mathcal{N}} + P_{q_i \bar{q}_j} \otimes f_{\bar{q}_j,\mathcal{N}}) + P_{q_i \mathrm{g}} \otimes f_{\mathrm{g},\mathcal{N}} \right], \\
\frac{\mathrm{d}f_{\bar{q}_i,\mathcal{N}}}{\mathrm{d}t} &= \frac{\alpha_{\mathrm{s}}(t)}{2\pi} \left[\sum_j (P_{\bar{q}_i q_j} \otimes f_{q_j,\mathcal{N}} + P_{\bar{q}_i \bar{q}_j} \otimes f_{\bar{q}_j,\mathcal{N}}) + P_{\bar{q}_i g} \otimes f_{\mathrm{g},\mathcal{N}} \right], \\
\frac{\mathrm{d}f_{\mathrm{g},\mathcal{N}}}{\mathrm{d}t} &= \frac{\alpha_{\mathrm{s}}(t)}{2\pi} \left[\sum_j (P_{\mathrm{g} q_j} \otimes f_{q_j,\mathcal{N}} + P_{\mathrm{g} \bar{q}_j} \otimes f_{\bar{q}_j,\mathcal{N}}) + P_{\mathrm{gg}} \otimes f_{\mathrm{g},\mathcal{N}} \right].
\end{aligned} \tag{3.2.96}$$

This coupled system of $2N_{\mathrm{f}} + 1$ equations is reduced by using charge-conjugation and flavour symmetry. The splitting functions are written in terms of *flavour-singlet* (S) and *flavour-non-singlet* (V) contributions,

$$\begin{aligned}
P_{q_i q_j} &= P_{\bar{q}_i \bar{q}_j} = \delta_{ij} P^{\mathrm{V}}_{\mathrm{qq}} + P^{\mathrm{S}}_{\mathrm{qq}}, \\
P_{q_i \bar{q}_j} &= P_{\bar{q}_i q_j} = \delta_{ij} P^{\mathrm{V}}_{\mathrm{q\bar{q}}} + P^{\mathrm{S}}_{\mathrm{q\bar{q}}}, \\
P_{q_i \mathrm{g}} &= P_{\bar{q}_i g} = P_{\mathrm{qg}}, \\
P_{\mathrm{g} q_j} &= P_{\mathrm{g} \bar{q}_j} = P_{\mathrm{gq}}.
\end{aligned} \tag{3.2.97}$$

Furthermore, we introduce *distribution functions with definite charge-conjugation parity*,

$$f^{\pm}_{q_i,\mathcal{N}} = f_{q_i,\mathcal{N}} \pm f_{\bar{q}_i,\mathcal{N}}, \tag{3.2.98}$$

[9]This corresponds to operator mixing in the operator-product expansion.

and flavour-symmetry properties, namely non-singlet quark distributions

$$
\begin{aligned}
f^{\mathrm{NS},-}_{q_i,\mathcal{N}} &= f^{-}_{q_i,\mathcal{N}}\,, \qquad i = 1,\dots,N_\mathrm{f} \\
f^{\mathrm{NS},+}_{k,\mathrm{q},\mathcal{N}} &= \sum_{i=1}^{k} f^{+}_{q_k,\mathcal{N}} - k f^{+}_{q_{k+1},\mathcal{N}}\,, \qquad k = 1,\dots,N_\mathrm{f}-1,
\end{aligned} \tag{3.2.99}
$$

and singlet quark distributions

$$
f^{\mathrm{S},+}_{\mathrm{q},\mathcal{N}} = \sum_{i=1}^{N_\mathrm{f}} f^{+}_{q_i,\mathcal{N}}. \tag{3.2.100}
$$

Then, the evolution equations decompose into

$$
\begin{aligned}
\frac{\mathrm{d}}{\mathrm{d}t} f^{\mathrm{NS},-}_{q_i,\mathcal{N}} &= \frac{\alpha_\mathrm{s}(t)}{2\pi}\left[(P^\mathrm{V}_{\mathrm{qq}} - P^\mathrm{V}_{\mathrm{q\bar{q}}}) \otimes f^{\mathrm{NS},-}_{q_i,\mathcal{N}} + (P^\mathrm{S}_{\mathrm{qq}} - P^\mathrm{S}_{\mathrm{q\bar{q}}}) \otimes \sum_j f^{\mathrm{NS},-}_{q_j,\mathcal{N}}\right], \\
\frac{\mathrm{d}}{\mathrm{d}t} f^{\mathrm{NS},+}_{k,\mathrm{q},\mathcal{N}} &= \frac{\alpha_\mathrm{s}(t)}{2\pi}\left[(P^\mathrm{V}_{\mathrm{qq}} + P^\mathrm{V}_{\mathrm{q\bar{q}}}) \otimes f^{\mathrm{NS},+}_{k,\mathrm{q},\mathcal{N}}\right], \\
\frac{\mathrm{d}}{\mathrm{d}t} f^{\mathrm{S},+}_{\mathrm{q},\mathcal{N}} &= \frac{\alpha_\mathrm{s}(t)}{2\pi}\left[\left\{P^\mathrm{V}_{\mathrm{qq}} + P^\mathrm{V}_{\mathrm{q\bar{q}}} + N_\mathrm{f}\left(P^\mathrm{S}_{\mathrm{qq}} + P^\mathrm{S}_{\mathrm{q\bar{q}}}\right)\right\} \otimes f^{\mathrm{S},+}_{\mathrm{q},\mathcal{N}}\right. \\
&\qquad \left. + 2N_\mathrm{f} P_{\mathrm{qg}} \otimes f_{\mathrm{g},\mathcal{N}}\right], \\
\frac{\mathrm{d}}{\mathrm{d}t} f_{\mathrm{g},\mathcal{N}} &= \frac{\alpha_\mathrm{s}(t)}{2\pi}\left[P_{\mathrm{gq}} f^{\mathrm{S},+}_{\mathrm{q},\mathcal{N}} + P_{\mathrm{gg}} \otimes f_{\mathrm{g},\mathcal{N}}\right].
\end{aligned} \tag{3.2.101}
$$

There are two separate equations for the non-singlet quark-distribution functions $f^{\mathrm{NS},-}_{q_i,\mathcal{N}}$ and $f^{\mathrm{NS},+}_{k,\mathrm{q},\mathcal{N}}$. The quark-singlet distribution function $f^{\mathrm{S},+}_{\mathrm{q},\mathcal{N}}$ is coupled to the gluon-distribution function $f_{\mathrm{g},\mathcal{N}}$.

The splitting functions $P_{ij}(x)$ are related to the probabilities of finding parton i with a momentum fraction x in parton j. They are (+)-distributions and fulfill the *sum rules* for quark-number conservation

$$
\int_0^1 \mathrm{d}x \sum_i (P_{q_i q_j}(x) - P_{\bar{q}_i q_j}(x)) = 0, \tag{3.2.102}
$$

and momentum conservation

$$
\begin{aligned}
\int_0^1 \mathrm{d}x\, x \left(\sum_i \left[P_{q_i q_j}(x) + P_{\bar{q}_i q_j}(x)\right] + P_{\mathrm{g}q_j}(x)\right) &= 0, \\
\int_0^1 \mathrm{d}x\, x \left(\sum_i \left[P_{q_i \mathrm{g}}(x) + P_{\bar{q}_i \mathrm{g}}(x)\right] + P_{\mathrm{gg}}(x)\right) &= 0.
\end{aligned} \tag{3.2.103}
$$

The integrals have to vanish since the lowest-order contribution, i.e. the simple parton-model contribution, saturates the sum rules.

In the *one-loop approximation* the corresponding kernels are flavour- and scheme-independent and have the form

$$\begin{aligned}
P_{\mathrm{q\bar{q}}}^{V,(0)}(x) &= 0, \qquad P_{\mathrm{qq}}^{S,(0)}(x) = 0, \qquad P_{\mathrm{q\bar{q}}}^{S,(0)}(x) = 0, \\
P_{\mathrm{qq}}^{V,(0)}(x) &= P_{\mathrm{qq}}^{(0)}(x) = C_{\mathrm{F}} \left(\frac{1+x^2}{1-x}\right)_+, \\
P_{\mathrm{qg}}^{(0)}(x) &= T_{\mathrm{F}} \left[x^2 + (1-x)^2\right], \\
P_{\mathrm{gq}}^{(0)}(x) &= C_{\mathrm{F}} \frac{1+(1-x)^2}{x}, \\
P_{\mathrm{gg}}^{(0)}(x) &= 2C_{\mathrm{A}} \left[x\left(\frac{1}{1-x}\right)_+ + \frac{1-x}{x} + x(1-x)\right] \\
&\quad + \left(\frac{11}{6}C_{\mathrm{A}} - \frac{2}{3}N_{\mathrm{f}}T_{\mathrm{F}}\right)\delta(1-x).
\end{aligned} \tag{3.2.104}$$

Consequently, the non-singlet evolution equations are diagonal and have the same kernel $P_{\mathrm{qq}}^{(0)}(x)$. This kernel also enters the singlet equation. The QCD splitting functions differ from the QED case (2.8.117) only by the group-theoretical factors and the contribution of the gluon self-couplings. At two-loop order the singlet contributions $P_{\mathrm{qq}}^{\mathrm{S},(1)}$ and $P_{\mathrm{q\bar{q}}}^{\mathrm{S},(1)}$ are non-zero but obey the relation $P_{\mathrm{qq}}^{\mathrm{S},(1)} = P_{\mathrm{q\bar{q}}}^{\mathrm{S},(1)}$. Therefore, the non-singlet evolution equations still are diagonal.

As the strong fine-structure constant α_{s} is not small in the complete Q^2 region, higher-order approximations give a significant contribution. The *two-loop splitting functions* are known [Cu80, Fu80a, Fu82]. A complete second-order calculation for deep-inelastic scattering was performed in Refs. [Ma90, vN91, Zi91, Zi92] both in the $\overline{\mathrm{MS}}$ and the DIS (deep-inelastic scattering) scheme including non-leading terms.[10] In this calculation also the one-loop corrections to the Wilson coefficients [Ba78, Fl79] enter.

The evolution equations (3.2.88), (3.2.96) [or (3.2.101)] and (3.2.104) describe the QCD corrections to deep-inelastic lepton–nucleon scattering in a

[10] The DIS scheme is defined such that the simple parton model is exact at the renormalization point. The conversion formulas between the different schemes are given in Sect. 3.3.2.

concentrated form. The solution of the evolution equations can be approximated by suitable expressions with free parameters for $f_{q,\mathcal{N}}(x,t)$, $f_{\bar{q},\mathcal{N}}(x,t)$, and $f_{g,\mathcal{N}}(x,t)$. These have the typical form

$$f(x,t) = C(t)x^{a(t)}(1-x)^{b(t)}[1 + c(t)\sqrt{x} + d(t)x],$$
$$t = \log\left(\frac{\log Q^2/\Lambda^2}{\log Q_0^2/\Lambda^2}\right), \tag{3.2.105}$$

where C, a, b, c, and d are expanded in power series in t. Fitting these parameters to experiment, while respecting the QCD sum rules, one obtains phenomenologically useful, Q^2-dependent *parton distribution functions* (cf. [Gl98, La00, Ma00] and references therein).

3.2.4 Experimental tests of QCD

A huge amount of experimental results for deep-inelastic lepton-nucleon scattering is available from SLAC, CERN, Fermilab, and HERA. They cover a wide range of x and Q^2 values, namely

$$0.00002 \le x \le 0.85, \qquad 1\,\text{GeV}^2 \le Q^2 \le 30000\,\text{GeV}^2,$$

and have attained a very high degree of accuracy. A detailed discussion of these results and the corresponding theoretical QCD calculations is not possible in this book. Instead, we restrict ourselves to the discussion of some important results in order to give an impression of the quality of the applications of QCD in this field.

A comparison between the experimental data of the *nucleon structure function* $F_2(x,Q^2)$ and the solution of the evolution equation in next-to-leading order with parton distribution functions fitted at the initial scale $Q^2 = 4\,\text{GeV}^2$, and using a fixed value of the strong fine-structure constant $\alpha_s(M_Z^2) = 0.118$ is shown in Fig. 3.4 [Fl99]. The data agree impressively with the QCD prediction. This is also the case for high energies in the small x region where no deviation of the next-to-leading order QCD prediction is seen. Note, however, that the figure does not show the scaling violations and the comparably large experimental errors at large x. For a review see Ref. [St95], for a compilation of recent data Ref. [PDG00].

Combining the results of these experiments in a suitable way, it is possible to determine the *parton distribution functions* $f_{i,\mathcal{N}}(x,Q^2)$. Results are available for the valence-quark densities $f_{u,P}(x,Q^2)$ and $f_{d,P}(x,Q^2)$, but also for

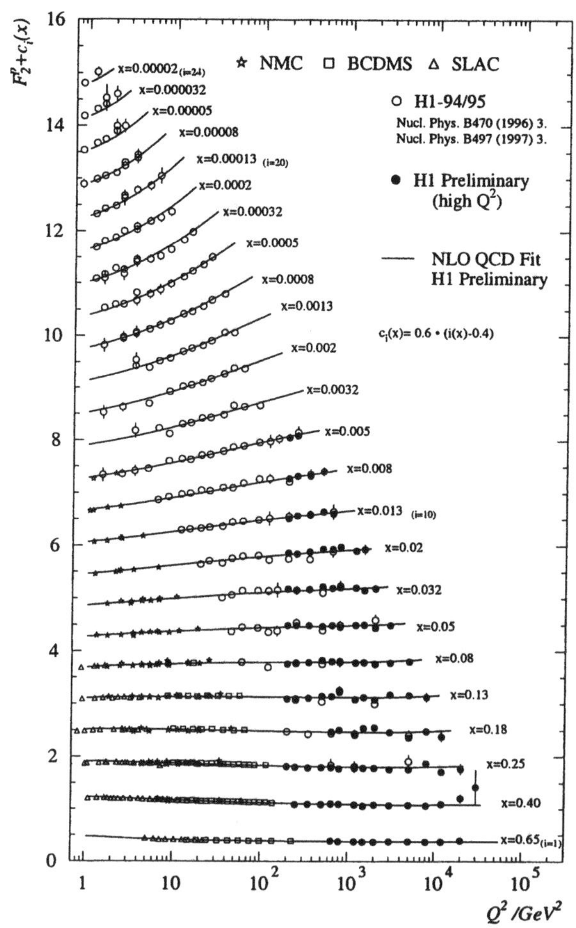

Fig. 3.4 The proton structure function $F_2(x, Q^2)$ for $1\,\mathrm{GeV}^2 \leq Q^2 \leq 30\,000\,\mathrm{GeV}^2$ and $0.00002 \leq x \leq 0.65$. The line is a next-to-leading order QCD fit (from Ref. [Fl99]). For display purposes $c(x) = 0.6(i(x) - 0.4)$, $i(x) = 1, \ldots, 24$ is added to F_2 each time the value of x is increased.

antiquark densities and the distribution of strange and charmed quarks in the proton. The *gluon distribution function* $f_{\mathrm{g,P}}(x, Q^2)$ dominates at small x although it enters the experimental results only in next-to-leading order.

As an example, the results on the parton distribution functions from the MRST collaboration [Ma98] are shown in Fig. 3.5. The valence quarks u and d are most important at $x > 0.1$. The sea quarks contribute at smaller values of x, and the gluon dominates for $x < 0.1$. The probability of carrying

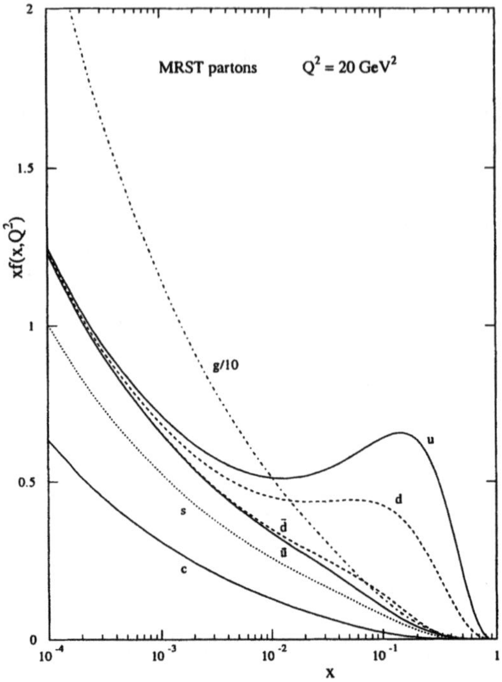

Fig. 3.5 The *parton distribution functions* $f_{i,\mathrm{P}}(x, Q^2)$, denoted by the parton symbols (u, d, ..., g), for $Q^2 = 20\,\mathrm{GeV}^2$ (from Ref. [Ma98])

a large momentum fraction x decreases for the single parton. Therefore, the parton distribution functions must decrease at large x with increasing Q^2. Because of momentum conservation they increase for small x.

The *longitudinal structure function* $F_\mathrm{L} = F_2 - 2xF_1$ vanishes in the quark–parton model (cf. Sect. 1.5.3.1). Owing to higher-order corrections it is no longer equal to zero in QCD but is given by

$$F_\mathrm{L}^{\gamma\gamma\mathcal{N}}(x, Q^2) = \frac{\alpha_\mathrm{s}(Q^2)}{2\pi} \int_x^1 \frac{\mathrm{d}z}{z} \left(\frac{x}{z}\right)^2 \left[\frac{8}{3} F_2^{\gamma\gamma\mathcal{N}}(z, Q^2) + 4 \sum_q Q_q^2 \left(1 - \frac{x}{z}\right) z f_{\mathrm{g},\mathcal{N}}(z, Q^2)\right] \tag{3.2.106}$$

the at the one-loop level [Al78, Gl78].

The $\mathcal{O}(\alpha_s^2)$ corrections have been calculated in Ref. [vN91, Zi91, Zi92, Mo00]. The measurements of the ratio $R = F_L/2xF_1$ are in agreement with the QCD prediction [Ar97, Ya00].

3.2.4.1 The QCD scale parameter Λ

The evolution of the structure functions with Q^2 depends on the *scale parameter* Λ_{QCD}. Its value in the $\overline{\text{MS}}$ scheme, taking into account four quark flavours and three-loop corrections, is [Sa99]

$$\Lambda^{(4)}_{\overline{\text{MS}}} = 282.7 \pm 25.1(\text{stat.}) \pm 24.5(\text{syst.})\,\text{MeV}, \tag{3.2.107}$$

corresponding to

$$\alpha_s(M_Z^2) = 0.1172 \pm 0.0017(\text{stat.}) \pm 0.0017(\text{syst.}). \tag{3.2.108}$$

3.2.4.2 Sum rules

We have presented the different *parton-model sum rules* and their tests already in Sect. 1.5. The two-loop QCD corrections have been calculated to most of them and are necessary to find agreement between the predictions and experiment.

3.2.4.3 Polarized deep-inelastic scattering

Results for polarized deep-inelastic electron–nucleon scattering have become available recently. The measured quantity is the nucleon spin asymmetry[11]

$$A(x, Q^2) = \frac{d\sigma^{\uparrow\downarrow}(x, Q^2) - d\sigma^{\uparrow\uparrow}(x, Q^2)}{d\sigma^{\uparrow\downarrow}(x, Q^2) + d\sigma^{\uparrow\uparrow}(x, Q^2)}, \tag{3.2.109}$$

which depends on the *spin-dependent structure functions* $g_{1,2}(x, Q^2)$. These are defined by decomposing the hadronic tensor for polarized deep-inelastic

[11] The first arrow denotes the lepton spin, the second the nucleon spin.

lepton–nucleon scattering as (we restrict ourselves here to the photon contribution)

$$\begin{aligned} W_{\mu\nu}(p,q) = & -g^{\mathrm{T}}_{\mu\nu}F_1(x,Q^2) + \frac{p^{\mathrm{T}}_{\mu}p^{\mathrm{T}}_{\nu}}{pq}F_2(x,Q^2) \\ & - \frac{M}{pq}\mathrm{i}\varepsilon_{\mu\nu\rho\sigma}q^{\rho}\left[s^{\sigma}g_1(x,Q^2) + \left(s^{\sigma} - \frac{sq}{pq}p^{\sigma}\right)g_2(x,Q^2)\right], \end{aligned} \tag{3.2.110}$$

where s^{σ} denotes the spin vector of the nucleon with the properties $sp = 0$ and $s^2 = -1$.

The nucleon spin is composed of the spins of the valence quarks, the sea quarks, the gluons, and the orbital angular momenta. In the naïve parton model only the contribution of the valence quark spin is taken into account. Then, the structure functions $g_{1,2}$ depend on the difference of the distribution functions for quarks with spin up and spin down $\Delta_{q,\mathcal{N}} = f^{\uparrow}_{q,\mathcal{N}} - f^{\downarrow}_{q,\mathcal{N}}$. For electromagnetic deep-inelastic scattering this gives

$$g_1^{\gamma\gamma\mathcal{N}}(x) = \frac{1}{2}\sum_q Q_q^2(\Delta_{q,\mathcal{N}}(x) + \Delta_{\bar{q},\mathcal{N}}(x)), \quad g_2^{\gamma\gamma\mathcal{N}}(x) = 0. \tag{3.2.111}$$

The QCD corrections to this parton-model result can be obtained from evolution equations for $\Delta_{q,\mathcal{N}}(x,Q^2)$ and $\Delta_{\mathrm{g},\mathcal{N}}(x,Q^2)$ of the form (3.2.88) with the *lowest-order polarized splitting functions* [Al77]

$$\begin{aligned} \Delta P^{(0)}_{\mathrm{qq}}(x) &= P_{\mathrm{qq}}(x) = C_{\mathrm{F}}\left(\frac{1+x^2}{1-x}\right)_+, \\ \Delta P^{(0)}_{\mathrm{qg}}(x) &= T_{\mathrm{F}}(2x-1), \\ \Delta P^{(0)}_{\mathrm{gq}}(x) &= C_{\mathrm{F}}(2-x), \\ \Delta P^{(0)}_{\mathrm{gg}}(x) &= C_{\mathrm{A}}\left(\left(\frac{2}{1-x}\right)_+ - 4x + 2\right) + \delta(1-x)\frac{11C_{\mathrm{A}} - 4N_{\mathrm{f}}T_{\mathrm{F}}}{6}. \end{aligned} \tag{3.2.112}$$

Taking input data and an ansatz as in (3.2.105), predictions for the Q^2 dependence of $g_1(x,Q^2)$ can be made (again $C_{\mathrm{F}} = 4/3$, $T_{\mathrm{F}} = 1/2$, and $C_{\mathrm{A}} = 3$ for QCD). The *two-loop polarized splitting functions* are also known [Me96, Vo96].

From the experimental data it turns out that approximately only one third of the nucleon spin is carried by the valence quarks [Kr99] (spin crisis [Le88]).

However, because of the axial anomaly (cf. Sect. 2.7), the spin densities of the corresponding constituent and parton quarks do not have to be equal. The possible contribution of the gluons to the quark spin function may cure this defect [Al97].

The spin structure of the nucleons is also tested by the *Bjorken sum rule* for the spin asymmetry [Bj66],

$$S_{\mathrm{Bj}} = \int_0^1 \mathrm{d}x \left[g_1^{\mathrm{P}}(x, Q^2) - g_1^{\mathrm{N}}(x, Q^2)\right] = \frac{1}{6}\left|\frac{g_{\mathrm{A}}}{g_{\mathrm{V}}}\right|\left(1 - \frac{\alpha_{\mathrm{s}}}{\pi}\right), \quad (3.2.113)$$

where g_{V} and g_{A} are the vector and axial-vector couplings of the nucleon. The Bjorken sum rule is based on isospin symmetry and current algebra. In the formula above also the leading QCD correction is included. The measured value shows that the Bjorken sum rule is satisfied within present errors. From this, it is possible to derive, taking into account the next-to-leading order (NLO) QCD corrections, a value for the strong coupling constant $\bar{\alpha}_{\mathrm{s}}(M_{\mathrm{Z}}^2) = 0.120 + 0.010 - 0.008$ [Al97] which is in good agreement with other results.

Summing up this short survey on experimental tests of QCD in deep-inelastic scattering, there is no doubt that QCD is confirmed as a non-abelian gauge theory with gauge group SU(3). Gluonic radiative corrections, including the effects of the NLO, are needed for an adequate interpretation of the data. The methods used in this section, namely the operator-product expansion and the evolution equations, allow for the decomposition of the amplitudes into parts that can be treated by perturbation theory and non-perturbative parts.

The OPE gives results for the moments of the structure functions. This method is restricted to processes which are described by local current operators. Other hard-scattering processes can be treated using the factorization theorems and the evolution equations that have been derived for QCD. They are the basis for perturbative QCD. This material is presented in the next section.

3.3 Perturbative Quantum Chromodynamics

The operator-product expansion is a method suitable for dealing with deep-inelastic lepton–nucleon scattering. However, it cannot be used for many other hard-scattering processes such as the inclusive production of hadrons, the Drell–Yan process, or jet production in e^+e^- annihilation. Therefore, it is important to provide the techniques for a direct application of the QCD perturbation series to these processes.

An important property of QCD is *factorization* of processes into short-distance and long-distance parts. The short-distance part is a hard parton-scattering function, which is calculable in perturbation theory. It does not depend on the external hadrons. It is IR-safe, i.e. it does not involve IR or mass singularities. After renormalization it is UV-finite and depends on the renormalization scale. The long-distance parts, the parton distribution functions and the *fragmentation functions*, are independent of the hard-scattering process, IR-sensitive, and contain the properties of the external hadrons. Both parts are put together at the factorization scale like in the parton model (for a review see e.g. [St95]). The detailed rules needed are derived for deep-inelastic lepton–nucleon scattering in Sect. 3.3.1. A prescription for the factorization of hard-scattering processes is formulated in Sect. 3.3.2. This approach is compared with the operator-product expansion in Sect. 3.3.3. As applications, fermion-pair production (Sect. 3.3.4) and jet production (Sect. 3.3.5) are considered.

3.3.1 One-loop corrections to the parton model

First, we perform a perturbative calculation of the QCD corrections of order α_s to the parton model for the case of lepton–nucleon scattering. The relevant diagrams have been shown in Fig. 3.2 on page 436. As can be read off from (1.5.30), the calculation of the quark–lepton scattering in lowest order (α_s^0) yields

$$\begin{aligned} \hat{F}_2^{qV}(\hat{x}, Q^2) &= G_v^{qV}\hat{x}\delta(1-\hat{x}) \qquad \text{with} \quad \hat{x} = \frac{Q^2}{2\hat{p}q}, \\ \hat{F}_L^{qV}(\hat{x}, Q^2) &= \hat{F}_2^{qV}(\hat{x}, Q^2) - 2\hat{x}\hat{F}_1^{qV}(\hat{x}, Q^2) = 0 \end{aligned} \tag{3.3.1}$$

for the parity-conserving quark structure functions. Variables referring to partons are indicated by a hat, and $G_v^{qV} = G_v^{qVV}$ is the product of coupling constants of the vector boson (1.5.11) for $V = V'$, the case that we are considering here.

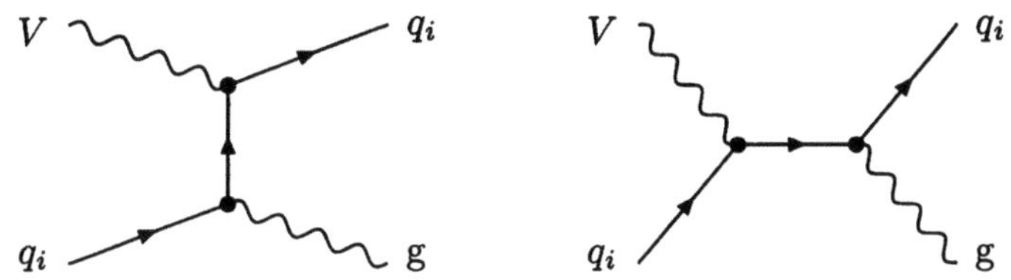

Fig. 3.6 Feynman diagrams for $V + q_i \to q_i + \mathrm{g}$

The Feynman diagrams of Fig. 3.6 for gluon bremsstrahlung off quarks $V(q) + q_i(\hat{p}) \to q_i(\hat{p}') + \mathrm{g}(k)$ ($q^2 = -Q^2$, $\hat{p}^2 = \hat{p}'^2 = k^2 = 0$) are identical with those of QED up to the colour matrix λ^a. The corresponding invariant matrix element for the vector coupling v_q^V of massless quarks and gluons reads

$$\mathcal{M}_\mu = ev_q^V \bar{u}(\hat{p}') g_\mathrm{s} \frac{\lambda^a}{2} \left(\gamma^\alpha \frac{\hat{\not{p}} + \not{q}}{\hat{s}} \gamma_\mu + \gamma_\mu \frac{\hat{\not{p}} - \not{k}}{\hat{t}} \gamma^\alpha \right) u(\hat{p}) \varepsilon_\alpha(k), \quad (3.3.2)$$

where ε_α is the polarization vector of the gluon ($\varepsilon^2 = -1$, $\varepsilon k = 0$), and

$$\hat{s} = (\hat{p} + q)^2 = \frac{1 - \hat{x}}{\hat{x}} Q^2, \qquad \hat{t} = (\hat{p} - k)^2 = -\frac{Q^2}{2\hat{x}}(1 - \cos\vartheta),$$
$$\hat{u} = (q - k)^2 = -\frac{Q^2}{2\hat{x}}(1 + \cos\vartheta), \qquad \hat{s} + \hat{t} + \hat{u} = -Q^2. \quad (3.3.3)$$

Our interest lies in the unpolarized, parity-conserving part of the cross section summed over gluon colours and integrated over the gluon momentum. Using the polarization sum in Feynman gauge $\sum_\mathrm{pol} \varepsilon_\alpha^* \varepsilon_\beta = -g_{\alpha\beta}$, the following expression results for the invariant matrix element squared:

$$\mathcal{M}_{\mu\nu} = -\frac{1}{2} \sum_\text{spins} \frac{1}{3} \sum_\text{colour} \mathcal{M}_\mu^* \mathcal{M}_\nu = -\frac{1}{2} e^2 (v_q^V)^2 g_\mathrm{s}^2 C_\mathrm{F}$$
$$\times \operatorname{Tr}\left(\frac{\gamma_\mu (\hat{\not{p}} + \not{q}) \gamma_\alpha \hat{\not{p}}' \gamma^\alpha (\hat{\not{p}} + \not{q}) \gamma_\nu \hat{\not{p}}}{\hat{s}^2} + \frac{\gamma^\alpha (\hat{\not{p}} - \not{k}) \gamma_\mu \hat{\not{p}}' \gamma_\nu (\hat{\not{p}} - \not{k}) \gamma_\alpha \hat{\not{p}}}{\hat{t}^2} \right.$$
$$\left. + \frac{\gamma_\mu (\hat{\not{p}} + \not{q}) \gamma_\alpha \hat{\not{p}}' \gamma_\nu (\hat{\not{p}} - \not{k}) \gamma^\alpha \hat{\not{p}} + \gamma^\alpha (\hat{\not{p}} - \not{k}) \gamma_\mu \hat{\not{p}}' \gamma_\alpha (\hat{\not{p}} + \not{q}) \gamma_\nu \hat{\not{p}}}{\hat{s}\hat{t}} \right). \quad (3.3.4)$$

We regularize possible IR divergences by calculating in $D = 4 + 2\epsilon$ dimensions. Then, we have to use the D-dimensional two-particle massless phase space,

$$\mathrm{d}\Gamma_2^D = \mu^{(4-D)}\delta^{(D)}(p + q - p' - k)\frac{\mathrm{d}^D p'}{(2\pi)^{D-1}}\delta(p'^2)\frac{\mathrm{d}^D k}{(2\pi)^{D-1}}\delta(k^2),$$

and the relation between the differential "cross section" and the invariant matrix element squared,

$$\frac{\mathrm{d}\hat{\sigma}_{\mu\nu}^{Vq\to qg}}{\mathrm{d}\Omega} = \frac{1}{64\pi^2\hat{s}}\frac{p'}{p}\mathcal{M}_{\mu\nu}\frac{1}{\Gamma(1+\epsilon)}\left(\frac{p'^2\sin^2\vartheta}{4\pi\mu^2}\right)^{\epsilon}, \tag{3.3.5}$$

where p and p' are the initial- and final-state three-momenta in Vq CM system, ϑ the scattering angle, and μ the scale parameter of dimensional regularization (cf. Sect. 2.5.2.1). We have the relations

$$p = \frac{\hat{s} + Q^2}{2\sqrt{\hat{s}}}, \qquad p' = \frac{\sqrt{\hat{s}}}{2}, \qquad \frac{1}{\hat{s}}\frac{p'}{p} = \frac{\hat{x}}{Q^2}. \tag{3.3.6}$$

In order to obtain the contribution to the structure functions we relate the "total cross section" $\hat{\sigma}_{\mu\nu}$ to the hadronic tensor [cf. (1.2.24), (1.5.9) and (3.2.1)],

$$e^2\hat{W}_{\mu\nu} = \frac{q\hat{p}}{\pi}\hat{\sigma}_{\mu\nu} = \frac{1}{2\pi}\hat{s}\frac{p}{p'}\hat{\sigma}_{\mu\nu} = \frac{1}{\pi}\frac{Q^2}{2\hat{x}}\hat{\sigma}_{\mu\nu}, \tag{3.3.7}$$

and project out the structure functions like in (3.2.8),

$$\begin{aligned}\frac{1}{\hat{x}}\hat{F}_2 &= \frac{1}{1+\epsilon}\left(-g^{\mu\nu} + 2\hat{x}\frac{\hat{p}^\mu\hat{p}^\nu}{\hat{p}q}(3+2\epsilon)\right)\hat{W}_{\mu\nu}, \\ \hat{F}_\mathrm{L} &= 4\hat{x}^2\frac{\hat{p}^\mu\hat{p}^\nu}{\hat{p}q}\hat{W}_{\mu\nu}.\end{aligned} \tag{3.3.8}$$

The contractions of (3.3.4) give, adding the contribution of the axial coupling with (1.5.11)

$$\begin{aligned}g_{\mu\nu}\mathcal{M}^{\mu\nu} &= -4e^2 G_v^{qV} g_\mathrm{s}^2 C_\mathrm{F}(1+\epsilon)\left[\frac{\hat{s}}{\hat{t}} - \frac{2Q^2\hat{u}}{\hat{s}\hat{t}} + \frac{\hat{t}}{\hat{s}} + \epsilon\frac{(\hat{s}+\hat{t})^2}{\hat{s}\hat{t}}\right], \\ \hat{p}_\mu\hat{p}_\nu\mathcal{M}^{\mu\nu} &= -4e^2 G_v^{qV} g_\mathrm{s}^2 C_\mathrm{F}(1+\epsilon)\left[\frac{\hat{u}}{2}\right].\end{aligned} \tag{3.3.9}$$

The D-dimensional integral for the total cross section reads ($p'^2 = (1 - \hat{x})Q^2/4\hat{x}$)

$$\hat{\sigma}^{Vq\to qg}_{g,\mu\nu}(\hat{x}, Q^2) = \frac{1}{64\pi^2}\frac{\hat{x}}{Q^2}\frac{1}{\Gamma(1+\epsilon)}\left(\frac{(1-\hat{x})Q^2}{16\pi\mu^2\hat{x}}\right)^{\epsilon} \times 2\pi\int_0^{\pi} d\vartheta\,(\sin\vartheta)^{1+2\epsilon}\mathcal{M}_{\mu\nu}. \tag{3.3.10}$$

Rewriting this integral in terms of the momentum transfer $\hat{t} = -Q^2(1 - \cos\vartheta)/2\hat{x}$ instead of the scattering angle gives

$$\hat{\sigma}^{Vq\to qg}_{g,\mu\nu}(\hat{x}, Q^2) = \frac{1}{16\pi}\left(\frac{\hat{x}}{Q^2}\right)^2\frac{1}{\Gamma(1+\epsilon)}\left(\frac{1-\hat{x}}{4\pi\mu^2}\right)^{\epsilon} \times \int_{-Q^2/\hat{x}}^{0} d\hat{t}\,(-\hat{t})^{\epsilon}\left(1+\frac{\hat{x}\hat{t}}{Q^2}\right)^{\epsilon}\mathcal{M}_{\mu\nu}(\hat{t}). \tag{3.3.11}$$

From the $\hat{t}$-pole terms of $\mathcal{M}_{\mu\nu}(\hat{t})$, (3.3.9), the mass singularity, i.e. logarithmically divergent contributions to the quark structure functions, emerges. Since $\hat{t} = -2E_qE_g(1 - \cos\vartheta)$, small $\hat{t}$ corresponds either to $E_g \approx 0$, i.e. *radiation of soft gluons*, or to $\vartheta \approx 0$, i.e. *collinear radiation of gluons.*

Contracting (3.3.11) with $g^{\mu\nu}$, performing the integral, and expressing everything by the scaling variable $\hat{x}$ leads to

$$-g^{\mu\nu}\hat{\sigma}^{Vq\to qg}_{g,\mu\nu}(\hat{x}, Q^2) = 4\pi\alpha G^{qV}_{v}\,\alpha_s C_F\frac{\hat{x}}{Q^2}\left(\frac{(1-\hat{x})Q^2}{4\pi\mu^2\hat{x}}\right)^{\epsilon}\frac{\Gamma(1+\epsilon)}{\Gamma(1+2\epsilon)} \times\left[\Gamma(\epsilon)\frac{1+\hat{x}^2}{1-\hat{x}} - \frac{3}{2}\frac{1}{1-\hat{x}} + 3 - \hat{x} - \frac{7}{2}\epsilon\frac{1}{1-\hat{x}}\right] + \mathcal{O}(\epsilon). \tag{3.3.12}$$

Similarly the contraction with $p^\mu p^\nu$ yields

$$p^\mu p^\nu\hat{\sigma}^{Vq\to qg}_{g,\mu\nu} = 4\pi\alpha G^{qV}_{v}\,\alpha_s C_F\frac{\hat{x}}{Q^2}\frac{Q^2}{4x} + \mathcal{O}(\epsilon). \tag{3.3.13}$$

Using (3.3.7) and (3.3.8) we obtain for the *gluon-bremsstrahlung* contribution to the quark structure functions

$$\hat{F}^{Vq,(1)}_{2}(\hat{x}, Q^2)_{\mathrm{real}} = G^{qV}_{v}\frac{\alpha_s}{2\pi}C_F\hat{x}\left(\frac{(1-\hat{x})Q^2}{4\pi\mu^2\hat{x}}\right)^{\epsilon}\frac{\Gamma(1+\epsilon)}{\Gamma(1+2\epsilon)} \times\left[\frac{1+\hat{x}^2}{1-\hat{x}}\Gamma(\epsilon) - \frac{3}{2}\frac{1}{1-\hat{x}} + 3 + 2\hat{x} - \frac{7}{2}\epsilon\frac{1}{1-\hat{x}}\right] + \mathcal{O}(\epsilon),$$

$$\hat{F}^{Vq,(1)}_{L}(\hat{x}, Q^2)_{\mathrm{real}} = G^{qV}_{v}\frac{\alpha_s}{2\pi}C_F 2\hat{x}^2 + \mathcal{O}(\epsilon). \tag{3.3.14}$$

where we have omitted terms that give a vanishing contribution for $\epsilon \to 0$. This "real" part of $\hat{F}_2^{Vq}$ is Q^2-dependent and IR-singular. For the complete expansion of (3.3.14) in ϵ, the following identity for distributions is useful:

$$x^{-\epsilon}(1-x)^{-1+\epsilon} = \frac{1}{\epsilon}\delta(1-x) + \left(\frac{1}{1-x}\right)_+ + \epsilon\left(\frac{\log(1-x)}{1-x}\right)_+ - \epsilon\frac{\log(x)}{1-x} + \mathcal{O}(\epsilon^2). \quad (3.3.15)$$

The contribution of real-gluon emission graphs finally reads

$$\begin{aligned}\hat{F}_2^{Vq,(1)}(\hat{x}, Q^2)_{\text{real}} &= G_v^{qV}\frac{\alpha_s}{2\pi}C_F\hat{x}\left(\frac{Q^2}{4\pi\mu^2}\right)^{\epsilon}\frac{\Gamma(1+\epsilon)}{\Gamma(1+2\epsilon)} \\ &\times\left[\frac{2}{\varepsilon^2}\delta(1-\hat{x}) + \frac{1}{\varepsilon}\left(\frac{1+\hat{x}^2}{1-\hat{x}}\right)_+ + (1+\hat{x}^2)\left(\frac{\log(1-\hat{x})}{1-\hat{x}}\right)_+\right. \\ &\left.-\frac{3}{2}\left(\frac{1}{1-\hat{x}}\right)_+ + \frac{7}{2}\delta(1-\hat{x}) - \frac{1+\hat{x}^2}{1-\hat{x}}\log\hat{x} + 2\hat{x} + 3\right]. \quad (3.3.16)\end{aligned}$$

The contribution of the internal radiative corrections depicted in Fig. 3.2d on page 436 is also IR-singular but proportional to $\delta((\hat{p}+q)^2) = \delta(1-\hat{x})/2\hat{p}q$. The detailed calculation proceeds like in Sect. 2.8.3.2 and shows indeed that the singularities at $\hat{x} = 1$ of the real and virtual corrections cancel.

The total result for the corrections in order α_s for the *quark structure function* in the dimensional regularization scheme reads [Al79, Hu81]

$$\begin{aligned}\hat{F}_2^{Vq,(1)}(\hat{x}, Q^2) &= G_v^{qV}\frac{\alpha_s}{2\pi}C_F\hat{x}\left[\left(\frac{1+\hat{x}^2}{1-\hat{x}}\right)_+\left(\frac{1}{\epsilon} + \gamma + \log\frac{Q^2}{4\pi\mu^2}\right)\right. \\ &+ (1+\hat{x}^2)\left(\frac{\log(1-\hat{x})}{1-\hat{x}}\right)_+ - \frac{3}{2}\left(\frac{1}{1-\hat{x}}\right)_+ \\ &\left.-\left(\frac{\pi^2}{3} + \frac{9}{2}\right)\delta(1-\hat{x}) - \frac{1+\hat{x}^2}{1-\hat{x}}\log\hat{x} + 2\hat{x} + 3\right] \\ &= G_v^{qV}\frac{\alpha_s}{2\pi}C_F\hat{x}\left[\frac{1+\hat{x}^2}{1-\hat{x}}\left(\frac{1}{\epsilon} + \gamma + \log\frac{Q^2}{4\pi\mu^2}\right)\right. \\ &\left.+\frac{1+\hat{x}^2}{1-\hat{x}}\left(\log\frac{1-\hat{x}}{\hat{x}} - \frac{3}{4}\right) + \frac{9+5\hat{x}}{4}\right]_+ \\ &= G_v^{qV}\frac{\alpha_s}{2\pi}\hat{x}\left[P_{qq}^{(0)}(\hat{x})\left(\frac{1}{\epsilon} + \gamma + \log\frac{Q^2}{4\pi\mu^2}\right) + \hat{\Phi}_2^{Vq}(\hat{x})\right] \quad (3.3.17)\end{aligned}$$

with the finite, scheme-dependent *coefficient function* $\hat{\Phi}_2^{Vq}(\hat{x})$ and the *splitting function* [cf. (3.2.90)]

$$P_{\rm qq}^{(0)}(\hat{x}) = C_{\rm F}\left(\frac{1+\hat{x}^2}{1-\hat{x}}\right)_+ . \tag{3.3.18}$$

The quark structure functions are not directly measurable quantities reflecting the fact that the quarks are not free particles but confined in the hadrons. On this level we cannot expect cancellation of the collinear singularities. Therefore in one-loop order, $\hat{F}_2^{Vq,(1),{\rm NS}}(\hat{x},Q^2)$ still contains the *collinear singularity* $P_{\rm qq}^{(0)}(\hat{x})(1/\epsilon+\gamma+\log(Q^2/\mu^2))$. Moreover, it is not expressed by ordinary functions but by (+)-distributions which were defined in (2.8.81).

A second reaction mechanism in $\mathcal{O}(\alpha_{\rm s})$ is quark–antiquark production in gluon–vector-boson scattering, $V+{\rm g}\to q_i+\bar{q}_i$. The calculation gives

$$\begin{aligned}
\hat{F}_2^{V{\rm g},(1)}(\hat{x},Q^2) &= 2\sum_q G_v^{qV}\hat{x}\frac{\alpha_{\rm s}}{2\pi}\left[P_{\rm qg}^{(0)}(\hat{x})\left(\frac{1}{\epsilon}+\gamma+\log\frac{Q^2}{4\pi\mu^2}\right)\right. \\
&\quad \left. + P_{\rm qg}^{(0)}(\hat{x})\log\frac{1-\hat{x}}{\hat{x}} - T_{\rm F}(6\hat{x}^2-6\hat{x})\right] \\
&= 2\sum_q G_v^{qV}\hat{x}\frac{\alpha_{\rm s}}{2\pi}\left[P_{\rm qg}^{(0)}(\hat{x})\left(\frac{1}{\epsilon}+\gamma+\log\frac{Q^2}{4\pi\mu^2}\right)+\hat{\Phi}_2^{V{\rm g}}(\hat{x})\right]
\end{aligned} \tag{3.3.19}$$

for the *gluon structure function* in the dimensional regularization scheme [Al79, Hu81] with the *splitting function* [cf. (3.2.104)]

$$P_{\rm qg}^{(0)}(z) = T_{\rm F}(z^2+(1-z)^2). \tag{3.3.20}$$

and the *coefficient function* $\hat{\Phi}_2^{V{\rm g}}(\hat{x})$ [Ba78]. The QCD corrections to the quark and gluon structure functions are flavour-independent at one-loop order.

The physical process is inclusive lepton–hadron scattering. The hadron structure functions are built up from the parton structure functions and the parton distribution functions. The singularities of both have to compensate each other in order to give finite and unique hadron structure functions. The mechanism of this compensation is discussed in the next section.

3.3.2 Factorization

In order to obtain the *nucleon structure function* $F_2^{V\mathcal{N}}(x, Q^2)$ we generalize the parton-model procedure (1.5.31). There, the *parton structure function* (3.3.1) $\hat{F}_2^{Vq,(0)}(\hat{x}) = G_v^{qV} \hat{x}\delta(1-\hat{x})$ is multiplied with the *parton distribution function* $f_{q,\mathcal{N}}(z)$, the probability of finding the corresponding parton with four-momentum $\hat{p} = zp$ in the nucleon. The condition $\hat{x} = Q^2/2\hat{p}q = Q^2/2zpq = x/z$ is imposed with help of the δ-function $\delta(x - \hat{x}z)$. Finally, the resulting expression is integrated over $\hat{x}$ and z. This *factorization* into a short-distance and a long-distance part was proven, for instance, for deep-inelastic scattering and the Drell-Yan process [Co89] and is the basis of perturbative QCD. The first factor can be calculated in perturbation theory, whereas the second one has to be taken from experiment.

In perturbative QCD, like in QED [cf. (2.8.115)], factorization is a *multiplicative convolution*,

$$\begin{aligned} \frac{1}{x}F_2^{V\mathcal{N}}(x, Q^2) &= \int_0^1 \mathrm{d}z \int_0^1 \mathrm{d}\hat{x}\, \delta(x - \hat{x}z) \\ &\times \left(\sum_q \frac{1}{\hat{x}} \hat{F}_2^{Vq}(\hat{x}, Q^2) \Big(f_{q,\mathcal{N}}(z) + f_{\bar{q},\mathcal{N}}(z) \Big) + \frac{1}{\hat{x}} \hat{F}_2^{V\mathrm{g}}(\hat{x}, Q^2) f_{\mathrm{g},\mathcal{N}}(z) \right) \\ &= \left(\left[\sum_q \hat{\mathcal{F}}_2^{Vq} \otimes (f_{q,\mathcal{N}} + f_{\bar{q},\mathcal{N}}) \right] + \hat{\mathcal{F}}_2^{V\mathrm{g}} \otimes f_{\mathrm{g},\mathcal{N}} \right) (x, Q^2), \qquad (3.3.21) \end{aligned}$$

where the sum runs over all quark flavours, and $\mathcal{F}_2(x) = F_2(x)/x$.

An important property of factorization is that the hard-scattering parts, the parton-scattering cross sections—in deep-inelastic scattering the *parton structure functions parton!structure function* $\hat{F}^{Vi}(\hat{x}, Q^2)$—depend only on the interacting vector boson V and the parton i but not on the external hadrons. They are calculated in form of loop corrections to the Born approximation and depend on the strong coupling constant g_s, i.e. α_s, the renormalization scale $\mathcal{M}$, and the hard-scattering variable Q^2. They are *IR-safe*, i.e. free of soft-gluon singularities.

On the other hand, the long-distance parts, the parton distribution functions $f_{i,\mathcal{N}}(z)$, depend on the external hadron and the interacting parton but not on the specific interacting vector boson, i.e. they are universal. They are independent of the hard-scattering variable Q^2. Since they contain the long-distance effects they cannot be calculated in perturbation theory and contain all the IR sensitivity of the original cross section.

The separation of the cross section (3.3.21) into a long-distance part and a short-distance part is not unique. It is possible to shift finite contributions from the long-distance to the short-distance part and vice versa. Therefore, the separation requires the definition of a soft–hard scale parameter or factorization scale $\mathcal{M}_{\rm SH}$ relative to which soft and hard is defined and a prescription for the distribution of the finite parts. These two points have to be fixed by a *factorization scheme.*

In order to make this more explicit, we rewrite the $\mathcal{O}(\alpha_{\rm s})$ expression for $F_2^{V\mathcal{N}}(x, Q^2)$ as

$$\begin{aligned}\frac{1}{x}F_2^{V\mathcal{N}}(x,Q^2) = \sum_q G_v^{qV} \int_x^1 \frac{\mathrm{d}z}{z} \Bigg\{ \Bigg(\delta\left(1-\frac{x}{z}\right) \\ + \frac{\alpha_{\rm s}}{2\pi}\left[P_{\rm qq}^{(0)}\left(\frac{x}{z}\right)\left(\frac{1}{\epsilon}+\gamma+\log\frac{Q^2}{4\pi\mu^2}\right)+\hat{\Phi}_2^{Vq}\left(\frac{x}{z}\right)\right]\Bigg) \\ \times\left(f_{q,\mathcal{N}}^{(0)}(z)+f_{\bar{q},\mathcal{N}}^{(0)}(z)\right) \\ + 2\frac{\alpha_{\rm s}}{2\pi}\left[P_{\rm qg}^{(0)}\left(\frac{x}{z}\right)\left(\frac{1}{\epsilon}+\gamma+\log\frac{Q^2}{4\pi\mu^2}\right)+\hat{\Phi}_2^{V{\rm g}}\left(\frac{x}{z}\right)\right] f_{{\rm g},\mathcal{N}}^{(0)}(z)\Bigg\},\end{aligned} \tag{3.3.22}$$

where $f_{i,\mathcal{N}}^{(0)}(z)$ are bare parton distribution functions. The mass-singular term $\propto 1/\epsilon$ has to be, the other parts may be absorbed into $f_{i,\mathcal{N}}^{(0)}(z)$ leaving finite but scheme-dependent, *"renormalized" parton structure and parton distribution functions.* The absorption of the mass-singular term can be considered as a kind of renormalization of the parton distribution functions. In this way the measured nucleon structure function leads to scheme-dependent parton distribution functions. Therefore, all manipulations have to be performed in the same scheme.

The following factorization schemes are mainly used in the applications of perturbative QCD.

In the $\overline{\rm MS}$ *scheme*, the leading-logarithmic term is shifted to the parton distribution function. The hadron structure function of the $\overline{\rm MS}$ scheme is obtained as convolution of these parton structure functions with parton distribution functions $f_{i,\mathcal{N}}^{\overline{\mathrm{MS}}}$ just like in (3.3.21),

$$\begin{aligned}\mathcal{F}_2^{V\mathcal{N}}(x,Q^2) = \sum_q G_v^{qV} \int_x^1 \frac{\mathrm{d}z}{z} \Bigg\{ \Bigg[\delta\left(1-\frac{x}{z}\right) \\ + \frac{\alpha_\mathrm{s}}{2\pi}\left[P_{\mathrm{qq}}^{(0)}\left(\frac{x}{z}\right)\log\frac{Q^2}{\mathcal{M}_{\mathrm{SH}}^2} + \hat{\Phi}_2^{Vq}\left(\frac{x}{z}\right)\right]\Bigg] \\ \times \left(f_{q,\mathcal{N}}^{\overline{\mathrm{MS}}}(z,Q^2) + f_{\bar{q},\mathcal{N}}^{\overline{\mathrm{MS}}}(z,Q^2)\right) \\ + 2\frac{\alpha_\mathrm{s}}{2\pi}\left[P_{\mathrm{qg}}^{(0)}\left(\frac{x}{z}\right)\log\frac{Q^2}{\mathcal{M}_{\mathrm{SH}}^2} + \hat{\Phi}_2^{V\mathrm{g}}\left(\frac{x}{z}\right)\right] f_{\mathrm{g},\mathcal{N}}^{\overline{\mathrm{MS}}}(z,Q^2)\Bigg\} \\ = \sum_q \hat{\mathcal{F}}_{2,\overline{\mathrm{MS}}}^{Vq} \otimes \left(f_{q,\mathcal{N}}^{\overline{\mathrm{MS}}} + f_{\bar{q},\mathcal{N}}^{\overline{\mathrm{MS}}}\right) + \hat{\mathcal{F}}_{2,\overline{\mathrm{MS}}}^{V\mathrm{g}} \otimes f_{\mathrm{g},\mathcal{N}}^{\overline{\mathrm{MS}}}. \end{aligned} \tag{3.3.23}$$

The $\overline{\mathrm{MS}}$ parton distribution functions obey the equation

$$\begin{aligned} f_{q,\mathcal{N}}^{\overline{\mathrm{MS}}}(z,\mathcal{M}_{\mathrm{SH}}^2) = f_{q,\mathcal{N}}^{(0)}(z) + \frac{\alpha_\mathrm{s}}{2\pi}\left(\frac{1}{\epsilon}+\gamma+\log\frac{\mathcal{M}_{\mathrm{SH}}^2}{4\pi\mu^2}\right) \\ \times \int_z^1 \frac{\mathrm{d}y}{y}\left[P_{\mathrm{qq}}^{(0)}\left(\frac{z}{y}\right) f_{q,\mathcal{N}}^{(0)}(y) + P_{\mathrm{qg}}^{(0)}\left(\frac{z}{y}\right) f_{\mathrm{g},\mathcal{N}}^{(0)}(y)\right], \end{aligned} \tag{3.3.24}$$

and similarly for $f_{\bar{q},\mathcal{N}}^{\overline{\mathrm{MS}}}(z,Q^2)$. From (3.3.23) it is possible to determine the parton distribution functions $f_{q,\mathcal{N}}^{\overline{\mathrm{MS}}}(z,\mathcal{M}_{\mathrm{SH}}^2)$ experimentally. QCD then predicts the Q^2 dependence of the nucleon structure functions in $\mathcal{O}(\alpha_\mathrm{s})$. Often the soft–hard scale parameter $\mathcal{M}_{\mathrm{SH}}$ is identified with the renormalization scale parameter $\mathcal{M}$, i.e. $\mathcal{M}_{\mathrm{SH}} = \mathcal{M}$.

In the DIS *scheme*, all corrections to $F_2^{Vq}(x,Q^2)$ are absorbed into the quark distribution functions at $Q^2 = \mathcal{M}_{\mathrm{SH}}^2$,

$$\begin{aligned}\mathcal{F}_2^{V\mathcal{N}}(x,Q^2) = \sum_q G_v^{qV}\left(f_{q,\mathcal{N}}^{\mathrm{DIS}}(x,Q^2) + f_{\bar{q},\mathcal{N}}^{\mathrm{DIS}}(x,Q^2)\right) \\ + \sum_q G_v^{qV}\frac{\alpha_\mathrm{s}}{2\pi}\log\frac{Q^2}{\mathcal{M}_{\mathrm{SH}}^2}\int_x^1\frac{\mathrm{d}z}{z} \\ \times\Bigg[P_{\mathrm{qq}}^{(0)}\left(\frac{x}{z}\right)\left(f_{q,\mathcal{N}}^{\mathrm{DIS}}(z,Q^2) + f_{\bar{q},\mathcal{N}}^{\mathrm{DIS}}(z,Q^2)\right) \\ + 2P_{\mathrm{qg}}^{(0)}\left(\frac{x}{z}\right) f_{\mathrm{g},\mathcal{N}}^{\mathrm{DIS}}(z,Q^2)\Bigg] \end{aligned} \tag{3.3.25}$$

with the DIS parton distribution functions

$$
\begin{aligned}
f_{q,\mathcal{N}}^{\mathrm{DIS}}(z,\mathcal{M}_{\mathrm{SH}}^2) &= f_{q,\mathcal{N}}^{(0)}(z) + \frac{\alpha_{\mathrm{s}}}{2\pi}\int_z^1 \frac{\mathrm{d}y}{y} \\
&\times \left\{ \left[P_{\mathrm{qq}}^{(0)}\left(\frac{z}{y}\right)\left(\frac{1}{\epsilon}+\gamma+\log\frac{\mathcal{M}_{\mathrm{SH}}^2}{4\pi\mu^2}\right) + \hat{\Phi}_2^{Vq}\left(\frac{z}{y}\right)\right] f_{q,\mathcal{N}}^{(0)}(z) \right. \\
&\left. + \left[P_{\mathrm{qg}}^{(0)}\left(\frac{z}{y}\right)\left(\frac{1}{\epsilon}+\gamma+\log\frac{\mathcal{M}_{\mathrm{SH}}^2}{4\pi\mu^2}\right) + \hat{\Phi}_2^{V\mathrm{g}}\left(\frac{z}{y}\right)\right] f_{\mathrm{g},\mathcal{N}}^{(0)}(z) \right\},
\end{aligned}
\tag{3.3.26}
$$

and similarly for $f_{\bar{q},\mathcal{N}}^{\mathrm{DIS}}$. The relation between the two schemes reads

$$
f_{q,\mathcal{N}}^{\mathrm{DIS}} = f_{q,\mathcal{N}}^{\overline{\mathrm{MS}}} + \frac{\alpha_{\mathrm{s}}}{2\pi}\left\{ \hat{\Phi}_2^{Vq} \otimes f_{q,\mathcal{N}}^{\overline{\mathrm{MS}}} + \hat{\Phi}_2^{V\mathrm{g}} \otimes f_{\mathrm{g},\mathcal{N}}^{\overline{\mathrm{MS}}} \right\}
\tag{3.3.27}
$$

at $Q^2 = \mathcal{M}_{\mathrm{SH}}^2$ and analogously for the antiquark distribution function.

The hadron structure functions $F_1(\hat{x}, Q^2)$ and $F_3(\hat{x}, Q^2)$ including the finite parts can be treated similarly. [Yn99, El96].

The parton distribution functions are determined from experiment using (3.3.21). Since the parton structure functions have to be fixed within a factorization scheme, the parton distribution functions are factorization-scheme-dependent.

3.3.3 Factorization and the operator-product expansion

3.3.3.1 Moments of the nucleon structure function

The multiplicative convolution (3.3.21) of the quark structure functions with the quark distribution functions can be split up into factors by a *Mellin transformation.* This yields for the moments of the structure functions

$$
\begin{aligned}
F_{2,n}^{V\mathcal{N}}(Q^2) &= \int_0^1 \mathrm{d}x\, x^{n-2} F_2^{V\mathcal{N}}(x,Q^2) \\
&= \int_0^1 \mathrm{d}x\, x^{n-2}\left[\sum_q \hat{F}_2^{Vq} \otimes \left(f_{q,\mathcal{N}}^{(0)} + f_{\bar{q},\mathcal{N}}^{(0)}\right) + \hat{F}_2^{V\mathrm{g}} \otimes f_{\mathrm{g},\mathcal{N}}^{(0)}\right](x,Q^2) \\
&= \sum_i \hat{F}_{2,n}^{Vi}(Q^2) f_{i,\mathcal{N},n}^{(0)}, \qquad i = \{q, \bar{q}, \mathrm{g}\}.
\end{aligned}
\tag{3.3.28}
$$

We considered in Sect. 3.2.2 the OPE for the leading-logarithmic (LL) approximation of the moments of the structure function. Therefore, we take the corresponding approximation of the one-loop *parton structure functions* [cf. (3.3.17) and (3.3.19)]

$$\hat{F}_{2,\mathrm{LL}}^{Vi}(\hat{x}, Q^2) = \sum_j G_v^{iV} \hat{x} \Bigg[\delta_{ij}\delta(1-\hat{x}) + \frac{\alpha_s}{2\pi} P_{ij}^{(0)}(\hat{x}) \left(\frac{1}{\epsilon} + \gamma + \log \frac{Q^2}{4\pi\mu^2} \right) \Bigg] . \qquad (3.3.29)$$

An explicit expression for the moments can be calculated from (3.3.29) by using the connection (3.2.95) between splitting function $P_{ij}^{(0)}(x)$ and anomalous dimensions $\gamma_{n,0}^{ij}$. Factorizing at $\mathcal{M}_{\mathrm{SH}}$ as described in Sect. 3.3.2 yields

$$F_{2,\mathrm{LL},n}^{V\mathcal{N}} = \sum_{ij} G_v^{iV} \left[\delta_{ij} - \frac{\alpha_s(\mathcal{M})}{8\pi} \gamma_{n,0}^{ij} \log \frac{Q^2}{\mathcal{M}_{\mathrm{SH}}^2} \right] f_{j,\mathcal{N},n}(\mathcal{M}_{\mathrm{SH}}^2) . \qquad (3.3.30)$$

3.3.3.2 Summation of leading higher-order terms

The order-α_s calculations already reproduce some of the results previously obtained using operator-product expansion and renormalization-group formalism, summarized by the evolution equations. Working in higher orders of α_s, quite a lot of Feynman diagrams contribute. Thus, the interesting question is which diagrams dominate in the Bjorken limit. An answer was given by V.N. Gribov and L.N. Lipatov [Gr72] for abelian vector-gluon theories, and later extended to non-abelian gauge theories like QCD [Ll78, Am78, Fo97]. In an appropriate gauge, *generalized ladder diagrams* with renormalized propagators and vertices give the *leading-logarithmic contributions*.[12] Unitarity relates the hadronic tensor $\hat{W}_{\mu\nu}$ to the forward Comp-

[12] We present only a few details for the photonic case; for further information we refer to the quoted literature.

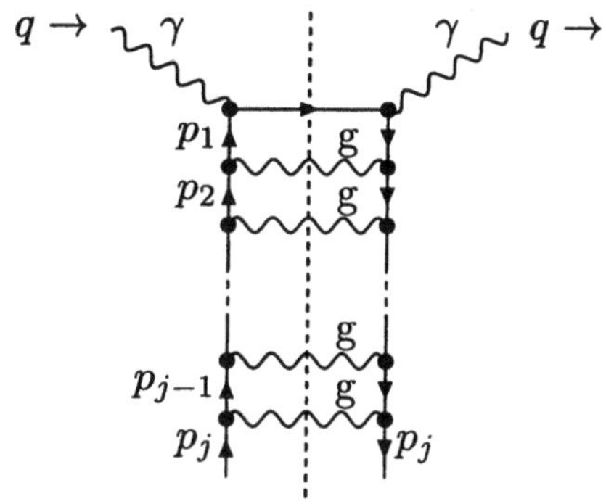

Fig. 3.7 Ladder diagram for photon–quark scattering

ton scattering amplitude of virtual photons off quarks,

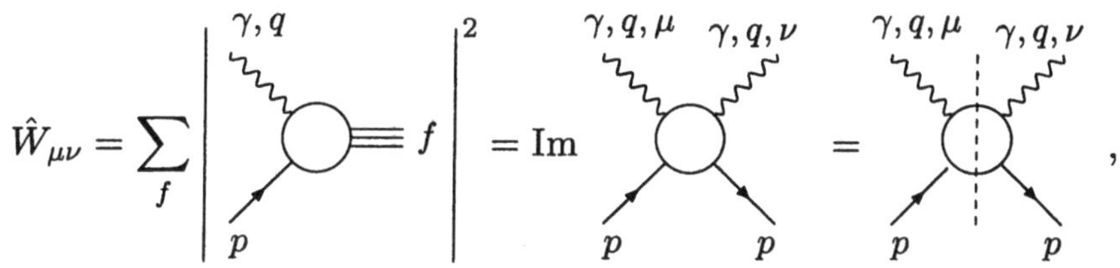

(3.3.31)

where the dashed line indicates cutting, i.e. the corresponding propagators are taken on-shell. In formulas, the relation reads

$$W_{\mu\nu}(x, Q^2) = \frac{1}{2\pi}\mathrm{Im}\{T_{\mu\nu}(x, Q^2)\}. \tag{3.3.32}$$

The ladder-like gluon-exchange diagrams with bare propagators and vertices (an example is shown in Fig. 3.7) can be investigated by iteration. The relevant quantity is the square of the amplitude $\hat{T}$ for forward-scattering of virtual photons off quarks. The change in the corresponding expression occurring when enlarging the ladder from j rungs to $j+1$ rungs by the emission of an additional gluon can be calculated in a similar way as in Sect. 3.3.1 for the emission of one gluon. In the leading-logarithmic approximation only the terms with the t-channel pole are relevant. A similar calculation has been performed in Sect. 2.8.3.1. One obtains the following recursion formula for the square of the amplitude $\hat{T}$ for forward-scattering of virtual photons off quarks:

$$\hat{T}_{j+1}(\hat{p}_{j+1}, q) = \int_{-Q^2}^{-Q_0^2} \frac{\bar{\alpha}_s(\hat{t}_j)}{2\pi} \frac{d\hat{t}_j}{\hat{t}_j} \int_0^1 \frac{dz_j}{z_j} P_{qq}^{(0)}(z_j) \hat{T}_j(\hat{p}_j, q). \quad (3.3.33)$$

Here, for simplicity we regularize the mass singularities by a cut-off parameter Q_0. Dressed vertices and propagators are included by taking the running coupling constant $\bar{\alpha}_s(\hat{t}_j) = 4\pi/(\beta_0 \log(-\hat{t}_j/\Lambda^2))$. The variables are defined as

$$\hat{t}_j = \hat{p}_j^2, \qquad p_j = z_j p_{j+1} \quad (3.3.34)$$

and obey

$$-Q^2 = q^2 \leq t_1 \leq t_2 \leq \ldots \leq t_j \leq -Q_0^2. \quad (3.3.35)$$

The IR singularities of the real and virtual corrections compensate each other. Inserting the explicit expression for the running coupling constant yields

$$\hat{T}_{j+1}(\hat{p}_{j+1}, q) = \int_{-Q^2}^{-Q_0^2} \frac{d\hat{t}_j}{\hat{t}_j} \frac{2}{\beta_0 \log \frac{-\hat{t}_j}{\Lambda^2}} \int_0^1 \frac{dz_j}{z_j} P_{qq}^{(0)}(z_j) \hat{T}_j(\hat{p}_j, q). \quad (3.3.36)$$

This is the desired iteration formula for the dominant terms. Owing to the relations (3.3.32) and (3.2.8) it also holds for the corresponding structure functions.

The term of zeroth order is the contribution of the naïve parton model, $\hat{F}_{2,0}^{Vq} = G_v^{qV} \hat{x}\delta(1-\hat{x})$ [cf. (3.3.1)]. Using this expression as starting point for the iteration, an explicit representation of $\hat{F}_{2,j}^{Vq}$ can be achieved,

$$\begin{aligned} \frac{1}{\hat{x}} \hat{F}_{2,j}^{Vq}(\hat{p}, q) = {} & G_v^{qV} \left(\frac{2}{\beta_0}\right)^j \\ & \times \int_{-Q^2}^{-Q_0^2} \frac{d\hat{t}_1}{\hat{t}_1} \frac{1}{\log \frac{-\hat{t}_1}{\Lambda^2}} \cdots \int_{\hat{t}_{j-1}}^{-Q_0^2} \frac{d\hat{t}_2}{\hat{t}_2} \frac{1}{\log \frac{-\hat{t}_2}{\Lambda^2}} \\ & \times \int_0^1 \frac{dz_1}{z_1} P_{qq}^{(0)}(z_1) \cdots \int_0^1 \frac{dz_j}{z_j} P_{qq}^{(0)}(z_j) \\ & \times \delta\left(1 - \frac{\hat{x}}{z_1 \cdots z_{j-1} z_j}\right) \end{aligned} \quad (3.3.37)$$

with

$$\hat{x} = \frac{Q^2}{2\hat{p}q} \approx \frac{Q^2}{2\hat{p}_j q} z_j \approx \ldots \approx \frac{Q^2}{2\hat{p}_1 q} z_j z_{j-1} \ldots z_1. \tag{3.3.38}$$

The $\hat{t}$ integrals can be evaluated in the following way:

$$\begin{aligned}
&\int_{-Q^2}^{-Q_0^2} \frac{\mathrm{d}\hat{t}_1}{\hat{t}_1} \frac{1}{\log \frac{-\hat{t}_1}{\Lambda^2}} \int_{\hat{t}_1}^{-Q_0^2} \frac{\mathrm{d}\hat{t}_2}{\hat{t}_2} \frac{1}{\log \frac{-\hat{t}_2}{\Lambda^2}} \\
&= \int_{-Q^2}^{-Q_0^2} \frac{\mathrm{d}\hat{t}_1}{\hat{t}_1} \frac{1}{\log \frac{-\hat{t}_1}{\Lambda^2}} \left[\log\log \frac{Q_0^2}{\Lambda^2} - \log\log \frac{-\hat{t}_1}{\Lambda^2} \right] \\
&= \frac{1}{2} \left[\log\log \frac{Q_0^2}{\Lambda^2} \right]^2 - \log\log \frac{Q_0^2}{\Lambda^2} \log\log \frac{Q^2}{\Lambda^2} + \frac{1}{2} \left[\log\log \frac{Q^2}{\Lambda^2} \right]^2 \\
&= \frac{1}{2} \left[\log \frac{\log(Q_0^2/\Lambda^2)}{\log(Q^2/\Lambda^2)} \right]^2 .
\end{aligned}$$

The result of the j-fold integration is

$$\frac{1}{j!} \left[\log \frac{\log(Q^2/\Lambda^2)}{\log(Q_0^2/\Lambda^2)} \right]^j .$$

The z integrals again can be decoupled by considering Mellin moments,

$$\begin{aligned}
\hat{F}^{Vq}_{2,j,n}(Q^2) &= \int_0^1 \mathrm{d}\hat{x}\; \hat{x}^{n-2} \hat{F}^{Vq}_{2,j}(\hat{x}, Q^2) \\
&= G_v^{qV} \frac{1}{j!} \left[\frac{2}{\beta_0} \log \frac{\log(Q^2/\Lambda^2)}{\log(Q_0^2/\Lambda^2)} \int_0^1 \mathrm{d}y\; y^{n-1} P^{(0)}_{\mathrm{qq}}(y) \right]^j \\
&= G_v^{qV} \frac{1}{j!} \left[d^{\mathrm{NS}}_{n,0} \log \frac{\log(Q^2/\Lambda^2)}{\log(Q_0^2/\Lambda^2)} \right]^j ,
\end{aligned} \tag{3.3.39}$$

where (3.2.89) and $d^{\mathrm{NS}}_{n,0} = -\gamma_{n,0}/2\beta_0$ have been inserted. Summation over the number of rungs finally leads to

$$\sum_{j=0}^{\infty} \hat{F}^{Vq}_{2,j,n}(Q^2) = G_v^{qV} \left[\frac{\log(Q^2/\Lambda^2)}{\log(Q_0^2/\Lambda^2)} \right]^{d^{\mathrm{NS}}_{n,0}} \tag{3.3.40}$$

for the ladder diagrams. This expression can be factorized into the mass singularity and a regular part. The LL result for the *moments of the non-singlet nucleon structure function* follows with (3.3.28),

$$\begin{aligned} F_{2,n,\mathrm{LL}}^{V\mathcal{N},\mathrm{NS}}(Q^2) &= \sum_q G_v^{qV} \left[\log\frac{Q^2}{\Lambda^2}\right]^{d_{n,0}^{\mathrm{NS}}} \left[\log\frac{Q_0^2}{\Lambda^2}\right]^{-d_{n,0}^{\mathrm{NS}}} f_{q,\mathcal{N},n}^{(0)} \\ &= \sum_q G_v^{qV} \left[\log\frac{Q^2}{\Lambda^2}\right]^{d_{n,0}^{\mathrm{NS}}} f_{q,\mathcal{N},n}(\Lambda). \end{aligned} \tag{3.3.41}$$

Using (3.1.5), it can be seen that this result is the same as the one obtained with the RG equation and the Wilson expansion in (3.2.66). It consists of the products of the moments of the quark structure functions and the moments of the quark distribution functions. Consequently, from the summation of the leading-logarithmic terms of generalized ladder diagrams using the running strong coupling constant the results of the OPE arise for deep-inelastic lepton–nucleon scattering.

Let us summarize the most important results. The dominant QCD process is the collinear radiation of hard gluons. The associated mass singularities factorize. Formally, this corresponds in (3.3.41) to the replacement of the bare Wilson coefficients $C_n = Q_q^2$, occurring also in (3.2.42), by the renormalized, mass-singularity-free ones $C_n(\alpha_s)$ from (3.2.64). This picture of hard QCD processes can be extended to cover other reactions and higher-order calculations.

3.3.4 Lepton-pair production in hadron–hadron scattering

The ideas and methods presented in the preceding section for deep-inelastic lepton–nucleon scattering are the starting point for applying QCD to other *hard-scattering processes.* Perturbative QCD justifies the universality of the parton model and permits calculations of radiative corrections for each particular process [Do80]. The rules for dealing with hard-scattering processes in higher orders of QCD can be formulated as follows. Based on factorization and the parton model, the cross section is composed of the parton-model cross sections $\hat{\sigma}$ for parton–lepton or parton–parton scattering, convoluted with universal Q^2-dependent quark distribution functions $f_{q,h}(x, Q^2)$, antiquark distribution functions $f_{\bar{q},h}(x, Q^2)$, gluon distribution functions

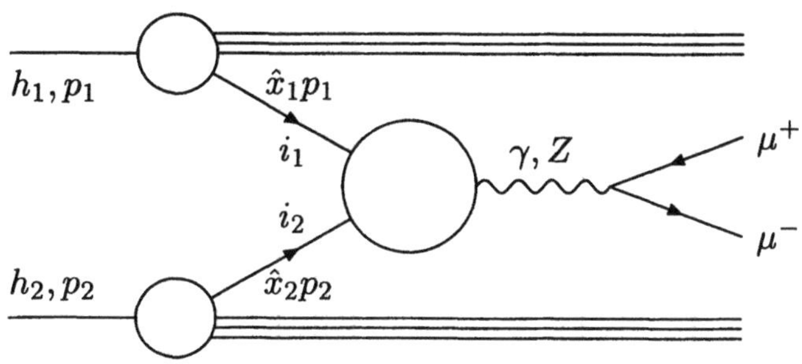

Fig. 3.8 The parton-model diagram for the Drell–Yan process

$f_{g,h}(x,Q^2)$, or parton *fragmentation functions* $D_{h,q}(z,Q^2)$, $D_{h,\bar{q}}(z,Q^2)$, and $D_{h,g}(z,Q^2)$.

This prescription is successfully applied to lepton-pair and jet production in hadron–hadron scattering, jet production and inclusive hadron production in e^+e^- annihilation and lepton–hadron scattering, and others. For illustration we present some typical examples: the Drell–Yan process in Sect. 3.3.4.1, jet production in Sect. 3.3.5, and the total cross section in e^+e^- annihilation in Sect. 3.3.6.

3.3.4.1 QCD corrections to the Drell–Yan cross section

The first example is lepton-pair ($e^+e^-, \mu^+\mu^-, \mu^+\nu_\mu$, etc.) production in hadron–hadron scattering, the so-called *Drell–Yan process* (cf. Fig. 3.8) [Dr70, Po77, Sa78]

$$h_1(p_1) + h_2(p_2) \to V + X \to l_1(k_1) + l_2(k_2) + X, \qquad V = \gamma,\ \mathrm{Z},\ \mathrm{W}.$$

The parton momenta are $\hat{x}_1 p_1$ and $\hat{x}_2 p_2$ with $p_1^2 = p_2^2 = 0$. With the Mandelstam variables

$$\begin{aligned} s &= (p_1+p_2)^2 = 2p_1p_2, \\ Q^2 &= (k_1+k_2)^2 = (\hat{x}_1p_1+\hat{x}_2p_2)^2 = \hat{x}_1\hat{x}_2 2p_1p_2 = \hat{x}_1\hat{x}_2 s, \end{aligned} \qquad (3.3.42)$$

the parton-model cross section reads

$$\begin{aligned} \frac{\mathrm{d}\sigma^V_{h_1h_2,\mathrm{pm}}(p_1,p_2,Q^2)}{\mathrm{d}Q^2} &= \sum_{q_1,\bar{q}_2} \int_0^1 \mathrm{d}\hat{x}_1 \int_0^1 \mathrm{d}\hat{x}_2 \, \frac{\mathrm{d}\hat{\sigma}^V_{q_1\bar{q}_2}(\hat{x}_1p_1,\hat{x}_2p_2)}{\mathrm{d}Q^2} \\ &\quad \times \left[f_{q_1,h_1}(\hat{x}_1) f_{\bar{q}_2,h_2}(\hat{x}_2) + f_{\bar{q}_2,h_1}(\hat{x}_1) f_{q_1,h_2}(\hat{x}_2) \right]. \end{aligned} \qquad (3.3.43)$$

In the electromagnetic case, $q\bar{q} \to \gamma^*$ ($q_1 = q_2$), we have

$$\frac{\mathrm{d}\hat{\sigma}^{\gamma}_{q\bar{q}}(\hat{x}_1 p_1, \hat{x}_2 p_2)}{\mathrm{d}Q^2} = \frac{4\pi\alpha^2}{3N_{\mathrm{C}}Q^2} Q_q^2 \delta(Q^2 - \hat{x}_1 \hat{x}_2 s), \tag{3.3.44}$$

yielding

$$\frac{\mathrm{d}\sigma^{\gamma}_{h_1 h_2,\mathrm{pm}}(s, Q^2)}{\mathrm{d}Q^2} = \frac{4\pi\alpha^2}{3N_{\mathrm{C}}Q^2} \sum_q Q_q^2 \int_0^1 \mathrm{d}\hat{x}_1 \int_0^1 \mathrm{d}\hat{x}_2\, \delta(Q^2 - \hat{x}_1 \hat{x}_2 s)$$
$$\times \left[f_{q,h_1}(\hat{x}_1) f_{\bar{q},h_2}(\hat{x}_2) + f_{\bar{q},h_1}(\hat{x}_1) f_{q,h_2}(\hat{x}_2) \right], \tag{3.3.45}$$

where the factor $1/N_{\mathrm{C}} = N_{\mathrm{C}}/N_{\mathrm{C}}^2$ results from the colour factor N_{C} and the average over the colours in the initial state. The Z and W bosons were discovered in $\mathrm{P\bar{P}}$ scattering. In the Z-boson case, the quark cross section is diagonal in the quark flavours and reads

$$\frac{\mathrm{d}\hat{\sigma}^{\mathrm{Z}}_{q\bar{q}}(\hat{x}_1 p_1, \hat{x}_2 p_2)}{\mathrm{d}Q^2} = \frac{4\pi\alpha^2}{3N_{\mathrm{C}}Q^2} (v_l^2 + a_l^2)(v_q^2 + a_q^2) \delta(Q^2 - \hat{x}_1 \hat{x}_2 s)$$
$$\times \frac{(Q^2)^2}{(Q^2 - M_{\mathrm{Z}}^2)^2 + \Gamma_{\mathrm{Z}}^2 M_{\mathrm{Z}}^2} \tag{3.3.46}$$

with the couplings (1.4.38) and the total width Γ_{Z} given in (1.4.37) and (1.4.39). A more detailed discussion has to include the γ–Z-interference terms.

The cross section for W production is built up in a similar way,

$$\frac{\mathrm{d}\hat{\sigma}^{\mathrm{W}}_{q\bar{q}'}(\hat{x}_1 p_1, \hat{x}_2 p_2)}{\mathrm{d}Q^2} = \frac{4\pi\alpha^2}{3N_{\mathrm{C}}Q^2} \frac{|\mathbf{V}_{qq'}|^2}{16 s_{\mathrm{W}}^4} \delta(Q^2 - \hat{x}_1 \hat{x}_2 s)$$
$$\times \frac{(Q^2)^2}{(Q^2 - M_{\mathrm{W}}^2)^2 + \Gamma_{\mathrm{W}}^2 M_{\mathrm{W}}^2}, \tag{3.3.47}$$

with the total width Γ_{W} from (1.4.40) and (1.4.41).

For h_1 and h_2 being nucleons, the quark distribution functions measured in deep-inelastic lepton–nucleon scattering can be used to make *absolute predictions* [Gl78] based on these parton-model formulas. However, these are by a factor $K = 1.5$–2 lower than the experimental results [Co82, Be85, Al87, An87]. This difference is due to the higher-order corrections, which are important since α_s is not small.

As already mentioned in Sect. 3.3.2, factorization was proven to be valid for the Drell–Yan process (cf. Ref. [Co89] and references therein). The hard-scattering part is $q+\bar{q}\to V$, the long-distance part is hidden in the quark distribution functions $f_{q,h}$ and $f_{\bar{q},h}$. The *virtual one-loop corrections* to $q+\bar{q}\to V$ are vertex and self-energy diagrams, the real ones are due to *gluon bremsstrahlung*, $q+\bar{q}\to V+\mathrm{g}$. The calculation of these contributions is closely connected to those for deep-inelastic scattering, the difference is that the momentum of the virtual boson V is time-like for lepton-pair production, whereas it is space-like in deep-inelastic scattering. In one-loop order, there is also a contribution from the gluons in the hadrons, namely the processes $q+\mathrm{g}\to q+V$ and $\bar{q}+\mathrm{g}\to\bar{q}+V$. Therefore, also the gluon distribution function $f_{\mathrm{g},h}$ enters the result.

As in deep-inelastic scattering, *dimensional regularization* with $D=4+2\epsilon$ is used to handle the collinear singularities. Then, the parton distribution functions of the corresponding results for deep-inelastic scattering can be used. With $z=Q^2/\hat{x}_1\hat{x}_2 s$, the one-loop result for the hard part of the electromagnetic production reads [Al79, Hu81]

$$\begin{aligned}\frac{\mathrm{d}\hat{\sigma}_{q\bar{q}}^{\gamma,(1)}(z,Q^2)}{\mathrm{d}Q^2} &= \frac{4\pi\alpha^2}{3N_\mathrm{C}Q^2}Q_q^2\frac{\alpha_\mathrm{s}}{\pi}\frac{1}{s}\left[\left(\frac{1}{\epsilon}+\gamma+\log\frac{Q^2}{4\pi\mu^2}\right)P_{\mathrm{qq}}^{(0)}(z)\right.\\ &\qquad\left.+\,\hat{\Phi}_{\mathrm{DY},q\bar{q}}(z)\right],\\ \frac{\mathrm{d}\hat{\sigma}_{q\mathrm{g}}^{\gamma,(1)}(z,Q^2)}{\mathrm{d}Q^2} &= \frac{4\pi\alpha^2}{3N_\mathrm{C}Q^2}Q_q^2\frac{\alpha_\mathrm{s}}{2\pi}\frac{1}{s}\left[\left(\frac{1}{\epsilon}+\gamma+\log\frac{Q^2}{4\pi\mu^2}\right)P_{\mathrm{qg}}^{(0)}(z)\right.\\ &\qquad\left.+\,\hat{\Phi}_{\mathrm{DY},q\mathrm{g}}(z)\right],\end{aligned}\tag{3.3.48}$$

where $P_{\mathrm{qq}}^{(0)}$ and $P_{\mathrm{qg}}^{(0)}$ are given again by (3.2.104). In the $\overline{\mathrm{MS}}$ *scheme*, the finite, scheme-dependent parts read

$$\begin{aligned}\hat{\Phi}_{\mathrm{DY},q\bar{q}}^{\overline{\mathrm{MS}}}(z) &= C_\mathrm{F}\left[\left(\frac{\pi^2}{3}-4\right)\delta(1-z)+4\left(\frac{\log(1-z)}{1-z}\right)_+\right.\\ &\qquad\left.-\,2(1+z)\log(1-z)-\frac{1+z^2}{1-z}\log z\right],\\ \hat{\Phi}_{\mathrm{DY},q\mathrm{g}}^{\overline{\mathrm{MS}}}(z) &= T_\mathrm{F}\left[(1-2z+2z^2)\log\frac{(1-z)^2}{z}+\frac{3+2z-3z^2}{2}\right].\end{aligned}\tag{3.3.49}$$

These equations together with the corresponding parton distribution functions provide one-loop-improved absolute predictions for the lepton-pair-production cross section. It is important to use the same scheme for the parton cross sections and for the parton distribution functions, e.g. in both cases the $\overline{\mathrm{MS}}$ or the DIS scheme. The one-loop corrections are not small and therefore essential for a good agreement with the experimental results. This is even more the case in the *DIS scheme* which is related to the $\overline{\mathrm{MS}}$ scheme by

$$\hat{\Phi}^{\mathrm{DIS}}_{\mathrm{DY},q\bar{q}}(z) = \hat{\Phi}^{\mathrm{DIS}}_{\mathrm{DY},\bar{q}q}(z) = \hat{\Phi}^{\overline{\mathrm{MS}}}_{\mathrm{DY},q\bar{q}}(z) - \hat{\Phi}^{V\mathrm{q}}_{2}(z), \tag{3.3.50}$$

with $\hat{\Phi}^{V\mathrm{q}}_{2}(z)$ from (3.3.17) and similarly for $\hat{\Phi}^{\mathrm{DIS}}_{\mathrm{DY},g\bar{q}}(z)$. In this scheme one obtains very large vertex corrections, i.e. a large factor $1 + 2\pi\alpha_{\mathrm{s}}C_{\mathrm{F}}/3$ multiplying the $\delta(1-z)$ function. For a typical value of $\alpha_{\mathrm{s}} \approx 0.3$, this gives a correction factor of ≈ 1.8. It was proposed to cure the effect of these large vertex corrections by exponentiation of the soft-photon contribution [Pa80, Cu80a]. The remaining correction then is small. Therefore exponentiation of the soft-radiation contribution is essential in the DIS scheme.

Full two-loop calculations for deep-inelastic scattering and lepton-pair production exist (cf. [Ma90, Ha91, vN92] and references therein). They are smaller than the one-loop corrections and lead to a very good agreement with the experimental results. This is important for W and Z production and the determination of the scale parameter Λ of QCD.

3.3.5 Jet cross sections

3.3.5.1 $e^+e^- \to$ jets in the parton model

Production of jet pairs in e^+e^- annihilation [El76, De78, Wu84, Kr84] was discussed in the framework of the naïve parton model in Sect. 1.5.4. The basic process is the production of a quark–antiquark pair, $e^+e^- \to (\gamma^*, Z) \to q\bar{q}$. Taking into account for simplicity only the annihilation via an off-shell photon γ^*, we get the parton-model cross section [cf. (1.2.45)]

$$\frac{\mathrm{d}\sigma^{\mathrm{2jet}}_{\mathrm{pm}}(s,\theta)}{\mathrm{d}\Omega} = \frac{\alpha^2}{4s} N_{\mathrm{C}} \sum_q Q_q^2 (1+\cos^2\theta) \tag{3.3.51}$$

and the total cross section [cf. (1.2.46)]

$$\sigma_{\rm pm}^{\rm 2jet}(s) = \frac{4\pi\alpha^2}{3s} N_{\rm C} \sum_q Q_q^2. \tag{3.3.52}$$

Each primary quark hadronizes completely into a bunch of hadrons, a jet.

3.3.5.2 Jets in QCD

In QCD, this simple picture has to be extended and refined to include besides quark jets also gluon jets, radiative corrections, details of hadronization, quark decays, etc. Moreover, a jet definition is needed which leads to IR-safe observables and allows for factorization into hard, perturbatively treatable parts and soft parts including, in particular, hadronization.

In lowest order ($\alpha_{\rm s}^0$) of QCD, i.e. in the parton model, two-jet production results from the quark–vector-boson vertex $V^* \to q\bar{q}$. In order $\alpha_{\rm s}$, besides one-loop virtual QCD corrections to the quark–vector-boson vertex, real three-jet events occur. The primary process in this case is *gluon bremsstrahlung* $V^* \to q\bar{q}{\rm g}$ followed by hadronization, leading to three jets if the primary quarks and gluon have high energies.

We discuss the primary process of the final state, the decay of the virtual vector boson into a quark pair and a gluon. The matrix element of this process is given by the same Feynman diagrams as gluon bremsstrahlung in deep-inelastic scattering or the Drell–Yan process. However, the virtual vector boson is time-like in the jet case ($q^2 = s > 0$), and the kinematics is that of a decay process, $V^*(q) \to q(p_q)\bar{q}(p_{\bar{q}}){\rm g}(k)$, where the momenta are indicated in parentheses.

Using the *Mandelstam variables*

$$\begin{aligned}
s_1 &= (q - p_q)^2 = (k + p_{\bar{q}})^2 = s - 2\sqrt{s}E_q = 2E_{\rm g}E_{\bar{q}}(1 - \cos\theta_{{\rm g}\bar{q}}), \\
s_2 &= (q - p_{\bar{q}})^2 = (k + p_q)^2 = s - 2\sqrt{s}E_{\bar{q}} = 2E_{\rm g}E_q(1 - \cos\theta_{\rm g}q), \\
s_3 &= (q - k)^2 = (p_q + p_{\bar{q}})^2 = s - 2\sqrt{s}E_{\rm g} = 2E_qE_{\bar{q}}(1 - \cos\theta_{q\bar{q}}), \\
s &= s_1 + s_2 + s_3, \quad 0 \le s_{1,2,3} \le s, \quad s_1 + s_2 \le s,
\end{aligned} \tag{3.3.53}$$

one obtains, after performing the angular integrations of the three-particle phase space, the differential gluon-bremsstrahlung cross section

$$
\begin{aligned}
s^2 \frac{d\sigma(e^+e^- \to q\bar{q}g)}{ds_1 ds_2} &= \frac{4\pi\alpha^2}{3s} N_C \sum_q Q_q^2 \frac{\alpha_s}{2\pi} C_F \frac{(s-s_1)^2 + (s-s_2)^2}{s_1 s_2} \\
&= \sigma_{pm}^{2jet}(s) \frac{\alpha_s}{2\pi} C_F \frac{(s-s_1)^2 + (s-s_2)^2}{s_1 s_2}. \qquad (3.3.54)
\end{aligned}
$$

We omit the contribution of the Z boson for simplicity. For $s_1 = 0$ or $s_2 = 0$, IR divergences occur. They correspond, as can be seen from (3.3.53), to soft and collinear gluon emission and can be regularized dimensionally.

Because of the IR singularity, the differential cross section for $e^+e^- \to q\bar{q}g$ rises steeply for $s_{1,2} \to 0$, becoming eventually bigger than the one of the basic parton process $e^+e^- \to q\bar{q}$. This region of phase space corresponds to soft and collinear gluons which do not hadronize into observable jets. Therefore, that part of the cross section does not lead to three-jet events. A cut-off in phase space is needed to define a finite, measurable jet cross section. This is free of singularities but cut-off dependent. Under the assumption that hadronization is a soft process one identifies the cut-off in the parton cross section with a corresponding cut-off for the hadrons which form the jets.

From the several existing definitions of jet observables we discuss here only two: the JADE algorithm [Be88, Be92] and the Sterman–Weinberg approach [St77].

3.3.5.3 The JADE algorithm

In the *JADE algorithm* one defines variables

$$
y_{ij} = \frac{2E_i E_j}{s}(1 - \cos\theta_{ij}), \qquad (3.3.55)
$$

where E_i are the energies of the observed hadrons and θ_{ij} the angles between the pairs (ij). Since pairs with small y_{ij} have either small energies or are almost collinear, the hadrons are grouped into jets according to the following iterative procedure: the pair of hadrons with the smallest y_{ij} is combined into one object with four-momentum $p^\mu = p_i^\mu + p_j^\mu$. Then a new list of y's is calculated and the combining procedure of pairs of objects is iterated until every remaining y is bigger than a cut-off value $y_c > 0$. Thus, the hadronic final state is decomposed into jets depending on y_c. This procedure is IR-safe since soft and collinear hadrons are combined directly into jets.

Introducing the cut-off y_c in the parton phase space by

$$\frac{s_{1,2}}{s} \geq y_c, \qquad \frac{s - s_1 - s_2}{s} = \frac{2p_q p_{\bar{q}}}{s} \geq y_c, \qquad y_c \leq \frac{1}{3}, \tag{3.3.56}$$

yields the total cross section for $e^+e^- \to q\bar{q}g$ [Be92]

$$\begin{aligned}
\sigma^{3\text{jet}}(s, y_c) &= \sigma^{2\text{jet}}_{\text{pm}}(s) \frac{\alpha_s}{2\pi} C_F \\
&\times \int_{y_c s}^{s-2y_c s} ds_1 \int_{y_c s}^{s-s_1-y_c s} ds_2 \, \frac{(s-s_1)^2 + (s-s_2)^2}{s_1 s_2} \\
&= \sigma^{2\text{jet}}_{\text{pm}}(s) \frac{\alpha_s}{\pi} C_F \left[2\,\text{Li}_2\left(\frac{y_c}{1-y_c}\right) + \log^2\left(\frac{y_c}{1-y_c}\right) - \frac{\pi^2}{6} \right. \\
&\left. + \frac{3}{2}(1-2y_c) \log\left(\frac{y_c}{1-2y_c}\right) + (1-3y_c)\left(\frac{5}{4} + \frac{3}{4} y_c\right) \right],
\end{aligned} \tag{3.3.57}$$

where $\text{Li}_2(z)$ is the dilogarithm (2.5.77). For not too small values of y_c, this defines an acceptable three-jet cross section.

The JADE algorithm has the tendency to create spurious jets. This is avoided by the so-called *Durham algorithm* [Ca91] where instead of y_{ij}, (3.3.55), the variable

$$y^{\text{Dur}}_{ij} = \frac{2\min(E_i^2, E_j^2)}{s} (1 - \cos\theta_{ij}) \tag{3.3.58}$$

is used.

The two-jet cross section to order α_s is obtained by adding the virtual corrections and the gluon-bremsstrahlung cross section with the cuts $s_{1,2,3} \leq s y_c$ to the parton-model result

$$\sigma^{2\text{jet}}(s, y_c) = \sigma^{2\text{jet}}_{\text{pm}}(s) \left(1 + \frac{3\alpha_s}{4\pi} C_F\right) - \sigma^{3\text{jet}}(s, y_c). \tag{3.3.59}$$

The first term is taken from the result for the total cross section of the annihilation into hadrons [cf. (3.3.67)], which is calculated in Sect. 3.3.6,

$$\sigma(e^+e^- \to \text{hadrons}) = \sigma_{\text{pm}}(e^+e^- \to \text{hadrons}) \left(1 + \frac{3\alpha_s}{4\pi} C_F\right). \tag{3.3.60}$$

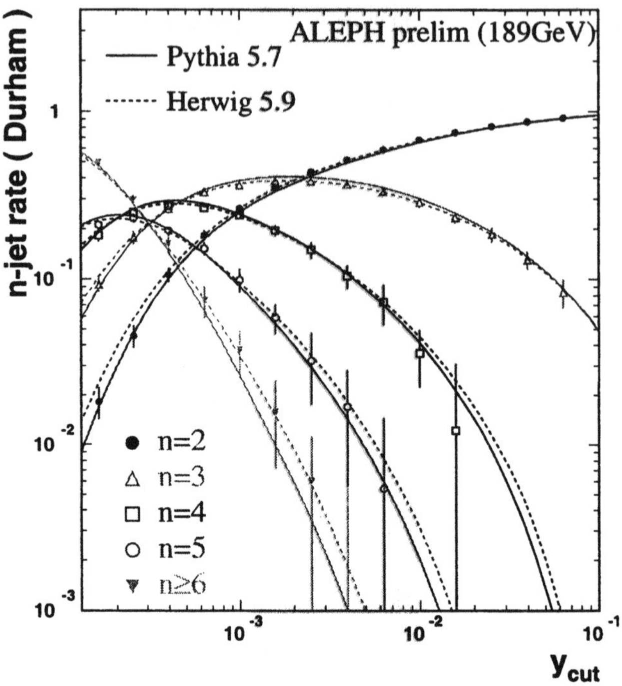

Fig. 3.9 y_c dependence of the classification into multijet ($n = 2, 3, 4, 5, \geq 6$) events (from Ref. [AL99])

In lowest order, the ratio of three-jet to two-jet events is given by α_s. Therefore, it allows—dependent on the renormalization scheme and the soft-hard scale—for the determination of $\Lambda_{\overline{MS}}$.

In a similar way, multijet cross sections, for example in order α_s^2 four-jet cross sections resulting from $q\bar{q}q\bar{q}$ and $q\bar{q}gg$ final states, have been calculated [Al79a, Al80, Kr84]. The four-jet cross section depends directly on the triple-gluon vertex. Therefore, it allows to test the non-abelian character of QCD. The experimental results clearly rule out an abelian gauge theory of the strong interaction.

A QCD analysis of the y_c dependence of the classification of the events into multijets by the ALEPH collaboration at 189 GeV shows good agreement between experiment and theory as described by the Monte-Carlo generators PYTHIA and HERWIG (Fig. 3.9).

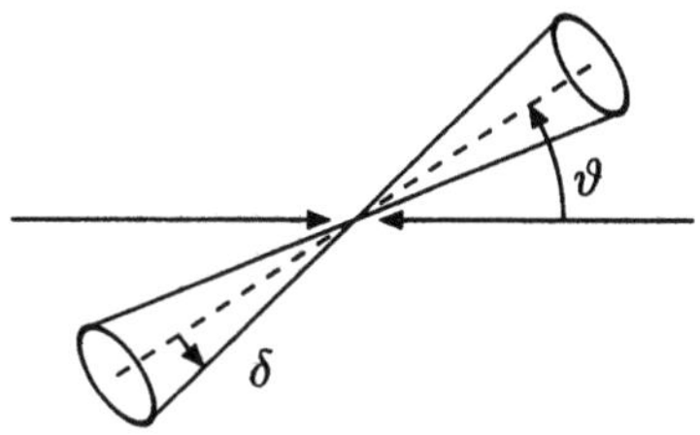

Fig. 3.10 Sterman–Weinberg definition of a jet

At large values of the cut-off $y_{\rm c}$ most events are interpreted as two broad jets, for smaller $y_{\rm c}$ these split into narrower multijet topologies. For $y_{\rm c} \lesssim 0.005$ the multijet events become as frequent as the two-jet ones, indicating a breakdown of this classification.

A more detailed analysis compares the shapes of the jets with the QCD predictions. For this purpose, IR-safe, calculable shape variables such as thrust, sphericity, moments, etc. have been defined. For details we refer to [El96].

3.3.5.4 The Sterman–Weinberg approach

Following the Bloch–Nordsieck procedure, in the *Sterman–Weinberg approach* jet cross sections are defined as inclusive cross sections like in QED (cf. Sects. 2.8.2.3 and Sect. 2.8.3.9), where the bremsstrahlung process is added to the virtual corrections in order to obtain IR-finite, mass-singularity-free quantities. The final-state hadrons are considered as soft if their energy E_i is small compared with the total energy $\sqrt{s}$, $\varepsilon = E_i/\sqrt{s} \ll 1$. All particles that lie in a cone with half-angle aperture $\delta \ll 1$ are collected into a jet (see Fig. 3.10). The jet cross section then depends on these cuts ε and δ. Including the QCD corrections the Sterman–Weinberg two-jet cross section [St77] reads

$$\frac{\mathrm{d}\sigma^{\rm 2jet}_{\rm phys}}{\mathrm{d}\Omega} = \frac{\mathrm{d}\sigma^{\rm 2jet}_{\rm pm}}{\mathrm{d}\Omega}\left[1 - \frac{\alpha_{\rm s}}{\pi}C_{\rm F}\left(4\log\delta\log 2\varepsilon + 3\log\delta + \frac{\pi^2}{3} - \frac{5}{2} + \mathcal{O}(\varepsilon,\delta)\right)\right]. \tag{3.3.61}$$

The Sterman–Weinberg approach is based on the simple picture of cones but not suited to analyse details of multijet events. Therefore, nowadays the (modified) JADE algorithm is usually used in the experimental analysis and in theoretical calculations.

3.3.5.5 Parton showers and hadronization

Up to now, we have considered only the primary steps in the formation of jets, namely the production of a quark–antiquark pair including the effects of gluon bremsstrahlung and virtual radiative corrections. A more realistic approach has to consider the successive production of *parton showers* and, after the soft energy scale is reached, *hadronization.* The full calculation of parton cascades is, although it is a hard-scattering process, too complicated to be performed analytically. But it is possible to treat the most important contributions: collinear emission and soft radiation. The collinear emission is formulated with evolution equations with kernels like in deep-inelastic scattering. Multiple soft-gluon radiation leads, like in QED, to exponentiation, yielding the so-called Sudakov form factors. Hadronization is a soft process and is described by models such as independent fragmentation, string models, and cluster models. All these calculations are implemented in computer simulations, i.e. Monte Carlo programs, finally producing QCD event generators. For details we refer to the literature [El96].

3.3.6 Total hadronic cross section

The parton-model cross section for the annihilation of electron–positron pairs into hadrons was discussed in Sect. 1.5.4.1. The basic mechanism is annihilation into a virtual photon or Z boson followed by the production of quark–antiquark pairs. These pairs radiate gluons and further quark pairs, and finally hadronize.

Here, we are mainly interested in the QCD effects and therefore treat only the photonic contribution to the cross section. It is similar to the point-like cross section of $\sigma(\mathrm{e^+e^-} \to \mu^+\mu^-)$. Therefore, the ratio (1.5.46),

$$R(s) = \frac{\sigma(\mathrm{e^+e^-} \to \text{hadrons})}{\sigma(\mathrm{e^+e^-} \to \mu^+\mu^-)}, \qquad s = E_{\mathrm{CM}}^2, \tag{3.3.62}$$

is given in lowest order, i.e. in the parton model with massless quarks, by

$$R_{\mathrm{pm}}(s) = N_{\mathrm{C}} \sum_q Q_q^2, \tag{3.3.63}$$

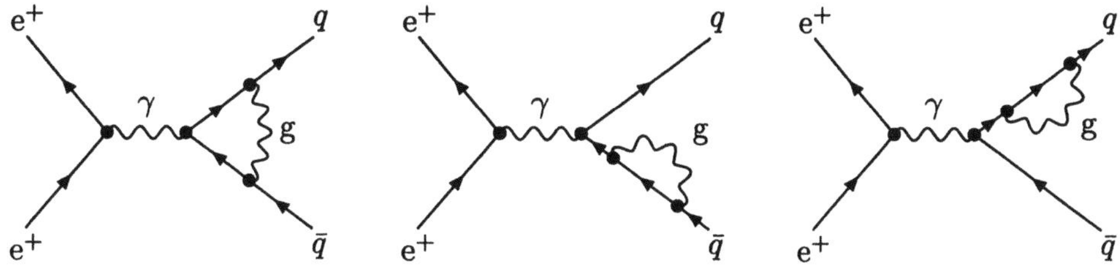

Fig. 3.11 Feynman diagrams for the one-loop corrections to $R(s)$

where only those quarks contribute which can be produced at the energy considered.

The QCD calculation of $R(s)$ can proceed in two ways. In the first way, the jet cross sections are integrated to obtain the total hadronic cross section. The second way is based on the fact that the photonic cross section is determined by the matrix elements of the electromagnetic current operator between the vacuum and hadron states (cf. Sect. 1.2.1.9). Therefore, it is related to the *hadronic tensor* $H_{\mu\nu}$, i.e. the vacuum expectation value of the *current–current commutator*,

$$H_{\mu\nu}(q) = \frac{1}{4\pi} \int \mathrm{d}^4x \, \mathrm{e}^{\mathrm{i}qx} \langle 0 | [j_\mu^{\mathrm{em}}(x), j_\nu^{\mathrm{em}}(0)] | 0 \rangle, \tag{3.3.64}$$

where $q^2 = s$ now is time-like, in contrast to deep-inelastic scattering. The tensor $H_{\mu\nu}$ can be treated like $W_{\mu\nu}$ with renormalization-group techniques.

The diagrams for the QCD corrections to $R(s)$ in order α_s are shown in Fig. 3.11. The virtual corrections are UV-finite after renormalization but IR- and mass-divergent. The total cross section is the most inclusive cross section and therefore, according to the *KLN theorem*, free of mass divergences.[13] Indeed, adding the real gluon bremsstrahlung contribution, the result is IR-finite. The calculation of these contributions is the same as in QED; only a colour factor C_F has to be added. The detailed calculations for both cases can be found in Ref. [Fi89]. The results in the dimensional regularization scheme read

$$\frac{\delta R^{(1)}_{\mathrm{virt}}}{R_{\mathrm{pm}}} = \frac{\alpha_s}{\pi} C_F \left(\frac{4\pi\mu^2}{s}\right)^{-\epsilon} \frac{1}{\Gamma(1+\epsilon)} \left[-\frac{1}{\epsilon^2} + \frac{3}{2\epsilon} - 4 + \frac{\pi^2}{2} \right] + \mathcal{O}(\epsilon),$$

[13] There are no QCD corrections to the initial state.

$$\frac{\delta R^{(1)}_{\text{real}}}{R_{\text{pm}}} = \frac{\alpha_{\text{s}}}{\pi} C_{\text{F}} \left(\frac{4\pi\mu^2}{s}\right)^{-\epsilon} \frac{1}{\Gamma(1+\epsilon)} \left[\frac{1}{\epsilon^2} - \frac{3}{2\epsilon} + \frac{19}{4} - \frac{\pi^2}{2}\right] + \mathcal{O}(\epsilon). \tag{3.3.65}$$

The sum of these contribution is finite and very simple,

$$\frac{\delta R^{(1)}}{R_{\text{pm}}} = \frac{3\alpha_s}{4\pi} C_{\text{F}}. \tag{3.3.66}$$

The application of the renormalization-group equation leads to the replacement of α_{s} by the running coupling $\bar{\alpha}_{\text{s}}$. The reason is that the anomalous dimensions of the conserved electromagnetic current vanish. Therefore, the final result, including the first-order leading corrections, reads [Ap73, Ze73]

$$\begin{aligned} R(s) &= N_{\text{C}} \sum_q Q_q^2 \left[1 + \frac{3\bar{\alpha}_{\text{s}}(s)}{4\pi} C_{\text{F}}\right] + \mathcal{O}(\bar{\alpha}_{\text{s}}^2) \\ &\overset{\text{SU(3)}}{\longrightarrow} 3 \sum_q Q_q^2 \left[1 + \frac{\bar{\alpha}_{\text{s}}(s)}{\pi}\right] + \mathcal{O}(\bar{\alpha}_{\text{s}}^2). \end{aligned} \tag{3.3.67}$$

The two-loop contributions in the $\overline{\text{MS}}$ scheme have been calculated in Ref. [Ch79],

$$\begin{aligned} \delta R^{(2)} &= \left(\frac{\bar{\alpha}_{\text{s}}(s)}{\pi}\right)^2 \left[-\frac{3}{32}(C_{\text{F}})^2 + \left(\frac{123}{32} - \frac{11}{4}\zeta(3)\right) C_{\text{F}} C_{\text{A}} \right. \\ &\quad \left. + \left(-\frac{11}{8} + \zeta(3)\right) C_{\text{F}} T_{\text{F}} N_{\text{f}}\right] \\ &\overset{\text{SU(3)}}{\longrightarrow} \left(\frac{\bar{\alpha}_{\text{s}}(s)}{\pi}\right)^2 \left[\frac{365}{24} - 11\zeta(3) + \left(-\frac{11}{12} + \frac{2}{3}\zeta(3)\right) N_{\text{f}}\right] \\ &\approx \left(\frac{\bar{\alpha}_{\text{s}}(s)}{\pi}\right)^2 (1.986 - 0.115 N_{\text{f}}) \end{aligned} \tag{3.3.68}$$

with $\zeta(3) = \sum_{n=1}^{\infty} n^{-3} \approx 1.202$. Also the three-loop corrections are known [Go91, Su91].

When comparing with experimental results, the threshold regions have to be omitted. The number of quark flavours N_{f} in $\bar{\alpha}_{\text{s}}$ has to correspond to those quarks that can be pair-produced at the corresponding energy. Under these conditions good agreement between theory and experiment is found as can be seen from Fig. 1.13.

With these results we conclude this short survey of perturbative QCD. Methods for dealing with hard-scattering processes beyond leading order are firmly established. The same applies for the treatment of the effects of even higher orders and of the processes that occur during the hadronization of quark and gluon jets.[14] A consistent picture of these processes, including higher orders in α_s exists and is tested in many details by experiments. All these show good agreement between experiment and theory. One possibility to get an impression of the validity of QCD is to compare the values of α_s obtained from different experiments. This is shown in Sect. 3.8.

3.4 Heavy-quark effective theory (HQET)

The values of the effective quark masses range from several MeV for the light quarks to 4 GeV for the bottom quark and 174 GeV for the top quark [PDG00]. The masses of the heavy quarks are large compared with the QCD scale parameter $\Lambda_{\overline{MS}} \approx 250\,\text{MeV}$, and in many cases with the energies of the considered process. Consequently, in these applications the effects of heavy quarks Q can formally be described in the limit $m_Q \to \infty$, i.e. in the heavy-quark effective theory (HQET) [Is92, Ne94, Ma97, Ne98a].

In atomic and molecular physics the properties of bound states of light with heavy particles are studied. In atomic physics, the nucleus may be replaced in very good approximation by a static source, the Coulomb potential. The effects of the finiteness of the nuclear mass, the Rydberg corrections, are very small. The Born–Oppenheimer method in molecular physics is based on the fact that the movement of the nuclei is much slower than the one of the electrons. As a consequence of the small velocity of the heavy particles, the spin-dependent effects are suppressed since spin interactions are relativistic effects and proportional to the inverse mass.

The situation in HQET is very similar. In lowest order, the heavy quark acts like a static source of the strong interaction with a strength independent of the mass of the heavy quark m_Q. Therefore, there is no difference between the interactions of different heavy quarks. This leads to an approximate symmetry, the *heavy-flavour symmetry*, with the corresponding

[14] An overview of the application of QCD to high-energy scattering processes is given in [El96].

heavy-flavour group $SU(N_{hf})_{hf}$ [Is89], where N_{hf} is the number of heavy flavours.

Moreover, because heavy quarks move slowly, a non-relativistic treatment is possible in a first approximation. The interaction is independent of the spin orientation of the heavy quark. This *spin symmetry* [Gr90, Ge90] gets corrections from relativistic spin-dependent effects which are proportional to $1/m_Q$ and thus small.

Both symmetries may be combined into the *flavour–spin symmetry* with associated symmetry group $SU(2N_{hf})_{hf}$.

3.4.1 The Lagrangian of HQET

Our starting point of the treatment of heavy quarks is the path-integral representation of QCD (2.4.16). The HQET approximation is based on an idea similar to the *Foldy–Wouthuysen expansion* of the Dirac equation. The small components of the Dirac field of the (slow) heavy quark are integrated out in the path integral. The resulting non-local effective theory is expanded into inverse powers of the heavy-quark mass m_Q yielding terms of the order $1/m_Q$, $1/m_Q^2$, ...

HQET has been applied to exclusive and inclusive weak decays of heavy flavours. As in all applications of perturbative QCD, a separation of short- and long-distance parts is necessary. Physical quantities have to be independent of the corresponding scale parameter. This is guaranteed by the renormalization-group equations of QCD.

3.4.1.1 Path-integral derivation

We write the generating functional of the Green functions of QCD in a form in which the heavy-quark field $Q(x)$ with mass m_Q and its source η are given explicitly, whereas all other fields are collected into Φ_λ with action S_λ and source λ,

$$Z\{\eta,\bar{\eta},\lambda\} = \int \mathcal{D}[Q]\,\mathcal{D}[\bar{Q}]\,\mathcal{D}[\Phi_\lambda] e^{i\left(S\{\eta,\bar{\eta}\}+S_\lambda+\int d^4x\,\Phi_\lambda\lambda\right)},$$

where

$$S\{\eta,\bar\eta\} = \int \mathrm{d}^4x \left[\bar Q(\mathrm{i}\not{D} - m_Q)Q + \bar\eta Q - \bar Q\eta\right],$$
$$D^\mu = \partial^\mu - \mathrm{i}g_\mathrm{s}A^\mu, \qquad A^\mu = A^{a,\mu}\frac{\lambda^a}{2}. \tag{3.4.1}$$

For the applications to heavy hadrons the momentum p_Q^μ of the heavy quark is decomposed as $p_Q^\mu = m_Q v^\mu + k^\mu$, where v^μ is the four-velocity of the quark ($v^0 > 0$, $v^2 = 1$), and k^μ describes the off-shell part of the momentum of the heavy quark inside the hadron.

In order to derive the heavy-quark approximation, the heavy-quark field $Q(x)$ is decomposed into its large $h_v^+(x)$ and small $h_v^-(x)$ components using the projectors $P_v^\pm = (1 \pm \not{v})/2$ with $(P_v^\pm)^2 = P_v^\pm$ and $P_v^+ P_v^- = 0$. In addition, the phase factor of the free movement of the heavy component is separated off

$$h_v^+ = \mathrm{e}^{\mathrm{i}m_Q vx} P_v^+ Q, \qquad h_v^- = \mathrm{e}^{\mathrm{i}m_Q vx} P_v^- Q. \tag{3.4.2}$$

The covariant derivative D^μ is split into a longitudinal and transverse part,

$$D^\mu = v^\mu(vD) + D_\perp^\mu, \qquad v_\mu D_\perp^\mu = 0, \qquad \{\not{D}_\perp, \not{v}\} = 0. \tag{3.4.3}$$

Inserting this into (3.4.1) and denoting the sources for the large and small components by ρ^+ and ρ^-, respectively, yields

$$S\{\rho^+,\rho^-,\bar\rho^+,\bar\rho^-\} = \int \mathrm{d}^4x \Big[\bar h_v^+ \mathrm{i}vD h_v^+ - \bar h_v^-(2m_Q + \mathrm{i}vD)h_v^-$$
$$+ \bar h_v^+ \mathrm{i}\not{D}_\perp h_v^- + \bar h_v^- \mathrm{i}\not{D}_\perp h_v^+ + \bar\rho^+ h_v^+ - \bar h_v^+ \rho^+ + \bar\rho^- h_v^- - \bar h_v^- \rho^-\Big] \tag{3.4.4}$$

for the action of the heavy quark. The small component h_v^- of the heavy-quark field with the "mass term" $2m_Q$ is now integrated out using the gaussian fermionic functional-integral formula (2.2.62). After switching off the sources, $\rho^- = \bar\rho^- = 0$, and performing the algebraic completion of the quadratic form, this results in

$$Z\{\rho^+,\bar\rho^+,\lambda\} = \int \mathcal{D}[h^+]\,\mathcal{D}[\bar h^+]\,\mathcal{D}[\Phi_\lambda]\,\det(2m_Q + \mathrm{i}vD)$$
$$\times \exp\left[\mathrm{i}\left(S_h\{\rho^+\} + S_\lambda + \int \mathrm{d}^4x\,\Phi_\lambda\lambda\right)\right] \tag{3.4.5}$$

with

$$S_h\{\rho^+,\bar{\rho}^+\} = \int \mathrm{d}^4x \Big[\bar{h}_v^+ \mathrm{i}vD h_v^+ - \bar{h}_v^+ \not{D}_\perp \frac{1}{2m_Q + \mathrm{i}vD - \mathrm{i}\varepsilon} \not{D}_\perp h_v^+ \\ + \bar{\rho}^+ h^+ - \bar{h}^+ \rho^+\Big] . \tag{3.4.6}$$

The relation between the small and large components is

$$h_v^- = \frac{1}{2m_Q + \mathrm{i}vD - \mathrm{i}\varepsilon} \mathrm{i}\not{D}_\perp h_v^+ . \tag{3.4.7}$$

Equations (3.4.5) and (3.4.6) are still exact. However, they contain the fermion determinant and a non-local interaction.

The fermion determinant comprises the effects of closed loops of the small components of the heavy-quark fields. As it stands, it is highly divergent. The constant part of the divergent determinant can be absorbed by dividing Z through the determinant of free quark fields $\det(2m_Q + \mathrm{i}v\partial)$ resulting in

$$\frac{\det(2m_Q + \mathrm{i}vD)}{\det(2m_Q + \mathrm{i}v\partial)} = \det\left(1 + \frac{g_s vA}{2m_Q + \mathrm{i}v\partial}\right) \\ = \exp\left(\mathrm{Tr}\log\left(1 + \frac{g_s vA}{2m_Q + \mathrm{i}v\partial}\right)\right) . \tag{3.4.8}$$

Since the generators of the colour group $SU(3)_C$ are traceless, $\mathrm{Tr}(A) = 0$, the physical effects are of order $1/m_Q^2$. In a first approximation they are usually neglected.

3.4.1.2 Heavy-mass expansion

The non-locality of the interaction has its origin in the fact that we have integrated out structures with an extension of the Compton wavelength $1/m_Q$. At tree level, the higher-order terms in an expansion in $1/m_Q$ are obtained by expanding the heavy quark propagator in (3.4.6) into a geometrical series,

$$\frac{1}{2m_Q + \mathrm{i}vD - \mathrm{i}\varepsilon} = \frac{1}{2m_Q} - \frac{\mathrm{i}vD}{(2m_Q)^2} + \ldots , \tag{3.4.9}$$

resulting in the Lagrangian

$$\mathcal{L}_{\text{eff}} = \bar{h}_v^+ \mathrm{i}vD h_v^+ - \frac{1}{2m_Q}\bar{h}_v^+(\not{D}_\perp)^2 h_v^+ + \frac{\mathrm{i}}{4m_Q^2}\bar{h}_v^+ \not{D}_\perp vD \not{D}_\perp h_v^+ + \dots . \tag{3.4.10}$$

The physical content of the leading terms of $\mathcal{L}_{\text{eff}}$ can be seen by going into the rest frame of the quark, $v^\mu = (1, \mathbf{0})$,

$$\mathcal{L}_{\text{eff}} = \bar{h}_v^+ \mathrm{i}D^0 h_v^+ - \frac{1}{2m_Q}\bar{h}_v^+(\boldsymbol{\gamma D})(\boldsymbol{\gamma D})h_v^+ + \dots . \tag{3.4.11}$$

Varying $\mathcal{L}_{\text{eff}}$ with respect to $\bar{h}_v^+$ gives the Schrödinger equation for the heavy-quark field

$$\mathrm{i}\partial^0 h_v^+ = -\frac{1}{2m_Q}(\nabla - \mathrm{i}g_{\mathrm{s}}\mathbf{A})^2 h_v^+ - g_{\mathrm{s}}A^0 h_v^+ + \frac{g_{\mathrm{s}}}{2m_Q}\boldsymbol{\Sigma}\mathbf{B}h_v^+, \tag{3.4.12}$$

where $\mathbf{B} = \nabla \times \mathbf{A}$ is the chromo-magnetic field and $\boldsymbol{\Sigma}$ is the Dirac spin matrix [cf. (A.1.21)]. Equation (3.4.12) is the correct non-relativistic approximation including the interaction of the chromo-magnetic field with the quark spin.

In order $1/m_Q$, the HQET and the Foldy–Wouthuysen transformation are identical. They deviate formally at order $1/m_Q^2$ because they are based on different expansions of the heavy-quark fields. Taking (3.4.2) and (3.4.7) yields

$$Q(x) = \mathrm{e}^{-\mathrm{i}m_Q vx}\left(1 + \frac{1}{2m_Q + \mathrm{i}vD - \mathrm{i}\varepsilon}\mathrm{i}\not{D}_\perp\right) h_v^+(x), \tag{3.4.13}$$

i.e. a geometric series in $1/m_Q$ after expansion. On the other hand, the *Foldy–Wouthuysen transformation* works with products of unitary operators

$$Q(x) = \mathrm{e}^{-\mathrm{i}m_Q vx}\left(\dots\, \mathrm{e}^{\mathrm{i}S_3/8m_Q^3}\, \mathrm{e}^{\mathrm{i}S_2/4m_Q^2}\, \mathrm{e}^{\mathrm{i}S_1/2m_Q}\right) h_v^+(x). \tag{3.4.14}$$

They are constructed such that the small terms of the Dirac fields decouple from the large ones. Only the leading term $S_1 = \not{D}_\perp$ is identical with the leading term of the HQET expansion. The different representations of the quark field generate different expansions of the Lagrangian.

The first term of the HQET Lagrangian $\mathcal{L}_{\text{eff}}$ (3.4.10), the dimension-four operator, is unique and renormalizable. It gives rise to the following momentum-space *Feynman rules*:

- heavy-quark propagator: $\mathrm{i}P_v^-(vk+\mathrm{i}\varepsilon)^{-1}$, $P_v^- = (1-\not{v})/2$,
- heavy-quark–gluon vertex: $\mathrm{i}g_\mathrm{s}v^\mu T^a_{cc'}$.

With these Feynman rules field-theoretical corrections like self-energies, vertex functions, renormalization constants, and anomalous dimensions can be calculated in the usual way.

The higher-order terms in the $1/m_Q$ expansion consist of operators with dimension 5, 6, ... and therefore are not renormalizable. They occur in a finite number of insertions only and can be treated by matching the regularized, cut-off-dependent expression at a certain scale to the full theory.

3.4.2 Symmetries of HQET

The leading term of the effective Lagrangian (3.4.10), the static approximation of HQET, is invariant under flavour and spin transformations. These approximate symmetries lead to many useful relations between the masses, or the electromagnetic and weak properties of the heavy hadrons.

3.4.2.1 Heavy-flavour and spin symmetry

The Lagrangian of QCD approaches in the infinite-mass limit, $m_Q \to \infty$, the static Lagrangian, i.e. the first term in (3.4.10). This leads to symmetry relations for matrix elements involving hadrons with charm and bottom quarks.[15] The static Lagrangian for the (c, b)-quark system with spin orientations $(\uparrow,\downarrow)$,

$$\mathcal{L}_{\text{eff}}^{(0)} = \bar{c}_v^\uparrow \mathrm{i}vDc_v^\uparrow + \bar{b}_v^\uparrow \mathrm{i}vDb_v^\uparrow + \bar{c}_v^\downarrow \mathrm{i}vDc_v^\downarrow + \bar{b}_v^\downarrow \mathrm{i}vDb_v^\downarrow, \tag{3.4.15}$$

is invariant under spin rotations $\mathrm{SU}(2)_\text{spin}$ and $\mathrm{SU}(2)_\text{hf}$ flavour transformations

$$\begin{pmatrix} c'_v \\ b'_v \end{pmatrix} = U \begin{pmatrix} c_v \\ b_v \end{pmatrix}. \tag{3.4.16}$$

We consider mesons with one heavy quark Q and one light antiquark $\bar{q}$ ($\bar{q} = \bar{\mathrm{u}}, \ldots$) and denote the flavour part of their state vector by $|Q_v, \bar{q}\rangle$. The generator I_hf^- of the $\mathrm{SU}(2)_\text{hf}$ group then transforms a B-meson state into

[15] The top quark does not live long enough to form bound states.

a charmed one, $I^{-}_{\rm hf}|{\rm b}_v,\bar{q}\rangle = |{\rm c}_v,\bar{q}\rangle$, and similarly for the other generators. This leads to symmetry relations for matrix elements of the bilinear, current-type operators $\bar{h}'_{v'}(x)\Gamma h_v(x)$, where $\Gamma = \{\gamma_\mu, \gamma_\mu\gamma_5\}$. Thus, there are relations between the matrix elements $\langle Q'_{v'},\bar{q}'|\bar{h}'_{v'}\Gamma h_v|Q_v,\bar{q}\rangle$.

3.4.2.2 Spin symmetry and meson states

In contrast to full QCD, the static Lagrangian (3.4.15) is invariant under spin rotations. Consequently, the mesons with the same flavour content and in the same orbital state are degenerate in mass. In this approximation the pseudoscalar and the vector ground-state mesons differ only in the spin part of the wave function and are related by a spin-flip operation. These spin parts $\chi^{\rm PS}$ and $\chi^{\rm V}$ of the state vectors of the pseudoscalar and the vector mesons are in Dirac-matrix notation given by

$$\chi^{\rm PS} = \sqrt{M}(\gamma_5 P^-_v), \qquad \chi^{\rm V}(\varepsilon) = \sqrt{M}(\not{\varepsilon} P^-_v), \qquad P^-_v = \frac{1-\not{v}}{2}, \quad (3.4.17)$$

where $\varepsilon^\mu(\lambda)$ ($\varepsilon^2 = -1$, $\varepsilon v = 0$) is the polarization vector of the vector meson with mass M. The spin amplitudes $\chi_{A\alpha}$ carry a heavy-quark spinor index A and a light-quark spinor index α and are normalized according to

$$\mathrm{Tr}(\bar{\chi}\chi) = 2M, \qquad \bar{\chi} = \gamma^0\chi^\dagger\gamma^0. \quad (3.4.18)$$

The operator $S_v(\varepsilon) = \gamma_5\not{v}\not{\varepsilon}$, $S_v^2(\varepsilon) = 1$, flips the spin of the heavy quark $S_v(\varepsilon)\chi^{\rm PS} = \chi^{\rm V}(\varepsilon)$.

3.4.2.3 Flavour–spin symmetry

The static Lagrangian (3.4.15) is invariant under the flavour-spin group $\mathrm{SU}(2N_{\rm hf})_{\rm hf}$ which is generated by simultaneous flavour and spin transformations. Thereby pseudoscalar and vector mesons with charm and beauty are members of the adjoint representation of this group. This large approximate symmetry leads to many relations between matrix elements, some of which are discussed in Sect. 3.4.3.

3.4.2.4 Reparametrization invariance

The representation of the quark four-momentum by the four-velocity of the quark and k^μ is not unique. Other decompositions, i.e. replacements $v^\mu \to v^\mu + \delta v^\mu$ must lead to the same physical results. Moreover, a reparametrization of the effective quark mass, $m_Q \to m_Q + \delta m_Q$, does not change the predictions of HQET since the quark masses are, in contrast to the meson masses, no physical quantities. The invariance under these reparametrization transformations leads to exact relations in HQET [Fa92].

3.4.3 Applications of HQET

3.4.3.1 Masses of heavy mesons

As a consequence of the spin symmetry of HQET, one expects equal masses for the heavy pseudoscalar mesons and the vector mesons with the same flavour and orbital quantum numbers, $M_{\mathrm{D}} \approx M_{\mathrm{D}^*}$ and $M_{\mathrm{B}} \approx M_{\mathrm{B}^*}$. This is indeed approximately true [PDG00]: $M_{\mathrm{D}} = 1869\,\mathrm{MeV} \approx M_{\mathrm{D}^*} = 2007\,\mathrm{MeV}$ and $M_{\mathrm{B}} = 5279\,\mathrm{MeV} \approx M_{\mathrm{B}^*} = 5325\,\mathrm{MeV}$.

Moreover, since the spin interaction is proportional to $1/m_Q$ we expect

$$\begin{aligned}\Delta_{\mathrm{V,PS}} &= M_{\mathrm{V}}^2 - M_{\mathrm{PS}}^2 = (M_{\mathrm{V}} + M_{\mathrm{PS}})(M_{\mathrm{V}} - M_{\mathrm{PS}}) \\ &\approx 2m_Q(M_{\mathrm{V}} - M_{\mathrm{PS}}) \approx 2m_Q \frac{\text{constant}}{2m_Q} = \text{constant}. \end{aligned} \tag{3.4.19}$$

This relation is fulfilled experimentally: $\Delta_{\mathrm{B}^*,\mathrm{B}} \approx 0.49\,\mathrm{GeV}^2$, $\Delta_{\mathrm{D}^*,\mathrm{D}} \approx 0.53\,\mathrm{GeV}^2$, $\Delta_{\mathrm{K}^*,\mathrm{K}} \approx 0.55\,\mathrm{GeV}^2$, and even $\Delta_{\rho,\pi} \approx 0.57\,\mathrm{GeV}^2$. Similar results hold for heavy baryons.

3.4.3.2 Relations between current matrix elements

Electromagnetic and weak transitions of hadrons are obtained from the hadronic matrix elements of current operators, i.e. $\langle H_2, v_2, \lambda_2 | j^\mu(0) | H_1, v_1, \lambda_1 \rangle$, and $\langle H_2, v_2, \lambda_2 | j_5^\mu(0) | H_1, v_1, \lambda_1 \rangle$, where $|H, v, \lambda\rangle$ is the state vector of a heavy hadron H with four-velocity v^μ and helicity λ, and j^μ and j_5^μ are the vector and axial-vector parts of the hadronic charged currents (1.4.24).[16]

[16] We here use the symbol j in order to avoid confusion with the heavy quark field h_v.

For simplicity we discuss the current matrix elements between pseudoscalar and vector-meson states. Because of Lorentz invariance, they have the form

$$\begin{aligned}
\langle \mathrm{PS}_2, v_2|j^\mu(0)|\mathrm{PS}_1, v_1\rangle &= (v_1+v_2)^\mu F_+(q^2) + (v_1-v_2)^\mu F_-(q^2),\\
\langle \mathrm{V}_2, \lambda_2, v_2|j^\mu(0)|\mathrm{PS}_1, v_1\rangle &= \mathrm{i}\varepsilon^{\mu\nu\rho\sigma}\varepsilon^*_{2,\nu} v_{2,\rho} v_{1,\sigma} F_{\mathrm{V}}(q^2),\\
\langle \mathrm{V}_2, \lambda_2, v_2|j_5^\mu(0)|\mathrm{PS}_1, v_1\rangle &= \varepsilon_2^{\mu *} F_{\mathrm{A}_1}(q^2) \qquad (3.4.20)\\
&\quad + v_1^\mu(\varepsilon_2^* v_1) F_{\mathrm{A}_2}(q^2) + v_2^\mu(\varepsilon_2^* v_1) F_{\mathrm{A}_3}(q^2)
\end{aligned}$$

with $q^2 = (p_1-p_2)^2 = M_1^2 + M_2^2 - 2M_1M_2v_1v_2$. The form factors $F_i(q^2)$ contain the dynamics of the strong interaction.

We investigate the current matrix elements in the heavy-quark limit. Because of flavour–spin symmetry, they can be decomposed as

$$\langle H_2, v_2, \bar{q_2}|\bar{h}_{v_2}\Gamma h_{v_1}|H_1, v_1, \bar{q_1}\rangle = \mathrm{Tr}\Big(\bar{\chi}_{H_2}\Gamma\chi_{H_1}\Big)\,\xi(y), \qquad (3.4.21)$$

where $y = v_1v_2$. The *Isgur–Wise function* $\xi(v_1v_2)$ [Is89] corresponds to the form factor. It is a universal dynamical quantity which group-theoretically was obtained—as can be seen from above—after performing the spin- and flavour-symmetry decompositions by applying the Wigner–Eckart theorem and separating off Clebsch–Gordan coefficients.

In our case we take the amplitudes (3.4.17) for the pseudoscalar and vector mesons with $\Gamma = \gamma^\mu$ and $\Gamma = \gamma^\mu\gamma_5$ for the vector and axial-vector currents, respectively. Using (3.4.21) yields for instance for the PS $\to$ PS transition

$$\begin{aligned}
\langle \mathrm{PS}_2, v_2|j^\mu(0)|\mathrm{PS}_1, v_1\rangle &= \sqrt{M_2M_1}\,\mathrm{Tr}\left\{P_{v_2}^-\gamma_5\gamma^\mu\gamma_5P_{v_1}^-\right\}\,\xi(y)\\
&= -\frac{\sqrt{M_2M_1}}{4}\,\mathrm{Tr}\left\{(1-\not{v}_2)\gamma^\mu(1-\not{v}_1)\right\}\xi(y)\\
&= \sqrt{M_1M_2}(v_1^\mu + v_2^\mu)\,\xi(y). \qquad (3.4.22)
\end{aligned}$$

This gives the two relations

$$F_+(q^2) = \sqrt{M_1M_2}\,\xi(y), \qquad F_-(q^2) = 0. \qquad (3.4.23)$$

The results for the other form factors read

$$\begin{aligned}
&F_{\mathrm{V}} = \sqrt{M_1M_2}\,\xi(y), \qquad F_{\mathrm{A}_1} = \sqrt{M_1M_2}(1+v_1v_2)\,\xi(y),\\
&F_{\mathrm{A}_2} = 0, \qquad F_{\mathrm{A}_3} = -\sqrt{M_1M_2}\,\xi(y). \qquad (3.4.24)
\end{aligned}$$

The six form factors are reduced to one quantity, the Isgur–Wise function. For $v = v'$, i.e. $vv' = 1$ or $q^2 = 0$, the left-hand side of (3.4.22) is the matrix element of a conserved current associated with the heavy-flavour symmetry. The corresponding conserved charges simply transform the heavy quarks into each other and the diagonal charges count the number of heavy quarks. This fixes the normalization of the form factor F_+ in (3.4.20), which in turn implies that the Isgur–Wise function is normalized at the kinematic point $v = v'$ as $\xi(1) = 1$.

In this way one obtains symmetry relations between different decays which can be compared with experiment. For B decays into D and D* mesons plus a lepton pair it is possible to compare the following ratios of form factors with experiment [Du96]:

$$(1 + v_1 v_2)\frac{F_{\mathrm{V}}}{F_{\mathrm{A}_1}} = 1.18 \pm 0.32$$

$$(1 + v_1 v_2)\frac{-F_{\mathrm{A}_3} - (M_{\mathrm{B}}/M_{\mathrm{D}^*})F_{\mathrm{A}_2}}{F_{\mathrm{A}_1}} = 0.71 \pm 0.23. \qquad (3.4.25)$$

These ratios should be equal to one in the infinite-mass limit. The first ratio is in good agreement with theory; the second shows sensible deviations from 1. The reason is that those form factors that do not vanish in lowest order get only second-order corrections in $1/m_Q$, whereas F_{A_2} and F_- get first-order corrections. This is the content of *Luke's theorem* [Lu90].

HQET was successfully applied also to purely leptonic decays of heavy mesons, semi-leptonic decays into light mesons, baryon decays, inclusive decays, etc. For details we refer to reviews, e.g. [Ne94, Ma97, Ne98a].

3.4.3.3 Corrections to the leading-order results

For an adequate comparison with precise experiments, it is necessary to calculate the corrections to the results of the static Lagrangian $\mathcal{L}_{\mathrm{eff}}^{(0)}$, (3.4.15). There are two types of corrections. The first class are the recoil corrections. They originate from the heavy-mass expansion of the Dirac operator in $\mathcal{L}_{\mathrm{eff}}$ (3.4.10), and the expansion (3.4.13) of the heavy-quark fields $Q(x)$ into the large component fields. The expansion of the field operator and the corresponding corrections to the state vector of the heavy meson may be interpreted as corrections to the current operator. The expansion of $\mathcal{L}_{\mathrm{eff}}$ in

$1/m_Q$ leads to non-renormalizable operators. The matrix elements of these new operators give new reduced amplitudes.

The second class consists of the field-theoretical loop corrections. They can be calculated using the Feynman rules of QCD for the gluons and light quarks and those for the heavy quarks given in Sect. 3.4.1.2. In this way the renormalization constants and anomalous dimensions are determined, which enter the renormalization-group-improved results.

Moreover, in some cases it may be necessary to include electromagnetic radiative corrections. For details on the many applications of HQET we refer to the reviews [Is92, Ne94, Ma97, Ne98a].

3.5 Light quarks and chiral perturbation theory

The running coupling constant of QCD grows with decreasing energies. This leads to the confinement of quarks and gluons in hadrons and prohibits the application of perturbation theory in the region below 1 GeV. Low-energy hadron dynamics can be investigated by lattice QCD (cf. Sect. 2.10) or by the method of effective field theories. The latter is based on the fact that at low energies a description of the strong interaction directly in terms of the physical degrees of freedom, the light hadrons, e.g. the pseudoscalar mesons, is possible. A lot of phenomenologically successful work in low-energy hadron physics was done in the framework of current algebra [Ad68, Al73] long before QCD was established. This led to the development of effective field theories for low-energy hadron physics, the non-linear σ model, and chiral perturbation theory [We79, Ga84]. Among the many reviews of this field we mention only Refs. [Le91, Le94, Pi95, Le96].

3.5.1 Chiral symmetry of massless QCD

The bridge between the QCD Lagrangian (2.3.53) and the low-energy effective theories is built by the symmetries of the light-quark sector which appear if the masses of the N_f light quarks (u, d, s, ...) vanish, i.e. in the *chiral limit*.

We start by studying the possible symmetries of the quark mass term

$$\mathcal{L}_{\mathrm{QCD,m}} = -\bar{\psi} M_\mathrm{q} \psi = -\sum_q \bar{\psi}_q(x) m_q \psi_q(x). \tag{3.5.1}$$

The quark mass matrix $M_{\mathrm{q}} = \mathrm{diag}(m_{\mathrm{u}}, m_{\mathrm{d}}, m_{\mathrm{s}}, \ldots)$ in general allows for a global $(\mathrm{U}(1)_{\mathrm{V}})^{N_{\mathrm{f}}}$ symmetry $\psi_q \to \exp(\mathrm{i}\alpha^q)\psi_q$, $q = \mathrm{u}, \mathrm{d}, \mathrm{s}, \ldots$, with the conserved vector currents

$$j_q^\mu(x) = \bar{\psi}_q(x)\gamma^\mu\psi_q(x), \qquad \partial_\mu j_q^\mu(x) = 0. \tag{3.5.2}$$

If the N_{f} quark masses are equal, $\mathcal{L}_{\mathrm{QCD,m}}$ is invariant under a $\mathrm{SU}(N_{\mathrm{f}})_{\mathrm{V}}$ symmetry $\psi \to \exp(\mathrm{i}\alpha^a\lambda^a/2)\psi$ [λ^a are the $\mathrm{SU}(N)$ Gell-Mann matrices (1.2.51)], leading to the *conserved vector currents*

$$j_\mu^a(x) = \bar{\boldsymbol{\psi}}(x)\gamma_\mu\frac{\lambda^a}{2}\boldsymbol{\psi}(x), \qquad \partial^\mu j_\mu^a(x) = 0. \tag{3.5.3}$$

In the case of vanishing quark masses also chiral transformations $\psi \to \exp(\mathrm{i}\beta_a\gamma_5\lambda^a/2)\psi$ become a symmetry. As a consequence, we have in addition *conserved axial-vector currents*,

$$j_{5,\mu}^a(x) = \bar{\boldsymbol{\psi}}(x)\gamma_\mu\gamma_5\frac{\lambda^a}{2}\boldsymbol{\psi}(x), \qquad \partial^\mu j_{5,\mu}^a(x) = 0. \tag{3.5.4}$$

The flavour currents (3.5.3) and (3.5.4) as well as their charges form the basis of *current algebra*.

The charges of these currents, together with those of the singlet vector and axial-vector currents ($\mathbf{1}$ is the unit matrix in flavour space),

$$j_\mu^0(x) = \bar{\boldsymbol{\psi}}(x)\gamma_\mu\mathbf{1}\boldsymbol{\psi}(x), \qquad j_{5,\mu}^0(x) = \bar{\boldsymbol{\psi}}(x)\gamma_\mu\gamma_5\mathbf{1}\boldsymbol{\psi}(x), \tag{3.5.5}$$

generate the global group

$$\mathrm{SU}(N_{\mathrm{f}})_{\mathrm{V}} \times \mathrm{SU}(N_{\mathrm{f}})_{\mathrm{A}} \times \mathrm{U}(1)_{\mathrm{V}} \times \mathrm{U}(1)_{\mathrm{A}}$$

with dimension $2N_{\mathrm{f}}^2$.

In this context it is very convenient to consider besides the vector and axial-vector currents also chiral currents

$$j_{\mathrm{L},\mu}(x) = \frac{1}{2}[j_\mu(x) - j_{5,\mu}(x)], \qquad j_{\mathrm{R},\mu}(x) = \frac{1}{2}[j_\mu(x) + j_{5,\mu}(x)]. \tag{3.5.6}$$

3.5.1.1 Spontaneous breaking of chiral symmetry

This chiral symmetry is spontaneously broken,[17]

$$\mathrm{SU}(N_\mathrm{f})_\mathrm{V} \times \mathrm{SU}(N_\mathrm{f})_\mathrm{A} \times \mathrm{U}(1)_\mathrm{V} \times \mathrm{U}(1)_\mathrm{A} \longrightarrow \mathrm{SU}(N_\mathrm{f})_\mathrm{V} \times \mathrm{U}(1)_\mathrm{V},$$

to the subgroup of vector symmetries with dimension N_f^2. The breaking of $\mathrm{SU}(N_\mathrm{f})_\mathrm{A}$ generates a multiplet of $N_\mathrm{f}^2 - 1$ pseudoscalar *Goldstone bosons.* For the case of three massless quarks, these are eight massless pseudoscalar mesons. They are identified with the lightest pseudoscalars, π^+, π^0, π^-, K^+, K^0, $\bar{\mathrm{K}}^0$, K^-, and η_8. The breaking of chiral symmetry by the QCD quark-mass term generates their experimentally observed masses [cf. Sect. 3.5.4]. The global group $\mathrm{U}(1)_\mathrm{A}$ is not anomaly-free and therefore gives no massless Goldstone boson. [cf. Sect. 3.5.1.2]

The transition amplitude from the vacuum $|0\rangle$ to a pseudoscalar-meson state $|p, b\rangle$,

$$\begin{aligned} &\langle 0|j^a_{5,\mu}(0)|p,b\rangle = \langle 0|\bar\psi(0)\gamma_\mu\gamma_5\frac{\lambda^a}{2}\psi(0)|p,b\rangle = \mathrm{i}p_\mu f^a_{\mathrm{PS}}\delta^{ab}, \\ &a, b = 1, \ldots, N_\mathrm{f}^2 - 1, \end{aligned} \tag{3.5.7}$$

contains the *order parameter*, the pseudoscalar-meson decay constant f^a_{PS}, which is related to the *quark condensate* $\langle 0|\bar\psi(0)\psi(0)|0\rangle$ [cf. (3.5.25)]. In strict $\mathrm{SU}(N_\mathrm{f})_\mathrm{V}$ symmetry f^a_{PS} is independent of a.

The current matrix element (3.5.7) obeys as a consequence of (3.5.4) the Ward identity [cf. the *PCAC relation* (2.7.29)]

$$p_\mu\langle 0|j_5^{a,\mu}(0)|p,b\rangle = \mathrm{i}p^2 f^a_{\mathrm{PS}}\delta^{ab} = 0. \tag{3.5.8}$$

This implies $p^2 = 0$, i.e. the Goldstone bosons are massless.

[17] Details of spontaneous symmetry breaking are presented in Sect. 4.1.

3.5.1.2 The U(1)$_{\mathrm{A}}$ anomaly

The Ward identity of the flavour-singlet axial-vector current (3.5.5), the generator of $U(1)_{\mathrm{A}}$, contains an anomaly[18] [cf. (2.7.40)]

$$\partial^\mu j^0_{5,\mu} = -\frac{N_{\mathrm{f}} g_{\mathrm{s}}^2}{16\pi^2} \varepsilon_{\mu\nu\rho\sigma} \operatorname{Tr}(F^{\mu\nu} F^{\rho\sigma}), \tag{3.5.9}$$

where $F^{\mu\nu}$ is the gluon field-strength tensor. Consequently, the SU(N_{f}) singlet meson, the η', remains massive also in the case of massless light quarks (cf. Sect. 2.9.3.10).

3.5.2 Pion-pole dominance and effective low-energy theory

Low-energy physics deals with the S matrix and current matrix elements of the pseudoscalar mesons $|p, a\rangle$. In order to derive the relation between QCD and the effective low-energy theory we consider the axial-vector current Green functions

$$\langle 0| \mathrm{T}\, j^{a_1}_{5,\mu_1}(x_1)\, j^{a_2}_{5,\mu_2}(x_2) \cdots j^{a_n}_{5,\mu_n}(x_n)|0\rangle$$

and their generating functional

$$\begin{aligned} \exp(T_{\mathrm{c}}\{\mathrm{J}\}) = \sum_n \frac{\mathrm{i}^n}{n!} \int \mathrm{d}^4x_1 \cdots \mathrm{d}^4x_n\, J^{a_1,\mu_1}(x_1) \cdots J^{a_n,\mu_n}(x_n) \\ \times \langle 0|\mathrm{T} j^{a_1}_{5,\mu_1}(x_1) \cdots j^{a_n}_{5,\mu_n}(x_n)|0\rangle, \end{aligned} \tag{3.5.10}$$

where $J^a_\mu(x)$ denotes the sources of the currents.

The current two-point function is dominated at low energies by the poles of the pseudoscalar mesons (*pion-pole dominance*). The contribution of the continuum states begins at the two-particle threshold $p^2 = 4M^2_{\mathrm{PS}}$, i.e. at

[18] Gauge theories with anomalies are discussed in Sect. 2.7.2.

higher energies. We investigate the one-particle contribution by inserting a complete set of meson states $|q\rangle$ and using (3.5.7),[19]

$$\int \mathrm{d}^4x\, \mathrm{e}^{\mathrm{i}px} \langle 0|\mathrm{T} j^a_{5,\mu}(x) j^b_{5,\nu}(0)|0\rangle = \int \mathrm{d}^4x\, \mathrm{e}^{\mathrm{i}px} \int \frac{\mathrm{d}^3q}{(2\pi)^3 2E_q}$$

$$\times \Big(\theta(x^0)\langle 0|j^a_{5,\mu}(0)|q\rangle\langle q|j^b_{5,\nu}(0)|0\rangle \mathrm{e}^{-\mathrm{i}qx}$$

$$+ \theta(-x^0)\langle 0|j^b_{5,\nu}(0)|q\rangle\langle q|j^a_{5,\mu}(0)|0\rangle \mathrm{e}^{\mathrm{i}qx} \Big)$$

$$= \int \mathrm{d}^4x\, \mathrm{e}^{\mathrm{i}px} \int \frac{\mathrm{d}^3q}{(2\pi)^3 2E_q} \Big(\theta(x^0)\mathrm{e}^{-\mathrm{i}qx} + \theta(-x^0)\mathrm{e}^{\mathrm{i}qx} \Big) p_\mu p_\nu \delta^{ab} (f^a_{\mathrm{PS}})^2$$

$$= \mathrm{i} \frac{p_\mu p_\nu}{p^2 - (M^a_{\mathrm{PS}})^2 + \mathrm{i}\varepsilon} \delta^{ab} (f^a_{\mathrm{PS}})^2. \tag{3.5.11}$$

For the last step see (2.1.12)–(2.1.14).

In a similar way, other Green functions can be analysed, and vertices may be constructed that describe the effective low-energy interactions. Based on pion-pole dominance, the axial-vector currents can be replaced by the corresponding pseudoscalar meson fields,

$$j^a_{5,\mu}(x) = -f^a_{\mathrm{PS}} \partial_\mu \pi^a(x). \tag{3.5.12}$$

In fact, the derivative $\partial_\mu \pi^a(x)$ of the pseudoscalar fields transforms like an octet of axial-vector currents, i.e. it has the same quantum numbers as $j^a_{5,\mu}$, and can therefore be used as an interpolating field for the axial-vector currents. In this way, the effective low-energy interactions can be collected in an *effective Lagrangian* $\mathcal{L}_{\mathrm{eff}}(\pi, \partial\pi)$ for pseudoscalar meson fields. The effective Lagrangian must fulfill the condition

$$\exp(T_\mathrm{c}\{\mathbf{J}\}) = \frac{1}{Z} \int \mathcal{D}[\pi]\, \mathrm{e}^{\mathrm{i}\int \mathrm{d}^4x\, (\mathcal{L}_{\mathrm{eff}}(\pi,\partial\pi,\ldots) + \pi\mathbf{J})}, \tag{3.5.13}$$

and has to reproduce the symmetry properties of the current Green functions, Ward identities, and pion-pole dominance. The effective Lagrangian $\mathcal{L}_{\mathrm{eff}}$ can be expanded in a series in the number of the π fields,

$$\mathcal{L}_{\mathrm{eff}} = \mathcal{L}^{(2)}_{\mathrm{eff}} + \mathcal{L}^{(4)}_{\mathrm{eff}} + \ldots, \tag{3.5.14}$$

[19] The meson masses vanish in the case of strict chiral invariance. We write the following equation for finite masses of the pseudoscalar mesons since chiral invariance is obviously broken in nature.

which corresponds to an expansion in powers of the (small) momenta. As a consequence of current conservation, this expansion starts with [We79, Le91, Le94]

$$\mathcal{L}_{\text{eff}}^{(2)} = \frac{1}{2}\partial_\mu \pi^a \partial^\mu \pi^a. \tag{3.5.15}$$

No Lorentz- and flavour-invariant π^3 terms can be constructed. Therefore, the interaction terms start with the quartic expressions $\mathcal{L}_{\text{eff}}^{(4)}$ in the π fields.

A detailed analysis [Ga85] shows that the effective Lagrangian is best expressed in terms of the exponentials of the π fields. This leads to the non-linear σ model. The linear and non-linear σ models were discussed as effective low-energy models for the theory of the strong interaction already before QCD was proposed.

3.5.3 The non-linear σ model

In the preceding sections we have seen that at low energies chiral massless QCD is governed by axial-vector-current Green functions and pion dominance and further structured by Ward identities and analyticity. This structure can be formulated as an *effective theory* in the form of the *non-linear σ model* [Ga85]. The fields are written in the form of an exponential,

$$U(x) = \exp\left(\frac{\mathrm{i}\Phi(x)}{f_{\mathrm{PS}}}\right), \tag{3.5.16}$$

where the octet of the light pseudoscalar mesons is arranged in a 3×3 matrix,

$$\Phi(x) = \pi^a(x)\lambda^a = \begin{pmatrix} \pi_3 + \frac{\eta_8}{\sqrt{3}} & \sqrt{2}\pi^+ & \sqrt{2}K^+ \\ \sqrt{2}\pi^- & -\pi_3 + \frac{\eta_8}{\sqrt{3}} & \sqrt{2}K^0 \\ \sqrt{2}K^- & \sqrt{2}\bar{K}^0 & -\frac{2\eta_8}{\sqrt{3}} \end{pmatrix}. \tag{3.5.17}$$

The exponential U transforms linearly under chiral transformations. With the unitary $\mathrm{SU}(N_\mathrm{f})$ matrices V_L and V_R this transformation reads

$$U'(x) = V_\mathrm{L} U(x) V_\mathrm{R}^\dagger. \tag{3.5.18}$$

The Lagrangian of the non-linear σ model is constructed from $U(x)$ in such a way that it reproduces (3.5.15) for the pseudoscalar mesons,

$$\mathcal{L}_\sigma = \frac{f_{\mathrm{PS}}^2}{4} \operatorname{Tr}[(\partial_\mu U)(\partial^\mu U^\dagger)]. \tag{3.5.19}$$

Expanding the exponential functions in the Lagrangian $\mathcal{L}_\sigma$ we get, besides the bilinear term $\mathcal{L}_\sigma^{(2)}$ (3.5.15), higher-order terms. The quartic coupling

$$\mathcal{L}_\sigma^{(4)} = \frac{1}{12 f_{\mathrm{PS}}^2} \operatorname{Tr}\{[\Phi(\partial_\mu \Phi) - (\partial_\mu \Phi)\Phi][\Phi(\partial^\mu \Phi) - (\partial_\mu \Phi)\Phi]\} \tag{3.5.20}$$

as well as all other couplings are of derivative type. The effective Lagrangian (3.5.19) gives the correct expression $\mathcal{M}(\pi\pi \to \pi\pi) = t/f_{\mathrm{PS}}^2$ for the elastic low-energy $\pi\pi$ scattering amplitude (t is the momentum transfer squared) [We66]. Expanding further, similar results can be calculated for multi-pion scattering.

Noether's theorem leads to the following conserved vector and axial-vector currents [Le91]:

$$\begin{aligned} j_\mu^a(x) &= \mathrm{i}\frac{f_{\mathrm{PS}}^2}{4} \operatorname{Tr}(\lambda^a [\partial_\mu U, U^\dagger]), \\ j_{5,\mu}^a(x) &= \mathrm{i}\frac{f_{\mathrm{PS}}^2}{4} \operatorname{Tr}(\lambda^a \{\partial_\mu U, U^\dagger\}). \end{aligned} \tag{3.5.21}$$

3.5.4 Breaking of chiral invariance

Chiral invariance is the basis of the connection of massless QCD with the non-linear σ model. However in reality, the masses of the light quarks do not vanish. Chiral symmetry is broken by the QCD quark-mass term (3.5.1) which induces a corresponding term in the non-linear σ model. A suitable ansatz is

$$\mathcal{L}_{\sigma,\mathrm{sb}} = \frac{f_{\mathrm{PS}}^2}{4} 2B_0 \operatorname{Tr}[M_{\mathrm{q}}(U + U^\dagger)] \tag{3.5.22}$$

with a free constant B_0 and the mass matrix M_{q} of the light quarks.

The complete Lagrangian of the *non-linear massive σ model* now reads [Ga85]

$$\mathcal{L}_{\sigma,\mathrm{m}} = \frac{f_{\mathrm{PS}}^2}{4}\left(\mathrm{Tr}[(\partial_\mu U)(\partial^\mu U^\dagger)] + 2B_0\,\mathrm{Tr}[M_{\mathrm{q}}(U+U^\dagger)]\right). \qquad (3.5.23)$$

It contains two parameters, the decay constant f_{PS} and the constant B_0 of the symmetry-breaking term. Comparing the vacuum expectation values of (3.5.1) and (3.5.22) yields

$$\begin{aligned} &m_{\mathrm{u}}\langle 0|\overline{\psi}_{\mathrm{u}}\psi_{\mathrm{u}}|0\rangle + m_{\mathrm{d}}\langle 0|\overline{\psi}_{\mathrm{d}}\psi_{\mathrm{d}}|0\rangle + m_{\mathrm{s}}\langle 0|\overline{\psi}_{\mathrm{s}}\psi_{\mathrm{s}}|0\rangle \\ &\quad = -f_{\mathrm{PS}}^2 B_0\,(m_{\mathrm{u}} + m_{\mathrm{d}} + m_{\mathrm{s}})\,. \end{aligned} \qquad (3.5.24)$$

This relates B_0 to the value of the *quark condensate*. Assuming approximate flavour symmetry of the quark condensate, (3.5.24) simplifies to

$$\langle 0|\overline{\psi}_q\psi_q|0\rangle = -f_{\mathrm{PS}}^2 B_0 \qquad \text{no sum over } q. \qquad (3.5.25)$$

The symmetry-breaking quark-mass terms contribute not only to the meson masses but also to decay and scattering amplitudes. The low-energy expression for the $\pi\pi$ scattering amplitude, for instance, is changed to $\mathcal{M}(\pi\pi \to \pi\pi) = (t - M_\pi^2)/f_{\mathrm{PS}}^2$.

The non-linear σ model is the simplest low-energy approximation of QCD. Instead of quarks and gluons it contains the pseudoscalar mesons which indeed dominate hadron physics at energies below the ρ-meson mass. The tree graphs resulting from (3.5.23) give the simplest approximation to current matrix elements, scattering amplitudes, and masses of the pseudoscalar mesons. A general, complete set of terms with four derivatives respecting chiral invariance and Lorentz invariance and including the quark masses was constructed by Gasser and Leutwyler [Ga85],

$$\mathcal{L}_{\sigma,\mathrm{m}}^{(4)} = \sum_{i=1}^{8} L_i P_i, \qquad (3.5.26)$$

where the L_i are free parameters and

$$\begin{aligned} P_1 &= (\mathrm{Tr}[(\partial_\mu U)(\partial^\mu U^\dagger)])^2, \\ P_2 &= \mathrm{Tr}[(\partial_\mu U)(\partial_\nu U^\dagger)]\ \mathrm{Tr}[(\partial^\mu U)(\partial^\nu U^\dagger)], \end{aligned}$$

$$\begin{aligned}
P_3 &= \mathrm{Tr}[(\partial_\mu U)(\partial^\mu U^\dagger)(\partial_\nu U)(\partial^\nu U^\dagger)],\\
P_4 &= 2B_0\,\mathrm{Tr}[(\partial_\mu U)(\partial^\mu U^\dagger)]\,\mathrm{Tr}[M_\mathrm{q}(U+U^\dagger)],\\
P_5 &= 2B_0\,\mathrm{Tr}[(\partial_\mu U)(\partial^\mu U^\dagger)(M_\mathrm{q}U+U^\dagger M_\mathrm{q})],\\
P_6 &= 4B_0^2(\mathrm{Tr}[M_\mathrm{q}(U+U^\dagger)])^2,\\
P_7 &= 4B_0^2(\mathrm{Tr}[M_\mathrm{q}(U-U^\dagger)])^2,\\
P_8 &= 4B_0^2\,\mathrm{Tr}[M_\mathrm{q}U^\dagger M_\mathrm{q}U^\dagger + UM_\mathrm{q}UM_\mathrm{q}].
\end{aligned} \tag{3.5.27}$$

If there are less than three flavours, only P_1, P_2, P_4, P_6, and P_7 are independent.

In order p^4, the axial anomaly also contributes to the effective Lagrangian [Le91],

$$\mathcal{L}_\mathrm{WZ} = f(U)\varepsilon_{\mu\nu\rho\sigma}(\partial^\mu U)(\partial^\nu U)(\partial^\rho U)(\partial^\sigma U), \tag{3.5.28}$$

where the function $f(U)$ is determined by the Wess–Zumino consistency condition (cf. Sect. 2.7.2.3). Physical effects of this term can be seen, for instance, in the two-photon decays of the neutral pseudoscalar mesons, $\pi^0 \to \gamma\gamma$, $\eta \to \gamma\gamma$ [Bi93].

The non-linear σ model is non-renormalizable. Since each loop integration gives two additional powers of the momenta, the one-loop corrections can be written as higher-derivative terms. Therefore, the divergences can be absorbed in the unknown coefficients L_i leaving renormalized couplings L_i^ren which can be fitted to experiment. Besides this, the loops give rise to logarithmic tadpole terms of the type

$$A(M^2) = \frac{1}{32\pi^2}\frac{M^2}{f_\mathrm{PS}^2}\log\frac{M^2}{\mu^2}, \tag{3.5.29}$$

where μ is the parameter of dimensional regularization (cf. Sect. 2.5.2.1 on page 208). As in every non-renormalizable theory the number of undetermined parameters increases with the number of loops. This reduces the predictive power of the higher-order corrections considerably.

The coefficients L_i^ren were determined from experiment, for instance, in Ref. [Bi94]. They are typically of the order of 10^{-3} and produce effects of 2% at $t = M_\pi^2$ and of 20% at $t = M_\mathrm{K}^2$.

3.5.5 Applications of chiral perturbation theory

3.5.5.1 Masses of the pseudoscalar mesons

Expanding the exponential of the pseudoscalar fields $U(x)$ in the symmetry-breaking term (3.5.22) to second order and leaving out the constant term, gives the mass term of the pseudoscalar mesons,

$$\begin{aligned}\mathcal{L}_{\mathrm{M}} &= \frac{f_{\mathrm{PS}}^2}{4} 2B_0 \operatorname{Tr}\left(-\frac{1}{f_{\mathrm{PS}}^2} M_{\mathrm{q}}\lambda^a\lambda^b\right)\pi^a(x)\pi^b(x) \\ &= -\frac{B_0}{2}\operatorname{Tr}\left(M_{\mathrm{q}}\lambda^a\lambda^b\right)\pi^a(x)\pi^b(x) \\ &= -B_0\left[m_{\mathrm{u}}\left(\frac{\pi_3^2}{2}+\pi^+\pi^-+K^+K^-+\frac{\eta_8^2}{6}+\frac{\pi_3\eta_8}{\sqrt{3}}\right)\right. \\ &\quad + m_{\mathrm{d}}\left(\frac{\pi_3^2}{2}+\pi^+\pi^-+\bar{K}^0K^0+\frac{\eta_8^2}{6}-\frac{\pi_3\eta_8}{\sqrt{3}}\right) \\ &\quad \left. + m_{\mathrm{s}}\left(K^+K^-+\bar{K}^0K^0+\frac{2\eta_8^2}{3}\right)\right].\end{aligned} \tag{3.5.30}$$

After diagonalization of the $(\pi_3\eta_8)$ mixing term, we find for the masses

$$\begin{aligned}M_{\pi^\pm}^2 &= (m_{\mathrm{u}}+m_{\mathrm{d}})B_0, \\ M_{\mathrm{K}^\pm}^2 &= (m_{\mathrm{u}}+m_{\mathrm{s}})B_0, \qquad M_{\mathrm{K}^0}^2 = (m_{\mathrm{d}}+m_{\mathrm{s}})B_0, \\ M_{\pi^0}^2 &= \left(m_{\mathrm{u}}+m_{\mathrm{d}}-\frac{(m_{\mathrm{u}}-m_{\mathrm{d}})^2}{2(2m_{\mathrm{s}}-m_{\mathrm{u}}-m_{\mathrm{d}})}\right)B_0+\mathcal{O}\left((m_{\mathrm{u}}-m_{\mathrm{d}})^3\right), \\ M_\eta^2 &= \left(\frac{m_{\mathrm{u}}+m_{\mathrm{d}}+4m_{\mathrm{s}}}{3}+\frac{(m_{\mathrm{u}}-m_{\mathrm{d}})^2}{2(2m_{\mathrm{s}}-m_{\mathrm{u}}-m_{\mathrm{d}})}\right)B_0 \\ &\quad +\mathcal{O}\left((m_{\mathrm{u}}-m_{\mathrm{d}})^3\right)\end{aligned} \tag{3.5.31}$$

with the $(\pi_3\eta_8)$ mixing angle

$$\tan 2\delta = \sqrt{3}\frac{m_{\mathrm{d}}-m_{\mathrm{u}}}{2m_{\mathrm{s}}-m_{\mathrm{u}}-m_{\mathrm{d}}}. \tag{3.5.32}$$

From (3.5.31) the sum rule

$$3(M_\eta^2+M_{\pi^0}^2)-2(M_{\pi^+}^2+M_{\mathrm{K}^0}^2+M_{\mathrm{K}^+}^2)=0 \tag{3.5.33}$$

follows, or, neglecting isospin breaking, the Gell-Mann–Okubo mass formula

$$3M_\eta^2 + M_\pi^2 - 4M_{\rm K}^2 = 0, \tag{3.5.34}$$

which is in good agreement with experiment. For the ratios of the quark masses one obtains from (3.5.31)

$$\begin{aligned} \frac{m_{\rm u}}{m_{\rm d}} &= \frac{M_{\pi^+}^2 + M_{\rm K^+}^2 - M_{\rm K^0}^2}{M_{\pi^+}^2 - M_{\rm K^+}^2 + M_{\rm K^0}^2}, \\ \frac{m_{\rm u}}{m_{\rm s}} &= \frac{M_{\pi^+}^2 + M_{\rm K^+}^2 - M_{\rm K^0}^2}{-M_{\pi^+}^2 + M_{\rm K^+}^2 + M_{\rm K^0}^2} \end{aligned} \tag{3.5.35}$$

up to terms of order $(m_{\rm u} - m_{\rm d})^2$. The value of B_0 can be determined only if one of the quark masses is known. A detailed analysis including higher-order terms in the effective Lagrangian and loop corrections was performed by Gasser and Leutwyler [Ga85, Le96a]. They obtain for the ratios of the quark masses the values

$$\frac{m_{\rm u}}{m_{\rm d}} = 0.553 \pm 0.043, \qquad \frac{m_{\rm s}}{m_{\rm d}} = 18.9 \pm 0.8. \tag{3.5.36}$$

Taking from QCD sum rules $m_{\rm s}(1\,{\rm GeV}) = 175 \pm 25\,{\rm MeV}$ [Bi95] for the running mass of the strange quark at 1 GeV, gives for the running masses $m_{\rm u}(1\,{\rm GeV}) = 5.1 \pm 0.9\,{\rm MeV}$ and $m_{\rm d}(1\,{\rm GeV}) = 9.3 \pm 1.4\,{\rm MeV}$. The value of the quark condensate follows from (3.5.25) as $\langle 0|\bar\psi\psi|0\rangle(1\,{\rm GeV}) = -(235 \pm 20\,{\rm MeV})^3$. This value is in agreement with the results obtained from sum rules (2.9.161) and quenched lattice approximation (2.9.162) given in Sect. 2.9.3.7.

3.5.5.2 Decay constants

Many precise experimental results have been derived in low-energy hadron physics for decay constants, electromagnetic and weak form factors, scattering amplitudes, non-leptonic decays, etc. On the one hand they can be used to determine the couplings $L_i^{\rm ren}$, on the other hand they allow for tests of chiral perturbation theory including contributions of loops and of $\mathcal{L}_{\sigma,\rm m}^{(4)}$.

As a typical case we present the expressions for the π and K decay constants in the isospin limit ($m_\mathrm{u} = m_\mathrm{d}$) [Ga85],

$$\begin{aligned} f_\pi &= f_\mathrm{PS}\left(1 + \frac{4M_\pi^2}{f_\mathrm{PS}^2} L_5^\mathrm{ren} + \frac{8M_\mathrm{K}^2 + 4M_\pi^2}{f_\mathrm{PS}^2} L_4^\mathrm{ren} - 2A(M_\pi^2) - A(M_\mathrm{K}^2)\right), \\ f_\mathrm{K} &= f_\mathrm{PS}\left(1 + \frac{4M_\mathrm{K}^2}{f_\mathrm{PS}^2} L_5^\mathrm{ren} + \frac{8M_\mathrm{K}^2 + 4M_\pi^2}{f_\mathrm{PS}^2} L_4^\mathrm{ren}\right. \\ &\quad \left. - \frac{3}{4}A(M_\pi^2) - \frac{3}{2}A(M_\mathrm{K}^2) - \frac{3}{4}A(M_\eta^2)\right), \end{aligned} \tag{3.5.37}$$

where $A(M^2)$ is the tadpole term (3.5.29). Taking the ratio f_π/f_K eliminates L_4^ren, giving up to logarithmic terms

$$\frac{f_\pi}{f_\mathrm{K}} = 1 - \frac{4(M_\mathrm{K}^2 - M_\pi^2)}{f_\mathrm{PS}^2} L_5^\mathrm{ren}. \tag{3.5.38}$$

Performing a complete analysis and using the value $f_\mathrm{K}/f_\pi = 1.22 \pm 0.01$ [Le84] yields $L_5^\mathrm{ren} = (1.4 \pm 0.5) \times 10^{-3}$ [Bi94]. This value is very close to the result of phenomenological considerations, i.e. pole dominance of scalar mesons, $L_5^\mathrm{ren} = f_\mathrm{PS}^2/4M_\mathrm{S}^2$, where $M_\mathrm{S} = 980\,\mathrm{MeV}$ is the mass of the corresponding scalar meson, the $f_0(980)$ meson. The value for f_η can be predicted from a formula similar to (3.5.37) as $f_\eta/f_\pi = 1.30 \pm 0.05$ [Ga85].

As already mentioned, many results have been obtained for observables in meson physics. All of them are in agreement with experiment. Extensions including assumptions like vector-meson dominance have been successfully applied not only to low-energy meson physics but also to baryon physics.

3.6 Results of lattice approximation of QCD

In Sect. 2.10 we discussed the lattice approximation of gauge theories. Applied to QCD, the main results are due to numerical methods which make use of powerful computers. In order to get physically relevant results, one has to overcome several problems. In spite of the remarkable progress in the theoretical understanding and great computational efforts, these are only solved incompletely. We repeat some of the sources of these problems:

- extrapolation to the continuum limit,

- extrapolation to the infinite-volume limit,
- quenched approximation of full QCD with dynamical quarks,
- extrapolation to the chiral limit, and
- algorithmic sources of systematic errors.

The related uncertainties lead to sizable systematic errors. Thus, we have results which in the best cases have errors of 10–20%. With this precaution, we discuss some topics which are typical for non-perturbative QCD: the spectrum of the light hadrons (Sect. 3.6.1), the glue balls (Sect. 3.6.2), and the relation between non-perturbative and perturbative QCD (Sect. 3.6.3). We refer to the literature for lattice results on matrix elements [Ku00] and structure functions [Pe00].

3.6.1 The hadron spectrum

Since the quark model of the hadrons (Sect. 1.3) played an important rôle in establishing QCD, the calculation of hadron masses from first principles was always a main goal of lattice QCD [Ha81a, We82]. However, the available computer power allows for large enough lattices only in the *quenched approximation* where the effects of quark loops are neglected. Nevertheless, it is hoped that these results reproduce some qualitative features of the physical hadron spectrum. But the error of this approximation is not really understood.

3.6.1.1 The parameters of the simulations

For the quenched approximation, the biggest lattices used have typically $N_{\rm s}^3 \times N_{\rm t} = 16^3 \times 32$ to $48^3 \times 64$ points. In the recent calculations (1994–1998) [Go97, Yo98], the coupling constant $\beta = 6/g_0^2$ (2.10.12) varies between $\beta = 5.7$ and $\beta = 6.5$.

In Table 3.1 we collect the results of some simulations with a variety of β values. The lattice constant a is determined from the mass of the ρ meson ($m_\rho = 770\,\text{MeV}$) in calculations with zero quark mass. We compare the input β with the calculated value $\beta_{\rm calc}$ following from formula (2.10.78). The agreement of both β values gives an impression of *perturbative asymptotic scaling.* For the calculation $\Lambda_{\overline{\rm MS}} = 170\,\text{MeV}$ was assumed, which corresponds according to (2.10.82) to $\Lambda_{\rm L} = 6\,\text{MeV}$. With all the uncertainties about this

β		$N_s^3 \times N_t$	a (fm)	size (fm)	β_{calc}
5.70	W	$24^3 \times 48$	0.13	3.4	5.68
6.00	W	$24^3 \times 32$	0.10	2.0	5.89
6.20	S		0.075		6.19
6.20	W	$24^3 \times 48$	0.070	1.6	6.19
6.25	W	$48^3 \times 84$	0.063	3.03	6.27
6.50	W		0.05		6.53
6.50	S	$48^3 \times 64$	0.047	2.6	6.53

Tab. 3.1 Quenched lattice simulations of QCD for Wilson fermions (W) and staggered fermions (S). Further notations are explained in the text.

fictitious *quenched* Λ *value* with quark number $N_q = 0$, this seems to be a reasonable value (cf. the physical $\Lambda_{\overline{MS}} = 208 + 25 - 23$ MeV of Sect. 3.1.2).

The finite-size effect (cf. Sect. 2.10.5) was checked by comparing the results in volumes of different physical sizes. For the "measurement" of the masses of the ρ meson and of the nucleon the corrections are less than 2% for spatial sizes $La = 2$ fm, and negligible for sizes above $La = 3$ fm (cf. Table 3.1).

The finite-size effects become severe for light particles (Sect. 2.10.5). Besides a large spatial volume, the measurement of small masses $m = \xi/a$ requires a large extension in time direction [cf. (2.10.18)]. The requirement of a large physical volume with small lattice constants exceeds the power of present-day computers. In the present calculations the quarks get a mass which makes the pion massive, and therefore $m_\pi/m_\rho \geq 0.5$. This renders the extrapolation to the chiral limit $m_q \to 0$, and therefore to $m_\pi \to 0$, particularly unreliable (cf. Sect. 3.6.1.4).

3.6.1.2 Spectrum of hadron masses

Hadron masses calculated in the quenched approximation with Wilson fermions are shown in Table 3.2. We took the recent values of the CP-PACS collaboration [Ao00]. They are extrapolated to the continuum.

In order to fix the bare quark masses $m_{u,0}$, $m_{d,0}$, and $m_{s,0}$, and the lattice scale Λ_L, the values of m_π, m_ρ, and either m_K or m_Φ are taken as an input.

	π	ρ	K	K*	Φ			
exp.	140	768	498	896	1019			
	input	input	input	858	957			
	input	input	553	889	input			
	N	Λ	Σ	Ξ	Δ	Σ^*	Ξ^*	Ω
exp.	940	1116	1193	1315	1232	1384	1532	1672
	878	1019	1117	1201	1257	1359	1459	1561
	878	1060	1176	1288	1257	1388	1517	1647

Tab. 3.2 Lattice results for the hadron spectrum using besides m_π, m_ρ, and m_{K^*} alternatively m_Φ (line 3) or m_K (line 4) as input (from [Ao00])

The errors of the whole procedure are estimated to 5–10% for the mesons and around 12% for the baryons. Within these errors, the baryon masses are the same for either m_K or m_Φ as input. A comparison with the experimental results (Table 3.2) indicates that the quenched lattice approximation can describe the experimental spectrum with errors in the 10% region.

3.6.1.3 Hyperfine splitting

Lattice QCD allows in principle for the determination of the masses of light quarks [Uk93]. The renormalization-group invariant masses belong to the fundamental parameters of the standard model. Therefore, they are of great importance.

The hyperfine splitting in the K system, i.e. the mass difference $m_{K^*} - m_K$, may be used to determine the mass m_s of the strange quark since in this difference an additive quark-mass renormalization drops out. The calculation proceeds by fixing the bare parameters from physics and then transforming them into the running mass of the $\overline{\text{MS}}$ scheme. We have shown in Table 3.2 how numerical simulations describe the hyperfine splitting. From this we gain the bare mass $m_{s,0}$.

For the transition from the bare lattice perturbation theory to the $\overline{\text{MS}}$ scheme in the continuum, we have to use arguments similar to those for α_s in Sect. 2.10.4.3. Similarly to (2.10.75), we get [Ma97a]

$$\overline{m}_s(\mu) = \widetilde{m}\left\{1 + \frac{g_0^2}{4\pi}\left(-\frac{2}{\pi}\log(a\mu) + \log\widetilde{C}_m\right)\right\}. \tag{3.6.1}$$

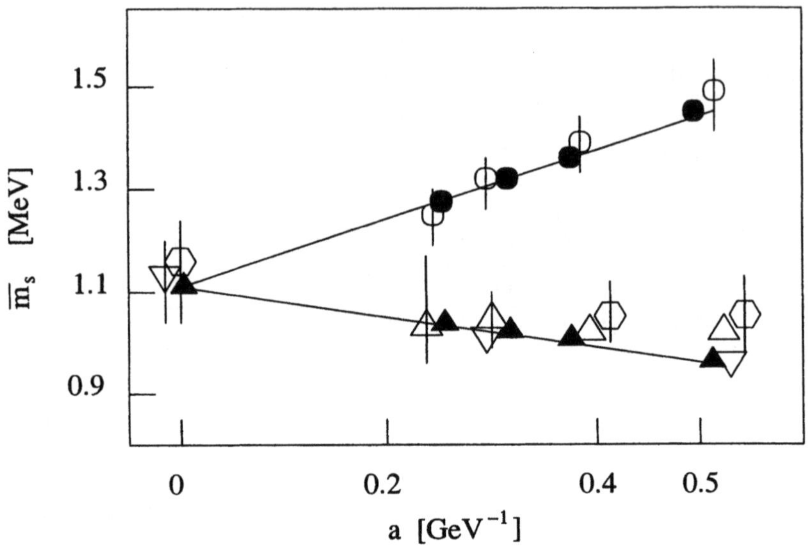

Fig. 3.12 Different simulations for the determination of $\overline{m}_s(\mu)$ for $\mu = 2\,\text{GeV}$ (based on Ref. [Yo98]). Circles refer to the perturbative scheme, triangles to the PCAC scheme.

Here a is the lattice constant, μ the $\overline{\text{MS}}$ scale parameter, and $\widetilde{C}_m$ a calculable constant which depends on the details of the lattice regularization. For the Wilson action, $\widetilde{C}_m = 1.67$. The mass $\tilde{m}$ is given by $\tilde{m} = m_0$ in one-loop lattice perturbation theory and by $\tilde{m} = m_0/(1 - g_0^2/12)$ for the mean-field-improved bare mass [Le93a]. A similar expression holds for a non-perturbative improved bare mass. In Fig. 3.12 we compare $\overline{m}_s$ for different simulations [Yo98].

Another method for the determination of $\overline{m}_s$ uses a lattice version of the *PCAC relation* (2.7.29) [Lu99],

$$m_K^2 f_K = Z_P(\overline{m}_u + \overline{m}_s)\langle 0|\bar{u}\gamma_5 s|K^+\rangle, \tag{3.6.2}$$

where $m_K^2 f_K$ is taken from experiment, and the renormalization constant Z_P can be determined by lattice perturbation theory as above or by non-perturbative methods [Ca99]. We represent this determination also in Fig. 3.12.

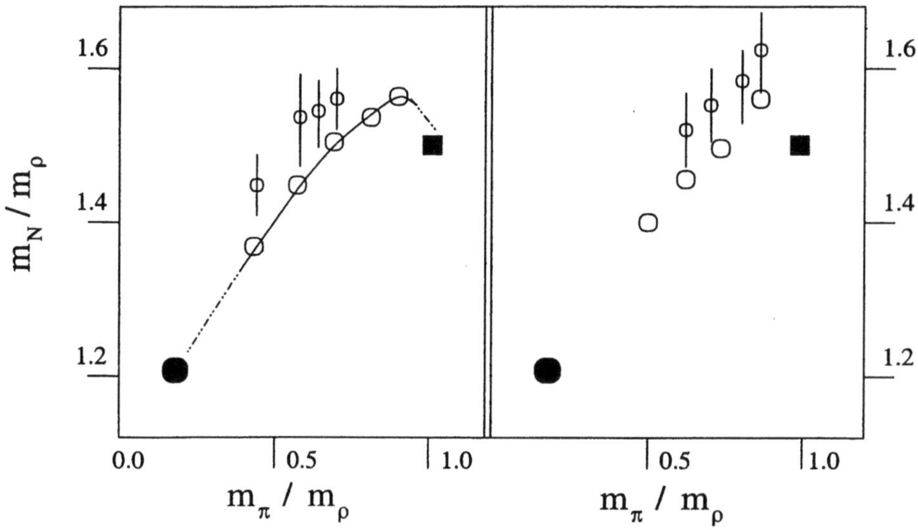

Fig. 3.13 Edinburgh plot: quenched approximation with staggered fermions, $\beta = 6.15$ (open large circles on the left-hand side); full QCD with staggered fermions, two flavours, $\beta = 5.60$ (small circles with error bars on the left-hand side); quenched approximation with Wilson fermions, $\beta = 6.17$ (open circles on the right-hand side); dynamical Wilson fermions, $\beta = 5.5$ (small circles with error bars on the right-hand side) (based on Ref. [Go97])

The result as derived from Fig. 3.12 is $\overline{m}_s(2\,\text{GeV}) = (110\text{–}130)\,\text{MeV}$ in quenched approximation. A determination based on (3.6.2) including an estimate of all errors gives $\overline{m}_s(2\,\text{GeV}) = 97(4)\,\text{MeV}$ [Ga00] (cf. Sect. 3.1.3).

3.6.1.4 The Edinburgh plot

The Edinburgh plot shows the mass ratio of the pion and the ρ meson m_π/m_ρ versus the mass ratio of the nucleon and the ρ meson m_N/m_ρ. This plot is particularly well-suited to demonstrate the progress and the difficulties of the lattice approximation of the hadron spectrum.

In Fig. 3.13 we present the Edinburgh plot extracted from [Go97]. The different points of the simulations refer to different bare quark masses. The filled square is the result of a strong-coupling approximation yielding $m_\mathrm{N}/m_\rho = 1.5\,m_\pi/m_\rho$. This means that all mesons consist of two quarks and all baryons of three quarks. The filled circles represent the experimental values $m_\pi/m_\rho = 0.18$ and $m_\mathrm{N}/m_\rho = 1.22$.

As a relation between mass ratios, the Edinburgh plot does not depend on a determination of the lattice constant. For quenched staggered fermions, simulations with different β allow to perform the continuum limit. The result is given by the line through the points for $\beta = 6.15$. The dashed line is a theoretical continuation based on a phenomenological quark model. Figure 3.13 also shows that the simulations connect the regions of the strong-coupling approximation with the physical point to which the quenched approximation allows to extrapolate.

The simulations with dynamical quarks follow in general the line of the quenched approximation. However, the large computational errors do not allow for definite conclusions. (For quenched approximations, the errors lie within the size of the symbols.)

The approach to the chiral limit, and hence to the physical point, is still poor. For Wilson fermions we have $m_\pi/m_\rho > 0.6$. For staggered fermions it is somewhat better. There are some indications that improvement of Wilson fermions will make some progress in this respect.

3.6.2 Glue balls

Glue balls are a new feature of SU(3) gauge theory [Fr72a]. In contrast to the linear Maxwell equations of the electromagnetic field, the gluon fields have self-interaction. This leads to bound states of gluons, the *glue balls.* Their masses can be determined by the lattice approximation of SU(3) gauge theories [Is82]. However, in full QCD the coupling to quark fields introduces mixing with quarkonium states. This is one of the reasons why this important prediction of QCD is not yet definitely confirmed by experiment.

3.6.2.1 Quantum numbers of glue balls

In the limit where the glue balls may be considered as free relativistic particles, they possess quantum numbers [Mi89] like mass, spin, parity, charge parity, and internal quantum numbers (cf. Sect. 1.2.1.1). In QCD, the internal quantum numbers are determined by the quark content. Since glue balls are essentially bound states of pure gluons, they have the flavour quantum numbers of the vacuum.

Lattice approximation violates rotational symmetry. Therefore, the definition of spin by the irreducible representations of the rotation group SO(3)

O_h	J^{PC}	$m/\sqrt{\sigma}$	m_G(MeV)
A_1^{++}	0^{++}	3.52 (12)	1550 ± 50
E^{++}	2^{++}	5.25 (25)	2270 ± 100
T_2^{++}		5.07 (17)	
A_1^{-+}	0^{-+}	5.3 (6)	2330 ± 250

Tab. 3.3 The glue-ball spectrum (from Ref. [Ba93])

fails, and SO(3) has to be substituted by the cubic group. There are five irreducible representations O_h of the finite symmetry group of the cube, A_1, A_2, E, T_1, and T_2 with dimensions 1, 1, 2, 3, and 3, respectively. If we restrict the rotations to those of the cubic group, the spin-0 representation of SO(3) becomes A_1, spin 1 gets T_1, spin 2 corresponds to $E \oplus T_2$, and spin 3 contains A_2. Rotation symmetry is restored in the continuum limit of lattice approximation. Consequently, for a spin-2 glue ball there are two mass-degenerate states in the lattice representations E and T_2. This example indicates the procedure for the determination of spins in lattice approximation. Parity P and charge-conjugation parity C can be determined on the lattice as in the continuum.

There are combinations of quantum numbers that cannot be formed by pure $q\bar{q}$ states: 0^{+-}, 1^{-+}, 2^{+-}, ... (cf. Sect. 1.3.1.2). Such spin exotics are particularly interesting candidates for glue balls. However, there might be mixing of glue balls with $q\bar{q}$ states with vacuum quantum numbers, leading to *hybrids*, which show similar features.

3.6.2.2 Masses of glue balls

Simulations of the type discussed in Sect. 3.6.1.1 allow to calculate the mass spectrum of the glue balls in quenched approximation. We give in Table 3.3 the results for glue balls with a mass below 3 GeV (from Ref. [Ba93]).

The simulations range from $\beta = 5.7$ to $\beta = 6.4$. They allow for a good extrapolation to the continuum for the spin states 0^{++} and 2^{++}. In particular, the lattice states E^{++} and T_2^{++} are represented by mass-degenerate states. As we have discussed above, this is a signal for the restoration of rotational invariance. The *string constant* $\kappa = (440\,\text{MeV})^2$ (cf. Sects. 3.7.1.1 and 3.7.3) is used for the transition to the physical mass scale. This value is in good agreement with the scale determined by m_ρ.

According to Ref. [Ba93], there are indications for states above 3 GeV. But there are no exotic states below 3 GeV.

Do glueballs as a new feature of gauge theories shed some light on their non-perturbative structure as discussed in Sect. 2.9.3? It might be worthwhile to mention that there exists a treatment of glueballs by an effective Hamiltonian which is based on the vacuum tunnelling picture [vB96, vH97]. A result of this method for the mass ratio is $m(2^{++})/m(0^{++}) \approx 1.5$, in good agreement with the lattice result of Table 3.3.

3.6.2.3 Comparison with experiment

Quenched lattice approximation gives increasing evidence for a glue-ball mass of $m_{\mathrm{G}} \approx 1600\,\mathrm{MeV}$. How does this compare with experiments? There are three 0^{++} resonances in this mass region, $f_0(1370)$, $f_0(1500)$, and $f_j(1710)$ with $j = 0$ or 2. In this energy region one expects besides the glue-ball state $|\mathrm{G}\rangle$, the quark states $|q\bar{q}(8)\rangle = |\mathrm{u\bar{u}} + \mathrm{d\bar{d}} - 2\mathrm{s\bar{s}}\rangle/\sqrt{6}$ and $|q\bar{q}(1)\rangle = |\mathrm{u\bar{u}} + \mathrm{d\bar{d}} + \mathrm{s\bar{s}}\rangle/\sqrt{3}$ (cf. Sect. 1.3.1.1). For the f_0 mesons one expects mixing,

$$f_0(y) = \alpha(y)|\mathrm{G}\rangle + \beta(y)|q\bar{q}(1)\rangle + \gamma(y)|q\bar{q}(8)\rangle,$$
$$y = 1370,\ 1500,\ 1710. \tag{3.6.3}$$

So far, there seems to be no agreement on the mixing pattern. However, there is good experimental evidence that the glue ball fits in such a scheme [Cl98, We97].

3.6.3 The connection between long- and short-distance physics: non-perturbative renormalization group

A fundamental problem of QCD is the relation between non-perturbative physics at long distances and asymptotic freedom at short distances. Of course, lattice approximation and perturbation theory approximate the same formal Lagrangian. But having in mind the different steps in lattice approximation (Sect. 2.10), including the transition to the continuum on the one hand and the unsolved problem of convergence of renormalized perturbation theory on the other (cf. Sect. 1.6.2.2), one would like to see that both schemes match smoothly.

3.6.3.1 The α_s program

This problem is pursued by the α_s program [Lu93]. The aim of this program is to calculate the running coupling $\bar{\alpha}_s(M_Z^2)$ with good precision from some low-energy data at the hadronic scale. Since the perturbative coupling $\bar{\alpha}_s(Q^2)$ is not a good choice for the description of low-energy hadron data, other definitions of the strong coupling are used. The determination of $\bar{\alpha}_s(Q^2)$ from the lattice approximation involves three steps:

(1) Define a suitable coupling constant α_h together with quark masses m_h and measure their values on the lattice.

(2) Set the scale at which α_h and m_h take their values.

(3) Convert α_h to $\bar{\alpha}_s(Q^2)$ and m_h to $\overline{m}_s$.

In the following we discuss these different points.

3.6.3.2 The input from long-distance physics

As low-energy input data we may take, for instance, the values of different masses, e.g. the mass of the lightest glue ball, the spin splitting of quarkonia, or the potential between heavy quarks. The corresponding lattice approximation sets the scale Λ_h at which α_h is determined from these data. One possibility for the low energy input is to take the force $F_{q\bar{q}}(r)$ between a $q\bar{q}$ pair at a certain distance r_0. The reference scale r_0 has been chosen such that

$$r_0^2 F_{q\bar{q}}(r_0) = 1.65, \tag{3.6.4}$$

giving $r_0 \simeq 0.5\,\text{fm}$ for typical phenomenological quarkonium potentials (cf. Sect. 1.3.3.2) [So94]. If we put $F_{q\bar{q}} = \kappa$ at $r_0 = 0.5\,\text{fm}$ for a pure confinement potential (1.3.17), then we get $\kappa = 1.65 \times r_0^{-2} = 1.3\,\text{GeV}\,\text{fm}^{-1}$. The difference to a realistic value $\kappa = 1\,\text{GeV}\,\text{fm}^{-1}$ can be attributed to the gluon-exchange part (1.3.16).

Since there are not yet good lattice data with dynamical quarks, we have to restrict ourselves essentially to the quenched approximation. In this case we do not have to consider quark masses and their scale dependence.

For the transition to the continuum, asymptotic scaling of the bare coupling constant g_0, or $\beta_L = 6/g_0^2$, is used. In particular, we recall Table 3.1 and

Sect. 3.6.1.1. In order to get α_s with good precision, we have to improve perturbation theory. One of these attempts goes back to Parisi [Pa81], who noted that the poor convergence of bare perturbation theory may be due to the appearance of tadpole diagrams. These diagrams can be re-summed by replacing the bare coupling g_0 by $g_P = g_0/(1 - g_0^2/6 + \ldots)$. This is one example of the choice of a suitable coupling α_h which possibly leads to a better approach in lattice constant a to the physical continuum value than those of β_L in Table 3.1. Another possibility is the Symanzik–Lüscher improvement program (Sect. 2.10.6.6).

3.6.3.3 Non-perturbative evolution

The conversion of α_P, or in general α_h, to $\bar{\alpha}_s$ cannot be approached directly in a controlled manner with high precision. As we have seen in Sect. 3.6.1.1, the typical lattice approach to the continuum involves a scale of order $0.1\,\mathrm{fm} \simeq (2\,\mathrm{GeV})^{-1}$ whereas $\bar{\alpha}_s$ is reliably measured at the Z-boson mass, $\bar{\alpha}_s(M_Z^2) \approx 0.12$. Such a large scale difference cannot be covered by numerical simulation on a single lattice, since this would require a size of 1000^4 lattice points. There are schemes to overcome this problem. These are based on finite-size techniques and the non-perturbative renormalization group.

It is beyond the scope of this book to describe these methods in all technical details. Instead we refer to the work of M. Lüscher [Lu99] and present directly the results obtained for $\bar{\alpha}_s(M_Z^2)$ and $\Lambda_{\overline{\mathrm{MS}}}$ from low-energy data.

3.6.3.4 Results

Starting from (3.6.4) as low-energy input, the non-perturbative evolution for the quenched approximation results in $\Lambda_{\overline{\mathrm{MS}}} = 0.602(48)/r_0$. As explained above, one takes $r_0 = 0.5\,\mathrm{fm}$ from charmonium physics giving

$$\Lambda_{\overline{\mathrm{MS}}} = (238 \pm 19)\,\mathrm{MeV}, \tag{3.6.5}$$

or by the two-loop formula (3.1.16) for $N_q = 0$

$$\bar{\alpha}_s(M_Z^2) = 0.079. \tag{3.6.6}$$

We may compare this with the result from asymptotic scaling of the bare coupling constant in Sect. 3.6.1.1, $\Lambda_{\overline{\mathrm{MS}}} = 170\,\mathrm{MeV}$ or $\bar{\alpha}_s(M_Z^2) = 0.075$.

From heavy quark physics [Sh97] one gets $\Lambda_{\overline{\rm MS}} = 350\,\mathrm{MeV}$ or $\bar{\alpha}_s(M_Z^2) = 0.084$. This result is based on the determination of $\alpha_P(8.2\,\mathrm{GeV})$ from spin splitting, and the relation $\bar{\alpha}_s(Q^2) = \alpha_P(\mathrm{e}^{5/3}Q^2)(1 + 2\alpha_P/\pi + C_2(N_q)\alpha_P^2)$, where $C_2(N_q)$ lies between 0 and 1. Non-perturbative evolution to M_Z is not used.

The quenched value $\bar{\alpha}_s(M_Z^2) = 0.079$ is significantly lower than the measured perturbative one, $\bar{\alpha}_s(M_Z^2) = 0.1185 \pm 0.0020$ [PDG00]. However, sea quarks affect the evolution of the coupling, and they also influence the low-energy reference scale. As an example, the quenched value $\Lambda_{\overline{\rm MS}} = (238 \pm 19)\,\mathrm{MeV}$ yields $\bar{\alpha}_s(M_Z^2) = 0.119$ for five flavours.

A result like (3.6.5), (3.6.6) for an unquenched simulation is still missing. Nevertheless, there are estimates in full QCD using the incomplete data with dynamical quarks to extrapolate to the physical flavours and masses. Using heavy quark data as input one gets $\bar{\alpha}_s(M_Z^2) = 0.118$ or $\bar{\alpha}_s(M_Z^2) = 0.116$ for $N_q = 5$ [Sh97]. However, these values have not the quality of the quenched result (3.6.5). For the determination of $\bar{\alpha}_s(M_Z^2)$ in full QCD one has to take into account the running of the masses $\bar{m}(Q^2)$. Similar methods as explained above lead to some recent progress (cf. Sect. 3.6.1.3).

Finally, we want to emphasize again the significance of this section. If unquenched results would reach the same controlled precision of the quenched ones, this would put potentially the treatment of non-perturbative effects on the same level as perturbative QCD.

3.7 Quark confinement

The simplest physical explanation of quark confinement can be given by describing the quark interaction by means of an attractive potential which increases as distances become larger, the *confinement potential.* Such a model was discussed in the treatment of quarkonia in Sect. 1.3.3.1. In this section we investigate the question whether the concept of a confinement potential has any significance in QCD. Most probably, the discussion of this basic question will determine the treatment of the long range structure of QCD.

3.7.1 The Wilson criterion

When defining a potential within QCD, the relativistic nature of the interaction together with gauge invariance gives rise to several effects which must be dealt with first of all. In a relativistic field theory—and this includes QCD—the interaction is always linked with the creation of virtual and real particles [Ak65]. This type of physics is not described by a potential. Therefore, the first step in formulating the potential concept for the confinement problem is to consider infinite-mass quarks in QCD. This freezes all kinetic degrees of freedom of the quarks, and there is no quark-pair creation. The colour charges of the $\mathrm{SU}(3)_{\mathrm{C}}$ triplet quarks are merely static sources and the finiteness of the propagation of the gluon interaction is not significant. In this approximation, the existence of a confinement potential between static triplet and antitriplet colour charges is a property of pure gluon dynamics.

When considering the existence of a confinement potential in pure gauge theory as evidence for confinement, it should always be remembered that this potential refers to the interaction of static quarks. The physical confinement problem is only fully solved if it can be shown that the existence of a confinement potential guarantees that low-mass dynamical quarks cannot be produced as free particles. We start this discussion with a two-dimensional model which shows the effects of such an interaction.

3.7.1.1 Definition of the confining potential

K. Wilson [Wi74] defined a potential between a quark q and an antiquark $\bar{q}$ in a colour-singlet state by considering the quantum-mechanical vacuum expectation value $W(C)$ for the path C of a heavy static $q\bar{q}$ pair from x to y,

$$W(C) = \langle \mathrm{Tr}\, U(C) \rangle = \frac{1}{Z} \int \mathcal{D}[\mathbf{A}]\, \mathrm{Tr}\, U(C) \exp(-S_{\mathrm{E}}\{\mathbf{A}\}). \tag{3.7.1}$$

Here the *Wilson loop*,

$$U(C) = \mathrm{P} \exp\left(ig \int_C \mathrm{d}s^\mu A_\mu^a(s) \frac{\lambda^a}{2} \right) \tag{3.7.2}$$

represents the finite parallel displacement along a path C as in (2.3.36). As the antiquark $\bar{q}$ at y has the opposite colour charge of the quark q at x, the colour displacement connecting these sources is a closed line (cf. Fig. 3.14). The trace of $U(C)$ occurs since the $q\bar{q}$ pair is in a colour-singlet state. The quantum-mechanical expectation value of $U(C)$ is calculated using the euclidean path-integral formula (2.2.67).

The gauge invariance of $W(C)$ follows directly from the gauge dependence of the colour displacement (2.3.35) along a closed path,

$$\operatorname{Tr} U(C) \to \operatorname{Tr}[g(x)U(C)g^{-1}(x)] = \operatorname{Tr} U(C). \tag{3.7.3}$$

To calculate the potential, C is taken as a planar, closed curve like in Fig. 3.14. Then,

$$V(r) = -\lim_{T\to\infty} \frac{\log W(C(r,T))}{T} \tag{3.7.4}$$

can be interpreted as a static potential $V(r)$ between a quark–antiquark pair in a colour-singlet state. This gauge-invariant definition of a potential is justified in the following section.

The interpretation of the Wilson integral (3.7.1) as a potential, (3.7.4), provides the *Wilson criterion for quark confinement.* If the euclidean expectation value $W(C)$ for similarly shaped paths C decreases exponentially with the minimal surface $\mathrm{Ar}(C) = rT$ enclosed by C,

$$W(C) \propto \exp(-\kappa \mathrm{Ar}(C)) \qquad \text{for large paths } C, \tag{3.7.5}$$

this is regarded as the criterion for confinement, because a confinement potential

$$V(r) \underset{r\to\infty}{\sim} \kappa r, \tag{3.7.6}$$

follows according to (3.7.4) for large rectangular paths C. The constant κ is called *string constant*. The reasons for this are described in Sect. 3.7.3. The string constant κ has dimension $(\mathrm{mass})^2$ and is a renormalization-group-invariant quantity which cannot be calculated by perturbation theory.

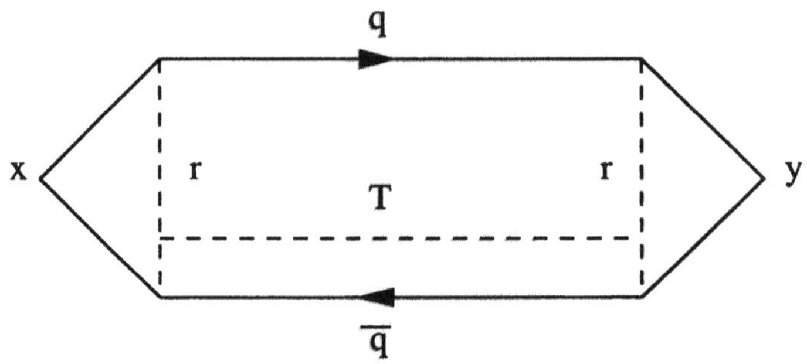

Fig. 3.14 Wilson loop $C(r,T)$ for the definition of a gauge-invariant static quark–antiquark potential

3.7.1.2 The Wilson definition of the Coulomb potential

In order to give a better idea of the details of the gauge-invariant potential definition by K. Wilson, we consider (3.7.1) and (3.7.2) in the framework of QED. For this purpose, the pure electromagnetic gauge-field Lagrangian $\mathcal{L}_{\text{gauge}}$ (2.3.11) is supplemented by a gauge-fixing term $\mathcal{L}_{\text{fix}} = (\partial^\mu A_\mu)^2/2$. After Wick rotation, i.e. continuation to the euclidean time $x_4 = \mathrm{i}x_0$, $A_4 = \mathrm{i}A_0$, we obtain the action

$$\begin{aligned} S_{\mathrm{E}}\{\mathbf{A}\} &= \int \mathrm{d}^4x \left[\frac{1}{4}(\partial_\mu A_\nu - \partial_\nu A_\mu)(\partial^\mu A^\nu - \partial^\nu A^\mu) + \frac{1}{2}(\partial^\mu A_\mu)^2\right] \\ &= \frac{1}{2}\int \mathrm{d}^4x\,(\partial_\mu A_\nu)(\partial^\mu A^\nu). \end{aligned} \tag{3.7.7}$$

Hence, the Wilson integral (3.7.1) becomes

$$\begin{aligned} W(C) = \frac{1}{Z}\int \mathcal{D}[\mathbf{A}] \exp\left(\mathrm{i}e\int_C \mathrm{d}s^\mu A_\mu(s)\right) \\ \times \exp\left(-\frac{1}{2}\int \mathrm{d}^4x\,(\partial^\mu A^\nu)(\partial_\mu A_\nu)\right). \end{aligned} \tag{3.7.8}$$

This is a gaussian integral as (2.2.11). It can be evaluated like the example (2.2.51)–(2.2.55) but with $\int \mathrm{d}^4x\, J(x)\psi(x)$ replaced by $e\int_C \mathrm{d}s^\mu A_\mu(s)$. The result is

$$W(C) = \exp\left(-\frac{e^2}{2}\int_C \mathrm{d}s^\mu \int_C \mathrm{d}t^\nu\, S_{\mu\nu}(s-t)\right). \tag{3.7.9}$$

Here, $S_{\mu\nu}(x)$ is the *euclidean Green function* of the electromagnetic field, also called the *Schwinger function*, in Feynman gauge,

$$S_{\mu\nu}(x) = -\frac{1}{4\pi^2}\delta_{\mu\nu}\frac{1}{x^2} = \frac{\delta_{\mu\nu}}{(2\pi)^4}\int \mathrm{d}^4p\,\frac{\mathrm{e}^{\mathrm{i}px}}{p^2},$$
$$\Box S_{\mu\nu}(x-y) = -\delta_{\mu\nu}\delta^{(4)}(x-y). \tag{3.7.10}$$

The two-point Green function [cf. (2.1.22)]

$$G_{\mu\nu}(x,y) = -g_{\mu\nu}\mathrm{i}\Delta_{\mathrm{F}}(x-y),$$

and the Schwinger function $S_{\mu\nu}(x-y) = \delta_{\mu\nu}S(x-y)$ are related by analytic continuation,

$$S(x-y) = \mathrm{i}\Delta_{\mathrm{F}}(x-y)|^{x_0=-\mathrm{i}x_4}_{y_0=-\mathrm{i}y_4}, \qquad \tilde{S}(p) = -\Delta_{\mathrm{F}}(p)|_{p_0=-\mathrm{i}p_4}. \tag{3.7.11}$$

In order to evaluate the Wilson integral (3.7.9) further, we choose a rectangle with sides of lengths r and T (cf. Fig. 3.14) for C. Hence, with (3.7.10), we have

$$\begin{aligned}\int_C \mathrm{d}s^\mu \int_C \mathrm{d}t^\nu\, S_{\mu\nu}(s-t) = &-\frac{1}{2\pi^2}\int_0^T \mathrm{d}t' \int_0^T \mathrm{d}t''\frac{1}{r^2+(t'-t'')^2}\\ &-\frac{1}{2\pi^2}\int_0^r \mathrm{d}r' \int_0^r \mathrm{d}r''\frac{1}{T^2+(r'-r'')^2}\\ &+\frac{1}{2\pi^2}\int_0^T \mathrm{d}t' \int_0^T \mathrm{d}t''\frac{1}{(t'-t'')^2}\\ &+\frac{1}{2\pi^2}\int_0^r \mathrm{d}r' \int_0^r \mathrm{d}r''\frac{1}{(r'-r'')^2}.\end{aligned}$$

Because of the factor $\delta_{\mu\nu}$ in $S_{\mu\nu}$, only integrations between parallel sides give a contribution. Those integrals in which t and s run over opposite (same) sides are shown in the first (last) two lines. Elementary integration gives

$$\int_0^T\int_0^T \frac{\mathrm{d}t'\mathrm{d}t''}{r^2+(t'-t'')^2} = \frac{2T}{r}\arctan\frac{T}{r} - \log\left(1+\frac{T^2}{r^2}\right). \tag{3.7.12}$$

The integrals over the same side diverge. Systematic discussion of these divergences would involve gauge-invariant regularization as in Sect. 2.5.1.

For the following consideration a simpler regularization is sufficient, which deforms the contour around the singularity,

$$\begin{aligned}\text{reg}\int_0^T\int_0^T \frac{dt'dt''}{(t'-t'')^2} &= 2\int_0^{T-\varepsilon} dt' \int_{t'+\varepsilon}^{T} dt'' \frac{1}{(t'-t'')^2} \\ &= \frac{2}{\varepsilon}(T-\varepsilon) + 2\log\frac{\varepsilon}{T}. \end{aligned} \tag{3.7.13}$$

Adding all contributions gives

$$\begin{aligned} W(C) = \exp\Big\{ \frac{e^2}{2\pi^2}\Big[&\frac{T}{r}\arctan\frac{T}{r} + \frac{r}{T}\arctan\frac{r}{T} \\ &- \frac{1}{2}\log\left(\left(1+\frac{T^2}{r^2}\right)\left(1+\frac{r^2}{T^2}\right)\right) - \frac{1}{\varepsilon}(T+r) + 2 - \log\frac{\varepsilon^2}{rT}\Big]\Big\} \end{aligned} \tag{3.7.14}$$

for the Wilson integral. Hence, according to (3.7.4), the electric potential between opposite charges is

$$V(r) = -\lim_{T\to\infty}\frac{\log W(C)}{T} = -\frac{e^2}{4\pi r} + \frac{e^2}{2\pi^2\varepsilon}, \tag{3.7.15}$$

i.e. the known Coulomb potential. The singular constant term $\propto \varepsilon^{-1}$ represents the self-energy of a regularized point charge. It is subtracted by the renormalization process. The logarithmically divergent term $\sim \log(\varepsilon^2/rT)$ does not appear for paths C with a continuous tangent [Do80a].

In a general R_ξ gauge, a term $((\xi-1)/16\pi^2)\partial_\mu\partial_\nu \log x^2/\partial^2$ has to be added to $S_{\mu\nu}$. A path integral along closed loops with respect to such gradient terms vanishes after regularization. Thus, the Wilson integral is independent of the gauge fixing.

3.7.2 Quark confinement in strong-coupling approximation

In Sect. 2.10.2 we introduced the basics of the strong-coupling approximation on the lattice. The most important example is the application to the Wilson loop $W(C)$. Based on the Osterwalder–Seiler theorem [Os78], there is a convergent strong-coupling expansion of $W(C)$ on the infinite lattice for small β. This factorizes the Boltzmann factor $\exp(-S_{\mathrm{E}}\{\mathbf{A}\})$ by character expansion leading to effective coupling constants $v(k)$, where k denotes the irreducible representations of SU(3). The expectation value of an observable is expanded in strong-coupling lattice graphs G with a given order in $v(k)$. We have shown this for the Wilson loop $W(\mathcal{C})$ in a simple model, the Z_2 gauge theory in two dimensions, in Sect. 2.10.2.1.

For quark triplets, triality-flux conservation can be used to show that the lowest-order, non-vanishing contribution to the Wilson loop $W(\mathcal{C})$ comes from the graph $G^{(m)}$ which has $\mathcal{C}$ as boundary, $\partial G^{(m)} = \mathcal{C}$, and which consists of the smallest number m of plaquettes which can fill $\mathcal{C}$ [Wi74]. According to the Wilson criterion (3.7.5) we have quark confinement in lowest-order strong-coupling approximation. In this case $v(k) = v(3) = v$ and the contribution of $G^{(m)}$ is

$$W(\mathcal{C}) = \text{constant} \times v^m = \exp\left(\frac{1}{a^2}\text{Ar}[\mathcal{C}]\log v\right). \tag{3.7.16}$$

The leading approximation of the *string constant* can be read off from this: $a^2\kappa = -\log v \approx \log 3g_0^2$ with the lattice constant a, the expansion $v = (g_0^{-2} + g_0^{-4}/2 + \ldots)/3$, and $\beta = 6/g_0^2$.

Terms of higher order of the strong-coupling approximation, described in Sect. 2.10.2, give the result

$$\begin{aligned} a^2\kappa = -\Big(&\log v + 4v^4 + 12v^5 - 10v^6 - 36v^7 + \frac{391}{2}v^8 \\ &+ \frac{1131}{10}v^9 + \frac{2550837}{5120}v^{10} + \ldots\Big). \end{aligned} \tag{3.7.17}$$

Details and the relevant graphs can be found in the literature [Dr83, Mu81].

We emphasize again, that the strong-coupling expansion gives a convergent series expansion for the Wilson loop on the lattice if the coupling constant is sufficiently large, i.e. for small β. Quark confinement is proven in this way under the conditions given. This gave rise to the hope that a consistent application of QCD can describe the phenomenon of quark confinement. However, for a complete proof of confinement, the non-trivial transition to the continuum must be analysed. This does not yet exist for the strong-coupling approximation (cf. Sect. 2.10.4.3).

3.7.3 The string picture

The validity of the Wilson criterion (3.7.5) would imply that pure gluon dynamics describes a potential between a static quark–antiquark pair that increases linearly with increasing distance. This type of potential is very different from the Coulomb potential, which appears in QED and arises also in perturbative QCD from one-gluon exchange. Thus, there must be a dynamical process which cannot be described by perturbation theory and which causes confinement in QCD. One physical picture for this stems from the theory of superconductivity and is called the *chromo-electric Meissner effect* [tH76b, Ma76].

3.7.3.1 The Abrikosov-Nielson-Olesen string

The *Abrikosov–Nielsen–Olesen string* [Ab57, Ni73] is a field-theoretical model for a linear potential between two magnetic monopoles with opposite magnetic charges in a superconductor of second kind. We start with the relativistic *Ginzburg–Landau Lagrangian* [Gi50] for a superconductor of second kind,

$$\mathcal{L} = [(\partial^\mu + \mathrm{i}eA^\mu)\Phi^*][(\partial_\mu - \mathrm{i}eA_\mu)\Phi] - \frac{1}{4}F_{\mu\nu}F^{\mu\nu} - \frac{\lambda}{4}\left(|\Phi|^2 - \frac{\mu^2}{\lambda}\right)^2 \tag{3.7.18}$$

with $\lambda > 0$ and $\mu^2 > 0$. This Lagrangian describes the gauge-invariant coupling of the electromagnetic field A_μ with a self-interacting, scalar, complex field Φ. In superconductivity, $|\Phi|^2$ describes an order parameter which is proportional to the number of superconducting electrons, i.e. Cooper pairs. The equations of motion for (3.7.18), the Ginzburg–Landau equations

$$\begin{aligned} (\partial_\mu - \mathrm{i}eA_\mu)^2\Phi &= \frac{\mu^2}{2}\Phi - \frac{\lambda}{2}|\Phi|^2\Phi, \\ \partial^\nu F_{\nu\mu} = -ej_\mu &= -\mathrm{i}e(\Phi^*\partial_\mu\Phi - \Phi\partial_\mu\Phi^*) - 2e^2A_\mu\Phi^*\Phi, \end{aligned} \tag{3.7.19}$$

give a macroscopic approximation [Go58, Pa69, La80] of the atomistic BCS theory [Ba57]. They are applicable in situations where the variations in Φ are small compared with the coherence length ξ, $|\partial\Phi| < \xi^{-1}$, i.e. predominantly for temperatures in the critical range.

Non-abelian generalizations of such gauge theories play an important rôle in the description of the weak interaction (cf. Sect. 4.1).

In this section static, cylindrically-symmetric solutions of (3.7.19) are considered. We use cylindrical coordinates ρ, θ, z, and t: $x_1 = \rho\cos\theta$, $x_2 = \rho\sin\theta$, $x_3 = x_z$, $x_4 = t$ and correspondingly for A_μ: A_ρ, A_θ, A_z, and A_t. In these coordinates we make the ansatz

$$\begin{aligned} \Phi(x) &= \mathrm{e}^{\mathrm{i}n\theta} G(\rho), \qquad n = 0, \pm 1, \pm 2, \dots\,, \\ A_\theta(x) &= A(\rho), \qquad A_t = A_\rho = A_z = 0. \end{aligned} \tag{3.7.20}$$

This implies that the non-vanishing element of the field-strength tensor, the magnetic induction in z direction $B(\rho)$, has the form

$$-F_{12}(x) = B(\rho) = \frac{1}{\rho}\frac{\mathrm{d}}{\mathrm{d}\rho}[\rho A(\rho)], \tag{3.7.21}$$

and the field equations reduce to

$$\left[-\frac{1}{\rho}\frac{\mathrm{d}}{\mathrm{d}\rho}\rho\frac{\mathrm{d}}{\mathrm{d}\rho} + \left(eA(\rho) - \frac{n}{\rho}\right)^2 - \frac{\mu^2}{2} + \frac{\lambda}{2}G^2(\rho)\right] G(\rho) = 0, \tag{3.7.22}$$

$$\frac{\mathrm{d}}{\mathrm{d}\rho}\frac{1}{\rho}\frac{\mathrm{d}}{\mathrm{d}\rho}[\rho A(\rho)] = e j_\theta(\rho) = 2e\left(eA(\rho) - \frac{n}{\rho}\right) G^2(\rho). \tag{3.7.23}$$

The energy of such a field configuration is obtained from the Lagrangian (3.7.18). In *temporal gauge*, $A_t = 0$, the Hamiltonian has the form

$$\begin{aligned} H &= \int \mathrm{d}^3x \left[\frac{\mathbf{B}^2 + \mathbf{E}^2}{2} + |\partial_t\Phi|^2 + |(\nabla - \mathrm{i}e\mathbf{A})\Phi|^2 + \frac{\lambda}{4}\left(|\Phi|^2 - \frac{\mu^2}{\lambda}\right)^2\right] \\ &= 2\pi \int \mathrm{d}z \int_0^\infty \rho\,\mathrm{d}\rho \left[\frac{1}{2\rho^2}\left(\frac{\mathrm{d}}{\mathrm{d}\rho}[\rho A(\rho)]\right)^2 + \left(\frac{\mathrm{d}}{\mathrm{d}\rho}G(\rho)\right)^2 \right. \\ &\qquad \left. + \left(eA(\rho) - \frac{n}{\rho}\right)^2 + \frac{\lambda}{4}\left(G^2(\rho) - \frac{\mu^2}{\lambda}\right)^2\right] \\ &\equiv \int \mathrm{d}z\, K\{A, G\}, \end{aligned} \tag{3.7.24}$$

where K represents the energy per cylinder length. The solution of the field equations (3.7.22) and (3.7.23) with lowest energy, i.e. with $K = 0$, the classical vacuum solution, is

$$A(\rho) = \frac{n}{e\,\rho}, \qquad G(\rho) = \frac{\mu}{\sqrt{\lambda}}. \tag{3.7.25}$$

Transformation of this solution into cartesian coordinates shows that the associated potential

$$A^\mu(x) = \frac{n}{e}\,\partial^\mu\left(\arccos\frac{x_1}{\sqrt{x_1^2+x_2^2}}\right) \tag{3.7.26}$$

is a pure gauge. Accordingly, $\Phi(x)$ is obtained by the corresponding gauge transformation of the constant $\mu/\sqrt{\lambda}$,

$$\Phi(x) = \mathrm{e}^{in\theta}\frac{\mu}{\sqrt{\lambda}}. \tag{3.7.27}$$

Now, we consider other solutions with finite energy. As K is made up of positive expressions, these must disappear individually for $\rho \to \infty$. Equation (3.7.24) shows that $A(\rho)$ and $G(\rho)$ thus approach the vacuum solution for $\rho \to \infty$. Therefore, these solutions can be written as

$$A(\rho) = \frac{n}{e\rho} + \bar{A}(\rho), \qquad G(\rho) = \frac{\mu}{\sqrt{\lambda}} + \bar{G}(\rho). \tag{3.7.28}$$

Inserting this into (3.7.22) and (3.7.23), we can read off the asymptotic behaviour for $\rho \to \infty$,

$$\bar{A}(\rho) \sim \rho^{-1/2}\exp\left(-\sqrt{\frac{2}{\lambda}}e\mu\rho\right), \qquad \bar{G}(\rho) \sim \rho^{-1/2}\mathrm{e}^{-\mu\rho}. \tag{3.7.29}$$

Similarly as in (2.9.5), we apply Stokes' theorem to the magnetic flux $\Phi_B(\mathcal{K})$ through a large circular surface $\mathcal{K}$ with boundary $\partial\mathcal{K}$,

$$\Phi_B(\mathcal{K}) = \int_{\mathcal{K}} \mathbf{B}\,\mathrm{d}\mathbf{f} = \int_{\partial\mathcal{K}} A_\mu(s)\,\mathrm{d}s^\mu = \lim_{\rho\to\infty}\rho\int_0^{2\pi} A_\theta\,\mathrm{d}\theta = \frac{2\pi}{e}n. \tag{3.7.30}$$

Because of the strong decrease of $\bar{A}(\rho)$, only the asymptotic vacuum solution contributes at the boundary. The fact that n is an integer follows from the uniqueness of $\Phi(x)$ according to (3.7.20). This means that the *magnetic flux is quantized* by the integer *topological quantum number* n [Lo50, Ab57].

As an integer valued functional of A and Φ, the flux $\Phi_B(\mathcal{K})$ can be deformed continuously without changing its value. The boundary condition for $\rho \to 0$,

$$A(\rho) \underset{\rho\to 0}{\sim} \text{constant}, \qquad \text{or} \quad \bar{A}(\rho) \underset{\rho\to 0}{\sim} -\frac{n}{e\rho}, \tag{3.7.31}$$

follows from $\int_{\mathcal{K}} \mathbf{B}\mathrm{d}\mathbf{f} = -2\pi(\rho A)|_{\rho=0}$. This condition establishes the form of a solution by a power series around $\rho = 0$. As $A(\rho) \sim$ constant according to (3.7.31), and since the repulsive potential $\sim n^2/\rho^2$ in (3.7.22) acts like the centrifugal potential in the Schrödinger equation, the form of $A(\rho)$ and $G(\rho)$ for small ρ is

$$A(\rho) = C + \dots, \qquad G(\rho) = C_n \rho^n + \dots . \tag{3.7.32}$$

This implies for the magnetic field strength, using (3.7.21) and (3.7.23)

$$\begin{aligned} B(\rho) &= B(0) + \int_0^\rho \mathrm{d}\rho' \frac{\mathrm{d}}{\mathrm{d}\rho'} B(\rho') = B(0) + \int_0^\rho \mathrm{d}\rho' \frac{\mathrm{d}}{\mathrm{d}\rho'} \frac{1}{\rho'} \frac{\mathrm{d}}{\mathrm{d}\rho'} [\rho' A(\rho')] \\ &= B(0) + \int_0^\rho \mathrm{d}\rho'\, 2e \left(eA - \frac{n}{\rho'} \right) G^2(\rho'). \end{aligned} \tag{3.7.33}$$

Inserting the expansions (3.7.32), this gives

$$B(\rho) = B(0) - eC_n^2 \rho^{2n} + \dots . \tag{3.7.34}$$

In general, the complete solution of the non-linear equation (3.7.22) cannot be given analytically. However, there is a limiting case, $\sqrt{2/\lambda}e\mu \gg 2\mu$, for which $\Phi(x)$ gets constant, $\Phi_0 = \sqrt{2\mu/\lambda} =$ finite, with $2\mu/\lambda^2 \to 0$, the *London limit*. In this case (3.7.23) can be reduced to Bessel's differential equation. The final result contains the modified Bessel function K_0,

$$B(\rho) = \frac{2\mu^2 e}{\lambda} K_0 \left(e\mu \sqrt{\frac{2}{\lambda}} \rho \right) . \tag{3.7.35}$$

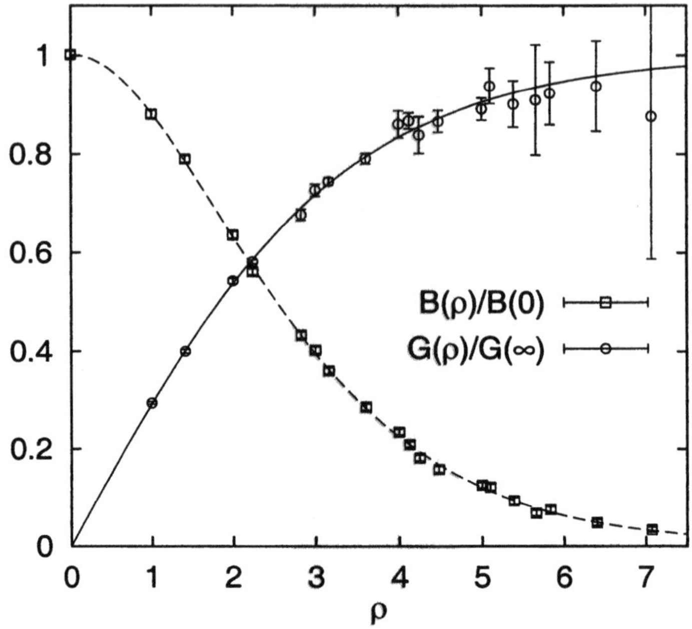

Fig. 3.15 The Nielsen–Olesen string: $B(\rho)/B(0)$ (dashed line) and $G(\rho)/G(\infty)$ (solid line) as a function of ρ in lattice units $\sim 0.08\,\mathrm{fm}$

The *string constant*, i.e. the energy per cylinder length, becomes

$$\kappa = 2\pi \int_0^\infty \rho\, \mathrm{d}\rho \frac{1}{2} B^2(\rho) = \frac{\pi}{\lambda}\mu^2 \tag{3.7.36}$$

in this limit. A numerical solution is given in Fig. 3.15. The parameters are taken from a lattice simulation [Ba98a] which we discuss in detail in Sect. 3.7.3.3. It incorporates the features discussed here.

3.7.3.2 The chromo-electric Meissner effect

The classical calculation of the magnetic string of Sect. 3.7.3.1 is an interesting intuitive model for quark confinement. The following features are worthwhile to mention:[20]

[20]Some of them like the order parameter and the mass generation occur also in the weakly interacting Electroweak Standard Model.

- The undisturbed superconducting medium is characterized by the value of the order parameter $|\phi| = \mu/\sqrt{\lambda}$, the classical vacuum value of $\phi(x)$.
- In the medium, a magnetic flux tube, i.e. a magnetic string, is formed between very distant magnetic poles. The property of a superconductor of expelling a magnetic field (Meissner effect) leads in this case to a magnetic induction $B(\rho)$ being concentrated in a string. As a magnetic field cannot penetrate a superconductor, the latter is regarded as a perfect diamagnet with permeability $\mu_{\text{magn}} = 0$.
- The constant, finite energy per string length [K in (3.7.24)] gives a linear potential when the monopoles are very distant.
- The magnetic flux is quantized, (3.7.30), and so are the magnetic charges of the monopoles. Quantization of the classical magnetic flux in units of $2\pi/e$ shows the importance of topological quantum numbers for the non-perturbative structure of this mechanism described here.
- The magnetic flux is stabilized by a circular current $j_\theta = \mathrm{d}B(\rho)/\mathrm{d}\rho$ [cf. (3.7.21) and (3.7.23].
- The medium resumes its behaviour as a normal conductor, $|\phi| \sim 0$, at the centre of the magnetic string.
- The *penetration depth* of the magnetic field is $\sqrt{\lambda/2}/(e\mu)$ [cf. (3.7.29)]. In dynamical solutions, this corresponds to the Compton wave length of a massive "photon" (cf. Higgs mechanism in Sect. 4.1.2.1).
- The displaced scalar field is restored to the vacuum value with a coherence length μ^{-1} (3.7.29). In perturbation theory of the Higgs model this is related to the Higgs-boson mass $M = \sqrt{2}\mu$ (Sect. 4.1.2.1).

For QCD we consider the Abrikosov–Nielsen–Olesen string as a classical model for the Wilson criterion of quark confinement. Because now we have a string between electric quark charges and not between magnetic monopoles, we have to exchange the electromagnetic fields by a *duality transformation*, $\mathbf{E} \to \mathbf{B}$, $\mathbf{B} \to -\mathbf{E}$. The gluon field between the colour charges of a separated static quark–antiquark pair is compressed to a string in the QCD vacuum. Similar to the Meissner effect, the QCD vacuum acts as a confinement medium (cf. Sect. 3.7.4) in the Abrikosov–Nielsen–Olesen string. This confinement mechanism is called the *chromo-electric Meissner effect* [tH81, tH82]. The constant field energy per unit length of the string gives the

	magnetic Abrikosov–Nielsen–Olesen string	chromo-electric gluon string
sources	magnetic monopoles with opposite magnetic charges	non-commutative charges of quarks and antiquarks
string of	magnetic flux between magnetic charges	chromo-electric flux between colour charges
embedded in	perfect diamagnet $\mu_{\text{magn}} = 0$ (superconductor)	confinement vacuum $\mu_{\text{el}} = 0,\ \varepsilon = 1/\mu_{\text{el}}$ (dielectric constant)
produced by	magnetic Meissner effect	chromo-electric Meissner effect
internal phases	normal conducting	asymptotically free

Tab. 3.4 The analogy between magnetic and chromo-electric strings

linear potential required by the Wilson confinement criterion. In Table 3.4 the analogy is conceptionally carried further.

In order to be more explicit, we discuss the modifications of Sect. 3.7.3.1 for the description of a confining string. We consider the string between $q\bar{q}$ sources on the 3 axis at $z \to \pm\infty$. In the scheme described in Table 3.4, magnetic and electric fields are exchanged like in a duality transformation. Therefore, we have to substitute:

(a) $B \to -E_3$, the electric component in 3 direction;

(b) $\Phi_B(\mathcal{K}) \to \Phi_{\text{el}} = \int_{\mathcal{K}} E_3\,\text{df}$, (3.7.30);

(c) $2\pi/e \to 1$, because we give unit charge to the quarks. This substitution corresponds to the Dirac condition $e_{\text{el}}\, e_{\text{mg}} = 2\pi N$, N integer, in a theory with electric and magnetic charges [Di31, Di48, tH81];

(d) $\sqrt{\lambda/2}/(e\mu) \to \delta$, the dual penetration length, (3.7.29);

(e) $\sqrt{2}/\mu \to \xi$, the dual coherence length, (3.7.29);

(f) $A_\theta \to \tilde{A}_\theta$. In order to get dynamical equations, one has to introduce dual potentials $\tilde{F}_{\mu\nu} = \partial_\mu \tilde{A}_\nu - \partial_\nu \tilde{A}_\mu$;

(g) $j_\theta \to \tilde{k}_\theta$, the magnetic current;

(h) $G(\rho) \to \tilde{G}(\rho) \equiv |\tilde{\Phi}(x)|$, (3.7.20), where $\tilde{\Phi}(x)$ is an effective description of the confining vacuum.

Up to now we have assumed uncritically that we can use the duality concept in non-abelian gauge theories like in abelian theories. One hypothesis could justify this assumption, the conjecture of *abelian dominance*. This assumes that there is a gauge in which fields related to an abelian subgroup of SU(N) dominate the long-distance physics. We make some comments on this assumption in Sect. 3.7.4.1.

We want to mention two effects not described by this picture of confinement:

i) The string thickness increases with increasing string length [Lu80b] in quantum mechanics.

ii) The centre of the gauge group plays an important rôle in a non-abelian gauge theory. The colour charges of static quarks have a triality which differs from zero. In a pure gauge theory, these cannot be screened by the triality-zero colour charges of the gluons. A quark charge is always a source of a chromo-electric flux which can end only in a colour charge of opposite triality, e.g. an antiquark. As mentioned above, the chromo-electric Meissner effect concentrates this flux into a string and thus produces a confinement potential. Hence, the representation of the centre of the colour group, characterized by triality, is important in the dynamics of quark confinement.

3.7.3.3 Lattice approximation of the string picture

There are first attempts to demonstrate string formation between static $q\bar{q}$ pairs in lattice approximation of pure gauge theories [Ba95c]. Figure 3.16 shows such a string-like configuration of the action density in a pure SU(2) gauge theory. For the string picture, only the difference of the vacuum expectation values (vevs) of observables calculated from the action with and without static sources are relevant. The lattice approximation of the action density, i.e. the Lagrangian at a fixed time, follows the line of derivation of the Wilson action in Sect. 2.10.1.2. We consider two sources q and $\bar{q}$ at spatial distance R. The action density on the lattice $\sigma_R(\mathbf{n})$ consists of a chromo-electric part $\mathcal{E}_R(\mathbf{n})$ and a chromo-magnetic part $\mathcal{B}_R(\mathbf{n})$,

$$\sigma_R(\mathbf{n}) = \frac{1}{2}[\mathcal{E}_R(\mathbf{n}) - \mathcal{B}_R(\mathbf{n})] \tag{3.7.37}$$

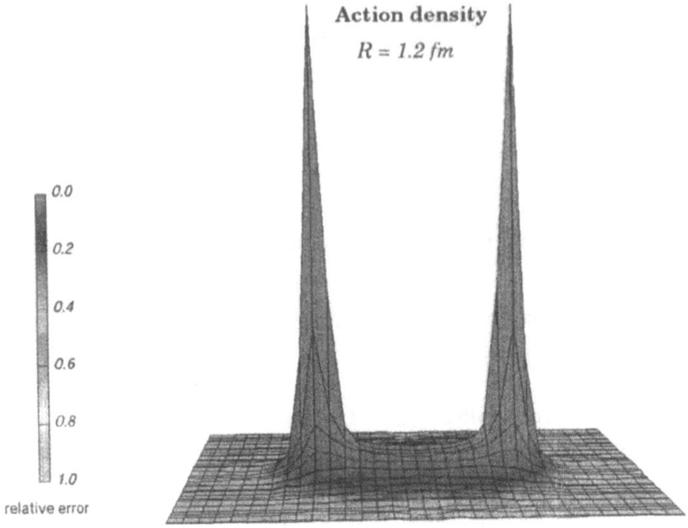

Fig. 3.16 A SU(2) string in lattice approximation [Ba95c]: between the static $q\bar{q}$ sources at a distance $R = 1.2\,\text{fm}$ the action density develops a string-like configuration. Close to the $q\bar{q}$ sources the action density becomes singular.

with

$$\begin{aligned}\mathcal{E}_R(\mathbf{n}) &= \langle 0, R|\mathbf{E}^2(\mathbf{n})|0, R\rangle - \langle 0|\mathbf{E}^2(\mathbf{n})|0\rangle \equiv \langle|\mathbf{E}^2(\mathbf{n})|\rangle_{|0,R\rangle-|0\rangle},\\ \mathcal{B}_R(\mathbf{n}) &= \langle 0, R|\mathbf{B}^2(\mathbf{n})|0, R\rangle - \langle 0|\mathbf{B}^2(\mathbf{n})|0\rangle \equiv \langle|\mathbf{B}^2(\mathbf{n})|\rangle_{|0,R\rangle-|0\rangle}.\end{aligned} \tag{3.7.38}$$

The sign conventions in (3.7.37) correspond to the action density $\mathcal{L} = \frac{1}{2}(\mathbf{E}^2 - \mathbf{B}^2)$ in the continuum with Minkowski metric. The lowest energy state with these sources, i.e. the vacuum in the presence of the $q\bar{q}$ sources, is denoted by $|0, R\rangle$.

We have seen that vevs of static quantities, like masses m in Sect. 2.10.1.4, or $V(r)$ in Sect. 3.7.1.1, can be obtained as a $T \to \infty$ limit. In this sense, vevs of operators $\langle 0, R|\mathcal{O}(\mathbf{n})|0, R\rangle$ can be obtained as a $T \to \infty$ limit of correlations of $\mathcal{O}(\mathbf{n})$ with Wilson loops [Ba95c]

$$\langle \mathcal{O}(\mathbf{n}, T)\rangle_W \equiv \frac{\langle \mathcal{O}(\mathbf{n}, T/2) W(R,T)\rangle - \langle \mathcal{O}(\mathbf{n}, T/2)\rangle\langle W(R,T)\rangle}{\langle W(R,T)\rangle}. \tag{3.7.39}$$

The Wilson loop $W(R,T)$ stands for the string of the $q\bar{q}$ pairs. The correlation (3.7.39) yields directly the difference of the vevs with and without the $q\bar{q}$ sources. As an example, we consider the calculation of $\mathcal{E}_R(\mathbf{n})$ and $\mathcal{B}_R(\mathbf{n})$. Similar as in (2.10.12), we get

$$U_{\mu\nu} \equiv \frac{1}{2}\operatorname{Tr}[U(\partial\mathcal{P})] = 1 - \frac{a^4}{2\beta}F^b_{\mu\nu}F^b_{\mu\nu} + \dots \qquad \text{no sum} \qquad (3.7.40)$$

for the oriented plaquette $\mathcal{P} = [x;\mu,\nu]$. Inserting $U_{\mu\nu}$ in (3.7.39), we obtain in the limit $T \to \infty$

$$\frac{2\beta}{a^4}\langle U_{i4}(\mathbf{n},T)\rangle_W \underset{T\to\infty}{\longrightarrow} \langle E_i^2(\mathbf{n})\rangle_{|0,R\rangle-|0\rangle}, \qquad (3.7.41)$$

$$-\frac{2\beta}{a^4}\langle U_{ij}(\mathbf{n},T)\rangle_W \underset{T\to\infty}{\longrightarrow} \langle B_k^2(\mathbf{n})\rangle_{|0,R\rangle-|0\rangle} \quad i,j,k \text{ cyclic.} \qquad (3.7.42)$$

[Note that $E_i^2 = E_i^a E_i^a = 2\operatorname{Tr}(E_i^a T^a E_i^b T^b)$]. The expressions for the field operators $E_i(\mathbf{n})$ and $B_k(\mathbf{n})$ can be found in a similar way, based on the geometric interpretation of Sect. 2.10.1.1.

The lattice approximation of such extended structures encounters many difficult problems, which we suppressed in our cursory presentation above.

- One has to compromise between a large physical lattice extension aN_s and a small a for the transition to the continuum (N_s is given by the computer power).
- In order that $W(R,T)$ describes the $q\bar{q}$ string well, the ground state overlap must be increased by suitably smeared operators. Corrections for finite T in the limit $T \to \infty$ have to be introduced.
- One has to reduce the noise in the combined heat-bath and over-relaxation procedure in the algorithm for updating the gauge fields (cf. Sect. 2.10.3.5).
- There are the general lattice problems like the transition to the continuum, finite size effects, etc. The treatment of a string in an $N_s \times N_t = 16^4$, 32^4, or $48^3 \times 64$ lattice for $\beta = 2.50$, 2.65, or 2.74 requires in the limit $T \to \infty$ ten thousands of measurements [Ba95c].

All these efforts lead to the picture shown in Fig. 3.16 together with the following results.

1) There is a static potential $V(r)$ which for $r \geq 0.5\,\mathrm{fm}$ can be well approximated by

$$V(r) = \kappa r - \frac{\pi}{12}\frac{1}{r}. \tag{3.7.43}$$

This result is obtained by fitting the potential in lattice units for different values of β, and transforming the different fits with help of the canonical string tension $\sqrt{\kappa} = 440\,\mathrm{MeV}$ into physical units. For different β, asymptotic scaling is approached reasonably well. A constant term is omitted in the lattice fits. Because of the infinite self-energy of the static sources [cf. the last term in (3.7.15)], the lattice approximation is not valid in these regions.

2) The universal constant with the value $\pi/12$ follows from the solution of a classical string equation [Lu80b]. It is the second term of an asymptotic solution for large r. The lattice approximation confirms this result. [For $r \ll 0.5\,\mathrm{fm}$ one expects that the Coulomb constant approaches the running coupling constant value $4\alpha(r)/3$].

3) According to (3.6.4), the reference scale is $r_0 \approx 0.5\,\mathrm{fm}$. Under the assumptions

   (i) that the quenched approximation leads approximately to the same potential (3.7.43) for SU(2) and SU(3) (this is confirmed by lattice calculation) and

   (ii) that the relation between quenched and unquenched approximation described in Sect. 3.6.3.4 is correct,

   the potential (3.7.43) is in agreement with the perturbative coupling constant $\alpha_{\overline{\mathrm{MS}}} \approx 0.12$.

4) The transverse width of the flux distribution in the plane perpendicular to the string halfway between the sources is described by Fig. 3.16 with arguments transformed by duality [Ba98]. The result of the "lattice measurement" of the penetration length δ and the coherence length ξ is

$$\delta = 0.16(2)\,\mathrm{fm}, \qquad \xi = 0.27(3)\,\mathrm{fm}. \tag{3.7.44}$$

These results are valid for separations R between 1 fm and 2 fm.

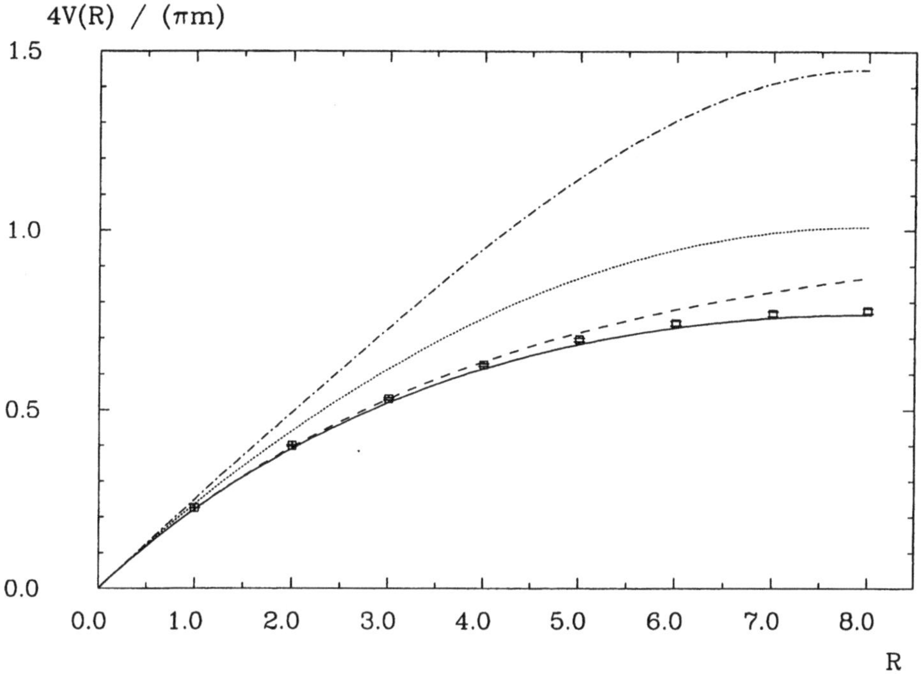

Fig. 3.17 Screened potential of the Schwinger model

3.7.3.4 Quark confinement with dynamical quarks

The Wilson criterion for confinement applies only for pure gauge theories. In QCD with dynamical quarks, colour charges of quarks and antiquarks can be screened by quark-pair creation. The physical states should be the conventional, colourless hadrons which can be considered as bound states of $q\bar{q}$ and qqq in the simple quark model. In the string picture there is *string breaking*. This occurs at those distances R where the string of the $Q\bar{Q}$ source has enough energy so that it can decay into a pair of $q\bar{Q}$ and $\bar{q}Q$ mesons.

String breaking is not yet established in the lattice approximation of SU(2) and SU(3) gauge theories with dynamical quarks. All the problems of observing extended structures discussed in the last section occur even more in unquenched approximations [Sc00]. We show string breaking in a (1+1)-dimensional QED model with "photons" coupled to massless fermions (*Schwinger model* [Sc62]). Since in (1+1) dimensions the Coulomb potential is $V(r) = e^2|r|/2$, we have confinement according to the Wilson criterion.

For large $r > 0$, the Coulomb potential is screened by the $q\bar{q}$ pairs. Figure 3.17 is based on an exact calculation of the model in a finite volume [Di92, Di94]. Its unquenched lattice approximation is based on staggered fermions treated by a hybrid Monte Carlo algorithm. Figure 3.17 shows the different steps:

(i) Coulomb potential in a finite volume (dot-dashed line);

(ii) Coulomb potential in a finite volume with topological sectors omitted (dotted line);

(iii) Screened potential in an infinite volume (dashed line);

(iv) Screened potential in a finite volume (solid line);

(v) Lattice approximation with $\beta = (ea)^{-2} = 10$, where a is the lattice constant and e the coupling constant in the (1+1)-dimensional Coulomb potential with dimension of a mass (squares).

Lattice approximations of such low-dimensional models may give valuable hints on what one expects in more realistic situations (cf. Sect. 2.9.1.1 and Sect. 2.9.2.1).

3.7.4 Long-range correlations of the QCD vacuum

As an outlook on recent developments of some old ideas [tH81, tH82] on non-perturbative properties of gauge theories we discuss again the dual Meissner effect (Sect. 3.7.3.2). As already stated, we consider the combination of lattice approximation with the intuitive pictures of topological properties derived from the semi-classical approximation as the appropriate method for non-perturbative problems. The progress in this program is due to the fact that computers now allow for the treatment of volumes of sizes $\approx (2$–$3\,\mathrm{fm})^4$ with a resolution 0.05–0.1 fm (cf. Sect. 3.6.1.1). However, there are still many unsolved shortcomings of this approach. We mention only two.

- It is not yet finally established how much the intuitive picture depends on the chosen gauge.
- In lattice approximation the study of long-range effects depends on the cooling (Sect. 2.10.3.5). The effect of cooling is not yet completely understood.

Thus, the non-perturbative theory of QCD is not yet firmly established.

3.7.4.1 Maximal abelian gauge and the chromo-electric Meissner effect

Applying the duality concept of Sect. 3.7.3.2 to the Ginzburg–Landau effective Lagrangian (3.7.18), we get the (euclidean) Lagrangian of the *dual gauge theory*

$$\mathcal{L} = \frac{1}{4} G_{\mu\nu} G^{\mu\nu} + \frac{1}{2} (D_\mu \Phi)^* (D^\mu \Phi) + \frac{\lambda}{4} \left(|\Phi|^2 - \frac{\mu^2}{\lambda} \right)^2 \tag{3.7.45}$$

with

$$(D^\mu \Phi) = (\partial_\mu + \mathrm{i} g_{\mathrm{mg}} c_\mu) \Phi, \quad G_{\mu\nu} = \partial_\mu c_\nu - \partial_\nu c_\mu. \tag{3.7.46}$$

How can one understand *dual gauge theory* as an effective gauge theory? This question is approached by the conjecture of *abelian dominance* and the *condensation of magnetic monopoles* [tH81, tH82]. We illustrate the main concepts of this procedure for SU(2).

The meaning of *maximal abelian gauge* is simple: a regular SU(2) potential $\mathcal{A}_\mu(x) = g\, A^b_\mu(x) \tau^b/2$ is diagonalized by a gauge transformation as far as possible. To this end, the expression

$$R\{\mathbf{A}\} = \int \mathrm{d}^4 x \, [A^1_\mu(x) A^{1,\mu}(x) + A^2_\mu(x) A^{2,\mu}(x)] \tag{3.7.47}$$

with the components of the gauge field corresponding to non-diagonal Pauli matrices is minimized by a gauge transformation $\omega(x)$,

$$\mathcal{A}^\omega_\mu(x) \equiv \omega(x) \mathcal{A}_\mu(x) \omega^{-1}(x) - \mathrm{i}\omega(x) \partial_\mu \omega^{-1}(x) \equiv \bar{\mathcal{A}}_\mu + \mathcal{M}_\mu, \tag{3.7.48}$$

where $\bar{\mathcal{A}}_\mu$ is defined as the homogeneous transformation of $\mathcal{A}_\mu$ and $\mathcal{M}_\mu$ as the inhomogeneous part. The condition for a local extremum of the functional $R\{\mathbf{A}\}$ is

$$(\partial_\mu \pm \mathrm{i} g A^3_\mu(x)) A^\mp_\mu(x) = 0, \qquad A^\pm_\mu(x) = A^1_\mu(x) \pm \mathrm{i} A^2_\mu(x). \tag{3.7.49}$$

In this gauge, we define the abelian field $a_\mu(x)$ and its field strength $f_{\mu\nu}(x)$ as

$$a_\mu(x) = \frac{1}{g}\,\mathrm{Tr}[\tau^3 A^\omega_\mu(x)], \tag{3.7.50}$$

$$f_{\mu\nu}(x) = \partial_\mu a_\nu(x) - \partial_\nu a_\mu(x) = \partial_\mu \bar{A}^3_\nu - \partial_\nu \bar{A}^3_\mu + \partial_\mu M^3_\nu - \partial_\nu M^3_\mu. \tag{3.7.51}$$

The definition of the abelian potential $a_\mu(x)$ still admits gauge transformations of an U(1) subgroup of SU(2), $g(x) = \exp(-\mathrm{i}\,\theta(x)\,\tau_3/2)$,

$$a'_\mu(x) = a_\mu(x) + \frac{1}{g}\partial_\mu\theta(x). \tag{3.7.52}$$

Thus, $a_\mu(x)$ plays the rôle of an abelian gauge field. The fields $A^{\omega,\pm}_\mu(x)$ transform as charged fields (cf. Sect. 2.3.2.5),

$$A'^{\omega,\pm}_\mu(x) = \mathrm{e}^{\mp\mathrm{i}\theta} A^{\omega,\pm}_\mu(x). \tag{3.7.53}$$

According to (3.7.48), the first term in (3.7.51) originates from a gauge-rotated regular vector potential $\bar{A}^3_\mu$. By the abelian Bianchi identity, this part of the dual field strength $\tilde{f}_{\mu\nu} = \varepsilon_{\mu\nu\rho\sigma} f^{\rho\sigma}/2$ is divergence-less. Only the term involving M_μ might be singular and contribute to a non-vanishing, magnetic, conserved current [Kr87]

$$k^\mu = \frac{1}{4\pi}\partial_\nu \tilde{f}^{\mu\nu}, \qquad \partial_\mu k^\mu = 0, \tag{3.7.54}$$

as

$$\begin{aligned} k_\mu &= \frac{1}{8\pi g}\varepsilon_{\mu\nu\rho\sigma}\,\mathrm{Tr}(\tau^3\partial^\nu(\partial^\rho \mathcal{M}^\sigma - \partial^\sigma \mathcal{M}^\rho)) \\ &= -\frac{\mathrm{i}}{8\pi g}\varepsilon_{\mu\nu\rho\sigma}\,\mathrm{Tr}(\tau^3\partial^\nu[\mathcal{M}^\rho, \mathcal{M}^\sigma]). \end{aligned} \tag{3.7.55}$$

In the last step we used the explicit form of $\mathcal{M}_\mu = -\mathrm{i}\omega\partial_\mu\omega^{-1}$. A detailed analysis shows, that M^3_μ becomes singular at the position of a magnetic charge, i.e. of a magnetic monopole [tH81, Su95, Br97].

The conserved current $k_\mu(x)$ generates a symmetry $\mathrm{U(1)_{mg}}$. However, one assumes that this symmetry is spontaneously broken (cf. Sect. 4.1.2) by a *condensation of the magnetic monopoles,*

$$\langle k_\mu(x) k_\nu(y)\rangle = g_{\mathrm{mg}}^2 \delta_{\mu\nu}\delta^4(x-y) f(x) + \dots . \tag{3.7.56}$$

In a translation-invariant field theory, $f(x)$ is constant and describes a mass in tree-graph approximation. Phenomenologically, one can describe the condensed monopoles by a field $\tilde{\Phi}(x)$. Then, the conjecture of abelian dominance and monopole condensation suggests a possible interpretation of the basic fields of dual SU(2), $G_{\mu\nu}(x)$, $c_\mu(x)$, and $\Phi(x)$ [cf. (3.7.46)] by the identification

$$G_{\mu\nu}(x) = \tilde{f}_{\mu\nu}(x), \qquad c_\mu(x) = a_\mu(x), \qquad \Phi(x) = \tilde{\Phi}(x). \tag{3.7.57}$$

The coupling constants of dual SU(2) are λ, μ, and g_{mg} according to (3.7.45) and (3.7.46). The magnetic coupling g_{mg} satisfies together with the ("electric") coupling constant g the Dirac quantization condition [Di31, Di48], $g\,g_{\mathrm{mg}} = 2\pi N$, N integer. The results (3.7.44) for the penetration length $\delta = 0.16(2)\,\mathrm{fm}$ and the coherence length $\xi = 0.27(3)\,\mathrm{fm}$ of the lattice analysis of the dual Abrikosov–Nielsen–Olesen string (Sect. 3.7.3.3) allow for $N = 1$, $g = 1$, $g_{\mathrm{mg}} = 2\pi$ for an approximate determination of the coupling constants:

$$\begin{aligned} \mu &= \frac{\sqrt{2}}{\xi} \approx 5.23\,\mathrm{fm}^{-1}, \qquad \lambda = 8\pi^2\delta^2\mu^2 \approx 55.28, \\ \langle\Phi\rangle^2 &= \frac{\mu^2}{\lambda} \approx 0.5\,\mathrm{fm}^{-2}. \end{aligned} \tag{3.7.58}$$

In a more systematic derivation of dual SU(2) one integrates the path integral over the massive charged fields $A^{\pm}(x)$ [Ko98]. If one takes into account the gauge fixing by the BRS procedure (Sect. 2.4.3.1), one can show that $\alpha_{\mathrm{mg}} = g_{\mathrm{mg}}^2/4\pi$ satisfies the renormalization-group equation in one-loop approximation. Thus, $\alpha_{\mathrm{mg}}(Q^2)$ should run consistently with the Dirac quantization. Unfortunately, there are no Q^2 regions where both α_{mg} and α are small.

The maximal abelian gauge can be approximated on the lattice [Kr87]. Lattice calculations for the string tension show evidence for the importance of the diagonal gauge-field elements, i.e. for abelian dominance. They indicate also the existence of condensates of magnetic monopoles [Ha98b].

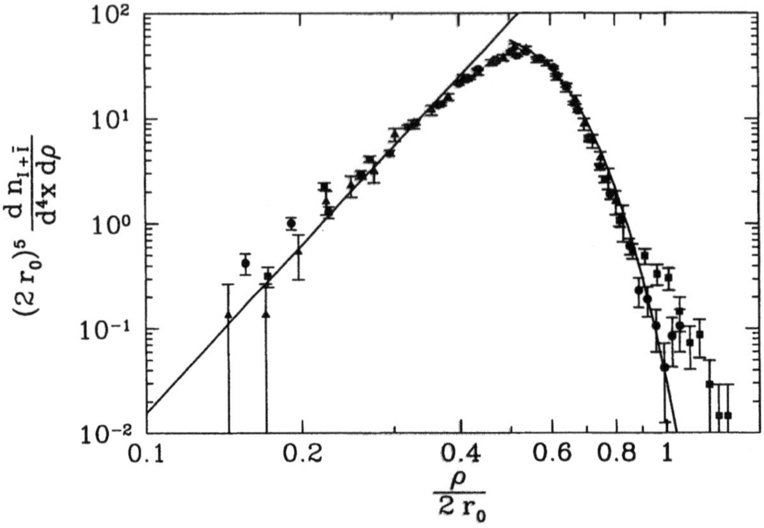

Fig. 3.18 Instanton distribution: the points represent lattice results, the straight line to the left gives the result of perturbative semi-classical approximation, and the line to the right gives a fit to dual gauge theory (from Ref. [Ri99]).

3.7.4.2 Lattice results for instantons

Topological configurations seem to play an important rôle in the understanding of non-perturbative aspects of gauge theories. Figure 3.18 illustrates, how instantons fit in this picture [Ri99]. It shows the density distribution of instantons $D_{\mathrm{I}}(\rho, \alpha(\mu_0))$ (2.9.121) of the size ρ (called λ in Sect. 2.9.3) measured by a cooled lattice approximation [Sm98].

For small instantons, i.e. $\rho \leq 0.3\,\mathrm{fm}$, 't Hooft's semi-classical approximation (Sect. 2.9.3.2) works well. The straight line gives the result of the two-loop approximation of (2.9.121). It is an absolute prediction of $D_{\mathrm{I}}(\rho, \alpha(\mu_0))$ without any fitting parameters. The $\Lambda_{\overline{MS}}$ parameter is determined by the non-perturbative evolution of Sect. 3.6.3.4: $\Lambda_{\overline{MS}} = (238 \pm 19)\,\mathrm{MeV}$. The parameter $r_0 \approx 0.5\,\mathrm{fm}$ was introduced in this context.

We speculate that the large instantons are screened with a penetration length of the dual Meissner effect. An extension of 't Hooft's calculation

(Sect. 2.9.3.2) contains such an effect. The result is an additional contribution in the action, $\Delta S = 4\pi^2\rho^2\langle\Phi\rangle^2$. We include such a contribution in the density distribution $D_{\mathrm{I}}(\rho)$ [Sh99]. A fit for $\rho \geq 0.6\,\mathrm{fm}$ with the ansatz

$$D_{\mathrm{I}}(\rho) = D_{\mathrm{I,pert}}(\rho)\mathrm{e}^{-4\pi^2\rho^2\langle\Phi\rangle^2} = \text{constant} \times \rho^6\mathrm{e}^{-4\pi^2\rho^2\langle\Phi\rangle^2} \qquad (3.7.59)$$

gives $\langle\Phi\rangle^2 = 0.39\,\mathrm{fm}^{-2}$. It is included in Fig. 3.18 as a line interpolating the lattice points. It might be compared with the result (3.7.58) from the SU(2) parameters (3.7.44), $\langle\Phi\rangle^2 = \mu^2/\lambda \approx 0.5\,\mathrm{fm}^{-2}$. In view of the comparison of SU(2) with SU(3), and all the other approximations mentioned above, this seems to support these considerations.

With this interpretation, Fig. 3.18 shows how the instanton interpolates between the perturbative (quenched) SU(3) QCD and the long-range part described by the dual gauge theory.

3.7.4.3 Experimental consequences

What are the experimental consequences of this complicated picture of the non-perturbative QCD vacuum? We mention only three examples.

There is the prediction that for high physical temperatures (cf. Sect. 2.2.4.4) a *deconfining phase transition* takes place [Co75a]. This means that quarks are no longer confined at high temperatures but form a plasma of approximately free quarks and gluons interacting according to asymptotically free QCD. Lattice calculation gives $T_{\mathrm{c}} \approx 270\,\mathrm{MeV}$ for the transition temperature in quenched approximation and $T_{\mathrm{c}} \approx 170 - 190\,\mathrm{MeV}$ for 3-flavour QCD [Ka00]. According to the picture above, the magnetic-monopole condensation is responsible for quark confinement. Indeed, lattice calculations show a sharp decrease of the order parameter of magnetic-monopole condensation at the transition temperature [Di99]. Experimentally, one tries to generate a *quark–gluon plasma* by high-energy heavy-ion collisions [Ap99, Sa00a]. In the theories about the development of the early universe, this phase transition plays also an important rôle [Ko90].

Many "QCD-inspired" phenomenological models on non-perturbative features of elementary particle physics are based on dual gauge theories, i.e. on dual QCD. We mention only the model of the *stochastic vacuum* [Do88]. This model tries to explain the main features of high-energy hadron–hadron scattering and photon–hadron scattering by non-perturbative QCD [Do94,

Be99]. It relates the long-distance correlation function of the field strength to dual QCD [Ba98].

Since instantons and their derivates play such an important rôle in non-perturbative gauge theories, it would be of utmost importance to see experimental signals of their actions [Ri00].

3.8 A test of QCD: the running of α_s

Twenty-five years after it was proposed, QCD has become the generally accepted theory of the strong interaction. Since then, effective methods for the calculation of physical quantities have been worked out. For hard-scattering processes a consistent perturbation theory is available. In most cases two-loop calculations exist, sometimes there are even three-loop results, and for the β-function a fourth-order calculation was performed. Methods for the treatment of QCD in the strong-interaction region have been developed. These include the strong-coupling expansion and the lattice approximation of QCD which has found many applications. The investigation of the topological properties of QCD has let to a deeper understanding of this non-abelian gauge theory. The physical meaning of the corresponding non-perturbative instanton solutions still has to be investigated.

The accuracy of these perturbative and non-perturbative calculations is in most cases adequate for a comparison with the experimental results. We have shown this for the perturbative calculations for deep-inelastic scattering, lepton-pair production, the total hadronic cross section in e^+e^- annihilation, and jet production. The non-perturbative methods were applied to low-energy hadron physics. In this way the masses and other properties of mesons, glue balls, and baryons were obtained from the description of these states as bound states of quarks, antiquarks, and gluons. Based on QCD, models for heavy-quark physics on the one hand, low-energy effective theories in the form of chiral perturbation theory on the other hand, and many more were developed.

A huge amount of detailed experimental and theoretical results is available. We conclude this discussion of QCD by considering the results for one of the fundamental constants of nature, the strong coupling constant g_s, or $\alpha_s = g_s^2/4\pi$. This enters all QCD results for physical quantities and in this sense has to be universal, i.e. the same in hard-scattering processes, bound state properties, and low-energy scattering. Using the renormalization group

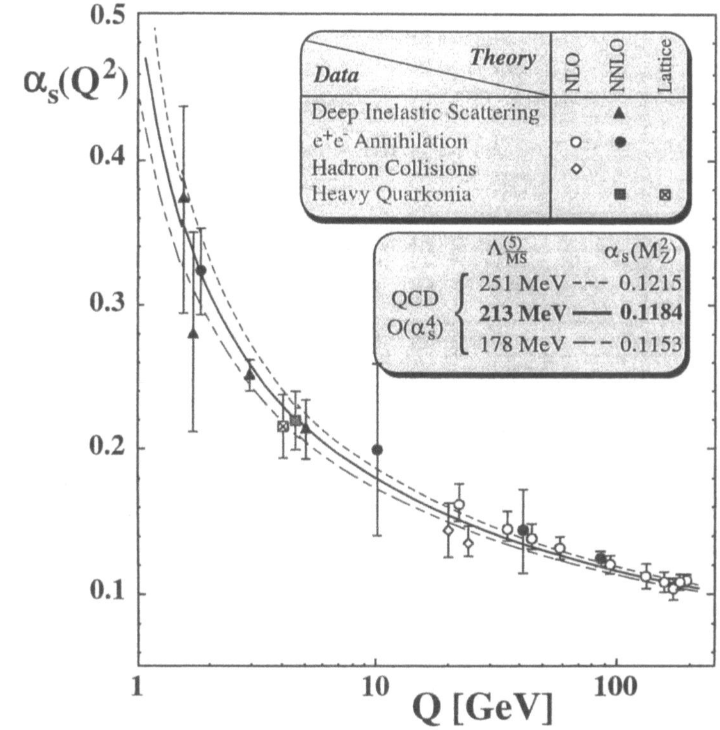

Fig. 3.19 Summary of experimental determinations of $\alpha_s(Q^2)$ (from Ref. [Be00]) [(N)NLO = (next-to-)next-to-leading order]

the value of strong coupling is usually given at M_Z since the most precise determinations come from this region. Instead of $\alpha_s(M_Z^2)$ in many cases the *asymptotic scale parameter* Λ of QCD (cf. Sect. 3.1) is given.

In Fig. 3.19, a summary of the measurements of the *running coupling* $\alpha_s(Q^2)$ together with a QCD fit in the $\overline{\text{MS}}$ scheme for five flavours is shown. Over a region from 2 GeV up to 200 GeV an impressive agreement between theory and experiment is found. The best fit gives the values [Be00]

$$\Lambda^{(5)}_{\overline{\text{MS}}} = 213 \begin{matrix} +38 \\ -35 \end{matrix} \text{ MeV}, \qquad \alpha_s(M_Z^2) = 0.1184 \pm 0.0031. \tag{3.8.1}$$

As a by-product, the number of colours is determined as $N_C = 3.03 \pm 0.12$, via the dependence of the β-function on N_C.

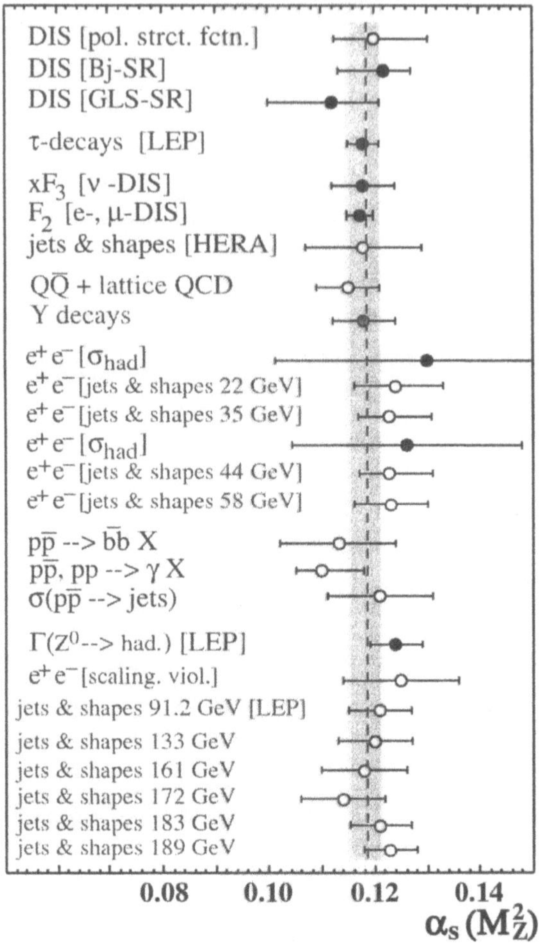

Fig. 3.20 Summary of experimental determinations of $\alpha_s(M_Z^2)$ (from Ref. [Be00])

A comparison between different measurements of $\alpha_s(M_Z^2)$ is shown in Fig. 3.20. The dashed line represents the statistical expectation value of $\alpha_s(M_Z^2)$, the grey band corresponds to the one-standard deviation range. This illustrates the impressive agreement of QCD with data from many different experiments.

References to Chapter 3

[Ab57] A.A. Abrikosov, *Sov. Phys. JETP* **5** (1957) 1174.

[Ab98] P. Abreu et al. (DELPHI Coll.), *Phys. Lett.* **B418** (1998) 430.

[Ad68] S.L. Adler and R.F. Dashen, *Current Algebras and Applications to Particle Physics,* Frontiers in physics Vol. 30, (W.A. Benjamin, Inc., New York, 1968).

[Ak65] S.Ø. Aks, *J. Math. Phys.* **6** (1965) 516.

[Al73] V. de Alfaro, S. Fubini, G. Furlan and C. Rossetti, *Currents in Hadron Physics*, (North Holland, Amsterdam, 1973).

[Al77] G. Altarelli and G. Parisi, *Nucl. Phys.* **B126** (1977) 298.

[Al78] G. Altarelli and G. Martinelli, *Phys. Lett.* **76B** (1978) 89.

[Al79] G. Altarelli, R.K. Ellis and G. Martinelli, *Nucl. Phys.* **B157** (1979) 461.

[Al79a] A. Ali et al., *Phys. Lett.* **B82** (1979) 285 and *Nucl. Phys.* **B167** (1980) 454.

[Al80] A. Ali, E. Pietarinen, G. Kramer and J. Willrodt, *Phys. Lett.* **93B** (1980) 155.

[Al87] C. Albajar et al. (UA1 Coll.), *Phys. Lett.* **B198** (1987) 271.

[Al97] G. Altarelli, R.B. Ball, St. Forte and G. Ridolfi, *Nucl. Phys.* **B496** (1997) 337.

[AL99] ALEPH Coll., CERN-OPEN-99-291, CERN-ALEPH-99-023.

[Am78] D. Amati, R. Petronzio and G. Veneziano, *Nucl. Phys.* **B140** (1978) 54 and **B146** (1978) 29.

[An87] R. Ansari et al. (UA2 Coll.), *Phys. Lett.* **B194** (1987) 158.

[Ao00] S. Aoki et al. (CP-PACS Coll.), *Phys. Rev. Lett.* **84** (2000) 238.

[Ap73] T. Appelquist and H. Georgi, *Phys. Rev.* **D8** (1973) 4000.

[Ap99] H. Appelshauser et al. (NA49 Coll.), *Phys. Lett.* **B459** (1999) 679.

[Ar97] M. Arnedo et al. (NMC Coll.), *Nucl. Phys.* **B483** (1997) 3.

[Ba57] J. Bardeen, L.N. Cooper and J.R. Schrieffer, *Phys. Rev.* **106** (1957) 162 and **108** (1957) 1175.

[Ba78] W.A. Bardeen, A.J. Buras, D.W. Duke and T. Muta, *Phys. Rev.* **D18** (1978) 3998.

[Ba93] G.S. Bali et al (UKQCD Coll.), *Phys. Lett.* **B309** (1993) 378.

[Ba95c] G.S. Bali, K. Schilling and C. Schlichter, *Phys. Rev.* **D51** (1995) 5165.

[Ba98] M. Baker, N. Brambilla, H.G. Dosch and A. Vairo, *Phys. Rev.* **D58** (1998) 034010.

[Ba98a] G.S. Bali, K. Schilling and C. Schlichter, *Prog. Theor. Phys. Suppl.* **131** (1998) 645.

[Be83] W. Bernreuther, *Ann. Phys.* **151** (1983) 127.

[Be85] B. Betev et al., *Z. Phys.* **C28** (1985) 9 and 15.

[Be88] S. Bethke et al. (JADE Coll.), *Phys. Lett.* **B213** (1988) 235.

[Be92] S. Bethke, Z. Kunszt, D.E. Soper and W.J. Stirling, *Nucl. Phys.* **B370** (1992) 310, **E***Nucl. Phys.* **B523** (1998) 681.

[Be99] E.R. Berger and O. Nachtmann, *Eur. Phys. J.* **C7** (1999) 459.

[Be00] S. Bethke, *J. Phys.* **G26** (2000) R27.

[Bi93] J. Bijnens, *Int. J. Mod. Phys.* **8A** (1993) 3045.

[Bi94] J. Bijnens, G. Ecker and J. Gasser, *Chiral Perturbation Theory* in *The 2nd DAΦNE Phys. Handbook*, eds. L. Maiani, G. Pancheri and N. Paver (Frascati, 1994) p. 125.

[Bi95] J. Bijnens, J. Prades and E. de Rafael, *Phys. Lett.* **B348** (1995) 226.

[Bj69] J.D. Bjorken, *Phys. Rev.* **179** (1969) 1547.

[Bj66] J.D. Bjorken *Phys. Rev.* **148** (1966) 1467 and *Phys. Rev.* **D1** (1970) 1376.

[Br97] R.C. Brower, K.N. Orginos and Ch.-I Tan, *Phys. Rev.* **D55** (1997) 6313.

[Br99] A. Brandenburg et al., *Phys. Lett.* **B468** (1999) 168.

[Ca91] S. Catani, Yu.L. Dokshitzer, M. Olsson, G. Turnok and B.R. Webber, *Phys. Lett.* **B269** (1991) 432.

[Ca99] S. Capitani et al. (ALPHA Coll.), *Nucl. Phys.* **B544** (1999) 669.

[Ch79] K.G. Chetyrkin, A.L. Kataev and F.V. Tkachov, *Phys. Lett.* **85B** (1979) 277.

[Ch97] K.G. Chetyrkin, *Phys. Lett.* **B404** (1997) 161.

[Cl98] F.E. Close, *Nucl. Phys. B (Proc. Suppl.)* **63** (1998) 28.

[Co73] S. Coleman and E. Weinberg, *Phys. Rev.* **D7** (1973) 1888.

[Co75a] J.C. Collins and M.J. Perry, *Phys. Rev. Lett.* **34** (1975) 1353.

[Co82] B. Cox, *Dimuon Production in Hadronic Interactions* in *Proceedings of the 21st International Conference on High Energy Physics, J. Phys. Suppl. (France) 43*, eds. P. Petiau and M. Porneuf (Paris, 1982) p. C3-140.

[Co84] J.C. Collins, *Renormalization*, (Cambridge University Press, Cambridge, 1984).

[Co89] J.C. Collins, D.E. Soper and G. Sterman, *Factorization of Hard Processes in QCD* in *Perturbative Quantum Chromodynamics*, ed. A.H. Mueller (Singapore, 1989) p. 1.

[Cu80] G. Curci, W. Furmanski and R. Petronzio, *Nucl. Phys.* **B175** (1980) 27.

[Cu80a] G. Curci and M. Greco, *Phys. Lett.* **92B** (1980) 175.

[De78] A. de Rujula, J. Ellis, E.G. Floratos and M.K. Gaillard, *Nucl. Phys.* **B138** (1978) 387.

[Di31] P.A.M. Dirac, *Proc. Roy. Soc.* **A133** (1931) 60.

[Di48] P.A.M. Dirac, *Phys. Rev.* **74** (1948) 817.

[Di74] J.A. Dixon and J.C. Taylor, *Nucl. Phys.* **B78** (1974) 552.

[Di92] H. Dilger, *Phys. Lett.* **B294** (1992) 263.

[Di94] H. Dilger and H. Joos, *Nucl. Phys. B (Proc. Suppl.)* **B34** (1994) 195.

[Di99] A. Di Giacomo et al., *Nucl. Phys. B (Proc. Suppl.)* **B74** (1999) 405.

[Do77] Yu.L. Dokshitzer, *Sov. Phys. JETP* **46** (1977) 641.

[Do80] Yu.L. Dokshitzer, D.I. Dyakonov and S.I. Troyan, *Phys. Rept.* **58** (1980) 269.

[Do80a] V.S. Dotsenko and S.N. Vergeles, *Nucl. Phys.* **B169** (1980) 527.

[Do88] H.G. Dosch and Yu.A. Simonov, *Phys. Lett.* **B205** (1988) 339.

[Do94] H.G. Dosch, E. Ferreira and A. Krämer, *Phys. Rev.* **D50** (1994) 1992.

[Dr70] S.D. Drell and T.M. Yan, *Phys. Rev. Lett.* **24** (1970) 855.

[Dr83] J.-M. Drouffe and J.-B. Zuber, *Phys. Rept.* **102** (1983) 1.

[Du96] J.E. Duboscq et al. (CLEO Coll.), *Phys. Rev. Lett.* **76** (1996) 3898.

[El76] J. Ellis, M.K. Gaillard and G.G. Ross, *Nucl. Phys.* **B111** (1976) 253.

[El96] R.K. Ellis, W.J. Stirling and B.R. Webber, *QCD and Collider Physics*, (Cambridge University Press, Cambridge, 1996).

[Fa92] A. Falk, M. Neubert and M. Luke, *Nucl. Phys.* **B388** (1992) 363.

[Fi89] R.D. Field, *Applications of Perturbative QCD*, Frontieres in Physics Vol. 77, (Addison-Wesley, Redwood City, Calif., 1989).

[Fl77] E.G. Floratos, D.A. Ross and C.T. Sachrajda, *Nucl. Phys.* **B129** (1977) 66.

[Fl79] E.G. Floratos, D.A. Ross and C.T. Sachrajda, *Nucl. Phys.* **B152** (1979) 493.

[Fl99] M. Fleischer (H1 Coll.), *Measurement of Neutral and Charged Current Cross Sections at High Q^2* in *Proceedings of the 29th International Conference on High Energy Physics*, eds. A. Astbury, D. Axen and J. Robinson (Singapore, etc., 1999) p. 803.

[Fo97] J.R. Forshaw and D.A. Ross, *Quantum Chromodynamics and the Pomeron*, (Cambridge University Press, Cambridge, 1997).

[Fr72a] H. Fritzsch and M. Gell-Mann, *Current Algebra: Quarks and what else?* in *Proceedings of the XVIth International Conference on High Energy Physics*, eds. J.D. Jackson and A. Roberts (Chicago, 1972) p. 135.

[Fr73] H. Fritzsch, M. Gell-Mann and H. Leutwyler, *Phys. Lett.* **47B** (1973) 365.

[Fu80a] W. Furmanski and R. Petronzio, *Phys. Lett.* **97B** (1980) 437.

[Fu82] W. Furmanski and R. Petronzio, *Z. Phys.* **C11** (1982) 293.

[Ga84] J. Gasser and H. Leutwyler, *Ann. Phys.* **158** (1984) 142.

[Ga85] J. Gasser and H. Leutwyler, *Nucl. Phys.* **B250** (1985) 465.

[Ga00] J. Garden et al. (ALPHA and UKQCD Coll.), *Nucl. Phys.* **B571** (2000) 237.

[Ge72] M. Gell-Mann, *Quarks* in *Elementary Particle Physics, Multiparticle Aspects,* Acta Physica Austriaca, Suppl. IX, ed. P. Urban (Wien, New York, 1972) p. 733.

[Ge90] H. Georgi, *Phys. Lett.* **B240** (1990) 447.

[Gi50] V.L. Ginzburg and L.D. Landau, *JETP (USSR)* **20** (1950) 1064.

[Gl78] M. Glück and E. Reya, *Nucl. Phys.* **B145** (1978) 24.

[Gl98] M. Glück, E. Reya and A. Vogt, *Eur. Phys. J.* **C5** (1998) 461.

[Go58] L.P. Gor'kov, *Sov. Phys. JETP* **7** (1958) 505.

[Go79] A. Gonzales-Arroyo, C. Lopez and F.J. Yndurain, *Nucl. Phys.* **B153** (1979) 161;
A. Gonzales-Arroyo and C. Lopez, *Nucl. Phys.* **B166** (1980) 429.

[Go91] S.G. Gorishny, A.L. Kataev and S.A. Larin, *Phys. Lett.* **B259** (1991) 144.

[Go97] S. Gottlieb, *Nucl. Phys. B (Proc. Suppl.)* **53** (1997) 155.

[Gr64] O.W. Greenberg, *Phys. Rev. Lett.* **13** (1964) 598.

[Gr72] V.N. Gribov and L.N. Lipatov, *Sov. J. Nucl. Phys.* **15** (1972) 438 and 675.

[Gr73a] D.J. Gross and F. Wilczek, *Phys. Rev.* **D8** (1973) 3633 and **D9** (1974) 980.

[Gr90] B. Grinstein, *Nucl. Phys.* **B339** (1990) 253.

[Ha65] M.Y. Han and Y. Nambu, *Phys. Rev.* **139B** (1965) 1006.

[Ha81a] H. Hamber and G. Parisi, *Phys. Rev. Lett.* **47** (1981) 1792.

[Ha91] R. Hamberg, W.L. van Neerven and T. Matsuura, *Nucl. Phys.* **B359** (1991) 343.

[Ha92a] R. Hamberg and W.L. van Neerven, *Nucl. Phys.* **B379** (1992) 143.

[Ha98b] A. Hart and M. Teper, *Phys. Rev.* **D58** (1998) 014504.

[Hu81] B. Humpert and W.L. van Neerven, *Nucl. Phys.* **B184** (1981) 225.

[Is82] K. Ishikawa, M. Teper and G. Schierholz, *Phys. Lett.* **110B** (1982) 399.

[Is89] N. Isgur and M. Wise, *Phys. Lett.* **232B** (1989) 113 and **237B** (1989) 527.

[Is92] N. Isgur and M. Wise, *Heavy Quark Symmetry* in *Heavy Flavours*, eds. A. Buras and M. Lindner (World Scientific, 1992) p. 234.

[Jo76] S.D. Joglekar and B.W. Lee, *Ann. Phys.* **97** (1976) 160.

[Ka69] L.P. Kadanoff, *Phys. Rev. Lett.* **23** (1969) 1430.

[Ka00] F. Karsch, *Nucl. Phys. B (Proc. Suppl.)* **B83-84** (2000) 14.

[Ko90] E.W. Kolb and M.S. Turner, *The Early Universe*, (Addison Wesley, Redwood City, 1990).

[Ko98] K.-I. Kondo, *Phys. Rev.* **D57** (1998) 7467.

[Kr84] G. Kramer, *Theory of Jets in Electron-Positron Annihilation*, Springer Tracts in Modern Physics 102, (Springer, Berlin, etc., 1984).

[Kr87] A.S. Kronfeld, G. Schierholz and U.-J. Wiese, *Nucl. Phys.* **B293** (1987) 461.

[Kr99] A.D. Krisch, *Summary of the 13th International Symposium on High Energy Spin Physics* in *Proceedings of the 13th International Symposium on High Energy Spin Physics*, eds. N.E. Tyurin et al. (Singapore, etc., 1999) p. 268.

[Ku00] Y. Kuramashi, *Nucl. Phys. B (Proc. Suppl.)* **B83-84** (2000) 24.

[La80] L.D. Landau, E.M. Lifschitz and L.P. Pitaevskii, *Statistical Physics, Part 2, Sect. 45*, (Oxford, Pergamon Press, 1980).

[La97] S.A. Larin, P. Nogueira, T. van Ritbergen and J.A.M. Vermaseren, *Nucl. Phys.* **B492** (1997) 338.

[La00] H.L. Lai et al. (CTEQ Coll.), *Eur. Phys. J.* **C12** (2000) 375.

[Le84] H. Leutwyler and M. Roos, *Z. Phys.* **C25** (1984) 91.

[Le88] E. Leader and M. Anselmino, *Z. Phys.* **C41** (1988) 239.

[Le91] H. Leutwyler, *Chiral Effective Lagrangians* in *Recent Aspects of Quantum Fields*, eds. H. Mitter and H. Gausterer (Berlin, etc., 1991) p. 1.

[Le93a] P. Lepage and P.B. Mackenzie, *Phys. Rev.* **D48** (1993) 2250.

[Le94] H. Leutwyler, *Ann. Phys.* **235** (1994) 165.

[Le96] H. Leutwyler, *Light Quark Effective Theory* in *Effective Theories and Fundamental Interactions*, Proceedings of the International School of Subnuclear Physics, Erice, 1996, ed. A. Zichichi (Singapore, 1996) p. 53.

[Le96a] H. Leutwyler, *Phys. Lett.* **B378** (1996) 313.

[Li75] L.N. Lipatov, *Sov. J. Nucl. Phys.* **20** (1975) 95.

[Ll78] C.H. Llewellyn Smith, *Jets and QCD* in *Facts and Prospects in Gauge Theories*, Acta Physica Austriaca, Suppl. XIX, ed. P. Urban (Wien, New York, 1978) p. 331.

[Lo50] F. London, *Superfluids, Vol. 1*, (New York, John Wiley & Sons, 1950).

[Lu80b] M. Lüscher, K. Symanzik and P. Weisz, *Nucl. Phys.* **B173** (1980) 365.

[Lu90] M. Luke, *Phys. Lett.* **B252** (1990) 447.

[Lu93] M. Lüscher, R. Sommer, P. Weisz and U. Wolff, *Nucl. Phys.* **B389** (1993) 247.

[Lu99] M. Lüscher, *Advanced Lattice QCD* in *Probing the Standard Model of Particle Interactions*, Les Houches 1997, eds. R. Gupta et al. (Amsterdam, 1999) p. 229.

[Ma76] S. Mandelstam, *Phys. Rept.* **23** (1976) 307.

[Ma84] W. Marciano, *Phys. Rev.* **D29** (1984) 580.

[Ma90] T. Matsuura, R. Hamberg and W.L. van Neerven, *Nucl. Phys.* **B345** (1990) 345.

[Ma97] T. Mannel, *Rept. Prog. Phys.* **60** (1997) 1113.

[Ma97a] P.B. Mackenzie, *Nucl. Phys. B (Proc. Suppl.)* **53** (1997) 23.

[Ma98] A.D. Martin, R.G. Roberts, W.J. Stirling and R.S. Thorne, *Eur. Phys. J.* **C4** (1998) 463.

[Ma00] A.D. Martin, R.G. Roberts, W.J. Stirling and R.S. Thorne, *Eur. Phys. J.* **C14** (2000) 133.

[Me96] R. Mertig and W.L. van Neerven, *Z. Phys.* **C70** (1996) 637.

[Mi89] C. Michael and M. Teper, *Nucl. Phys.* **B314** (1989) 347.

[Mo00] S. Moch and J.A.M. Vermaseren, *Nucl. Phys.* **B573** (2000) 853.

[Mu81] G. Münster, *Nucl. Phys.* **B180** (1981) 23.

[Na81] O. Nachtmann and W. Wetzel, *Nucl. Phys.* **B187** (1981) 333.

[Na82] S. Narison, *Phys. Rept.* **84** (1982) 263.

[Ne94] M. Neubert, *Phys. Rept.* **245** (1994) 259.

[Ne98a] M. Neubert, *B decays and the Heavy Quark Expansion* in *Heavy Flavours II*, eds. A.J. Buras and M. Lindner (World Scientific, 1998) p. 239.

[Ni73] H.B. Nielsen and P. Olesen, *Nucl. Phys.* **B61** (1973) 45.

[Os78] K. Osterwalder and E. Seiler, *Ann. Phys.* **110** (1978) 440.

[Pa69] R.D. Parks (ed.), *Superconductivity*, (New York, etc., Marcel Dekker, 1969).

[Pa80] G. Parisi, *Phys. Lett.* **90B** (1980) 295.

[Pa81] G. Parisi, *Recent Progress in Gauge Theories* in *High-Energy Physics 1980 (XXth Int. Conf. Madison)*, eds. L. Durand and L.G. Pondrom (New York, 1981) p. 1531.

[PDG00] D.E. Groom et al. (Particle Data Group), *Eur. Phys. J.* **C15** (2000) 1.

[Pe00] R. Petronzio, *Nucl. Phys. B (Proc. Suppl.)* **B83-84** (2000) 136.

[Pi95] A. Pich, *Rept. Prog. Phys.* **58** (1995) 563.

[Po73] H.D. Politzer, *Phys. Rev. Lett.* **30** (1973) 1346.

[Po74] H.D. Politzer, *Phys. Rev.* **14C** (1974) 129.

[Po77] H.D. Politzer, *Nucl. Phys.* **B129** (1977) 301.

[Ri97] T. van Ritbergen, J.A.M Vermaseren and S.A. Larin, *Phys. Lett.* **B400** (1997) 379.

[Ri99] A. Ringwald and F. Schrempp, *Phys. Lett.* **B459** (1999) 249.

[Ri00] A. Ringwald and F. Schrempp, *Theory and Phenomenology of Instantons at HERA* in *New Trends in HERA Physics*, eds. G. Grindhammer, B. Kniehl and G. Kramer (Berlin, 2000) p. 203.

[Sa78] C.T. Sachrajda, *Phys. Lett.* **73B** (1978) 185 and **76B** (1978) 100.

[Sa99] J. Santiago and F.J. Yndurain, *Nucl. Phys.* **B563** (1999) 45.

[Sa00a] H. Satz, *Rept. Prog. Phys.* **63** (2000) 1511.

[Sc62] J. Schwinger, *Phys. Rev.* **125** (1962) 397, **128** (1962) 2425.

[Sc00] K. Schilling, *Nucl. Phys. B (Proc. Suppl.)* **B83-84** (2000) 140.

[Sh97] J. Shigemitsu, *Nucl. Phys. B (Proc. Suppl.)* **53** (1997) 16.

[Sh99] E. Shuryak, hep-ph/9909458.

[Sm98] D.A. Smith and M.J. Teper (UKQCD Coll.), *Phys. Rev.* **D58** (1998) 014505.

[So94] R. Sommer, *Nucl. Phys.* **B411** (1994) 839.

[St77] G. Sterman and S. Weinberg, *Phys. Rev. Lett.* **39** (1977) 1436.

[St95] G. Sterman et al., *Rev. Mod. Phys.* **67** (1995) 67.

[Su91] L.R. Surguladze and M.A. Samuel, *Phys. Rev. Lett.* **66** (1991) 560.

[Su95] H. Suganuma, S. Sasaki and H. Toki, *Nucl. Phys.* **B435** (1995) 207.

[Ta81] E. Tarrach, *Nucl. Phys.* **B183** (1981) 384.

[Ta82] O.V. Tarasov, Dubna report JINR-P2-82-900 (1982).

[tH76b] G. 't Hooft, *Gauge Theories with Unified Weak, Electromagnetic and Strong Interactions* in *High Energy Physics*, ed. A. Zichichi (Bologna, 1976) p. 1225.

[tH81] G. 't Hooft, *Nucl. Phys.* **B190** (1981) 455.

[tH82] G. 't Hooft, *Physica Scripta* **25** (1982) 133.

[tH85] G. 't Hooft, *Nucl. Phys.* **B254** (1985) 11.

[Uk93] A. Ukawa, *Nucl. Phys. B (Proc. Suppl.)* **30** (1993) 3.

[vB96] P. van Baal, *Global Issues in Gauge Fixing* in *Non-pertubative Approaches to Quantum Chromodynamics*, ed. D. Diakonov (Gatchina, 1996) p. 4.

[vH97] B. van den Heuvel, *Nucl. Phys.* **B488** (1997) 282.

[vN91] W.L. van Neerven and E.B. Zijlstra, *Phys. Lett.* **B272** (1991) 127.

[vN92] W.L. van Neerven and E.B. Zijlstra, *Nucl. Phys.* **B382** (1992) 11.

[vN00] W.L. van Neerven and A. Vogt, *Nucl. Phys.* **B568** (2000) 263.

[Ve97] J.A.M Vermaseren, S.A. Larin and T. van Ritbergen, *Phys. Lett.* **B405** (1997) 327.

[Vo96] W. Vogelsang, *Nucl. Phys.* **B475** (1996) 47.

[We66] S. Weinberg, *Phys. Rev. Lett.* **17** (1966) 616.

[We79] S. Weinberg, *Physica* **A96** (1979) 327.

[We82] D. Weingarten, *Phys. Lett.* **109B** (1982) 57.

[We97] D. Weingarten, *Nucl. Phys. B (Proc. Suppl.)* **53** (1997) 232.

[Wi69] K.G. Wilson, *Phys. Rev.* **179** (1969) 1499.

[Wi74] K.G. Wilson, *Phys. Rev.* **D10** (1974) 2445.

[Wu84] S.L. Wu, *Phys. Rept.* **C107** (1984) 59.

[Ya00] U.K. Yang and A. Bodek, *Eur. Phys. J.* **C13** (2000) 241.

[Yn99] F.J. Yndurain, *The Theory of Quark and Gluon Interactions*, (Springer Verlag, Berlin, etc., 1999) (third edition).

[Yo98] T. Yoshié, *Nucl. Phys. B (Proc. Suppl.)* **63** (1998) 3.

[Ze73] A. Zee, *Phys. Rev.* **D8** (1973) 4038.

[Zi73] W. Zimmermann, *Ann. Phys.* **77** (1973) 536 and 570.

[Zi91] E.B. Zijlstra and W.L. van Neerven, *Phys. Lett.* **B273** (1991) 476 and **B297** (1992) 377.

[Zi92] E.B. Zijlstra and W.L. van Neerven, *Nucl. Phys.* **B383** (1992) 525.

4 Gauge theories of the electroweak interaction

As discussed in Sect. 1.4, the electroweak interactions of fermions exhibit a symmetry under the group $SU(2)_w \times U(1)_Y$. Promoting this symmetry to a gauge symmetry yields a gauge theory of the electroweak interaction. However, in contrast to the gauge bosons of the electromagnetic and of the strong interaction, the electroweak gauge bosons are massive. Therefore, the formulation of a gauge theory for the electroweak interaction requires a new concept, *spontaneous symmetry breaking*. By applying this concept to gauge theories we arrive at the only known form for a renormalizable quantum field theory for the weak interaction with massive gauge bosons.

Upon combining the $SU(2)_w \times U(1)_Y$ gauge theory with spontaneous symmetry breaking we construct the Standard Model of the electroweak interaction [We67, Sa68, Gl70]. This theory is in accordance with all known experimental facts about the electroweak interaction.[1] As a remnant of spontaneous symmetry breaking it predicts a physical scalar boson, the Higgs boson [Hi64, En64, Gu64, Ki67], which has not been detected experimentally so far. However, its mass can be confined to a limited range by comparing the Electroweak Standard Model with precision data.

In Sect. 4.1 we explain spontaneous breaking of global and local symmetries as well as the quantization and renormalization of quantum field theories with spontaneously broken symmetries. In Sect. 4.2 we introduce the Lagrangian of the Standard Model of the electroweak interaction. Simple applications of this model are given in Sect. 4.3. Its quantization and renormalization are performed in Sects. 4.4 and 4.5, respectively. Finally, the electroweak radiative corrections for various processes are discussed in Sect. 4.6.

[1] Neutrino oscillations can in principle be described within the Standard Model by adding right-handed neutrinos (cf. Sect. 5).

4.1 Spontaneous symmetry breaking

Exact pure gauge symmetry demands that all gauge fields are massless. On the other hand, we know from experiment that the electroweak interaction is mediated, apart from the photon, by massive vector particles. Explicit mass terms $M^2 A_\mu A^\mu/2$ must not be added to the Lagrangian as they violate gauge invariance. This, however, is crucial for the Slavnov–Taylor identities (Sect. 2.4.3) which are needed for the proof of renormalizability and unitarity as outlined in Sect. 2.5.4. The only known possibility for introducing masses without violating gauge invariance is the *Higgs mechanism* which is based on spontaneous symmetry breaking. Thereby, gauge fields are coupled with additional, postulated scalar fields which, owing to their self-interactions, acquire asymmetric non-vanishing *vacuum expectation values* (vevs), despite the gauge symmetry of the Lagrangian. These scalar fields are also necessary to guarantee tree-level unitarity in a theory with massive vector bosons [Ll73, Co74a]. Some of these scalar fields, the so-called *Higgs fields*, give rise to physically observable scalar particles.

In order to define the meaning of *spontaneous symmetry breaking*[2] we consider a Lagrangian that is invariant under a group of transformations G. There are two alternatives for the ground states $|0\rangle$, i.e. the states with lowest energy:

1. Only one unique ground state $|0\rangle$ exists; in this case, $|0\rangle$ must be invariant under transformations with respect to G.
2. There is a set of degenerate ground states which transform into one another under the group G. All these ground states are physically equivalent. However, each of them selects a direction in representation space. Choosing one of them as the ground state of the theory, the G symmetry is said to be spontaneously broken.

Systems with a finite number of degrees of freedom usually possess a unique ground state. The possibility of spontaneous symmetry breaking is typical for theories involving an infinite number of degrees of freedom. The phenomenon is, independently of its application for mass generation in gauge field theories, of distinct physical significance. For instance, the spontaneous breaking of chiral symmetry has been discussed in Sect. 3.5.1.1

[2]The phrase was coined by M. Baker and S.L. Glashow [Ba62].

A well-known example is the infinitely extended ferromagnet below the Curie temperature [He28]: although the interaction of the magnetic moments is symmetric under rotations, in the ground state all moments are aligned in an arbitrary but fixed direction, which causes a spontaneous magnetization $\mathbf{M} \neq \mathbf{0}$. Conversely, rotation symmetry is not spontaneously broken in the phase above the Curie temperature, and thus $\mathbf{M} = \mathbf{0}$. This example exhibits two characteristic features which occur in all the models described below:

1. A symmetric phase and a phase with spontaneous symmetry breaking can (separated by a phase transition) occur in the same system depending on the physical boundary conditions (the temperature of the ferromagnet, in the example).
2. In contrast to the symmetric phase, the phase with spontaneous symmetry breaking is characterized by a non-vanishing *order parameter*; for the ferromagnet this is the magnetization $\mathbf{M}$. In quantized models these are vacuum expectation values.

The Higgs mechanism has already been discussed on the classical level using the example of the Nielsen–Olesen model during discussions on the chromoelectric Meissner effect in QCD (Sects. 3.7.3.1 and 3.7.3.2). After a preparatory section on the breaking of a global symmetry (Sect. 4.1.1), the concept will be extended to include non-abelian gauge symmetry (Sect. 4.1.2). The Higgs mechanism is used to explain dynamical mass generation in the Glashow–Salam–Weinberg model for the electroweak interaction in Sect. 4.2.

An experimentally interesting question is whether Higgs fields are fundamental fields corresponding to physically detectable particles or not (cf. Sect. 4.3.2).

4.1.1 Spontaneous breaking of a global symmetry

Because of the importance of spontaneous symmetry breaking its dynamics is first described using a global symmetry as an example.

4.1.1.1 Goldstone model

The *Goldstone model* [Go61] describes the self-interaction of a complex field $\phi(x)$. Its Lagrangian

$$\mathcal{L} = (\partial^\mu \phi)^\dagger \partial_\mu \phi - V(\phi) \tag{4.1.1}$$

with

$$V(\phi) = \frac{\lambda}{4}\left(|\phi|^2 - \frac{\mu^2}{\lambda}\right)^2 = \frac{\lambda}{4}|\phi|^4 - \frac{\mu^2}{2}|\phi^2| + \frac{\mu^4}{4\lambda} \tag{4.1.2}$$

is invariant under global phase transformations

$$\phi(x) \to e^{i\theta}\phi(x), \qquad \phi^\dagger(x) \to \phi^\dagger(x)e^{-i\theta}, \qquad \theta = \text{constant}. \tag{4.1.3}$$

The coupling constant λ is positive, otherwise the energy would not be bounded from below. On the other hand, there are two possibilities for the sign of μ^2 in (4.1.2):

1. $\mu^2 < 0$: the term involving μ^2 is a mass term, and the field ϕ has the mass $M^2 = -\mu^2$; the potential $V(\phi)$ has a unique, classical vacuum (= field configuration with lowest energy), $\phi_0(x) \equiv 0$.

2. $\mu^2 > 0$: in this case, (4.1.1) describes a phase with spontaneously broken symmetry. The constant fields

$$\phi_0(x) \equiv v e^{i\theta}, \quad v = \sqrt{\frac{\mu^2}{\lambda}}, \quad \theta = \text{constant}, \quad 0 \le \theta < 2\pi, \tag{4.1.4}$$

minimize the potential (cf. Fig. 4.1) and therefore describe configurations of lowest energy. The modulus v of ϕ_0 is the *order parameter*, the phase θ is arbitrary and can be changed by a global gauge transformation (4.1.3).

In order to formulate a perturbation theory for $\mu^2 > 0$ we have to expand the scalar field about one of the ground states (4.1.4). We use the parametrization

$$\phi(x) = \left(v + \frac{1}{\sqrt{2}}\eta(x)\right) \exp\left[i\left(\theta + \frac{\xi(x)}{\sqrt{2}v}\right)\right] \tag{4.1.5}$$

in which $\eta(x)$ and $\xi(x)$ are real fields describing small deviations of the modulus and the phase of the field ϕ from the ground state values (4.1.4). The choice of one of the ground states, i.e. of a particular phase θ, leads

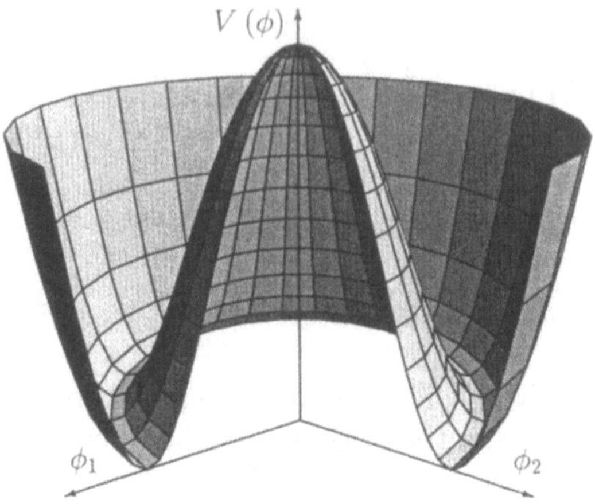

Fig. 4.1 Potential of the Goldstone model in the phase with spontaneous symmetry breaking $\mu^2 > 0$

to *spontaneous symmetry breaking.* Inserting (4.1.5), the Lagrangian (4.1.1) reads

$$\begin{aligned}\mathcal{L} = &\frac{1}{2}(\partial_\mu \xi)\partial^\mu \xi + \frac{1}{2}(\partial_\mu \eta)\partial^\mu \eta - \frac{1}{2}\mu^2\eta^2 - \frac{\lambda v}{2\sqrt{2}}\eta^3 - \frac{\lambda}{16}\eta^4 \\ &+ \frac{1}{2}(\partial_\mu \xi)\partial^\mu \xi \left(\frac{\sqrt{2\lambda}}{\mu}\eta + \frac{\lambda}{2\mu^2}\eta^2 \right), \end{aligned} \tag{4.1.6}$$

and we find:

1. $\eta(x)$ is a scalar field with mass μ; it describes the radial oscillations of ϕ around the vev v (cf. Fig. 4.1). The finite mass results from the non-vanishing radial curvature of the potential.

2. $\xi(x)$ is a massless, scalar field, the *Goldstone field.* It is massless, as a result of the vanishing curvature along the potential minima. The Goldstone field $\xi(x)$ describes the associated normal mode with eigenfrequency zero (cf. Fig. 4.1).

The Goldstone model exhibits a general property of theories with spontaneous breaking of a continuous global symmetry, namely the occurrence of massless, scalar field excitations, the *Goldstone bosons.*

4.1.1.2 Generalization to non-abelian symmetry groups and the Goldstone theorem

The gauge group of the Goldstone model is the commutative group of phase transformations (4.1.3), the group U(1). Its extension to non-abelian groups G is obvious. We consider a multiplet of n real scalar fields $\boldsymbol{\phi}(x) = (\phi_i(x)) = (\phi_1(x), \ldots, \phi_n(x))$ which transforms under a real representation T^a $[(T^a_{ij})^* = -T^a_{ij}, a = 1, \ldots, \dim G]$ of the group G as

$$\begin{aligned} \phi_i(x) &\to [\exp(\mathrm{i}\theta^a T^a)]_{ij}\phi_j(x), \qquad \theta^a = \text{constant}, \\ \delta\phi_i(x) &= \mathrm{i}T^a_{ij}\phi_j(x)\delta\theta^a \end{aligned} \tag{4.1.7}$$

[cf. (1.2.83) and (1.2.84)]. The potential $V(\boldsymbol{\phi})$ of the Lagrangian

$$\mathcal{L} = \frac{1}{2}(\partial^\mu\phi_i)\partial_\mu\phi_i - V(\boldsymbol{\phi}) \tag{4.1.8}$$

is required to be invariant under the gauge transformations (4.1.7),

$$0 = \delta V = \frac{\partial V}{\partial\phi_i}\delta\phi_i = \mathrm{i}\frac{\partial V}{\partial\phi_i}T^a_{ij}\phi_j\delta\theta^a. \tag{4.1.9}$$

Since $\delta\theta^a$ is arbitrary, this implies

$$\frac{\partial V}{\partial\phi_i}T^a_{ij}\phi_j = 0, \qquad a = 1, \ldots, \dim G. \tag{4.1.10}$$

Differentiation with respect to ϕ_k yields

$$\frac{\partial^2 V}{\partial\phi_k\partial\phi_i}T^a_{ij}\phi_j + \frac{\partial V}{\partial\phi_i}T^a_{ik} = 0. \tag{4.1.11}$$

As the example of the Goldstone model showed, spontaneous symmetry breaking occurs when $V(\boldsymbol{\phi})$ has a minimum for $\boldsymbol{\phi}(x) \equiv \mathbf{v} = (v_i) \neq 0$, i.e.

$$\left.\frac{\partial V}{\partial\phi_i}\right|_{\boldsymbol{\phi}=\mathbf{v}} = 0, \qquad \left.\frac{\partial^2 V}{\partial\phi_k\partial\phi_i}\right|_{\boldsymbol{\phi}=\mathbf{v}} = (M^2)_{ki}, \quad M^2 \geq 0. \tag{4.1.12}$$

The symmetric matrix M^2 formed from the second derivatives of V is positive semi-definite in the minimum, i.e. it has non-negative eigenvalues.

Because of the invariance of $V(\phi)$ under the transformations (4.1.7), also $[\exp(\mathrm{i}\theta^a T^a)]_{ij} v_j$ is a field configuration with minimal energy.

In perturbation theory, deviations from the vacuum field configuration $\phi(x) = \mathbf{v} =$ constant are considered in analogy to the Goldstone model, and hence fields $\tilde{\phi}_i(x) = \phi_i(x) - v_i$ are introduced. Expansion of the Lagrangian (4.1.8) in a power series in $\tilde{\phi}_i$ up to second order yields

$$\mathcal{L} = \frac{1}{2}[\partial_\mu \tilde{\phi}_i(x)]\partial^\mu \tilde{\phi}_i(x) - \frac{1}{2}(M^2)_{ki}\tilde{\phi}_k(x)\tilde{\phi}_i(x) + \dots . \tag{4.1.13}$$

The appearance of $(M^2)_{ik}$ in the bilinear terms provides the physical interpretation of $(M^2)_{ik}$ as the matrix of mass squares for the fields $\tilde{\phi}_i(x)$. In order to obtain information on the eigenvalues of M^2 we put $\phi_i(x) \equiv v_i$ in (4.1.11) resulting in

$$(M^2)_{ki} T^a_{ij} v_j = 0, \qquad a = 1, \dots, \dim G. \tag{4.1.14}$$

In general, $\mathbf{v}$ is invariant under a subgroup H of G, the stability group. If we choose an appropriate basis for the generators T^a of G we can write the generators of the Lie algebra of H as $\hat{T}^{\hat{a}} \equiv T^{\hat{a}}$, $\hat{a} = 1, \dots, \dim H$, and the invariance of $\mathbf{v}$ under transformations from H means

$$\hat{T}^{\hat{a}}_{ij} v_j = 0, \qquad \hat{a} = 1, \dots, \dim H. \tag{4.1.15}$$

The remaining linearly independent generators, $\check{T}^{\check{a}} \equiv T^{\hat{a}}$, $\check{a} = \dim H + 1, \dots, \dim G$, which span the coset space G/H, obey

$$\check{T}^{\check{a}}_{ij} v_j \neq 0, \qquad \check{a} = \dim H + 1, \dots, \dim G, \tag{4.1.16}$$

As a result, $\dim H$ of the $\dim G$ equations (4.1.14) are trivially satisfied. However, the remaining $\dim G - \dim H$ equations force eigenvalues zero for M^2. The result of this consideration is the *Goldstone theorem* [Go61, Go62, Bl63]: the matrix of mass squares M^2 has $\dim G - \dim H$ eigenvalues zero, one for each broken generator. The associated field excitations are called *Goldstone fields* and result from spontaneous symmetry breaking. According to (4.1.14), the Goldstone fields are linear combinations of $\check{T}^{\check{a}}_{ij} v_j$ The remaining eigenvalues of M^2 are positive.

This result can be directly generalized to complex unitary representations by noting the *relation between complex and real representations.* Any multiplet of complex fields ψ that transforms according to some complex representation of G with hermitian generators T^a can be rewritten in terms of a multiplet of real fields

$$\phi = \begin{pmatrix} \operatorname{Re}\psi \\ \operatorname{Im}\psi \end{pmatrix} \tag{4.1.17}$$

that transforms according to a real representation with the generators

$$\mathrm{i}\, t^a = \begin{pmatrix} -\operatorname{Im} T^a & -\operatorname{Re} T^a \\ \operatorname{Re} T^a & -\operatorname{Im} T^a \end{pmatrix}. \tag{4.1.18}$$

4.1.1.3 Linear σ model

An important example for a spontaneously broken non-abelian symmetry is provided by the linear σ model. The fields ϕ_i form N-component, real vectors which are transformed according to the group SO(N), the rotation group in N dimensions. The invariant Lagrangian can be written as [cf. (4.1.1) and (4.1.2)]

$$\mathcal{L} = \frac{1}{2}(\partial_\mu \phi_i)\partial^\mu \phi_i - V(\phi), \qquad V(\phi) = \frac{\lambda}{4}\left(\phi_i\phi_i - \frac{\mu^2}{\lambda}\right)^2. \tag{4.1.19}$$

The classical configuration with lowest energy is a constant field $\phi_0(x) \equiv \mathbf{v}$ that minimizes the potential $V(\phi)$

$$\phi_{i0}\phi_{i0} = v_i v_i = \frac{\mu^2}{\lambda} = v^2 > 0, \tag{4.1.20}$$

i.e. the fields that minimize V form a $(N-1)$-dimensional sphere with radius v in N-dimensional euclidean ϕ space. It is conventional to choose the field that points into the Nth direction

$$\phi_0 = \mathbf{v} = (0, \ldots, 0, v). \tag{4.1.21}$$

as ground state. This is invariant under the orthogonal transformations of the first $N-1$ coordinates, i.e. under the subgroup $H = \mathrm{SO}(N-1)$. Hence $\dim G = N(N-1)/2$, $\dim H = (N-1)(N-2)/2$, and $\dim G - \dim H = N-1$.

According to the Goldstone theorem, there are $N-1$ massless fields $\boldsymbol{\pi} = (\pi_1, \ldots, \pi_{N-1})$ and one massive field σ. Writing $\boldsymbol{\phi} = (\boldsymbol{\pi}, \sigma + v)$, the Lagrangian (4.1.19) is expressed as

$$\begin{aligned}\mathcal{L} = &\frac{1}{2}(\partial_\mu \sigma)\partial^\mu \sigma - \frac{1}{2}(2\mu^2)\sigma^2 + \frac{1}{2}(\partial_\mu \boldsymbol{\pi})\partial^\mu \boldsymbol{\pi} \\ &- \frac{\lambda}{4}(\boldsymbol{\pi}^2 + \sigma^2)^2 - \lambda v \sigma (\boldsymbol{\pi}^2 + \sigma^2).\end{aligned} \tag{4.1.22}$$

The mass of the field σ is $\sqrt{2}\mu$.

For $N = 4$, the group SO(4) is locally isomorphic to the group SU(2)×SU(2). The chiral SU(2) × SU(2) symmetry is the classical example of a broken symmetry in elementary particle physics (cf. Sect. 3.5.1). The quanta of the three Goldstone fields can be interpreted as pions and those of the massive field as σ meson. This can be formulated within the so-called σ model (cf. Sect. 3.5.3).

4.1.1.4 Goldstone theorem for quantized theories

So far, the models were considered as classical field theories. They represent the lowest order of a semi-classical approximation for the corresponding quantized theories according to Sect. 2.2.5.3. We now show that the Goldstone theorem is also valid for the *quantized model.* This could be proven by analysing the matrix elements of the charge operator which generates the internal, spontaneously broken symmetry [Gi64, Be74a]. Instead we use the path-integral quantization (Sect. 2.4) [Jo64] and consider the generating functional of general Green functions $[\mathbf{J} = (J_i),\ \boldsymbol{\phi} = (\phi_i)]$,

$$\begin{aligned}T\{\mathbf{J}\} &= Z\{\mathbf{J}\}/Z\{0\}, \\ Z\{\mathbf{J}\} &= \int \mathcal{D}[\boldsymbol{\phi}] \exp\left(\mathrm{i} \int \mathrm{d}^4 x \, [\mathcal{L} + J_i(x)\phi_i(x)]\right)\end{aligned} \tag{4.1.23}$$

with $\mathcal{L}$ from (4.1.8). We perform an infinitesimal gauge transformation in the path integral $Z\{\mathbf{J}\}$ in the same way as for the derivation of the Ward identities in Sects. 2.2.6 and 2.4.3.2, $\phi_i(x) \to \phi_i(x) + \delta\phi_i(x)$ with $\delta\phi_i(x)$ according to (4.1.7). As a result of the gauge invariance of $\mathcal{L}$ and $\mathcal{D}[\boldsymbol{\phi}]$, we obtain

$$\int \mathrm{d}^4 x \, J_i(x) T^a_{ik} \frac{\delta T\{\mathbf{J}\}}{\mathrm{i}\delta J_k(x)} = 0, \tag{4.1.24}$$

or, in terms of the generating functional of connected Green functions $T_c\{\mathbf{J}\} = \log T\{\mathbf{J}\}$ [cf. (2.2.31)],

$$\int \mathrm{d}^4x\, J_i(x) T^a_{ik} \frac{\delta T_c\{\mathbf{J}\}}{\mathrm{i}\delta J_k(x)} = 0. \tag{4.1.25}$$

The Legendre transformation to the generating functional of the vertex functions [cf. (2.2.35)–(2.2.36)] has the form

$$\begin{aligned} \mathrm{i}\Gamma\{\boldsymbol{\phi}\} &= -\mathrm{i} \int \mathrm{d}^4x\, J_i(x)\phi_i(x) + T_c\{\mathbf{J}\}, \\ \phi_i(x) &= \frac{\delta T_c\{\mathbf{J}\}}{\mathrm{i}\delta J_i(x)}, \qquad J_i(x) = -\frac{\delta\Gamma\{\boldsymbol{\phi}\}}{\delta\phi_i(x)}, \end{aligned} \tag{4.1.26}$$

and (4.1.25) results in

$$\int \mathrm{d}^4x\, \frac{\delta\Gamma\{\boldsymbol{\phi}\}}{\delta\phi_i(x)} T^a_{ik}\phi_k(x) = 0. \tag{4.1.27}$$

For spontaneous symmetry breaking, the field remains constant but differs from zero in the ground state, i.e.

$$v_i = \langle 0|\phi_i(x)|0\rangle = \left.\frac{\delta T_c\{\mathbf{J}\}}{\mathrm{i}\delta J_i(x)}\right|_{\mathbf{J}\equiv 0} \neq 0 \tag{4.1.28}$$

for some i, or equivalently

$$\left.\frac{\delta\Gamma\{\boldsymbol{\phi}\}}{\delta\phi_i(x)}\right|_{\boldsymbol{\phi}\equiv\mathbf{v}} = 0. \tag{4.1.29}$$

The last equation is a consequence of (4.1.26) and (4.1.28) because for $\mathbf{J} \equiv 0$, $\boldsymbol{\phi}$ takes the value $\mathbf{v}$ and vice versa. The inverse propagator of the field ϕ then follows from (2.2.41),

$$-\left[\frac{\delta^2 T_c\{\mathbf{J}\}}{\mathrm{i}\delta J_i(x)\mathrm{i}\delta J_k(y)}\right]^{-1}\Bigg|_{\mathbf{J}\equiv 0} = \left.\frac{\mathrm{i}\delta^2\Gamma\{\boldsymbol{\phi}\}}{\delta\phi_i(x)\phi_k(y)}\right|_{\boldsymbol{\phi}=\mathbf{v}} = -(G^{-1})_{ik}(x-y). \tag{4.1.30}$$

Differentiating (4.1.27) with respect to ϕ_k, putting ϕ_i equal to v_i, and performing the transformation to momentum space yields with (4.1.29) and (4.1.30)

$$(G^{-1})_{li}(p=0)T^a_{ik}v_k = 0, \tag{4.1.31}$$

i.e. the same formula as in the classical approximation (4.1.14) which was used as the starting point for the proof of the Goldstone theorem. This means that massless Goldstone fields occur also in quantized theories with spontaneously broken global symmetry.

4.1.1.5 Renormalization of the σ model

The Lagrangian of the σ model is renormalizable in the sense of power counting. It turns out that the renormalization of the spontaneously broken case is related to the renormalization of the symmetric case (cf. [Le72b] and references therein).

Consider the potential in (4.1.19). The case with spontaneous symmetry breaking, $\mu^2 > 0$, can be obtained from the symmetric case, $\mu^2 < 0$, by a continuous variation of μ^2. Unfortunately this procedure involves a transition through the singular point $\mu^2 = 0$, in which physical quantities are not analytic in μ^2. This problem can be circumvented by introducing a linear symmetry-breaking term $c_i\phi_i$ in the Lagrangian. After the inclusion of this term, the Goldstone phase can be reached from the symmetric phase by varying μ^2 and $\mathbf{c} = (c_i)$ without passing through singular points.

Consider now the Lagrangian (4.1.19) plus a linear breaking term

$$\mathcal{L}^c = \mathcal{L} + c_i\phi_i(x). \tag{4.1.32}$$

In order to generate the correct perturbation theory we introduce the shifted fields

$$\tilde{\boldsymbol{\phi}} = \boldsymbol{\phi} - \mathbf{v}, \qquad \mathbf{v} = \langle 0|\boldsymbol{\phi}|0\rangle. \tag{4.1.33}$$

The generating functional of Green functions for the fields $\tilde{\phi}_i$ in the theory with symmetry breaking reads

$$T^c\{\mathbf{J}\} = \exp\left(T_c^c\{\mathbf{J}\}\right) = \\ = \frac{\int \mathcal{D}[\tilde{\boldsymbol{\phi}}] \exp\left(\mathrm{i}\int \mathrm{d}^4x\, [\mathcal{L}(\tilde{\boldsymbol{\phi}}+\mathbf{v}) + (J_i + c_i)\tilde{\phi}_i]\right)}{\int \mathcal{D}[\tilde{\boldsymbol{\phi}}] \exp\left(\mathrm{i}\int \mathrm{d}^4x\, [\mathcal{L}(\tilde{\boldsymbol{\phi}}+\mathbf{v}) + c_i\tilde{\phi}_i]\right)}, \tag{4.1.34}$$

where $T_c^c\{\mathbf{J}\}$ is the generating functional of connected Green functions, and the denominator ensures $T_c^c\{0\} = 0$. Performing a substitution $\tilde{\boldsymbol{\phi}} \to \boldsymbol{\phi} - \mathbf{v}$ yields

$$\exp\left(T_c^c\{\mathbf{J}\}\right) = \frac{\int \mathcal{D}[\boldsymbol{\phi}] \exp\left(\mathrm{i}\int \mathrm{d}^4x\, [\mathcal{L}(\boldsymbol{\phi}) + (J_i + c_i)(\phi_i - v_i)]\right)}{\int \mathcal{D}[\boldsymbol{\phi}] \exp\left(\mathrm{i}\int \mathrm{d}^4x\, [\mathcal{L}(\boldsymbol{\phi}) + c_i(\phi_i - v_i)]\right)}. \tag{4.1.35}$$

This allows us to express $T_c^c\{\mathbf{J}\}$ in terms of the corresponding functional for $\mathbf{c} = 0$, i.e. $T_c\{\mathbf{J}\} = \log T\{\mathbf{J}\}$ with $T\{\mathbf{J}\}$ from (4.1.23),

$$T_c^c\{\mathbf{J}\} = T_c\{\mathbf{J}+\mathbf{c}\} - T_c\{\mathbf{c}\} - \mathrm{i}\int \mathrm{d}^4x\, J_i v_i. \tag{4.1.36}$$

Note that v_i is determined by the condition that the derivative of T_c^c with respect to $\mathbf{J}$ vanishes at $\mathbf{J} \equiv 0$,

$$\langle 0|\tilde{\phi}_i(x)|0\rangle = \left.\frac{\delta T_c^c\{\mathbf{J}\}}{\mathrm{i}\delta J_i(x)}\right|_{\mathbf{J}\equiv 0} = 0, \tag{4.1.37}$$

i.e.

$$v_i = \left.\frac{\delta T_c\{\mathbf{J}\}}{\mathrm{i}\delta J_i(x)}\right|_{\mathbf{J}\equiv 0}. \tag{4.1.38}$$

A Legendre transformation yields

$$\mathrm{i}\Gamma^c\{\tilde{\boldsymbol{\phi}}\} = T_c^c\{\mathbf{J}\} - \mathrm{i}\int \mathrm{d}^4x\, J_i\tilde{\phi}_i \\ = T_c\{\mathbf{J}+\mathbf{c}\} - T_c\{\mathbf{c}\} - \mathrm{i}\int \mathrm{d}^4x\, J_i(\tilde{\phi}_i + v_i) \tag{4.1.39}$$

for the corresponding vertex functional with

$$\tilde{\phi}_i = \frac{\delta T^c_{\rm c}\{\mathbf{J}\}}{\mathrm{i}\delta J_i(x)} = \frac{\delta T_{\rm c}\{\mathbf{J}+\mathbf{c}\}}{\mathrm{i}\delta J_i(x)} - v_i. \tag{4.1.40}$$

Using (4.1.26) we find

$$\Gamma^c\{\tilde{\boldsymbol{\phi}}\} = \Gamma\{\tilde{\boldsymbol{\phi}}+\mathbf{v}\} - \Gamma\{\mathbf{v}\} + \int \mathrm{d}^4x\, c_i\tilde{\phi}_i \tag{4.1.41}$$

for the vertex functional with

$$c_i = -\left.\frac{\delta\Gamma\{\boldsymbol{\phi}\}}{\delta\phi_i(x)}\right|_{\boldsymbol{\phi}(x)\equiv\mathbf{v}}. \tag{4.1.42}$$

Expanding (4.1.41) in $\tilde{\boldsymbol{\phi}}$ leads to ($n > 1$)

$$\Gamma^c_{i_1\ldots i_n}(x_1,\ldots,x_n) = \tag{4.1.43}$$

$$\sum_{m=0}^{\infty}\frac{1}{m!}\int\left(\prod_{k=1}^{m}\mathrm{d}^4y_k\, v_{j_k}\right)\Gamma_{i_1\ldots i_n j_1\ldots j_m}(x_1,\ldots,x_n,y_1,\ldots y_m),$$

i.e. the vertex functions for the case with broken symmetry are obtained by re-summing vertex functions in the symmetric theory with external legs contracted with vevs. In this way, in particular, mass terms in the two-point functions result from the vertex functions with more external legs. Note that Γ^c are Green functions for fluctuations around the asymmetric vacuum.

In terms of Γ^c the identity (4.1.27) reads

$$\int \mathrm{d}^4x\left[\frac{\delta\Gamma^c\{\tilde{\boldsymbol{\phi}}\}}{\delta\tilde{\phi}_i(x)} - c_i\right]T^a_{ik}\left[\tilde{\phi}_k(x)+v_k\right] = 0. \tag{4.1.44}$$

Differentiating with respect to $\tilde{\phi}_j$, putting $\tilde{\boldsymbol{\phi}} = 0$, using

$$\left.\frac{\delta\Gamma^c\{\tilde{\boldsymbol{\phi}}\}}{\delta\tilde{\phi}_i(x)}\right|_{\tilde{\boldsymbol{\phi}}=0} = 0, \tag{4.1.45}$$

i.e. the vanishing of the $\tilde{\phi}$-field one-point function or *tadpole*, and transforming to momentum space yields

$$T^a_{ik}\left[v_k\Gamma^{c,\tilde{\phi}\tilde{\phi}}_{ji}(0) - c_i\delta_{kj}\right] = T^a_{ik}\left[v_k\Gamma^{c,\tilde{\phi}\tilde{\phi}}_{ji}(0) + c_k\delta_{ji}\right] = 0 \qquad (4.1.46)$$

for the inverse propagator $\Gamma^{c,\tilde{\phi}\tilde{\phi}}_{ij}(p^2)$ at zero momentum, where we used the antisymmetry of the matrices T^a_{ik} in the last equality. This generalizes (4.1.31) to $\mathbf{c} \neq 0$. Because $\Gamma^{c,\tilde{\phi}\tilde{\phi}}_{ij}$ is proportional to δ_{ij} in the space transverse to $\mathbf{c}$, where the symmetry is unbroken, (4.1.46) implies that $\mathbf{c}$ and $\mathbf{v}$ are collinear. One can define a mass parameter for the (with respect to $\mathbf{c}$) transverse states according to

$$\Gamma^{c,\tilde{\phi}\tilde{\phi}}_{\mathrm{T}}(0) = -m^2_{\mathrm{T}}, \qquad \mathbf{c} = m^2_{\mathrm{T}}\mathbf{v}. \qquad (4.1.47)$$

For $\mathbf{c} = 0$ this implies $m^2_{\mathrm{T}} = 0$, i.e. the Goldstone theorem, which is thus proven to all orders in perturbation theory.

Equation (4.1.41) expresses the vertex functional of the theory with broken symmetry in terms of the vertex functional of the symmetric theory. Thus, the renormalization of the symmetric theory implies the renormalization of the broken one. The renormalized functional of the symmetric theory is related to the bare one as

$$\Gamma_{\mathrm{R}}(\boldsymbol{\phi}, -\mu^2, \lambda) = \Gamma(\boldsymbol{\phi}_0, -\mu_0^2, \lambda_0) = \Gamma(Z^{1/2}\boldsymbol{\phi}, -\mu^2 - \delta\mu^2, Z_\lambda\lambda). \qquad (4.1.48)$$

Then (4.1.41) implies for the renormalized functional of the broken case

$$\begin{aligned}\Gamma^c_{\mathrm{R}}(\tilde{\boldsymbol{\phi}}, -\mu^2, \lambda, \mathbf{c}, \mathbf{v}) &= \Gamma^c(\tilde{\boldsymbol{\phi}}_0, -\mu_0^2, \lambda_0, \mathbf{c}_0, \mathbf{v}_0) \qquad (4.1.49)\\ &= \Gamma^c(Z^{1/2}\tilde{\boldsymbol{\phi}}, -\mu^2 - \delta\mu^2, Z_\lambda\lambda, Z^{-1/2}\mathbf{c}, Z^{1/2}\mathbf{v}).\end{aligned}$$

In addition to the renormalization of the parameters of the symmetric theory a renormalization of $\mathbf{v}$ and $\mathbf{c}$ is necessary. The rescaling of $\mathbf{c}$ ensures the invariance of the term $c_i\phi_i$ under renormalization.

Equation (4.1.49) allows us to discuss the renormalization of the case with spontaneous symmetry breaking. By starting with $\mu^2 < 0$, $\mathbf{c} = 0$, we first continue in $\mathbf{c}$ to $\mathbf{c} \neq 0$, then for constant $\mathbf{c} \neq 0$ in μ^2 to $\mu^2 > 0$ and finally in $\mathbf{c}$ to $\mathbf{c} = 0$. In this way (4.1.49) is analytically continued from $\mu^2 < 0$ to $\mu^2 > 0$ without hitting the singularity at $\mu^2 = 0$. Thus, we arrive at the result that the spontaneously broken theory can be renormalized by the same mass, coupling, and field renormalization constants as the symmetric theory supplemented by a renormalization of the vev.

A power-counting argument shows that even the values of the coupling and field renormalization constants can be taken over from the symmetric theory, and only the mass counterterm has to be adjusted to renormalize the broken theory [Le74]. In order to ensure the validity of (4.1.45) the counterterm $\delta\mu^2$ has to be fixed appropriately.

By introducing finite renormalizations one can relax the relations on the renormalization constants in (4.1.49) while keeping (4.1.45). This is, for instance, useful if one wants to fix the renormalization of μ^2 at the mass shell of a physical particle. We note that the validity of (4.1.45) is not necessary. However, if it does not hold, the presence of tadpole contributions considerably increases the number of graphs contributing to a given Green function.

There is a different approach where the vev $\mathbf{v}$ is not renormalized [Co84]. Then, the vertex functions are no longer finite, but the connected Green functions apart from the one-point function still are. This can be directly seen from (4.1.36) by putting $\mathbf{c} = 0$ in the spontaneously broken phase ($\mathbf{v} \neq 0$)

$$T^c_{\mathrm{c}}\{\mathbf{J}\}|_{\mathbf{c}=0} = T_{\mathrm{c}}\{\mathbf{J}\} - \mathrm{i}\int \mathrm{d}^4x J_i v_i. \tag{4.1.50}$$

With the exception of the tadpole, the connected Green functions in the spontaneously broken phase are equal to those in the symmetric phase. Thus, the counterterms of the symmetric phase render also the connected Green functions in the spontaneously broken phase finite. Since the tadpole does not enter S-matrix elements, this approach can be used equally well for the calculation of physical observables. Note that this approach requires to take into account all tadpoles explicitly. We finally mention that in the same way one could impose any renormalization condition for the tadpole without changing the S matrix.

4.1.2 Spontaneous breaking of a gauge symmetry

4.1.2.1 The Higgs mechanism

In theories with local symmetry, in which the gauge fields are minimally coupled to scalar fields $\boldsymbol{\phi} = (\phi_i)$, spontaneous symmetry breaking implies that the massless Goldstone bosons disappear from the physical spectrum and some of the gauge fields become massive. This mass-generation mechanism is called *Higgs mechanism* [Hi64, En64, Gu64, Ki67, Be74a]. The massive scalar fields that remain in the physical spectrum are called *Higgs fields.*

We investigate the Higgs mechanism using the gauged version of the Lagrangian (4.1.8). It describes a non-abelian gauge theory with a multiplet of n scalar fields coupled to the gauge fields,

$$\begin{aligned}\mathcal{L} &= -\frac{1}{4}F^a_{\mu\nu}F^{a,\mu\nu} + \frac{1}{2}(D_\mu\phi_i)(D^\mu\phi_i) - V(\phi),\\ D_\mu &= \partial_\mu - \mathrm{i}gA^a_\mu T^a.\end{aligned} \tag{4.1.51}$$

The Lagrangian $\mathcal{L}$ is invariant under local gauge transformations. The gauge fields transform according to the adjoint representation of the gauge group G [cf. (2.3.32)], the scalar multiplet $\boldsymbol{\phi}$ according to a real representation [cf. (4.1.7)] with generators T^a. The results can straightforwardly be generalized to complex Higgs representations. A specific example is provided by the gauged linear σ model, where the gauge group is SO(N), the scalar fields transform according to the defining representation of this group ($n = N$), and the scalar potential is given in (4.1.19).

We assume that the potential is chosen in such a way that the symmetry is spontaneously broken, i.e. the potential is minimized by the constant field $\boldsymbol{\phi} \equiv \mathbf{v} \neq 0$ with $\mathbf{v}^2 = v^2$. In the σ model for $\mu^2 > 0$ the minimum is given by (4.1.21). In this case, the stability group, i.e. the subgroup H the symmetry of which is preserved, is SO($N-1$). As at the end of Sect. 4.1.1.2, the generators of G involve the $\dim H$ generators of H, $\hat{T}^{\hat{a}}$, $\hat{a} = 1, \ldots, \dim H$, which leave the ground state invariant (4.1.15), and the $\dim G - \dim H$ generators of the coset space G/H, $\check{T}^{\check{a}}$, $\check{a} = \dim H + 1, \ldots, \dim G$, which obey (4.1.16).

Similarly to the Goldstone model (4.1.5), small excitations from the ground state are parametrized by

$$\boldsymbol{\phi}(x) = \exp\left(\frac{\mathrm{i}\check{T}^{\check{a}}\xi^{\check{a}}(x)}{v}\right)(\mathbf{v} + \boldsymbol{\eta}(x)), \quad \check{a} = 1, \ldots, \dim G - \dim H. \tag{4.1.52}$$

The $\dim G - \dim H$ Goldstone fields $\xi^{\check{a}}(x)$ appear in the exponential, and the vector $\boldsymbol{\eta}(x)$ involves only $n - [\dim G - \dim H]$ non-vanishing components. Since the Goldstone modes $\xi^{\check{a}}$ are linear combinations of $T^{\check{a}}_{ij}v_j$

(cf. Sect. 4.1.1.2), the condition that $\boldsymbol{\eta}(x)$ involves no Goldstone modes reads

$$\eta_i^\dagger(x)\check{T}_{ij}^{\check{a}}v_j = 0. \tag{4.1.53}$$

For the group SO(N), $\dim G - \dim H = N(N-1)/2 - (N-1)(N-2)/2 = N-1$, and $\boldsymbol{\eta}(x)$ involves only one finite component, $\boldsymbol{\eta}(x) = (0, \ldots, 0, \eta(x))$.

In contrast to the models considered so far, in gauged models the Goldstone fields $\xi^{\check{a}}(x)$ can be regarded as parameters of local gauge transformations $\exp(\mathrm{i}\check{T}^{\check{a}}\xi^{\check{a}}(x)/v)$ and thus have no physical significance. As a result of the gauge symmetry of the Lagrangian (4.1.51), the Goldstone degrees of freedom can be eliminated by carrying out the gauge transformation $\exp(-\mathrm{i}\check{T}^{\check{a}}\xi^{\check{a}}(x)/v)$. In this gauge, the Higgs field $\boldsymbol{\phi}(x)$ in (4.1.52) is given by

$$\boldsymbol{\phi}(x) = (\mathbf{v} + \boldsymbol{\eta}(x)) \,. \tag{4.1.54}$$

So far, gauges have always been fixed by specifying the form of the gauge field, e.g. by requiring $\partial^\mu A_\mu^a(x) = 0$. However, any other fields that do not transform trivially can be used for fixing the gauge as well. The requirement that $\boldsymbol{\phi}(x)$ has the form (4.1.54), or equivalently the condition (4.1.53), fixes a gauge, which is in fact a *unitary gauge* because the unphysical degrees of freedom $\xi(x)$ do not occur, and therefore the unitarity of the S matrix is evident (cf. Sect. 2.5.4.5). In the gauge (4.1.54), the Lagrangian (4.1.51) reads

$$\begin{aligned}\mathcal{L} = -\frac{1}{4}F^{a,\mu\nu}F_{\mu\nu}^a &+ \frac{1}{2}M_{\check{a}\check{b}}^2 A_\mu^{\check{a}} A^{\check{b},\mu} + \frac{1}{2}(\partial_\mu \eta_i)(\partial^\mu \eta_i) \\ &+ \text{mass terms for } \boldsymbol{\eta}(x) \\ &+ \text{terms involving 3 or 4 fields.}\end{aligned} \tag{4.1.55}$$

The gauge-boson mass term results from inserting the vev $\mathbf{v}$ into the gauge-invariant kinetic-energy term of the scalar fields,

$$\begin{aligned}\frac{1}{2}(D_\mu v_i)(D^\mu v_i) &= \frac{1}{2}g^2(\mathrm{i}\check{T}_{ij}^{\check{a}}v_j)(\mathrm{i}\check{T}_{ik}^{\check{b}}v_k)A_\mu^{\check{a}}A^{\check{b},\mu} \\ &= \frac{1}{2}g^2(v_j\check{T}_{ji}^{\check{a}}\check{T}_{ik}^{\check{b}}v_k)A_\mu^{\check{a}}A^{\check{b},\mu} = \frac{1}{2}M_{\check{a}\check{b}}^2 A_\mu^{\check{a}}A^{\check{b},\mu},\end{aligned} \tag{4.1.56}$$

where we used (4.1.15) and the antisymmetry of the matrices T^a for real representations. For complex representations we obtain, using (4.1.18), the matrix of gauge-boson masses squared

$$M^2_{\check{a}\check{b}} = g^2(\check{T}^{\check{a}}\mathbf{v})^\dagger_i(\check{T}^{\check{b}}\mathbf{v})_i = \frac{g^2}{2}\mathbf{v}^\dagger[\check{T}^{\check{a}},\check{T}^{\check{b}}]_+\mathbf{v}. \tag{4.1.57}$$

From (4.1.57) it can be easily seen that $M^2_{\check{a}\check{b}}$ is real and positive-definite. Its rank is $\dim G - \dim H$. Consequently, all gauge bosons associated with generators in G/H become massive. In the σ model, the gauge-boson mass matrix is given by

$$M^2_{\check{a}\check{b}} = g^2v^2(\check{T}^{\check{a}}\check{T}^{\check{b}})_{nn} = g^2v^2\delta_{\check{a}\check{b}} \qquad \text{(no sum over } n\text{)}. \tag{4.1.58}$$

From this one concludes:

1. The $\dim G - \dim H$ [$= N-1$ for SO(N)] Goldstone fields $\xi^{\check{a}}(x)$ have disappeared. No Goldstone bosons appear in spontaneous symmetry breaking of gauge theories.

2. The $n - [\dim G - \dim H]$ scalar fields in $\eta(x)$ remain as in general massive fields; their quanta are called *Higgs particles.* In the σ model, one finds one *Higgs field* $\eta(x)$ with mass $M = \mu\sqrt{2}$.

3. $\dim G - \dim H$ [$= N-1$ for SO(N)] vector fields have become massive with $M \propto gv$ [$M = gv$ in the σ model] and thus have longitudinal, physical polarization states, in addition to the two transverse polarization states of massless vector fields [cf. (2.1.20)–(2.1.22)].

4. The remaining $\dim H$ [$= (N-1)(N-2)/2$ for SO(N)] vector fields have zero mass.

The comparison of the degrees of freedom described by $\mathcal{L}$ according to (4.1.51), or gauge-equivalent (4.1.55), gives the picture shown in Table 4.1 for the general case, and for the σ model in brackets. The transmutation of Goldstone modes into longitudinal polarization states of massive vector bosons is known as the *Higgs mechanism.*

	$\mathcal{L}$ as in (4.1.51)	$\mathcal{L}$ as in (4.1.55)
massless gauge fields	$\dim G$ $\left[= \frac{N(N-1)}{2}\right]$	$\dim H$ $\left[= \frac{(N-1)(N-2)}{2}\right]$
physical states	$2 \dim G$	$2 \dim H$
massive gauge fields	0	$\dim G - \dim H$ $[= N - 1]$
physical states	0	$3\,(\dim G - \dim H)$
Higgs fields	n $[= N]$	$n - [\dim G - \dim H]$ $[= 1]$
all physical states	$2 \dim G + n$	$2 \dim G + n$

Tab. 4.1 Degrees of freedom in the symmetric and spontaneously broken Lagrangians [$G = \mathrm{SO}(N)$, $H = \mathrm{SO}(N-1)$]

4.1.2.2 Quantization of spontaneously broken gauge theories

The gauge theory defined by (4.1.51) can be quantized using the path-integral formalism according to the procedure outlined in Sect. 2.4. This requires the fixing of a gauge (Sect. 2.4.1.2) and the introduction of Faddeev–Popov fields (Sects. 2.4.1.3).

The unitary gauge used in (4.1.55) gives the canonical form (2.1.22) of the vector-boson propagators. As unphysical scalar bosons are not present, the unitarity of the S matrix is manifest. However, Green functions are not renormalizable in this gauge. Divergences do not cancel until the S matrix is calculated [We72, Le72a, Ap72]. Moreover, the definition of the finite part of the S matrix may not be unique under certain circumstances [Ja72, Ba72a]. Thus, the choice of a renormalizable gauge (in the sense of Sects. 2.5.1.3 and 2.5.1.4) is preferable, just as it was for the case without symmetry breaking.

In principle, (2.4.11) and (2.4.17) can be used as gauge-fixing term, but this is often unsuitable. Instead one prefers generalized forms, e.g. the Fradkin–Tyutin gauges [Fr74] or the *'t Hooft gauges* [tH71].

As a 't Hooft gauge is employed in the next section, we discuss it in some detail. Consider first the gauged σ model (4.1.51) with the scalar potential of (4.1.19) for the abelian gauge group SO(2). With $\phi = (\phi_1, v + \eta)$ and $T = \tau_2$ [cf. (A.1.11)] the gauge-invariant Lagrangian takes the form

$$\begin{aligned}\mathcal{L} = &-\frac{1}{4}F^{a,\mu\nu}F^a_{\mu\nu} + \frac{M^2}{2}A_\mu A^\mu + \frac{1}{2}(\partial_\mu\eta)(\partial^\mu\eta) - \mu^2\eta^2 \\ &+ \frac{1}{2}(\partial_\mu\phi_1)(\partial^\mu\phi_1) - MA^\mu(\partial_\mu\phi_1) \\ &+ \text{terms involving 3 or 4 fields}\end{aligned} \tag{4.1.59}$$

with $M = gv$. In the 't Hooft gauge, the gauge-fixing part of the Lagrangian, $\mathcal{L}_{\text{fix}}$, is chosen in such a way that all bilinear terms that involve two different fields, i.e. $-MA^\mu(\partial_\mu\phi_1)$ in the example, cancel in the Lagrangian (up to irrelevant total derivatives). For the Lagrangian (4.1.59) this is achieved by choosing

$$C[A_\mu(x), \phi_1(x)] = \partial^\mu A_\mu(x) + \xi M\phi_1(x), \tag{4.1.60}$$

$$\mathcal{L}_{\text{fix}} = -\frac{1}{2\xi}(C[A_\mu(x), \phi_1(x)])^2 = -\frac{1}{2\xi}\left(\partial^\mu A_\mu(x) + \xi M\phi_1(x)\right)^2.$$

Thus, the mass matrix of $\mathcal{L} + \mathcal{L}_{\text{fix}}$ is diagonal in the fields. The *would-be Goldstone field* ϕ_1 acquires a non-vanishing mass $\sqrt{\xi}M$, and its propagator is given by

$$\mathrm{i}\Delta^{\phi_1} = \frac{\mathrm{i}}{k^2 - \xi M^2 + \mathrm{i}\varepsilon}. \tag{4.1.61}$$

The *gauge-field propagator* has the form

$$\mathrm{i}\Delta^A_{\mu\nu} = \mathrm{i}\frac{-g_{\mu\nu} + k_\mu k_\nu(1-\xi)/(k^2 - \xi M^2)}{k^2 - M^2 + \mathrm{i}\varepsilon}. \tag{4.1.62}$$

The Lagrangian for the Faddeev–Popov ghost fields $\mathcal{L}_{\text{ghost}}$ is calculated from $C^a[A^\mu(x), \phi_1(x)]$ according to (2.4.17) taking into account the gauge variation of the field $\phi_1(x)$

$$\delta\phi_1 = g(v + \eta)\delta\theta. \tag{4.1.63}$$

We obtain

$$\mathcal{L}_{\text{ghost}} = -\int \mathrm{d}^4x\, \bar{u}(x)\left[\partial_\mu\partial^\mu + \xi M g(v+\eta)\right] u(x). \qquad (4.1.64)$$

Thus, also the Faddeev–Popov ghost acquires a mass $\sqrt{\xi}M$ and is no longer free in the abelian theory. Note that in the 't Hooft gauge, all boson propagators behave as $1/k^2$ for large k^2, and the theory is renormalizable according to power counting.

Consider now a more general gauge theory with spontaneous symmetry breaking as given for example by (4.1.51). We assume that the gauge group is simple and hence has a single coupling constant and that the scalar fields $\boldsymbol{\phi}$ belong to a real representation. Scalar fields with vanishing vev are obtained by the shift $\boldsymbol{\phi} = \tilde{\boldsymbol{\phi}} + \mathbf{v}$ with $\langle\tilde{\boldsymbol{\phi}}\rangle = 0$. The shifted fields $\tilde{\boldsymbol{\phi}}_i$ involve Goldstone fields χ_i, i.e. linear combinations of the $T^a v_i$ and other linearly independent scalar fields characterized by (4.1.53). By introducing the fields $\tilde{\boldsymbol{\phi}}$ in the part of the Lagrangian (4.1.51) that describes the interaction of the scalar fields with the gauge field, we find the following bilinear mixing terms between gauge and scalar fields

$$-\mathrm{i}g(\partial^\mu\tilde{\phi}_i)A^{\check{a}}_\mu(\check{T}^{\check{a}}v)_i = -\mathrm{i}g(\partial^\mu\chi_i)A^{\check{a}}_\mu(\check{T}^{\check{a}}v)_i. \qquad (4.1.65)$$

Because of (4.1.53), the mixing terms (4.1.65) involve only would-be Goldstone fields χ_i and are thus absent in the unitary gauge.

The mixing terms (4.1.65) can be eliminated by introducing a *'t Hooft gauge-fixing term* of the form

$$\mathcal{L}_{\text{fix}} = -\frac{1}{2\xi}\left(\partial^\mu A^a_\mu + \mathrm{i}\xi g\chi_i\check{T}^a_{ij}v_j\right)^2. \qquad (4.1.66)$$

The term quadratic in the scalar fields χ_i yields a mass matrix for the would-be Goldstone fields,

$$\frac{1}{2}(M^2_\chi)_{ij}\chi_i\chi_j = -\frac{1}{2}\xi g^2(\chi_i\check{T}^{\check{a}}_{ik}v_k)(v_l\check{T}^{\check{a}}_{lj}\chi_j). \qquad (4.1.67)$$

Note that the gauge-fixing term (4.1.66) does not modify the masses of the physical scalar fields. The masses of the would-be Goldstone bosons are $\sqrt{\xi}$

times the non-zero vector-boson masses. This can be seen by inspecting an eigenvector $x^{\check{a}}$ of $(M^2)_{\check{a}\check{b}}$ with eigenvalue M^2,

$$(M^2)_{\check{a}\check{b}}x^{\check{b}} = -g^2(\check{T}^{\check{a}}v)_i(\check{T}^{\check{b}}v)_i x^{\check{b}} = M^2 x^{\check{a}}. \tag{4.1.68}$$

Then, $x^{\check{b}}(T^{\check{b}}v)_i$ is an eigenvector of $(M^2_\chi)_{ij}$ with eigenvalue $\sqrt{\xi}M^2$,

$$\begin{aligned}(M^2_\chi)_{ij}x^{\check{b}}(\check{T}^{\check{b}}v)_j &= -\xi g^2(\check{T}^{\check{a}}v)_i(\check{T}^{\check{a}}v)_j x^{\check{b}}(\check{T}^{\check{b}}v)_j \\ &= \xi(M^2)_{\check{a}\check{b}}x^{\check{b}}(\check{T}^{\check{a}}v)_i = \xi M^2 x^{\check{a}}(\check{T}^{\check{a}}v)_i.\end{aligned} \tag{4.1.69}$$

The Faddeev–Popov Lagrangian is given by

$$\mathcal{L}_{\text{ghost}} = -\int \mathrm{d}^4x\, \bar{u}^a(x) \left(\partial^\mu D^{ab}_\mu - \xi g^2 v_i T^a_{ij} T^b_{jk}(v_k + \tilde{\phi}_k)\right) u^b(x). \tag{4.1.70}$$

The ghost-mass matrix reads

$$(M^2_{\text{ghost}})_{\check{a}\check{b}} = \xi g^2 v_i \check{T}^{\check{a}}_{ij}\check{T}^{\check{b}}_{jk} v_k = \xi(M^2)_{\check{a}\check{b}}, \tag{4.1.71}$$

and thus is proportional to the gauge-boson mass matrix (4.1.57). The fact that the masses of the Faddeev–Popov ghosts and would-be Goldstone bosons are equal to those of the longitudinal gauge-bosons is crucial for the cancellation of the poles of these unphysical fields in the S matrix.

For finite ξ, all boson propagators behave as $1/k^2$ for large k^2, i.e. the 't Hooft gauge is renormalizable. For $\xi \to \infty$, the gauge-boson propagators agree with the propagator (1.4.13) in the unitary gauge. Consequently, the unitary gauge is a limiting case of the 't Hooft gauge (4.1.66). The masses of the would-be Goldstone bosons and of the Faddeev–Popov ghosts associated with massive gauge bosons diverge like $\sqrt{\xi}$ for large ξ. Hence, in this limit these fields decouple from the S matrix. As a consequence, the proof of the unitarity of the S matrix simplifies as discussed in Sect. 2.5.4.5.

4.1.2.3 Renormalization of spontaneously broken gauge theories

The renormalizability of spontaneously broken gauge theories [tH71, Le74] can be proven in a way similar to the renormalizability of the σ model (cf. Sect. 4.1.1.5). We restrict ourselves to the discussion of the gauged σ model. The complete Lagrangian is given by (4.1.51) and (4.1.19) supplemented by gauge-fixing and Faddeev–Popov ghost terms. We assume that the gauge-fixing functional is linear in the fields, as for example in the 't Hooft gauge (4.1.66).

We set $\mu^2 > 0$, so that the field $\boldsymbol{\phi}$ acquires a non-vanishing vev $\langle \boldsymbol{\phi} \rangle = \mathbf{v}$. We want to show that the counterterms for the symmetric theory ($\mu^2 < 0$) plus a renormalization of the vev are sufficient to render the broken theory ($\mu^2 > 0$) finite. Since the point $\mu^2 = 0$ is not analytic (a phase transition occurs at this point) we cannot simply continue μ^2 from positive to negative values.

We use the same strategy as in Sect. 4.1.1.5 and introduce a constant external source coupled to the field $\boldsymbol{\phi}$ as in (4.1.32). Then, (4.1.41) holds here as well. From Sect. 2.5.4.2 we know how to renormalize the vertex functional of the symmetric theory:

$$\begin{aligned}\Gamma_{\mathrm{R}}(\boldsymbol{\phi}, \mathbf{A}, \ldots, g, -\mu^2, \lambda) &= \Gamma(\boldsymbol{\phi}_0, \mathbf{A}_0, \ldots, g_0, -\mu_0^2, \lambda_0) \\ &= \Gamma(Z^{1/2}\boldsymbol{\phi}, Z_A^{1/2}\mathbf{A}, \ldots, Z_g g, -\mu^2 - \delta\mu^2, Z_\lambda \lambda). \qquad (4.1.72)\end{aligned}$$

The dots represent unphysical fields and gauge parameters. The renormalization of the gauge parameters is determined (in linear gauges) in such a way that the complete gauge-fixing term is not renormalized. Equation (4.1.41) tells us that the broken theory is renormalized as

$$\begin{aligned}\Gamma^{\mathrm{c}}_{\mathrm{R}}(\tilde{\boldsymbol{\phi}}, \mathbf{A}, \ldots, g, -\mu^2, \lambda, \mathbf{c}, \mathbf{v}) &= \Gamma^{\mathrm{c}}(\tilde{\boldsymbol{\phi}}_0, \mathbf{A}_0, \ldots, g_0, -\mu_0^2, \lambda_0, \mathbf{c}_0 . \mathbf{v}_0) \\ &= \Gamma^{\mathrm{c}}(Z^{1/2}\tilde{\boldsymbol{\phi}}, Z_A^{1/2}\mathbf{A}, \ldots, Z_g g, -\mu^2 - \delta\mu^2, Z_\lambda \lambda, Z^{-1/2}\mathbf{c}, Z^{1/2}\mathbf{v}). \\ & \qquad (4.1.73)\end{aligned}$$

Upon analytic continuation this result holds also for the case of purely spontaneous symmetry breaking, i.e. $\mathbf{c} = 0$ and $\mu^2 > 0$.

The corresponding normalization conditions for the various vertex functions can be deduced from the identity (4.1.41) and the normalization conditions of the symmetric functions. Once we know how to renormalize spontaneously broken theories in terms of symmetric counterterms we can perform additional finite renormalizations in order to conform with more physical renormalization conditions. One can, for example, demand that the coupling constant is fixed as the value of a vertex function at some on-shell point, or that the renormalized particle masses are equal to the physical masses.

According to (4.1.73) the vev is renormalized in the same way as the Higgs field itself. It is, however, convenient to perform a further finite renormalization of the vev in order to introduce the true vev of the interacting Higgs field even if the other counterterms are already fixed by physical renormalization conditions. Requiring that the renormalized vev is the true one, i.e. given by the actual minimum of the effective Higgs potential, is equivalent to the requirement that the *renormalized tadpole*, i.e. the term in the effective potential linear in the Higgs field, vanishes. In this scheme, tadpole contributions do not have to be taken into account in practical calculations. Instead of a counterterm for the vev one can directly introduce a tadpole counterterm.

Similarly to the σ model (Sect. 4.1.1.5), also spontaneously broken gauge theories can be renormalized without renormalizing the vevs [Ap73a]. While the S matrix and the connected Green functions, apart from the tadpole, are finite in this approach, the vertex functions are still divergent.

4.1.2.4 General theory with spontaneous symmetry breaking

A non-abelian gauge theory for spinor and scalar fields interacting with each other via massive intermediate vector bosons is a suitable dynamical concept for describing the electroweak phenomena. The Higgs mechanism can be used for constructing a renormalizable theory of this kind: the vevs of some of the scalar fields generate the masses of the gauge bosons by spontaneous symmetry breaking. The construction of the general, gauge-invariant part of the Lagrangian of such a theory proceeds as follows:

1. Choose a gauge group G. As the gauge fields always transform according to the adjoint representation of G, the number of gauge fields $W^a = (W^+, W^-, Z^0, \gamma, ...)$ is fixed. The gauge-field couplings are given by the structure constants f^{abc}.

2. Choose representations L^a, R^a, and P^a of G under which left- and right-handed fermions $\psi_i = (\nu_e, e, \nu_\mu, \mu, \nu_\tau, \tau \ldots, u_c, d_c, c_c, s_c, t_c, b_c \ldots)$

(c denotes the colour index) and, for definiteness real, scalars ϕ_r transform, respectively,

$$[L^a, L^b] = \mathrm{i}f^{abc}L^c, \qquad [R^a, R^b] = \mathrm{i}f^{abc}R^c,$$
$$[P^a, P^b] = \mathrm{i}f^{abc}P^c. \tag{4.1.74}$$

When choosing the fermion representations, care must be taken that no γ_5 anomalies can occur (cf. Sect. 2.7.2.2).

3. The most general form of the gauge-invariant, local, renormalizable Lagrangian, which also includes self-couplings $V(\phi)$ of the scalar fields and a coupling between scalars and fermions, is given by[3]

$$\begin{aligned}\mathcal{L} = &-\frac{1}{4}F^a_{\mu\nu}F^{a,\mu\nu} + \bar{\psi}_i(\mathrm{i}\gamma_\mu D^\mu_{ij} - M_{ij})\psi_j \\ &+\frac{1}{2}\left(D_{rs,\mu}\phi_s\right)\left(D^\mu_{rs'}\phi_{s'}\right) - V(\phi) \\ &+ g\bar{\psi}_i\phi_r\left(X^{\mathrm{L},r}_{ij}\frac{1-\gamma_5}{2} + X^{\mathrm{R},r}_{ij}\frac{1+\gamma_5}{2}\right)\psi_j\end{aligned} \tag{4.1.75}$$

with

$$\begin{aligned}F^a_{\mu\nu} &= \partial_\mu W^a_\nu - \partial_\nu W^a_\mu + gf^{abc}W^b_\mu W^c_\nu, \\ D^\mu_{ij}\psi_j &= \left[\delta_{ij}\partial^\mu - \mathrm{i}gW^{a,\mu}\left(L^a_{ij}\frac{1-\gamma_5}{2} + R^a_{ij}\frac{1+\gamma_5}{2}\right)\right]\psi_j, \\ D^\mu_{rs}\phi_s &= \left[\delta_{rs}\partial^\mu - \mathrm{i}gW^{a,\mu}P^a_{rs}\right]\phi_s.\end{aligned} \tag{4.1.76}$$

The gauge-invariant potential $V(\phi)$ is a polynomial of the fourth degree at most in ϕ [cf. (4.1.19)]. Higher powers in the fields must not occur in the other coupling terms, too. Otherwise coupling constants with negative mass dimension would destroy the renormalizability of the theory (cf. Sect. 2.5.1.3 and Table 2.1). The G symmetry can be spontaneously broken and a Higgs–Kibble mechanism introduced by a suitable choice of $V(\phi)$. The gauge bosons of the symmetry group $H \subset G$, which leave a classical vacuum vector $\mathbf{v}$ invariant, remain massless, and the other gauge bosons become massive (cf. Sect. 4.1.2.1). The gauge symmetry is said to be broken spontaneously from $G \to H$. The $\bar{\psi}\phi\psi$ coupling term is gauge-invariant if (in matrix notation)

$$[L^a, X^{\mathrm{L},r}] = X^{\mathrm{L},s}P^a_{sr}, \qquad [R^a, X^{\mathrm{R},r}] = X^{\mathrm{R},s}P^a_{sr} \tag{4.1.77}$$

[3]For the choice of the coupling constants, see the note given at the end of Sect. 2.3.3.3, (2.3.52).

applies.

The phenomenological analysis of the electroweak interaction in Sect. 1.4 has shown that the appropriate gauge group of the *weak isospin* and *weak hypercharge* is $G = \mathrm{SU}(2)_{\mathrm{W}} \times \mathrm{U}(1)_{\mathrm{Y}}$, and hence four gauge fields W_μ^+, W_μ^-, Z_μ, and A_μ are needed as the basis for a "minimal" theory. More complicated models based on the above scheme must contain this as a substructure.

We summarize the important points of the concept of *spontaneous symmetry breaking*, introduced in this section. The Lagrangian (4.1.55) is gauge-invariant. The currents coupled to vector fields are conserved. Only the necessity to choose one of the physically equivalent ground states as basis for the perturbation theory hides the gauge invariance. The gauge transformation law, expressed in the shifted fields $\tilde{\phi}$, becomes inhomogeneous. Spontaneous symmetry breaking introduces mass terms for gauge bosons. Since gauge invariance is not broken but only hidden, the Higgs mechanism can be used to formulate a *renormalizable theory with massive vector fields.*

4.2 The Standard Model of the Electroweak Interaction

In Sect. 2.3.3.5 we argued that the theory of the electroweak interaction is a non-abelian gauge theory with the gauge group $\mathrm{SU}(2)_{\mathrm{W}} \times \mathrm{U}(1)_Y$. The masses of the gauge bosons can be introduced by spontaneous symmetry breaking as described in Sect. 4.1.2. In this section we describe the *Electroweak Standard Model* (EWSM) [We67, Sa68, Gl70]. It realizes in a minimal way the unification of the electromagnetic with the charged and neutral weak interactions by postulating the $\mathrm{SU}(2)_{\mathrm{W}} \times \mathrm{U}(1)_Y$ gauge group of weak isospin and hypercharge and mass generation by a complex doublet of scalar fields.

We formulate the classical Lagrangian of the EWSM in its symmetric form and then pass to the representation in terms of physical fields and parameters so that its physical content becomes visible.

4.2.1 The Lagrangian of the Electroweak Standard Model

We start by formulating the Lagrangian of the EWSM in a form in which the symmetry structure is manifest.

As the gauge group is not simple but rather a product of SU(2) and U(1), electromagnetic and weak interactions are combined by two *gauge coupling constants*, g_2 for the *weak isospin* $\mathrm{SU}(2)_\mathrm{w}$ and g_1 for the *weak hypercharge* $\mathrm{U}(1)_Y$. We denote the three generators of weak isospin by I^a_w, the generator of weak hypercharge by Y_w. The generator of the electric charge is defined via the *Gell-Mann–Nishijima relation*

$$Q = I^3_\mathrm{w} + \frac{Y_\mathrm{w}}{2}. \tag{4.2.1}$$

4.2.1.1 Gauge part

According to the dimension of the gauge group $\mathrm{SU}(2)_\mathrm{w} \times \mathrm{U}(1)_Y$ there are four gauge fields which transform according to the adjoint representation. The gauge fields belonging to the weak-isospin group $\mathrm{SU}(2)_\mathrm{w}$ are called $\mathrm{W}^a_\mu(x)$, $a = 1, 2, 3$, and the one belonging to the weak-hypercharge group $\mathrm{U}(1)_Y$ is called $B_\mu(x)$. The Lagrangian for the gauge fields reads (cf. Sect. 2.3.3)

$$\mathcal{L}_\mathrm{gauge} = -\frac{1}{4} F_{1,\mu\nu}(x) F_1^{\mu\nu}(x) - \frac{1}{4} F^a_{2,\mu\nu}(x) F_2^{a,\mu\nu}(x). \tag{4.2.2}$$

The gauge field strengths are defined as

$$\begin{aligned} F_{1,\mu\nu}(x) &= \partial_\mu B_\nu(x) - \partial_\nu B_\mu(x), \\ F^a_{2,\mu\nu}(x) &= \partial_\mu W^a_\nu(x) - \partial_\nu W^a_\mu(x) + g_2 \varepsilon^{abc} W^b_\mu(x) W^c_\nu(x), \end{aligned} \tag{4.2.3}$$

where ε_{abc} are the totally antisymmetric structure constants of SU(2) (cf. Sect. 1.2.2).

The gauge interaction of the matter fields is determined by the covariant derivative

$$D_\mu = \left(\partial_\mu - \mathrm{i} g_2 I^a_\mathrm{w} W^a_\mu + \mathrm{i} g_1 \frac{Y_\mathrm{w}}{2} B_\mu \right). \tag{4.2.4}$$

4.2.1.2 Fermionic part

According to the structure of the fermion–gauge-boson interaction discussed in Sect. 1.4, the fermions are arranged into isospin doublets $(\psi'_{i,+}, \psi'_{i,-})$, where the indices + and − (represented by σ in the following) distinguish the two members of the doublets. The index i runs over the three lepton doublets (1.4.6) and the three quark doublets (1.4.23) including the three different colours so that there are in fact nine quark doublets. The colour index is suppressed in the following. The prime at the fermion fields indicates eigenstates of the electroweak interaction, i.e. the covariant derivatives are diagonal with respect to the generation indices in this basis. These states are not necessarily mass eigenstates.

The *left-handed fields* $\psi'^{\mathrm{L}}(x) = \frac{1}{2}(1-\gamma_5)\psi'(x)$ in these doublets are associated with the fundamental two-dimensional representation of the weak-isospin group $\mathrm{SU}(2)_{\mathrm{W}}$, $\Psi_i'^{\mathrm{L}}(x) = (\psi_{i,+}'^{\mathrm{L}}, \psi_{i,-}'^{\mathrm{L}})$, with representation matrices $I^a_{\mathrm{W}} = \tau^a/2$, where τ^a are the Pauli matrices (A.1.11). Each *right-handed field* $\psi'^{\mathrm{R}}(x) = \frac{1}{2}(1+\gamma_5)\psi'(x)$ is put into a one-dimensional trivial representation of $\mathrm{SU}(2)_{\mathrm{W}}$ ($I^a_{\mathrm{W}} = 0$). Consequently, the term involving g_2 in the covariant derivative is absent for right-handed fields. Note that in the minimal Standard Model no right-handed neutrinos are included.[4] The weak hypercharge of the fermions is fixed in such a way that the correct electric charge results from the Gell-Mann–Nishijima formula (4.2.1), leading to

$$Y^{\mathrm{R}}_{i,+} = Y^{\mathrm{L}}_i + 1 = 2Q_{i,+}, \qquad Y^{\mathrm{R}}_{i,-} = Y^{\mathrm{L}}_i - 1 = 2Q_{i,-} \tag{4.2.5}$$

for the weak hypercharges of right- and left-handed fermion fields. The fermionic part of the Lagrangian reads

$$\mathcal{L}_{\mathrm{F}} = \sum_i \overline{\Psi}_i'^{\mathrm{L}} \mathrm{i}\not{D} \Psi_i'^{\mathrm{L}} + \sum_{i,\sigma} \overline{\psi}_{i,\sigma}'^{\mathrm{R}} \mathrm{i}\not{D} \psi_{i,\sigma}'^{\mathrm{R}}. \tag{4.2.6}$$

Because left- and right-handed fields transform according to different representations of the symmetry group $\mathrm{SU}(2)_{\mathrm{W}} \times \mathrm{U}(1)_Y$, the theory is chiral, and explicit mass terms for the fermions are forbidden.

[4]Right-handed neutrinos can be easily included in the Standard Model without affecting its basic structure. This becomes necessary in order to allow for the description of the recent experimental results on neutrino oscillations.

It is often convenient to denote the left-handed lepton and quark doublets by

$$L'^{\mathrm{L}}_i = \begin{pmatrix} \nu'^{\mathrm{L}}_i \\ l'^{\mathrm{L}}_i \end{pmatrix}, \qquad Q'^{\mathrm{L}}_i = \begin{pmatrix} u'^{\mathrm{L}}_i \\ d'^{\mathrm{L}}_i \end{pmatrix}, \qquad i = 1,2,3, \tag{4.2.7}$$

respectively, and the right-handed fermion singlets by l'^{R}_i, u'^{R}_i, and d'^{R}_i. Here ν, l, u, and d represent neutrino, charged-lepton, up-type-quark, and down-type-quark fields, respectively, and i is the generation index ($i = 1,2,3$). In this notation the fermionic part of the Lagrangian has the form

$$\mathcal{L}_{\mathrm{F}} = \sum_i \left(\bar{L}'^{\mathrm{L}}_i \mathrm{i}\not{D} L'^{\mathrm{L}}_i + \bar{Q}'^{\mathrm{L}}_i \mathrm{i}\not{D} Q'^{\mathrm{L}}_i\right)$$
$$+ \sum_i \left(\bar{l}'^{\mathrm{R}}_i \mathrm{i}\not{D} l'^{\mathrm{R}}_i + \bar{u}'^{\mathrm{R}}_i \mathrm{i}\not{D} u'^{\mathrm{R}}_i + \bar{d}'^{\mathrm{R}}_i \mathrm{i}\not{D} d'^{\mathrm{R}}_i\right). \tag{4.2.8}$$

4.2.1.3 Higgs part

In order to introduce mass terms for the weak gauge bosons we use the *Higgs mechanism* to break the $\mathrm{SU}(2)_{\mathrm{w}} \times \mathrm{U}(1)_Y$ symmetry in such a way that the electromagnetic symmetry $\mathrm{U}(1)_{\mathrm{em}}$ is the remaining symmetry (cf. Sect. 4.1.2). To this end, we postulate a weak-isospin doublet ($I^a_{\mathrm{w}} = \tau^a/2$) of two complex scalars fields $\Phi(x)$, the Higgs doublet, which couples in a gauge-invariant way to the vector bosons. In order to allow for a electrically neutral component with non-vanishing vev, its hypercharge must be $Y_\Phi = \pm 1$. By convention we choose the + sign, so that $\Phi(x) = (\phi^+(x), \phi^0(x))$. The choice of a weak-isospin doublet with hypercharge $Y_\Phi = \pm 1$ also allows for *Yukawa couplings*, i.e. couplings between scalar and fermion fields, that are necessary for the generation of fermion masses via spontaneous symmetry breaking. The Higgs part of the Lagrangian reads

$$\mathcal{L}_{\mathrm{H}} = (D_\mu \Phi)^\dagger (D^\mu \Phi) - V(\Phi)$$
$$- \sum_{i,j} (g_{ij} \bar{\Psi}^{\mathrm{L}}_i \psi^{\mathrm{R}}_{j,-} \Phi + \tilde{g}_{ij} \bar{\Psi}^{\mathrm{L}}_i \psi^{\mathrm{R}}_{j,+} \Phi^{\mathrm{c}} + \mathrm{h.c.}), \tag{4.2.9}$$

where

$$V(\Phi) = \frac{\lambda}{4}\left(\Phi^\dagger \Phi\right)^2 - \mu^2 \Phi^\dagger \Phi, \qquad \mu^2, \lambda > 0, \tag{4.2.10}$$

is the gauge-invariant self-interaction of the scalar doublet with coupling λ and mass parameter μ^2, and $\Phi^{\mathrm{c}}(x)$ is the charge-conjugate Higgs field $\Phi^{\mathrm{c}}(x) = \mathrm{i}\tau^2\Phi^* = (\phi^{0*}(x), -\phi^-(x))$. Note that the Higgs-doublet field allows for Yukawa interactions to both up-type and down-type right-handed fermion fields. The Higgs self-interaction is chosen in such a way that the Higgs field acquires a non-vanishing vacuum expectation value. This vacuum expectation value leads to mass terms for the gauge bosons and—arising from the Yukawa couplings—also for the fermions. This is explained in detail in Sect. 4.2.2.2.

In the notation of (4.2.8) the Higgs Lagrangian reads

$$\begin{aligned}\mathcal{L}_{\mathrm{H}} &= (D_\mu\Phi)^\dagger (D^\mu\Phi) - V(\Phi) \\ &\quad - \sum_{i,j}\left(\bar{L}_i^{\prime\mathrm{L}} G_{ij}^l l_j^{\prime\mathrm{R}}\Phi + \bar{Q}_i^{\prime\mathrm{L}} G_{ij}^u u_j^{\prime\mathrm{R}}\Phi^{\mathrm{c}} + \bar{Q}_i^{\prime\mathrm{L}} G_{ij}^d d_j^{\prime\mathrm{R}}\Phi + \mathrm{h.c.}\right),\end{aligned} \tag{4.2.11}$$

where G^f denote Yukawa coupling matrices.

The classical Lagrangian $\mathcal{L}_{\mathrm{cl}}$ of the GSW theory is the sum of the individual terms (4.2.2), (4.2.8) and (4.2.9) [or (4.2.11)],

$$\mathcal{L}_{\mathrm{cl}} = \mathcal{L}_{\mathrm{gauge}} + \mathcal{L}_{\mathrm{F}} + \mathcal{L}_{\mathrm{H}}. \tag{4.2.12}$$

It is invariant under the infinitesimal *gauge transformations*,

$$\begin{aligned}W_\mu^a(x) &\to W_\mu^a(x) + \frac{1}{g_2}\partial_\mu\delta\theta^a + \varepsilon^{abc}W_\mu^b\delta\theta^c, \\ B_\mu(x) &\to B_\mu(x) + \frac{1}{g_1}\partial_\mu\delta\theta^Y, \\ \Phi(x) &\to \left[1 - \mathrm{i}\frac{1}{2}\delta\theta^Y(x) + \mathrm{i}\frac{\tau^a}{2}\delta\theta^a(x)\right]\Phi(x), \\ \Psi_i^{\prime\mathrm{L}}(x) &\to \left[1 - \mathrm{i}\frac{Y_i^{\mathrm{L}}}{2}\delta\theta^Y(x) + \mathrm{i}\frac{\tau^a}{2}\delta\theta^a(x)\right]\Psi_i^{\prime\mathrm{L}}(x), \\ \psi_i^{\prime\sigma,\mathrm{R}}(x) &\to \left[1 - \mathrm{i}\frac{Y_{i,\sigma}^{\mathrm{R}}}{2}\delta\theta^Y(x)\right]\psi_i^{\prime\sigma,\mathrm{R}}(x),\end{aligned} \tag{4.2.13}$$

where θ^a and θ^Y are the parameters of the gauge transformations corresponding to the groups $\mathrm{SU}(2)_{\mathrm{w}}$ and $\mathrm{U}(1)_Y$, respectively.

The Lagrangian $\mathcal{L}_{\mathrm{cl}}$ in the above form depends on the two gauge couplings g_2 and g_1, the Yukawa couplings G_{ij}^f, and the parameters μ^2 and λ of the scalar potential.

4.2.2 The Lagrangian in the physical basis

The gauge symmetry forbids explicit mass terms in the Lagrangian $\mathcal{L}_{\text{cl}}$, apart from a mass term for the scalar doublet Φ. In order to introduce gauge-boson and fermion masses in the EWSM we use spontaneous symmetry breaking, i.e. the Higgs–Kibble mechanism [Hi64]. In order to extract the physical content of the Lagrangian in the spontaneously broken form, we rewrite it in terms of physical parameters and fields corresponding to mass and charge eigenstates.

4.2.2.1 Spontaneous symmetry breaking in the EWSM

The scalar-field self-interaction is chosen in such a way that spontaneous symmetry breaking and Higgs–Kibble mechanism take place as specified in Sect. 4.1.2. The scalar potential $V(\Phi)$, (4.2.10), has a minimum for

$$|\langle\Phi\rangle|^2 = \frac{2\mu^2}{\lambda} \equiv \frac{v^2}{2} \neq 0. \tag{4.2.14}$$

Out of this three-dimensional subspace we choose a vector Φ_0 that is annihilated by the electric charge operator Q,

$$Q\Phi_0 = \left(\frac{\tau^3}{2} + \frac{1}{2}Y_\Phi\right)\Phi_0 = \begin{pmatrix} 1 & 0 \\ 0 & 0 \end{pmatrix}\begin{pmatrix} \phi_{01} \\ \phi_{02} \end{pmatrix} = 0, \tag{4.2.15}$$

so that the remaining (unbroken) symmetry is the one of electromagnetic gauge transformations $\mathrm{U}(1)_{\mathrm{em}}$. Note that Φ_0 is neither invariant under $\mathrm{SU}(2)_{\mathrm{w}}$ nor under $\mathrm{U}(1)_Y$. The conditions (4.2.14) and (4.2.15) have (up to a phase) a unique solution,

$$\Phi_0 = \begin{pmatrix} 0 \\ v/\sqrt{2} \end{pmatrix}. \tag{4.2.16}$$

In order to expand the Lagrangian about this classical ground state we write the Higgs-doublet field as

$$\Phi(x) = \begin{pmatrix} \phi^+(x) \\ \frac{1}{\sqrt{2}}[v + \eta(x) + \mathrm{i}\chi(x)] \end{pmatrix}, \qquad \phi^-(x) = [\phi^+(x)]^\dagger, \tag{4.2.17}$$

where $\eta(x)$, $\chi(x)$, and $\phi^{\pm}(x)$ have vanishing vacuum expectation value. The fields $\phi^{+}(x)$, $\phi^{-}(x)$, and $\chi(x)$, the *would-be Goldstone fields*, are unphysical degrees of freedom and can be eliminated by transition to the *unitary gauge* according to (4.1.54). We will see that $\eta(x)$ is a massive physical *Higgs field*, that gives rise to the physical *Higgs boson*.

4.2.2.2 Physical fields and parameters

The physical content of the EWSM can be extracted most easily in the *unitary gauge*, where $\phi^{\pm} = \chi = 0$ (cf. Sect. 4.1.2.1). In this gauge no unphysical degrees of freedom occur and the physical degrees of freedom can be classified as eigenstates of electric charge and mass.

Since the gauge bosons have vanishing hypercharge, the corresponding charge eigenstates can be found as eigenstates of the (adjoint-representation) matrix $(I^3_{\mathrm{W}})_{ab} = -\mathrm{i}\varepsilon_{3ab}$,

$$
\begin{aligned}
W^{\pm}_{\mu}(x) &= \frac{1}{\sqrt{2}}[W^{1}_{\mu}(x) \mp \mathrm{i}W^{2}_{\mu}(x)], \qquad\qquad (4.2.18)\\
QW^{\pm}_{\mu}(x) &= I^3_{\mathrm{W}}W^{\pm}_{\mu}(x) = \pm W^{\pm}_{\mu}(x), \qquad QW^{3}_{\mu}(x) = QB_{\mu}(x) = 0.
\end{aligned}
$$

Thus, we have one positively charged gauge boson, W^{+}, one negatively charged one, W^{-}, and two neutral ones, W^{3} and B.

In order to determine the mass eigenstates we have to inspect the Higgs part of the Lagrangian. Using (4.2.17) and (4.2.18) and putting $\phi^{\pm} = \chi = 0$, it reads

$$
\begin{aligned}
\mathcal{L}_{\mathrm{H}} = {}& \frac{1}{4}v^2 g_2^2 W^{-}_{\mu}W^{+\mu} + \frac{1}{8}v^2(g_2 W^3_{\mu} + g_1 B_{\mu})(g_2 W^{3\mu} + g_1 B^{\mu})\\
&+ \frac{1}{2}(\partial_{\mu}\eta)(\partial^{\mu}\eta) - \mu^2\eta^2 + \text{trilinear and quadrilinear terms}\\
&- \frac{1}{\sqrt{2}}(v+\eta)\sum_{i,j}\left(\bar{l}'^{\mathrm{L}}_{i}G^{l}_{ij}l'^{\mathrm{R}}_{j} + \bar{u}'^{\mathrm{L}}_{i}G^{u}_{ij}u'^{\mathrm{R}}_{j} + \bar{d}'^{\mathrm{L}}_{i}G^{d}_{ij}d'^{\mathrm{R}}_{j} + \text{h.c.}\right).
\end{aligned}
\qquad (4.2.19)
$$

From (4.2.19) we can deduce the following properties of the physical states:

1. The scalar, neutral Higgs field $\eta(x)$ has the mass

$$
M_{\mathrm{H}} = \sqrt{2}\mu. \qquad (4.2.20)
$$

2. The *masses of the vector fields* result from the bilinear form

$$\frac{1}{4}g_2^2 W_\mu^- W^{+\mu} v^2 + \frac{1}{8}(g_2 W_\mu^3 + g_1 B_\mu)^2 v^2.$$

The mass of the charged vector boson $W^\pm$ can be directly read off as

$$M_{\mathrm{W}} = \frac{g_2}{2} v. \tag{4.2.21}$$

In order to determine the masses of the neutral vector bosons we have to diagonalize the corresponding mass matrix

$$(M^0)^2 = \frac{1}{4}\begin{pmatrix} g_1^2 & g_1 g_2 \\ g_1 g_2 & g_2^2 \end{pmatrix} v^2. \tag{4.2.22}$$

As the determinant of $(M^0)^2$ vanishes, one of the two eigenvalues is zero. This reflects the fact that owing to the unbroken $\mathrm{U}(1)_{\mathrm{em}}$ symmetry one of the neutral vector fields—the photon—remains massless. The second (non-vanishing) eigenvalue is directly obtained from the trace of $(M^0)^2$ and provides the *mass for the heavy neutral vector boson*, the Z boson,

$$M_{\mathrm{Z}} = \frac{1}{2}\sqrt{g_1^2 + g_2^2}\, v. \tag{4.2.23}$$

3. The mass eigenstates of the neutral vector bosons are obtained by an orthogonal transformation

$$\begin{pmatrix} A_\mu \\ Z_\mu \end{pmatrix} = \begin{pmatrix} c_{\mathrm{W}} & -s_{\mathrm{W}} \\ s_{\mathrm{W}} & c_{\mathrm{W}} \end{pmatrix} \begin{pmatrix} B_\mu \\ W_\mu^3 \end{pmatrix} \tag{4.2.24}$$

with

$$\begin{aligned} c_{\mathrm{W}} \equiv \cos\theta_{\mathrm{W}} &= \frac{g_2}{\sqrt{g_1^2 + g_2^2}} = \frac{M_{\mathrm{W}}}{M_{\mathrm{Z}}}, \\ s_{\mathrm{W}} \equiv \sin\theta_{\mathrm{W}} &= \frac{g_1}{\sqrt{g_1^2 + g_2^2}}, \end{aligned} \tag{4.2.25}$$

and the *weak mixing angle* or *Weinberg angle* θ_{W}.

4. The Yukawa interaction term of the Lagrangian (4.2.11) provides mass terms for the fermions with the following mass matrices for leptons, up-type quarks, and down-type quarks

$$M_{ij}^l = \frac{1}{\sqrt{2}} G_{ij}^l v, \quad M_{ij}^u = \frac{1}{\sqrt{2}} G_{ij}^d v, \quad M_{ij}^d = \frac{1}{\sqrt{2}} G_{ij}^d v. \tag{4.2.26}$$

They can be diagonalized by a bi-unitary transformation with $U_{ik}^{f,\mathrm{L}}$ and $U_{ik}^{f,\mathrm{R}}$ for left-handed and right-handed fermion fields, respectively, resulting in the fermion mass eigenstates,

$$f_i^{\mathrm{L}} = \sum_k U_{ik}^{f,\mathrm{L}} f'^{\mathrm{L}}_k, \qquad f_i^{\mathrm{R}} = \sum_k U_{ik}^{f,\mathrm{R}} f'^{\mathrm{R}}_k, \tag{4.2.27}$$

and the *fermion masses*,

$$m_{f,i} = \frac{1}{\sqrt{2}} \sum_{k,m} U_{ik}^{f,\mathrm{L}} G_{km}^{f} U_{mi}^{f,\mathrm{R}\dagger} v. \tag{4.2.28}$$

Owing to the absence of right-handed neutrinos in $\mathcal{L}_{\mathrm{cl}}$, no gauge-invariant Yukawa interaction term exists that can provide masses for the neutrinos. Consequently, all neutrinos remain massless.

5. After diagonalization of the fermion mass matrices also the interaction of the Higgs field $\eta(x)$ with fermions is diagonal and the associated couplings m_f/v are proportional to the fermion masses.

In terms of the physical fields (4.2.18), (4.2.24), and (4.2.27), the fermion–gauge-boson interaction reads

$$\begin{aligned}
\mathcal{L}_{\mathrm{F,int}} = & \sum_{f=l,\nu,u,d} \sum_i [-g_2 s_{\mathrm{W}} Q_f \bar{f}_i \gamma^\mu f_i] A_\mu \\
& + \sum_{f=l,\nu,u,d} \sum_i \frac{g_2}{c_{\mathrm{W}}} [I^3_{\mathrm{W},f} \bar{f}_i^{\mathrm{L}} \gamma^\mu f_i^{\mathrm{L}} - s_{\mathrm{W}}^2 Q_f \bar{f}_i \gamma^\mu f_i] Z_\mu \\
& + \sum_{i,j} \frac{g_2}{\sqrt{2}} [\bar{u}_i^{\mathrm{L}} \gamma^\mu (U^{u,\mathrm{L}} U^{d,\mathrm{L}\dagger})_{ij} d_j^{\mathrm{L}} W_\mu^+ + \bar{d}_i^{\mathrm{L}} \gamma^\mu (U^{d,\mathrm{L}} U^{u,\mathrm{L}\dagger})_{ij} u_j^{\mathrm{L}} W_\mu^-] \\
& + \sum_{i,j} \frac{g_2}{\sqrt{2}} [\bar{\nu}_i^{\mathrm{L}} \gamma^\mu (U^{\nu,\mathrm{L}} U^{l,\mathrm{L}\dagger})_{ij} l_j^{\mathrm{L}} W_\mu^+ + \bar{l}_i^{\mathrm{L}} \gamma^\mu (U_{ij}^{l,\mathrm{L}} U^{\nu,\mathrm{L}\dagger}) \nu_j^{\mathrm{L}} W_\mu^-],
\end{aligned} \tag{4.2.29}$$

where we used $g_1 c_{\mathrm{W}} = g_2 s_{\mathrm{W}}$ from (4.2.25). Furthermore, $I^3_{\mathrm{W},f}$ and Q_f are the third component of the weak isospin and the charge of the fermion f, respectively. Inspection of (4.2.29) yields:

1. By requiring that the *photon–fermion couplings* equal those in QED,

$$\mathcal{L}_{\mathrm{QED}} = \sum_{f=l,\nu,u,d} \sum_i [\bar{f}_i (\mathrm{i}\partial\!\!\!/ - m_f) f_i - e Q_f \bar{f}_i \gamma^\mu f_i A_\mu],$$

we can relate the coupling constants g_1 and g_2 to the *elementary charge*,

$$\sqrt{4\pi\alpha} = e = g_2 s_\mathrm{W} = g_1 c_\mathrm{W} = \frac{g_1 g_2}{\sqrt{g_1^2 + g_2^2}}, \qquad \alpha \approx \frac{1}{137}. \quad (4.2.30)$$

2. Owing to their unitarity, the matrices $U_{ik}^{f,\mathrm{L}}$ and $U_{ik}^{f,\mathrm{R}}$ drop out of the interaction terms between fermions and neutral gauge bosons. Consequently, there are no flavour-changing neutral currents at tree level. However, in the quark–W-boson interaction a non-trivial matrix remains, the unitary *quark-mixing matrix* (cf. Sect. 1.4.2.2):

$$\mathbf{V} = U^{u,\mathrm{L}} U^{d,\mathrm{L}\dagger}. \quad (4.2.31)$$

Since all neutrinos are massless and thus degenerate in mass with each other, we can choose $U_{ik}^{\nu,\mathrm{L}}$ freely and eliminate the mixing matrix in the lepton–W-boson interaction by a suitable choice of the neutrino fields,

$$\nu_i^\mathrm{L} = \sum_k U_{ik}^{l,\mathrm{L}} \nu'^{\mathrm{L}}_k. \quad (4.2.32)$$

Thus, the physical neutrino states are fixed in such a way that their charged-current interaction is diagonal.

Note that massive neutrinos with non-degenerate masses would give rise to a non-trivial mixing matrix for leptons. On the other hand, the quark-mixing matrix would be trivial for degenerate masses of the up-type or down-type quarks.

As a consequence of spontaneous symmetry breaking we got the masses of the physical Higgs boson M_H, of the charged and neutral gauge bosons M_W and M_Z, of the fermions m_f, and the quark-mixing matrix $\mathbf{V}$. In addition we could identify the massless gauge boson with the photon, having the correct coupling to fermions proportional to e. This is the only remaining coupling parameter.

4.2.2.3 The Standard Model Lagrangian in terms of physical fields and parameters

We can now rewrite the complete classical Lagrangian $\mathcal{L}_\mathrm{cl}$ of the EWSM in terms of the physical fields, i.e. the mass and charge eigenstates A_μ, Z_μ,

$W_\mu^\pm$, η, l, ν, u, and d, the would-be Goldstone fields $\phi^\pm$ and χ, and the physical parameters e, M_W, M_Z, M_H, m_f, and $\mathbf{V}$,

$$\begin{aligned}
\mathcal{L}_\mathrm{cl} = & \sum_{f=l,\nu,u,d} \sum_i [\bar{f}_i(\mathrm{i}\partial\!\!\!/ - m_f)f_i - eQ_f\bar{f}_i\gamma^\mu f_i A_\mu] \\
& + \sum_{f=l,\nu,u,d} \sum_i \frac{e}{s_\mathrm{W}c_\mathrm{W}}[I^3_{\mathrm{W},f}\bar{f}_i^\mathrm{L}\gamma^\mu f_i^\mathrm{L} - s_\mathrm{W}^2 Q_f \bar{f}_i\gamma^\mu f_i]Z_\mu \\
& + \sum_{i,j} \frac{e}{\sqrt{2}s_\mathrm{W}}[\bar{u}_i^\mathrm{L}\gamma^\mu \mathbf{V}_{ij} d_j^\mathrm{L} W_\mu^+ + \bar{d}_i^\mathrm{L}\gamma^\mu \mathbf{V}_{ij}^\dagger u_j^\mathrm{L} W_\mu^-] \\
& + \sum_i \frac{e}{\sqrt{2}s_\mathrm{W}}[\bar{\nu}_i^\mathrm{L}\gamma^\mu l_i^\mathrm{L} W_\mu^+ + \bar{l}_i^\mathrm{L}\gamma^\mu \nu_i^\mathrm{L} W_\mu^-] \\
& - \frac{1}{4}\left|\partial_\mu A_\nu - \partial_\nu A_\mu - \mathrm{i}e(W_\mu^- W_\nu^+ - W_\nu^- W_\mu^+)\right|^2 \\
& - \frac{1}{4}\left|\partial_\mu Z_\nu - \partial_\nu Z_\mu + \mathrm{i}e\frac{c_\mathrm{W}}{s_\mathrm{W}}(W_\mu^- W_\nu^+ - W_\nu^- W_\mu^+)\right|^2 \\
& - \frac{1}{2}\Big|\partial_\mu W_\nu^+ - \partial_\nu W_\mu^+ - \mathrm{i}e(W_\mu^+ A_\nu - W_\nu^+ A_\mu) \\
& \qquad + \mathrm{i}e\frac{c_\mathrm{W}}{s_\mathrm{W}}(W_\mu^+ Z_\nu - W_\nu^+ Z_\mu)\Big|^2 \\
& + \frac{1}{2}\left|\partial_\mu(\eta + \mathrm{i}\chi) - \mathrm{i}\frac{e}{s_\mathrm{W}}W_\mu^-\phi^+ + \mathrm{i}M_\mathrm{Z}Z_\mu + \mathrm{i}\frac{e}{2c_\mathrm{W}s_\mathrm{W}}Z_\mu(\eta + \mathrm{i}\chi)\right|^2 \\
& + \Big|\partial_\mu\phi^+ + \mathrm{i}eA_\mu\phi^+ - \mathrm{i}e\frac{c_\mathrm{W}^2 - s_\mathrm{W}^2}{2c_\mathrm{W}s_\mathrm{W}}Z_\mu\phi^+ - \mathrm{i}M_\mathrm{W}W_\mu^+ \\
& \qquad - \mathrm{i}\frac{e}{2s_\mathrm{W}}W_\mu^+(\eta + \mathrm{i}\chi)\Big|^2 \\
& - \frac{1}{2}M_\mathrm{H}^2\eta^2 - e\frac{M_\mathrm{H}^2}{2s_\mathrm{W}M_\mathrm{W}}\eta\left(\phi^-\phi^+ + \frac{1}{2}|\eta + \mathrm{i}\chi|^2\right) \\
& - e^2\frac{M_\mathrm{H}^2}{8s_\mathrm{W}^2M_\mathrm{W}^2}\left(\phi^-\phi^+ + \frac{1}{2}|\eta + \mathrm{i}\chi|^2\right)^2 \\
& - \sum_{f=l,\nu,u,d} \sum_i e\frac{m_{f,i}}{2s_\mathrm{W}M_\mathrm{W}}(\bar{f}_i f_i\eta - 2I^3_{\mathrm{W},f}\mathrm{i}\bar{f}_i\gamma_5 f_i\chi) \\
& + \sum_{i,j} \frac{e}{\sqrt{2}s_\mathrm{W}}\left(\frac{m_{u,i}}{M_\mathrm{W}}\left(\bar{u}_i^\mathrm{R}\mathbf{V}_{ij}d_j^\mathrm{L}\phi^+ + \bar{d}_i^\mathrm{L}\mathbf{V}_{ij}^\dagger u_j^\mathrm{R}\phi^-\right)\right.
\end{aligned}$$

$$-\frac{m_{d,j}}{M_\mathrm{W}}\left(\bar{u}_i^\mathrm{L}\mathbf{V}_{ij}d_j^\mathrm{R}\phi^+ + \bar{d}_i^\mathrm{R}\mathbf{V}_{ij}^\dagger u_j^\mathrm{L}\phi^-\right)\Bigg)$$
$$-\sum_i \frac{e}{\sqrt{2}s_\mathrm{W}}\frac{m_{l,j}}{M_\mathrm{W}}\left(\bar{\nu}_i^\mathrm{L}l_i^\mathrm{R}\phi^+ + \bar{l}_i^\mathrm{R}\nu_i^\mathrm{L}\phi^-\right). \tag{4.2.33}$$

The physical content of (4.2.33) is the following:

1. The first four lines constitute the free fermion Lagrangian and the couplings of the fermions with the gauge bosons. These couplings agree with the ones given in Sect. 1.4.2.5. Thus, the *muon decay constant* is in lowest order given by [cf. (1.4.18)]
$$\frac{4G_\mu}{\sqrt{2}} = \frac{e^2}{2s_\mathrm{W}^2 M_\mathrm{W}^2} = \frac{2}{v^2}, \tag{4.2.34}$$
i.e. it is directly related to the vacuum expectation value v. The experimental value for G_μ yields $v \approx 246\,\mathrm{GeV}$.
2. Lines 5–8 contain besides the kinetic energies of the gauge bosons their non-abelian gauge couplings, which depend only on e and s_W.
3. Lines 9–11 originate from the gauge-invariant Higgs–gauge-boson interaction. Besides the kinetic terms for the Higgs bosons they contain the gauge-boson mass terms and bilinear terms that give rise to tree-level mixing between gauge bosons and would-be Goldstone bosons.
4. Lines 12 and 13 describe the Higgs-boson self-interactions. This part depends on the Higgs-boson mass, a parameter which is presently not yet known experimentally.
5. The Yukawa couplings in the last 4 lines, however, are fixed by $v = 2s_\mathrm{W}M_\mathrm{W}/e$ and m_f only.

The Lagrangian (4.2.33) is invariant under gauge transformations that follow from (4.2.13) by inserting the physical fields and parameters,

$$\begin{aligned}
A_\mu &\to A_\mu + \partial_\mu\delta\theta^A + \mathrm{i}e[W_\mu^+\delta\theta^- - W_\mu^-\delta\theta^+],\\
Z_\mu &\to Z_\mu + \partial_\mu\delta\theta^Z - \mathrm{i}e\frac{c_\mathrm{W}}{s_\mathrm{W}}[W_\mu^+\delta\theta^- - W_\mu^-\delta\theta^+],\\
W_\mu^\pm(x) &\to W_\mu^\pm(x) + \partial_\mu\delta\theta^\pm\\
&\quad \mp\mathrm{i}\frac{e}{s_\mathrm{W}}\left[W_\mu^\pm\left(s_\mathrm{W}\delta\theta^A - c_\mathrm{W}\delta\theta^Z\right) - (s_\mathrm{W}A_\mu - c_\mathrm{W}Z_\mu)\delta\theta^\pm\right],
\end{aligned}$$

$$
\begin{aligned}
\eta &\to \eta + \frac{e}{2s_{\mathrm{W}}c_{\mathrm{W}}}\chi\delta\theta^Z + \frac{\mathrm{i}e}{2s_{\mathrm{W}}}[\phi^+\delta\theta^- - \phi^-\delta\theta^+], \\
\chi &\to \chi - \frac{e}{2s_{\mathrm{W}}c_{\mathrm{W}}}[v+\eta]\delta\theta^Z + \frac{e}{2s_{\mathrm{W}}}[\phi^+\delta\theta^- + \phi^-\delta\theta^+], \\
\phi^\pm &\to \phi^\pm \mp \mathrm{i}e\phi^\pm\left[\delta\theta^A + \frac{s_{\mathrm{W}}^2 - c_{\mathrm{W}}^2}{2c_{\mathrm{W}}s_{\mathrm{W}}}\delta\theta^Z\right] \pm \frac{\mathrm{i}e}{2s_{\mathrm{W}}}[v+\eta\pm\mathrm{i}\chi]\delta\theta^\pm, \\
\psi^{\mathrm{L}}_{i,\pm} &\to \psi^{\mathrm{L}}_{i,\pm} - \mathrm{i}e\left[Q_{i,\pm}\delta\theta^A + \frac{s_{\mathrm{W}}}{c_{\mathrm{W}}}\left(Q_{i,\pm} \mp \frac{1}{2s_{\mathrm{W}}^2}\right)\delta\theta^Z\right]\psi^{\mathrm{L}}_{i,\pm} \\
&\quad + \mathrm{i}\frac{e}{\sqrt{2}s_{\mathrm{W}}}\delta\theta^\pm v^\pm_{ij}\psi^{\mathrm{L}}_{j,\mp}, \\
\psi^{\mathrm{R}}_{i,\sigma} &\to \psi^{\mathrm{R}}_{i,\sigma} - \mathrm{i}eQ_{i,\sigma}\left(\delta\theta^A + \frac{s_{\mathrm{W}}}{c_{\mathrm{W}}}\delta\theta^Z\right)\psi^{\mathrm{R}}_{i,\sigma},
\end{aligned}
\tag{4.2.35}
$$

where

$$
\delta\theta^\pm = \frac{1}{g_2}(\delta\theta^1 \mp \mathrm{i}\delta\theta^2), \tag{4.2.36}
$$

$$
\delta\theta^A = \frac{1}{g_1}c_{\mathrm{W}}\delta\theta^Y - \frac{1}{g_2}s_{\mathrm{W}}\delta\theta^3, \qquad \delta\theta^Z = \frac{1}{g_2}c_{\mathrm{W}}\delta\theta^3 + \frac{1}{g_1}s_{\mathrm{W}}\delta\theta^Y,
$$

$v^+_{ij} = \mathbf{V}_{ij}$, $v^-_{ij} = \mathbf{V}^\dagger_{ij}$ for quarks, and $v^\pm_{ij} = \delta_{ij}$ for leptons.

After adding a gauge-fixing term for the photon as in QED [cf. (2.4.11) and (2.4.23)], the Lagrangian (4.2.33) is suitable for tree-level calculations. The would-be Goldstone fields $\phi^\pm$ and χ can be eliminated by a gauge transformation. This yields $\mathcal{L}_{\mathrm{cl}}$ in the unitary gauge which is directly obtained from (4.2.33) by putting $\phi^\pm = \chi = 0$. The corresponding Feynman rules can be obtained from those given in the App. A.2 by leaving out all terms involving the would-be Goldstone fields $\phi^\pm$ and χ and the Faddeev–Popov ghosts u^{A}, u^{Z}, and $u^\pm$, omitting in the remaining terms all counterterms δZ, and using the massive *gauge-boson propagators in the unitary gauge* (1.4.13) $(V = W, Z)$,

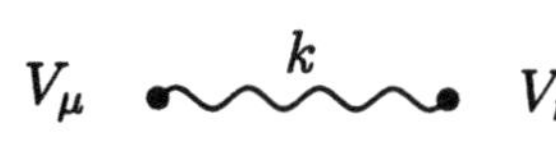

$$
= \mathrm{i}\frac{-g_{\mu\nu} + k_\mu k_\nu/M_V^2}{k^2 - M_V^2}. \tag{4.2.37}
$$

If we choose the parameter v in (4.2.17) in such a way that it does not satisfy (4.2.14), we get additional contributions to the Lagrangian (4.2.33) arising from the Higgs potential,

$$\begin{aligned}\delta\mathcal{L}_t = t\eta &+ \frac{te}{2s_{\mathrm{w}}M_{\mathrm{W}}}\left(\phi^-\phi^+ + \frac{1}{2}\chi^2\right) \\ &- e\frac{te}{4s_{\mathrm{w}}^2M_{\mathrm{W}}^2}\eta\left(\phi^-\phi^+ + \frac{1}{2}|\eta + \mathrm{i}\chi|^2\right) \\ &- e^2\frac{te}{16s_{\mathrm{w}}^3M_{\mathrm{W}}^3}\left(\phi^-\phi^+ + \frac{1}{2}|\eta + \mathrm{i}\chi|^2\right)^2 ,\end{aligned} \tag{4.2.38}$$

with the *tadpole*

$$t = v\left(\mu^2 - \frac{1}{4}\lambda v^2\right). \tag{4.2.39}$$

From this equation it is manifest that the correct vacuum expectation value v corresponds to the vanishing of the tadpole t.

4.3 Simple applications of the Electroweak Standard Model

We have already discussed some applications of the EWSM in Sect. 1.4. We considered, in particular, muon decay, neutrino–electron scattering, Z and W decays, and Z-boson production in e^+e^- annihilation. In all these reactions only the fermion–gauge-boson interaction of the EWSM enters in lowest-order perturbation theory, which is exactly the one considered in Sect. 1.4. In this section we add examples that are sensitive to other sectors of the EWSM. In order to illustrate the effects of the non-abelian gauge couplings, we consider W-pair production in e^+e^- collisions. As examples for reactions involving the Higgs boson and its interactions we discuss decays of the Higgs boson and production of the Higgs boson in e^+e^- collisions via $e^+e^- \to \mathrm{HZ}$.

4.3.1 W-pair production in e^+e^- annihilation

The process $e^+e^- \to W^+W^-$ allows to study directly the properties of the W-boson, i.e. its mass, its total and partial widths, and its couplings in the clean environment of e^+e^- colliders. One important aspect is the investigation of the non-abelian gauge interaction. On the other hand, the W-pair production process in the LEP2 region can be used for an independent direct measurement of the W-boson mass at the level of 0.05% [Ku96]. The present experimental error is 46 MeV [Gu00]. This in turn gives rise to a stringent precision test of the EWSM (cf. Sect. 4.6.5).

We first settle our conventions. Because W-pair production requires a centre-of-mass (CM) energy greater than twice the W-boson mass, we can safely neglect the electron mass. The momenta and helicities of the incoming $e^\pm$ are denoted by $p_\pm$ and $\sigma_\pm$, respectively, those of the outgoing $W^\pm$ by $k_\pm$ and $\lambda_\pm$. In the CM system of the e^+e^- pair, the momenta read

$$p_\pm^\mu = E(1,0,0,\pm 1), \qquad k_\pm^\mu = E(1,\pm\beta\sin\vartheta,0,\pm\beta\cos\vartheta) \tag{4.3.1}$$

with E denoting the beam energy, ϑ the scattering angle between the e^+ and the W^+, and $\beta = \sqrt{1 - M_W^2/E^2}$ the velocity of the W bosons in the CM frame. The *Mandelstam variables* are given by

$$\begin{aligned} s &= (p_+ + p_-)^2 = (k_+ + k_-)^2 = 4E^2, \\ t &= (p_+ - k_+)^2 = (p_- - k_-)^2 = M_W^2 - 2E^2 + 2E^2\beta\cos\vartheta, \\ u &= (p_+ - k_-)^2 = (p_- - k_+)^2 = M_W^2 - 2E^2 - 2E^2\beta\cos\vartheta. \end{aligned} \tag{4.3.2}$$

In order to define helicity amplitudes, we need to introduce the corresponding polarization vectors for the W bosons,[5]

$$\begin{aligned} \varepsilon_\pm^{*\mu}(k_\pm,+1) &= \frac{1}{\sqrt{2}}(0,\pm\cos\vartheta,-\mathrm{i},\mp\sin\vartheta), \\ \varepsilon_\pm^{*\mu}(k_\pm,-1) &= \frac{1}{\sqrt{2}}(0,\pm\cos\vartheta,\mathrm{i},\mp\sin\vartheta), \\ \varepsilon_\pm^{*\mu}(k_\pm,\ \ 0) &= \frac{E}{M_W}(\beta,\pm\sin\vartheta,0,\pm\cos\vartheta) \\ &= \frac{1}{M_W}k_\pm^\mu - \frac{M_W}{E(1+\beta)}(1,\mp\sin\vartheta,0,\mp\cos\vartheta). \end{aligned} \tag{4.3.3}$$

[5]Note that the helicity of massive W bosons is not Lorentz-invariant. We define it in the CM system.

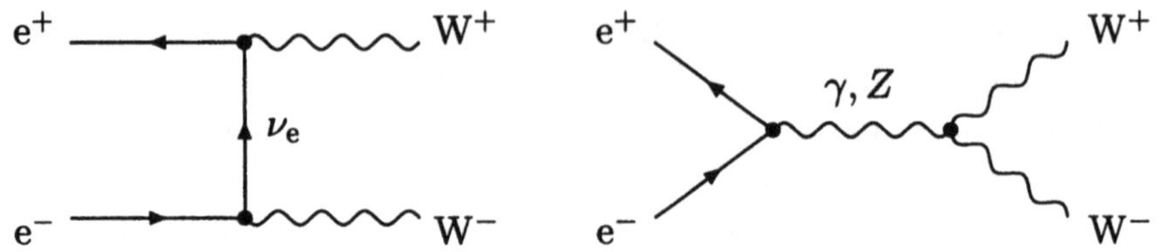

Fig. 4.2 Lowest-order diagrams for $e^+e^- \to W^+W^-$

Since we neglect the electron mass, the invariant matrix element is only non-zero for $\sigma_- = -\sigma_+ = \sigma$. Right-handed electrons are labelled by $\sigma = +$ and left-handed ones by $\sigma = -$.

At tree level only the three diagrams shown in Fig. 4.2 contribute, if we neglect a Higgs-exchange diagram which is suppressed by a factor m_e/M_W originating from the Higgs–electron coupling.[6] The t-channel ν_e-exchange diagram contributes only for left-handed electrons, whereas the s-channel diagrams containing the non-abelian triple gauge coupling contribute for both helicities. Introducing the basic matrix elements

$$
\begin{aligned}
\mathcal{M}_t^{\pm} &= \bar{v}(p_+)\not{\varepsilon}_+(\not{k}_+ - \not{p}_+)\not{\varepsilon}_- \frac{1\pm\gamma_5}{2} u(p_-), \\
\mathcal{M}_s^{\pm} &= \bar{v}(p_+)[2\not{\varepsilon}_+(\varepsilon_- k_+) - 2\not{\varepsilon}_-(\varepsilon_+ k_-) \\
&\quad - (\not{k}_+ - \not{k}_-)(\varepsilon_+\varepsilon_-)] \frac{1\pm\gamma_5}{2} u(p_-),
\end{aligned} \tag{4.3.4}
$$

we can write the lowest-order amplitude as

$$
\begin{aligned}
\mathcal{M}_{\mathrm{Born}}(\sigma, \lambda_+, \lambda_-, s, t) &= \frac{e^2}{2s_W^2}\frac{1}{t}\mathcal{M}_t^{\sigma}\delta_{\sigma-} + e^2\frac{1}{s}\mathcal{M}_s^{\sigma} - \frac{c_W}{s_W} g_e^{\sigma}\frac{1}{s-M_Z^2}\mathcal{M}_s^{\sigma} \\
&= \frac{e^2}{2s_W^2}\left[\frac{1}{t}\mathcal{M}_t^{\sigma} + \frac{1}{s-M_Z^2}\mathcal{M}_s^{\sigma}\right]\delta_{\sigma-} - e^2\left[\frac{M_Z^2}{s(s-M_Z^2)}\right]\mathcal{M}_s^{\sigma}
\end{aligned} \tag{4.3.5}
$$

with g_e^{σ} defined in (A.2.22). In the second line we have inserted the explicit form of the Z-boson–fermion couplings and rearranged the contributions of the individual diagrams into two gauge-invariant subsets, one of which

[6] Actually, the suppression factor is $(m_e/M_W)^2$, where the second factor results from $\bar{v}(p_+)u(p_-)$.

$\sqrt{s}$ [GeV]	σ^{Born}	$\sigma_{-}^{\mathrm{Born}}$	$\sigma_{+}^{\mathrm{Born}}$	$\sigma_{\mathrm{T}}^{\mathrm{Born}}$	$\sigma_{\mathrm{L}}^{\mathrm{Born}}$	$\sigma_{\mathrm{M}}^{\mathrm{Born}}$
165.0	10.66	21.27	0.05	5.19	1.34	4.13
176.0	16.47	32.74	0.21	8.84	1.79	5.84
190.0	18.12	35.92	0.32	10.76	1.65	5.70
250.0	15.40	30.52	0.27	11.77	0.75	2.87
500.0	6.71	13.37	0.05	6.34	0.11	0.25
1000.0	2.51	5.00	0.01	2.46	0.03	0.02

Tab. 4.2 Integrated lowest-order cross section in pb for different polarizations and different CM energies

involves the charged-current coupling constant $e/(\sqrt{2}s_{\mathrm{W}})$ and the other the electromagnetic coupling constant e.

According to (1.2.26) the unpolarized cross section is given by

$$\left(\frac{\mathrm{d}\sigma}{\mathrm{d}\Omega}\right)_{\mathrm{Born}} = \frac{\beta}{64\pi^2 s}\frac{1}{4}\sum_{\sigma,\lambda_+,\lambda_-}|\mathcal{M}_{\mathrm{Born}}(\sigma,\lambda_+,\lambda_-,s,t)|^2 . \tag{4.3.6}$$

Integration over the solid angle yields the total cross section [Al77a]

$$\begin{aligned}\sigma_{\mathrm{Born}} = \frac{\pi\alpha^2}{2s_{\mathrm{W}}^4}\frac{\beta}{s}\Bigg\{&\left[1+\frac{2M_{\mathrm{W}}^2}{s}+\frac{2M_{\mathrm{W}}^4}{s^2}\right]\frac{1}{\beta}\log\frac{1+\beta}{1-\beta}-\frac{5}{4}\\ &+\frac{M_{\mathrm{Z}}^2(1-2s_{\mathrm{W}}^2)}{s-M_{\mathrm{Z}}^2}\left[2\frac{M_{\mathrm{W}}^2}{s}\left(2+\frac{M_{\mathrm{W}}^2}{s}\right)\frac{1}{\beta}\log\frac{1+\beta}{1-\beta}\right.\\ &\left.-\frac{s}{12M_{\mathrm{W}}^2}-\frac{5}{3}-\frac{M_{\mathrm{W}}^2}{s}\right]\\ &+\frac{M_{\mathrm{Z}}^4(1-4s_{\mathrm{W}}^2+8s_{\mathrm{W}}^4)\beta^2}{48(s-M_{\mathrm{Z}}^2)^2}\left[\frac{s^2}{M_{\mathrm{W}}^4}+\frac{20s}{M_{\mathrm{W}}^2}+12\right]\Bigg\}.\end{aligned} \tag{4.3.7}$$

In Table 4.2 we list the cross section σ for different CM energies and different polarizations of the electrons and W bosons for $M_{\mathrm{W}} = 80.30\,\mathrm{GeV}$ and $M_{\mathrm{Z}} = 91.187\,\mathrm{GeV}$. The cases of two transverse, one transverse and one longitudinal, and two longitudinal W bosons are denoted by T, M, and L, respectively. Positrons are assumed to be unpolarized.

Since the longitudinal polarization vectors are proportional to E/M_W [cf. (4.3.3)], the contributions of individual Feynman diagrams to the invariant matrix element grow with energy for longitudinal W bosons. Using $\varepsilon_\pm^{*\mu} \sim k_\pm^\mu/M_\mathrm{W}$, $E \gg M_\mathrm{W}$, we obtain

$$\begin{aligned}\mathcal{M}_t^\pm &= -\frac{t}{M_\mathrm{W}^2}\bar{v}(p_+)\not{k}_+\frac{1\pm\gamma_5}{2}u(p_-),\\ \mathcal{M}_s^\pm &= \frac{s-2M_\mathrm{W}^2}{M_\mathrm{W}^2}\bar{v}(p_+)\not{k}_+\frac{1\pm\gamma_5}{2}u(p_-)\end{aligned} \tag{4.3.8}$$

for the matrix elements involving two longitudinal gauge bosons. The absolute square of the spinor chain $\bar{v}(p_+)\not{k}_+(1\pm\gamma_5)u(p_-)$ can be evaluated as a trace,

$$\begin{aligned}\left|\bar{v}(p_+)\not{k}_+\frac{1\pm\gamma_5}{2}u(p_-)\right|^2 &= \mathrm{Tr}\left\{\not{p}_+\not{k}_+\frac{1\pm\gamma_5}{2}\not{p}_-\not{k}_+\frac{1\pm\gamma_5}{2}\right\}\\ &= \frac{s^2}{4}\beta^2\sin^2\vartheta,\end{aligned} \tag{4.3.9}$$

resulting in

$$\bar{v}(p_+)\not{k}_+\frac{1\pm\gamma_5}{2}u(p_-) = \frac{s}{2}\beta\sin\vartheta\,\mathrm{e}^{\mathrm{i}\phi}, \tag{4.3.10}$$

where ϕ is a convention-dependent phase which we set to zero. Putting everything together we obtain for two longitudinal W bosons in the high-energy limit

$$\frac{\mathcal{M}_t^\sigma}{t} \sim -\frac{\mathcal{M}_s^\sigma}{s} \sim -\frac{s}{2M_\mathrm{W}^2}\sin\vartheta, \qquad s, |t| \gg M_\mathrm{W}^2. \tag{4.3.11}$$

Consequently, the cross section resulting from the t-channel diagram alone grows with s in the high-energy limit

$$\left(\frac{\mathrm{d}\sigma}{\mathrm{d}\Omega}\right)^{\mathrm{LL}}_{\mathrm{Born},t} \sim \frac{\alpha^2}{16s}\frac{s^2}{16s_\mathrm{W}^4 M_\mathrm{W}^4}\sin^2\vartheta, \qquad s, |t| \gg M_\mathrm{W}^2, \tag{4.3.12}$$

and thus eventually violates unitarity.

In the EWSM, the gauge symmetry guarantees that the s-channel diagrams exactly cancel the bad high-energy behaviour of the t-channel diagram so

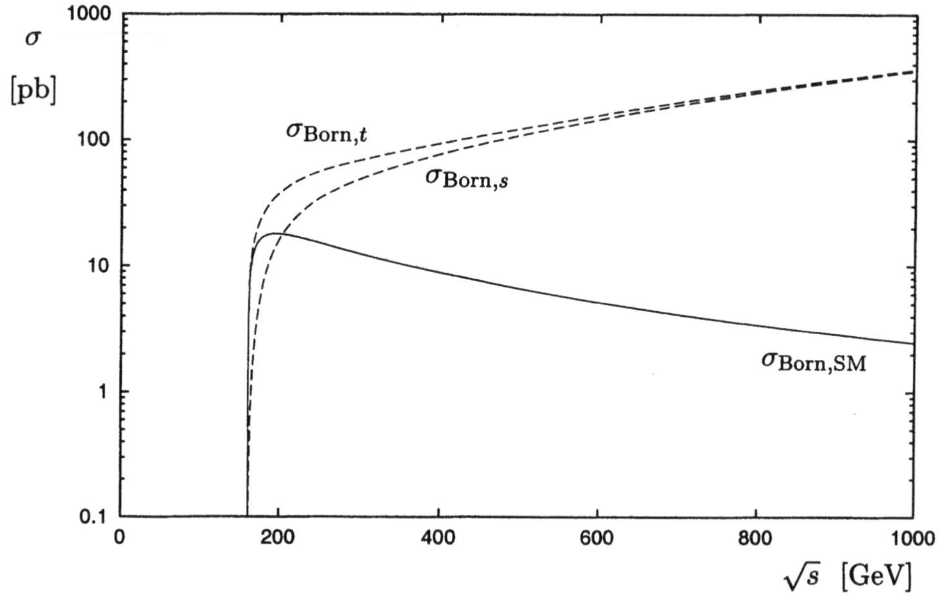

Fig. 4.3 Gauge cancellations in the total cross section for W-pair production: the individual curves correspond to the Born cross sections arising from the s-channel ($\sigma_{\mathrm{Born},s}$) and t-channel ($\sigma_{\mathrm{Born},t}$) diagrams alone, and to the complete EWSM cross section ($\sigma_{\mathrm{Born,SM}}$).

that the complete lowest-order matrix element for longitudinal W bosons approaches a constant at high energies. The cross section in the EWSM

$$\left(\frac{\mathrm{d}\sigma}{\mathrm{d}\Omega}\right)^{\mathrm{LL}}_{\mathrm{Born}} \sim \frac{\alpha^2}{16s}\frac{1+4s_{\mathrm{W}}^4}{16s_{\mathrm{W}}^4c_{\mathrm{W}}^4}\sin^2\vartheta, \qquad s \gg M_{\mathrm{W}}^2 \tag{4.3.13}$$

drops proportional to $1/s$ in the high-energy limit ($s \gg M_{\mathrm{W}}^2$). The cancellations between the individual Feynman diagrams are illustrated in Fig. 4.3 for the total cross section. They reach one order of magnitude at 400 GeV and two orders at 1 TeV.

We have found that the cross section $\sigma(\mathrm{e}^+\mathrm{e}^- \to \mathrm{W}_{\mathrm{L}}^+\mathrm{W}_{\mathrm{L}}^-)$ within the EWSM respects unitarity at lowest-order. This is an example of a general property of gauge theories: while tree-level unitarity is violated in processes involving longitudinal gauge bosons in general theories, it is guaranteed in gauge theories. In fact, the requirement of tree-level unitarity of all physical processes applied to a general quantum field theory involving vector fields enforces a gauge theory [Ll73, Co74a].

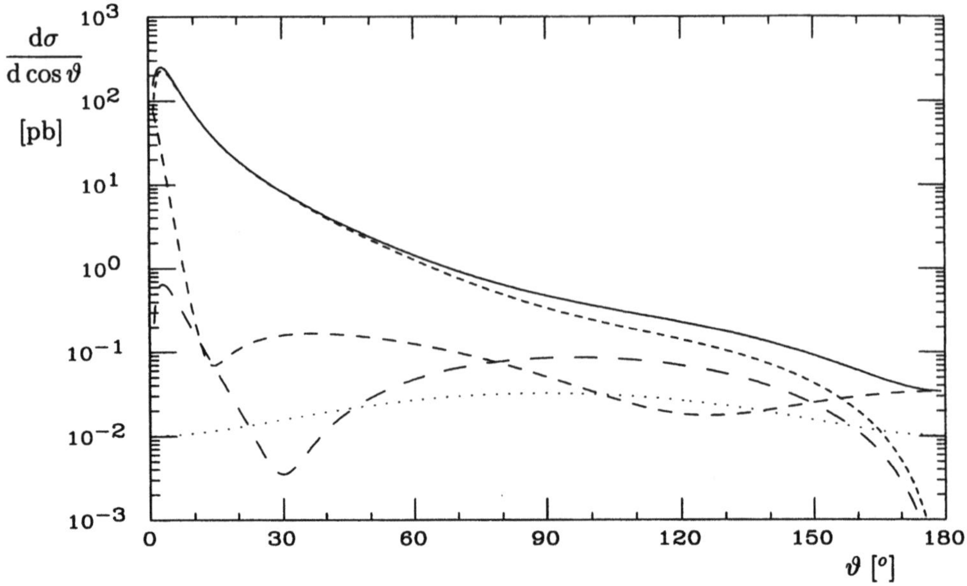

Fig. 4.4 Lowest-order differential cross section at $\sqrt{s} = 500\,\text{GeV}$ (The meaning of the curves is explained in the text)

After the gauge cancellations, the t-channel contributions dominate the EWSM cross section at high energies owing to a large coupling constant and the $1/t$ pole. However, the t-channel contributions are absent or suppressed for all polarized cross sections except for transverse W bosons and left-handed electrons. Thus, this polarization dominates the total cross section at high energies and all others are smaller by at least a factor $30 \log(s/M_W^2)$. These features are illustrated by Fig. 4.4, which shows the differential cross section for various polarizations at $\sqrt{s} = 500\,\text{GeV}$. The solid curve corresponds to unpolarized W bosons and unpolarized electrons, the dotted one to unpolarized W bosons and right-handed electrons. The dashed curves represent unpolarized electrons and, in order of increasing dash length, the cases of two transverse, one transverse and one longitudinal, and two transverse W bosons, respectively. The situation at $\sqrt{s} = 190\,\text{GeV}$ is illustrated in Fig. 4.5. Even at this energy a t-channel enhancement in the forward direction of about a factor 20 is present.

The threshold behaviour is important for the determination of the W-

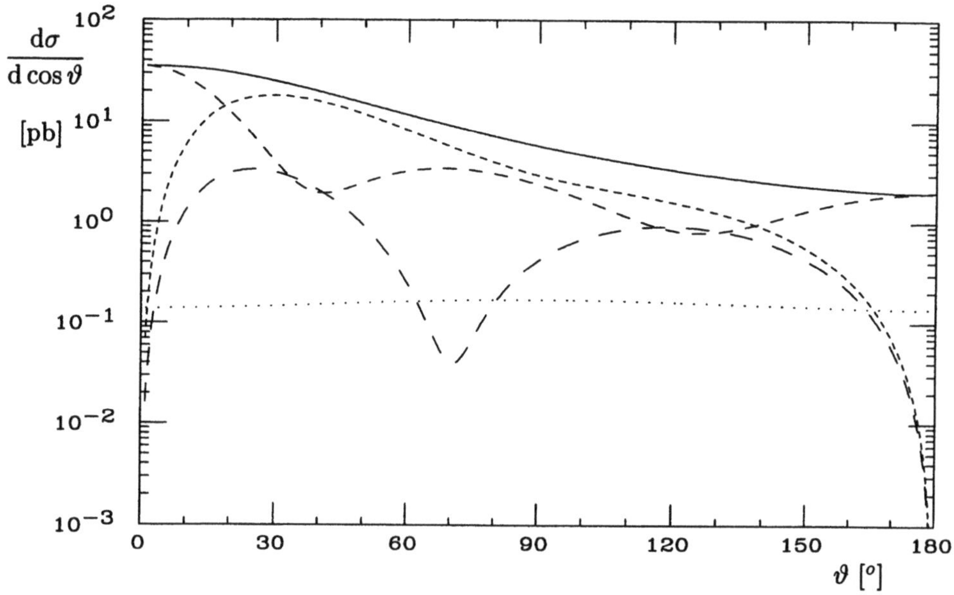

Fig. 4.5 Lowest-order differential cross section at $\sqrt{s} = 190\,\mathrm{GeV}$ (same conventions as in Fig. 4.4)

boson mass [Ku96] from the measurement of the W-pair-production cross section close to threshold. For small β, the matrix elements behave as

$$\frac{\mathcal{M}_s^\sigma}{s} \propto \beta, \qquad \frac{\mathcal{M}_t^\sigma}{t} \propto 1 \tag{4.3.14}$$

for fixed scattering angles. Consequently, the s-channel matrix elements vanish at threshold and the t-channel graph dominates in the threshold region. For $\beta \ll 1$ the differential cross section for unpolarized beams and W bosons is given by

$$\left(\frac{\mathrm{d}\sigma}{\mathrm{d}\Omega}\right)_{\mathrm{Born}} \approx \frac{\alpha^2}{s}\frac{1}{4s_\mathrm{W}^4}\beta\left[1 + 4\beta\cos\vartheta\,\frac{3c_\mathrm{W}^2-1}{4c_\mathrm{W}^2-1} + \mathcal{O}(\beta^2)\right], \tag{4.3.15}$$

where the (angular-independent) leading term $\propto \beta$ originates only from the t-channel diagram. In the total cross section

$$\sigma_{\mathrm{Born}} \approx \frac{\pi\alpha^2}{s}\frac{1}{4s_\mathrm{W}^4}4\beta + \mathcal{O}(\beta^3) \tag{4.3.16}$$

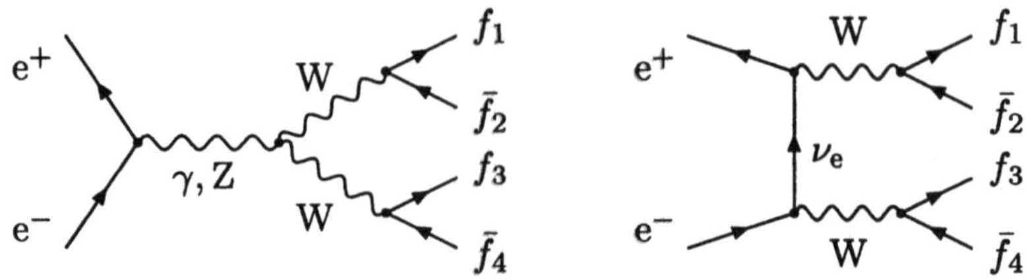

Fig. 4.6 Doubly-resonant lowest-order diagrams for $e^+e^- \to WW \to 4f$

all terms $\propto \beta^2$ drop out and the s-channel and the s–t-interference contribution are proportional to β^3. These properties can be derived from electron-helicity conservation, CP invariance, and the orthogonality of different partial waves [Be94].

The total cross section close to threshold is completely determined by the t-channel diagram and thus practically independent of possible non-standard physics contributions, such as anomalous gauge-boson couplings, which affect mainly the s channel. The normalization of the cross section in the threshold region is determined by the well known W-boson–fermion coupling. Its energy dependence is governed by the linear rise in β. This is even true after the inclusion of finite-width effects because the next-to-leading β^3 term becomes sizeable only several Γ_W above threshold. This fact allows for a largely model-independent determination of the W-boson mass from the W-pair-production threshold. In fact, the W-boson mass has been measured in this way at LEP with the result $M_W = 80.40 \pm 0.22\,\text{GeV}$ [LEP97].

As discussed already in Sect. 1.4.2.4, the W bosons are unstable and decay rapidly. The approximation of stable W bosons is not adequate, and one has to include the effects of the finite width of the W bosons. Thus, instead of the process $e^+e^- \to W^+W^-$ one has to consider $e^+e^- \to W^+W^- \to 4f$, i.e. a four-fermion final state [Be94]. The corresponding Feynman diagrams are shown in Fig. 4.6. The decay of the W bosons allows for an alternative method for the measurement of the W-boson mass: one measures the four-momenta of the final-state fermions and reconstructs from those the invariant masses of the fermion pairs. Then, the W-boson mass can be obtained by a fit to the invariant-mass distributions of the fermion pairs. The present experimental result from LEP2 for the W-boson mass reads $M_W = 80.427 \pm 0.046\,\text{GeV}$ [Gu00].

The cross section resulting from the Feynman diagrams in Fig. 4.6 is not gauge-invariant. This is due to the fact that subprocesses other than W-pair production contribute to the four-fermion final state. Therefore, all Feynman diagrams for $e^+e^- \to 4f$ have to be included in order to obtain the gauge-invariant cross section that is observed experimentally.

However, the W-pair-production diagrams dominate the cross section in regions of phase space where on-shell W-pair-production is allowed owing to the enhancement by the two resonant W-boson propagators. By subtracting the small contribution of the non-W-pair-production diagrams from the experimentally measured cross section one can define an off-shell W-pair-production cross section, conventionally called CC03 cross section [Ba96]. This cross section is theoretically defined from the W-pair-production or CC03 diagrams in Fig. 4.6, calculated in the Feynman gauge. Figure 4.7 shows the experimental data points for this cross section as given by the LEP Electroweak Working Group [LEP00] for the Summer 2000 conferences together with theoretical calculations. There is good agreement between data and theoretical predictions at the per-cent level.

To reach this agreement, the complete $\mathcal{O}(\alpha)$ radiative corrections to $e^+e^- \to WW \to 4f$ had to be included in the theoretical prediction. In order to obtain a gauge-invariant result these corrections have been calculated in the double-pole approximation. In this approximation only the leading term in an expansion of the matrix elements about the poles of the two resonant propagators is taken into account. The theoretical prediction including these and further relevant higher-order effects are provided by the Monte Carlo programs YFSWW3 [Ja98] and RACOONWW [De00a]. Below a CM energy of 170 GeV, where the double-pole approximation is no more applicable, the prediction in Figure 4.7 is continued by the program GENTLE [Ba93a], which does not include the non-universal electroweak corrections. The version of GENTLE used in this figure is tuned to reproduce the prediction of RACOONWW and YFSWW3 on the total cross section at LEP2 within a few per mill (see Ref. [Gr00]).

While the treatment of the four-fermion process is necessary for a theoretically adequate description, the essential features of W-pair production discussed above in the on-shell approximation are not affected by these refinements.

4.3.2 Production and decay of the Higgs boson

In the EWSM, the fundamental particles (leptons, quarks, and weak gauge bosons) acquire masses through their interactions with a scalar field with

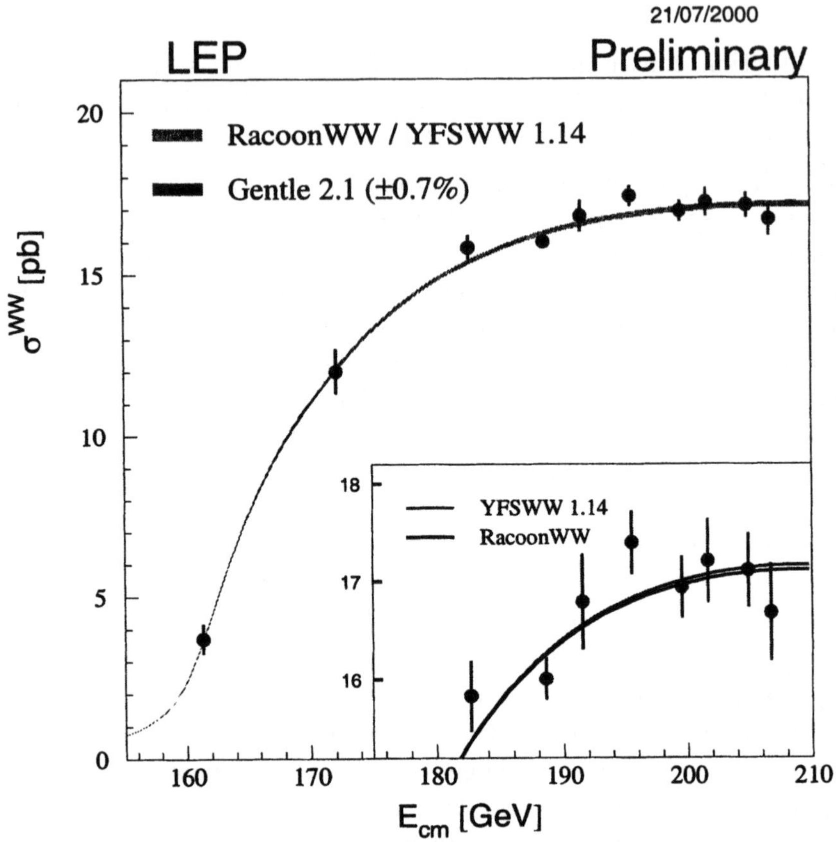

Fig. 4.7 Total WW production cross section at LEP2 as given by the LEPEWWG [LEP00]

non-zero vacuum expectation value. The experimentally observable consequence of this mechanism is the neutral scalar Higgs particle with its couplings to fermions and gauge bosons. So far, however, there is no experimental evidence for its existence. Within the EWSM the mass of the Higgs boson is a free parameter that is not directly related to known experimental quantities.

4.3.2.1 Higgs-boson mass

For a large Higgs-boson mass, the Higgs sector becomes strongly interacting and instead of the Higgs boson many resonances should show up in the TeV range. The physical content of the EWSM changes drastically. Therefore, it is interesting to derive limits on the mass of the Higgs boson by requiring that the EWSM does not become strongly interacting. These limits do not apply to the model in general but only to its perturbative realization.

The strength of the Higgs self-interaction is determined by the Higgs-boson mass itself. For instance, the coupling between four Higgs bosons is given by $g_{\mathrm{HHHH}} = 3e^2 M_{\mathrm{H}}^2/4s_{\mathrm{w}}^2 M_{\mathrm{W}}^2 = 3M_{\mathrm{H}}^2/2v^2$. Requiring $g_{\mathrm{HHHH}}/4\pi < 1$ implies [Ve77, Le77]

$$M_{\mathrm{H}} \lesssim \sqrt{8\pi/3}v \sim 700\,\mathrm{GeV}. \tag{4.3.17}$$

A similar limit can be obtained by considering the running of the quartic Higgs coupling $\lambda(\mu) = 2M_{\mathrm{H}}^2(\mu)/v^2$ [Ca79, Li86]. Solving the renormalization group equation for the quartic coupling

$$\frac{\mathrm{d}\lambda}{\mathrm{d}\log\mu^2} = \frac{3}{8\pi^2}(\lambda^2 + \lambda g_{\mathrm{t}}^2 - g_{\mathrm{t}}^4) + \mathcal{O}(e^2) + \mathcal{O}(g_f^2), \tag{4.3.18}$$

where $g_{\mathrm{t}} = \sqrt{2}m_{\mathrm{t}}/v$ denotes the Yukawa coupling of the top quark, one finds that the solution develops a Landau pole. Thus, the theory cannot be valid for arbitrary high energy scales unless the renormalized quartic coupling vanishes. This phenomenon is known as triviality. Assuming that the theory is valid only below a certain scale Λ and requiring that the Landau pole does not occur below this scale implies an upper limit on the Higgs-boson mass for a given top-quark mass. In fact, this limit is close to the limit that results from the requirement that the system does not become strongly interacting, i.e. $\lambda(\mu)/4\pi < 1$ below the scale Λ. The obtained limits become larger with decreasing Λ. The condition $M_{\mathrm{H}} < \Lambda$ sets an upper limit on the Higgs-boson mass of about 700 GeV. This renormalization group argument has been confirmed in lattice analyses [Ha87, Lu88, Go93].

Similar limits can be derived from the requirement of tree-level unitarity in longitudinal gauge-boson-scattering processes.

A lower limit on the Higgs-boson mass of 113.3 GeV has been obtained from direct experimental searches [Ig00]. Confronting precision experiments with the EWSM including radiative corrections favours a light Higgs boson. A typical upper limit from such an analysis is $M_{\mathrm{H}} < 170\,\mathrm{GeV}$ at 95% CL [Gu00] (cf. Sect. 4.6.5).

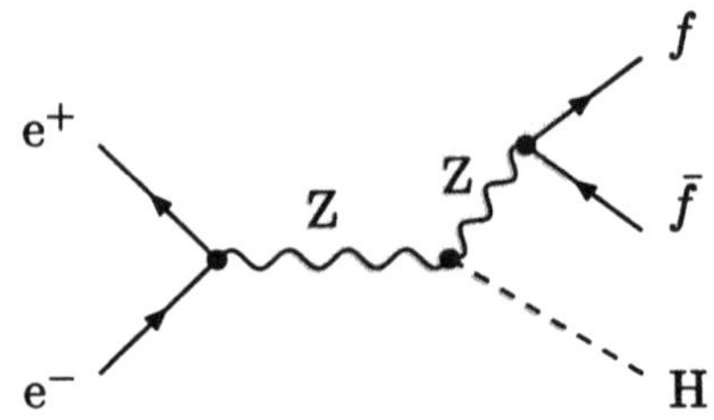

Fig. 4.8 Lowest-order diagram for $e^+e^- \to Hf\bar{f}$

4.3.2.2 Higgs-boson production

We have seen in Sect. 4.2 that the couplings of the Higgs boson to other particles are proportional to the masses of those particles. Therefore, light particles essentially play no rôle in reactions involving the Higgs boson.[7] In particular, the direct production of the Higgs boson via e^+e^- or $q\bar{q}$ annihilation is practically excluded. The Higgs boson can be most easily produced via its couplings to the weak gauge bosons or to the top quark.

The present lower limit on the Higgs-boson mass M_H of the order of 113 GeV originates from direct searches at LEP using the process $e^+e^- \to Hf\bar{f}$. The relevant Feynman diagram is shown in Fig. 4.8. For $E_{CM} \approx M_Z^2$, the cross section for this process is enhanced, because the Z boson that couples to e^+e^- becomes resonant. This allowed to search for a Higgs boson with a mass below 70 GeV. For larger Higgs-boson masses the Higgs signal cannot be separated from the background. However, for $E_{CM} > M_H + M_Z$ the second Z-boson can be real, and the corresponding cross section becomes again large. In fact, this process would be the dominating source of Higgs bosons for energies below ~ 500 GeV, should the Higgs-boson mass fall into this range.

We consider the Higgs-strahlung or Bjorken process

$$e^+(p_+,\sigma_+) + e^-(p_-,\sigma_-) \to Z(k_Z,\lambda) + H(k_H) \tag{4.3.19}$$

in some detail. Here $\sigma_\pm = +1/2, -1/2$ and $\lambda = +1, 0, -1$ denote the helicities of the incoming leptons and the outgoing Z boson. Neglecting the

[7]An exception is given by the gluons, which couple to the Higgs field via fermion loops involving the strong coupling.

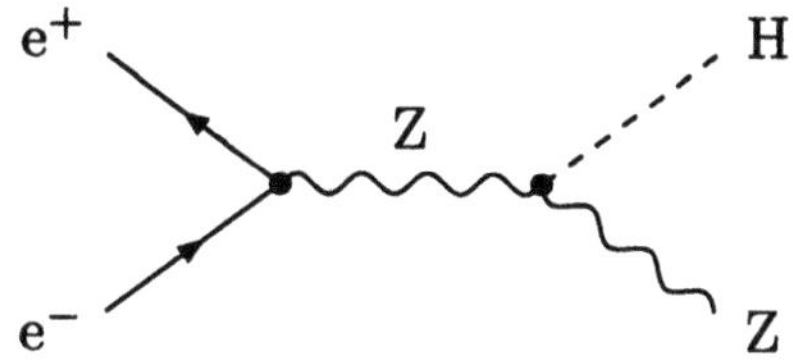

Fig. 4.9 Lowest-order diagram for $e^+e^- \to HZ$

electron mass, which is small with respect to the energies involved, the momenta read

$$\begin{aligned} p_\pm^\mu &= E(1,0,0,\pm 1), \\ k_{Z,H} &= E\left(1 \pm \frac{M_Z^2 - M_H^2}{s}, \pm\beta\sin\vartheta, 0, \pm\beta\cos\vartheta\right) \end{aligned} \tag{4.3.20}$$

in the centre-of-mass (CM) system, where E denotes the beam energy, $s = 4E^2$ the CM energy squared, ϑ the scattering angle between the e^+ and the Z boson, and

$$\beta = \frac{1}{s}\sqrt{[s-(M_Z+M_H)^2][s-(M_H-M_Z)^2]} \tag{4.3.21}$$

the velocity of the outgoing particles in the CM system.

The helicities of the Z boson are defined via the transverse ($\pm$) and longitudinal (0) polarization vectors

$$\begin{aligned} \varepsilon^{*\mu}(\pm) &= \frac{1}{\sqrt{2}}(0, \cos\vartheta, \mp i, -\sin\vartheta), \\ \varepsilon^{*\mu}(0) &= \frac{E}{M_Z}\left(\beta, \frac{s+M_Z^2-M_H^2}{s}\sin\vartheta, 0, \frac{s+M_Z^2-M_H^2}{s}\cos\vartheta\right). \end{aligned} \tag{4.3.22}$$

Since we neglect the electron mass we can omit all Feynman graphs containing a coupling of a (physical or unphysical) Higgs particle to an electron, as they contain an explicit suppression factor m_e/M_W. Then, only one Feynman diagram is left at lowest order (Fig. 4.9). The corresponding invariant amplitude reads

$$\mathcal{M}_{\text{Born}}\left(\pm\frac{1}{2},\lambda\right) = e g_{\text{e}}^{\pm}\frac{M_{\text{Z}}}{s_{\text{W}}c_{\text{W}}}\frac{1}{s-M_{\text{Z}}^2}\bar{v}(p_+)\not{\varepsilon}^*(\lambda)\frac{1\pm\gamma_5}{2}u(p_-), \tag{4.3.23}$$

where the coupling of the Z-boson to left- and right-handed electrons $g_{\text{e}}^{\pm}$ is given in (A.2.22).

From the amplitude (4.3.23) the differential polarized cross sections result as

$$\left(\frac{\mathrm{d}\sigma}{\mathrm{d}\Omega}\right)_{\text{Born}}\left(\pm\frac{1}{2},\lambda\right) = \frac{\alpha^2}{4}\frac{\beta M_{\text{Z}}^2}{(s-M_{\text{Z}}^2)^2}\frac{(g_{\text{e}}^{\pm})^2}{s_{\text{W}}^2c_{\text{W}}^2} \tag{4.3.24}$$
$$\times\begin{cases}\dfrac{1+\cos^2\vartheta}{2} & \text{for}\quad \lambda=\pm1\\[2ex] \left(\dfrac{s\beta^2}{4M_{\text{Z}}^2}+1\right)\sin^2\vartheta & \text{for}\quad \lambda=0.\end{cases}$$

The unpolarized cross section is obtained by summing over the polarizations of the Z boson and averaging over the fermion polarizations

$$\left(\frac{\mathrm{d}\sigma}{\mathrm{d}\Omega}\right)_{\text{Born}} = \frac{\alpha^2}{8}\frac{\beta M_{\text{Z}}^2}{(s-M_{\text{Z}}^2)^2}\frac{(g_{\text{e}}^{+})^2+(g_{\text{e}}^{-})^2}{s_{\text{W}}^2c_{\text{W}}^2}\left(\frac{s\beta^2}{8M_{\text{Z}}^2}\sin^2\vartheta+1\right). \tag{4.3.25}$$

For high energies and well above threshold, the longitudinal gauge bosons dominate the cross section owing to the factor s/M_{Z}^2 originating from the longitudinal polarization vector. While the cross section for transverse Z bosons drops like $1/s^2$ for high energies, the one for longitudinal Z bosons behaves as $1/s$. The $\sin^2\vartheta$ angular behaviour of the longitudinal cross sections is characteristic for the production of a scalar particle.

By integrating (4.3.24) and (4.3.25) over the production angle ϑ, we obtain the total polarized cross sections

$$\sigma_{\text{Born}}\left(\pm\frac{1}{2},\lambda\right) = \alpha^2\pi\frac{\beta M_{\text{Z}}^2}{(s-M_{\text{Z}}^2)^2}\frac{(g_{\text{e}}^{\pm})^2}{s_{\text{W}}^2c_{\text{W}}^2}\begin{cases}\dfrac{2}{3} & \text{for}\quad \lambda=\pm1\\[2ex] \dfrac{2}{3}+\dfrac{s\beta^2}{6M_{\text{Z}}^2} & \text{for}\quad \lambda=0,\end{cases} \tag{4.3.26}$$

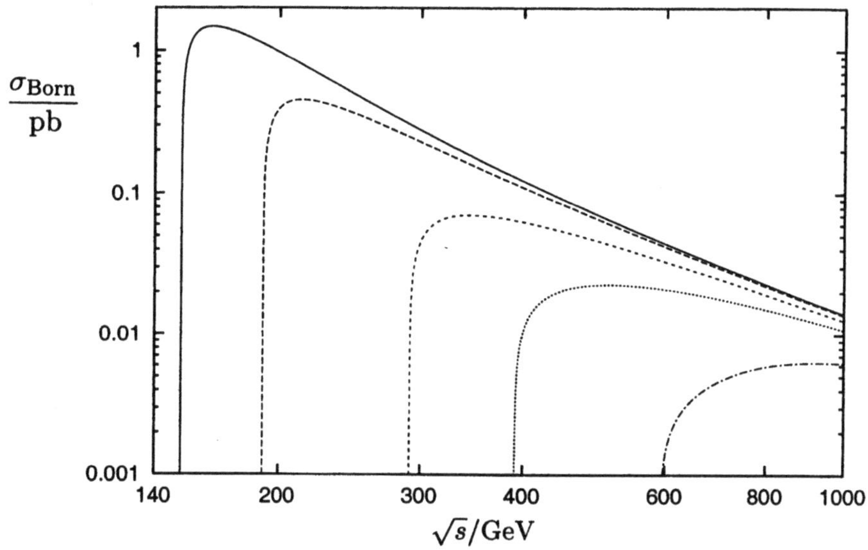

Fig. 4.10 Lowest-order cross section for different Higgs-boson masses $M_{\mathrm{H}} = 60$, 100, 200, 300, and 500 GeV

and the total unpolarized cross section

$$\sigma_{\text{Born}} = \frac{\alpha^2\pi}{2} \frac{\beta M_{\mathrm{Z}}^2}{(s - M_{\mathrm{Z}}^2)^2} \frac{(g_{\mathrm{e}}^+)^2 + (g_{\mathrm{e}}^-)^2}{s_{\mathrm{w}}^2 c_{\mathrm{w}}^2} \left(1 + \frac{s\beta^2}{12M_{\mathrm{Z}}^2}\right). \qquad (4.3.27)$$

The total cross section for different Higgs-boson masses is shown in Fig. 4.10. For small Higgs-boson masses it is about 0.5–1 pb at $\sqrt{s} = 200\,\text{GeV}$, and about 50 fb at $\sqrt{s} = 500\,\text{GeV}$. The unpolarized cross section drops proportional to $1/s$ at high energies.

At energies above about 500 GeV, the WW-fusion process, $\mathrm{e^+e^-} \to \nu_{\mathrm{e}}\bar{\nu}_{\mathrm{e}}\mathrm{H}$, becomes more important for Higgs-boson production since the corresponding cross section,

$$\sigma(\mathrm{e^+e^-} \to \nu_{\mathrm{e}}\bar{\nu}_{\mathrm{e}}\mathrm{H}) = \\ = \frac{\alpha^3}{16M_{\mathrm{W}}^2 s_{\mathrm{w}}^6} \left[\left(1 + \frac{M_{\mathrm{H}}^2}{s}\right) \log\frac{s}{M_{\mathrm{H}}^2} - 2\left(1 - \frac{M_{\mathrm{H}}^2}{s}\right)\right], \qquad (4.3.28)$$

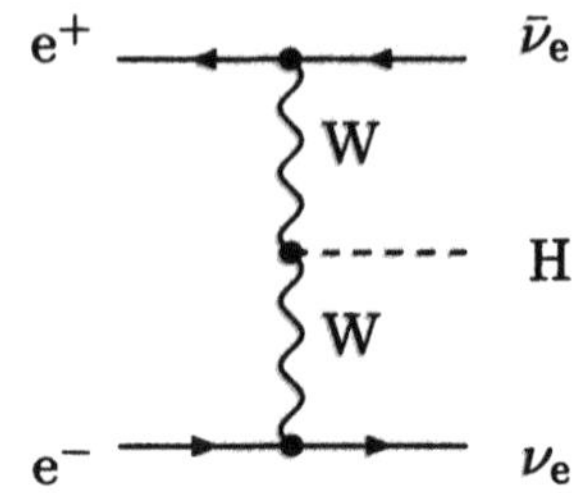

Fig. 4.11 Lowest-order diagram for $e^+e^- \to \nu_e\bar{\nu}_e H$

rises logarithmically with the energy. The cross section for ZZ fusion is about one order of magnitude smaller owing to the smaller couplings to the Z boson.

4.3.2.3 Higgs-boson decays

Since the couplings of the Higgs boson to other particles are proportional to the masses of those particles, the Higgs boson will predominantly decay into the heaviest particle that is kinematically allowed. If $M_H \lesssim 140\,\text{GeV}$, the Higgs-boson decays mainly into fermions, in particular, into bottom quarks. The corresponding partial decay width is

$$\Gamma(H \to f\bar{f}) = \frac{\alpha}{8s_w^2} M_H N_C^f \frac{m_f^2}{M_W^2}\left(1 - \frac{4m_f^2}{M_H^2}\right)^{3/2} \tag{4.3.29}$$

with the colour factor $N_C^f = 1$ for leptons and 3 for quarks.

For larger Higgs-boson masses the decays into one real and one virtual gauge boson become important. If $M_H > 2M_W$, the Higgs boson can decay into a pair of real W bosons and for $M_H > 2M_Z$ also into real Z bosons. The corresponding partial decay widths

$$\begin{aligned}\Gamma(H \to WW) &= \frac{\alpha}{16s_w^2}\frac{M_H^3}{M_W^2}\left(1 - \frac{4M_W^2}{M_H^2}\right)^{1/2}\left(1 - \frac{4M_W^2}{M_H^2} + 12\frac{M_W^4}{M_H^4}\right),\\ \Gamma(H \to ZZ) &= \frac{\alpha}{32s_w^2}\frac{M_H^3}{M_W^2}\left(1 - \frac{4M_Z^2}{M_H^2}\right)^{1/2}\left(1 - \frac{4M_Z^2}{M_H^2} + 12\frac{M_Z^4}{M_H^4}\right)\end{aligned} \tag{4.3.30}$$

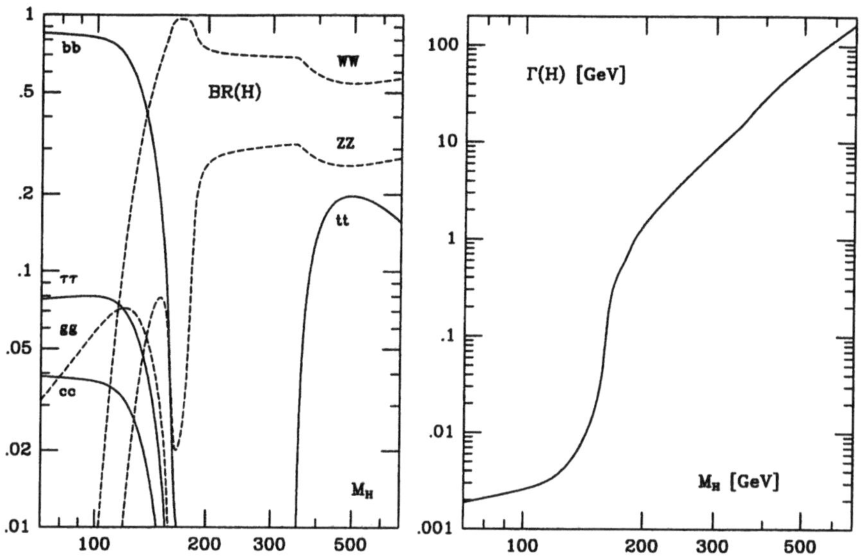

Fig. 4.12 Branching ratios of the important Higgs-boson decay modes and total Higgs-boson decay width (from Ref. [Dj96])

dominate for $M_{\mathrm{H}} \gtrsim 2M_{\mathrm{W}}$ even if the decay channel $\mathrm{H} \to \mathrm{t}\bar{\mathrm{t}}$ opens.

For $M_{\mathrm{H}} \gg M_{\mathrm{W}}$ the dominant contribution to the decay comes from longitudinal gauge bosons (with polarization vector proportional to $M_{\mathrm{H}}/M_{\mathrm{W}}$), and the total decay width is approximately given by

$$\Gamma_{\mathrm{H}} \approx \Gamma(\mathrm{H} \to \mathrm{WW}) + \Gamma(\mathrm{H} \to \mathrm{ZZ}) \approx \frac{3\alpha}{32 s_{\mathrm{w}}^2} M_{\mathrm{H}} \frac{M_{\mathrm{H}}^2}{M_{\mathrm{W}}^2}. \tag{4.3.31}$$

For

$$M_{\mathrm{H}} \approx \frac{8 s_{\mathrm{w}} M_{\mathrm{W}}}{\sqrt{6\alpha}} = \sqrt{\frac{32\pi}{3}}\, v \approx 1.4\,\mathrm{TeV} \tag{4.3.32}$$

the width becomes larger than the mass, and the physical interpretation of the Higgs boson as a particle becomes doubtful.

The total decay width of the Higgs boson and the branching ratios are shown in Fig. 4.12 as a function of the Higgs-boson mass. Apart from the decay modes discussed above, also the decays into a pair of gluons, $\mathrm{H} \to gg$, or photons, $\mathrm{H} \to \gamma\gamma$, are relevant for intermediate Higgs-boson masses.

Although these decays proceed via loop diagrams, the corresponding rates are non-negligible because the coupling of the Higgs boson to the virtual particles is not suppressed and, moreover, the gluons couple with the strong coupling constant.

4.4 Quantization of the Electroweak Standard Model

4.4.1 Gauge fixing and Faddeev–Popov fields

The quantization of the EWSM requires, as the quantization of all non-abelian gauge theories (cf. Sect. 2.4), the introduction of a *gauge-fixing term* and of *Faddeev–Popov fields.* For higher-order calculations and renormalization it is preferable to choose a renormalizable gauge, i.e. a gauge in which the gauge-boson propagators behave as $1/k^2$ for large k^2. Spontaneous symmetry breaking introduces mixing terms $V_\mu \, \partial^\mu \phi$ between the gauge-boson fields and the would-be Goldstone fields [cf. (4.2.33)], which give rise to non-diagonal propagators. As explained in Sect. 4.1.2.2, one can choose a renormalizable gauge in which these mixing terms are absent. To this end, we introduce a gauge-fixing term of the form

$$\mathcal{L}_{\text{fix}} = -\frac{1}{2\xi_A}(C^{\text{A}})^2 - \frac{1}{2\xi_Z}(C^{\text{Z}})^2 - \frac{1}{\xi_W}C^+C^- \tag{4.4.1}$$

with linear gauge-fixing operators

$$\begin{aligned} C^{\pm} &= \partial^\mu W^{\pm}_\mu \mp \mathrm{i} M_{\text{W}} \xi'_W \phi^{\pm}, \\ C^{\text{Z}} &= \partial^\mu Z_\mu - M_{\text{Z}} \xi'_Z \chi, \\ C^{\text{A}} &= \partial^\mu A_\mu. \end{aligned} \tag{4.4.2}$$

This general linear gauge contains five independent gauge parameters ξ_a, $a = A, Z, \pm$, and ξ'_a, $a = Z, \pm$, where $\xi^{(\prime)}_{\pm} \equiv \xi^{(\prime)}_W$.

For $\xi'_W = \xi_W$ and $\xi'_Z = \xi_Z$ the terms involving the would-be Goldstone fields in (4.4.2) cancel the mixing terms $V_\mu \, \partial^\mu \phi$ in the classical Lagrangian (4.2.33) up to irrelevant total derivatives. This gauge is called *'t Hooft gauge* and is used in the following if not stated otherwise.

The terms quadratic in the gauge fields allow for well-defined *gauge-boson propagators* [cf. (4.1.62)],

$$\mathrm{i}\Delta^{V^a}_{\mu\nu}(k) = \mathrm{i}\frac{-g_{\mu\nu} + k_\mu k_\nu/k^2}{k^2 - M_a^2} - \mathrm{i}\frac{k_\mu k_\nu}{k^2}\frac{\xi_a}{k^2 - \xi_a M_a^2}, \tag{4.4.3}$$

where $a = A, Z, \pm$, $V^Z \equiv Z$, $V^\pm \equiv W^\pm$, and $M_\pm = M_\mathrm{W}$. The transverse parts, which describe the physical degrees of freedom, are independent of the gauge parameters whereas the longitudinal parts are gauge-dependent.

The terms quadratic in the *would-be Goldstone fields* generate mass terms for those fields,

$$-\xi_W M_\mathrm{W}^2 \phi^+\phi^- - \frac{1}{2}\xi_Z M_\mathrm{Z}^2 \chi^2, \tag{4.4.4}$$

so that the corresponding propagators read

$$\mathrm{i}\Delta^{\phi^a}(k) = \frac{\mathrm{i}}{k^2 - \xi_a M_a^2} \tag{4.4.5}$$

with $a = Z, \pm$, $\phi^Z \equiv \chi$.

According to the general results of Sect. 2.4.1, the Faddeev–Popov ghost-field Lagrangian is obtained as follows $(a, b, c \in \{A, Z, \pm\})$

$$\mathcal{L}_\mathrm{ghost} = -\int \mathrm{d}^4z\, \mathrm{d}^4y\, \bar{u}^a(x)\left(\frac{\delta C^a(x)}{\delta V^c_\nu(z)}\frac{\delta V^c_\nu(z)}{\delta\theta^b(y)} + \frac{\delta C^a(x)}{\delta\phi^c(z)}\frac{\delta\phi^c(z)}{\delta\theta^b(y)}\right)u^b(y). \tag{4.4.6}$$

Inserting the gauge-fixing terms (4.4.2) and using the gauge transformations (4.2.35), we find up to total derivative terms

$$\begin{aligned}\mathcal{L}_\mathrm{ghost} = \Big\{ &-\bar{u}^+(\partial^\mu\partial_\mu + \xi_W M_\mathrm{W}^2)u^+ + \mathrm{i}e(\partial^\mu\bar{u}^+)\left(A_\mu - \frac{c_\mathrm{W}}{s_\mathrm{W}}Z_\mu\right)u^+ \\ &- \mathrm{i}e(\partial^\mu\bar{u}^+)W^+_\mu\left(u^A - \frac{c_\mathrm{W}}{s_\mathrm{W}}u^Z\right) \\ &- eM_\mathrm{W}\xi_W\bar{u}^+\left[\frac{1}{2s_\mathrm{W}}(H + \mathrm{i}\chi)u^+ - \phi^+\left(u^A - \frac{c_\mathrm{W}^2 - s_\mathrm{W}^2}{2c_\mathrm{W}s_\mathrm{W}}u^Z\right)\right] \\ &- (u^+ \to u^-, W^+ \to W^-, \phi^+ \to \phi^-, \mathrm{i} \to -\mathrm{i})\Big\}\end{aligned}$$

$$
\begin{aligned}
&- \bar{u}^Z (\partial^\mu \partial_\mu + \xi_Z M_Z^2) u^Z - \mathrm{i}e\frac{c_W}{s_W}(\partial^\mu \bar{u}^Z)\left(W_\mu^+ u^- - W_\mu^- u^+\right) \\
&- eM_Z \xi_Z \bar{u}^Z \left[\frac{1}{2c_W s_W}\eta u^Z - \frac{1}{2s_W}\left(\phi^+ u^- + \phi^- u^+\right)\right] \\
&- \bar{u}^A \partial^\mu \partial_\mu u^A + \mathrm{i}e(\partial^\mu \bar{u}^A)\left(W_\mu^+ u^- - W_\mu^- u^+\right). \qquad (4.4.7)
\end{aligned}
$$

The ghosts are coupled to the gauge fields and to the would-be Goldstone fields. From the bilinear terms we obtain the *Faddeev–Popov ghost propagators* [cf. (4.1.62)],

$$
\mathrm{i}\Delta^{u^a}(k) = \frac{\mathrm{i}}{k^2 - \xi_a M_a^2}. \qquad (4.4.8)
$$

The Faddeev–Popov ghost field u^A is massless. The masses of the other ghost fields coincide with those of the scalar gauge bosons [$\varepsilon(k) \propto k$] and the would-be Goldstone bosons (cf. Sect. 4.1.2.2). This is necessary to allow for the cancellation of the corresponding unphysical poles in S-matrix elements. The gauge dependence of these masses reflects the unphysical nature of the corresponding fields.

The form of the propagators (4.4.3), (4.4.5), and (4.4.8) ensures that the 't Hooft gauge leads to a renormalizable theory for finite values of the gauge parameters ξ_a. For $\xi_a \to \infty$ the masses of the unphysical degrees of freedom (except for u^A) tend to infinity. As a consequence, the unphysical degrees of freedom decouple (u^A becomes a free field) and we are left with the theory in the (non-renormalizable) unitary gauge.

For $\xi_a = 1$, the *'t Hooft–Feynman gauge*, pole positions and residues of the longitudinal parts of the gauge-boson propagators coincide with those of the transverse parts, so that the gauge-boson propagators take the simple form

$$
\mathrm{i}\Delta_{\mu\nu}^{V^a}(k) = \frac{-\mathrm{i}g_{\mu\nu}}{k^2 - M_a^2}. \qquad (4.4.9)
$$

In this gauge, the masses of the unphysical fields equal those of the corresponding physical fields and the $k_\mu k_\nu$ parts of the gauge-boson propagators are absent. Therefore, the 't Hooft–Feynman gauge is most convenient for practical higher-order calculations.

Adding up all terms (4.2.33), (4.4.1) and (4.4.7) we obtain the complete Lagrangian of the EWSM suitable for higher-order calculations,

$$\mathcal{L}_{\mathrm{GSW}} = \mathcal{L}_{\mathrm{cl}} + \mathcal{L}_{\mathrm{fix}} + \mathcal{L}_{\mathrm{ghost}}. \tag{4.4.10}$$

The corresponding Feynman rules are given in App. A.2.

4.4.2 BRS symmetry and physical fields

The gauge-fixing and Faddeev–Popov terms destroy the ordinary gauge invariance of $\mathcal{L}_{\mathrm{cl}}$. However, as discussed in Sect. 2.4.3 for a general gauge theory, $\mathcal{L}_{\mathrm{GSW}}$ is invariant under the *Becchi–Rouet–Stora (BRS) transformations* [Be74]. The transformations for the gauge-boson, scalar, and fermion fields are obtained from (4.2.35) by the replacement $\delta\theta^a = \delta\lambda u^a$, $a = A, Z, \pm$, where $u^a(x)$ are Faddeev–Popov ghost fields, and $\delta\lambda$ is an infinitesimal anticommuting constant. Using the BRS operator s, introduced in Sect. 2.4.3.1, these transformations read

$$\begin{aligned}
sA_\mu &= \partial_\mu u^A + \mathrm{i}e[W_\mu^+ u^- - W_\mu^- u^+],\\
sZ_\mu &= \partial_\mu u^Z - \mathrm{i}e\frac{c_{\mathrm{W}}}{s_{\mathrm{W}}}[W_\mu^+ u^- - W_\mu^- u^+],\\
sW_\mu^\pm &= \partial_\mu u^\pm \mp \mathrm{i}e\left[W_\mu^\pm\left(u^A - \frac{c_{\mathrm{W}}}{s_{\mathrm{W}}}u^Z\right) - \left(A_\mu - \frac{c_{\mathrm{W}}}{s_{\mathrm{W}}}Z_\mu\right)u^\pm\right],\\
s\eta &= \frac{e}{2s_{\mathrm{W}}c_{\mathrm{W}}}\chi u^Z + \frac{\mathrm{i}e}{2s_{\mathrm{W}}}[\phi^+ u^- - \phi^- u^+],\\
s\chi &= -\frac{e}{2s_{\mathrm{W}}c_{\mathrm{W}}}[v+\eta]u^Z + \frac{e}{2s_{\mathrm{W}}}[\phi^+ u^- + \phi^- u^+],\\
s\phi^\pm &= \mp\mathrm{i}e\phi^\pm\left[u^A + \frac{s_{\mathrm{W}}^2 - c_{\mathrm{W}}^2}{2c_{\mathrm{W}}s_{\mathrm{W}}}u^Z\right] \pm \frac{\mathrm{i}e}{2s_{\mathrm{W}}}[v+\eta\pm\mathrm{i}\chi]u^\pm,\\
s\psi_{i,\pm}^{\mathrm{L}} &= -\mathrm{i}e\left[Q_{i,\pm}u^A + \frac{s_{\mathrm{W}}}{c_{\mathrm{W}}}\left(Q_{i,\pm} \mp \frac{1}{2s_{\mathrm{W}}^2}\right)u^Z\right]\psi_{i,\pm}^{\mathrm{L}}\\
&\quad + \mathrm{i}\frac{e}{\sqrt{2}s_{\mathrm{W}}}u^\pm v_{ij}^\pm \psi_{j,\mp}^{\mathrm{L}},\\
s\psi_{i,\pm}^{\mathrm{R}} &= -\mathrm{i}eQ_{i,\pm}\left(u^A + \frac{s_{\mathrm{W}}}{c_{\mathrm{W}}}u^Z\right)\psi_{i,\pm}^{\mathrm{R}}
\end{aligned} \tag{4.4.11}$$

with $v_{ij}^\pm$ defined after (4.2.36).

The BRS transformation of the ghost fields in the symmetric basis is directly obtained from (2.4.35),

$$\delta_{\mathrm{BRS}} u^a = -\delta\lambda \frac{1}{2} g \varepsilon^{abc} u^b u^c, \qquad a, b, c \in \{1, 2, 3\},$$
$$\delta_{\mathrm{BRS}} u^B = 0. \tag{4.4.12}$$

Translating this into the basis of mass and charge eigenstates

$$u^{\pm} = \frac{1}{\sqrt{2}}(u^1 \mp \mathrm{i} u^2),$$
$$u^Z = c_{\mathrm{W}} u^3 + s_{\mathrm{W}} u^B, \qquad u^A = -s_{\mathrm{W}} u^3 + c_{\mathrm{W}} u^B, \tag{4.4.13}$$

results in

$$\delta_{\mathrm{BRS}} u^{\pm} = \pm\delta\lambda\, \mathrm{i} \frac{e}{s_{\mathrm{W}}} u^{\pm}(s_{\mathrm{W}} u^A - c_{\mathrm{W}} u^Z) \equiv \delta\lambda s u^{\pm},$$
$$\delta_{\mathrm{BRS}} u^Z = -\delta\lambda\, \mathrm{i} e \frac{c_{\mathrm{W}}}{s_{\mathrm{W}}} u^- u^+ \equiv \delta\lambda s u^Z,$$
$$\delta_{\mathrm{BRS}} u^A = \delta\lambda\, \mathrm{i} e u^- u^+ \equiv \delta\lambda s u^A. \tag{4.4.14}$$

The BRS transformation of the antighost fields is determined by the gauge-fixing terms $[\bar{u}^{\pm} = (\bar{u}^1 \pm \mathrm{i}\bar{u}^2)/\sqrt{2}]$,

$$\delta_{\mathrm{BRS}} \bar{u}^{\pm} = -\delta\lambda \frac{1}{\xi_W} C^{\mp} \equiv \delta\lambda s \bar{u}^{\pm},$$
$$\delta_{\mathrm{BRS}} \bar{u}^Z = -\delta\lambda \frac{1}{\xi_Z} C^Z \equiv \delta\lambda s \bar{u}^Z,$$
$$\delta_{\mathrm{BRS}} \bar{u}^A = -\delta\lambda \frac{1}{\xi_A} C^A \equiv \delta\lambda s \bar{u}^A. \tag{4.4.15}$$

In order to get BRS transformations that are independent of the gauge-fixing term and strictly nilpotent, one introduces auxiliary fields B^a, $a = A, Z, \pm$, the so-called *Nakanishi–Lautrup fields* [Na66a, La67], and rewrites the gauge-fixing Lagrangian as (cf. Sect. 2.4.3.1)

$$\mathcal{L}_{\mathrm{fix}} = B^+ C^- + B^- C^+ + B^Z C^Z + B^A C^A$$
$$+ \xi_W B^+ B^- + \frac{\xi_Z}{2} B^Z B^Z + \frac{\xi_A}{2} B^A B^A \tag{4.4.16}$$

with the gauge-fixing operators from (4.4.2). The BRS transformations of the antighost fields are replaced by

$$\begin{aligned}\delta_{\mathrm{BRS}}\bar{u}^a &= \delta\lambda B^a \equiv \delta\lambda s\bar{u}^a, \quad a = \pm, Z, A \\ \delta_{\mathrm{BRS}}B^a &= 0 \equiv \delta\lambda s B^a. \end{aligned} \tag{4.4.17}$$

As argued in Sect. 2.5.4.3, the *physical degrees of freedom* of a gauge theory can be extracted by considering the BRS transformations of the asymptotic fields which are related to those of the free fields. We do not go through the formalism but consider directly the BRS transformations of the free fields,

$$\begin{aligned} sA_\mu &= \partial_\mu u^{\mathrm{A}}, \qquad sZ_\mu = \partial_\mu u^{\mathrm{Z}}, \qquad sW_\mu^\pm = \partial_\mu u^\pm, \\ s\eta &= 0, \qquad s\chi = -M_{\mathrm{Z}}u^{\mathrm{Z}}, \qquad s\phi^\pm = \pm \mathrm{i} M_{\mathrm{W}} u^\pm, \\ s\psi_{\mathrm{L}}^{i,\pm} &= 0, \qquad s\psi_{\mathrm{R}}^{i,\pm} = 0, \\ su^a &= 0, \qquad s\bar{u}^a = B^a, \qquad sB^a = 0, \qquad a = \pm, Z, A. \end{aligned} \tag{4.4.18}$$

The physical degrees of freedom of the photon field are as discussed in Sect. 2.5.4.3 (pages 245–246) for the general non-abelian massless gauge field. For massive gauge fields, the situation is as follows. The scalar component of the massive gauge fields,[8] the would-be-Goldstone fields χ and ϕ, and the antighost fields $\bar{u}^a$ are not annihilated and thus correspond to unphysical degrees of freedom. The ghost fields and the auxiliary field B^a are BRS transformations and consequently involve only zero-norm states that decouple from the physical S matrix. Thus, the physical states are the three degrees of freedom of the massive gauge-boson fields, the two transverse photon polarizations, the Higgs field, and the fermion fields.

Here we see again the *quartet mechanism* at work. The unphysical fields form quartets, $(\bar{u}^\pm, \phi^\pm, u^\pm, B^\pm)$, $(\bar{u}^Z, \chi, u^Z, B^Z)$, and $(\bar{u}^A, A_{\mathrm{L}}^\mu, u^A, B^A)$. The quartet involving the longitudinal mode of the photon field A_{L}^μ is equivalent to the quartets in an unbroken gauge theory and has already been discussed in Sect. 2.5.4.3 on page 245. For the quartets of the broken gauge theory, the equations of motion for the B fields, $B^a = -C^a/\xi_a$, tell us that also

[8] The scalar polarization of vector fields is generally defined by $\varepsilon(k)k \neq 0$. While for massless vector fields this implies $\varepsilon(k) \propto (|\boldsymbol{k}|, -\mathbf{k})$, for massive vector fields it entails $\varepsilon(k) \propto k$. Note that for massless vector fields $\varepsilon(k) \propto k$ corresponds to the longitudinal polarization.

the C^a, which are linear combinations of the would-be Goldstone fields and the scalar components of the associated massive gauge fields, are zero-norm states. The linearly independent combinations of these fields are unphysical owing to their non-vanishing BRS transformation. Instead of this linear combinations we have written for simplicity the would-be Goldstone fields in the quartets.

4.4.3 Slavnov–Taylor identities

As shown in Sect. 2.4.3, the BRS symmetry gives rise to relations between different Green functions, the Ward identities or Slavnov–Taylor identities. The generic form of these identities (2.4.62) can be directly transferred to the Glashow–Salam–Weinberg theory:

$$0 = \frac{\delta_{\mathrm{BRS}}}{\delta\lambda}\Big\langle T\prod_l \Psi_{I_l}\Big\rangle, \tag{4.4.19}$$

where Ψ_I represents anyone of the fields V_μ^a, u^a, $\bar{u}^a$ $(a = A, Z, \pm)$, ϕ^a $(a = Z, \pm)$, η, $\psi_{i,\pm}^{\{\mathrm{L,R}\}}$, or $\bar{\psi}_{i,\pm}^{\{\mathrm{L,R}\}}$. Here, we use the formalism without B fields.

Consider the identities that follow from the BRS invariance of Green functions involving one antighost field and arbitrary other fields. Using (4.4.19) we obtain

$$\begin{aligned} 0 &= \frac{\delta_{\mathrm{BRS}}}{\delta\lambda}\Big\langle T\bar{u}^a\prod_l \Psi_{I_l}\Big\rangle \\ &= -\Big\langle T\frac{1}{\xi_a}C^a\prod_l \Psi_{I_l}\Big\rangle \\ &\quad +\sum_k \sigma_k\Big\langle T\bar{u}^a\Big(\prod_{l<k}\Psi_{I_l}\Big)(s\Psi_{I_k})\prod_{m>k}\Psi_{I_m}\Big\rangle = 0, \end{aligned} \tag{4.4.20}$$

where σ_k stands for the sign that results from anticommuting $\delta\lambda$ to the left and the sum $\sum_k$ extends over all fields in $\prod_l \Psi_{I_l}$.

If all the fields Ψ_{I_l} are physical and on shell, their BRS variations vanish [cf. (2.5.160)], and we find (omitting terms that vanish after truncation)

$$0 = \Big\langle TC^a\prod_l \Psi_{I_l}^{\mathrm{as,phys}}\Big\rangle. \tag{4.4.21}$$

For the case with additional gauge-fixing terms we obtain in the same way (no sum over a)

$$\begin{aligned} 0 &= \frac{\delta_{\mathrm{BRS}}}{\delta\lambda}\Big\langle T\bar{u}^a\Big(\prod_m C^{a_m}\Big)\prod_l \Psi_{I_l}^{\mathrm{as,phys}}\Big\rangle \\ &= -\Big\langle T\frac{1}{\xi_a}C^a\Big(\prod_m C^{a_m}\Big)\prod_l \Psi_{I_l}^{\mathrm{as,phys}}\Big\rangle \\ &\quad - \sum_k \Big\langle T\bar{u}^a(sC^{a_k})\Big(\prod_{m\neq k} C^{a_m}\Big)\prod_l \Psi_{I_l}^{\mathrm{as,phys}}\Big\rangle = 0. \end{aligned} \tag{4.4.22}$$

Using the equation of motion for the antighost fields [cf. (2.4.63)]

$$\mathrm{i}\Big\langle T\frac{\delta}{\delta\bar{u}^a(x)}X\Big\rangle = -\Big\langle T\Big(sC^a(x)\Big)X\Big\rangle, \tag{4.4.23}$$

where X stands for any product of fields, and the fact that neither C^{a_m} nor $\Psi_{I_l}^{\mathrm{as,phys}}$ in (4.4.22) involve antighost fields, we arrive at

$$\begin{aligned} &\Big\langle T\frac{1}{\xi_a}C^a(x)\Big(\prod_m C^{a_m}(x_m)\Big)\prod_l \Psi_{I_l}^{\mathrm{as,phys}}\Big\rangle \\ &\quad = -\mathrm{i}\sum_k \delta^{aa_k}\delta^{(4)}(x-x_k)\Big\langle T\Big(\prod_{m\neq k} C^{a_m}(x_m)\Big)\prod_l \Psi_{I_l}^{\mathrm{as,phys}}\Big\rangle. \end{aligned} \tag{4.4.24}$$

The terms on the right-hand side involve δ-functions and thus correspond to disconnected parts. Thus, we obtain for connected Green functions

$$\Big\langle T\frac{1}{\xi_a}C^a(x)\prod_m C^{a_m}(x_m)\prod_l \Psi_{I_l}^{\mathrm{as,phys}}\Big\rangle_{\mathrm{c}} = 0. \tag{4.4.25}$$

Without physical fields $\Psi_{I_l}^{\mathrm{as,phys}}$ and for one extra gauge-fixing term, equation (4.4.24) reduces to

$$\frac{1}{\xi_a}\Big\langle TC^a(x)C^b(y)\Big\rangle = -\mathrm{i}\delta^{ab}\delta^{(4)}(x-y). \tag{4.4.26}$$

Inserting the gauge-fixing functions (4.4.2) and going to momentum space we find the following non-trivial relations for the propagators of the charged bosons

$$\begin{aligned} k^\mu k^\nu G^{W^+W^-}_{\mu\nu}(k) &- \xi_W M_W k^\mu G^{W^+\phi^-}_\mu(k) \\ &- \xi_W M_W k^\mu G^{\phi^+W^-}_\mu(k) + \xi_W^2 M_W^2 G^{\phi^+\phi^-}(k) = -i\xi_W, \end{aligned} \tag{4.4.27}$$

and for the neutral boson propagators

$$\begin{aligned} k^\mu k^\nu G^{ZZ}_{\mu\nu}(k) - 2i\xi_Z M_Z k^\mu G^{Z\chi}_\mu(k) + \xi_Z^2 M_Z^2 G^{\chi\chi}(k) &= -i\xi_Z, \\ k^\mu k^\nu G^{AZ}_{\mu\nu}(k) - i\xi_Z M_Z k^\mu G^{A\chi}_\mu(k) &= 0, \\ k^\mu k^\nu G^{AA}_{\mu\nu}(k) &= -i\xi_A. \end{aligned} \tag{4.4.28}$$

According to the conventions of App. A.1.3.2 the field labels indicate the outgoing fields. The momentum arguments denote the incoming momentum of the particle corresponding to the first upper index. These relations between the longitudinal parts of the gauge-boson propagators, the gauge-field–would-be-Goldstone-field mixing propagators, and the propagators of the would-be-Goldstone bosons are exact and of course fulfilled in lowest order. They show that the unphysical parts of these propagators are not independent and, in particular, guarantee that their poles coincide. The relation for the photon propagator ensures that its longitudinal part gets no higher-order corrections.

4.4.4 Lee identities

Just as in the Yang–Mills theory discussed in Sect. 2.4.3.4 also in the Standard Model of the electroweak interaction *Lee identities* can be derived from the BRS invariance of the effective action. For linear *gauge-fixing functionals* of the form ($a = A, Z, \pm$)

$$C^a\{\mathbf{V},\phi;x\} = \sum_{b=A,Z,\pm} \Phi^{ab}_\mu V^{b,\mu}(x) + \sum_{b=Z,\pm} \Phi^{ab}_\phi \phi^b(x), \tag{4.4.29}$$

like (4.4.2), the identities corresponding to (2.4.77) and (2.4.78) read

$$0 = \int \mathrm{d}^4x \Bigg[\sum_{a=A,Z,\pm} \left(\frac{\delta\tilde{\Gamma}}{\delta V^a_\mu} \frac{\delta\tilde{\Gamma}}{\delta K_V^{a,\mu}} + \frac{\delta\tilde{\Gamma}}{\delta u^a} \frac{\delta\tilde{\Gamma}}{\delta K_u^a} \right) + \sum_{a=Z,\pm} \left(\frac{\delta\tilde{\Gamma}}{\delta \phi^a} \frac{\delta\tilde{\Gamma}}{\delta K_\phi^a} \right) + \frac{\delta\tilde{\Gamma}}{\delta \eta} \frac{\delta\tilde{\Gamma}}{\delta K_\eta} + \sum_i \left(\frac{\delta\tilde{\Gamma}}{\delta K^i_{\bar{\psi}}} \frac{\delta\tilde{\Gamma}}{\delta \psi_i} + \frac{\delta\tilde{\Gamma}}{\delta \bar{\psi}_i} \frac{\delta\tilde{\Gamma}}{\delta K^i_\psi} \right) \Bigg], \tag{4.4.30}$$

$$0 = \sum_{b=A,Z,\pm} \left(\Phi_V^{ab,\mu} \frac{\delta\tilde{\Gamma}}{\delta K_V^{b,\mu}} \right) + \sum_{b=Z,\pm} \left(\Phi_\phi^{ab} \frac{\delta\tilde{\Gamma}}{\delta K_\phi^b} \right) + \frac{\delta\tilde{\Gamma}}{\delta \bar{u}^a}, \tag{4.4.31}$$

where $\tilde{\Gamma}$ is the vertex functional with the gauge-fixing terms subtracted [cf. (2.4.76)], and i runs over all fermion fields. The terms involving the sources K of the BRS-transformed fields [cf. (2.4.53)] read

$$\mathcal{L}_{\text{BRS-sources}} = \sum_{a=A,Z,\pm} \left(K_V^{a,\mu} s A^a_\mu + K_u^a s u^a \right) + \sum_{a=Z,\pm} K_\phi^a s \phi^a + K_\eta s\eta + K^i_\psi s \bar{\psi}_i + (s\bar{\psi}_i) K^i_{\bar{\psi}}, \tag{4.4.32}$$

and the sources for the charged gauge-boson, ghost, and would-be Goldstone fields are defined as $K^\pm = (K^1 \pm \mathrm{i}K^2)/\sqrt{2}$.

Classically, the *physical fields* and their properties can be directly derived from the Lagrangian. In the quantized gauge theory these properties can be extracted from the Lee identities as follows.

We investigate the consequences of the Lee identities for the two-point functions in the charged-boson sector following Ref. [Ao80]. The relevant fields are the charged gauge-bosons fields $W^\pm$, the charged would-be Goldstone-boson fields $\phi^\pm$, and the charged ghost and antighost fields $u^\pm$ and $\bar{u}^\pm$. Differentiating (4.4.30) with respect to $W^\pm_\mu$ and $u^\mp$, putting the sources equal to zero, and transforming to momentum space yields

$$\tilde{\Gamma}^{W^\pm W^\mp}_{\mu\nu}(k,-k)\tilde{\Gamma}^{K_V^\mp u^\mp,\nu}(k,-k) + \tilde{\Gamma}^{W^\pm \phi^\mp}_{\mu}(k,-k)\tilde{\Gamma}^{K_\phi^\mp u^\mp}(k,-k) = 0, \tag{4.4.33}$$

where for instance $\tilde{\Gamma}^{K_V^\mp u^\mp,\nu}(k,-k)$ is the Fourier transform of

$$\tilde{\Gamma}^{K_V^\mp u^\mp,\nu}(x,y) = \frac{\delta\tilde{\Gamma}}{\delta K^\mp_{V,\nu}(x) \delta u^\mp(y)}. \tag{4.4.34}$$

All other terms vanish because of ghost-number and charge conservation. Similarly, differentiation of (4.4.30) with respect to $\phi^{\pm}$ and $u^{\mp}$ yields

$$\tilde{\Gamma}_{\nu}^{\phi^{\pm}W^{\mp}}(k,-k)\tilde{\Gamma}^{K_V^{\mp}u^{\mp},\nu}(k,-k)+\tilde{\Gamma}^{\phi^{\pm}\phi^{\mp}}(k,-k)\tilde{\Gamma}^{K_{\phi}^{\mp}u^{\mp}}(k,-k)=0. \tag{4.4.35}$$

In order to proceed we have to decompose the vertex functions into covariants and invariant functions. For the W and ϕ vertex functions we write

$$\begin{aligned}\Gamma_{\mu\nu}^{W^{\pm}W^{\mp}}(k,-k)&=g_{\mu\nu}^{\mathrm{T}}\Gamma_{\mathrm{T}}^{WW}(k^2)+g_{\mu\nu}^{\mathrm{L}}\Gamma_{\mathrm{L}}^{WW}(k^2),\\ \Gamma_{\mu}^{W^{\pm}\phi^{\mp}}(k,-k)&=-\Gamma_{\mu}^{\phi^{\mp}W^{\pm}}(k,-k)=k_{\mu}\Gamma_{\mathrm{L}}^{W^{\pm}\phi^{\mp}}(k^2),\end{aligned} \tag{4.4.36}$$

with the projectors on transverse and longitudinal parts

$$g_{\mu\nu}^{\mathrm{T}}=\left(g_{\mu\nu}-\frac{k_{\mu}k_{\nu}}{k^2}\right),\qquad g_{\mu\nu}^{\mathrm{L}}=\frac{k_{\mu}k_{\nu}}{k^2}. \tag{4.4.37}$$

In addition we have

$$\tilde{\Gamma}_{\nu}^{K_V^{\pm}u^{\pm}}(k,-k)=k_{\nu}\tilde{\Gamma}_{\mathrm{L}}^{K_V^{\pm}u^{\pm}}(k^2). \tag{4.4.38}$$

In the absence of absorptive parts (imaginary parts that arise from loop integrals), the effective action is hermitian and we obtain the further relations[9]

$$\begin{aligned}\tilde{\Gamma}_{\mathrm{L}}^{W^{+}\phi^{-}}(k,-k)&=-\left[\tilde{\Gamma}_{\mathrm{L}}^{W^{-}\phi^{+}}(k,-k)\right]^{*}\equiv\tilde{\Gamma}_{\mathrm{L}}^{W\phi}(k,-k),\\ \tilde{\Gamma}_{\mathrm{L}}^{K_V^{+}u^{+}}(k,-k)&=\left[\tilde{\Gamma}_{\mathrm{L}}^{K_V^{-}u^{-}}(k,-k)\right]^{*}\equiv\tilde{\Gamma}_{\mathrm{L}}^{K_W u}(k,-k),\\ \tilde{\Gamma}^{K_{\phi}^{+}u^{+}}(k,-k)&=-\left[\tilde{\Gamma}^{K_{\phi}^{-}u^{-}}(k,-k)\right]^{*}\equiv\tilde{\Gamma}^{K_{\phi}u}(k,-k).\end{aligned} \tag{4.4.39}$$

If we require CP invariance in addition, these vertex functions are purely real. Since the CP violation resulting from the quark mixing matrix in the EWSM does not affect the two-point functions at the one-loop level, these are real in the general case. Therefore, we assume no CP violation in two-point functions in the following discussion.

[9] In the presence of absorptive parts these relations still hold if the complex conjugation is accompanied by a multiplication of the absorptive parts with −1.

Inserting everything into (4.4.33) and (4.4.35), we arrive at

$$\begin{aligned} &\tilde{\Gamma}_{\mathrm{L}}^{WW}\tilde{\Gamma}_{\mathrm{L}}^{K_W u} - \tilde{\Gamma}_{\mathrm{L}}^{W\phi}\tilde{\Gamma}^{K_\phi u} = 0, \\ &k^2\tilde{\Gamma}_{\mathrm{L}}^{W\phi}\tilde{\Gamma}_{\mathrm{L}}^{K_W u} - \tilde{\Gamma}^{\phi\phi}\tilde{\Gamma}^{K_\phi u} = 0, \end{aligned} \tag{4.4.40}$$

and by eliminating the ghost contributions

$$\tilde{\Gamma}_{\mathrm{L}}^{WW}\tilde{\Gamma}^{\phi\phi} = k^2(\tilde{\Gamma}_{\mathrm{L}}^{W\phi})^2. \tag{4.4.41}$$

The inverse propagators Γ^{IJ} are obtained by adding the contributions of the gauge-fixing Lagrangian to $\tilde{\Gamma}^{IJ}$. Using the gauge-fixing terms (4.4.1), we obtain for the inverse propagator matrix

$$\begin{aligned} \Gamma^{\pm}_{(\mu\nu)} &= \begin{pmatrix} \Gamma_{\mu\nu}^{W^\pm W^\mp} & \Gamma_{\mu}^{W^\pm \phi^\mp} \\ \Gamma_{\nu}^{\phi^\pm W^\mp} & \Gamma^{\phi^\pm\phi^\mp} \end{pmatrix} \\ &= \begin{pmatrix} g_{\mu\nu}^{\mathrm{T}}\Gamma_{\mathrm{T}}^{WW} + g_{\mu\nu}^{\mathrm{L}}\left(\tilde{\Gamma}_{\mathrm{L}}^{WW} - \frac{1}{\xi_W}k^2\right) & k_\mu\left(\pm\tilde{\Gamma}_{\mathrm{L}}^{W\phi} \mp M_{\mathrm{W}}\frac{\xi'_W}{\xi_W}\right) \\ k_\nu\left(\pm\tilde{\Gamma}_{\mathrm{L}}^{W\phi} \mp M_{\mathrm{W}}\frac{\xi'_W}{\xi_W}\right) & \tilde{\Gamma}^{\phi\phi} - M_{\mathrm{W}}^2\frac{(\xi'_W)^2}{\xi_W} \end{pmatrix}. \end{aligned} \tag{4.4.42}$$

By choosing the gauge parameter[10]

$$\xi'_W = \frac{\xi_W}{M_{\mathrm{W}}}\tilde{\Gamma}_{\mathrm{L}}^{W\phi}(k_0^2), \tag{4.4.43}$$

this matrix becomes diagonal for $k^2 = k_0^2$. Selecting k_0^2 so that

$$\Gamma_{\mathrm{L}}^{WW}(k_0^2) = \tilde{\Gamma}_{\mathrm{L}}^{WW}(k_0^2) - \frac{1}{\xi_W}k_0^2 = 0, \tag{4.4.44}$$

and using (4.4.41), we obtain

$$\begin{aligned} \Gamma^{\phi\phi}(k_0^2) &= \tilde{\Gamma}^{\phi\phi}(k_0^2) - M_{\mathrm{W}}^2\frac{(\xi'_W)^2}{\xi_W} = \tilde{\Gamma}^{\phi\phi}(k_0^2)\left(1 - \frac{\xi_W}{k_0^2}\tilde{\Gamma}_{\mathrm{L}}^{WW}(k_0^2)\right) \\ &= -\frac{\xi_W}{k_0^2}\tilde{\Gamma}^{\phi\phi}(k_0^2)\Gamma_{\mathrm{L}}^{WW}(k_0^2) = 0. \end{aligned} \tag{4.4.45}$$

[10] In other gauges the matrix can be diagonalized by a redefinition of the fields of the form $W_\mu \to aW_\mu + b\partial_\mu\phi$. In the presence of absorptive parts only the real part of $\Gamma^\pm$ can be made diagonal with real ξ'_W.

Thus, the inverse propagators for the ϕ field and the longitudinal component of the W-boson field vanish at the same position, i.e. the poles of the corresponding propagators coincide.

Next we determine the pole of the propagator of the charged Faddeev–Popov ghost fields. In the gauge (4.4.2) the identity (4.4.31) for $a = \pm$ reads

$$0 = \partial^\mu \frac{\delta \tilde{\Gamma}}{\delta K_V^{\pm,\mu}} \mp \mathrm{i}\xi'_W M_\mathrm{W} \frac{\delta \tilde{\Gamma}}{\delta K_\phi^{\pm}} + \frac{\delta \tilde{\Gamma}}{\delta \bar{u}^{\pm}}. \tag{4.4.46}$$

Differentiating with respect to $u^\pm$ and putting the sources to zero implies after transforming to momentum space

$$0 = \mathrm{i}k^\mu \tilde{\Gamma}_\mu^{K_V^\pm u^\pm}(k,-k) \mp \mathrm{i}\xi'_W M_\mathrm{W} \tilde{\Gamma}^{K_\phi^\pm u^\pm}(k,-k) + \Gamma^{\bar{u}^\pm u^\pm}(k,-k). \tag{4.4.47}$$

Inserting (4.4.38) and using (4.4.40) and (4.4.43), we find

$$\begin{aligned} \Gamma^{\bar{u}^\pm u^\pm}(k^2) &= \mp \mathrm{i}k^2 \tilde{\Gamma}_\mathrm{L}^{K_W u}(k^2) \pm \mathrm{i}\xi'_W M_\mathrm{W} \tilde{\Gamma}^{K_\phi u}(k^2) \\ &= \mp \mathrm{i}k^2 \tilde{\Gamma}_\mathrm{L}^{K_W u}(k^2) \left(1 - \xi_W \frac{\tilde{\Gamma}^{W\phi}(k_0^2)}{k^2} \frac{\tilde{\Gamma}_\mathrm{L}^{WW}(k^2)}{\tilde{\Gamma}^{W\phi}(k^2)}\right) \end{aligned} \tag{4.4.48}$$

for the ghost propagator and thus owing to (4.4.44)

$$\Gamma^{\bar{u}^\pm u^\pm}(k_0^2) = 0, \tag{4.4.49}$$

i.e. also the poles of the charged Faddeev–Popov ghost propagators coincide with the one of the longitudinal part of the W-boson propagator.

Thus, the propagators of the four states, the scalar component of the $\mathrm{W}^\pm$ boson, the would-be Goldstone boson $\phi^\pm$, the ghost $u^\pm$, and the antighost $\bar{u}^\pm$, have a pole in a common position. The fact that all these poles coincide is necessary for the decoupling of these unphysical excitations from the S matrix. These quartet particles can appear in the physical subspace only in a zero-norm combination and hence cannot be observed.

We now turn to the two-point functions of the neutral bosons. In the absence of CP-violating effects, the two-point functions of the Higgs field do not mix with those of the other bosons. Equation (4.4.30) gives rise to the following relations [corresponding to (4.4.40)]:

$$\sum_{a=A,Z} \tilde{\Gamma}_{\mathrm{L}}^{V^b V^a} \tilde{\Gamma}_{\mathrm{L}}^{K_V^a u^c} + \tilde{\Gamma}_{\mathrm{L}}^{V^b \chi} \tilde{\Gamma}^{K_\chi u^c} = 0, \tag{4.4.50}$$

$$k^2 \sum_{a=A,Z} \tilde{\Gamma}_{\mathrm{L}}^{V^a \chi} \tilde{\Gamma}_{\mathrm{L}}^{K_V^a u^c} - \tilde{\Gamma}^{\chi\chi} \tilde{\Gamma}^{K_\chi u^c} = 0, \tag{4.4.51}$$

where $b, c \in \{A, Z\}$. Interpreting $\tilde{\Gamma}_{\mathrm{L}}^{V^b V^a}$ and $\tilde{\Gamma}_{\mathrm{L}}^{K_V^b u^a}$ as matrices with indices b and a, (4.4.50) implies

$$\det \tilde{\Gamma}_{\mathrm{L}}^{VV} \det \tilde{\Gamma}_{\mathrm{L}}^{K_V u} = -\det(\tilde{\Gamma}_{\mathrm{L}}^{V\chi} \tilde{\Gamma}^{K_\chi u}) = 0. \tag{4.4.52}$$

Since $\det \tilde{\Gamma}_{\mathrm{L}}^{K_V u} \neq 0$ at tree level and thus in perturbation theory, this means

$$\det \tilde{\Gamma}_{\mathrm{L}}^{VV} = 0, \tag{4.4.53}$$

i.e. $\tilde{\Gamma}_{\mathrm{L}}^{VV}$ has one vanishing eigenvalue. This is just a consequence of the fact that there is only one neutral Goldstone boson that can give mass to the neutral gauge bosons.

The Lee identities involve only the longitudinal parts of the gauge-boson propagators. In order to gain information on the physical, transverse parts we need further input. At least within perturbation theory, one-particle irreducible Green functions do not involve any poles. Together with the covariant decomposition of gauge-boson two-point functions

$$\Gamma_{\mu\nu}^{V^a V^b}(k, -k) = g_{\mu\nu}^{\mathrm{T}} \Gamma_{\mathrm{T}}^{V^a V^b} + g_{\mu\nu}^{\mathrm{L}} \Gamma_{\mathrm{L}}^{V^a V^b} \tag{4.4.54}$$

this entails

$$\tilde{\Gamma}_{\mathrm{T}}^{V^a V^b}(0) = \tilde{\Gamma}_{\mathrm{L}}^{V^a V^b}(0), \tag{4.4.55}$$

and consequently

$$\det \tilde{\Gamma}_{\mathrm{T}}^{VV}(0) = \det \tilde{\Gamma}_{\mathrm{L}}^{VV}(0) = 0. \tag{4.4.56}$$

Therefore, also the matrix of the transverse parts of the neutral-boson two-point-functions has one eigenvalue zero. The corresponding massless gauge boson is identified with the photon. The physical, transverse photon field is defined in such a way that it does not mix with the Z-boson field at zero momentum transfer. Then, the transverse parts and owing to (4.4.55) also the longitudinal parts of the photon two-point function and of the photon–Z-boson mixing vanish for $k^2 = 0$,

$$\tilde{\Gamma}_{\mathrm{T}}^{AA}(0) = \tilde{\Gamma}_{\mathrm{L}}^{AA}(0) = 0, \qquad \tilde{\Gamma}_{\mathrm{T}}^{AZ}(0) = \tilde{\Gamma}_{\mathrm{L}}^{AZ}(0) = 0. \tag{4.4.57}$$

Inserting this into (4.4.50) for $b = A$ and $c = Z$, and observing that $\tilde{\Gamma}^{K_\chi u^Z}(0) \neq 0$ in lowest order and thus in perturbation theory, we find

$$\tilde{\Gamma}_{\mathrm{L}}^{A\chi}(0) = 0. \tag{4.4.58}$$

Thus, the on-shell photon decouples from all the other fields.

By analysing the identities (4.4.50) and (4.4.51) further, one can deduce in a similar way as for the charged bosons that the zeros of the Z-boson and χ-boson two-point functions coincide. To this end, it is convenient to choose the gauge-fixing function

$$C^{\mathrm{Z}} = \partial^\mu Z_\mu + \xi''_Z \partial^\mu A_\mu - M_Z \xi'_Z \chi \tag{4.4.59}$$

and to fix the gauge parameters ξ'_Z and ξ''_Z in such a way that $\Gamma_{\mathrm{L}}^{A\chi}(k_0^2) = 0$ and $\Gamma_{\mathrm{L}}^{Z\chi}(k_0^2) = 0$ at the value of k_0^2 for which $\Gamma_{\mathrm{L}}^{ZZ}(k_0^2) = 0$.

Finally, we consider the two-point functions of the neutral ghosts. We start from the identity (4.4.31) in the gauge (4.4.1) for $a = A, Z$, where A and Z are the physical photon and Z-boson fields,

$$\begin{aligned} 0 &= \partial^\mu \frac{\delta \tilde{\Gamma}}{\delta K_V^{A,\mu}} + \frac{\delta \tilde{\Gamma}}{\delta \bar{u}^A}, \\ 0 &= \partial^\mu \frac{\delta \tilde{\Gamma}}{\delta K_V^{Z,\mu}} - \xi'_Z M_Z \frac{\delta \tilde{\Gamma}}{\delta K_\chi} + \frac{\delta \tilde{\Gamma}}{\delta \bar{u}^Z}. \end{aligned} \tag{4.4.60}$$

Differentiating this with respect to u^a, $a = A, Z$, gives rise to

$$0 = \mathrm{i}k^\mu \tilde{\Gamma}_\mu^{K_V^A u^a}(k,-k) + \Gamma^{\bar{u}^A u^a}(k,-k), \tag{4.4.61}$$

$$0 = \mathrm{i}k^\mu \tilde{\Gamma}_\mu^{K_V^Z u^a}(k,-k) - \xi'_Z M_Z \tilde{\Gamma}^{K_\chi u^a}(k,-k) + \Gamma^{\bar{u}^Z u^a}(k,-k). \tag{4.4.62}$$

After inserting the covariant decomposition

$$\Gamma_{\mu}^{K_V^b u^a}(k,-k) = k_\mu \Gamma_{\mathrm{L}}^{K_V^b u^a}(k^2), \tag{4.4.63}$$

(4.4.61) turns to

$$\Gamma^{\bar{u}^A u^a}(k^2) = -\mathrm{i}k^2 \tilde{\Gamma}_{\mathrm{L}}^{K_V^A u^a}(k^2), \tag{4.4.64}$$

i.e.

$$\Gamma^{\bar{u}^A u^A}(0) = 0 = \Gamma^{\bar{u}^A u^Z}(0). \tag{4.4.65}$$

The photon ghost is massless.

We can define the photon ghost field u^A in such a way that it decouples from the Z ghost at $k^2 = 0$, i.e.

$$\Gamma^{\bar{u}^Z u^A}(0) = 0. \tag{4.4.66}$$

Then, as a consequence of (4.4.62) and (4.4.50) for $b = Z$ and $c = A$

$$\Gamma^{K_\chi u^A}(0) = 0 = \Gamma_{\mathrm{L}}^{K_V^Z u^A}(0), \tag{4.4.67}$$

which is used later.

It can further be shown that the Z-ghost two-point function has a zero at the same location as the longitudinal Z-boson propagator.

All this is consistent with the fact that the contributions of scalar Z bosons ($\varepsilon^\mu \propto k^\mu$), χ bosons, u^Z, and $\bar{u}^Z$ as well as the contributions of longitudinal photons ($\varepsilon^\mu \propto k^\mu$), scalar photons [$\varepsilon^\mu \propto (k^0, -\mathbf{k})$], u^A, and $\bar{u}^A$ cancel in S-matrix elements (cf. Sects. 2.5.4.3 and 4.4.2). As discussed above, there are two quartets of unphysical modes in the neutral sector.

4.4.5 The background-field method for the Electroweak Standard Model

4.4.5.1 The gauge-invariant effective action

The *background-field method* discussed in Sect. 2.4.4 can be directly applied to the Glashow–Salam–Weinberg model [De95]. To this end, the gauge fields are split as in (2.4.107) into background fields (denoted with a caret) and quantum fields,

$$W_\mu^a \to \hat{W}_\mu^a + W_\mu^a, \qquad B_\mu \to \hat{B}_\mu + B_\mu. \tag{4.4.68}$$

For the fermion fields no splitting is necessary, as discussed at the end of Sect. 2.4.4.2. On the other hand, we have to split the Higgs-doublet field since we want to include it in the gauge-fixing term in order to eliminate tree-level mixing between the gauge bosons and the corresponding would-be Goldstone bosons. This is done in such a way that the background Higgs field $\hat{\Phi}$ has the usual non-vanishing vacuum expectation value v while the one of the quantum Higgs field Φ is zero,

$$\begin{aligned}\hat{\Phi}(x) &= \begin{pmatrix} \hat{\phi}^+(x) \\ \frac{1}{\sqrt{2}}[v + \hat{\eta}(x) + \mathrm{i}\hat{\chi}(x)] \end{pmatrix}, \\ \Phi(x) &= \begin{pmatrix} \phi^+(x) \\ \frac{1}{\sqrt{2}}[\eta(x) + \mathrm{i}\chi(x)] \end{pmatrix}. \end{aligned} \tag{4.4.69}$$

In order to retain background-field gauge invariance, we must introduce a *gauge-fixing functional* that is covariant with respect to gauge transformations of the background fields. A suitable ansatz is [Ei89]

$$\begin{aligned}\mathcal{L}_{\mathrm{GF,BFM}} = &-\frac{1}{2\xi_W}\Big[(\delta^{ac}\partial_\mu + g_2\varepsilon^{abc}\hat{W}_\mu^b)W^{c,\mu} \\ &\qquad - \mathrm{i}g_2\xi_W'\frac{1}{2}(\hat{\Phi}_i^\dagger\tau_{ij}^a\Phi_j - \Phi_i^\dagger\tau_{ij}^a\hat{\Phi}_j)\Big]^2 \\ &-\frac{1}{2\xi_B}\Big[\partial_\mu B^\mu + \mathrm{i}g_1\xi_B'\frac{1}{2}(\hat{\Phi}_i^\dagger\Phi_i - \Phi_i^\dagger\hat{\Phi}_i)\Big]^2 . \end{aligned} \tag{4.4.70}$$

Background-field gauge invariance requires that the background gauge fields appear only within a covariant derivative in the gauge-fixing term and that the terms in brackets transform according to the adjoint representation of the gauge group. Moreover, only four independent gauge parameters are allowed,[11] ξ_W and ξ'_W for $\mathrm{SU(2)_w}$ and ξ_B and ξ'_B for $\mathrm{U(1)}_Y$. Mixing between gauge bosons and the corresponding would-be Goldstone bosons is eliminated by choosing $\xi'_W = \xi_W$ and $\xi'_B = \xi_B$, resulting in the *'t Hooft gauge* for the background-field formalism. In order to avoid tree-level mixing between the quantum photon and Z-boson fields, we set $\xi = \xi_W = \xi_B$. Note that (4.4.70) translates to the conventional gauge-fixing term (4.4.1) for $\xi_W = \xi_Z = \xi_A$ upon replacing the background Higgs field by its vacuum expectation value and omitting the background $\mathrm{SU(2)_w}$ triplet field $\hat{W}^a_\mu$.

The Faddeev–Popov Lagrangian can be derived as usual, and physical (background and quantum) fields can be introduced as in Sects. 4.2.2.2 and 4.4.2. More details and a complete list of Feynman rules can be found in Ref. [De95].

4.4.5.2 Background-field Ward identities

The most important property of the background-field formalism is the invariance of the background-field effective action under background-field gauge transformations with associated group parameters $\hat{\theta}^a$ [cf. (2.4.121)],

$$\frac{\delta\Gamma}{\delta\hat{\theta}^a} = 0, \qquad a = A, Z, \pm. \tag{4.4.71}$$

This invariance implies linear identities for the vertex functions that are precisely the *Ward identities* related to the classical Lagrangian. For the two-point functions involving neutral bosons these identities read

$$k^\mu \Gamma^{\hat{A}\hat{A}}_{\mu\nu}(k) = 0, \qquad k^\mu \Gamma^{\hat{A}\hat{Z}}_{\mu\nu}(k) = 0, \tag{4.4.72}$$

$$k^\mu \Gamma^{\hat{A}\hat{\chi}}_{\mu}(k) = 0, \qquad k^\mu \Gamma^{\hat{Z}\hat{Z}}_{\mu\nu}(k) - \mathrm{i}M_\mathrm{Z}\Gamma^{\hat{\chi}\hat{Z}}_{\nu}(k) = 0, \tag{4.4.73}$$

$$k^\mu \Gamma^{\hat{Z}\hat{\chi}}_{\mu}(k) - \mathrm{i}M_\mathrm{Z}\Gamma^{\hat{\chi}\hat{\chi}}(k) + \frac{\mathrm{i}e}{2s_\mathrm{w}c_\mathrm{w}}\Gamma^{\hat{\eta}}(0) = 0, \tag{4.4.74}$$

$$k^\mu \Gamma^{\hat{W}^\pm\hat{W}^\mp}_{\mu\nu}(k) \mp M_\mathrm{W}\Gamma^{\hat{\phi}^\pm\hat{W}^\mp}_{\nu}(k) = 0, \tag{4.4.75}$$

$$k^\mu \Gamma^{\hat{W}^\pm\hat{\phi}^\mp}_{\mu}(k) \mp M_\mathrm{W}\Gamma^{\hat{\phi}^\pm\hat{\phi}^\mp}(k) \pm \frac{e}{2s_\mathrm{w}}\Gamma^{\hat{\eta}}(0) = 0. \tag{4.4.76}$$

[11] This follows also in the conventional formalism by demanding rigid, i.e. global, $\mathrm{SU(2)_w} \times \mathrm{U(1)}_Y$ invariance of the complete Lagrangian $\mathcal{L}_\mathrm{GSW}$ (4.4.10).

Since the vertex functions are one-particle irreducible, they contain no tadpole contributions, and the tadpoles appear explicitly as $\Gamma^{\hat{\eta}}(0)$ in the Ward identities.

For the photon–fermion and the photon–W-boson vertices, QED-like Ward identities are found,

$$k^\mu \Gamma^{\hat{A}\bar{f}f}_\mu(k,\bar{p},p) = -eQ_f\left[\Gamma^{\bar{f}f}(\bar{p}) - \Gamma^{\bar{f}f}(-p)\right], \tag{4.4.77}$$

$$k^\mu \Gamma^{\hat{A}\hat{W}^+\hat{W}^-}_{\mu\rho\sigma}(k,k_+,k_-) = e\left[\Gamma^{\hat{W}^+\hat{W}^-}_{\rho\sigma}(k_+) - \Gamma^{\hat{W}^+\hat{W}^-}_{\rho\sigma}(-k_-)\right],$$

$$k_+^\rho \Gamma^{\hat{A}\hat{W}^+\hat{W}^-}_{\mu\rho\sigma}(k,k_+,k_-) - M_W \Gamma^{\hat{A}\hat{\phi}^+\hat{W}^-}_{\mu\sigma}(k,k_+,k_-) =$$

$$e\left[\Gamma^{\hat{W}^+\hat{W}^-}_{\mu\sigma}(-k_-) - \Gamma^{\hat{A}\hat{A}}_{\mu\sigma}(k) + \frac{c_W}{s_W}\Gamma^{\hat{A}\hat{Z}}_{\mu\sigma}(k)\right]. \tag{4.4.78}$$

From (4.4.77) the universality of the electric charge in the EWSM can be derived in the same way as for QED (cf. Sect. 2.5.1.6).

4.4.5.3 Connected Green functions

The construction of connected background-field Green functions and of the S matrix has been described in Sect. 2.4.4.3. In order to define

$$\hat{\Gamma}^{\text{full}} = \hat{\Gamma} + \int d^4x\, \mathcal{L}^{\text{BF}}_{\text{GF}}, \tag{4.4.79}$$

we choose the usual *'t Hooft gauge-fixing term*

$$\mathcal{L}^{\text{BF}}_{\text{GF}} = -\frac{1}{2\hat{\xi}_A}\left(\partial_\mu \hat{A}^\mu\right)^2 - \frac{1}{2\hat{\xi}_Z}\left(\partial_\mu \hat{Z}^\mu - \hat{\xi}_Z M_Z \hat{\chi}\right)^2$$
$$- \frac{1}{\hat{\xi}_W}\left|\partial_\mu \hat{W}^{+,\mu} - \mathrm{i}\hat{\xi}_W M_W \hat{\phi}^+\right|^2 \tag{4.4.80}$$

for the background fields with the background-gauge-fixing parameters $\hat{\xi}_A$, $\hat{\xi}_Z$, and $\hat{\xi}_W$. Then, the generating functional of connected background-field Green functions, $\hat{T}_c$, is obtained from $\hat{\Gamma}^{\text{full}}$ by a Legendre transformation [De96],

$$\hat{T}_c\{J_{\hat{F}}, J_f, J_{\bar{f}}\} = \mathrm{i}\hat{\Gamma}^{\text{full}}\{\hat{F}, f, \bar{f}\}$$
$$+ \mathrm{i}\int d^4x\left[\sum_{\hat{F}} J_{\hat{F}}\hat{F} + \sum_f (J_{\bar{f}} f - \bar{f} J_f)\right], \tag{4.4.81}$$

where $\hat{F} = \hat{A}, \hat{Z}, \hat{W}^+, \hat{W}^-, \hat{\eta}, \hat{\chi}, \hat{\phi}^+, \hat{\phi}^-$, and

$$J_{\hat{F}} = -\frac{\delta \hat{\Gamma}^{\text{full}}}{\delta \hat{F}}, \qquad J_f = \frac{\delta \hat{\Gamma}^{\text{full}}}{\delta f}, \qquad J_{\bar{f}} = \frac{\delta \hat{\Gamma}^{\text{full}}}{\delta \bar{f}}. \tag{4.4.82}$$

As a consequence, the one-particle reducible Green functions and S-matrix elements are composed as in the conventional formalism from a tree structure of vertex functions. While the vertices in these trees are directly given by the background-field vertex functions, the propagators are determined as the inverse of the two-point vertex functions resulting from $\hat{\Gamma}^{\text{full}}$. Consequently, these propagators are exact to the given order of perturbation theory, i.e. the *Dyson re-summation* of self-energy corrections is included by construction. Of course, one can expand the propagators and recover the ordinary perturbative expansion.

The Ward identities (4.4.71) for the effective action $\hat{\Gamma}$, which generates the vertex functions, translate into Ward identities for the functional $\hat{T}_{\text{c}}$, which generates the connected Green functions. These Ward identities were explicitly derived in Ref. [De96]. The two-point functions involving neutral gauge bosons, for instance, obey,

$$\begin{aligned}
&k^\mu G^{\hat{A}\hat{A}}_{\text{c},\mu\nu}(k) = \frac{-\mathrm{i}\hat{\xi}_A k_\nu}{k^2}, \qquad k^\mu G^{\hat{A}\hat{Z}}_{\text{c},\mu\nu}(k) = 0, \\
&k^\mu G^{\hat{Z}\hat{Z}}_{\text{c},\mu\nu}(k) + \mathrm{i}\hat{\xi}_Z M_Z G^{\hat{\chi}\hat{Z}}_{\text{c},\nu}(k) = \frac{-\mathrm{i}\hat{\xi}_Z k_\nu}{k^2 - \hat{\xi}_Z M_Z^2}, \\
&k^\mu G^{\hat{Z}\hat{\chi}}_{\text{c},\mu}(k) + \mathrm{i}\hat{\xi}_Z M_Z G^{\hat{\chi}\hat{\chi}}_{\text{c}}(k) = \frac{-\hat{\xi}_Z M_Z}{k^2 - \hat{\xi}_Z M_Z^2}.
\end{aligned} \tag{4.4.83}$$

For the photon–fermion vertex we find

$$-\frac{\mathrm{i}}{\hat{\xi}_A} k^2 k^\mu G^{\hat{A}f\bar{f}}_{\text{c},\mu}(k,p,\bar{p}) = -eQ_f \left[G^{f\bar{f}}_{\text{c}}(p) - G^{f\bar{f}}_{\text{c}}(-\bar{p}) \right], \tag{4.4.84}$$

i.e. the same Ward identity as in QED (2.4.88).

After truncating the Green functions and putting fields on shell, many terms drop out in the Ward identities. For example, we find the following identities

$$\begin{aligned}
&k^\nu G^{\hat{A}\dots}_{\text{trunc},\nu} = 0, \qquad k^\nu G^{\hat{Z}\dots}_{\text{trunc},\nu} = \mathrm{i}M_Z G^{\hat{\chi}\dots}_{\text{trunc}}, \\
&k^\nu G^{\hat{W}^\pm\dots}_{\text{trunc},\nu} = \pm M_W G^{\hat{\phi}^\pm\dots}_{\text{trunc}},
\end{aligned} \tag{4.4.85}$$

where the ellipses stand for any on-shell fields. The first of these identities expresses electromagnetic current conservation, the others imply the *Goldstone-boson equivalence theorem* (cf. Sect. 4.4.7), as discussed in Ref. [De96].

The background-field Ward identities for connected Green functions hold exactly even in finite orders of perturbation theory if the re-summed propagators are used to connect the vertex functions [De96]. It is only necessary to calculate the inverse propagators in the same order of perturbation theory as all the other vertex functions. This property is due to the linearity of the background-field Ward identities for vertex functions. In contrast, the Slavnov–Taylor identities in general hold for connected Green functions in a given order of perturbation theory only if all contributions, including the propagators, are expanded up to this order.

The background-field method can be used to simplify calculations in the EWSM. This is mainly due to the fact that the gauge fixing of the background fields is totally unrelated to the gauge fixing of the quantum fields. This freedom can be used to choose a particularly suitable background gauge, e.g. the unitary gauge, in order to reduce the number of Feynman diagrams considerably. Moreover, in the 't Hooft–Feynman gauge ($\xi = 1$) for the quantum fields, many vertices simplify with respect to the conventional formalism [De95].

4.4.6 Charge universality in the Electroweak Standard Model

In QED, the charges of the elementary fields are products of the charge quantum numbers and the elementary charge e. This fact is built into the Lagrangian and guaranteed by gauge invariance. While higher-order corrections affect the elementary charge, i.e. the electromagnetic coupling constant, they do not modify the ratios of the physical charges of the particles, which are defined as the on-shell couplings of the charged particles to the photon. The fact that (apart from the charge quantum numbers) the on-shell coupling constant of charged particles to the photon is independent of the species of the charged particles is called *charge universality*. It turns out that it holds in the same way in the EWSM.

In QED, charge universality is easily derived from the identities (2.4.102) or (2.4.104), which hold not only for the electron but for arbitrary charged fermions. Similar identities are valid for electrically charged scalar bosons and vector bosons that couple gauge-invariantly to the photon. Sandwiching (2.4.104) between two spinors and using the on-shell renormalization condition (2.5.37), which can be written as [note that $\Gamma^{\psi\bar{\psi}} = -\Gamma^{\bar{\psi}\psi}$]

$$\left[\frac{\partial}{\partial p_\mu}\Gamma^{\psi\bar{\psi}}(p,-p)\right]u(p) = \gamma^\mu u(p), \tag{4.4.86}$$

yields

$$\bar{u}(p)\Gamma^{A\psi\bar{\psi}}_\mu(0,p,-p)u(p) = -eQ_f\bar{u}(p)\gamma^\mu u(p) = -2eQ_f p_\mu, \tag{4.4.87}$$

i.e. the on-shell coupling e is independent of the fermion species (and more generally of the matter species). The identity (2.4.102) is a direct consequence of the fact that the photon couples to the conserved current that generates the U(1) gauge transformations.

In the EWSM the situation is different, as the current to which the photon couples is only conserved up to unphysical terms. This renders the proof of charge universality non-trivial.

Following [Ao80], we start by differentiating the Lee identity (4.4.30) with respect to $\bar{\psi}_l$, ψ_k, and u^A, where the charges of the fermion fields are equal $Q_k = Q_l$. After transformation to momentum space we obtain

$$\begin{aligned}
0 = &- \sum_{a=A,Z} \tilde{\Gamma}^{V^a\bar{\psi}_l\psi_k}_\mu(k,\bar{p},p)\tilde{\Gamma}^{K^a_V u^A,\mu}(-k,k) \\
&- \tilde{\Gamma}^{\chi\bar{\psi}_l\psi_k}(k,\bar{p},p)\tilde{\Gamma}^{K_\chi u^A}(-k,k) \\
&+ \tilde{\Gamma}^{\bar{\psi}_l K^i_{\bar{\psi}} u^A}(\bar{p},p,k)\tilde{\Gamma}^{\bar{\psi}_i\psi_k}(-p,p) + \tilde{\Gamma}^{\bar{\psi}_l\psi_i}(\bar{p},-\bar{p})\tilde{\Gamma}^{K^i_\psi\psi_k u^A}(\bar{p},p,k).
\end{aligned} \tag{4.4.88}$$

We insert the covariant decomposition (4.4.38), differentiate with respect to k^μ for p^μ fixed and $\bar{p}^\mu = -p^\mu - k^\mu$, put $k_\mu = 0$, and sandwich everything between spinors $\bar{u}(p)\ldots u(p)$. From the first term of (4.4.88) only the part in which the derivative acts on the explicit k^μ in (4.4.38) contributes. Moreover, the AZ mixing contribution vanishes owing to (4.4.67). The second term drops out, because $\tilde{\Gamma}^{K^a_\chi u^A}$ vanishes at $k^2 = 0$ [cf. (4.4.67)], and its derivative with respect to k^μ is proportional to k^μ. The third term drops out because $\tilde{\Gamma}^{\bar{\psi}_i\psi_k}(-p,p)$ vanishes when projected on $u(p)$, i.e. on the mass shell. For the same reason, the fourth term contributes only if the derivative acts on $\tilde{\Gamma}^{\bar{\psi}_l\psi_i}(\bar{p},-\bar{p})$. Thus, we arrive at the following result

$$\begin{aligned}
&\bar{u}(p)\tilde{\Gamma}^{A\bar{\psi}_l\psi_k}_{\mu}(0,-p,p)u(p)\,\tilde{\Gamma}^{K^A_V u^A}_{\mathrm{L}}(0) \\
&\quad = -\bar{u}(p)\left[\frac{\partial}{\partial p^\mu}\tilde{\Gamma}^{\bar{\psi}_l\psi_i}(-p,p)\right]\tilde{\Gamma}^{K^i_\psi\psi_k u^A}(-p,p,0)u(p) \\
&\quad = -\bar{u}(p)\gamma_\mu\tilde{\Gamma}^{K^l_\psi\psi_k u^A}(-p,p,0)u(p) \\
&\quad = -\frac{p_\mu}{m_l}\bar{u}(p)\tilde{\Gamma}^{K^l_\psi\psi_k u^A}(-p,p,0)u(p),
\end{aligned} \tag{4.4.89}$$

where we used (4.4.86) and the fact that owing to Lorentz invariance the expression in the last but one line must be proportional to p_μ. From (4.4.64) we obtain $[\Gamma^{\bar{u}u}=-\Gamma^{u\bar{u}}]$

$$\tilde{\Gamma}^{K^A_V u^A}_{\mathrm{L}}(0) = \mathrm{i}\,\frac{\partial}{\partial k^2}\Gamma^{\bar{u}^A u^A}(k^2)\bigg|_{k^2=0} = -\mathrm{i}, \tag{4.4.90}$$

where we assumed, without affecting the argument, that the residue of the photon ghost propagator is equal to one. Inserting this into (4.4.89) we find

$$\bar{u}(p)\tilde{\Gamma}^{A\bar{\psi}_l\psi_k}_{\mu}(0,-p,p)u(p) = -\mathrm{i}\frac{p_\mu}{m_l}\bar{u}(p)\tilde{\Gamma}^{K^l_\psi\psi_k u^A}(-p,p,0)u(p). \tag{4.4.91}$$

By fixing $\bar{p}$ instead of p in the above derivation one obtains instead

$$\bar{u}(p)\tilde{\Gamma}^{A\bar{\psi}_l\psi_k}_{\mu}(0,-p,p)u(p) = \mathrm{i}\frac{p_\mu}{m_k}\bar{u}(p)\tilde{\Gamma}^{\bar{\psi}_l K^k_{\bar{\psi}} u^A}(-p,p,0)u(p). \tag{4.4.92}$$

In QED, the ghosts are free and the factor $\tilde{\Gamma}^{K^l_{\bar{\psi}}\bar{\psi}_k u^A}(-p,p,0)$ is given by its tree-level contribution which can be read off directly from the BRS transformation

$$\tilde{\Gamma}^{K^l_\psi\psi_k u^A}(-p,p,0) = -\tilde{\Gamma}^{\bar{\psi}_l K^k_{\bar{\psi}} u^A}(-p,p,0) = \mathrm{i}eQ_l\delta_{lk}, \tag{4.4.93}$$

and (4.4.91) reduces to (4.4.87). In the Electroweak Standard Model the vertex function $\tilde{\Gamma}^{K^l_\psi\psi_k u^A}(-p,p,0)$ plays the rôle of the fermion charge.

For the WWA vertex function we can derive a similar relation

$$\begin{aligned}
&\varepsilon^{*\nu}(k)\tilde{\Gamma}^{AW^+W^-}_{\mu\nu\rho}(0,-k,k)\varepsilon^{\rho}(k) \\
&= 2\mathrm{i}k_{\mu}\varepsilon^{*,\nu}(k)\tilde{\Gamma}^{K^+_V W^+ u^A}_{\rho\nu}(k,-k,0)\varepsilon^{\rho}(k) \\
&= -2\mathrm{i}k_{\mu}\varepsilon^{*,\nu}(k)\tilde{\Gamma}^{K^-_V W^- u^A}_{\nu\rho}(-k,k,0)\varepsilon^{\rho}(k),
\end{aligned} \tag{4.4.94}$$

and obviously $\tilde{\Gamma}^{K^{\pm}_V W^{\pm} u^A}_{\nu\rho}(\pm k,\mp k,0)$ is related to the charge of the W boson.

A relation between $\tilde{\Gamma}^{K^l_{\psi}\psi_k u^A}(-p,p,0)$ and $\tilde{\Gamma}^{K^{\pm}_V W^{\mp} u^A}_{\nu\rho}(\pm k,\mp k,0)$ can be obtained by differentiating (4.4.30) with respect to $\bar{\psi}_l$, ψ_k, u^A, and $W^{\pm,\nu}$, where the charges of the fermion fields fulfill $Q_k - Q_l \pm 1 = 0$. After transforming to momentum space, setting the momentum corresponding to u^A equal to zero, putting the other external fields on shell, and multiplying the fermion legs with spinors and the W leg with a polarization vector, we find the following non-vanishing terms

$$\begin{aligned}
0 = -\bar{u}(\bar{p})\Big[&\tilde{\Gamma}^{W^{\pm}\bar{\psi}_l\psi_k}_{\mu}(k_{\pm},\bar{p},p)\tilde{\Gamma}^{K^{\pm}_V W^{\pm} u^A,\mu\nu}(-k_{\pm},k_{\pm},0) \\
&+\tilde{\Gamma}^{W^{\pm}\bar{\psi}_l\psi_i,\nu}(k_{\pm},\bar{p},p)\tilde{\Gamma}^{K^i_{\psi}\psi_k u^A}(-p,p,0) \\
&+\tilde{\Gamma}^{\bar{\psi}_l K^i_{\bar{\psi}} u^A}(\bar{p},-\bar{p},0)\tilde{\Gamma}^{W^{\pm}\bar{\psi}_i\psi_k,\nu}(k_{\pm},\bar{p},p)\Big]u(p)\varepsilon^{(*)}_{\nu}(k_{\pm}).
\end{aligned} \tag{4.4.95}$$

The contributions of $\phi^{\pm}$ dropped out when projecting on the polarization vector $\varepsilon_{\nu}(k_{\pm})$. Equation (4.4.95) supports the interpretation of the quantities $\Gamma^{K^i_{\psi}\psi_k u^A}(-p,p,0)$ and $\tilde{\Gamma}^{K^{\pm}_V W^{\pm} u^A,\mu\nu}(-k_{\pm},k_{\pm},0)$ as charges and is consistent with

$$\begin{aligned}
\tilde{\Gamma}^{K^{\pm}_V W^{\pm} u^A,\mu\nu}(-k,k,0) &= \pm\mathrm{i}e' g_{\mu\nu}, \\
\tilde{\Gamma}^{K^l_{\psi}\psi_k u^A}(-p,p,0) = -\tilde{\Gamma}^{\bar{\psi}_l K^k_{\bar{\psi}} u^A}(-p,p,0) &= \mathrm{i}e' Q_l \delta_{lk},
\end{aligned} \tag{4.4.96}$$

with some constant e'. Unlike in the case of a symmetric gauge theory with a simple gauge group, in the EWSM the Lee identities are not sufficient to prove this result.

Fortunately, the functions appearing in (4.4.96) can be easily evaluated in the *Landau gauge*, i.e. for $\xi_a = \xi'_a = 0$. In this gauge, all gauge-boson propagators are purely transverse, the photon ghost u^A couples only to gauge bosons, and the corresponding coupling is proportional to the momentum of the antighost $\bar{u}^A$. Therefore, any diagram of the form

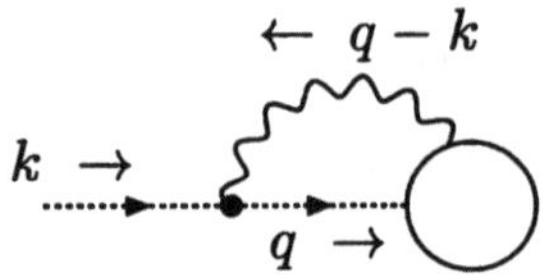

(4.4.97)

vanishes for $k = 0$, because the momentum q is contracted with the purely transverse propagator. As a consequence, the photon ghost with zero momentum does not interact and the vertex functions (4.4.96) are given by their tree-level (and counterterm) contributions. These are directly obtained from the BRS transformations of the fields (4.4.11),

$$\begin{aligned}
\tilde{\Gamma}^{K^l_\psi \psi^k u^A}(-p,p,0) &= -\tilde{\Gamma}^{\bar{\psi}_l K^k_{\bar{\psi}} u^A}(-p,p,0) = \mathrm{i}eQ_l\delta_{lk}, \\
\tilde{\Gamma}^{K^\pm_V W^\pm u^A}_{\mu\nu}(-p,p,0) &= \pm\mathrm{i}eg_{\mu\nu}.
\end{aligned} \tag{4.4.98}$$

This shows that the coupling of the photon to charged particles is described by a universal coupling, i.e. the universality of the electric charge.

4.4.7 The Goldstone-boson equivalence theorem

An important consequence of the Slavnov–Taylor identities in spontaneously broken gauge theories is the *Goldstone-boson equivalence theorem* [Co74a, Ch85, Go86]. It states that the amplitudes for reactions involving high-energetic, longitudinal vector bosons are asymptotically proportional to the amplitudes where these are replaced by their associated would-be Goldstone bosons. The existence of non-trivial proportionality factors resulting from higher-order corrections has been realized in Ref. [Ya88]. The practical virtue of the Goldstone-boson equivalence theorem is twofold. First of all, it facilitates the calculation of cross sections for reactions with longitudinal vector bosons at high energies, as the amplitudes for external scalars are much easier to evaluate. On the other hand, it might allow to derive information on the mechanism of spontaneous symmetry breaking from the experimental study of longitudinal vector bosons.

We demonstrate the derivation of the equivalence theorem for the case of one longitudinal W boson as an example. We start with the Ward identity (4.4.21), i.e.

$$0 = \left\langle T(\partial^\mu W^\mp_\mu \pm \mathrm{i}M_\mathrm{W}\xi_W\phi^\mp)\prod_l \Psi^{\mathrm{as,phys}}_{I_l}\right\rangle. \tag{4.4.99}$$

In order to obtain a relation for S-matrix elements, we have to truncate the external legs, put them on shell, and restrict to physical fields. This is most conveniently done in momentum space, where (4.4.99) reads

$$0 = \left\langle T(k^\mu W^\mp_\mu(k) \pm M_\mathrm{W}\xi_W\phi^\mp(k))\prod_l \Psi^{\mathrm{as,phys}}_{I_l}(k_l)\right\rangle, \tag{4.4.100}$$

and k and k_l denote incoming momenta. The truncation of the physical legs is straightforward. The truncation of the gauge-fixing leg is complicated by the fact that the W-boson field is contracted with a derivative. Owing to the mixing between the W and ϕ bosons we have to introduce the (full) propagator matrix

$$\begin{aligned} G^\mp_{(\mu\nu)} &= \begin{pmatrix} G^{W^\mp W^\pm}_{\mu\nu} & G^{W^\mp\phi^\pm}_{\mu} \\ G^{\phi^\mp W^\pm}_{\nu} & G^{\phi^\mp\phi^\pm} \end{pmatrix} \\ &= \begin{pmatrix} g^\mathrm{T}_{\mu\nu}G^{WW}_\mathrm{T} + g^\mathrm{L}_{\mu\nu}G^{WW}_\mathrm{L} & (\pm k_\mu G^{W\phi}_\mathrm{L}) \\ (\pm k_\nu G^{W\phi}_\mathrm{L}) & G^{\phi\phi} \end{pmatrix}, \end{aligned} \tag{4.4.101}$$

where the projectors on transverse and longitudinal parts are defined in (4.4.37), and k is the incoming momentum of fields corresponding to the first indices. Making the propagators in the gauge-fixing legs explicit, (4.4.100) can be written as (underlining indicates truncation)

$$\begin{aligned} 0 &= (k^\mu, \pm\xi_W M_\mathrm{W})G^\mp_{(\mu\nu)} \begin{pmatrix} \left\langle T\underline{W}^{\pm,\nu}\prod_l \Psi^{\mathrm{as,phys}}_{I_l}\right\rangle \\ \left\langle T\underline{\phi}^{\pm}\prod_l \Psi^{\mathrm{as,phys}}_{I_l}\right\rangle \end{pmatrix} \\ &= \left(k^\nu(G^{WW}_\mathrm{L} + \xi_W M_\mathrm{W} G^{W\phi}_\mathrm{L}), \pm(k^2 G^{W\phi}_\mathrm{L} + \xi_W M_\mathrm{W} G^{\phi\phi})\right) \\ &\quad\times \begin{pmatrix} \left\langle T\underline{W}^{\pm}_\nu\prod_l \Psi^{\mathrm{as,phys}}_{I_l}\right\rangle \\ \left\langle T\underline{\phi}^{\pm})\prod_l \Psi^{\mathrm{as,phys}}_{I_l}\right\rangle \end{pmatrix}. \end{aligned}$$

This yields

$$k^\nu \Big\langle T \underline{W}_\nu^\pm \prod_l \Psi_{I_l}^{\mathrm{as,phys}} \Big\rangle = \pm M_\mathrm{W} A^\pm(k^2) \Big\langle T \underline{\phi}^\pm \prod_l \Psi_{I_l}^{\mathrm{as,phys}} \Big\rangle \quad (4.4.102)$$

with

$$A^\pm(k^2) = -\frac{k^2 G_\mathrm{L}^{W\phi} + \xi_W M_\mathrm{W} G^{\phi\phi}}{M_\mathrm{W}(G_\mathrm{L}^{WW} + \xi_W M_\mathrm{W} G_\mathrm{L}^{W\phi})}. \quad (4.4.103)$$

Equation (4.4.102) provides a relation between a truncated Green function involving an unphysical scalar W-boson with the one involving the associated would-be Goldstone boson instead.

By using the Ward identity (4.4.27) and the relation [cf. (2.2.41)]

$$G^\mp_{(\mu\lambda)} \Gamma^{\pm,(\lambda\nu)} = \mathrm{i} \begin{pmatrix} \delta_\mu^\nu & 0 \\ 0 & 1 \end{pmatrix}, \quad (4.4.104)$$

between the propagator matrix $G^\pm$ and the matrix of the corresponding vertex functions (4.4.42), the factor $A^\pm(k^2)$ can be expressed as

$$A^\pm(k^2) = \frac{k^2/\xi_W + \Gamma_\mathrm{L}^{WW}}{M_\mathrm{W}(M_\mathrm{W} + \Gamma_\mathrm{L}^{W\phi})} = \frac{\tilde{\Gamma}_\mathrm{L}^{WW}}{M_\mathrm{W}\tilde{\Gamma}_\mathrm{L}^{W\phi}} \quad (4.4.105)$$

and thus involves no explicit gauge dependence. In lowest order it reduces to 1. In one-loop approximation it can be expressed in terms of self-energies as

$$A^\pm(k^2) = 1 - \frac{\Sigma_\mathrm{L}^{WW}(k^2)}{M_\mathrm{W}^2} - \frac{\Sigma^{W\phi}(k^2)}{M_\mathrm{W}}. \quad (4.4.106)$$

The self-energies Σ_L^{WW} and $\Sigma^{W\phi}$ are defined by splitting off the tree-level contributions, i.e.

$$\tilde{\Gamma}_\mathrm{L}^{WW} = M_\mathrm{W}^2 - \Sigma_\mathrm{L}^{WW}, \qquad \tilde{\Gamma}_\mathrm{L}^{W\phi} = M_\mathrm{W} + \Sigma^{W\phi}. \quad (4.4.107)$$

In the background-field method $A^\pm(k^2)$ is equal to one in all orders of perturbation theory owing to the Ward identity (4.4.75) [De96].

In order to derive the equivalence theorem from (4.4.102), we have to consider the longitudinal polarization vector of a massive vector boson. For a boson with momentum $k^\mu = (k^0, \mathbf{k})$ and mass M it reads

$$\varepsilon_{\mathrm{L}}^\mu(k) = \left(\frac{|\mathbf{k}|}{M}, \frac{\mathbf{k}}{|\mathbf{k}|}\frac{k^0}{M}\right). \tag{4.4.108}$$

For energies large compared with the vector-boson mass it becomes parallel to k^μ, i.e.

$$\varepsilon_{\mathrm{L}}^\mu(k) = \frac{k^\mu}{M_{\mathrm{W}}} + \mathcal{O}\left(\frac{M_{\mathrm{W}}}{k^0}\right). \tag{4.4.109}$$

This relation allows to replace the unphysical scalar polarization vector k_μ/M_{W} in (4.4.102) by the physical longitudinal one.

According to the *reduction formula* (cf. Sect. 2.1.3) the S-matrix element is obtained from the truncated Green function as

$$\langle \dots |S| W_{\mathrm{L}}^\pm(k) \dots \rangle = R_W^{1/2} \varepsilon_{\mathrm{L}}^\mu \langle T \underline{W_\mu^\pm}(k) \dots \rangle, \tag{4.4.110}$$

where the dots stand for the physical fields, and $R_W^{1/2}$ is the wave-function normalization constant of the W boson. Inserting (4.4.109) and using (4.4.102) this results in

$$\langle \dots |S| W_{\mathrm{L}}^\pm(k) \dots \rangle = \pm R_W^{1/2} A^\pm(M_{\mathrm{W}}^2) \langle T \underline{\phi}^\pm(k) \dots \rangle + \mathcal{O}\left(\frac{M_{\mathrm{W}}}{k^0}\right). \tag{4.4.111}$$

In the case of the Z boson one has to take care of the mixing with the photon. Nevertheless, via the use of the Ward identities (4.4.21) for the photon and the Z boson and (4.4.28) one can derive a relation equivalent to (4.4.111),

$$\langle \dots |S| Z_{\mathrm{L}}(k) \dots \rangle = \mathrm{i} R_Z^{1/2} A^0(M_{\mathrm{Z}}^2) \langle T \underline{\chi}(k) \dots \rangle + \mathcal{O}\left(\frac{M_{\mathrm{Z}}}{k^0}\right), \tag{4.4.112}$$

with

$$A^0(k^2) = \frac{\mathrm{i}k^2 \Delta^{Z\chi} - \xi_Z M_{\mathrm{Z}} \Delta^{\chi\chi}}{M_{\mathrm{Z}}(\Delta_{\mathrm{L}}^{ZZ} - \mathrm{i}\xi_Z M_{\mathrm{Z}} \Delta^{Z\chi})} = \frac{k^2/\xi_Z + \Gamma_{\mathrm{L}}^{ZZ}}{M_{\mathrm{Z}}(M_{\mathrm{Z}} - \mathrm{i}\Gamma^{Z\chi})} = \frac{\mathrm{i}\tilde{\Gamma}_{\mathrm{L}}^{ZZ}}{M_{\mathrm{Z}}\tilde{\Gamma}^{Z\chi}}. \tag{4.4.113}$$

The invariant functions related to the Z–χ propagator matrix and two-point functions are defined analogously to those of the W and χ bosons. In lowest order the factor $A^0(k^2)$ equals 1. In one-loop approximation it can be expressed in terms of self-energies as

$$A^0(k^2) = 1 - \frac{\Sigma_{\mathrm{L}}^{ZZ}(k^2)}{M_{\mathrm{Z}}^2} + \mathrm{i}\frac{\Sigma^{Z\chi}(k^2)}{M_{\mathrm{Z}}} \tag{4.4.114}$$

with

$$\tilde{\Gamma}_{\mathrm{L}}^{ZZ} = M_{\mathrm{Z}}^2 - \Sigma_{\mathrm{L}}^{ZZ}, \qquad \tilde{\Gamma}_{\mathrm{L}}^{Z\chi} = \mathrm{i}M_{\mathrm{Z}} + \Sigma^{Z\chi}. \tag{4.4.115}$$

In the background-field formalism this holds in all orders owing to (4.4.73).

For more external longitudinal vector bosons similar relations can be derived. For the case of several longitudinal W bosons, (4.4.25) yields

$$\Big\langle T\Big(\prod_n \Big[\varepsilon_{\mathrm{L}}^{\mu}(k_n)\underline{W}_{\mu}^{\pm}(k_n) \mp A^{\pm}\underline{\phi}^{\pm}(k_n)\Big]\Big)\prod_l \Psi_{I_l}^{\mathrm{as,phys}}\Big\rangle = \mathcal{O}\left(\frac{M_{\mathrm{W}}}{k_n^0}\right) \tag{4.4.116}$$

after truncation of the external legs, putting them on-shell, and replacing their momenta by the corresponding longitudinal polarization vectors.

By starting from Green functions in which some of the gauge-fixing functions are replaced by physical W bosons we arrive at relations that differ from (4.4.116) only by the absence of some of the ϕ fields. By combining all these relations we finally obtain

$$\begin{aligned}&\Big(\prod_n \varepsilon_{\mathrm{L}}^{\mu}(k_n)\Big)\Big\langle T\Big(\prod_n \underline{W}_{\mu}^{\pm}(k_n)\Big)\prod_l \Psi_{I_l}^{\mathrm{as,phys}}\Big\rangle \\ &\quad = \Big(\prod_n \Big[\pm A^{\pm}(k_n^2)\Big]\Big)\Big\langle T\Big(\prod_n \underline{\phi}^{\pm}(k_n)\Big)\prod_l \Psi_{I_l}^{\mathrm{as,phys}}\Big\rangle + \mathcal{O}\left(\frac{M_{\mathrm{W}}}{k_n^0}\right),\end{aligned} \tag{4.4.117}$$

or for S-matrix elements ($\{n\} = \{n_1\} \cup \{n_2\}$)

$$\begin{aligned}&\Big\langle \prod_{n_1} W_{\mathrm{L}}^{\mp}(k_{n_1})\ldots|S|\prod_{n_2} W_{\mathrm{L}}^{\pm}(k_{n_2})\ldots\Big\rangle \\ &\quad = \Big(\prod_n \Big[\pm R_W^{1/2}A^{\pm}(k_n^2)\Big]\Big)\Big\langle T\Big(\prod_n \underline{\phi}^{\pm}(k_n)\Big)\ldots\Big\rangle + \mathcal{O}\left(\frac{M_{\mathrm{W}}}{k_n^0}\right).\end{aligned} \tag{4.4.118}$$

If we include longitudinal Z bosons, on the right-hand side for each of them a factor $\mathrm{i}R_Z^{1/2}A^0(k^2)$ and the corresponding χ field appear.

Equation (4.4.118) and the corresponding generalization involving Z bosons constitutes the *Goldstone-boson equivalence theorem.* It relates S-matrix elements involving longitudinal gauge bosons to matrix elements involving the corresponding would-be-Goldstone-boson fields in the high-energy limit.

4.5 Renormalization of the Electroweak Standard Model

The EWSM is a non-abelian gauge theory with spontaneous symmetry breaking and as such a renormalizable theory. All infinities can be absorbed into counterterms that are generated by a redefinition of the free parameters and a renormalization of the fields and/or the external wave functions.

In order to define a *renormalization scheme* one has to choose a set of independent parameters. One possibility is to start from the Lagrangian in its symmetric form (4.2.2) and (4.2.8)–(4.2.12). As discussed in Sects. 4.1.1.5 and 4.1.2.3, one needs the *counterterms* of the symmetric theory plus a renormalization of the vacuum expectation value (vev) of the Higgs field.

For practical calculations it is more convenient to fix the counterterms by renormalization conditions on the physical parameters. Since there is a one-to-one correspondence between the physical parameters and the parameters of the symmetric Lagrangian, this approach also yields finite quantities. It has become customary to require that the renormalized masses of the gauge bosons, of the Higgs boson, and of the fermions equal the physical masses and that the renormalized electric charge equals the one measured in the *Thomson limit*, i.e. in the limit of low-energy Compton scattering of on-shell particles. This is the so-called *on-shell renormalization scheme* which was originally proposed by Ross and Taylor [Ro73], then worked out and used by many authors [Si80, Ba80a, Fl81, Sa81, Ao80, Bo86, De93]. The on-shell scheme is distinguished by the fact that all parameters have a direct physical meaning and can be measured directly in suitable experiments. This is evident for the particle masses.[12] It is also true for the definition of the electric

[12]This is not the case for the masses of the light quarks owing to the presence of the strong interaction. However, their contributions can often be expressed in terms of directly measurable quantities.

charge, because in the Thomson limit of Compton scattering all higher-order corrections vanish, and the lowest-order QED cross section becomes exact in all orders of perturbation theory [Th41, Di97]. The renormalization of the parameters together with the unambiguous renormalization of the external wave functions [cf. (2.1.34)] is sufficient to obtain finite S-matrix elements.

In order to obtain finite Green functions, field renormalization is required in addition. When renormalizing directly the Lagrangian in its symmetric form, it is sufficient to introduce only one field renormalization constant for each multiplet. This minimal field renormalization [Bo86] requires one renormalization constant for the triplet of $\mathrm{SU}(2)_{\mathrm{w}}$ gauge fields, one for the $\mathrm{U}(1)_Y$ gauge field, one for the complex Higgs-doublet field, one for every left-handed fermion doublet, and one for every right-handed fermion singlet. In this scheme, additional wave-function renormalizations have to be applied when calculating S-matrix elements.

Another possibility is to renormalize directly the physical fields [Ao80, De93]. Because fields that carry the same charge and spin in general mix, the field renormalization constants take the form of matrices. This allows us to perform the field renormalization such that the renormalized fields do not mix on shell and that no further wave-function renormalization is required.

In this section we sometimes make use of CP conservation. We note that the CP violation resulting from the quark-mixing matrix in the EWSM does not affect the two-point functions and three-point functions at the one-loop level, because at least four quark-mixing-matrix elements are required to construct CP violating quantities (cf. Sect. 1.4.2.3). Since the renormalization relies only on the two-point functions and three-point functions, our discussion is general at the one-loop level.

4.5.1 The renormalization transformation in the on-shell scheme

We perform the renormalization directly for the physical parameters and fields, i.e. the fields that correspond to definite charge and mass.

The bare parameters are split into *renormalized parameters* and *counter-terms* as follows (bare quantities are denoted by an index 0):

$$M_{\mathrm{W},0}^2 = M_{\mathrm{W}}^2 + \delta M_{\mathrm{W}}^2, \qquad M_{\mathrm{Z},0}^2 = M_{\mathrm{Z}}^2 + \delta M_{\mathrm{Z}}^2,$$

$$
\begin{aligned}
M_{\mathrm{H},0}^2 &= M_\mathrm{H}^2 + \delta M_\mathrm{H}^2,\\
m_{f,i,0} &= m_{f,i} + \delta m_{f,i},\\
V_{ij,0} &= V_{ij} + \delta V_{ij},\\
e_0 &= Z_e e \;=\; (1+\delta Z_e)e \;=\; e + \delta e.
\end{aligned}
\tag{4.5.1}
$$

Here M_W, M_Z, M_H, and $m_{f,i}$ are the physical masses of the corresponding particles defined by the poles of the propagators.

In addition we perform a renormalization of the tadpole

$$
t_0 = t + \delta t,
\tag{4.5.2}
$$

which corresponds to a renormalization of the vev of the Higgs-doublet field (cf. Sect. 4.1.2.3).

We introduce field renormalization of a general form using renormalization matrices for fields with equal quantum numbers. We assume that the Higgs field does not mix with other fields, which holds if CP conservation is required. The bare physical fields are split as

$$
\begin{aligned}
W_0^\pm &= Z_W^{1/2} W^\pm = \left(1 + \frac{1}{2}\delta Z_W\right) W^\pm,\\
\begin{pmatrix} Z_0 \\ A_0 \end{pmatrix} &= \begin{pmatrix} Z_{ZZ}^{1/2} & Z_{ZA}^{1/2} \\ Z_{AZ}^{1/2} & Z_{AA}^{1/2} \end{pmatrix} \begin{pmatrix} Z \\ A \end{pmatrix} = \begin{pmatrix} 1+\frac{1}{2}\delta Z_{ZZ} & \frac{1}{2}\delta Z_{ZA} \\ \frac{1}{2}\delta Z_{AZ} & 1+\frac{1}{2}\delta Z_{AA} \end{pmatrix} \begin{pmatrix} Z \\ A \end{pmatrix},\\
\eta_0 &= Z_\eta^{1/2}\eta = \left(1 + \frac{1}{2}\delta Z_\eta\right)\eta,\\
f_{i,0}^\mathrm{L} &= Z_{ij}^{1/2,f,\mathrm{L}} f_j^\mathrm{L} = \left(\delta_{ij} + \frac{1}{2}\delta Z_{ij}^{f,\mathrm{L}}\right) f_j^\mathrm{L},\\
f_{i,0}^\mathrm{R} &= Z_{ij}^{1/2,f,\mathrm{R}} f_j^\mathrm{R} = \left(\delta_{ij} + \frac{1}{2}\delta Z_{ij}^{f,\mathrm{R}}\right) f_j^\mathrm{R},
\end{aligned}
\tag{4.5.3}
$$

where the last expressions in each line are valid in first-order approximation. This general transformation allows us to define the fields in such a way that their quanta are mass eigenstates also in the presence of higher-order corrections.

The counterterms introduced so far are sufficient to render physical S-matrix elements and Green functions of physical particles finite. For a complete renormalization of the EWSM we need the renormalization of the unphysical sector in addition. This consists of the field renormalization of the would-be Goldstone fields,

$$\begin{aligned}\chi_0 &= Z_\chi^{1/2}\chi = \left(1+\frac{1}{2}\delta Z_\chi\right)\chi,\\ \phi_0^\pm &= Z_\phi^{1/2}\phi^\pm = \left(1+\frac{1}{2}\delta Z_\phi\right)\phi^\pm,\end{aligned} \tag{4.5.4}$$

of the Faddeev–Popov ghost fields,

$$u_0^\pm = \tilde{Z}_\pm u^\pm = \left(1+\delta\tilde{Z}_\pm\right)u^\pm, \tag{4.5.5}$$

$$\begin{pmatrix} u_0^Z \\ u_0^A \end{pmatrix} = \begin{pmatrix} \tilde{Z}_{ZZ} & \tilde{Z}_{ZA} \\ \tilde{Z}_{AZ} & \tilde{Z}_{AA} \end{pmatrix}\begin{pmatrix} u^Z \\ u^A \end{pmatrix} = \begin{pmatrix} 1+\delta\tilde{Z}_{ZZ} & \delta\tilde{Z}_{ZA} \\ \delta\tilde{Z}_{AZ} & 1+\delta\tilde{Z}_{AA} \end{pmatrix}\begin{pmatrix} u^Z \\ u^A \end{pmatrix},$$

and of the renormalization of the gauge parameters,

$$\xi_{0,W}^{(\prime)} = Z_{\xi_W^{(\prime)}}\xi_W^{(\prime)}, \qquad \xi_{0,Z}^{(\prime)} = Z_{\xi_Z^{(\prime)}}\xi_W^{(\prime)}, \qquad \xi_{0,A} = Z_{\xi_A}\xi_A. \tag{4.5.6}$$

The Faddeev–Popov antighost fields $\bar{u}$ are not renormalized [cf. (2.5.150)].

By choosing $\xi'_W = \xi_W$ and $\xi'_Z = \xi_Z$, the would-be Goldstone bosons decouple from the scalar gauge bosons, and the poles of their propagators coincide with the physical masses in lowest order. In order to have these features at higher orders, one has to renormalize ξ' and ξ independently.

By inserting the renormalization transformations into the bare Lagrangian $\mathcal{L}_0$ and writing $Z = 1+\delta Z$ for the multiplicative renormalization constants (matrices) we can split $\mathcal{L}_0$ [cf. (2.5.4)]

$$\mathcal{L}_0 = \mathcal{L} + \mathcal{L}_{\text{ct}}. \tag{4.5.7}$$

The basic Lagrangian $\mathcal{L}$ has the same form as $\mathcal{L}_0$ but with unrenormalized parameters and fields replaced by renormalized ones. The counterterm Lagrangian $\mathcal{L}_{\text{ct}}$ yields the counterterms. If not stated otherwise, we restrict ourselves to the one-loop corrections and consistently neglect terms of order $(\delta Z)^2$ in the following.

4.5.2 Renormalization conditions

The renormalization constants are fixed by imposing *renormalization conditions.* These consist of three sets: the conditions that define the renormalized physical parameters, those that define the renormalized physical fields, and those that fix the renormalization in the unphysical sector. While only the first set is relevant for physical quantities (S-matrix elements), a clever choice of the second set allows us not only to eliminate the explicit wave-function renormalization of the external particles, but also to simplify the explicit form of the renormalization conditions for the physical parameters considerably. The choice of the third set determines the form of the renormalized Slavnov–Taylor identities.

4.5.2.1 Tadpole renormalization

As discussed in Sect. 4.1.2.3, it its convenient to require the vanishing of the renormalized tadpole, i.e. the vanishing of the renormalized Higgs-field one-point vertex function,

$$\Gamma^{\eta}_{\mathrm{R}} = \Gamma^{\eta} + \delta t = 0. \tag{4.5.8}$$

As a consequence of this condition no Feynman diagrams involving tadpoles as subdiagrams need to be considered in actual calculations. However, δt enters some of the counterterms. We assume in the following that the *tadpole renormalization* (4.5.8) is performed. Otherwise, extra tadpole terms would appear in the renormalization conditions discussed below.

4.5.2.2 Mass renormalization and renormalization of physical fields

The renormalized mass parameters of the physical particles are fixed by the requirement that they are equal to the physical masses, i.e. to the real parts of the poles of the corresponding propagators. These poles are equivalent to the zeros of the one-particle irreducible two-point functions projected on physical states, i.e. on polarization vectors $\varepsilon_\mu(k)$ for gauge bosons and on spinors $u(p)$ and $v(p)$ for fermions. In case of mass matrices these conditions have to be fulfilled by the corresponding eigenvalues resulting in complicated expressions. These can be considerably simplified by requiring simultaneously the on-shell conditions for the field renormalization matrices [Ao80, De93]. The latter conditions require that close to its pole each propagator is given by its lowest-order expression with the bare mass replaced by the renormalized mass. As a consequence, an on-shell particle does not mix with others, and its propagator has residue one, i.e. the quanta of the renormalized fields are mass eigenstates. Thus, we arrive at the following renormalization conditions for the two-point functions for on-shell external physical fields:

$$
\begin{aligned}
\widetilde{\mathrm{Re}}\, \Gamma^{VV'^\dagger}_{\mathrm{R},\mu\nu}(k)\varepsilon^\nu(k)\Big|_{k^2=M_V^2} &= 0, \quad V,V' = W,Z,A, \\
\lim_{k^2\to M_V^2} \frac{1}{k^2-M_V^2}\, \widetilde{\mathrm{Re}}\, \Gamma^{VV^\dagger}_{\mathrm{R},\mu\nu}(k)\varepsilon^\nu(k) &= -\varepsilon_\mu(k), \\
\widetilde{\mathrm{Re}}\, \Gamma^{\eta\eta}_{\mathrm{R}}(k)\Big|_{k^2=M_\mathrm{H}^2} &= 0, \\
\lim_{k^2\to M_\mathrm{H}^2} \frac{1}{k^2-M_\mathrm{H}^2}\, \widetilde{\mathrm{Re}}\, \Gamma^{\eta\eta}_{\mathrm{R}}(k) &= 1, \\
\widetilde{\mathrm{Re}}\, \Gamma^{f\bar f}_{\mathrm{R},ij}(p)u_j(p)\Big|_{p^2=m_{f,j}^2} &= 0, \\
\lim_{p^2\to m_{f,i}^2} \frac{\not p + m_{f,i}}{p^2-m_{f,i}^2}\, \widetilde{\mathrm{Re}}\, \Gamma^{f\bar f}_{\mathrm{R},ii}(p)u_i(p) &= u_i(p).
\end{aligned}
\tag{4.5.9}
$$

Here $\widetilde{\mathrm{Re}}$ takes the real part of the loop integrals appearing in the self-energies but not of the elements of the quark-mixing matrix appearing there. Its purpose is to remove the absorptive parts. Owing to the required hermiticity of the renormalized Lagrangian, the counterterms can only affect the non-absorptive parts.

In order to proceed, we decompose the two-point functions into covariants. For the fermionic two-point functions we write,

$$
\begin{aligned}
\Gamma^{f\bar f}_{\mathrm{R},ij}(p) = {} & \not p \frac{1-\gamma_5}{2}\Gamma^{f,\mathrm{L}}_{ij}(p^2) + \not p \frac{1+\gamma_5}{2}\Gamma^{f,\mathrm{R}}_{ij}(p^2) \\
& + \frac{1-\gamma_5}{2}\Gamma^{f,\mathrm{l}}_{ij}(p^2) + \frac{1+\gamma_5}{2}\Gamma^{f,\mathrm{r}}_{ji}(p^2).
\end{aligned}
\tag{4.5.10}
$$

If we neglect absorptive parts, which are irrelevant for the renormalization, the hermiticity of the Lagrangian implies the hermiticity of the effective action. Therefore, the fermionic two-point functions must have the following symmetry

$$\Gamma^{f\bar{f}}_{\mathrm{R},ij}(p) = \gamma^0 \Gamma^{f\bar{f}\dagger}_{\mathrm{R},ij}(p)\gamma^0, \tag{4.5.11}$$

which implies

$$\begin{aligned} &\Gamma^{f,\mathrm{L}}_{ij}(p^2) = \Gamma^{f,\mathrm{L}*}_{ji}(p^2), \qquad \Gamma^{f,\mathrm{R}}_{ij}(p^2) = \Gamma^{f,\mathrm{R}*}_{ji}(p^2), \\ &\Gamma^{f,\mathrm{l}}_{ij}(p^2) = \Gamma^{f,\mathrm{r}*}_{ji}(p^2). \end{aligned} \tag{4.5.12}$$

Together with the covariant decomposition (4.4.36) for the gauge-boson two-point functions we obtain from (4.5.9)

$$\begin{aligned} \widetilde{\mathrm{Re}}\,\Gamma^{VV'\dagger}_{\mathrm{R,T}}(M_V^2) = 0, &\qquad \widetilde{\mathrm{Re}}\,\frac{\partial\Gamma^{VV\dagger}_{\mathrm{R,T}}(k^2)}{\partial k^2}\bigg|_{k^2=M_V^2} = -1, \\ \widetilde{\mathrm{Re}}\,\Gamma^{\eta\eta}_{\mathrm{R}}(M_\mathrm{H}^2) = 0, &\qquad \widetilde{\mathrm{Re}}\,\frac{\partial\Gamma^{\eta\eta}_{\mathrm{R}}(k^2)}{\partial k^2}\bigg|_{k^2=M_\mathrm{H}^2} = 1, \\ m_{f,j}\widetilde{\mathrm{Re}}\,\Gamma^{f,\mathrm{L}}_{\mathrm{R},ij}(m_{f,j}^2) &+ \widetilde{\mathrm{Re}}\,\Gamma^{f,\mathrm{r}}_{\mathrm{R},ij}(m_{f,j}^2) = 0, \\ m_{f,j}\widetilde{\mathrm{Re}}\,\Gamma^{f,\mathrm{R}}_{\mathrm{R},ij}(m_{f,j}^2) &+ \widetilde{\mathrm{Re}}\,\Gamma^{f,\mathrm{l}}_{\mathrm{R},ij}(m_{f,j}^2) = 0, \\ \widetilde{\mathrm{Re}}\bigg\{\Gamma^{f,\mathrm{R}}_{\mathrm{R},ii}(m_{f,i}^2) + \Gamma^{f,\mathrm{L}}_{\mathrm{R},ii}(m_{f,i}^2) &+ 2\frac{\partial}{\partial p^2}\Big[m_{f,i}^2\left(\Gamma^{f,\mathrm{R}}_{\mathrm{R},ii}(p^2) + \Gamma^{f,\mathrm{L}}_{\mathrm{R},ii}(p^2)\right) \\ &+ m_{f,i}\left(\Gamma^{f,\mathrm{r}}_{\mathrm{R},ii}(p^2) + \Gamma^{f,\mathrm{l}}_{\mathrm{R},ii}(p^2)\right)\Big]\bigg|_{p^2=m_{f,i}^2}\bigg\} = 2. \end{aligned} \tag{4.5.13}$$

Note that the condition $\Gamma^{AA}_{\mathrm{R,T}}(0) = 0$ is guaranteed by the Lee identities and gives no constraint on the counterterms.

4.5.2.3 Renormalization of the quark-mixing matrix

The mixing angles of the *quark-mixing matrix* V_{ij} are free parameters of the Electroweak Standard Model and thus have to be defined by renormalization conditions in higher-order calculations. The definition of the mixing angles depends on the parametrization (cf. Sect. 1.4.2.2). One could use definite physical observables to fix these parameters in a unique way. This, however, would destroy the symmetry of the renormalized quark-mixing matrix with respect to the quark generations.

In order to formulate a renormalization condition for the quark-mixing matrix that is independent of the parametrization and respects the symmetry, we reconsider the origin of the quark-mixing matrix. The bare quark-mixing matrix $V_{0,ij}$ is given by [cf. (4.2.31)]

$$V_{0,ij} = U^{u,\mathrm{L}}_{0,ik} U^{d,\mathrm{L}\dagger}_{0,kj}, \tag{4.5.14}$$

where the matrices $U_0^{f,\mathrm{L}}$ transform the bare weak-interaction eigenstates f'_0 to the bare mass eigenstates f_0,

$$U^{f,\mathrm{L}\dagger}_{ij} f^{\mathrm{L}}_{j,0} = f'^{\mathrm{L}}_{i,0}. \tag{4.5.15}$$

In the on-shell renormalization scheme, the higher-order mass eigenstates are related to the bare mass eigenstates through the field renormalization constants of the fermions,

$$f^{\mathrm{L}}_i = Z^{1/2,f,\mathrm{L}}_{ij} f^{\mathrm{L}}_{j,0}. \tag{4.5.16}$$

This suggests to define the renormalized quark-mixing matrix in analogy to the unrenormalized one through the rotation from the weak-interaction eigenstates to the renormalized mass eigenstates [De90]. Since the renormalized quark-mixing matrix should be unitary just as the bare one, only the unitary part of the field renormalization matrices can enter.

In the one-loop approximation the rotation contained in the fermion wave-function renormalization $1 + \frac{1}{2}\delta Z^{\mathrm{L}}$ is simply given by its antihermitian part

$$\delta Z^{f,\mathrm{L,AH}}_{ij} = \frac{1}{2}(\delta Z^{f,\mathrm{L}}_{ij} - \delta Z^{f,\mathrm{L}\dagger}_{ij}). \tag{4.5.17}$$

This leads us to define the renormalized quark-mixing matrix as

$$\begin{aligned} V_{ij} &= \left(\delta_{ik} + \frac{1}{2}\delta Z^{u,\mathrm{L,AH}\dagger}_{ik}\right) U^{u,\mathrm{L}}_{km} U^{d,\mathrm{L}\dagger}_{mn} \left(\delta_{nj} + \frac{1}{2}\delta Z^{d,\mathrm{L,AH}}_{nj}\right) \\ &= \left(\delta_{ik} + \frac{1}{2}\delta Z^{u,\mathrm{L,AH}\dagger}_{ik}\right) V_{0,kn} \left(\delta_{nj} + \frac{1}{2}\delta Z^{d,\mathrm{L,AH}}_{nj}\right) \\ &= V_{0,ij} + \frac{1}{2}\left(\delta Z^{u,\mathrm{L,AH}\dagger}_{ik} V_{0,kj} + V_{0,ik} Z^{d,\mathrm{L,AH}}_{kj}\right). \end{aligned} \tag{4.5.18}$$

It has been shown that this condition correctly cancels all one-loop divergences and that $V_{ij} = V_{0,ij}$ in the limit of degenerate up- or down-type quark masses [De90].

Using the fermion wave-function renormalization constants defined by the on-shell conditions (4.5.13), yields a gauge-dependent counterterm for the quark-mixing matrix and thus gauge-dependent S-matrix elements [Ga99]. Gauge-invariant results can be obtained by using a different set of fermion wave-function renormalization constants $\mathcal{Z}_{ij}^{1/2,f,\mathrm{L}}$ and $\mathcal{Z}_{ij}^{1/2,f,\mathrm{R}}$ for the renormalization of the quark-mixing matrix. These renormalization constants result from requiring the renormalization conditions (4.5.9) for the off-diagonal fermion two-point functions not on the mass shell, but at the symmetric point $p = 0$,

$$\begin{aligned}
\widetilde{\mathrm{Re}}\,\Gamma^{f\bar{f}}_{\mathrm{R},ij}(0) &= \widetilde{\mathrm{Re}}\,\Gamma^{f,\mathrm{l}}_{\mathrm{R},ij}(0)\frac{1-\gamma_5}{2} + \widetilde{\mathrm{Re}}\,\Gamma^{f,\mathrm{r}}_{\mathrm{R},ij}(0)\frac{1+\gamma_5}{2} = 0, \\
\frac{\partial}{\partial \not{p}}\widetilde{\mathrm{Re}}\,\Gamma^{f\bar{f}}_{\mathrm{R},ij}(p)\bigg|_{p=0} &= \widetilde{\mathrm{Re}}\,\Gamma^{f,\mathrm{L}}_{\mathrm{R},ij}(0)\frac{1-\gamma_5}{2} + \widetilde{\mathrm{Re}}\,\Gamma^{f,\mathrm{R}}_{\mathrm{R},ij}(0)\frac{1+\gamma_5}{2} = 0.
\end{aligned} \tag{4.5.19}$$

Since the right-handed off-diagonal fermion self-energies are finite and proportional to the fermion masses, one can also choose to disregard the renormalization condition for $\Gamma^{f,\mathrm{R}}_{\mathrm{R},ij}$ and set the right-handed off-diagonal renormalization constants $\mathcal{Z}_{ij}^{1/2,f,\mathrm{R}}$ to zero. Note that while the renormalization constants $\mathcal{Z}_{ij}^{1/2,f,\mathrm{R,L}}$ have to be used for the renormalization of the quark-mixing matrix, they must not be used for wave-function renormalization in S-matrix elements. There, the usual on-shell wave-function renormalization constants enter.

4.5.2.4 Charge Renormalization

The electrical charge is defined as the full $ee\gamma$ coupling for on-shell external particles in the *Thomson limit*, i.e. for vanishing photon momentum. The full $ee\gamma$ coupling consists of the corresponding vertex function and the wave-function renormalization constants. Owing to the photon–Z-boson mixing, also the eeZ coupling occurs in general. In the complete on-shell renormalization scheme the field renormalization is chosen in such a way that the on-shell photon does not mix with the Z boson and that all wave-function renormalization constants are trivial. Then, the *charge renormalization condition* takes the simple form

$$\bar{u}(p)\Gamma^{Ae\bar{e}}_{\mathrm{R},\mu}(p,p)u(p)\big|_{p^2=m_e^2} = e\bar{u}(p)\gamma_\mu u(p) \tag{4.5.20}$$

for the (truncated) vertex function

$\mathrm{i}\Gamma^{Ae\bar{e}}_{\mathrm{R},\mu}(p,p') =$ 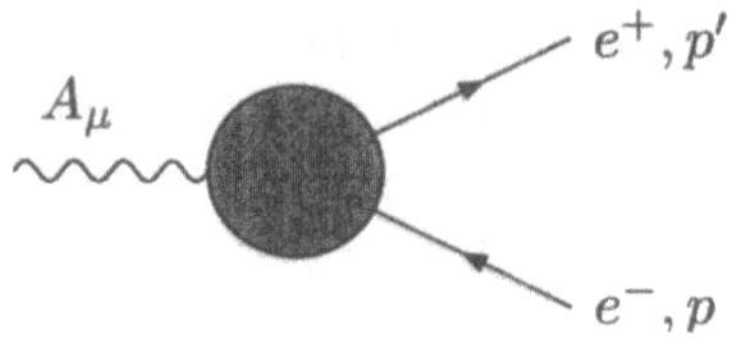 (4.5.21)

Owing to *charge universality* we could impose the above renormalization condition on any charged particle to obtain the same renormalized charge.

4.5.2.5 Renormalization of the unphysical sector

The renormalization of the unphysical fields and of the gauge parameters does not affect S-matrix elements. Therefore, the choice of the corresponding renormalization conditions is a pure matter of convenience. We fix the renormalization of the unphysical sector such that the Lee identities (4.4.30) and (4.4.31), i.e. (4.4.46) and (4.4.60), still hold in terms of renormalized fields. We do not give the explicit formulas but only sketch the procedure.

We have seen in Sect. 2.5.4.2 that in linear gauges the gauge-fixing term need not be renormalized. We therefore require that the gauge-fixing Lagrangian has the form (4.4.1) and (4.4.2) in terms of renormalized fields. This fixes the renormalization of the gauge parameters. The corresponding bare gauge-fixing functionals can be obtained with the help of the renormalization transformations in Sect. 4.5.1. From these bare gauge-fixing functionals the Faddeev–Popov Lagrangian is derived as usual and expressed in terms of renormalized fields.

In order to ensure the validity of (4.4.46) and (4.4.60) for renormalized fields, the renormalization of the sources $K_V^{a,\mu}$ must be equal to that of the Faddeev–Popov antighost fields, i.e. in our conventions $K_V^{a,\mu}$ is not renormalized. Moreover, the renormalization of the sources $K_\phi^\pm$ and K_χ has to be fixed appropriately.

The validity of (4.4.30) for renormalized fields can be guaranteed by renormalizing the other sources K of the BRS variations appropriately.

The field renormalization of the would-be Goldstone-boson fields and the Faddeev–Popov fields is still free and can be chosen in such a way that the corresponding propagators have residue equal to one.

4.5.3 Explicit form of renormalization constants

The renormalized vertex functions consist of the unrenormalized ones and the counterterms. The renormalization conditions allow to express the counterterms by the unrenormalized vertex functions at specific external momenta. We perform this task for the *renormalization constants* for the physical fields and parameters at the one-loop level.

The tadpole counterterm is fixed by

$$\delta t = -\Gamma^{\eta}(0). \tag{4.5.22}$$

The renormalized two-point functions entering the renormalization conditions for the physical fields are decomposed into lowest-order contributions, unrenormalized self-energies, and counterterms as

$$\begin{aligned}
\Gamma^{VV'^\dagger}_{\mathrm{R,T}}(k^2) &= -\delta_{VV'}(k^2 - M_V^2) - \Sigma^{VV'^\dagger}_{\mathrm{T}}(k^2) \\
&\quad - \left[\frac{1}{2}(k^2 - M_V^2)\delta Z_{VV'} + \frac{1}{2}(k^2 - M_{V'}^2)\delta Z_{V'V} - \delta_{VV'}\delta M_V^2\right], \\
\Gamma^{\eta\eta}_{\mathrm{R}}(k^2) &= (k^2 - M_{\mathrm{H}}^2) + \Sigma^{\eta}(k^2) + (k^2 - M_{\mathrm{H}}^2)\delta Z_{\eta} - \delta M_{\mathrm{H}}^2, \\
\Gamma^{f,\mathrm{L}}_{\mathrm{R},ij}(p^2) &= \delta_{ij} + \Sigma^{f,\mathrm{L}}_{ij}(p^2) + \frac{1}{2}\left(\delta Z^{f,\mathrm{L}}_{ij} + \delta Z^{f,\mathrm{L}\dagger}_{ij}\right), \\
\Gamma^{f,\mathrm{R}}_{\mathrm{R},ij}(p^2) &= \delta_{ij} + \Sigma^{f,\mathrm{R}}_{ij}(p^2) + \frac{1}{2}\left(\delta Z^{f,\mathrm{R}}_{ij} + \delta Z^{f,\mathrm{R}\dagger}_{ij}\right), \\
\Gamma^{f,\mathrm{l}}_{\mathrm{R},ij}(p^2) &= -m_{f,i}\delta_{ij} + \Sigma^{f,\mathrm{l}}_{ij}(p^2) \\
&\quad - \frac{1}{2}\left(m_{f,i}\delta Z^{f,\mathrm{L}}_{ij} + m_{f,j}\delta Z^{f,\mathrm{R}\dagger}_{ij}\right) - \delta_{ij}\delta m_{f,i}, \\
\Gamma^{f,\mathrm{r}}_{\mathrm{R},ij}(p^2) &= -m_{f,i}\delta_{ij} + \Sigma^{f,\mathrm{r}}_{ij}(p^2) \\
&\quad - \frac{1}{2}\left(m_{f,i}\delta Z^{f,\mathrm{R}}_{ij} + m_{f,j}\delta Z^{f,\mathrm{L}\dagger}_{ij}\right) - \delta_{ij}\delta m_{f,i},
\end{aligned} \tag{4.5.23}$$

where $VV' = \{WW, AA, AZ, ZZ\}$.

Inserting these equations into the renormalization conditions, we find for the mass counterterms

$$
\begin{aligned}
\delta M_{\mathrm{W}}^2 &= \widetilde{\mathrm{Re}}\,\Sigma_{\mathrm{T}}^{WW}(M_{\mathrm{W}}^2), \qquad \delta M_{\mathrm{Z}}^2 = \widetilde{\mathrm{Re}}\,\Sigma_{\mathrm{T}}^{ZZ}(M_{\mathrm{Z}}^2),\\
\delta M_{\mathrm{H}}^2 &= \widetilde{\mathrm{Re}}\,\Sigma^{\eta}(M_{\mathrm{H}}^2),\\
\delta m_{f,i} &= \frac{1}{2}\widetilde{\mathrm{Re}}\left[m_{f,i}\left(\Sigma_{ii}^{f,\mathrm{L}}(m_{f,i}^2)+\Sigma_{ii}^{f,\mathrm{R}}(m_{f,i}^2)\right)\right.\\
&\qquad \left. + \Sigma_{ii}^{f,\mathrm{l}}(m_{f,i}{}^2)+\Sigma_{ii}^{f,\mathrm{r}}(m_{f,i}{}^2)\right]. \qquad (4.5.24)
\end{aligned}
$$

The field renormalization constants of the boson fields are obtained as

$$
\begin{aligned}
\delta Z_{VV} &= -\widetilde{\mathrm{Re}}\,\frac{\partial\Sigma_{\mathrm{T}}^{VV}(k^2)}{\partial k^2}\bigg|_{k^2=M_V^2}, \qquad V=A,Z,W,\\
\delta Z_{AZ} &= -2\widetilde{\mathrm{Re}}\,\frac{\Sigma_{\mathrm{T}}^{AZ}(M_{\mathrm{Z}}^2)}{M_{\mathrm{Z}}^2}, \qquad \delta Z_{ZA} = 2\frac{\Sigma_{\mathrm{T}}^{AZ}(0)}{M_{\mathrm{Z}}^2},\\
\delta Z_{\eta} &= -\widetilde{\mathrm{Re}}\,\frac{\partial\Sigma^{\eta}(k^2)}{\partial k^2}\bigg|_{k^2=M_{\mathrm{H}}^2}, \qquad (4.5.25)
\end{aligned}
$$

and those of the fermion fields read

$$
\begin{aligned}
\delta Z_{ij}^{f,\mathrm{L}} &= \frac{2}{m_{f,i}^2-m_{f,j}^2}\,\widetilde{\mathrm{Re}}\left[m_{f,j}^2\Sigma_{ij}^{f,\mathrm{L}}(m_{f,j}^2)+m_{f,i}m_{f,j}\Sigma_{ij}^{f,\mathrm{R}}(m_{f,j}^2)\right.\\
&\qquad\left. + m_{f,i}\Sigma_{ij}^{f,\mathrm{l}}(m_{f,j}^2)+m_{f,j}\Sigma_{ij}^{f,\mathrm{r}}(m_{f,j}^2)\right], \qquad i\neq j,\\
\delta Z_{ij}^{f,\mathrm{R}} &= \frac{2}{m_{f,i}^2-m_{f,j}^2}\,\widetilde{\mathrm{Re}}\left[m_{f,j}^2\Sigma_{ij}^{f,\mathrm{R}}(m_{f,j}^2)+m_{f,i}m_{f,j}\Sigma_{ij}^{f,\mathrm{L}}(m_{f,j}^2)\right.\\
&\qquad\left. + m_{f,j}\Sigma_{ij}^{f,\mathrm{l}}(m_{f,j}^2)+m_{f,i}\Sigma_{ij}^{f,\mathrm{r}}(m_{f,j}^2)\right], \qquad i\neq j,\\
\delta Z_{ii}^{f,\mathrm{L}} &= -\widetilde{\mathrm{Re}}\,\Sigma_{ii}^{f,\mathrm{L}}(m_{f,i}^2)-m_{f,i}\frac{\partial}{\partial p^2}\widetilde{\mathrm{Re}}\left[m_{f,i}\left(\Sigma_{ii}^{f,\mathrm{L}}(p^2)+\Sigma_{ii}^{f,\mathrm{R}}(p^2)\right)\right.\\
&\qquad\left.\left. + \Sigma_{ii}^{f,\mathrm{l}}(p^2)+\Sigma_{ii}^{f,\mathrm{r}}(p^2)\right]\right|_{p^2=m_{f,i}^2},\\
\delta Z_{ii}^{f,\mathrm{R}} &= -\widetilde{\mathrm{Re}}\,\Sigma_{ii}^{f,\mathrm{R}}(m_{f,i}^2)-m_{f,i}\frac{\partial}{\partial p^2}\widetilde{\mathrm{Re}}\left[m_{f,i}\left(\Sigma_{ii}^{f,\mathrm{L}}(p^2)+\Sigma_{ii}^{f,\mathrm{R}}(p^2)\right)\right.\\
&\qquad\left.\left. + \Sigma_{ii}^{f,\mathrm{l}}(p^2)+\Sigma_{ii}^{f,\mathrm{r}}(p^2)\right]\right|_{p^2=m_{f,i}^2}. \qquad (4.5.26)
\end{aligned}
$$

Owing to (4.5.12),

$$\delta Z^{\dagger}_{ij} = \delta Z_{ij}(m_i^2 \leftrightarrow m_j^2), \tag{4.5.27}$$

and, in particular,

$$\delta Z^{\dagger}_{ii} = \delta Z_{ii}. \tag{4.5.28}$$

In the lepton sector we have $V_{ij} = \delta_{ij}$. Consequently, all (one-loop) lepton self-energies are diagonal and the off-diagonal lepton wave-function renormalization constants are zero. The same holds for the quark sector if one replaces the quark-mixing matrix by a unit matrix, as is often done in calculations of radiative corrections for high-energy processes.

The renormalization constant for the quark-mixing matrix V_{ij} can be directly read off from (4.5.18),

$$\delta V_{ij} = \frac{1}{4}\Big[(\delta \mathcal{Z}^{u,\mathrm{L}}_{ik} - \delta \mathcal{Z}^{u,\mathrm{L}\dagger}_{ik})V_{kj} - V_{ik}(\delta \mathcal{Z}^{d,\mathrm{L}}_{kj} - \delta \mathcal{Z}^{d,\mathrm{L}\dagger}_{kj})\Big]. \tag{4.5.29}$$

The non-diagonal fermion-field renormalization constants determined from (4.5.19) read

$$\begin{aligned}\delta \mathcal{Z}^{f,\mathrm{L,AH}}_{ik} = \frac{1}{m_{f,i}^2 - m_{f,k}^2}\Big[&(m_{f,i}^2 + m_{f,k}^2)\Sigma^{f,\mathrm{L}}_{ik}(0) + 2m_{f,i}m_{f,k}\Sigma^{f,\mathrm{R}}_{ik}(0)\\ &+ 2m_{f,i}\Sigma^{f,\mathrm{l}}_{ik}(0) + 2m_{f,k}\Sigma^{f,\mathrm{r}}_{ik}(0)\Big], \qquad i \neq j.\end{aligned} \tag{4.5.30}$$

Inserting these fermion-field renormalization constants yields

$$\begin{aligned}\delta V_{ij} = \widetilde{\mathrm{Re}}\Bigg\{&\frac{1}{m_{u,i}^2 - m_{u,k}^2}\Bigg[\frac{m_{u,k}^2 + m_{u,i}^2}{2}\Sigma^{u,\mathrm{L}}_{ik}(0) + m_{u,i}m_{u,k}\Sigma^{u,\mathrm{R}}_{ik}(0)\\ &+ m_{u,i}\Sigma^{u,\mathrm{l}}_{ik}(0) + m_{u,k}\Sigma^{u,\mathrm{r}}_{ik}(0)\Bigg]V_{kj}\\ &- V_{ik}\frac{1}{m_{d,k}^2 - m_{d,j}^2}\Bigg[\frac{m_{d,j}^2 + m_{d,k}^2}{2}\Sigma^{d,\mathrm{L}}_{kj}(0) + m_{d,k}m_{d,j}\Sigma^{d,\mathrm{R}}_{kj}(0)\\ &+ m_{d,k}\Sigma^{d,\mathrm{l}}_{kj}(0) + m_{d,j}\Sigma^{d,\mathrm{r}}_{kj}(0)\Bigg]\Bigg\}.\end{aligned} \tag{4.5.31}$$

If one chooses to set the right-handed off-diagonal renormalization constants to zero, the terms involving Σ^{R} have to be omitted.

It remains to fix the charge renormalization constant δZ_e. To be more general, we investigate the $ff\gamma$ vertex for arbitrary fermions f. The renormalized vertex function reads

$$\Gamma^{Aff}_{\mathrm{R},ij,\mu}(p,p') = -eQ_f\gamma_\mu\delta_{ij} + e\Lambda^{Aff}_{\mathrm{R},ij,\mu}(p,p'). \tag{4.5.32}$$

For on-shell external fermions it can be decomposed as ($k = p' - p$)

$$\begin{aligned}\Lambda^{Aff}_{\mathrm{R},ij,\mu}(p,p') = \delta_{ij}\Bigg(&\gamma_\mu\Lambda^f_{\mathrm{R,V}}(k^2) - \gamma_\mu\gamma_5\Lambda^f_{\mathrm{R,A}}(k^2) \\ &+ \frac{(p+p')_\mu}{2m_f}\Lambda^f_{\mathrm{R,S}}(k^2) + \frac{(p'-p)_\mu}{2m_f}\gamma_5\Lambda^f_{\mathrm{R,P}}(k^2)\Bigg).\end{aligned} \tag{4.5.33}$$

Expressing the renormalized quantities by the unrenormalized ones and the counterterms, and inserting this in the analogue of the renormalization condition (4.5.20) for arbitrary fermions, we find, using the Gordon identities (A.1.42),

$$\begin{aligned}0 &= \bar{u}(p)\Lambda^{Aff}_{\mathrm{R},ii,\mu}(p,p)u(p) \\ &= \bar{u}(p)\gamma_\mu u(p)\Bigg[-Q_f\left(\delta Z_e + \delta Z^{f,\mathrm{V}}_{ii} + \frac{1}{2}\delta Z_{AA}\right) + \Lambda^f_{\mathrm{V}}(0) + \Lambda^f_{\mathrm{S}}(0) \\ &\qquad + v_f\frac{1}{2}\delta Z_{ZA}\Bigg] \\ &\quad - \bar{u}(p)\gamma_\mu\gamma_5 u(p)\left[-Q_f\delta Z^{f,\mathrm{A}}_{ii} + \Lambda^f_{\mathrm{A}}(0) + a_f\frac{1}{2}\delta Z_{ZA}\right],\end{aligned} \tag{4.5.34}$$

where

$$\delta Z^{f,\mathrm{V}}_{ii} = \frac{1}{2}(\delta Z^{f,\mathrm{L}}_{ii} + \delta Z^{f,\mathrm{R}}_{ii}), \qquad \delta Z^{f,\mathrm{A}}_{ii} = \frac{1}{2}(\delta Z^{f,\mathrm{L}}_{ii} - \delta Z^{f,\mathrm{R}}_{ii}), \tag{4.5.35}$$

and v_f, a_f are the vector and axial-vector couplings of the Z-boson to the fermion f, explicitly given in (1.4.38) or (A.2.23). This yields, in fact, two conditions, namely

$$0 = -Q_f\left(\delta Z_e + \delta Z^{f,\mathrm{V}}_{ii} + \frac{1}{2}\delta Z_{AA}\right) + \Lambda^f_{\mathrm{V}}(0) + \Lambda^f_{\mathrm{S}}(0) + v_f\frac{1}{2}\delta Z_{ZA}, \tag{4.5.36}$$

$$0 = -Q_f\delta Z^{f,\mathrm{A}}_{ii} + \Lambda^f_{\mathrm{A}}(0) + a_f\frac{1}{2}\delta Z_{ZA}. \tag{4.5.37}$$

Equation (4.5.36) for $f = \mathrm{e}$ fixes the charge renormalization constant. Equation (4.5.37) should be fulfilled automatically.

In fact, similar to the QED Ward identity (2.4.104) there is an identity in the Electroweak Standard Model. At the one-loop level it reads

$$\begin{aligned}\bar{u}(p)\Lambda^{Aff}_{ii,\mu}(p,p)u(p) = &-Q_f\bar{u}(p)\left[\frac{\partial}{\partial p^\mu}\Sigma^f_{ii}(p)\right]u(p)\\ &- a_f\bar{u}(p)\gamma_\mu(1-\gamma_5)u(p)\frac{\Sigma^{AZ}_{\mathrm{T}}(0)}{M_{\mathrm{Z}}^2},\end{aligned} \tag{4.5.38}$$

as can be checked by direct calculation. Using

$$\begin{aligned}\bar{u}(p)\left[\frac{\partial}{\partial p^\mu}\Sigma^f_{ii}(p)\right]u(p) &= -\bar{u}(p)\gamma_\mu(\delta Z^{f,\mathrm{V}}_{ii} - \delta Z^{f,\mathrm{A}}_{ii}\gamma_5)u(p),\\ \Sigma^{AZ}_{\mathrm{T}}(0) &= \frac{1}{2}M_{\mathrm{Z}}^2\delta Z_{ZA},\end{aligned} \tag{4.5.39}$$

this results in (4.5.37) and

$$\Lambda^f_V(0) + \Lambda^f_S(0) - Q_f\delta Z^{f,\mathrm{V}}_{ii} + a_f\frac{1}{2}\delta Z_{ZA} = 0. \tag{4.5.40}$$

Inserting this in (4.5.36) we finally find (using $v_f - a_f = -Q_f s_{\mathrm{W}}/c_{\mathrm{W}}$)

$$\frac{\delta e}{e} = -\frac{1}{2}\delta Z_{AA} - \frac{s_{\mathrm{W}}}{c_{\mathrm{W}}}\frac{1}{2}\delta Z_{ZA} = \frac{1}{2}\Pi^{AA}(k^2)\bigg|_{k^2=0} - \frac{s_{\mathrm{W}}}{c_{\mathrm{W}}}\frac{\Sigma^{AZ}_{\mathrm{T}}(0)}{M_{\mathrm{Z}}^2}, \tag{4.5.41}$$

where we have introduced the photon vacuum polarization

$$\Pi^{AA}(k^2) = \frac{\Sigma^{AA}_{\mathrm{T}}(k^2)}{k^2}. \tag{4.5.42}$$

The result (4.5.41) is independent of the fermion species, reflecting electric charge universality; clearly it does not depend on a specific choice of field renormalization. Consequently, the analogue of (4.5.20) holds for arbitrary fermions f.

In the on-shell scheme the *weak mixing angle* is a derived quantity, i.e. just an abbreviation. Following Sirlin [Si80] we define it as

$$\sin^2\theta_{\mathrm{w}} \equiv s_{\mathrm{w}}^2 = 1 - \frac{M_{\mathrm{W}}^2}{M_{\mathrm{Z}}^2}, \tag{4.5.43}$$

using the renormalized gauge-boson masses. This definition is independent of a specific process and valid to all orders of perturbation theory. We note that the weak mixing angle is not a directly measurable quantity and (in higher orders) could be defined in different ways.

Since the auxiliary parameters s_{w}^2 and c_{w}^2 frequently appear, it is useful to introduce the corresponding counterterms

$$c_{\mathrm{w},0}^2 = c_{\mathrm{w}}^2 + \delta c_{\mathrm{w}}^2, \qquad s_{\mathrm{w},0}^2 = s_{\mathrm{w}}^2 + \delta s_{\mathrm{w}}^2. \tag{4.5.44}$$

Because of (4.5.43) these are directly related to the gauge-boson-mass counterterms. To one-loop order we obtain

$$\begin{aligned}\frac{\delta c_{\mathrm{w}}^2}{c_{\mathrm{w}}^2} &= \left(\frac{\delta M_{\mathrm{W}}^2}{M_{\mathrm{W}}^2} - \frac{\delta M_{\mathrm{Z}}^2}{M_{\mathrm{Z}}^2}\right) = \widetilde{\mathrm{Re}}\left(\frac{\Sigma_{\mathrm{T}}^{WW}(M_{\mathrm{W}}^2)}{M_{\mathrm{W}}^2} - \frac{\Sigma_{\mathrm{T}}^{ZZ}(M_{\mathrm{Z}}^2)}{M_{\mathrm{Z}}^2}\right),\\ \frac{\delta s_{\mathrm{w}}^2}{s_{\mathrm{w}}^2} &= -\frac{c_{\mathrm{w}}^2}{s_{\mathrm{w}}^2}\frac{\delta c_{\mathrm{w}}^2}{c_{\mathrm{w}}^2} = -\frac{c_{\mathrm{w}}^2}{s_{\mathrm{w}}^2}\widetilde{\mathrm{Re}}\left(\frac{\Sigma_{\mathrm{T}}^{WW}(M_{\mathrm{W}}^2)}{M_{\mathrm{W}}^2} - \frac{\Sigma_{\mathrm{T}}^{ZZ}(M_{\mathrm{Z}}^2)}{M_{\mathrm{Z}}^2}\right).\end{aligned} \tag{4.5.45}$$

4.5.4 Renormalization within the background-field method

In the *background-field formalism* for the Electroweak Standard Model (cf. Sect. 4.4.5) the parameters and quantum fields are renormalized just as in the conventional formalism (4.5.1)–(4.5.6). In addition, we have to renormalize the background fields,

$$\begin{aligned}\hat{W}_0^{\pm} &= Z_{\hat{W}}^{1/2}\hat{W}^{\pm} = \left(1 + \frac{1}{2}\delta Z_{\hat{W}}\right)\hat{W}^{\pm},\\ \begin{pmatrix}\hat{Z}_0\\ \hat{A}_0\end{pmatrix} &= \begin{pmatrix}Z_{\hat{Z}\hat{Z}}^{1/2} & Z_{\hat{Z}\hat{A}}^{1/2}\\ Z_{\hat{A}\hat{Z}}^{1/2} & Z_{\hat{A}\hat{A}}^{1/2}\end{pmatrix}\begin{pmatrix}\hat{Z}\\ \hat{A}\end{pmatrix} = \begin{pmatrix}1 + \frac{1}{2}\delta Z_{\hat{Z}\hat{Z}} & \frac{1}{2}\delta Z_{\hat{Z}\hat{A}}\\ \frac{1}{2}\delta Z_{\hat{A}\hat{Z}} & 1 + \frac{1}{2}\delta Z_{\hat{A}\hat{A}}\end{pmatrix}\begin{pmatrix}\hat{Z}\\ \hat{A}\end{pmatrix},\\ \hat{S}_0 &= Z_{\hat{S}}^{1/2}\hat{S} = (1 + \frac{1}{2}\delta Z_{\hat{S}})\hat{S}, \qquad \hat{S} = \hat{\eta}, \hat{\chi}, \hat{\phi}.\end{aligned} \tag{4.5.46}$$

These renormalization constants can be fixed in such a way that the renormalized effective action is invariant under background-field gauge transformations. This implies relations between the renormalization constants for the background fields and the parameter renormalization constants [De95]. At the one-loop level these read

$$\delta Z_{\hat{A}\hat{A}} = -2\delta Z_e, \qquad \delta Z_{\hat{Z}\hat{A}} = 0, \qquad \delta Z_{\hat{A}\hat{Z}} = 2\frac{c_{\mathrm{W}}}{s_{\mathrm{W}}}\frac{\delta c_{\mathrm{W}}^2}{c_{\mathrm{W}}^2},$$
$$\delta Z_{\hat{Z}\hat{Z}} = -2\delta Z_e - \frac{c_{\mathrm{W}}^2 - s_{\mathrm{W}}^2}{s_{\mathrm{W}}^2}\frac{\delta c_{\mathrm{W}}^2}{c_{\mathrm{W}}^2}, \qquad \delta Z_{\hat{W}} = -2\delta Z_e - \frac{c_{\mathrm{W}}^2}{s_{\mathrm{W}}^2}\frac{\delta c_{\mathrm{W}}^2}{c_{\mathrm{W}}^2},$$
$$\delta Z_{\hat{\eta}} = \delta Z_{\hat{\chi}} = \delta Z_{\hat{\phi}} = -2\delta Z_e - \frac{c_{\mathrm{W}}^2}{s_{\mathrm{W}}^2}\frac{\delta c_{\mathrm{W}}^2}{c_{\mathrm{W}}^2} + \frac{\delta M_{\mathrm{W}}^2}{M_{\mathrm{W}}^2}, \tag{4.5.47}$$

with δc_{W}^2 from (4.5.45). The relations (4.5.47) express the field renormalization constants of all gauge bosons and scalars completely in terms of the renormalization constants of the electric charge and the particle masses. In particular, note that we get the same relation as in QED (2.5.31) for the background-photon-field renormalization constant,

$$Z_{\hat{A}\hat{A}}^{-1/2} = Z_e. \tag{4.5.48}$$

In addition, the field renormalization constants of the fermions obey

$$\delta Z^{f_+,\mathrm{L}} = \delta Z^{f_-,\mathrm{L}}, \tag{4.5.49}$$

i.e. the renormalization constants for the two left-handed fermions in a doublet must be equal.

As the field renormalization constants are fixed by (4.5.47), in general the propagators acquire residues different from one, and different fields can mix on shell. This is similar to the minimal on-shell scheme of the conventional formalism [Bo86]. Therefore, when calculating S-matrix elements, one has to introduce (UV-finite) wave-function renormalization constants, which have been explicitly given in Ref. [De96] for the gauge fields. However, just as in QED, the on-shell definition of the electric charge together with gauge invariance automatically fixes the residue of the photon propagator to one.

As a consequence of the relations between the renormalization constants, the counterterm vertices of the background fields have a much simpler structure than the ones in the conventional formalism. In fact, all vertices originating from a separately gauge-invariant term in the Lagrangian acquire the same renormalization constants. The explicit form of the counterterm vertices at one-loop order has been given in Ref. [De95].

4.5.5 Mass renormalization for unstable particles

The renormalization conditions for the masses formulated in Sect. 4.5.2.2 are appropriate for stable particles. However, many of the particles in the EWSM, i.e. the Z boson, the W boson, the Higgs boson, the top quark, and most of the other fermions, are unstable. Since instability is a higher order effect, it can often be neglected. For precise predictions, however, it is necessary to take the instability into account.

We consider the definition of the W-boson mass in more detail. It turns out that the renormalization condition

$$\widetilde{\mathrm{Re}}\,\Gamma^{W^-W^+}_{\mathrm{R,T}}(M_{\mathrm{W}}^2) = 0 \tag{4.5.50}$$

leads to a renormalized mass parameter that is gauge-dependent beyond the one-loop approximation [Si91]. This implies that this mass parameter M_{W} cannot be interpreted as the physical mass of the W boson beyond one loop.

On the other hand, the pole of the W-boson propagator, or the zero of its inverse, in the complex plane is a gauge-independent quantity [Ga00a] and can be used to define the physical W-boson mass to all orders of perturbation theory. The renormalization condition then reads

$$\Gamma^{W^+W^-}_{\mathrm{R,T}}(\mu_{\mathrm{W}}) = 0, \tag{4.5.51}$$

where μ_{W} is the location of the complex pole of the W-boson propagator. The corresponding mass parameter $\overline{M}_{\mathrm{W}}$ and width parameter $\overline{\Gamma}_{\mathrm{W}}$ of the W boson are defined by

$$\mu_{\mathrm{W}} = \overline{M}_{\mathrm{W}}^2 - \mathrm{i}\overline{M}_{\mathrm{W}}\overline{\Gamma}_{\mathrm{W}}. \tag{4.5.52}$$

We use bars to distinguish these pole-scheme parameters from the parameters in the usual on-shell scheme.

In terms of unrenormalized quantities the renormalization condition (4.5.51) reads

$$\Gamma^{W^+W^-}_{\mathrm{T}}(\mu_{\mathrm{W}}) = -\left(\mu_{\mathrm{W}} - M_{\mathrm{W},0}^2 + \Sigma^{WW}_{\mathrm{T}}(\mu_{\mathrm{W}})\right) = 0. \tag{4.5.53}$$

Inserting (4.5.52) into (4.5.53) and expanding in $\bar{\Gamma}_W = \mathcal{O}(\alpha)$ yields

$$\begin{aligned} \overline{M}_W^2 - i\overline{M}_W\bar{\Gamma}_W &= M_{W,0}^2 - \Sigma_T^{WW}(\overline{M}_W^2 - i\overline{M}_W\bar{\Gamma}_W) \\ &= M_{W,0}^2 - \Sigma_T^{WW}(\overline{M}_W^2) + i\overline{M}_W\bar{\Gamma}_W\, \Sigma_T^{\prime WW}(\overline{M}_W^2) \\ &\quad + \frac{1}{2}(\overline{M}_W\bar{\Gamma}_W)^2\, \Sigma_T^{\prime\prime WW}(\overline{M}_W^2) + \mathcal{O}(\alpha^4), \end{aligned} \tag{4.5.54}$$

where $\Sigma_T^{\prime WW}(s)$ denotes the derivative of $\Sigma_T^{WW}(s)$ with respect to s. Separating into real and imaginary parts, we find

$$\begin{aligned} \overline{M}_W^2 &= M_{W,0}^2 - \mathrm{Re}\left\{\Sigma_T^{WW}(\overline{M}_W^2)\right\} \\ &\quad - \overline{M}_W\bar{\Gamma}_W\, \mathrm{Im}\left\{\Sigma_T^{\prime WW}(\overline{M}_W^2)\right\} + \mathcal{O}(\alpha^3), \\ \overline{M}_W\bar{\Gamma}_W &= \mathrm{Im}\left\{\Sigma_T^{WW}(\overline{M}_W^2)\right\} - \overline{M}_W\bar{\Gamma}_W\, \mathrm{Re}\left\{\Sigma_T^{\prime WW}(\overline{M}_W^2)\right\} \\ &\quad - \frac{1}{2}(\overline{M}_W\bar{\Gamma}_W)^2\, \mathrm{Im}\left\{\Sigma_T^{\prime\prime WW}(\overline{M}_W^2)\right\} + \mathcal{O}(\alpha^4). \end{aligned} \tag{4.5.55}$$

The width $\bar{\Gamma}_W$ can be eliminated from the right sides of (4.5.55) by iteration, resulting in

$$\begin{aligned} \overline{M}_W^2 &= M_{W,0}^2 - \mathrm{Re}\left\{\Sigma_T^{WW}(\overline{M}_W^2)\right\} \\ &\quad - \mathrm{Im}\left\{\Sigma_T^{WW}(\overline{M}_W^2)\right\} \mathrm{Im}\left\{\Sigma_T^{\prime WW}(\overline{M}_W^2)\right\} + \mathcal{O}(\alpha^3), \end{aligned} \tag{4.5.56}$$

$$\begin{aligned} \overline{M}_W\bar{\Gamma}_W &= \mathrm{Im}\left\{\Sigma_T^{WW}(\overline{M}_W^2)\right\} \Big[1 - \mathrm{Re}\left\{\Sigma_T^{\prime WW}(\overline{M}_W^2)\right\} \\ &\quad + \left(\mathrm{Re}\left\{\Sigma_T^{\prime WW}(\overline{M}_W^2)\right\}\right)^2 \\ &\quad - \frac{1}{2}\,\mathrm{Im}\left\{\Sigma_T^{WW}(\overline{M}_W^2)\right\} \mathrm{Im}\left\{\Sigma_T^{\prime\prime WW}(\overline{M}_W^2)\right\} + \mathcal{O}(\alpha^3)\Big]. \end{aligned} \tag{4.5.57}$$

On the other hand, we obtain

$$M_W^2 = M_{W,0}^2 - \mathrm{Re}\left\{\Sigma_T^{WW}(\overline{M}_W^2)\right\} \tag{4.5.58}$$

from (4.5.50) for the on-shell mass and by comparing this with (4.5.56) the relation

$$\overline{M}_W^2 = M_W^2 - \mathrm{Im}\left\{\Sigma_T^{WW}(\overline{M}_W^2)\right\} \mathrm{Im}\left\{\Sigma_T^{\prime WW}(\overline{M}_W^2)\right\} + \mathcal{O}(\alpha^3). \tag{4.5.59}$$

Thus, $\overline{M}_W^2$ and M_W^2 coincide at the one-loop level, but differ by two-loop corrections. Note that $\mathrm{Im}\left\{\Sigma_T^{\prime WW}(\overline{M}_W^2)\right\}$ is gauge-dependent.

The on-shell W-boson width is defined as

$$
\begin{aligned}
M_{\mathrm{W}}\Gamma_{\mathrm{W}} &= \frac{\mathrm{Im}\left\{\Sigma_{\mathrm{T}}^{WW}(M_{\mathrm{W}}^2)\right\}}{1+\mathrm{Re}\left\{\Sigma_{\mathrm{T}}^{\prime WW}(M_{\mathrm{W}}^2)\right\}} \\
&= \mathrm{Im}\left\{\Sigma_{\mathrm{T}}^{WW}(M_{\mathrm{W}}^2)\right\}\Big[1-\mathrm{Re}\left\{\Sigma_{\mathrm{T}}^{\prime WW}(M_{\mathrm{W}}^2)\right\} \\
&\qquad + \left(\mathrm{Re}\left\{\Sigma_{\mathrm{T}}^{\prime WW}(M_{\mathrm{W}}^2)\right\}\right)^2 + \mathcal{O}(\alpha^3)\Big].
\end{aligned}
\tag{4.5.60}
$$

Using (4.5.57) and (4.5.59) this can be related to the pole width as

$$
\begin{aligned}
\overline{M}_{\mathrm{W}}\overline{\Gamma}_{\mathrm{W}} = M_{\mathrm{W}}\Gamma_{\mathrm{W}} &- \mathrm{Im}\left\{\Sigma_{\mathrm{T}}^{WW}(M_{\mathrm{W}}^2)\right\}\left(\mathrm{Im}\left\{\Sigma_{\mathrm{T}}^{\prime WW}(M_{\mathrm{W}}^2)\right\}\right)^2 \\
&- \frac{1}{2}\left(\mathrm{Im}\left\{\Sigma_{\mathrm{T}}^{WW}(M_{\mathrm{W}}^2)\right\}\right)^2 \mathrm{Im}\left\{\Sigma_{\mathrm{T}}^{\prime\prime WW}(M_{\mathrm{W}}^2)\right\} + \mathcal{O}(\alpha^4).
\end{aligned}
\tag{4.5.61}
$$

We can simplify this further by taking into account that the W boson decays only into light fermions. Neglecting the fermion masses in Σ_{T}^{WW}, which is reasonable for $s \sim M_{\mathrm{W}}^2$, we have the relations

$$
\mathrm{Im}\left\{\Sigma_{\mathrm{T}}^{WW}(s)\right\} = \frac{\Gamma_{\mathrm{W}}}{M_{\mathrm{W}}}s, \qquad \mathrm{Im}\left\{\Sigma_{\mathrm{T}}^{\prime WW}(s)\right\} = \frac{\Gamma_{\mathrm{W}}}{M_{\mathrm{W}}}
\tag{4.5.62}
$$

in leading order. Then, (4.5.59) and (4.5.61) simplify to

$$
\begin{aligned}
\overline{M}_{\mathrm{W}}^2 &= M_{\mathrm{W}}^2 - \Gamma_{\mathrm{W}}^2 + \mathcal{O}(\alpha^3), \\
\frac{\overline{\Gamma}_{\mathrm{W}}}{\overline{M}_{\mathrm{W}}} &= \frac{\Gamma_{\mathrm{W}}}{M_{\mathrm{W}}} + \mathcal{O}(\alpha^4).
\end{aligned}
\tag{4.5.63}
$$

or

$$
\begin{aligned}
\overline{M}_{\mathrm{W}} &= M_{\mathrm{W}}\left(1-\frac{\Gamma_{\mathrm{W}}^2}{2M_{\mathrm{W}}^2}\right) + \mathcal{O}(\alpha^3), \\
\overline{\Gamma}_{\mathrm{W}} &= \Gamma_{\mathrm{W}}\left(1-\frac{\Gamma_{\mathrm{W}}^2}{2M_{\mathrm{W}}^2}\right) + \mathcal{O}(\alpha^4).
\end{aligned}
\tag{4.5.64}
$$

Inserting $M_{\mathrm{W}} \approx 81\,\mathrm{GeV}$ and $\Gamma_{\mathrm{W}} \approx 2.1\,\mathrm{GeV}$, we find

$$
\overline{M}_{\mathrm{W}} \approx M_{\mathrm{W}} - 27\,\mathrm{MeV}.
\tag{4.5.65}
$$

For the mass of the Z boson a similar analysis applies. There, the difference between the on-shell and the pole mass is given by [cf. Sect. 4.6.4.3 and, in particular, (4.6.77)]

$$\overline{M}_{\mathrm{Z}} \approx M_{\mathrm{Z}} - 34\,\mathrm{MeV}. \tag{4.5.66}$$

4.6 Electroweak radiative corrections

We have seen in Sect. 1.6 that in QED higher-order corrections are required in order to describe precision experiments adequately. The typical order of magnitude of one-loop corrections in QED at low energies is set by the loop-expansion parameter $\alpha/\pi \approx 0.0025$. At high energies the radiative corrections are enhanced by factors $\log(E/M)$, where M is a typical mass scale of the theory, in QED usually the electron mass. Moreover, phase-space cuts give rise to further enhancement factors, so that the QED corrections at high energies reach typically a few tens of per cent. In the EWSM the loop-expansion parameter is $\alpha/(\pi s_{\mathrm{W}}^2) \approx 0.01$ and additional enhancement factors appear owing to large top-quark and Higgs-boson masses. As a consequence, the typical order of magnitude of *electroweak radiative corrections* is 1% and can easily reach several per cent.

High-precision tests of a physical theory are important for various reasons. First of all they are necessary to determine the independent parameters accurately and to check the validity of the theory. In the case of a quantum field theory, high-precision tests are inevitable in order to check the quantum nature of the theory, since in lowest-order perturbation theory a quantum field theory is equivalent to a classical theory. Finally, the higher-order corrections depend on the complete structure of the theory including degrees of freedom that are not directly accessible at the experimentally available energy.

Especially during the last ten years, a large number of electroweak observables have been measured with high precision. Apart from low-energy measurements, which include the determination of the μ-decay constant and of the weak mixing angle from electron–neutrino or neutrino–nucleon scattering, the most prominent precision measurements are those of the Z-boson-pole observables and of the W-boson mass. Between 1989 and 1995 about 17 million Z bosons have been produced and analysed at the Large Electron–Positron Collider (LEP) at CERN. As a result, the mass of the Z boson, its

Fig. 4.13 Fermionic contributions to the gauge-boson self-energies

total and partial widths, and various Z-pole asymmetries have been measured with an accuracy at the per-mill level. The mass of the W boson has been determined at hadron colliders and at LEP2 with a precision of better than half a per mill, and this error will go down further in the future.

It is clear that an adequate theoretical prescription of such accurate experiments requires the inclusion of higher-order corrections. Even the complete one-loop corrections are not sufficiently precise for certain observables, and the leading two-loop effects have to be included [Ba95d].

In this section, we examine the most important one-loop radiative corrections. We give explicit results for the fermionic corrections to the gauge-boson self-energies and discuss the one-loop corrections to muon decay, the gauge-boson decays, and to the Z-pole observables.

4.6.1 Fermionic contributions to the gauge-boson self-energies

The fermionic one-loop contributions to the *gauge-boson self-energies* are given by the Feynman diagrams in Fig. 4.13. The *photon self-energy* diagram has already been computed in Sect. 2.5.3.1; the result is given in (2.5.93). The calculation of the other diagrams proceeds in a similar way, the only difference being the presence of axial-vector couplings and the two different fermion masses in the case of the W-boson self-energy. We give the results for the transverse parts of the self-energies [cf. (2.5.33) or (4.4.36)] for one doublet of fermions (leptons or quarks) with masses $m_\pm$, relative charges $Q_\pm$, and vector and axial-vector couplings to the Z boson $v_\pm$ and $a_\pm$, respectively,

$$\Sigma_{\mathrm{T}}^{AA}(q^2) = \frac{\alpha}{3\pi}\sum_{f=\pm} Q_f^2 H(q^2, m_f, m_f),$$

$$\Sigma_{\mathrm{T}}^{AZ}(q^2) = -\frac{\alpha}{3\pi}\sum_{f=\pm} Q_f v_f H(q^2, m_f, m_f),$$

$$\begin{aligned}\Sigma_{\mathrm{T}}^{ZZ}(q^2) &= \frac{\alpha}{3\pi}\sum_{f=\pm}\Big[(v_f^2+a_f^2)H(q^2,m_f,m_f)\\&\qquad - 6a_f^2m_f^2B_0(q^2,m_f,m_f)\Big],\\ \Sigma_{\mathrm{T}}^{WW}(q^2) &= \frac{\alpha}{3\pi}\frac{1}{4s_{\mathrm{w}}^2}\Big[H(q^2,m_+,m_-) - 3m_+m_-B_0(q^2,m_+,m_-)\Big],\end{aligned}\tag{4.6.1}$$

where

$$\begin{aligned}H(q^2,m,m) &= (q^2+2m^2)B_0(q^2;m,m) - 2m_f^2B_0(0;m,m) - \frac{q^2}{3},\\ H(q^2,m_+,m_-) &= \frac{3}{2}\Big[4B_{00}(q^2;m_+,m_-) - A_0(m_+) - A_0(m_-)\\&\qquad + \Big(q^2-(m_+-m_-)^2\Big)B_0(q^2;m_+,m_-)\Big],\end{aligned}\tag{4.6.2}$$

and A_0, B_0, and B_{00} are defined and evaluated in Sects. 2.5.2.1, 2.5.2.2, and 2.5.2.3. Summation over all fermion doublets, including the colour factors for quarks, yields the total fermionic contribution. For the photon self-energy and the photon–Z-boson mixing energy, this vanishes for $q^2 = 0$. As far as the photon self-energy is concerned, this follows from a Ward identity and holds also for the bosonic contributions. In the case of the photon–Z-boson mixing energy the bosonic contributions do not vanish for $q^2 = 0$ in the conventional formalism. In the background-field method they do vanish owing to the Ward identity (4.4.72).

For light fermions, i.e. $q^2 \gg m_f^2$, the above results simplify to

$$\begin{aligned}\Sigma_{\mathrm{T}}^{AA}(q^2) &= \frac{\alpha}{3\pi}\sum_{f=\pm}Q_f^2H(q^2,0,0),\\ \Sigma_{\mathrm{T}}^{AZ}(q^2) &= -\frac{\alpha}{3\pi}\sum_{f=\pm}Q_fv_fH(q^2,0,0),\\ \Sigma_{\mathrm{T}}^{ZZ}(q^2) &= \frac{\alpha}{3\pi}\sum_{f=\pm}(v_f^2+a_f^2)H(q^2,0,0),\\ \Sigma_{\mathrm{T}}^{WW}(q^2) &= \frac{\alpha}{3\pi}\frac{1}{4s_{\mathrm{w}}^2}H(q^2,0,0),\end{aligned}\tag{4.6.3}$$

where

$$\begin{aligned} H(q^2,0,0) &= q^2\left[B_0(q^2;0,0)-\frac{1}{3}\right] \\ &= q^2\left[\Delta-\log\frac{|q^2|}{\mu^2}+\frac{5}{3}+i\pi\theta(q^2)\right] \end{aligned} \tag{4.6.4}$$

with Δ and μ defined in (2.5.62) and after (2.5.50).

For a doublet of heavy fermions, $m_f^2 \gg q^2$, we find

$$\begin{aligned} \Sigma_{\mathrm{T}}^{AA}(q^2) &= \frac{\alpha}{3\pi}\sum_{f=\pm} Q_f^2 q^2\left[\Delta-\log\frac{m_f^2}{\mu^2}+\frac{q^2}{5m_f^2}\right]+q^2\mathcal{O}\left(\frac{q^4}{m_f^4}\right), \\ \Sigma_{\mathrm{T}}^{AZ}(q^2) &= -\frac{\alpha}{3\pi}\sum_{f=\pm} Q_f v_f q^2\left[\Delta-\log\frac{m_f^2}{\mu^2}+\frac{q^2}{5m_f^2}\right]+q^2\mathcal{O}\left(\frac{q^4}{m_f^4}\right), \\ \Sigma_{\mathrm{T}}^{ZZ}(q^2) &= -\frac{2\alpha}{\pi}\sum_{f=\pm} a_f^2 m_f^2\left[\Delta-\log\frac{m_f^2}{\mu^2}\right]+\mathcal{O}(q^2), \\ \Sigma_{\mathrm{T}}^{WW}(q^2) &= -\frac{\alpha}{4\pi}\frac{1}{2s_{\mathrm{W}}^2}\left\{m_+^2\left[\Delta-\log\frac{m_+^2}{\mu^2}\right]+m_-^2\left[\Delta-\log\frac{m_-^2}{\mu^2}\right]\right. \\ &\quad \left.+\frac{1}{2}(m_+^2+m_-^2)-\frac{m_+^2 m_-^2}{m_+^2-m_-^2}\log\frac{m_+^2}{m_-^2}\right\}+\mathcal{O}(q^2). \end{aligned} \tag{4.6.5}$$

For the self-energies of the neutral bosons, the above formulas hold also for single fermions if the sum over the doublet is omitted. In the case of the W-boson self-energy, we obtain for a doublet of one heavy and one light fermion, $m_+^2 \gg q^2 \gg m_-^2$,

$$\Sigma_{\mathrm{T}}^{WW}(q^2) = -\frac{\alpha}{4\pi}\frac{1}{2s_{\mathrm{W}}^2}m_+^2\left[\Delta-\log\frac{m_+^2}{\mu^2}+\frac{1}{2}\right]+\mathcal{O}(q^2). \tag{4.6.6}$$

4.6.2 Parameter relations in higher orders

4.6.2.1 Muon decay and Δr

The *muon lifetime* τ_μ is a very precisely known experimental quantity. Moreover, it can be calculated theoretically from first principles. Originally, this was done in the *Fermi model* with the result [Be56, Ki59]

$$\frac{1}{\tau_\mu} = \frac{G_\mu^2 m_\mu^5}{192\pi^3}\left(1 - 8\frac{m_\mathrm{e}^2}{m_\mu^2}\right)\left[1 + \frac{\alpha}{2\pi}\left(\frac{25}{4} - \pi^2\right)\right]. \tag{4.6.7}$$

The square bracket represents the QED corrections in the Fermi model. Equation (4.6.7) now serves as the *definition of the Fermi constant* G_μ in terms of the experimental muon lifetime

$$\tau_\mu^{\mathrm{exp}} = (2.19703 \pm 0.00004) \times 10^{-6}\,\mathrm{s}. \tag{4.6.8}$$

Recently, also the two-loop QED corrections to the muon lifetime have been calculated [Ri99b]. Including these in the definition of G_μ shifts its value from $G_\mu = 1.16639(1) \times 10^{-5}\,\mathrm{GeV}^{-2}$ to $G_\mu = 1.16637(1) \times 10^{-5}\,\mathrm{GeV}^{-2}$.

Comparing the result for τ_μ calculated in the EWSM with (4.6.7) allows to express G_μ in terms of the EWSM parameters. In lowest order this results in the relation (4.2.34). The radiative corrections to (4.2.34) consist of the W-boson self-energy, vertex and box diagrams, and counterterms.[13] Neglecting the electron and muon mass throughout, one obtains in first order

$$\begin{aligned} G_\mu &= \frac{\pi\alpha}{\sqrt{2}s_\mathrm{w}^2 M_\mathrm{W}^2}\left\{1 + 2\frac{\delta e}{e} - \frac{\delta s_\mathrm{w}^2}{s_\mathrm{w}^2} - \frac{\delta M_\mathrm{W}^2}{M_\mathrm{W}^2} + \frac{\Sigma_\mathrm{T}^{WW}(0)}{M_\mathrm{W}^2} + \delta_{\mathrm{vertex+box}}\right\} \\ &\equiv \frac{\pi\alpha}{\sqrt{2}s_\mathrm{w}^2 M_\mathrm{W}^2}(1 + \Delta r). \end{aligned} \tag{4.6.9}$$

The part that is already included in the Fermi-model QED corrections has been subtracted. The term $\delta_{\mathrm{vertex+box}}$ includes all remaining contributions from vertex and box diagrams. The quantity Δr is a finite combination of one-loop diagrams and counterterms and summarizes the *radiative corrections to muon decay* in the EWSM excluding those parts of the QED corrections already contained in (4.6.7) [Si80]. It can be calculated using

[13] We implicitly assume that the renormalized fermion propagators have residue one, so that no wave-function renormalizations are required.

the techniques and results given in Sect. 2.5.2. The W-boson self-energy and the counterterms have been defined in Sect. 4.5.3 and the contribution of the vertex and box diagrams is given by

$$\delta_{\text{vertex+box}} = \frac{\alpha}{\pi s_{\text{W}}^2}\left(\Delta - \log\frac{M_{\text{W}}^2}{\mu^2}\right) + \frac{\alpha}{4\pi s_{\text{W}}^2}\left(6 + \frac{7-4s_{\text{W}}^2}{2s_{\text{W}}^2}\log\frac{M_{\text{W}}^2}{M_{\text{Z}}^2}\right).$$

The UV-singular part can be expressed in terms of the photon–Z-boson mixing energy at zero momentum transfer

$$\frac{\alpha}{\pi s_{\text{W}}^2}\left(\Delta - \log\frac{M_{\text{W}}^2}{\mu^2}\right) = \frac{2}{s_{\text{W}}c_{\text{W}}}\frac{\Sigma_{\text{T}}^{AZ}(0)}{M_{\text{Z}}^2}. \tag{4.6.10}$$

Inserting these equations and the explicit expressions for the counterterms in the on-shell scheme, (4.5.24), (4.5.41), and (4.5.45), in (4.6.9) we obtain the one-loop result [Si80]

$$\begin{aligned}\Delta r = {} & \Pi^{AA}(0) + \frac{\Sigma_{\text{T}}^{WW}(0) - \operatorname{Re}\Sigma_{\text{T}}^{WW}(M_{\text{W}}^2)}{M_{\text{W}}^2} \\ & - \frac{c_{\text{W}}^2}{s_{\text{W}}^2}\operatorname{Re}\left(\frac{\Sigma_{\text{T}}^{ZZ}(M_{\text{Z}}^2)}{M_{\text{Z}}^2} - \frac{\Sigma_{\text{T}}^{WW}(M_{\text{W}}^2)}{M_{\text{W}}^2}\right) \\ & + 2\frac{c_{\text{W}}}{s_{\text{W}}}\frac{\Sigma_{\text{T}}^{AZ}(0)}{M_{\text{Z}}^2} + \frac{\alpha}{4\pi s_{\text{W}}^2}\left(6 + \frac{7-4s_{\text{W}}^2}{2s_{\text{W}}^2}\log\frac{M_{\text{W}}^2}{M_{\text{Z}}^2}\right) \\ = {} & \Delta r(\alpha, M_{\text{W}}, M_{\text{Z}}, M_{\text{H}}, m_{\text{t}}, m_f).\end{aligned} \tag{4.6.11}$$

Note that Δr depends (via the self-energies) on all parameters of the EWSM, in particular also on the top-quark mass and the experimentally unknown Higgs-boson mass.

4.6.2.2 Dominant contributions to Δr

It turns out that the dominant contributions to Δr originate from the (gauge-invariant) subset of $\mathcal{O}(\alpha)$ fermionic one-loop corrections. These corrections are given by the fermionic contributions to the gauge-boson self-energies in the first two lines of (4.6.11).

We first determine the contribution of a light fermion doublet ($m_f \ll M_{\mathrm{W}}$). From (4.6.3) and (4.6.4) we get

$$\begin{aligned}
\Pi^{AA}(0)\big|_{m_f\to 0} &= \frac{\partial \Sigma_{\mathrm{T}}^{AA}}{\partial q^2}\bigg|_{q^2=0,m_f\to 0} = \frac{\alpha}{3\pi}\sum_{f=\pm} Q_f^2\left[\Delta - \log\frac{m_f^2}{\mu^2}\right] \\
&= \frac{\alpha}{3\pi}\sum_{f=\pm} Q_f^2\left\{\left[\Delta - \log\frac{M_{\mathrm{Z}}^2}{\mu^2} + \frac{5}{3}\right] + \left[\log\frac{M_{\mathrm{Z}}^2}{m_f^2} - \frac{5}{3}\right]\right\}, \\
\frac{\Sigma_{\mathrm{T}}^{WW}(0)}{M_{\mathrm{W}}^2}\bigg|_{m_f=0} &= 0, \\
\frac{\Sigma_{\mathrm{T}}^{WW}(M_{\mathrm{W}}^2)}{M_{\mathrm{W}}^2}\bigg|_{m_f=0} &= \frac{\alpha}{3\pi}\frac{1}{4s_{\mathrm{w}}^2}\left[\Delta - \log\frac{M_{\mathrm{Z}}^2}{\mu^2} + \frac{5}{3}\right] - \frac{\alpha}{3\pi}\frac{1}{4s_{\mathrm{w}}^2}\log c_{\mathrm{w}}^2, \\
\frac{\Sigma_{\mathrm{T}}^{ZZ}(M_{\mathrm{Z}}^2)}{M_{\mathrm{Z}}^2}\bigg|_{m_f=0} &= \frac{\alpha}{3\pi}\sum_{f=\pm}(v_f^2 + a_f^2)\left[\Delta - \log\frac{M_{\mathrm{Z}}^2}{\mu^2} + \frac{5}{3}\right]. \qquad (4.6.12)
\end{aligned}$$

Inserting this into (4.6.11), all terms involving UV divergences (Δ) cancel [as can be seen by inserting the couplings v_f and a_f from (A.2.23)], and we arrive at

$$\begin{aligned}
\Delta r_{\text{light fermions}} &= \\
&= \frac{\alpha}{3\pi}\sum_{f=\pm}\left[Q_f^2\left(\log\frac{M_{\mathrm{Z}}^2}{m_f^2} - \frac{5}{3}\right) - \frac{c_{\mathrm{w}}^2 - s_{\mathrm{w}}^2}{8s_{\mathrm{w}}^4}\log c_{\mathrm{w}}^2\right] \qquad (4.6.13)
\end{aligned}$$

for the contribution of a light fermion doublet to Δr. For small fermion masses the logarithm $\log(M_{\mathrm{Z}}^2/m_f^2)$ in the first term gets large. This logarithm originates from the running of the electromagnetic coupling from $q^2 = 0$, where the scale is set by the small fermion masses m_f, to the electroweak scale which is typically given by $q^2 = M_{\mathrm{Z}}^2$. The large effect is due to the renormalization of the electromagnetic coupling at $q^2 = 0$. The *running of the electromagnetic coupling* can be more generally related to the UV-finite quantity

$$\Delta\alpha = \Pi^{AA}(0) - \mathrm{Re}\,\Pi^{AA}(M_{\mathrm{Z}}^2), \qquad (4.6.14)$$

which reproduces the first term in (4.6.13).

In particular, the leptonic contributions to $\Delta\alpha$, which fulfill $m_l \ll M_W$, read

$$\Delta\alpha_{\text{lep}} = \frac{\alpha}{3\pi} \sum_{l=e,\mu,\tau} Q_l^2 \left(\log \frac{M_Z^2}{m_l^2} - \frac{5}{3} \right) \approx 0.0314. \tag{4.6.15}$$

On the other hand, the contribution $\Delta\alpha_{\text{had}}^{(5)}$ of the five light quarks d, u, s, c, and b cannot be reliably calculated within perturbation theory because of the non-perturbative nature of the strong interaction at small momentum transfers. Instead, $\Delta\alpha_{\text{had}}^{(5)}$ can be derived directly from experimental data with the help of a subtracted *dispersion relation* [Ca61]

$$\Delta\alpha_{\text{had}}^{(5)} = -\frac{\alpha}{3\pi} M_Z^2 \,\text{Re} \left\{ \int_{4m_\pi^2}^{\infty} \text{d}s' \, \frac{R(s')}{s'(s' - M_Z^2 - \text{i}\varepsilon)} \right\}, \tag{4.6.16}$$

where

$$R(s) = \frac{\sigma(\text{e}^+\text{e}^- \to \gamma^* \to \text{hadrons})}{\sigma(\text{e}^+\text{e}^- \to \gamma^* \to \mu^+\mu^-)}. \tag{4.6.17}$$

Using e^+e^- data to calculate $R(s)$ for the energy range below 40 GeV and perturbative QCD for the high energy tail results in [Ei95, Bu95]

$$\Delta\alpha_{\text{had}}^{(5)}(s) = 0.0280 \pm 0.0007. \tag{4.6.18}$$

Summing up the leptonic contributions and those of the light quarks yields

$$\Delta\alpha = 0.0594 \pm 0.0007. \tag{4.6.19}$$

The contribution of heavy fermions ($m_f \gg M_W$) is small and vanishes in the limit of an infinite fermion mass. For the top quark we find, including the colour factor,

$$\Delta\alpha_t = -\frac{\alpha}{\pi} Q_t^2 \frac{1}{5} \frac{M_Z^2}{m_t^2}, \qquad m_t \gg M_W. \tag{4.6.20}$$

The top contribution does not change the value (4.6.19) within the given accuracy.

We turn to the case of a (hypothetical) heavy fermion doublet. Retaining only contributions proportional to m_f^2/M_W^2 in (4.6.5) and using a_f from (A.2.23), we get

$$\begin{aligned}
\Pi^{AA}(0) &= \frac{\alpha}{3\pi}\sum_{f=\pm} Q_f^2\left[\Delta - \log\frac{m_f^2}{\mu^2}\right] \sim 0,\\
\frac{\Sigma_T^{ZZ}(M_Z^2)}{M_Z^2} &\sim -\frac{\alpha}{4\pi}\frac{1}{2s_W^2}\sum_{f=\pm}\frac{m_f^2}{M_W^2}\left[\Delta - \log\frac{m_f^2}{\mu^2}\right],\\
\frac{\Sigma_T^{WW}(M_W^2)}{M_W^2} &\sim \frac{\Sigma_T^{WW}(0)}{M_W^2} \sim \frac{\Sigma_T^{ZZ}(M_Z^2)}{M_Z^2}\\
&\quad - \frac{\alpha}{4\pi}\frac{1}{4s_W^2 M_W^2}\left[m_+^2 + m_-^2 - \frac{2m_+^2 m_-^2}{(m_+^2 - m_-^2)}\log\frac{m_+^2}{m_-^2}\right],
\end{aligned} \tag{4.6.21}$$

and thus

$$\begin{aligned}
&\Delta r_{\text{heavy doublet}} =\\
&= -\frac{\alpha}{4\pi}\frac{c_W^2}{4s_W^4 M_W^2}\left[m_+^2 + m_-^2 - \frac{2m_+^2 m_-^2}{m_+^2 - m_-^2}\log\frac{m_+^2}{m_-^2}\right].
\end{aligned} \tag{4.6.22}$$

For equal masses, $m_+^2 = m_-^2$, $\Delta r_{\text{heavy doublet}}$ vanishes, i.e. there are no enhanced corrections from a heavy degenerate doublet. If on the other hand $m_+^2 \gg m_-^2$, (4.6.22) turns into

$$\begin{aligned}
&\Delta r_{\text{heavy doublet}} =\\
&= -\frac{\alpha}{4\pi}\frac{c_W^2}{4s_W^4 M_W^2}\left[m_+^2 + m_-^2 - 2m_-^2\log\frac{m_+^2}{m_-^2}\right], \qquad m_+^2 \gg m_-^2,
\end{aligned} \tag{4.6.23}$$

i.e. we find contributions to Δr that are enhanced by factors m_f^2/M_W^2. These effects are due to the breaking of the weak isospin symmetry.

The contribution of a doublet with a light and a heavy fermion to Δr is practically relevant for the top–bottom doublet. If we assume $m_+ \gg M_W \gg m_-$ and take only those terms into account that are enhanced by a factor m_+^2/M_W^2, we get the corresponding contribution directly from (4.6.23) by omitting the terms proportional to m_-^2/M_W^2

$$\Delta r_{\text{heavy/light doublet}} = \frac{\alpha}{4\pi}\frac{c_W^2}{4s_W^4}\frac{m_+^2}{M_W^2}. \tag{4.6.24}$$

The enhanced corrections originate exclusively from the quantity

$$\mathrm{Re}\left(\frac{\Sigma_{\mathrm{T}}^{ZZ}(M_{\mathrm{Z}}^2)}{M_{\mathrm{Z}}^2}-\frac{\Sigma_{\mathrm{T}}^{WW}(M_{\mathrm{W}}^2)}{M_{\mathrm{W}}^2}\right), \tag{4.6.25}$$

which in the on-shell renormalization scheme comes from the *renormalization of the weak mixing angle* (4.5.45).

The quantity (4.6.25) can be interpreted as a contribution to the ρ *parameter*. The ρ parameter is nowadays usually defined as

$$\rho=\frac{M_{\mathrm{W}}^2}{M_{\mathrm{Z}}^2\cos^2\theta_{\mathrm{w}}}. \tag{4.6.26}$$

According to (4.2.25) this is exactly equal to one at tree level and because of (4.5.43) also for the on-shell definition of the weak-mixing angle. However, originally the ρ parameter was defined as the ratio of the neutral- to the charged-current strength in neutrino scattering [Ro75]. Then, it contains the quantum correction

$$\mathrm{Re}\left(\frac{\Sigma_{\mathrm{T}}^{ZZ}(0)}{M_{\mathrm{Z}}^2}-\frac{\Sigma_{\mathrm{T}}^{WW}(0)}{M_{\mathrm{W}}^2}\right), \tag{4.6.27}$$

which is finite and equal to (4.6.25) when only heavy-fermion terms are considered.

The fact that (4.6.25) gets large in the presence of large mass splittings between the isospin partners in a doublet has been first observed by Veltman [Ve77]. The contribution of the top–bottom doublet reads, including the colour factor,

$$\Delta r_{\mathrm{t,b}}\sim-\frac{c_{\mathrm{w}}^2}{s_{\mathrm{w}}^2}\frac{3\alpha}{16\pi s_{\mathrm{w}}^2}\frac{m_{\mathrm{t}}^2}{M_{\mathrm{W}}^2}\equiv-\frac{c_{\mathrm{w}}^2}{s_{\mathrm{w}}^2}\Delta\rho,\qquad m_{\mathrm{t}}\gg M_{\mathrm{W}}, \tag{4.6.28}$$

and leads to a sizeable dependence of Δr on m_{t} (cf. Fig. 4.14).

As a result of the above analysis we have got a simple form for the leading terms in Δr,

$$\Delta r=\Delta\alpha-\frac{c_{\mathrm{w}}^2}{s_{\mathrm{w}}^2}\Delta\rho+\Delta r_{\mathrm{rem}}. \tag{4.6.29}$$

The correction to α, $\Delta\alpha$, gets large logarithmic contributions from the light fermions associated with the running of α. The correction to the *ρ parameter*, $\Delta\rho$, contains the leading quadratic corrections from a large top-quark mass. All other terms are collected in the remainder Δr_{rem}, which is typically of the order of ~ 0.01.

The remainder also contains terms logarithmic in the top-quark and Higgs-boson masses that are important. The complete contribution of a heavy top quark ($m_{\text{t}} \gg M_{\text{W}}$) to Δr reads

$$\Delta r_{\text{t}} = -\frac{\alpha}{16\pi s_{\text{W}}^2}\left[3\frac{c_{\text{W}}^2}{s_{\text{W}}^2}\frac{m_{\text{t}}^2}{M_{\text{W}}^2} + 2\left(\frac{c_{\text{W}}^2}{s_{\text{W}}^2} - \frac{1}{3}\right)\log\frac{m_{\text{t}}^2}{M_{\text{W}}^2} + \frac{4}{3}\log c_{\text{W}}^2 + \frac{c_{\text{W}}^2}{s_{\text{W}}^2} - \frac{7}{9}\right], \qquad (4.6.30)$$

and the contribution of a heavy Higgs boson ($M_{\text{H}} \gg M_{\text{W}}$)

$$\Delta r_{\text{H}} = \frac{\alpha}{16\pi s_{\text{W}}^2}\frac{11}{3}\left(\log\frac{M_{\text{H}}^2}{M_{\text{W}}^2} - \frac{5}{6}\right). \qquad (4.6.31)$$

Note that the dependence of Δr on the Higgs-boson mass is only logarithmic. The quantity Δr is visualized in Fig. 4.14 as a function of m_{t} and M_{H}. For comparison the experimental 1σ limits are also given (cf. Sect. 4.6.2.4).

4.6.2.3 Inclusion of leading higher-order effects

The size of the $\mathcal{O}(\alpha)$ result $\Delta r \sim 0.03$ indicates that the higher-order corrections are relevant for precision experiments $[(\Delta r)^2 \sim 0.001]$. In particular, the $\mathcal{O}(\alpha^2)$ terms associated with $\Delta\alpha \sim 0.06$ and $\Delta\rho \sim 0.01$ need to be included.

The *re-summation* of the leading logarithms contained in $\Delta\alpha$ can be inferred from the renormalization group [Ma79]. To this end, one considers the renormalization-group equation for the electromagnetic coupling in QED. Its solution is of the form given in (2.6.21) and contains the leading logarithms in re-summed form. Transferring this QED result to the EWSM implies that all orders in the leading-logarithmic correction $(\Delta\alpha)^n$ can be correctly taken into account by the replacement

$$\alpha(1+\Delta\alpha) \to \frac{\alpha}{1-\Delta\alpha} \equiv \alpha(M_{\text{Z}}^2) \qquad (4.6.32)$$

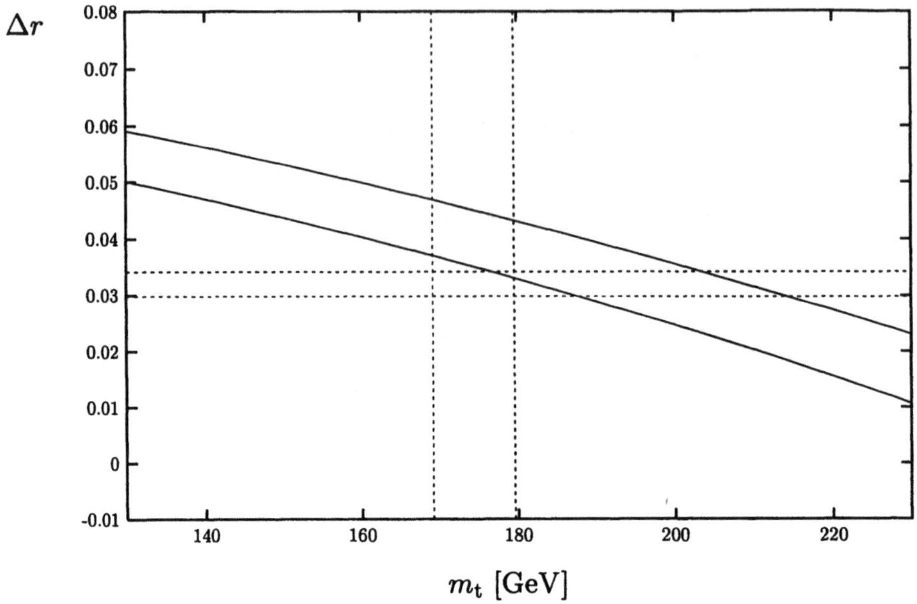

Fig. 4.14 Δr as a function of the top-quark mass for $103 \leq M_{\mathrm{H}} \leq 1000\,\mathrm{GeV}$ and $M_{\mathrm{Z}} = 91.1875\,\mathrm{GeV}$ (solid lines). The experimental 1σ bounds for $\Delta r = 0.0320 \pm 0.0022$ from $M_{\mathrm{W}} = 80.434 \pm 0.037$ and for $m_{\mathrm{t}} = 174.3 \pm 5.1\,\mathrm{GeV}$ are also indicated (dashed lines) [Ho00a]. (This plot is an update of Ref. [Ho00])

in the one-loop result. The non-leading QED terms of next order, i.e. terms of order $\alpha\Delta\alpha \sim \alpha^2 \log(m_f^2/s)$, are numerically not significant. Thus, in a situation where the large corrections are only due to $\Delta\alpha$, the re-summed form

$$G_\mu = \frac{\pi\alpha}{\sqrt{2}s_{\mathrm{w}}^2 M_{\mathrm{W}}^2} \frac{1}{1-\Delta r} \tag{4.6.33}$$

with Δr from (4.6.29) represents a good approximation to the full result.

The replacement (4.6.33) does not correctly re-sum the powers $(\Delta\rho)^n$ and has to be modified for large $\Delta\rho$ [Co89b]. A *re-summation* that is correct in the leading terms up to $\mathcal{O}(\alpha^2)$ is obtained by substituting

$$\alpha \to \alpha(M_{\mathrm{Z}}^2), \qquad s_{\mathrm{w}}^2 \to \bar{s}_{\mathrm{w}}^2 = s_{\mathrm{w}}^2 + c_{\mathrm{w}}^2 \Delta\rho \tag{4.6.34}$$

in the lowest-order relation (4.2.34) and adding the remainder,

$$G_\mu = \frac{\pi\alpha(M_Z^2)}{\sqrt{2}\bar{s}_W^2 M_W^2}(1 + \Delta r_{\text{rem}}). \tag{4.6.35}$$

This result can be refined [Co89b] by including irreducible two-loop corrections.

The QCD corrections to $\Delta\rho$ are known in $\mathcal{O}(\alpha_s)$ [Dj87] and $\mathcal{O}(\alpha_s^2)$ [Av94, Ch95]. These are universal corrections of order $G_\mu m_t^2 \alpha_s$ and $G_\mu m_t^2 \alpha_s^2$, respectively. The complete $\mathcal{O}(\alpha\alpha_s)$ corrections to the self-energies are available from perturbative calculations [Dj88, Kn90, Ha91a, Dj94] and by means of dispersion relations [Kn88, Kn92, Fa93]. The two-loop electroweak contributions to $\Delta\rho$ have been evaluated in the limit of a heavy top-quark mass in leading order (terms of order $G_\mu^2 m_t^4$) [Bi87, Ba92, Fl93] and sub-leading order (terms of order $G_\mu^2 m_t^2 M_Z^2$) [De96a, De97, De98]. Results based on the complete Feynman diagrams have been derived for the Higgs-mass dependence [Ba98b] and the complete fermionic two-loop contributions [Fr00a] to Δr. For an almost up to date review of higher-order corrections to Δr we refer to Ref. [Ho00].

A more detailed discussion of Δr, including theoretical uncertainties, can be found in Refs. [Bu89, Je91, Ba95d, Ho00].

4.6.2.4 Confronting theory with experiment

Equation (4.6.33) provides a correlation between the electroweak parameters. This correlation allows us to predict a value for the W-*boson mass* after the other parameters have been specified. Comparison of this prediction with the directly measured value of the M_W yields a test of the EWSM. Alternatively one can define the quantity Δr as a physical observable by

$$\Delta r = 1 - \frac{\pi\alpha}{\sqrt{2}G_\mu} \frac{1}{M_W^2(1 - M_W^2/M_Z^2)} \tag{4.6.36}$$

which can be determined through measurements of α, G_μ, M_Z and M_W/M_Z. Inserting the experimental values [Gu00] yields

$$\Delta r = 0.0320 \pm 0.0022, \tag{4.6.37}$$

i.e. Δr differs from zero by 14 standard deviations. If one takes into account only the running of the electromagnetic coupling in Δr, this still deviates from the experimental result (4.6.37) by 12 standard deviations. This provides an impressive confirmation of the presence of the electroweak radiative corrections.

Theoretically, Δr can be computed from α, G_μ, M_Z after specifying the masses m_t and M_H. The theoretical prediction for Δr as a function of the top-quark mass for two Higgs-boson masses is shown in Fig. 4.14 together with the experimental 1σ limits. As can be seen, the confrontation of the experimental value for Δr with the EWSM prediction constrains the *top-quark mass* to a value around 195 GeV, i.e. about 200 times the nucleon mass. As a remarkable success of the Standard Model this prediction does nicely agree with the direct measurement of the top-quark mass from the $p\bar{p}$ collider Tevatron at Fermilab, $m_t = 174.3 \pm 5.1\,\text{GeV}$ [PDG00].

4.6.3 Decay widths of the weak gauge bosons

4.6.3.1 The width of the Z boson

The Z boson dominantly decays into fermion–antifermion pairs. An accurate knowledge of the Z-boson width is required to describe the Z-resonance line shape appropriately. Furthermore, the partial widths for the decay of the Z boson into specific fermions f allow us, together with the Z-resonance asymmetries (cf. Sect. 4.6.4), to extract the weak coupling constants of the various types of fermions experimentally.

The partial widths $\Gamma(Z \to f\bar{f})$ and the total Z-boson decay width Γ_Z have been measured with an accuracy at the per-mill level at LEP1 [Gu00]. In order to match this accuracy, the inclusion of radiative corrections into the theoretical prediction is indispensable.

We consider the decay $Z(k) \to f(p)\bar{f}(\bar{p})$, where the particles carry the momenta indicated in parentheses. Because of Lorentz invariance and momentum conservation the corresponding matrix element must take the form

$$\begin{aligned}
\mathcal{M}^{Z\to f\bar{f}} = e\Big[&\bar{u}(p)\not{\varepsilon}(k)v(\bar{p})\, V_f(M_Z^2, m_f^2) \\
&- \bar{u}(p)\not{\varepsilon}(k)\gamma_5 v(\bar{p}))\, A_f(M_Z^2, m_f^2) \\
&+ \bar{u}(p)v(\bar{p})\, \varepsilon(k)(p-\bar{p})\, m_f S_f(M_Z^2, m_f^2) \\
&+ \bar{u}(p)\gamma_5 v(\bar{p})\, \varepsilon(k)(p-\bar{p})\, m_f P_f(M_Z^2, m_f^2)\Big],
\end{aligned} \tag{4.6.38}$$

where $\varepsilon(k)$ denotes the polarization vector of the Z boson, $\bar{u}(p)$ and $v(\bar{p})$ are the spinors of the fermion and antifermion, respectively, and V_f, A_f, S_f, and P_f are scalar coefficient functions.

Since all fermions in which the Z boson can decay are relatively light, their masses can be neglected in higher-order corrections in good approximation. Then, the last two terms in (4.6.38) can be dropped, and the matrix element can be written as

$$\mathcal{M}^{\mathrm{Z}\to f\bar{f}} = e\bar{u}(p)\not{\varepsilon}(k)\left[V_f(M_\mathrm{Z}^2) - A_f(M_\mathrm{Z}^2)\gamma_5\right]v(\bar{p}), \tag{4.6.39}$$

i.e. it has the same structure as the lowest-order matrix element, but with the lowest-order couplings v_f and a_f replaced by *effective couplings* $V_f(M_\mathrm{Z}^2)$ and $A_f(M_\mathrm{Z}^2)$. One can go even further and express the effective couplings in terms of an *effective mixing angle* θ_f and an effective normalization ρ_f,

$$\begin{aligned} A_f(M_\mathrm{Z}^2) &= \frac{\sqrt{\rho'_f}}{s_\mathrm{W}c_\mathrm{W}} I^3_{\mathrm{W},f} = \sqrt{\rho_f}\frac{1}{e}2\sqrt{\sqrt{2}G_\mu}M_\mathrm{Z} I^3_{\mathrm{W},f}, \\ V_f(M_\mathrm{Z}^2) &= \frac{\sqrt{\rho'_f}}{s_\mathrm{W}c_\mathrm{W}}(I^3_{\mathrm{W},f} - 2Q_f\sin^2\theta_f) \\ &= \sqrt{\rho_f}\frac{1}{e}2\sqrt{\sqrt{2}G_\mu}M_\mathrm{Z}(I^3_{\mathrm{W},f} - 2Q_f\sin^2\theta_f). \end{aligned} \tag{4.6.40}$$

For $\rho_f = 1$ and $\sin^2\theta_f = s_\mathrm{W}^2$ this yields the lowest-order expressions for the vector and axial-vector couplings in terms of G_μ. The parameter ρ'_f differs from ρ_f by the corrections to muon decay (cf. Sect. 4.6.2)

$$\rho_f = \frac{\rho'_f}{1 - \Delta r}. \tag{4.6.41}$$

Since the effective couplings (4.6.40) have the structure of the lowest-order couplings, the decay width, including higher-order corrections, can be directly read off from (1.4.37),

$$\begin{aligned} \Gamma(\mathrm{Z}\to f\bar{f}) &= N_\mathrm{C}^f\frac{\alpha M_\mathrm{Z}}{3}\left(V_f^2 + A_f^2\right) \\ &= N_\mathrm{C}^f\frac{\alpha M_\mathrm{Z}}{24c_\mathrm{W}^2s_\mathrm{W}^2}\rho'_f(1 - 4\sin^2\theta_f|Q_f| + 8\sin^4\theta_f|Q_f|^2) \\ &= N_\mathrm{C}^f\frac{G_\mu M_\mathrm{Z}^3}{12\pi\sqrt{2}}\rho_f(1 - 4\sin^2\theta_f|Q_f| + 8\sin^4\theta_f|Q_f|^2). \end{aligned} \tag{4.6.42}$$

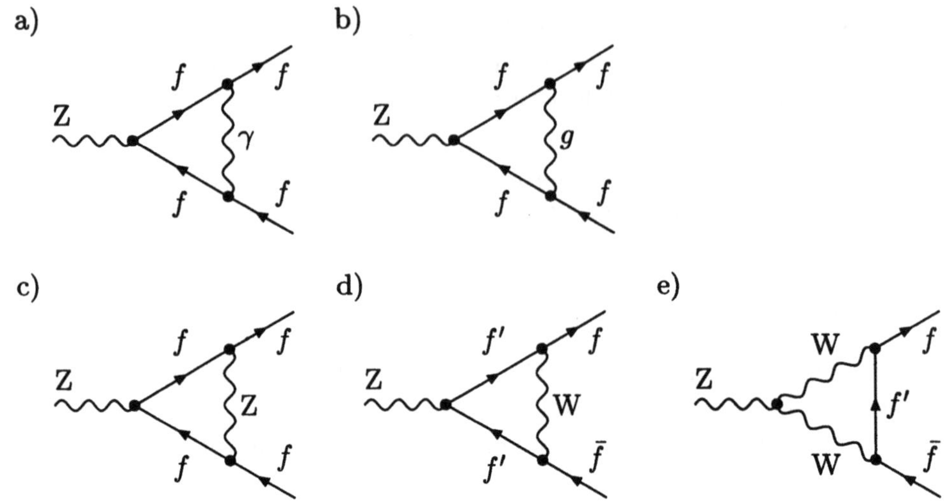

Fig. 4.15 Feynman diagrams for $Z \to f\bar{f}$

In one-loop approximation, the effective couplings can be determined from the Feynman diagrams shown in Fig. 4.15 together with the Zff vertex counterterms and possible wave-function renormalization constants. For vanishing fermion masses, the quark-mixing matrix can be chosen diagonal. The fermion f' denotes the isospin partner of the fermion f. If finite fermion masses are required, as it is necessary for the decay $Z \to b\bar{b}$ because of the presence of the virtual heavy top quark, additional diagrams have to be taken into account.

The one-loop corrections to the Z-boson decay can be split into the gauge-invariant subsets of QED, weak, and QCD corrections. The QED corrections involve all diagrams with virtual photons [diagram (a) of Fig. 4.15] together with the corresponding counterterms, the QCD corrections involve all diagrams with virtual gluons [diagram (b) of Fig. 4.15] together with the corresponding contributions to the fermion-field renormalization constants, and the weak corrections involve all other diagrams [diagrams (c), (d), and (e) of Fig. 4.15] and the corresponding counterterms. It is customary to include only the weak corrections into the effective couplings V_f and A_f or θ_f and ρ_f, and to specify the QED and QCD corrections separately.

Since the photonic and gluonic couplings do not involve γ_5, owing to parity invariance of these interactions, they contribute only to the normalization of the matrix element, but not to the relative size of the vector and axial vector form factors. The virtual QED and QCD corrections are IR-singular. In order to define a physical Z-boson decay width one has to include the real photonic and gluonic bremsstrahlung corrections, i.e. the decays $\mathrm{Z} \to f\bar{f}\gamma$ and $\mathrm{Z} \to f\bar{f}g$ in lowest order. After adding this contribution, integrated over the total photon or gluon phase space, one finds

$$\Gamma(\mathrm{Z} \to f\bar{f}) = N_{\mathrm{C}}^f \frac{G_\mu M_{\mathrm{Z}}^3}{12\pi\sqrt{2}} \rho_f (1 - 4\sin^2\theta_f |Q_f| + 8\sin^4\theta_f |Q_f|^2) \times (1 + \delta_{\mathrm{QED}}^f)(1 + \delta_{\mathrm{QCD}}^f) \tag{4.6.43}$$

for the one-loop corrected Z-boson decay width, where [Ma79a, Al80a, In80, Ch82a]

$$\delta_{\mathrm{QED}}^f = \frac{3\alpha}{4\pi} Q_f^2, \qquad \delta_{\mathrm{QCD}}^f = \begin{cases} \alpha_s(M_{\mathrm{Z}}^2)/\pi & \text{for quarks} \\ 0 & \text{for leptons.} \end{cases} \tag{4.6.44}$$

The weak one-loop corrections have been calculated for leptonic decays in Refs. [Si84, Je86] and for hadronic decays including the finite top-quark-mass effects in Refs. [Ak86, Be88a, Be88c].

As in the case of Δr, the dominant weak corrections to the Z-boson decay width originate from the running of the electromagnetic coupling and from corrections proportional to m_{t}^2. If we take only these corrections into account we can approximate

$$\rho_f \approx \rho = \frac{1}{1 - \Delta\rho}, \qquad \sin^2\theta_f \approx \bar{s}_{\mathrm{W}}^2, \tag{4.6.45}$$

with $\Delta\rho$ and $\bar{s}_{\mathrm{W}}^2$ from (4.6.28) and (4.6.34), respectively. Thus, we obtain for massless fermions

$$\Gamma(\mathrm{Z} \to f\bar{f}) \approx N_{\mathrm{C}}^f \frac{G_\mu M_{\mathrm{Z}}^3}{12\pi\sqrt{2}} \rho (1 - 4\bar{s}_{\mathrm{W}}^2 |Q_f| + 8\bar{s}_{\mathrm{W}}^4 |Q_f|^2) \times (1 + \delta_{\mathrm{QED}}^f)(1 + \delta_{\mathrm{QCD}}^f). \tag{4.6.46}$$

This reproduces the full one-loop results within 1% except for the $\mathrm{b\bar{b}}$ partial width, where additional corrections $\propto \alpha m_{\mathrm{t}}^2/M_{\mathrm{W}}^2$ occur, arising from virtual top quarks in the corrections to the $\mathrm{Zb\bar{b}}$ vertex [Co89b]. In the 't Hooft–Feynman gauge these enter exclusively through diagrams containing would-be Goldstone bosons coupled to the top quark. The leading terms are known

[Ak86] and can easily be incorporated in (4.6.46) by replacing the universal quantities ρ and $\bar{s}_{\mathrm{w}}^2$ by effective ones as

$$\rho \to \rho_{\mathrm{b}} = \rho\left(1 - \frac{4}{3}\Delta\rho\right), \qquad \bar{s}_{\mathrm{w}}^2 \to \sin^2\theta_{\mathrm{b}} = \bar{s}_{\mathrm{w}}^2\left(1 + \frac{2}{3}\Delta\rho\right) \tag{4.6.47}$$

with $\Delta\rho$ from (4.6.28). After this modification the approximation (4.6.46) describes also the partial width $\Gamma(\mathrm{Z} \to \mathrm{b\bar{b}})$ including the full one-loop electroweak corrections within 1%.

The status of the higher-order corrections to the Z-boson width and to the effective Z-boson–fermion coupling parameters ρ_f and $\sin^2\theta_f$ is similar as for Δr discussed in Sect. 4.6.2.3. Beyond the corrections of order $G_\mu^2 m_{\mathrm{t}}^4$, which are related to the ρ parameter, the sub-leading electroweak two-loop corrections of order $G_\mu^2 m_{\mathrm{t}}^2 M_{\mathrm{Z}}^2$ in the top-quark-mass expansion are known for the leptonic mixing angle $\sin^2\theta_l$ [De96a, De97, De98] and for ρ_l [De00]. Exact results based on the complete Feynman diagrams have been derived for the Higgs-mass dependence of the fermionic two-loop corrections to $\sin^2\theta_l$ [We98]. The non-universal contributions of order $G_\mu^2 m_{\mathrm{t}}^4$ to $\sin^2\theta_{\mathrm{b}}$ and ρ_b have been calculated in Refs. [Ba92, Fl93, De93a]. The QCD corrections to $\Gamma(Z \to q\bar{q})$ are known up to order α_{s}^3 and have been reviewed in Ref. [Ch96]. Finally the mixed QCD–electroweak corrections of order $\alpha\alpha_{\mathrm{s}}$ have been calculated in Ref. [Ha98c, Fl99a]. More details can also found in the recent review Ref. [Ho00].

4.6.3.2 The width of the W Boson

As the Z-boson width, also the W-boson width is dominated by decays into two fermions. The total width of the W boson is crucial for the shape of the W-pair production cross section in the threshold region. The leptonic partial widths allow to test lepton universality and the hadronic partial widths can serve to determine the elements of the quark-mixing matrix.

The radiative corrections to the W-boson decay can be calculated in the same way as for the Z-boson decay. Since the W boson couples only to left-handed fermions, only one form factor appears in the matrix element for vanishing fermion masses. The QCD corrections are given by (4.6.44). Because the W boson is charged, the QED corrections cannot be split off in a gauge-invariant way on the basis of Feynman diagrams. Thus, one arrives at the following result for the decay of a W-boson into massless fermions f_i and f'_j

$$\Gamma(\mathrm{W} \to f_i \bar{f}'_j) = N_{\mathrm{C}}^f \frac{\alpha}{6} \frac{M_{\mathrm{W}}}{2s_{\mathrm{W}}^2} |V_{ij}|^2 L_{f_i,f_j}(M_{\mathrm{W}}^2)(1 + \delta_{\mathrm{QCD}}), \tag{4.6.48}$$

where L_{f_i,f_j} includes the electroweak, i.e. electromagnetic and weak, corrections. Just as for the Z-boson width, the electroweak corrections can be incorporated into a new *effective weak mixing angle* by defining

$$\sin^2 \theta_{f_i,f_j} = \frac{s_{\mathrm{W}}^2}{L_{f_i,f_j}}. \tag{4.6.49}$$

The electroweak and QCD radiative corrections for decays into massless fermions ($m_f \ll M_{\mathrm{W}}$) have been calculated in Refs. [Al80a, Je86, Ba86]. The full one-loop electroweak and QCD corrections together with the complete *photonic and gluonic bremsstrahlung* were evaluated for arbitrary finite fermion masses in Ref. [De90a]. The bulk of the electroweak corrections can be obtained by expressing the lowest-order width in terms of $\alpha(M_{\mathrm{W}}^2)$ and $\bar{s}_{\mathrm{W}}^2$, or equivalently by G_μ and M_{W}. This leads to an improved Born approximation for the total and partial widths,

$$\Gamma^{\mathrm{W}l_i\nu_i} \approx \frac{G_\mu M_{\mathrm{W}}^3}{6\sqrt{2}\pi}, \qquad \Gamma^{\mathrm{W}q_iq'_j} \approx \frac{G_\mu M_{\mathrm{W}}^3}{2\sqrt{2}\pi} |V_{ij}|^2 \left(1 + \frac{\alpha_s}{\pi}\right),$$
$$\Gamma^{\mathrm{W}} \approx \frac{3G_\mu M_{\mathrm{W}}^3}{2\sqrt{2}\pi} \left(1 + \frac{2\alpha_s}{3\pi}\right). \tag{4.6.50}$$

These simple formulas reproduce the full one-loop corrected results including finite fermion masses with an accuracy of $\sim 0.5\%$ [De93].

4.6.4 Z-boson physics

The electroweak process that has been studied most intensively in recent years is $\mathrm{e}^+\mathrm{e}^- \to f\bar{f}$. It involves, as dominant subprocess, the production of a Z boson with subsequent decay into fermion–antifermion pairs. The Large Electron-Positron Collider (LEP) at CERN and the SLAC Linear Collider (SLC) have been dedicated many years to the investigation of this process. The high accuracy of these experiments renders this process the most prominent case for the relevance of electroweak radiative corrections. In the following we discuss the structure of these corrections.

4.6.4.1 Lowest-order results for $e^+e^- \to f\bar{f}$

We start out by summarizing the lowest-order predictions and the most important observables for $e^+e^- \to f\bar{f}$. The lowest-order matrix element was already given in (1.4.43). In contrast to Sect. 1.4.2.4, here we neglect the mass of the final-state fermion, but take into account the possibility of longitudinal polarization of the electron and positron. Denoting the degrees of longitudinal polarization of $e^\pm$ by $P^\pm$, the differential cross section can be written as

$$\left(\frac{d\sigma}{d\Omega}\right)_{\text{Born}} = \frac{\alpha^2}{4s} N_C^f [(1 - P^+P^-)X_{U,\text{Born}} + (P^+ - P^-)X_{L,\text{Born}}] \tag{4.6.51}$$

with

$$\begin{aligned} X_{U,\text{Born}} &= G_1(s)(1 + \cos^2\vartheta) + G_3(s)\,2\cos\vartheta, \\ X_{L,\text{Born}} &= H_1(s)(1 + \cos^2\vartheta) + H_3(s)\,2\cos\vartheta. \end{aligned} \tag{4.6.52}$$

The functions G_i were given in (1.4.45), and the functions H_i read

$$\begin{aligned} H_1(s) &= -2a_e v_f Q_f \operatorname{Re}\{\chi_Z(s)\} + 2v_e a_e (v_f^2 + a_f^2)|\chi_Z(s)|^2, \\ H_3(s) &= -2v_e a_f Q_f \operatorname{Re}\{\chi_Z(s)\} + 2(v_e^2 + a_e^2) v_f a_f |\chi_Z(s)|^2 \end{aligned} \tag{4.6.53}$$

with $\chi_Z(s)$ from (1.4.46) and v_f and a_f from (1.4.38).

From (4.6.51) various physical observables can be calculated. For unpolarized particles only $X_{U,0}$ contributes. The total cross section

$$\sigma_{\text{Born}}(e^+e^- \to f\bar{f}) = \int d\Omega \left(\frac{d\sigma}{d\Omega}\right)_{\text{Born}} = \frac{4\pi\alpha^2}{3s} N_C^f G_1(s) \tag{4.6.54}$$

involves only G_1. The *forward–backward asymmetry*, which is defined as

$$A_{FB}(e^+e^- \to f\bar{f}) = \frac{\sigma^F - \sigma^B}{\sigma^F + \sigma^B} \tag{4.6.55}$$

with

$$\sigma^F = \int_{\vartheta<\pi/2} d\Omega \frac{d\sigma}{d\Omega}, \qquad \sigma^B = \int_{\vartheta>\pi/2} d\Omega \frac{d\sigma}{d\Omega}, \tag{4.6.56}$$

measures the ratio of the C-odd to the C-even contributions in the cross section. In lowest order it is obtained as

$$A_{\mathrm{FB,Born}}(s) = \frac{3}{4}\frac{G_3(s)}{G_1(s)}. \tag{4.6.57}$$

More explicitly, the total lowest-order cross section can be written as

$$\begin{aligned}\sigma_{\mathrm{Born}}(\mathrm{e^+e^-} \to f\bar{f})(s) &= \frac{4\pi\alpha^2}{3s}N_{\mathrm{C}}^f \\ &\times\left\{\frac{(v_{\mathrm{e}}^2 + a_{\mathrm{e}}^2)(v_f^2 + a_f^2)s^2}{(s - M_{\mathrm{Z}}^2)^2 + M_{\mathrm{Z}}^2(\Gamma_{\mathrm{Z,Born}})^2}\left[1 + R_f\frac{s - M_{\mathrm{Z}}^2}{s}\right] + Q_f^2\right\},\end{aligned} \tag{4.6.58}$$

with

$$R_f = \frac{2Q_{\mathrm{e}}Q_f v_{\mathrm{e}} v_f}{(v_{\mathrm{e}}^2 + a_{\mathrm{e}}^2)(v_f^2 + a_f^2)}. \tag{4.6.59}$$

For Z physics we are interested in the energy region $E_{\mathrm{CMS}} \approx M_{\mathrm{Z}}$ where the Z-boson-exchange diagram dominates. The photon-exchange contributions are smaller by a factor $(\Gamma_{\mathrm{Z}}/M_{\mathrm{Z}})^2 \approx 1/1300$. The γ–Z interference vanishes on resonance and is suppressed by a factor $(s - M_{\mathrm{Z}}^2)/M_{\mathrm{Z}}^2$ nearby. The dominating Z-boson-exchange graph yields a simple Breit–Wigner cross section, which we can rewrite using the lowest-order results for the partial width of $Z \to f\bar{f}$ (1.4.37) as

$$\sigma_{f,\mathrm{Born}}(s) = 12\pi\frac{s}{M_{\mathrm{Z}}^2}\frac{\Gamma_{\mathrm{Born}}(\mathrm{Z} \to \mathrm{e^+e^-})\Gamma_{\mathrm{Born}}(\mathrm{Z} \to f\bar{f})}{(s - M_{\mathrm{Z}}^2)^2 + M_{\mathrm{Z}}^2(\Gamma_{\mathrm{Z,Born}})^2}. \tag{4.6.60}$$

The peak cross section $\sigma_{f,\mathrm{Born}}^{\mathrm{peak}}$ involves a product of branching ratios

$$\sigma_{f,\mathrm{Born}}^{\mathrm{peak}} = \sigma_{f,\mathrm{Born}}(M_{\mathrm{Z}}^2) = \frac{12\pi}{M_{\mathrm{Z}}^2}\frac{\Gamma_{\mathrm{Born}}(\mathrm{Z} \to \mathrm{e^+e^-})}{\Gamma_{\mathrm{Z,Born}}}\frac{\Gamma_{\mathrm{Born}}(\mathrm{Z} \to f\bar{f})}{\Gamma_{\mathrm{Z,Born}}}. \tag{4.6.61}$$

The on-resonance ($s = M_{\mathrm{Z}}^2$) forward–backward asymmetry is, in lowest order, given by (neglecting again the photon-exchange contributions and fermion masses)

$$A_{\mathrm{FB,Born}}^f(M_{\mathrm{Z}}^2) = \frac{3}{4}\mathcal{A}_{\mathrm{e}}\mathcal{A}_f \tag{4.6.62}$$

with

$$\mathcal{A}_f = \frac{2v_f a_f}{v_f^2 + a_f^2} = \frac{2|I^3_{\mathrm{w},f} - 4Q_f s_\mathrm{w}^2|}{1 + (2I^3_{\mathrm{w},f} - 4Q_f s_\mathrm{w}^2)^2}. \tag{4.6.63}$$

For polarized beams, additional asymmetries can be measured. The most important one is the left–right asymmetry

$$A_\mathrm{LR} = \frac{\sigma_\mathrm{L} - \sigma_\mathrm{R}}{\sigma_\mathrm{L} + \sigma_\mathrm{R}}, \tag{4.6.64}$$

where $\sigma_{\mathrm{L(R)}}$ denotes the integrated cross section for left- (right-)handed electrons, i.e. with $P_- = \pm 1$, and unpolarized positrons. On resonance it is given by

$$A_\mathrm{LR} = \mathcal{A}_\mathrm{e}. \tag{4.6.65}$$

In general, the quantity $\mathcal{A}_f$ determines all on-resonance asymmetries. It is a function of s_w^2 only. As a consequence, within the Electroweak Standard Model a measurement of an on-resonance asymmetry is equivalent to a *measurement of the weak mixing angle.* More generally, measurements of asymmetries determine ratios of vector and axial-vector couplings. Together with measurements of the partial widths, which determine $v_f^2 + a_f^2$, they allow to extract the couplings v_f and a_f. This offers the possibility for crucial tests of the EWSM.

4.6.4.2 Structure of radiative corrections

The radiative corrections to the cross section for $\mathrm{e}^+\mathrm{e}^- \to f\bar{f}$ consist of *virtual corrections*, i.e. loop diagrams, and real corrections, i.e. *bremsstrahlung* diagrams. These corrections can be divided into electromagnetic, weak, and QCD corrections in a gauge-invariant way.

The QED corrections consist of all those diagrams where an extra photon line is added to the lowest-order diagrams. While these corrections do not test the detailed structure of the electroweak theory, they are numerically very important and depend on the cuts applied to the emitted final-state photons.

The weak corrections are furnished by all other electroweak diagrams. They are sensitive to the complete structure of the electroweak model and depend, in particular, on the Higgs-boson and top-quark masses.

The QCD corrections appear in one-loop order only for quarks in the final state and involve all diagrams where additional real or virtual gluons are attached to the lowest-order diagrams. In higher orders, also gluonic corrections to quark loops in the boson self-energies show up. The QCD corrections depend on the running strong fine-structure constant $\bar{\alpha}_s$.

4.6.4.3 Weak corrections

The weak corrections consist of self-energy, vertex, and box corrections. The self-energy corrections are composed of the photon self-energy, the Z-boson self-energy, and the photon–Z-boson mixing energy. The vertex corrections involve the same contributions as in the case of the Z-boson decay and corrections to the $Af\bar{f}$ vertex. The fermion masses are neglected throughout. The contribution of the matrix element for $e^+e^- \to f\bar{f}$ from the one-loop weak corrections is obtained as

$$\begin{aligned}
\delta\mathcal{M} = e^2\Bigg\{ & -Q_e Q_f \frac{\Sigma^{AA}_{\mathrm{R,T}}(s)}{s^2}\, \bar{v}(p_+)\gamma^\mu u(p_-)\, \bar{u}(q_-)\gamma_\mu v(q_+) \\
& - \frac{\Sigma^{ZZ}_{\mathrm{R,T}}(s)}{(s-M_Z^2)^2}\, \bar{v}(p_+)\gamma^\mu(v_e - a_e\gamma_5)u(p_-)\, \bar{u}(q_-)\gamma_\mu(v_f - a_f\gamma_5)v(q_+) \\
& + \frac{\Sigma^{AZ}_{\mathrm{R,T}}(s)}{s(s-M_Z^2)} \Big[Q_f\, \bar{v}(p_+)\gamma^\mu(v_e - a_e\gamma_5)u(p_-)\, \bar{u}(q_-)\gamma_\mu v(q_+) \\
& \qquad + Q_e\, \bar{v}(p_+)\gamma^\mu u(p_-)\, \bar{u}(q_-)\gamma_\mu(v_f - a_f\gamma_5)v(q_+) \Big] \\
& + Q_e Q_f \frac{1}{s} \Big[\bar{v}(p_+)\gamma^\mu u(p_-)\, \bar{u}(q_-)\gamma_\mu(\delta V_f^\gamma - \delta A_f^\gamma\gamma_5)v(q_+) \\
& \qquad + \bar{v}(p_+)\gamma^\mu(\delta V_e^\gamma - \delta V_e^\gamma\gamma_5)u(p_-)\, \bar{u}(q_-)\gamma_\mu v(q_+) \Big] \\
& + \frac{1}{s-M_Z^2} \Big[\bar{v}(p_+)\gamma^\mu(v_e - a_e\gamma_5)u(p_-)\, \bar{u}(q_-)\gamma_\mu(\delta V_f^Z - \delta A_f^Z\gamma_5)v(q_+) \\
& \qquad + \bar{v}(p_+)\gamma^\mu(\delta V_e^Z - \delta V_e^Z\gamma_5)u(p_-)\, \bar{u}(q_-)\gamma_\mu(v_f - a_f\gamma_5)v(q_+) \Big] \Bigg\} \\
& + \delta\mathcal{M}_{\mathrm{box}}. \qquad (4.6.66)
\end{aligned}$$

At the Z-boson resonance, $s = M_Z^2$, the lowest-order Z-boson propagators diverge. In order to obtain a sensible matrix element one has to *re-sum the self-energy corrections* [Dy49]. The resulting propagators can be most easily obtained as the inverse of the corresponding two-point functions [cf. (2.2.42)]. The Z-boson propagator mixes with the photon propagator and the χ propagator. The transverse part, which enters in (4.6.66), is not affected by the χ propagator and can be considered independently of the longitudinal part. The relevant matrix of transverse parts of the renormalized two-point functions reads $(V, V' = \gamma, Z)$

$$\begin{aligned} \Gamma_{R,T}^{VV'}(s) &= i(G_{R,T}^{VV'}(s))^{-1} \\ &= -\begin{pmatrix} k^2 + \Sigma_{R,T}^{AA}(s) & \Sigma_{R,T}^{AZ}(s) \\ \Sigma_{R,T}^{AZ}(s) & k^2 - M_Z^2 + \Sigma_{R,T}^{ZZ}(s) \end{pmatrix} \end{aligned} \tag{4.6.67}$$

in terms of the renormalized self-energies. Inversion yields the re-summed *propagator matrix*

$$G_{R,T}^{VV'}(s) = \begin{pmatrix} G_{R,T}^{AA} & G_{R,T}^{AZ} \\ G_{R,T}^{AZ} & G_{R,T}^{ZZ} \end{pmatrix} \tag{4.6.68}$$

with

$$\begin{aligned} G_R^{AA}(s) &= \frac{i}{s + \Sigma_{R,T}^{AA}(s) - \frac{[\Sigma_{R,T}^{AZ}(s)]^2}{s - M_Z^2 + \Sigma_{R,T}^{ZZ}(s)}}, \\ G_R^{ZZ}(s) &= \frac{i}{s - M_Z^2 + \Sigma_{R,T}^{ZZ}(s) - \frac{[\Sigma_{R,T}^{AZ}(s)]^2}{s + \Sigma_{R,T}^{AA}(s)}}, \\ G_R^{AZ}(s) &= -\frac{\Sigma_{R,T}^{AZ}(s)}{s + \Sigma_{R,T}^{AA}(s)} G_R^{ZZ}(s). \end{aligned} \tag{4.6.69}$$

In the photon propagator, and away from the Z-boson resonance also in the other propagators, the higher-order terms $[\propto (\Sigma_{R,T}^{AZ})^2, \Sigma_{R,T}^{AA}\Sigma_{R,T}^{ZZ}]$ can be neglected. Around the Z-boson peak, however, the leading terms in the denominator of the Z-boson propagator cancel. In order to obtain nevertheless relative $\mathcal{O}(\alpha)$ accuracy one therefore has to include all terms of order $\mathcal{O}(\alpha^2)$ in the denominator, i.e. the terms involving $(\Sigma_{R,T}^{AZ})^2$ and the two-loop contributions to the Z-boson self-energy.

Performing the resummation (4.6.69) and including reducible higher-order corrections, we can write the matrix element for $e^+e^- \to f\bar{f}$ in the following way

$$
\begin{aligned}
\mathcal{M}(e^+e^- \to f\bar{f}) = {} & Q_e Q_f \bar{v}(p_+)\gamma^\mu [V_e^\gamma(s) - A_e^\gamma(s)\gamma_5] u(p_-) \\
& \times \frac{e^2}{s + \Sigma_{\mathrm{R,T}}^{AA}(s)} \bar{u}(q_-)\gamma_\mu [V_f^\gamma(s) - A_f^\gamma(s)\gamma_5] v(q_+) \\
& + \bar{v}(p_+)\gamma^\mu [V_e^Z(s) - V_e^Z(s)\gamma_5] u(p_-) \\
& \times \frac{e^2}{s - M_Z^2 + \Sigma_{\mathrm{R,T}}^{ZZ}(s) - \frac{[\Sigma_{\mathrm{R,T}}^{AZ}(s)]^2}{s + \Sigma_{\mathrm{R,T}}^{AA}(s)}} \\
& \times \bar{u}(q_-)\gamma_\mu [V_f^Z(s) - V_f^Z(s)\gamma_5] v(q_+) \\
& + \delta\mathcal{M}_{\mathrm{box}}
\end{aligned}
\tag{4.6.70}
$$

with the *effective couplings*

$$
\begin{aligned}
& V_f^\gamma(s) = 1 + \delta V_f^\gamma(s), \qquad A_f^\gamma(s) = \delta A_f^\gamma(s), \\
& V_f^Z(s) = v_f + \delta V_f^Z(s) + Q_f \frac{\Sigma_{\mathrm{R,T}}^{AZ}(s)}{s + \Sigma_{\mathrm{R,T}}^{AA}(s)}, \qquad A_f^Z(s) = a_f + \delta A_f^Z(s).
\end{aligned}
\tag{4.6.71}
$$

In the complete on-shell scheme we have

$$
\begin{aligned}
& \mathrm{Re}\{\Sigma_{\mathrm{R,T}}^{AZ}(M_Z^2)\} = 0, \qquad \mathrm{Re}\{\Sigma_{\mathrm{R,T}}^{ZZ}(M_Z^2)\} = 0, \\
& \left.\frac{\mathrm{d}\,\mathrm{Re}\{\Sigma_{\mathrm{R,T}}^{ZZ}(s)\}}{\mathrm{d}s}\right|_{s=M_Z^2} = 0,
\end{aligned}
\tag{4.6.72}
$$

and the *optical theorem* yields the relation

$$
M_Z\Gamma_Z = \mathrm{Im}\left\{\Sigma_{\mathrm{R,T}}^{ZZ}(M_Z^2)\right\}.
\tag{4.6.73}
$$

Using in addition the approximation

$$
\mathrm{Im}\left\{\Sigma_{\mathrm{R,T}}^{ZZ}(s)\right\} \approx \frac{s}{M_Z^2}\,\mathrm{Im}\left\{\Sigma_{\mathrm{R,T}}^{ZZ}(M_Z^2)\right\},
\tag{4.6.74}
$$

which is exact for massless fermions, we find for the denominator of the Z-boson propagator for $s \approx M_Z^2$

$$\begin{aligned} s - M_Z^2 + \Sigma_{R,T}^{ZZ}(s) &= s - M_Z^2 + \mathrm{i}\,\mathrm{Im}\,\Sigma_{R,T}^{ZZ}(s) + \mathcal{O}\left((s - M_Z^2)^2\right) \\ &\approx s - M_Z^2 + \mathrm{i}s\frac{\Gamma_Z}{M_Z}. \end{aligned} \tag{4.6.75}$$

For $s \approx M_Z^2$ this reproduces the Z-boson propagator introduced in (1.4.43). Note, however, that for the experimental accuracy of LEP this latter approximation is not allowed. The two forms of the propagator correspond to two different definitions of the Z-*boson mass* [Ba88]. In fact, a simple manipulation yields ($\gamma_Z = \Gamma_Z/M_Z$)

$$\begin{aligned} s - M_Z^2 + \mathrm{i}s\gamma_Z &= (1 + \mathrm{i}\gamma_Z)\left[s - \frac{M_Z^2 - \mathrm{i}M_Z\Gamma_Z}{1 + \gamma_Z^2}\right] \\ &= (1 + \mathrm{i}\gamma_Z)\left[s - (M_Z')^2 + \mathrm{i}M_Z'\Gamma_Z'\right] \end{aligned} \tag{4.6.76}$$

with

$$\begin{aligned} M_Z' &= \frac{M_Z}{\sqrt{1 + \gamma_Z^2}} \approx M_Z - \frac{\Gamma_Z^2}{2M_Z}, \\ \Gamma_Z' &= \frac{\Gamma_Z}{\sqrt{1 + \gamma_Z^2}} \approx \Gamma_Z - \frac{\Gamma_Z^3}{2M_Z^2}. \end{aligned} \tag{4.6.77}$$

The difference between M_Z and M_Z' is about 34 GeV. Note that M_Z' is equivalent to the pole mass $\overline{M}_Z$ introduced in Sect. 4.5.5 up to terms of order α^3.

Introducing the running electromagnetic coupling

$$e^2(s) = \frac{e^2}{1 + \Sigma_{R,T}^{AA}(s)/s} \tag{4.6.78}$$

and omitting the box diagrams (which give a contribution below 10^{-4} for $s \approx M_Z^2$), we can write the matrix element in the vicinity of the Z-boson resonance as

$$\begin{aligned}\mathcal{M}(\mathrm{e}^+\mathrm{e}^- \to f\bar{f}) &= Q_\mathrm{e} Q_f e^2(M_\mathrm{Z}^2)\bar{v}(p_+)\gamma^\mu[V_\mathrm{e}^\gamma(M_\mathrm{Z}^2) - A_\mathrm{e}^\gamma(M_\mathrm{Z}^2)\gamma_5]u(p_-)\\ &\times \frac{1}{s}\,\bar{u}(q_-)\gamma_\mu[V_f^\gamma(M_\mathrm{Z}^2) - A_f^\gamma(M_\mathrm{Z}^2)\gamma_5]v(q_+)\\ &+ \bar{v}(p_+)\gamma^\mu[V_\mathrm{e}^\mathrm{Z}(M_\mathrm{Z}^2) - V_\mathrm{e}^\mathrm{Z}(M_\mathrm{Z}^2)\gamma_5]u(p_-)\,\frac{e^2}{(s-M_\mathrm{Z}^2)^2 + \mathrm{i}s\frac{\Gamma_\mathrm{Z}}{M_\mathrm{Z}}}\\ &\times \bar{u}(q_-)\gamma_\mu[V_f^\mathrm{Z}(M_\mathrm{Z}^2) - V_f^\mathrm{Z}(M_\mathrm{Z}^2)\gamma_5]v(q_+). \qquad (4.6.79)\end{aligned}$$

Moreover, the photon-vertex form factors are small, $\delta V_f^\gamma(M_\mathrm{Z}^2) \approx A_f^\gamma(M_\mathrm{Z}^2) \approx 10^{-3}$. Thus, around the Z-boson resonance all electroweak corrections can be absorbed into the running Z-boson width and the effective fermion–Z-boson couplings to a good approximation. The matrix element including weak corrections has the form of the lowest-order matrix element (1.4.43).

Equation (4.6.79) provides a so-called *improved Born approximation*. It contains the most important radiative corrections and yet is very simple. In the resonance region it reproduces the complete $\mathcal{O}(\alpha)$ weak corrections with an accuracy of better than 1%. The cross section resulting from (4.6.79) takes the form of (4.6.58) with the couplings replaced by effective couplings and the constant width replaced by the running width,

$$\begin{aligned}\sigma_f(s) &= 12\pi\frac{s}{M_\mathrm{Z}^2}\frac{\Gamma(\mathrm{Z}\to\mathrm{e}^+\mathrm{e}^-)\Gamma(\mathrm{Z}\to f\bar{f})}{(s-M_\mathrm{Z}^2)^2 + \frac{s^2}{M_\mathrm{Z}^2}\Gamma_\mathrm{Z}^2}\left[1 + R_f\frac{s-M_\mathrm{Z}^2}{s}\right]\\ &+ \frac{4\pi\alpha^2(s)}{3s}Q_f^2 N_\mathrm{C}^f(1+\delta_\mathrm{QED}^f)(1+\delta_\mathrm{QCD}^f). \qquad (4.6.80)\end{aligned}$$

The γ–Z interference term has the form (4.6.59) but with the lowest order couplings replaced by the effective couplings (4.6.71). The QED and QCD corrections are the same as for the Z-decay width in (4.6.43).

The simplicity of (4.6.79) results from the fact that for $s \approx M_\mathrm{Z}^2$ the contributions involving a resonant propagator are dominant. These contributions factorize into the subprocesses of the production of the Z boson and its decay. As a consequence, only the vertex and self-energy corrections to the resonant matrix element are relevant if only moderate accuracy ($\sim 1\%$) is required.

The *Dyson re-summation of the self-energy insertions* brings the Z-boson width in the denominator and treats the leading higher-order corrections in a proper way. However, re-summation of the bosonic loops breaks gauge invariance. Therefore, actually only the fermionic loops, which contain the leading terms, are re-summed and the bosonic loops have to be understood as expanded to strict one-loop order. The re-summation of the bosonic loops in the 't Hooft–Feynman gauge does not yield significant differences but allows for a simple and compact notation.

The leading contributions to the effective couplings have been discussed in (4.6.40) and (4.6.45). In the case of $\mathrm{b\bar{b}}$ production the effective couplings depend also on the top-quark mass. In this case the effective parameters $\bar{s}_{\mathrm{w}}^2$ and ρ have to be modified as discussed in (4.6.47).

4.6.4.4 QED Corrections

The QED corrections are large around the Z-boson resonance and require a careful treatment including the leading higher-order contributions. They consist of virtual photonic corrections to the four-particle process $\mathrm{e^+e^-} \to f\bar{f}$ and the real bremsstrahlung corrections, i.e. the processes $\mathrm{e^+e^-} \to f\bar{f} + n\gamma$. In order to obtain IR-finite results, both contributions have to be combined. The QED corrections can be divided into final-state corrections, initial–final-state interference, and initial-state corrections. All of them depend on the cuts that are used to define the actual cross section. In the following we restrict ourselves to the discussion of the QED corrections to the total cross section [Be89]. The corrections to the asymmetries are similar but different in detail [Bo89].

The virtual electromagnetic and real soft-photonic corrections are essentially proportional to the lowest-order matrix elements. Also the emission of hard photons from the final states yields only a simple multiplicative correction to the cross section. However, if a hard photon is emitted from the initial state, the invariant mass of the virtual Z boson is changed and the resonance is shifted.

We first discuss the *final-state corrections*. If no cuts are applied, the total cross section gets the same small correction as the Z-boson width,

$$\Delta\sigma^{\mathrm{fin}} = \sigma_{\mathrm{Born}} \left(1 + \frac{3\alpha}{4\pi} Q_f^2\right) . \tag{4.6.81}$$

The results for the gluonic final-state corrections are obtained by replacing $3\alpha Q_f^2/(4\pi)$ by α_{s}/π.

For restrictive energy cuts $E_\gamma \leq \Delta E \ll E$, the final-state corrections to the integrated cross section can be found in the soft-photon approximation [Bo89]

$$\Delta\sigma^{\mathrm{fin}} = \sigma_{\mathrm{Born}}\frac{\alpha}{\pi}Q_f^2\left[\left(\log\frac{s}{m_f^2}-1\right)\left(2\log\frac{\Delta E}{E}+\frac{3}{2}\right)+\frac{\pi^2}{3}-\frac{1}{2}\right]. \tag{4.6.82}$$

In contrast to the initial-state and final-state QED corrections, the *initial–final-state interference* depends on the scattering angle and modifies the angular distribution. The leading angular-dependent term is of the form $\log(t/u)$. The initial–final-state interference is numerically small under many realistic experimental circumstances [Mo98]. Its contribution to the total on-resonance cross section is suppressed as $\alpha\Gamma_{\mathrm{Z}}/(\pi M_{\mathrm{Z}})$. This can be understood as follows: when the Z boson is produced close to its mass shell, the wave functions of the initial and final-state radiation are separated in space–time owing to the finite lifetime of the Z boson, so that their overlap is small.

The most important electromagnetic corrections are the *initial-state corrections.* The reason is that around the resonance the cross section has a rapid variation with energy. Emission of photons from the incoming charged particles reduces the energy that is available for the resonance formation. This changes drastically the resonance shape, builds up the radiative tail, decreases the peak cross section, and increases the observed width of the resonance peak.

The dominant initial-state corrections arise from leading logarithms of the form

$$\frac{\alpha}{\pi}\log\frac{s}{m_{\mathrm{e}}^2} \approx 0.06, \tag{4.6.83}$$

which originate from *collinear photon emission* from the incoming fermions. As has been demonstrated in Sect. 2.8.3, these collinear logarithms factorize, i.e. they can be associated to the incoming electron and positron, and can be described by distribution functions (cf. Sect. 2.8.3.7) [Ku85, Ni87, Be88a]. For the total cross section for $\mathrm{e^+e^-} \to$ anything, the factorization theorem reads [cf. (2.8.109)]

$$\sigma(s) = \int_0^1 \mathrm{d}x_+ \int_0^1 \mathrm{d}x_- f_{\mathrm{ee}}(x_+,Q^2) f_{\mathrm{ee}}(x_-,Q^2)\sigma_{\mathrm{Born}}(x_+x_-s), \tag{4.6.84}$$

where $\sigma_{\text{Born}}(s)$ is the lowest-order cross section and $\sigma(s)$ the cross section including the leading-logarithmic corrections. The *distribution function* f_{ee} describes mass-singular initial-state collinear radiation and represents the probability of finding in the parent particle $e^{\pm}$ at the scale Q^2 a particle $e^{\pm}$ with fraction x of longitudinal momentum. So $(1-x)$ denotes the fraction of energy carried away by collinear photon radiation. All large collinear logarithms are contained in the distribution functions, which depend, just like the hard-scattering cross section, on the mass-factorization scale Q^2. Equation (4.6.84) can be generalized to differential distributions.

The large logarithms in the structure functions can be expressed in terms of the *splitting functions* $P_{\text{ee}}^{(n)}(x)$ [Al77, Be89a] [cf. (2.8.107)],

$$\begin{aligned} f_{\text{ee}}(x, Q^2) = \delta(1-x) &+ \frac{\alpha}{2\pi} P_{\text{ee}}^{(0)}(x)\, L_{\text{e}} \\ &+ \frac{1}{2}\left(\frac{\alpha}{2\pi}\right)^2 \left[\left(P_{\text{ee}}^{(0)} \otimes P_{\text{ee}}^{(0)}\right)(x)\, L_{\text{e}}^2 + P_{\text{ee}}^{(1)}(x)\, L_{\text{e}}\right] \\ &+ \text{ higher-order terms,} \end{aligned} \tag{4.6.85}$$

with the convolution

$$(f \otimes g)(x) \equiv \int_0^1 \mathrm{d}x_1 \int_0^1 \mathrm{d}x_2\, f(x_1)\, g(x_2)\, \delta(x - x_1 x_2) \tag{4.6.86}$$

and the leading logarithm

$$L_{\text{e}} = \log \frac{Q^2}{m_{\text{e}}^2}. \tag{4.6.87}$$

The first-order splitting function $P_{\text{ee}}^{(0)}$ has been given in (2.8.82). It can be expressed as

$$\begin{aligned} P_{\text{ee}}^{(0)}(x) &= \left(\frac{1+x^2}{1-x}\right)_+ \\ &= \lim_{\varepsilon \to 0} \left\{\delta(1-x)\left[\frac{3}{2} + 2\log(\varepsilon)\right] + \theta(1-x-\varepsilon)\frac{1+x^2}{1-x}\right\}. \end{aligned} \tag{4.6.88}$$

The total cross-section (4.6.84) can be rewritten as

$$\sigma(s) = \int_0^1 \mathrm{d}x\, \Phi(x) \sigma_{\text{Born}}(xs) \tag{4.6.89}$$

with the *flux function*

$$\begin{aligned}\Phi(x) &= (f_{\text{ee}} \otimes f_{\text{ee}})(x) \\ &= \delta(1-x) + \frac{\alpha}{\pi} P_{\text{ee}}^{(0)}(x)\, L_{\text{e}} \\ &\quad + \frac{1}{2}\left(\frac{\alpha}{\pi}\right)^2 \left[\left(P_{\text{ee}}^{(0)} \otimes P_{\text{ee}}^{(0)}\right)(x)\, L_{\text{e}}^2 + \frac{1}{2} P_{\text{ee}}^{(1)}(x)\, L_{\text{e}}\right] \\ &\quad + \text{higher-order terms.}\end{aligned} \tag{4.6.90}$$

In the vicinity of the Z-boson resonance *soft photons* are dominating. Therefore, it is useful to improve the flux function (4.6.90) by re-summing the leading soft-photonic corrections. These originate from the $\log \varepsilon$ term in $P_{\text{ee}}^{(0)}$ and exponentiate as $\exp\{(2\alpha/\pi)\log\varepsilon\}$ according to [Ye61] [cf. (2.8.45) and (2.8.46)]. Together with these terms also the terms $2/(1-x)$ from the hard part of $P_{\text{ee}}^{(0)}$ should be re-summed in order to cancel the ε dependence. Therefore, the relevant soft-photon part of $P_{\text{ee}}^{(0)}$ reads

$$\begin{aligned}P_{\text{ee}}^{(0),\text{soft}}(x) &= \left(\frac{2}{1-x}\right)_+ \\ &= \lim_{\varepsilon\to 0}\left\{\delta(1-x)\,[2\log(\varepsilon)] + \theta(1-x-\varepsilon)\frac{2}{1-x}\right\}.\end{aligned} \tag{4.6.91}$$

Inserting this into (4.6.90), neglecting $P_{\text{ee}}^{(n)}$, $n \geq 1$, and extracting the leading terms in the soft-photon limit $x \to 1$ yields [Be88b]

$$\begin{aligned}\Phi^{\text{soft}}(x) &= \delta(1-x) + \lim_{\varepsilon\to 0}\sum_{n=1}^{\infty}\left(\frac{2\alpha}{\pi}L_{\text{e}}\right)^n \left[\delta(1-x)\frac{1}{n!}(\log\varepsilon)^n \right.\\ &\quad \left. + \theta(1-x-\varepsilon)\frac{1}{(n-1)!}\frac{\log^{n-1}(1-x)}{1-x}\right] \\ &= \lim_{\varepsilon\to 0}\left\{\delta(1-x)\exp\left(\frac{2\alpha}{\pi}L_{\text{e}}\log\varepsilon\right)\right. \\ &\quad \left. + \theta(1-x-\varepsilon)\frac{2\alpha}{\pi}L_{\text{e}}\frac{1}{1-x}\exp\left(\frac{2\alpha}{\pi}L_{\text{e}}\log(1-x)\right)\right\} \\ &= \frac{2\alpha}{\pi}L_{\text{e}}\frac{1}{1-x}\exp\left(\frac{2\alpha}{\pi}L_{\text{e}}\log(1-x)\right).\end{aligned} \tag{4.6.92}$$

Note that after re-summation ε can be put to zero. From the soft-photon limit [Ye61] one knows that instead of $(2\alpha/\pi)L_e$ in fact

$$\beta_e = \frac{2\alpha}{\pi}(L_e - 1) \tag{4.6.93}$$

should occur [cf. (4.6.82)]. Performing this replacement and including the non-soft-photonic $\mathcal{O}(\alpha)$ terms perturbatively, we arrive at

$$\Phi(x) = \beta_e(1-x)^{\beta_e - 1}(1 + \delta^{V+S}) + \delta^{H}. \tag{4.6.94}$$

In the quantities δ^{V+S} and δ^{H} in addition to the leading-logarithmic contributions derived from the splitting functions also the finite corrections obtained from a direct diagrammatic calculation for $e^+e^- \to Z^*$ can be included. In $\mathcal{O}(\alpha)$ these are given by

$$\begin{aligned} \delta^{V+S} &= \frac{\alpha}{\pi}\left(\frac{3}{2}L_e + \frac{\pi^2}{3} - 2\right) \\ \delta^{H} &= -\frac{\alpha}{\pi}(1+x)(L_e - 1). \end{aligned} \tag{4.6.95}$$

In the literature also the complete $\mathcal{O}(\alpha^2)$ terms for $e^+e^- \to Z^*$ [Be88b, Be89] and the leading-logarithmic terms of order $\mathcal{O}(\alpha^3)$ [Sk91, Ca92] are known.

The influence of the initial-state corrections on the Z-boson line shape is shown in Fig. 4.16. They lead to a reduction of the peak height of the resonance cross section by [Be89, Be90]

$$\Delta\sigma^{\text{peak}} \approx \sigma^{\text{peak},0}\left(\frac{\Gamma_Z}{M_Z}\right)^{\beta_e}(1 + \delta^{V+S}) \approx 0.26\,\sigma^{\text{peak}} \tag{4.6.96}$$

and to a shift of the peak position by

$$\Delta\sqrt{s_{\max}} \approx \frac{\beta_e \pi}{8}\Gamma_Z. \tag{4.6.97}$$

As a consequence the peak position and the Z-boson-mass are related by

$$\sqrt{s_{\max}} \approx M_Z + \frac{\beta_e\pi}{8}\Gamma_Z - \frac{\Gamma_Z^2}{4M_Z}, \tag{4.6.98}$$

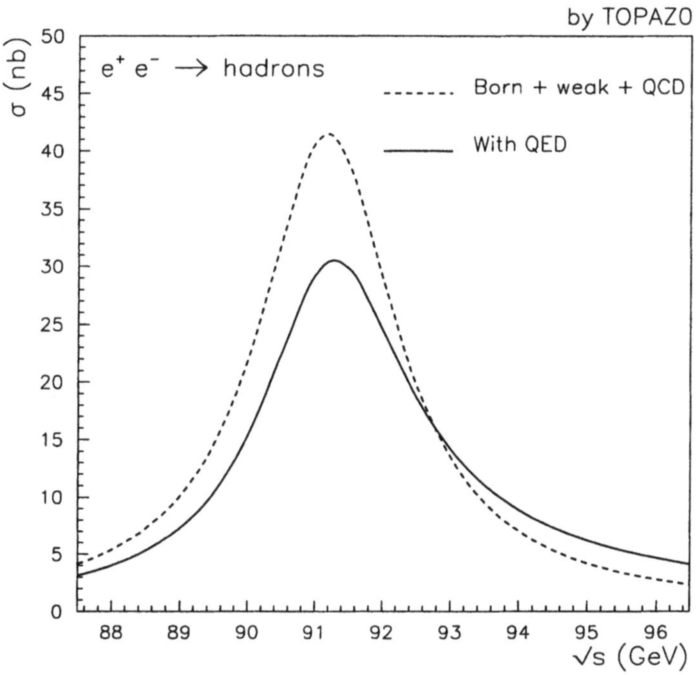

Fig. 4.16 The effect of QED corrections on the Z line shape of $e^+e^- \to$ hadrons. The dashed line is the cross section without QED corrections, the solid line is the QED corrected one (from Ref. [Mo98]).

where the last term results from the interplay of the factor s in the numerator and the factor s in front of Γ_Z in the denominator of (4.6.80).

The initial-state corrections to the *forward–backward asymmetry* can be obtained by a similar procedure. They are similarly large, because also the forward–backward asymmetry is a strongly varying function around the Z-boson resonance [Bo89].

We have not discussed *Bhabha scattering*, where besides the s-channel also t-channel exchange is present with corresponding radiative corrections. Bhabha scattering is not only interesting by itself but also because it is used as luminosity monitor. The structure of the associated weak radiative corrections is similar to those of $e^+e^- \to \mu^+\mu^-$, the electromagnetic corrections, however, are more complicated because of the t-channel exchange. Details can be found in Ref. [Ca89] and references therein.

4.6.4.5 Higher-order corrections

The precision experiments at LEP reached an accuracy of better than 10^{-3}. At this level not only the complete $\mathcal{O}(\alpha)$ corrections but also the leading higher-order corrections are required. The structure function formalism discussed in Sect. 4.6.4.4 can be used to incorporate the leading higher-order QED corrections, which are related to large logarithms of the form $[(\alpha/\pi)\log(M_Z/m_e)]^n$.

But also the non-photonic corrections involve enhanced terms that have to be taken into account in $\mathcal{O}(\alpha^2)$. The various contributions are summarized in [Ho00]. We shortly discuss their origin and structure.

Large higher-order weak effects originate from terms containing logarithms of the type $\log(M_Z/m_f)$ and from contributions that are enhanced by powers of m_t^2/M_W^2. The leading logarithms of $\mathcal{O}(\alpha^n \log^n(M_Z/m_f))$, which arise from the renormalization of α, can be included to all orders by introducing the running electromagnetic coupling constant (4.6.78). The corrections of the form $\propto (\alpha/\pi)^2(m_t^2/M_W^2)^2$ originating from the self-energies can be absorbed by using G_μ instead of α for charged-current interactions (cf. Sect. 4.6.2.3) or by parametrizing the Z-boson–fermion couplings by the effective parameters ρ and $\bar{s}_w^2$ [cf. Sect. 4.6.3.1, in particular (4.6.45)]. Further contributions of this type that originate from the Zbb vertex corrections can be included by suitable redefinitions of ρ_b and $\sin^2\theta_b$ (4.6.47). Moreover, all corrections to ρ_f and $\sin_f^\theta$ mentioned in Sect. 4.6.3.1 apply to $e^+e^- \to f\bar{f}$ as well.

Another class of enhanced two-loop corrections of order $\mathcal{O}(\alpha_s\alpha)$ arise from QCD corrections to electroweak one-loop diagrams. Further enhanced effects of order $\mathcal{O}(\alpha_s\alpha m_t^2/M_W^2)$ appear in QCD corrections to the top and/or bottom loops in self-energy corrections and in the electroweak corrections to the Zbb vertex. For a review on these corrections see Ref. [Ch96].

4.6.5 Precision tests of the electroweak interaction

A recent compilation of precision data from LEP, SLC, and other sources is given in Table 4.3 [Gu00]. It includes the measurements of the Z-boson mass M_Z, the total Z-boson width Γ_Z, ratios of partial widths, the peak cross

Observable	Data	Standard Model	Pull
M_Z [GeV]	91.1875 ± 0.0021	91.1874	+0.05
Γ_Z [GeV]	2.4952 ± 0.0023	2.4962	−0.43
$\sigma_{\text{had}}^{\text{peak}}$ [nb]	41.540 ± 0.037	41.480	+1.62
$\Gamma_Z^{\text{had}}/\Gamma_Z^{\ell\ell}$	20.767 ± 0.025	20.740	+1.08
$\Gamma_Z^{b\bar{b}}/\Gamma_Z^{\text{had}}$	0.21649 ± 0.00072	0.21577	+1.00
$\Gamma_Z^{c\bar{c}}/\Gamma_Z^{\text{had}}$	0.1683 ± 0.0038	0.1723	−1.04
A_{FB}^{ℓ}	0.01714 ± 0.00095	0.01642	+0.76
$\mathcal{A}_\tau$	0.1439 ± 0.0042	0.1480	−0.97
$\mathcal{A}_e$	0.1498 ± 0.0048	0.1480	+0.38
A_{FB}^{b}	0.0989 ± 0.0020	0.1037	−2.42
A_{FB}^{c}	0.0689 ± 0.0035	0.0741	−1.50
$\mathcal{A}_b$	0.922 ± 0.023	0.935	−0.55
$\mathcal{A}_c$	0.632 ± 0.026	0.668	−1.40
$\sin^2\theta_\ell$ ($\langle Q_{\text{FB}}\rangle$)	0.2321 ± 0.0010	0.23140	+0.7
$\sin^2\theta_\ell$ (A_{LR})	0.23098 ± 0.00026	0.23140	−1.62
M_W [GeV] (LEP)	80.427 ± 0.046	80.401	+0.55
M_W [GeV] (pp)	80.448 ± 0.062	80.401	+0.75
m_t [GeV]	174.3 ± 5.1	174.5	−0.04
$\Delta\alpha_{\text{had}}^{(5)}(M_Z)$ [GeV]	0.02804 ± 0.00065	0.02822	−0.28

Tab. 4.3 Summary of electroweak precision measurements (from Ref. [Gu00])

section, various on-resonance asymmetries (cf. Sect. 4.6.4.1), the effective leptonic mixing angle [cf. (4.6.40)], the W-boson mass, and the top-quark mass. The index ℓ stands for a generic lepton using lepton universality.

Column 3 of Table 4.3 contains the results of a Standard-Model fit to the precision data. Finally, the last column shows the deviations between the Standard-Model prediction and the experimental result in units of the experimental error. The agreement of the Standard Model with the data is impressive. No significant evidence for departures from the Standard-Model predictions has been found, and the deviations are as expected for a normal distribution of experimental measurements. We note that the inclusion of radiative corrections is crucial for the agreement. Without radiative corrections, or if only QED corrections were included, the theoretical predictions and the experimental results would differ by more than two standard deviations.

The predictions of the EWSM depend, besides the well-known parameters α, G_μ, M_Z, m_e, m_μ, and m_τ, also on $R(s)$ [cf. (4.6.17)], the strong fine-structure constant $\alpha_s(M_Z^2)$, the top-quark, and the Higgs-boson mass. The good agreement between Standard Model and data and the high precision allows to derive constraints on the top-quark mass, on the Higgs-boson mass, and on $\alpha_s(M_Z^2)$ from the precision data. Including all data in Table 4.3 one finds in addition to the parameters given in column 3 of that table

$$M_H = 62^{+53}_{-30}\,\text{GeV}, \qquad \alpha_s(M_Z^2) = 0.1183 \pm 0.0027. \tag{4.6.99}$$

This provides the best prediction of the Higgs-boson mass available. Note that the central value of this result is below the lower limit of 113.3 GeV [Ig00] from direct searches. The result for the strong coupling constant is in very good agreement with direct measurements [PDG00], $\alpha_s(M_Z^2) = 0.1185 \pm 0.0020$, and of similar precision.

The data can also be used within the Standard Model to determine the top-quark and W-boson masses indirectly and compare them to the direct measurements. To this end, one uses all data except those for M_W and m_t in the fit. In this way one obtains

$$M_W = 80.376 \pm 0.034\,\text{GeV}, \qquad m_t = 169^{+10}_{-8}\,\text{GeV}, \tag{4.6.100}$$

in good agreement with the direct measurements given in Table 4.3. This constitutes a striking success for the Standard Model at its quantum level and a remarkable confirmation of the precision tests of the electroweak interaction. The direct and indirect determinations are compared in Fig. 4.17 with the Standard Model prediction for various Higgs-boson masses. As can be seen, the indirect determinations prefer low M_H.

4.6.6 Status of the Electroweak Standard Model and perspectives

We have seen that the Electroweak Standard Model is in impressive agreement with the experimental data. The fermion sector and the gauge sector

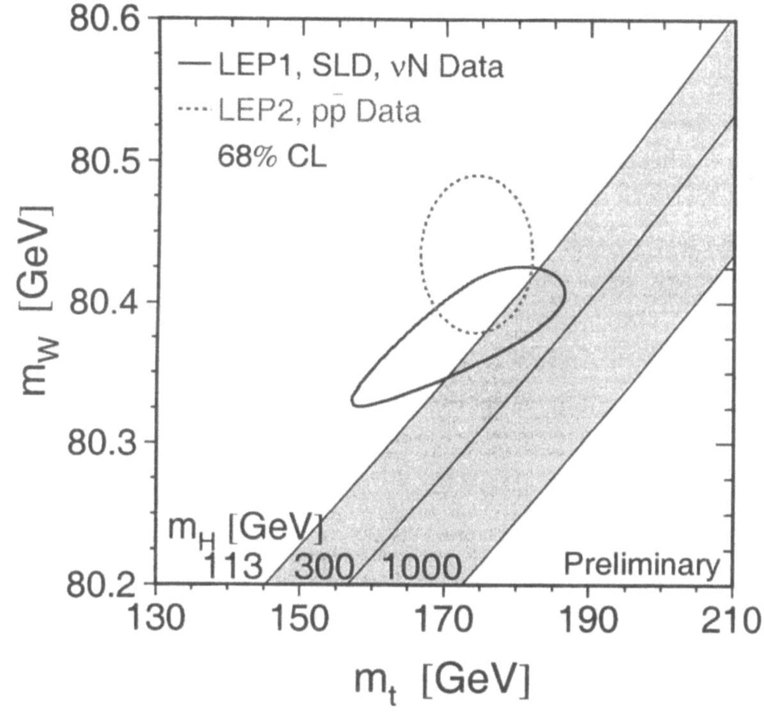

Fig. 4.17 Indirect (solid contour) and direct (dashed contour) determinations of M_W and m_t and the Standard-Model relationship for different values of the Higgs-boson mass (from Ref. [LEP00])

are experimentally established. However, so far there is no conclusive evidence for the Higgs sector.[14] The existence or non-existence of the Higgs boson will definitely be established with the help of the LHC, a proton–proton storage ring that will start operation in 2005.

Precision experiments have confirmed the Electroweak Standard Model at its quantum level, i.e. the electroweak radiative corrections including the genuine bosonic weak corrections have been experimentally verified. In order

[14]Recently an excess of events in Z-boson pair production has been found at LEP2 which could be interpreted as a signal of the Higgs boson [Ba00, Ac00].

to describe the high-precision data from LEP1 even two-loop corrections have to be included.

While other experiments cannot match this high precision at or below the per-mill level, many of the ongoing and future experiments will at least reach or exceed an accuracy of 1%. This requires to include one-loop radiative corrections. As an example we mention W-pair production in e^+e^- collisions, which has been investigated at LEP200 with an accuracy at the level of 1%. There, the complete electroweak one-loop corrections are needed for adequate theoretical predictions (cf. Sect. 4.3.1).

Since the electroweak radiative corrections grow logarithmically with increasing energy, their effect on physical observables will be more pronounced at future colliders. Therefore, in the future most tests of the Standard Model will have to rely on calculations for the corresponding radiative corrections.

References to Chapter 4

[Ac00] M. Acciari et al. (L3 coll.), *Phys. Lett.* **B495** (2000) 18.

[Ak86] A.A. Akhundov, D.Yu. Bardin and T. Riemann, *Nucl. Phys.* **B276** (1986) 1.

[Al77] G. Altarelli and G. Parisi, *Nucl. Phys.* **B126** (1977) 298.

[Al77a] W. Alles, C. Boyer and A. Buras, *Nucl. Phys.* **B119** (1977) 125.

[Al80a] D. Albert, W.J. Marciano, D. Wyler and Z. Parsa, *Nucl. Phys.* **B166** (1980) 460.

[Ao80] K.I. Aoki, Z. Hioki, R. Kawabe, M. Konuma and T. Muta, *Prog. Theor. Phys.* **64** (1980) 707, **65** (1981) 1001, and *Prog. Theor. Phys. Suppl.* **73** (1982) 1.

[Ap72] T. Appelquist and H.R. Quinn, *Phys. Lett.* **39B** (1972) 229.

[Ap73a] T. Appelquist, J. Carazzone, T. Goldman and H.R. Quinn, *Phys. Rev.* **D8** (1973) 1747.

[Av94] L. Avdeev, J. Fleischer, S.M Mikhailov and O. Tarasov, *Phys. Lett.* **B336** (1994) 560, E:**B349** (1995) 597.

[Ba62] M. Baker and S.L. Glashow, *Phys. Rev.* **128** (1962) 2462.

[Ba72a] I. Bars and M. Yoshimura, *Phys. Rev.* **D6** (1972) 374.

[Ba80a] D.Yu. Bardin, P.Ch. Christova and O.M. Federenko, *Nucl. Phys.* **175** (1980) 435 and **197** (1982) 1.

[Ba86] D.Yu. Bardin, S. Riemann and T. Riemann, *Z. Phys.* **C32** (1986) 121.

[Ba88] D.Yu. Bardin, A. Leike, T. Riemann and M. Sachwitz, *Phys. Lett.* **B206** (1988) 539.

[Ba92] R. Barbieri et al., *Phys. Lett.* **B288** (1992) 95, E:**B312** (1993) 511, and *Nucl. Phys.* **B409** (1993) 105.

[Ba93a] D. Bardin, M. Bilenky, A. Olchevski and T. Riemann, *Phys. Lett.* **B308** (1993) 403;
D. Bardin et al., *Comp. Phys. Commun.* **104** (1997) 161.

[Ba95d] D. Bardin, W. Hollik and G. Passarino (eds.), *Reports of the Working Group on Precision Calculations for the Z Resonance*, (CERN-95-03, Geneva, 1995).

[Ba96] D. Bardin, et al., *Event Generators for* WW *Physics* in *Physics at LEP 1*, CERN 96-01, Vol. 2, eds. G. Altarelli, T. Sjöstrand and F. Zwirner (Geneva, 1996) p. 3.

[Ba98b] S. Bauberger and G. Weiglein, *Phys. Lett.* **B419** (1998) 333.

[Ba00] R. Barate et al. (ALEPH coll.), *Phys. Lett.* **B495** (2000) 1.

[Be56] R.E. Behrends, R.J. Finkelnstein and A. Sirlin, *Phys. Rev.* **101** (1956) 866.

[Be74] C. Becchi, A. Rouet and R. Stora, *Phys. Lett.* **52B** (1974) 344 and *Commun. Math. Phys.* **42** (1975) 127.

[Be74a] J. Bernstein, *Rev. Mod. Phys.* **46** (1974) 1.

[Be88a] W. Beenakker and W. Hollik, *Z. Phys.* **C40** (1988) 141.

[Be88b] F.A. Berends, W.L. van Neerven and G.J.H. Burgers, *Nucl. Phys.* **297** (1988) 429; E:*Nucl. Phys.* **304** (1989) 921.

[Be88c] J. Bernabeu, A. Pich and A. Santamaria, *Phys. Lett.* **B200** (1988) 569.

[Be89] F.A. Berends et al., Z *Line Shape* in Z *Physics at LEP 1*, CERN 89-08, eds. G. Altarelli, R. Kleiss and C. Verzegnassi (Geneva, 1989) p. 89.

[Be89a] W. Beenakker, F.A. Berends and W.L. van Neerven, *Application of Renormalization Group Methods to Radiative Corrections* in *Radiative Corrections for* e^+e^- *Collisions*, ed. J.H. Kühn (Berlin, etc., 1989) p. 3.

[Be90] W. Beenakker, F.A. Berends and S.C. van der Marck, *Z. Phys.* **C46** (1990) 687.

[Be94] W. Beenakker and A. Denner, *Int. J. Mod. Phys.* **A9** (1994) 4837.

[Bi87] J.J. van der Bij and F. Hoogeveen, *Nucl. Phys.* **B283** (1987) 477.

[Bl63] A. Bludman and A. Klein, *Phys. Rev.* **131** (1963) 2364.

[Bo86] M. Böhm, W. Hollik and H. Spiesberger, *Fortschr. Phys.* **34** (1986) 687.

[Bo89] M. Böhm et al., *Forward–backward Asymmetries* in Z *Physics at LEP 1*, CERN 89-08, eds. G. Altarelli, R. Kleiss and C. Verzegnassi (Geneva, 1989) p. 203.

[Bu89] G. Burgers and F. Jegerlehner, *Δr, or the Relation between the Electroweak Couplings and the Weak Vector Boson Masses* in Z *Physics at LEP 1*, CERN 89-08, eds. G. Altarelli, R. Kleiss and C. Verzegnassi (Geneva, 1989) p. 7.

[Bu95] H. Burkhardt and B. Pietrzyk, *Phys. Lett.* **B356** (1995) 398.

[Ca61] N. Cabbibo and R. Gatto, *Phys. Rev.* **124** (1961) 1577.

[Ca79] N. Cabibbo, L. Maiani, G. Parisi and R. Petronzio, *Nucl. Phys.* **B158** (1979) 295.

[Ca89] M. Caffo et al., *Bhabha Scattering* in Z *Physics at LEP 1*, CERN 89-08, eds. G. Altarelli, R. Kleiss and C. Verzegnassi (Geneva, 1989) p. 171.

[Ca92] M. Cacciari, A. Deandrea, G. Montagna and O. Nicrosini, *Europhys. Lett.* **17** (1992) 123.

[Ch82a] T.H. Chang, K.J.F. Gaemers and W.L. van Neerven, *Nucl. Phys.* **B202** (1982) 407.

[Ch85] M.S. Chanowitz and M.K. Gaillard, *Nucl. Phys.* **B261** (1985) 379.

[Ch95] K.G. Chetyrkin, J.H. Kühn and M. Steinhauser, *Phys. Lett.* **B351** (1995) 331 and *Phys. Rev. Lett.* **75** (1995) 3820.

[Ch96] K.G. Chetyrkin, J.H. Kühn and A. Kwiatkowski, *Phys. Rept.* **277** (1996) 189 and *QCD Corrections to the* e^+e^- *Cross Section and the Z Boson Decay Rate* in *Reports of the Working Group on Precision Calculations for the* Z *Resonance,* CERN-95-03, eds. D. Bardin, W. Hollik and G. Passarino (Geneva, 1995) p. 175.

[Co74a] J.M. Cornwall, D.N. Levin and G. Tiktopoulos, *Phys. Rev.* **D10** (1974) 1145.

[Co84] J.C. Collins, *Renormalization*, (Cambridge University Press, Cambridge, 1984).

[Co89b] M. Consoli, W. Hollik and F. Jegerlehner, *Phys. Lett.* **B227** (1989) 167.

[De90] A. Denner and T. Sack, *Nucl. Phys.* **B347** (1990) 203.

[De90a] A. Denner and T. Sack, *Z. Phys.* **C46** (1990) 653.

[De93] A. Denner, *Fortschr. Phys.* **41** (1993) 307.

[De93a] A. Denner, W. Hollik and B. Lampe, *Z. Phys.* **C60** (1993) 93.

[De95] A. Denner, S. Dittmaier and G. Weiglein, *Nucl. Phys.* **B440** (1995) 95.

[De96] A. Denner and S. Dittmaier, *Phys. Rev.* **D54** (1996) 4499.

[De96a] G. Degrassi, P. Gambino and A. Vicini, *Phys. Lett.* **B383** (1996) 219.

[De97] G. Degrassi, P. Gambino and A. Sirlin, *Phys. Lett.* **B394** (1997) 188.

[De98] G. Degrassi, P. Gambino, M. Passera and A. Sirlin, *Phys. Lett.* **B418** (1998) 209.

[De00] G. Degrassi and P. Gambino, *Nucl. Phys.* **B567** (2000) 3.

[De00a] A. Denner, S. Dittmaier, M. Roth and D. Wackeroth, *Nucl. Phys.* **B587** (2000) 67.

[Di97] S. Dittmaier, *Phys. Lett.* **B409** (1997) 509.

[Dj87] A. Djouadi and C. Verzegnassi, *Phys. Rev.* **D20** (1979) 167.

[Dj88] A. Djouadi, *Nuovo Cimento* **A100** (1988) 357.

[Dj94] A. Djouadi and P. Gambino, *Phys. Rev.* **D49** (1994) 3499.

[Dj96] A. Djouadi, M. Spira and P.M. Zerwas, *Z. Phys.* **C70** (1996) 427.

[Dy49] F.J. Dyson, *Phys. Rev.* **75** (1949) 486 and 1736.

[Ei89] M.B. Einhorn and J. Wudka, *Phys. Rev.* **D39** (1989) 2758.

[Ei95] S. Eidelman and F. Jegerlehner, *Z. Phys.* **C67** (1995) 585.

[En64] F. Englert and R. Brout, *Phys. Rev. Lett.* **13** (1964) 321.

[Fa93] S. Fanchiotti, B.A. Kniehl and A. Sirlin, *Phys. Rev.* **D48** (1993) 307.

[Fl81] J. Fleischer and F. Jegerlehner, *Phys. Rev.* **D23** (1981) 2001.

[Fl93] J. Fleischer, F. Jegerlehner and O.V. Tarasov, *Phys. Lett.* **B319** (1993) 249 and *Phys. Rev.* **D51** (1995) 3820.

[Fl99a] J. Fleischer, F. Jegerlehner, M. Tentyukov and O. Veretin, *Phys. Lett.* **B459** (1999) 625.

[Fr74] E.S. Fradkin and I.V. Tyutin, *Revista del Nuovo Cim.* **4** (1974) 1.

[Fr00a] A. Freitas, W. Hollik, W. Walter and G. Weiglein, *Phys. Lett.* **B495** (2000) 338.

[Ga99] P. Gambino, P.A. Grassi and F. Madricardo, *Phys. Lett.* **B454** (1999) 98.

[Ga00a] P. Gambino and P.A. Grassi, *Phys. Rev.* **D62** (2000) 076002.

[Gi64] W. Gilbert, *Phys. Rev. Lett.* **12** (1964) 713.

[Gl70] S.L. Glashow, J. Iliopoulos and L. Maiani, *Phys. Rev.* **D2** (1970) 1285.

[Go61] J. Goldstone, *Nuovo Cimento* **19** (1961) 154.

[Go62] J. Goldstone, A. Salam and S. Weinberg, *Phys. Rev.* **127** (1962) 965.

[Go86] G.J. Gounaris, R. Kögerler and H. Neufeld, *Phys. Rev.* **D34** (1986) 3257.

[Go93] M. Göckeler, H. Kastrup, T. Neuhaus and F. Zimmermann, *Nucl. Phys.* **404** (1993) 517.

[Gr00] M. Grünewald et al., *Four Fermion Production in Electron Positron Collisions*, report of the four-fermion working group of the *LEP2 Monte Carlo workshop*, (CERN, Geneva, 2000), hep-ph/0005309.

[Gu64] G.S. Guralnik, C.R. Hagen and T.W.B. Kibble, *Phys. Rev. Lett.* **13** (1964) 585.

[Gu00] A. Gurtu, plenary talk given at the XXXth International Conference on High Energy Physics, July 27–August 2, 2000, Osaka, Japan.

[Ha87] A. Hasenfratz, T. Neuhaus, K. Jansen, H. Yoneyama and C.B. Lang, *Phys. Lett.* **B199** (1987) 531.

[Ha91a] F. Halzen and B.A. Kniehl, *Nucl. Phys.* **B353** (1991) 567.

[Ha98c] R. Harlander, T. Seidensticker and M. Steinhauser, *Phys. Lett.* **B426** (1998) 125.

[He28] W. Heisenberg, *Z. Physik* **49** (1928) 619.

[Hi64] P.W. Higgs, *Phys. Lett.* **12** (1964) 132, *Phys. Rev. Lett.* **13** (1964) 508, and *Phys. Rev.* **145** (1966) 1156.

[Ho00] W. Hollik and G. Duckeck, *Electroweak Precision Tests at LEP*, (Springer, Berlin, etc., 2000).

[Ho00a] W. Hollik, private communication.

[Ig00] P. Igo-Kemenes, plenary talk given at the XXXth International Conference on High Energy Physics, July 27–August 2, 2000, Osaka, Japan.

[In80] K. Inoue, A. Kakuto, H. Komatsu and S. Takeshita, *Prog. Theor. Phys.* **64** (1980) 1008.

[Ja72] R. Jackiw and S. Weinberg, *Phys. Rev.* **D5** (1972) 2396.

[Ja98] S. Jadach et al., *Phys. Lett.* **B417** (1998) 326, *Phys. Rev.* **D61** (2000) 113010 and hep-ph/0007012.

[Je86] F. Jegerlehner, *Z. Phys.* **C32** (1986) 425, E:*Z. Phys.* **C38** (1988) 519.

[Je91] F. Jegerlehner, *Testing the Standard Model* in *Renormalizing the Standard Model*, eds. M. Cvetic and P. Langacker (Singapore, etc., 1991) p. 476.

[Jo64] G. Jona-Lasinio, *Nuovo Cimento* **34** (1964) 1790.

[Ki59] T. Kinoshita and A. Sirlin, *Phys. Rev.* **113** (1959) 1652.

[Ki67] T.W.B. Kibble, *Phys. Rev.* **155** (1967) 1554.

[Kn88] B.A. Kniehl, J.H. Kühn and R.G. Stuart, *Phys. Lett.* **B214** (1988) 621.

[Kn90] B.A. Kniehl, *Nucl. Phys.* **B347** (1990) 86.

[Kn92] B.A. Kniehl and A. Sirlin, *Nucl. Phys.* **B371** (1992) 141 and *Phys. Rev.* **D47** (1993) 883.

[Ku85] E.A. Kuraev and V.S. Fadin, *Sov. J. Nucl. Phys.* **41** (1985) 466.

[Ku96] Z. Kunszt et al., *Determination of the Mass of the* W *Boson* in *Physics at LEP2,* CERN-96-01, eds. G. Altarelli, T. Sjöstrand and F. Zwirner (Geneva, 1996) p. 141.

[La67] B.E. Lautrup, *Mat. Fys. Medd. Kon. Dan. Vid.-Sel. Medd.* **35** (1967) 29.

[Le72a] S.Y. Lee, *Phys. Rev.* **D6** (1972) 1701 and 1803.

[Le72b] B.W. Lee, *Chiral Dynamics*, (Gordon and Breach Science Publishers, New York, 1972).

[Le74] B.W. Lee, *Phys. Rev.* **D9** (1974) 933.

[Le77] B.W. Lee, C. Quigg and H.B. Thacker, *Phys. Rev.* **D16** (1977) 1519.

[Ll73] C.H. Llewellyn Smith, *Phys. Lett.* **46B** (1973) 233.

[Li86] M. Lindner, *Z. Phys.* **C31** (1986) 295.

[Lu88] M. Lüscher and P. Weisz, *Phys. Lett.* **B212** (1988) 472.

[LEP97] The LEP Collaborations ALEPH, DELPHI, L3, OPAL, the LEP Electroweak Working Group and the SLD Heavy Flavour Group, CERN-PPE/97-154.

[LEP00] Homepage of the LEP Electroweak Working Group, http://lepewwg.web.cern.ch/LEPEWWG/.

[Ma79] W.J. Marciano, *Phys. Rev.* **D20** (1979) 274.

[Ma79a] W.J. Marciano and D. Wyler, *Z. Phys.* **C3** (1979) 181.

[Mo98] G. Montagna, O. Nicrosini and F. Piccinini, *Riv. del Nuovo Cimento* **21** (1998) 1.

[Na66a] N. Nakanashi, *Prog. Theor. Phys.* **35** (1966) 1111.

[Ni87] O. Nicrosini and L. Trentadue, *Phys. Lett.* **B196** (1987) 551.

[PDG00] D.E. Groom et al. (Particle Data Group), *Eur. Phys. J.* **C15** (2000) 1.

[Ri99b] T. van Ritbergen and R.G. Stuart, *Phys. Rev. Lett.* **82** (1999) 488.

[Ro75] D.A. Ross and M. Veltman, *Nucl. Phys.* **B95** (1975) 135.

[Ro73] D.A. Ross and J.C. Taylor, *Nucl. Phys.* **B51** (1973) 25.

[Sa68] A. Salam, *Weak and Electromagnetic Interactions* in *Elementary Particle Theory*, ed. N. Svartholm (Stockholm, 1968) p. 367.

[Sa81] S. Sakakibara, *Phys. Rev.* **D24** (1981) 1149.

[Si80] A. Sirlin, *Phys. Rev.* **D22** (1980) 971;
W.J. Marciano and A. Sirlin, *Phys. Rev.* **D22** (1980) 2695;
A. Sirlin and W.J. Marciano, *Nucl. Phys.* **B189** (1981) 442.

[Si84] A. Sirlin, *Phys. Rev.* **D29** (1984) 89.

[Si91] A. Sirlin, *Phys. Rev. Lett.* **67** (1991) 2127.

[Sk91] M. Skrzypek and S. Jadach, *Z. Phys.* **C49** (1991) 577.

[Th41] W. Thirring, *Phil. Mag.* **41** (1950) 233.

[tH71] G. 't Hooft, *Nucl. Phys.* **B33** (1971) 173 and **B35** (1971) 167.

[Ve77] M. Veltman, *Acta Phys. Pol.* **B8** (1977) 475.

[We67] S. Weinberg, *Phys. Rev. Lett.* **19** (1967) 1264.

[We72] S. Weinberg, *Phys. Rev. Lett.* **27** (1972) 1688.

[We98] G. Weiglein, *Acta Phys. Pol.* **B29** (1998) 2735.

[Ya88] Y.-P. Yao and C.-P. Yuan, *Phys. Rev.* **D38** (1988) 2237.

[Ye61] D.R. Yennie, S.C. Frautschi and H. Suura, *Ann. Phys.* **13** (1961) 379.

5 Extensions of the Standard Model

The Standard Model (SM) of particle physics successfully incorporates all known properties of the strong, weak, and electromagnetic interactions. The experimental evidence for neutrino oscillations [Fu98, Cl98a, Ma00a] can be accommodated within the SM by adding right-handed neutrinos. If the neutrinos are of Majorana type and violate lepton number, a Higgs singlet field is needed in addition. Apart from the neutrino oscillation phenomena, there is no compelling experimental result that contradicts the predictions of the Standard Model. However, many of the consequences of the theory have not yet been experimentally verified or remain to be worked out further, like the mass-generation mechanism or the strong-coupling regime of QCD.

Besides this, the SM leaves a lot of questions open. It is instructive to list the most prominent ones:

- The Standard Model involves not less than 19 free parameters. These consist of two gauge couplings, two gauge-boson masses, the mass of the Higgs boson, six quark and three lepton masses, three mixing angles and the phase parametrizing CP violation. In addition there is a further parameter θ_{QCD} which describes potential strong CP violation. For massive Dirac neutrinos six additional parameters appear, three neutrino masses and three parameters of the lepton-mixing matrix.[1] Can the number of free parameters be reduced?
- Why are there three different gauge couplings, α_s, $\alpha_w = \alpha/s_w^2$, and α? Why is the gauge group of the SM, G_{SM}, given by the direct product $SU(3)_C \times SU(2)_w \times U(1)_Y$?
- Why is the electric charge quantized (the hypercharge is arbitrary in the Standard Model), and why are lepton and quark charges related ($Q_e = 3Q_d = -3Q_u/2$)?

[1] If the right-handed neutrinos are Majorana particles, the phases related to those fields cannot be eliminated as for the quark fields, and three additional phases become relevant as compared to the case of massive Dirac neutrinos.

- What determines the matter multiplet structure of the Standard Model? Why do leptons and quarks appear in generations? What determines their number?

- Are leptons and quarks related in some way? The contributions of leptons and quarks to the chiral anomaly cancel each other, which is a necessary condition for the renormalizability of the Standard Model. Why is this the case?

- What determines the mass spectrum of the fermions and the form of the quark-mixing (and possibly the lepton-mixing) matrix? Note that most of the many free parameters of the Standard Model are related to the fermion masses.

- How can the Standard Model be combined with gravity?

There are several theoretical concepts that try to answer some of these questions. In the following, we describe *grand unified theories* (GUTs) and *supersymmetry* (SUSY).

5.1 Grand unified theories (GUTs)

GUTs [GUT] are strictly based on the concept of gauge theories. The essential additional idea is to describe the strong, weak, and electromagnetic interactions by a gauge theory with a single *simple* gauge group G. This immediately implies that only one gauge coupling g_{u} exists.

The choice of the gauge group G is restricted by the requirement that it must contain $G_{\mathrm{SM}} = \mathrm{SU}(3)_{\mathrm{C}} \times \mathrm{SU}(2)_{\mathrm{w}} \times \mathrm{U}(1)_{\mathrm{Y}}$ as a subgroup and it must allow for representations that can accommodate the known fermions with the correct G_{SM} quantum numbers. Since G_{SM} has rank 4, G must have at least rank 4. Moreover, G must admit complex representations since the SU(3)-triplet and SU(2)-doublet representations are complex. Finally, the fermion representations must be free of anomalies. Typical examples that fulfill these requirements are the groups SU(5) and SO(10) with rank 4 and 5, respectively, and the exceptional groups E_6, E_7, and E_8, where the index denotes the rank [GT].

Unification enlarges the dimension of the gauge group and thus gives rise to additional gauge bosons. Since the three gauge couplings differ at the presently experimentally accessible energies, grand unification can be realized only at very high energies, and the corresponding symmetry must be spontaneously broken. As a consequence additional Higgs bosons are required.

5.1.1 Unification of coupling constants

In the limit of unbroken G symmetry the unified strong, weak, and electromagnetic interactions are governed by the *minimal gauge-invariant coupling* of the charged fields to the gauge fields, i.e. by the covariant derivative

$$D_\mu = \partial_\mu - \mathrm{i} g_\mathrm{u} \sum_{a \in G} T^a A^a_\mu, \tag{5.1.1}$$

where T^a are the generators and A^a_μ the gauge fields of G. The covariant derivative has to contain the known interactions,

$$D_\mu = \partial_\mu - \mathrm{i} g_\mathrm{s} \sum_{a \in \mathrm{SU}(3)} T^a G^a_\mu - \mathrm{i} g_2 \sum_{a \in \mathrm{SU}(2)} T^a W^a_\mu + \mathrm{i} g_1 \frac{Y_\mathrm{w}}{2} B_\mu + \dots . \tag{5.1.2}$$

Comparing these expressions (in a suitable basis) immediately yields

$$g_\mathrm{u} = g_\mathrm{s} = g_2, \tag{5.1.3}$$

since all non-abelian generators are normalized in the same way. In order to find the proper relation for the hypercharge coupling g_1 we have to take into account the normalization of the abelian hypercharge generator $Y_\mathrm{w} = cT^a$, where T^a is one of the generators of G, e.g. T^{12} for SU(5). To this end, we assume that the fermions of one Standard-Model generation furnish complete multiplets of G and evaluate the sum of the squares of the hypercharges of all these fermions [cf. (5.1.24)]

$$\frac{1}{4} \mathrm{Tr}\{Y_\mathrm{w} Y_\mathrm{w}\} = \frac{10}{3} \stackrel{!}{=} c^2 \, \mathrm{Tr}\{T^a T^a\} = c^2 T(T^a), \qquad \text{no sum over } a. \tag{5.1.4}$$

On the other hand, the Dynkin index $T(T^a)$ [cf. (1.2.52)] can be determined from the SU(3) and SU(2) generators resulting in $T(T^a) = 2$ for a complete generation. Thus, we find $c = \sqrt{5/3}$ and

$$g_{\mathrm{u}} = \sqrt{\frac{5}{3}}\, g_1. \tag{5.1.5}$$

In the SM, the *weak hypercharge* generator is abelian and can have arbitrary eigenvalues. Within GUTs the hypercharge generator is a non-abelian generator and consequently its eigenvalues are quantized. This implies via the Gell-Mann–Nishijima relation the *quantization of the electric charge.* Moreover, the fact that leptons and quarks are in the same representation leads to a relation between the charges of leptons and quarks.

Using the relations (5.1.3) and (5.1.5), we find from (4.2.30) and (4.2.25)

$$e = \sqrt{\frac{3}{8}}\, g_{\mathrm{u}}, \qquad s_{\mathrm{W}}^2 = \frac{3}{8}. \tag{5.1.6}$$

The symmetry relations (5.1.3), (5.1.5), and (5.1.6) hold in a spontaneously broken gauge theory only above an energy scale m_{GUT} that is large with respect to the masses generated by the Higgs mechanism [Ge74a]. In order to compare with experimental values at low energies, the coupling constants have to be continued to low energies with help of the solutions of the RG equations [Ge74a, Bu78]. In one-loop approximation these read [cf. (2.6.21)]

$$\alpha_i^{-1}(\mu) = \alpha_i^{-1}(m_{\mathrm{GUT}}) - \frac{\beta_{0,i}}{2\pi}\log\left(\frac{m_{\mathrm{GUT}}}{\mu}\right), \qquad i = \mathrm{s}, 2, 1, \tag{5.1.7}$$

where $\alpha_i = g_i^2/(4\pi)$.

These solutions have been derived under the assumption that the β-functions are independent of the masses. However, they are generally valid also in effective theories for energies that are large compared to the masses of the excited degrees of freedom and small compared to the masses of the non-excited degrees of freedom [We60]. In this case, the one-loop coefficient of the β-function is given by

$$\beta_{0,i} = \frac{11}{3}C_{\mathrm{A}} - \frac{2}{3}T_{\mathrm{fer}} - \frac{1}{3}T_{\mathrm{H}}, \tag{5.1.8}$$

where C_A is the Casimir operator of the corresponding effective gauge group and T_{fer} and T_H the Dynkin indices of the effective (left-handed and right-handed) fermion multiplets and complex Higgs multiplets, respectively [cf. (1.2.56) and (1.2.52)].

If we assume that no new degrees of freedom occur between the electroweak scale and the GUT scale (hypothesis of "great desert"), then the running of the couplings between these scales is determined by the degrees of freedom of the SM with $N_g = 3$ fermion generations and $N_{Hd} = 1$ complex Higgs doublets, and we obtain

$$\begin{aligned} \beta_{0,s} &= 11 - \frac{4}{3}N_g = 7, \\ \beta_{0,2} &= \frac{22}{3} - \frac{4}{3}N_g - \frac{1}{6}N_{Hd} = \frac{19}{6}, \\ \beta_{0,1} &= 0 - \frac{20}{9}N_g - \frac{1}{6}N_{Hd} = -\frac{41}{6}. \end{aligned} \tag{5.1.9}$$

Identifying the electromagnetic coupling by $\alpha^{-1}(\mu) = \alpha_2^{-1}(\mu) + \alpha_1^{-1}(\mu)$, using $s_w^2(\mu) = \alpha(\mu)/\alpha_2(\mu)$, (5.1.5) and (5.1.3), we find from (5.1.7)

$$\begin{aligned} \alpha^{-1}(\mu) - \frac{8}{3}\alpha_s^{-1}(\mu) &= \frac{\frac{8}{3}\beta_{0,s} - \beta_{0,2} - \beta_{0,1}}{2\pi} \log\frac{m_{GUT}}{\mu} \\ &= \frac{66 + N_{Hd}}{6\pi}\log\frac{m_{GUT}}{\mu} = \frac{67}{6\pi}\log\frac{m_{GUT}}{\mu}, \\ s_w^2(\mu) &= \frac{3}{8} - \frac{\alpha(\mu)}{2\pi}\frac{5\beta_{0,2} - 3\beta_{0,1}}{8}\log\frac{m_{GUT}}{\mu} \\ &= \frac{3}{8} - \frac{\alpha(\mu)}{2\pi}\frac{110 - N_{Hd}}{24}\log\frac{m_{GUT}}{\mu} \\ &= \frac{3}{8} - \frac{\alpha(\mu)}{2\pi}\frac{109}{24}\log\frac{m_{GUT}}{\mu}. \end{aligned} \tag{5.1.10}$$

Eliminating m_{GUT} from these two equations yields

$$\begin{aligned} s_w^2(\mu) &= \frac{\beta_{0,s} - \beta_{0,2}}{\frac{8}{3}\beta_{0,s} - \beta_{0,2} - \beta_{0,1}} + \frac{\alpha(\mu)}{\alpha_s(\mu)}\frac{\frac{5}{3}\beta_{0,2} - \beta_{0,1}}{\frac{8}{3}\beta_{0,s} - \beta_{0,2} - \beta_{0,1}} \\ &= \frac{22 + N_{Hd}}{2(66 + N_{Hd})} + \frac{\alpha(\mu)}{\alpha_s(\mu)}\frac{110 - N_{Hd}}{3(66 + N_{Hd})} \\ &= \frac{23}{134} + \frac{\alpha(\mu)}{\alpha_s(\mu)}\frac{109}{201}. \end{aligned} \tag{5.1.11}$$

Because the fermions of one generation fill complete SU(5) multiplets, these predictions are independent of the number of fermion generations (at the one-loop level). They depend weakly on the number of Higgs doublets, i.e. the coefficients of the Higgs contributions are small. Equation (5.1.11) is the consistency condition for unification for a given set of fields and input parameters.

Using $\mu = M_Z$, $\alpha(M_Z) = 1/128$ and $\alpha_s(M_Z) = 0.12$ as input, (5.1.10) and (5.1.7) yield $m_{GUT} \approx 8 \times 10^{14}\,\text{GeV}$, $s_w^2(M_Z) \approx 0.207$, and $\alpha_s(m_{GUT}) \approx 1/42$. The value for $s_w^2(M_Z)$ is too small compared with the experimental value in the $\overline{\text{MS}}$ scheme $[s_w^2(M_Z)]_{exp} = 0.23117 \pm 0.00016$ [PDG00], which corresponds to the solution of the RG equation (5.1.7). In other words, the consistency condition (5.1.11) is not fulfilled. We note that our analysis is only at the one-loop level and neglects threshold effects and corrections from higher orders. Although these effects do not change the predictions significantly, they lead to uncertainties in the determination of m_{GUT}. A complete analysis is presented in the left-hand side of Fig. 5.2 on page 739. This figure shows the running of the gauge couplings starting from their experimental values at the electroweak scale. It is clearly seen, that the three curves do not intersect at one point.

This analysis is independent of the GUT group G and relies only on the fact that G is directly broken to G_{SM} at m_{GUT} and that no light particles other than those of the SM exist. Consequently, this analysis shows that more degrees of freedom have to be present between M_Z and m_{GUT} if grand unification is realized.

On the other hand, m_{GUT} is only a few orders of magnitude away from the *Planck scale*, i.e. the scale where gravitational interactions between particles become of order one,

$$m_{\text{Planck}} = (G_N/\hbar c)^{-1/2} \sim 1.2 \times 10^{19}\,\text{GeV} \tag{5.1.12}$$

with the gravitational constant $G_N = 6.707 \times 10^{-39}\,\hbar c\,(\text{GeV}/c^2)^{-2}$. This fact suggests the possibility that grand unification might be related to the unification of gravity with the forces of elementary particle physics.

5.1.2 Proton decay

Because leptons and quarks are members of the same multiplet of the group G, gauge bosons that mediate transitions between leptons and quarks must

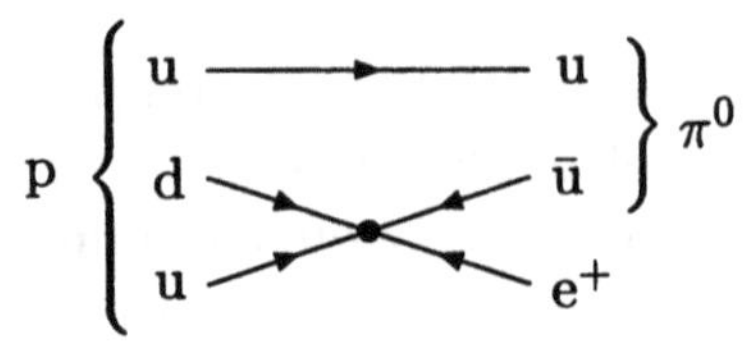

Fig. 5.1 Feynman diagram contributing to the proton decay $P \to e^+\pi^0$

exist. These lead to *baryon-number violation* and *proton decay.* Since these effects have not been observed so far, they must be very small, i.e. the gauge bosons that induce these transitions must be very heavy with masses of the order of m_{GUT}.

For low-energy processes, the interactions of the heavy gauge bosons can be described by effective *four-fermion interactions* [cf. (1.4.17)] [We79a, Wi79b] of the form [cf. (5.1.32), colour indices suppressed]

$$\mathcal{L}_{\mathrm{GUT}} = \frac{4G_{\mathrm{GUT}}}{\sqrt{2}} \left(\bar{u}^{\mathrm{c}}_{\mathrm{L}}\gamma^{\mu}u_{\mathrm{L}}\, \bar{e}^{\mathrm{c}}_{\mathrm{R}}\gamma_{\mu}d_{\mathrm{R}} + \ldots\right). \tag{5.1.13}$$

In the same way as in the transition from the GSW theory to the Fermi model, one obtains an effective Fermi coupling [cf. (1.4.18)]

$$\frac{G_{\mathrm{GUT}}}{\sqrt{2}} = \frac{g_{\mathrm{u}}^2}{8m_{\mathrm{GUT}}^2} = \frac{\pi\alpha(m_{\mathrm{GUT}}^2)}{2m_{\mathrm{GUT}}^2} \tag{5.1.14}$$

that determines the *proton lifetime*. A typical Feynman diagram for the decay $P \to e^+\pi^0$ is shown in Fig. 5.1. Analogously to the muon lifetime (4.6.7), the proton lifetime owing to the decay $P \to e^+\pi^0$ is given by

$$\tau_{P\to e^+\pi^0} \sim \frac{192\pi^3}{G_{\mathrm{GUT}}^2 m_{\mathrm{P}}^5} \sim \frac{m_{\mathrm{GUT}}^4}{[\alpha(m_{\mathrm{GUT}}^2)]^2 m_{\mathrm{P}}^5}. \tag{5.1.15}$$

Using the numbers obtained in the previous section leads to $\tau_{P\to e^+\pi^0} \sim 2 \times 10^{31}$ years, which is extremely large compared with the age of the universe ($\sim 10^{10}$ years) but already quite small compared with the experimental limit on the proton lifetime $\tau_{P\to e^+\pi^0} > 1.6 \times 10^{33}$ years [PDG00]. Note, however, that the theoretical value depends on the details of the GUT and, in particular, strongly on m_{GUT}.

The observation and the measurement of the proton decay are of utmost importance because this would provide a hint to a new energy range in which strong, weak, and electromagnetic interactions can be regarded as a universal interaction.

In addition, *baryon-number violation* is an essential ingredient for the explanation of the baryon asymmetry of the universe within the big-bang models [Sa67, Co93].

Violation of baryon number in proton decay is also theoretically satisfactory. Indeed, the replacement of global symmetries by local gauge invariance—possibly spontaneously broken—is one of the most essential aspects in the development of gauge theories. Although the simplest GUT based on the gauge group SU(5) still incorporates a global conservation law for the difference between baryon and lepton numbers, $B - L$, this is also replaced by a spontaneously broken local invariance when extended to SO(10) [Mo80].

5.1.3 Hierarchy (fine-tuning) problem

The appearance of the large scale m_{GUT} leads to a number of conceptual problems. The presence of two very different scales, the electroweak scale and the GUT scale is the origin of the so-called *hierarchy problem*

Consider a GUT extension of the Standard Model. It includes a Higgs sector that causes the spontaneous breaking of $\mathrm{SU}(2)_{\mathrm{W}} \times \mathrm{U}(1)_{\mathrm{Y}}$ and that is part of grand unification. In order to generate the observed masses of the weak gauge bosons, the Higgs field must have a negative mass-squared term, $-\mu^2$, of the order of $(100\,\mathrm{GeV})^2$. However, the mass of a scalar field receives additive renormalizations. In a theory with cut-off scale Λ, the corrections to μ^2 are of order Λ^2 and the renormalized mass can only be much smaller than Λ if the bare mass is fine-tuned in each order of perturbation theory to cancel the radiative corrections almost exactly. In GUTs this fine tuning has to be done to about 13 orders of magnitude, which looks rather unnatural. A possible solution of this problem is provided by supersymmetry (cf. Sect. 5.2).

5.1.4 SU(5) GUT

The choice of the gauge group G is restricted by the requirements listed at the beginning of Sect. 5.1. The smallest possible G has rank 4. Out of the nine possible rank-4 local Lie groups only two, namely $[\mathrm{SU}(3)]^2$ and SU(5), have complex representations. In $[\mathrm{SU}(3)]^2$ the generator of the electric charge does not allow for fractional charges and, being traceless, requires that the sum of all quark charges vanishes. Thus, SU(5) is the unique rank-4 candidate for a GUT [Ge74].

In the following we discuss grand unification in more detail using SU(5) as an example.

5.1.4.1 Gauge sector

The fundamental representation of SU(5) can be constructed in complete analogy to the fundamental representations of SU(2) and SU(3) in Sect. 1.2.2.2 resulting in $24 = \dim \mathrm{SU}(5)$ generalized 5×5 Gell-Mann matrices $\lambda^a_{(5)}$. We can choose the basis in such a way that the generators that involve non-zero entries only in the first three rows and columns can be associated with SU(3) and those that are non-vanishing only in the last two rows and columns with SU(2). Then, the remaining diagonal generator must correspond to hypercharge and the remaining twelve non-diagonal generators mix SU(3) and SU(2).

The local gauge invariance of SU(5) is mediated by 24 gauge fields which contain twelve additional gauge fields X_c, Y_c, $X_c^\dagger$, and $Y_c^\dagger$, $c = \mathrm{r, g, b}$, besides the eight gluons of QCD, G^a, and the four gauge bosons of the electroweak interaction, W^a and B. The gauge-field content of SU(5) can be summarized as

$$\mathcal{A}_\mu = g_\mathrm{u} A^a_\mu \frac{\lambda^a_{(5)}}{2} = g_\mathrm{u} \frac{1}{\sqrt{2}} \left(\begin{array}{ccc|cc} & & & X^\dagger_{\mathrm{r},\mu} & Y^\dagger_{\mathrm{r},\mu} \\ & \sqrt{2} G^a_\mu \frac{\lambda^a_{(3)}}{2} & & X^\dagger_{\mathrm{g},\mu} & Y^\dagger_{\mathrm{g},\mu} \\ & & & X^\dagger_{\mathrm{b},\mu} & Y^\dagger_{\mathrm{b},\mu} \\ \hline X_{\mathrm{r},\mu} & X_{\mathrm{g},\mu} & X_{\mathrm{b},\mu} & \multicolumn{2}{c}{\multirow{2}{*}{$\sqrt{2} W^a_\mu \frac{\sigma^a}{2}$}} \\ Y_{\mathrm{r},\mu} & Y_{\mathrm{g},\mu} & Y_{\mathrm{b},\mu} & & \end{array} \right)$$

$$- g_\mathrm{u} \frac{1}{2\sqrt{15}} \left(\begin{array}{ccc|cc} -2 & & & & \\ & -2 & & & 0 \\ & & -2 & & \\ \hline & & & +3 & \\ & 0 & & & +3 \end{array} \right) B_\mu . \qquad (5.1.16)$$

The coefficient $\lambda^{12}_{(5)}/2$ of $g_\mathrm{u} B_\mu$ is, up to normalization, equal to the hypercharge operator

$$\frac{\lambda^{12}_{(5)}}{2} = \sqrt{\frac{3}{5}} \frac{Y_\mathrm{w}}{2} . \qquad (5.1.17)$$

Consequently, we obtain

$$Y_{\mathrm{w}} = \frac{1}{3}\,\mathrm{diag}(-2,-2,-2,3,3) \tag{5.1.18}$$

and, via the *Gell-Mann–Nishijima relation* $Q = I^3_{\mathrm{w}} + Y_{\mathrm{w}}/2$, the charge operator

$$Q = \frac{1}{3}\,\mathrm{diag}(-1,-1,-1,3,0). \tag{5.1.19}$$

Using the definition of the adjoint representation [cf. (1.2.55)]

$$\left[\frac{\lambda^a_{(5)}}{2}, \mathcal{A}_\mu\right] = T^a_{\mathrm{adj}}\mathcal{A}_\mu, \tag{5.1.20}$$

one can derive the transformation properties of the gauge fields under the subgroups. The new gauge bosons X_c and Y_c transform as colour antitriplets and carry hypercharge 5/3. The pairs (X_c, Y_c) transform as three isospin doublets, and X_c and Y_c carry charge 4/3 and 1/3, respectively. Their antiparticles $X_c^\dagger$, and $Y_c^\dagger$ transform according to the complex-conjugate representations. The gauge bosons X_c and Y_c couple to leptons and quarks and are thus called *leptoquarks.* According to our analysis, the adjoint representation of SU(5) has the decomposition

$$\mathbf{24} = \underbrace{(\mathbf{8},\mathbf{1})_0}_{G^a_\mu} \oplus \underbrace{(\bar{\mathbf{3}},\mathbf{2})_{5/3}}_{X_c,Y_c} \oplus \underbrace{(\mathbf{3},\mathbf{2})_{-5/3}}_{X_c^\dagger,Y_c^\dagger} \oplus \underbrace{(\mathbf{1},\mathbf{3})_0}_{W^a_\mu} \oplus \underbrace{(\mathbf{1},\mathbf{1})_0}_{B_\mu} \tag{5.1.21}$$

under $\mathrm{SU}(3)_{\mathrm{C}} \times \mathrm{SU}(2)_{\mathrm{w}} \times \mathrm{U}(1)_{\mathrm{Y}}$, where the first entry in the parentheses indicates the SU(3) representation, the second entry the SU(2) representation, both labelled by their dimensions, and the index the weak hypercharge.

5.1.4.2 Fermion representation

In order to group the known fermions into SU(5) multiplets it is necessary to express all fermion fields in terms of left-handed fields only. This can be done by using the *charge conjugation matrix* $C = \mathrm{i}\gamma^2\gamma^0$, which relates the fields to their *charge conjugate fields*

$$\psi^{\mathrm{c}} = C\bar{\psi}^{\mathrm{T}}. \tag{5.1.22}$$

The properties of C are listed in (A.1.23) and (A.1.24). It can be easily seen that the charge conjugate of a left-handed field is right-handed and vice versa,

$$\begin{aligned}(\psi^{\mathrm{c}})_{\mathrm{L}} &= \frac{1}{2}(1-\gamma_5)\psi^{\mathrm{c}} = \frac{1}{2}(1-\gamma_5)C\bar{\psi}^{\mathrm{T}} \\ &= C\frac{1}{2}(1-\gamma_5^{\mathrm{T}})\bar{\psi}^{\mathrm{T}} = C\left[\bar{\psi}\frac{1}{2}(1-\gamma_5)\right]^{\mathrm{T}} = C(\bar{\psi}_{\mathrm{R}})^{\mathrm{T}}. \qquad (5.1.23)\end{aligned}$$

After replacing all right-handed fields by their charge-conjugate left-handed fields, we can enumerate the 15 fermion fields of the first generation (omitting right-handed neutrinos) in terms of their $\mathrm{SU}(3)_{\mathrm{C}} \times \mathrm{SU}(2)_{\mathrm{w}} \times \mathrm{U}(1)_{\mathrm{Y}}$ quantum numbers

$$\underbrace{(\mathbf{3},\mathbf{2})_{1/3}}_{(\mathrm{u_L},\mathrm{d_L})} \oplus \underbrace{(\bar{\mathbf{3}},\mathbf{1})_{2/3}}_{\mathrm{d_L^c}} \oplus \underbrace{(\bar{\mathbf{3}},\mathbf{1})_{-4/3}}_{\mathrm{u_L^c}} \oplus \underbrace{(\mathbf{1},\mathbf{2})_{-1}}_{(\nu_{\mathrm{L}},\mathrm{e_L})} \oplus \underbrace{(\mathbf{1},\mathbf{1})_{2}}_{\mathrm{e_L^c}} \qquad (5.1.24)$$

in the notation explained after (5.1.21).

Now we have to group these fields into representations of SU(5) (for the representation of SU(5) we refer to the literature [GT]). The 15-dimensional representation of SU(5) contains colour-sextet quarks and is thus not acceptable. Similarly, the 45-dimensional representation, which could possibly include all three fermion generations, contains colour octets and antisextets. On the other hand, the five-dimensional complex-conjugate fundamental representation and the 10-dimensional antisymmetric tensor representation have the decompositions

$$\begin{aligned}\bar{\mathbf{5}} &= (\bar{\mathbf{3}},\mathbf{1})_{2/3} \oplus (\mathbf{1},\mathbf{2})_{-1}, \\ \mathbf{10} &= (\mathbf{3},\mathbf{2})_{1/3} \oplus (\bar{\mathbf{3}},\mathbf{1})_{-4/3} \oplus (\mathbf{1},\mathbf{1})_{2}. \qquad (5.1.25)\end{aligned}$$

Comparison with (5.1.24) shows that the fermions of the first generation can be assigned to these multiplets as

$$\bar{\mathbf{5}} : \psi_{j,\mathrm{L}} = \begin{pmatrix} d_{\mathrm{r}}^{\mathrm{c}} \\ d_{\mathrm{g}}^{\mathrm{c}} \\ d_{\mathrm{b}}^{\mathrm{c}} \\ \hline e \\ -\nu_{\mathrm{e}} \end{pmatrix}_{\mathrm{L}} \qquad (5.1.26)$$

and

$$\mathbf{10}: \chi_{jk,\mathrm{L}} = \frac{1}{\sqrt{2}} \left(\begin{array}{ccc|cc} 0 & u_\mathrm{b}^\mathrm{c} & -u_\mathrm{g}^\mathrm{c} & -u_\mathrm{r} & -d_\mathrm{r} \\ -u_\mathrm{b}^\mathrm{c} & 0 & u_\mathrm{r}^\mathrm{c} & -u_\mathrm{g} & -d_\mathrm{g} \\ u_\mathrm{g}^\mathrm{c} & -u_\mathrm{r}^\mathrm{c} & 0 & -u_\mathrm{b} & -d_\mathrm{b} \\ \hline u_\mathrm{r} & u_\mathrm{g} & u_\mathrm{b} & 0 & -e^\mathrm{c} \\ d_\mathrm{r} & d_\mathrm{g} & d_\mathrm{b} & e^\mathrm{c} & 0 \end{array} \right)_\mathrm{L} . \tag{5.1.27}$$

The identification of $(e, -\nu_\mathrm{e})^\mathrm{T}$ as 2^* follows from the assignment of $(\nu_\mathrm{e}, e)^\mathrm{T}$ as 2 and the relation between these two SU(2) representations. The assignment of quantum numbers as given in (5.1.25) can be checked upon constructing the corresponding representation matrices from the fundamental representation via

$$\begin{aligned} (T^a\psi)^{(\bar{\mathbf{5}})} &= -\frac{\lambda^{*a}_{(5)}}{2}(\psi)^{(\bar{\mathbf{5}})}, \\ (T^a\chi)^{(\mathbf{10})} &= \frac{\lambda^{a}_{(5)}}{2}(\chi)^{(\mathbf{10})} + (\chi)^{(\mathbf{10})}\frac{\lambda^{\mathrm{T}a}_{(5)}}{2}. \end{aligned} \tag{5.1.28}$$

The other generations can be treated analogously if we suppress the complications related to the quark-mixing matrix.

The results of this classification are as follows:

- The multiplets (5.1.26) and (5.1.27) describe leptons and quarks on the same level. In this sense, SU(5) provides a realization of the idea of a lepton–quark symmetry.
- SU(5) symmetry yields a relation between lepton and quark charges and explains the fact that quarks carry third-integral charges. As a generator of SU(5) the charge operator (5.1.19) must be traceless. For the multiplet $\bar{\mathbf{5}}$ this yields $Q_\mathrm{e} + 3Q_{\mathrm{d}^\mathrm{c}} = 0$, i.e. $Q_\mathrm{d} = Q_\mathrm{e}/3$. Note that the factor 1/3 results from the three colours of the quarks.
- The fermion fields only fit into the SU(5) representations if the left-handed lepton doublet is accompanied by a right-handed SU(2) singlet of quarks. In this sense SU(5) predicts that right-handed quarks are SU(2) singlets.
- The representations $\bar{\mathbf{5}}$ and $\mathbf{10}$ have equal and opposite anomalies, and the SU(5) theory is anomaly-free.

- If right-handed neutrinos exist, they must be classified as SU(5) singlets.

The fermionic Lagrangian reads

$$\begin{aligned}\mathcal{L}_{\text{ferm}} &= \bar{\psi}\mathrm{i}\gamma_\mu D^\mu \psi + \mathrm{Tr}(\bar{\chi}\mathrm{i}\gamma_\mu D^\mu \chi) \\ &= \bar{\psi}_i \mathrm{i}\gamma_\mu (D^\mu \psi)_i + \bar{\chi}_{ji}\mathrm{i}\gamma_\mu (D^\mu \chi)_{ij} \\ &= \bar{\psi}_i \mathrm{i}\gamma_\mu (\delta_{ij}\partial^\mu + \mathrm{i}\mathcal{A}^{*\mu}_{ij})\psi_j + \bar{\chi}_{ji}\mathrm{i}\gamma_\mu (\delta_{ik}\partial^\mu - 2\mathrm{i}\mathcal{A}^\mu_{ik})\chi_{kj},\end{aligned} \tag{5.1.29}$$

where the covariant derivatives are given by

$$\begin{aligned}(D^\mu \psi)_i &= D^\mu_{ij}\psi_j = \partial^\mu \psi_i + \mathrm{i}\mathcal{A}^{*\mu}_{ij}\psi_j, \\ (D^\mu \chi)_{ij} &= \partial^\mu \chi_{ij} - \mathrm{i}(\mathcal{A}^\mu_{ik}\chi_{kj} + \mathcal{A}^\mu_{jk}\chi_{ik}).\end{aligned} \tag{5.1.30}$$

Upon inserting the explicit representations of the fermion fields (5.1.26) and (5.1.27) and of the gauge-boson fields (5.1.16) one obtains the usual fermionic interactions of the $SU(3)_C \times SU(2)_w \times U(1)_Y$ gauge theory plus the interactions of the new gauge bosons

$$\begin{aligned}\mathcal{L}_{\mathrm{I},XY} = &\frac{g_\mathrm{u}}{\sqrt{2}}\Big[X_i^{\dagger\mu}(\bar{d}_{i,\mathrm{R}}\gamma_\mu e^\mathrm{c}_\mathrm{R} + \bar{d}_{i,\mathrm{L}}\gamma_\mu e^\mathrm{c}_\mathrm{L} + \varepsilon_{ijk}\bar{u}^\mathrm{c}_{k,\mathrm{L}}\gamma_\mu u_{j,\mathrm{L}}) + \text{h.c.}\Big] \\ &- \frac{g_\mathrm{u}}{\sqrt{2}}\Big[Y_i^{\dagger\mu}(\bar{d}_{i,\mathrm{R}}\gamma_\mu \nu^\mathrm{c}_{\mathrm{e},\mathrm{R}} + \bar{u}_{i,\mathrm{L}}\gamma_\mu e^\mathrm{c}_\mathrm{L} + \varepsilon_{ijk}\bar{u}^\mathrm{c}_{j,\mathrm{L}}\gamma_\mu d_{k,\mathrm{L}}) + \text{h.c.}\Big].\end{aligned} \tag{5.1.31}$$

Note that the couplings of the X and Y bosons *violate baryon number B and lepton number L* but all of them conserve $B-L$ if the bosons X and Y carry $B - L = 2/3$.

At low energies, the Lagrangian (5.1.31) gives rise to an effective baryon-number-violating four-fermion interaction

$$\begin{aligned}\mathcal{L}_{\Delta B=1} = &\frac{g_\mathrm{u}^2}{2M_\mathrm{X}^2}\varepsilon_{ijk}\bar{u}^\mathrm{c}_{k,\mathrm{L}}\gamma^\mu u_{j,\mathrm{L}}\,(\bar{e}^\mathrm{c}_\mathrm{R}\gamma_\mu d_{i,\mathrm{R}} + \bar{e}^\mathrm{c}_\mathrm{L}\gamma_\mu d_{i,\mathrm{L}}) \\ &+ \frac{g_\mathrm{u}^2}{2M_\mathrm{Y}^2}\varepsilon_{ijk}\bar{u}^\mathrm{c}_{j,\mathrm{L}}\gamma^\mu d_{k,\mathrm{L}}\,(\bar{\nu}^\mathrm{c}_{\mathrm{e},\mathrm{R}}\gamma_\mu d_{i,\mathrm{R}} + \bar{e}^\mathrm{c}_\mathrm{L}\gamma_\mu u_{i,\mathrm{L}}) + \text{h.c.}.\end{aligned} \tag{5.1.32}$$

This Lagrangian leads, in particular, to proton decay. For example, the decay rate for $\mathrm{P} \to \mathrm{e}^+ + \pi^0$ is found to be (including leading-logarithmic QCD corrections) [La93]

$$\Gamma^{-1}(\mathrm{P} \to \mathrm{e}^+ + \pi^0) \approx 10^{31\pm 1}\,\text{years}, \tag{5.1.33}$$

which is too fast compared with the experimental limit [PDG00]

$$\Gamma^{-1}(\mathrm{P} \to \mathrm{e}^+ + \pi^0) > 1.6 \times 10^{33}\,\text{years}. \tag{5.1.34}$$

5.1.4.3 Spontaneous symmetry breaking

The non-observation of baryon-number violation requires the masses of the X and Y bosons to be very large. Therefore, symmetry breaking must occur in two steps

$$\mathrm{SU}(5) \underset{M_\mathrm{X}}{\rightarrow} \mathrm{SU}(3)_\mathrm{C} \times \mathrm{SU}(2)_\mathrm{w} \times \mathrm{U}(1)_\mathrm{Y} \underset{M_\mathrm{W}}{\rightarrow} \mathrm{SU}(3)_\mathrm{C} \times \mathrm{U}(1)_\mathrm{em}, \quad (5.1.35)$$

and at least two Higgs multiplets are required. The simplest choice is to use a scalar multiplet $\mathbf{\Sigma}$ transforming according to the 24-dimensional adjoint representation of SU(5)

$$\mathbf{\Sigma} = \frac{\lambda^a_{(5)}}{2} \Sigma^a \quad (5.1.36)$$

to break SU(5) down to the Standard-Model gauge group and a scalar quintuplet

$$\mathbf{H} = (H_\mathrm{r}, H_\mathrm{g}, H_\mathrm{b} \mid \phi^+, \phi^0)^\mathrm{T} \quad (5.1.37)$$

to break $\mathrm{SU}(2)_\mathrm{w} \times \mathrm{U}(1)_\mathrm{Y}$. The corresponding kinetic Lagrangian reads

$$\mathcal{L}_\mathrm{H,kin} = (D_\mu \mathbf{H})^\dagger (D^\mu \mathbf{H}) + \frac{1}{2} \mathrm{Tr}\{(D_\mu \mathbf{\Sigma})^\dagger (D^\mu \mathbf{\Sigma})\} \quad (5.1.38)$$

with

$$\begin{aligned} D^\mu \mathbf{H} &= \partial^\mu \mathbf{H} - \mathrm{i} \mathcal{A}^\mu \mathbf{H}, \\ D^\mu \mathbf{\Sigma} &= \partial^\mu \mathbf{\Sigma} - \mathrm{i}[\mathcal{A}^\mu, \mathbf{\Sigma}]. \end{aligned} \quad (5.1.39)$$

In order to break the SU(5) symmetry without affecting the SU(3) and SU(2) symmetry, we introduce a Higgs potential in such a way that the vacuum expectation value of $\mathbf{\Sigma}$ is given by

$$\langle \mathbf{\Sigma} \rangle = w \, \mathrm{diag}\left(1, 1, 1, -\frac{3}{2}, -\frac{3}{2}\right) \quad (5.1.40)$$

with w of the order of $m_{\rm GUT}$. Note that $\langle\boldsymbol{\Sigma}\rangle$ is proportional to the unit matrix in the SU(3) and SU(2) subspaces. The induced gauge-boson mass term reads

$$\begin{aligned}\mathcal{L}_{\rm mass,\boldsymbol{\Sigma}} &= \frac{1}{2}\,{\rm Tr}\{(D_\mu\langle\boldsymbol{\Sigma}\rangle)^\dagger(D^\mu\langle\boldsymbol{\Sigma}\rangle)\} = \frac{1}{2}\,{\rm Tr}\{[\mathcal{A}_\mu,\langle\boldsymbol{\Sigma}\rangle]^\dagger[\mathcal{A}^\mu,\langle\boldsymbol{\Sigma}\rangle]\}\\ &= \frac{25}{8}g_{\rm u}^2w^2(\mathbf{X}_\mu^\dagger\mathbf{X}^\mu+\mathbf{Y}_\mu^\dagger\mathbf{Y}^\mu),\end{aligned} \tag{5.1.41}$$

i.e. the X and Y bosons get masses

$$M_{\rm X} = M_{\rm Y} = \frac{5}{2\sqrt{2}}g_{\rm u}w, \tag{5.1.42}$$

and the other gauge bosons remain massless. The scalar multiplet $\boldsymbol{\Sigma}$ gives rise to 12 physical *Higgs bosons* corresponding to the representations $(\mathbf{8},\mathbf{1})_0$, $(\mathbf{1},\mathbf{3})_0$, and $(\mathbf{1},\mathbf{1})_0$ with masses of the order of w.

The potential for the field H is chosen to yield a vacuum expectation value of the form

$$\langle\mathbf{H}\rangle = \left(0,0,0\,\middle|\,0,\frac{v}{\sqrt{2}}\right)^{\rm T}, \tag{5.1.43}$$

where $v\approx 246\,$GeV [cf. (4.2.34)]. This leads to a gauge-boson mass term

$$\begin{aligned}\mathcal{L}_{\rm mass,H} &= (D_\mu\langle\mathbf{H}\rangle)^\dagger(D^\mu\langle\mathbf{H}\rangle)\\ &= \frac{1}{4}g_{\rm u}^2v^2\left[\mathbf{Y}_\mu^\dagger\mathbf{Y}^\mu+W_\mu^+W^{-\mu}+\frac{1}{2c_{\rm w}^2}Z_\mu Z^\mu\right],\end{aligned} \tag{5.1.44}$$

with ($c_{\rm w}^2=5/8$)

$$Z_\mu = c_{\rm w}\left(W_\mu^3+\sqrt{\frac{3}{5}}B_\mu\right) = c_{\rm w}W_\mu^3+s_{\rm w}B_\mu \tag{5.1.45}$$

and

$$M_{\rm W} = \frac{1}{2}g_{\rm u}v = c_{\rm w}M_{\rm Z}. \tag{5.1.46}$$

Out of the ten real fields of $\mathbf{H}$ only three are absorbed into the longitudinal modes of the massive weak vector bosons. In addition to the Standard-Model

Higgs boson, six additional physical particles remain which form two colour triplets with $Y_{\mathrm{w}} = -2/3$ and $Q = -1/3$. When choosing the Higgs potential appropriately, these triplet Higgs fields get masses of order w.

The fermion masses are generated via Yukawa couplings. Since there is no SU(5)-invariant Yukawa coupling involving a 24-plet, $\mathbf{\Sigma}$ does not couple to fermions and the fermion masses cannot be of the order of the GUT scale. On the other hand, Yukawa couplings involving $\mathbf{H}$ exist, e.g. for a single generation

$$\mathcal{L}_{\mathrm{Yuk}} = \left[\sqrt{2} g_d \bar{\psi}^{\mathrm{c}}_{i,\mathrm{R}} \chi_{ij,\mathrm{L}} H^{\dagger}_j + \frac{1}{2} g_u \varepsilon_{ijklm} \bar{\chi}^{\mathrm{c}}_{ij,\mathrm{R}} \chi_{kl,\mathrm{L}} H_m \right] + \mathrm{h.c.}. \tag{5.1.47}$$

Inserting the vacuum expectation value for $\mathbf{H}$ yields the fermion mass terms

$$\mathcal{L}_{\mathrm{mass,fer}} = -\frac{g_d v}{\sqrt{2}}(\bar{d}d + \bar{e}e) - 2\frac{g_u v}{\sqrt{2}}(\bar{u}u). \tag{5.1.48}$$

Thus, the fermions get masses of the order of the electroweak scale and, moreover, the masses of the electron and of the down quark are equal (at the unification scale). Similar equations hold for the other generations. After running the masses down to the electroweak scale, one obtains for the ratio of the bottom quark and the tau-lepton masses $m_{\mathrm{b}}/m_{\tau} \approx 3$ in fair agreement with the experimental value 2.4. The corresponding predictions for the light quark masses are worse and can only be explained by modifying the minimal SU(5) theory. Including other Higgs multiplets into the theory results in different mass relations.

5.1.4.4 Summary and critique of SU(5)

Grand unification based on SU(5) answers some of the questions listed at the beginning of this chapter. The strong, weak, and electromagnetic interactions are unified, the quantization of the electric charge and the fractional quark charges are explained, and the weak mixing angle is predicted. The other questions are not answered. The number of free parameters is even enlarged compared to the Standard Model, and the hierarchy problem becomes apparent. Most of these criticisms apply to all GUT theories.

A few weaknesses can be improved by extending the SU(5) symmetry to SO(10) [Fr75, Ge75]. One unattractive feature of the SU(5) classification

(5.1.26) and (5.1.27) is that particles of one generation are described by a reducible representation. In the SO(10) GUT, all fermions of a SM generation can be accommodated in a single irreducible 16-dimensional representation. In addition to the SM fermions a heavy neutral particle (a right-handed neutrino) is required in order to complete the 16-dimensional representation. The condition of anomaly freedom, which is satisfied for individual generations also finds a logical explanation in this model [Fr75].

We have seen that the predictions of the weak mixing angle and of the proton lifetime in minimal SU(5) disagree with experiment. A possible solution to change this predictions is to add more degrees of freedom [La93]. The failure of SU(5) cannot be resolved by adding additional families or light Higgs doublets. However, for example in SO(10) GUTs the contradictions can be avoided at the cost of predictive power. Probably, a more appealing approach is to extend the SU(5) GUT by requiring supersymmetry (cf. Sect. 5.2).

5.2 Supersymmetry

Supersymmetry [SUSY] is a symmetry that relates fermions and bosons. It is the only possible non-trivial extension of the Poincaré symmetry. Local supersymmetry includes naturally the algebra of general relativity. Moreover, supersymmetry provides a possible solution of the hierarchy problem of GUTs.

5.2.1 Supersymmetry algebra

In ordinary gauge theories (including GUTs) space–time symmetries and internal symmetries are independent of each other, i.e. the Lorentz algebra and the algebra of the inner symmetries decouple. It was proven by Coleman and Mandula [Co67] that this is necessarily the case in any four-dimensional quantum field theory with non-zero scattering amplitudes. The *Coleman–Mandula theorem* can be evaded by introducing fermionic symmetry generators, Q_α and $\bar{Q}_\alpha$ that obey anticommutation rather than commutation relations. These generators are Lorentz spinors and carry spin 1/2. Their anticommutator can be a spin-1 vector and is in fact given by the momentum operator,

$$\{Q_\alpha, \bar{Q}_\beta\} = 2\gamma^\mu_{\alpha\beta} P_\mu. \tag{5.2.1}$$

Together with the (anti-)commutation relations

$$\begin{aligned}
\{Q_\alpha, Q_\beta\} &= \{\bar{Q}_\alpha, \bar{Q}_\beta\} = 0, \\
[Q_\alpha, P^\mu] &= [\bar{Q}_\alpha, P^\mu] = 0, \qquad [P_\mu, P_\nu] = 0,
\end{aligned} \tag{5.2.2}$$

this is the simplest possible *supersymmetry algebra*. Because of the presence of anticommutators it is called a *graded algebra*. Since the algebra (5.2.1) and (5.2.2) relates the supersymmetry generators to those of the Poincaré group, a local supersymmetry implies a local Poincaré symmetry and thus gravity.

It is possible to build a finite set of different supersymmetries involving N spinorial charges ($N \leq 8$), the so-called N-extended supersymmetries. One can proof that these, together with the usual space–time and internal symmetries are the only possible symmetries of a physically reasonable theory [Ha75].

5.2.2 The chiral multiplet and the Wess–Zumino model

Since the generators Q_α and $\bar{Q}_\alpha$ carry spin, the representations of the supersymmetry algebra, the supermultiplets, must contain components with different spin. The simplest representation is the *chiral multiplet* consisting of four real scalar fields A, B, F, G and one *Majorana spinor*[2] ψ. Note that the number of fermionic degrees of freedom equals the number of bosonic degrees of freedom as required by supersymmetry. The *supersymmetry transformations* read

$$\begin{aligned}
\delta A &= [\mathrm{i}\bar{\varepsilon}Q, A] = \bar{\varepsilon}\psi, \\
\delta B &= [\mathrm{i}\bar{\varepsilon}Q, B] = -\mathrm{i}\bar{\varepsilon}\gamma_5\psi, \\
\delta \psi &= [\mathrm{i}\bar{\varepsilon}Q, \psi] = -\mathrm{i}\partial_\mu(A - \mathrm{i}\gamma_5 B)\gamma^\mu\varepsilon + (F + \mathrm{i}\gamma_5 G)\varepsilon, \\
\delta F &= [\mathrm{i}\bar{\varepsilon}Q, F] = -\mathrm{i}\bar{\varepsilon}\gamma^\mu\partial_\mu\psi, \\
\delta G &= [\mathrm{i}\bar{\varepsilon}Q, G] = -\bar{\varepsilon}\gamma^\mu\gamma_5\partial_\mu\psi,
\end{aligned} \tag{5.2.3}$$

where ε is a constant Grassmann Majorana spinor. It can be checked that repeated supersymmetry variations close linearly among themselves.

[2] A Majorana spinor is a spinor that equals its charge-conjugate, $\psi = \psi^c$ and thus involves 4 real degrees of freedom

The Lagrangian of the *Wess–Zumino model* [We74]

$$\begin{aligned}\mathcal{L} = &\frac{1}{2}(\partial_\mu A)(\partial^\mu A) + \frac{1}{2}(\partial_\mu B)(\partial^\mu B) + \frac{\mathrm{i}}{2}\bar{\psi}\partial\!\!\!/\psi + \frac{1}{2}F^2 + \frac{1}{2}G^2 \\ &+ m\left(FA - GB - \frac{1}{2}\bar{\psi}\psi\right) \\ &+ g[FA^2 - FB^2 - 2GAB - \bar{\psi}(A - \mathrm{i}\gamma_5 B)\psi]. \end{aligned} \tag{5.2.4}$$

is invariant under the transformations (5.2.3). The auxiliary fields F and G can be eliminated from the action by using their algebraic equations of motion. The resulting action does not have equal numbers of fermions and bosons but is still invariant under non-linear supersymmetry transformations. The corresponding algebra does not close off-shell but only upon the use of the equations of motion. Therefore, it is often more convenient to retain the auxiliary fields in order to maintain the complete off-shell linear algebra.

It is instructive to count the on-shell degrees of freedom of the action (5.2.4) to confirm the equality of the bosonic and fermionic degrees of freedom. Indeed, we find two bosonic degrees of freedom, A and B, and two fermionic degrees of freedom, the two on-shell components of the Majorana spinor ψ.

Note that the action (5.2.4) involves only one coupling and one mass. In general, all members of one supersymmetry multiplet have the same mass and the same coupling strength. Starting from the Lagrangian (5.2.4) perturbative calculations can be performed as usual.

The transformations (5.2.3) and the action (5.2.4) look rather weird. They become straightforward, when using the superspace formalism [Sa74].

5.2.3 Improved ultraviolet properties of supersymmetric theories

Supersymmetry links fermionic and bosonic degrees of freedom. This has some important consequences. First of all, supersymmetric theories exhibit an improved ultraviolet (UV) behaviour compared with non-supersymmetric theories. Because fermion and boson loops appear with opposite signs, the UV singularities cancel to a large extent. In particular, quadratic divergences are absent in supersymmetric quantum field theories [Ka82, Ni84]. Vice versa, the absence of quadratic divergences in an unrenormalized quantum field theory to all orders of perturbation theory implies that the theory is supersymmetric [In82, De83]. In the exact supersymmetric limit, the parameters of the superpotential, which describes the interactions between scalars and fermions [lines 2 and 3 in (5.2.4)], do not receive any (finite or infinite) corrections and thus need not to be renormalized [Gr79] (*non-renormalization theorem*). For instance, in the Wess–Zumino model neither mass nor coupling-constant renormalization is required, and all (logarithmic) divergences can be absorbed in *one* field renormalization constant. In general supersymmetric gauge theories, besides field renormalization, only a renormalization of the gauge coupling is needed [Ju83]. Theories with N-extended supersymmetry have even better UV properties. In $N = 2$ supersymmetric theories only the gauge coupling needs to be renormalized, and beyond one loop no divergences show up [Gr82]. By choosing the matter multiplets appropriately the $N = 2$ supersymmetric theories can be made completely finite to all orders of perturbation theory [Ho83, Ho83a, We85]. This is automatically ensured for the $N = 4$ supersymmetric gauge theory [So81, Br83, Ma83, Ho83a].

The non-renormalization of the superpotential provides a solution to the *hierarchy problem*. In the limit of exact supersymmetry there are no radiative corrections to any mass parameter, and the Higgs-boson mass can be fixed once and for all at tree level. Because so far no supersymmetric particle has been experimentally observed, supersymmetry must be broken. The supersymmetry breaking should introduce a splitting of the masses within supermultiplets but should not destroy important consequences of supersymmetry, like the cancellation of the quadratic divergences. In order to solve the hierarchy problem, one must ensure that the supersymmetry breaking does not spoil the non-renormalization theorem in a way that the Higgs-boson mass gets large radiative corrections. This can be achieved by allowing only soft supersymmetry-breaking terms, i.e. mass terms or local interactions with coupling of positive mass dimension (dimension-three and dimension-two operators) [Gi82a]. These terms induce only finite corrections to the parameters of the superpotential proportional to the mass parameters in the soft supersymmetry-breaking terms. Thus, in order to stabilize the weak scale, these masses should be below 1 TeV. As a consequence, the superpartners of the known particles should be within reach of the ongoing and proposed collider experiments.

5.2.4 The Minimal Supersymmetric Standard Model (MSSM)

Supersymmetry implies that for each fermionic degree of freedom a bosonic degree of freedom exists and vice versa. Consequently, in order to introduce supersymmetry in the Standard Model (SM) one has to add new degrees of freedom. It turns out, that none of the SM bosons can be a supersymmetric partner of one of the SM fermions, because of the differing quantum numbers. Therefore, one has to introduce a superpartner for each degree of freedom of the SM. The nomenclature is as follows: The fermionic partners of the bosons receive the name of the boson with the suffix *-ino*, and the bosonic partners of the SM fermions receive the name of the fermions with a prefix *s*. The scalar partners of the quarks and leptons are called *squarks* and *sleptons*, respectively, the fermionic partners of the gluons and electroweak gauge bosons (photon, W boson and Z boson) are the *gluinos* and *gauginos* (photino, wino and zino), and the fermionic partners of Higgs bosons are the *Higgsinos*. Moreover, one has to enlarge the Higgs sector of the SM by including a further Higgs doublet. This is necessary in order to ensure anomaly cancellation (the superpartners of the Higgs boson contribute to the anomaly) and to provide masses for all fermions without violating supersymmetry (supersymmetry does not allow for Yukawa couplings with both a Higgs doublet and its charge conjugate). In this way one arrives at the *Minimal Supersymmetric Standard Model* (MSSM) [Fa77, Ni84, Ha85]. The gauginos and Higgsinos mix and form neutral and charged mass eigenstates that are called *neutralinos* and *charginos*, respectively. There are 4 neutralinos which are represented by Majorana-spinor fields and 2 charginos represented by Dirac-spinor fields. This leads to the particle spectrum shown in Table 5.1.

The MSSM conserves *R parity*. This is a multiplicative quantum number, which is related to baryon number B, lepton number L, and spin s by

$$R = (-1)^{3(B-L)+2s}. \tag{5.2.5}$$

The ordinary particles have $R = 1$ and their superpartners have $R = -1$. The conservation of R parity implies that supersymmetric particles can only be produced in pairs and that the lightest supersymmetric particle (LSP) is stable.

From the calculational point of view, the MSSM is just an ordinary gauge theory with couplings restricted by supersymmetry. It can be evaluated in

standard particles		s	super partners		s
quarks	q	$\frac{1}{2}$	squarks	$\tilde{q}_1$, $\tilde{q}_2$	0
leptons	l, ν	$\frac{1}{2}$	sleptons	$\tilde{l}_1, \tilde{l}_2, \tilde{\nu}_1, , \tilde{\nu}_2$	0
gluons	g	1	gluinos	$\tilde{\mathrm{g}}$	$\frac{1}{2}$
gauge bosons	$\mathrm{W}^{\pm}$, Z, γ	1	neutralinos	$\tilde{\chi}_1^0$, $\tilde{\chi}_2^0$, $\tilde{\chi}_3^0$, $\tilde{\chi}_4^0$	$\frac{1}{2}$
Higgs bosons	h^0, H^0, A^0, $\mathrm{H}^{\pm}$	0	charginos	$\tilde{\chi}_1^{\pm}$, $\tilde{\chi}_2^{\pm}$	$\frac{1}{2}$

Tab. 5.1 Physical particle content of the MSSM (s denotes the spin of the particles.)

perturbation theory as usual. However, the usual regularization schemes like dimensional regularization violate supersymmetry. Therefore, extra finite counterterms have to be introduced to restore the SUSY Ward identities in higher orders.

One of the most important predictions of the MSSM is the fact that an upper limit on the mass of the lightest *Higgs boson* h^0 exists. In lowest order this limit is given by $M_{\mathrm{h}^0} < M_{\mathrm{Z}}$. Owing to large radiative corrections, the actual limit is $M_{\mathrm{h}^0} \lesssim 135\,\mathrm{GeV}$ (for a recent analysis, see [He99]). This prediction of the MSSM can be tested experimentally at the electron–positron and proton–proton colliders of the next generation.

Appropriate choices of the soft supersymmetry-breaking parameters allow to give masses to the supersymmetric particles that are high enough to evade experimental observation so far. However, in order to avoid the *hierarchy problem*, the masses of the supersymmetric particles should be of the order of the electroweak scale $v \approx 246\,\mathrm{GeV}$, i.e. not much larger than 1 TeV.

Another very interesting property of the MSSM is the possibility to introduce the electroweak symmetry breaking by radiative corrections. To this end, one incorporates universal scalar masses at the GUT scale in the supergravity inspired realization of the MSSM. When running these masses down to the electroweak scale, the mass squared of the Higgs boson that couples to the top quark gets large corrections owing to the large top-quark mass. In this way it can get negative at the electroweak scale and thus lead to spontaneous breaking of the electroweak symmetry.

The general MSSM involves more than one hundred free parameters, mainly masses, mixing angles, and phases. Moreover, many of these parameters have to be tuned in order to avoid conflict with experimental results, e.g.

on flavour-changing neutral currents and CP violation. The number of parameters can be reduced by using the concept of grand unification. In the GUT-related formulation, five parameters are sufficient to specify the supersymmetric particle sector.

5.2.5 Supersymmetric grand unification

In the MSSM additional fermions and scalars are present, and the coefficients (5.1.8) that determine the running of the gauge couplings change. Instead of (5.1.9) one obtains in the MSSM ($N_{\mathrm{g}} = 3$, $N_{\mathrm{Hd}} = 2$)

$$\begin{aligned} \beta_{0,\mathrm{s}} &= 9 - 2N_{\mathrm{g}} = 3, \\ \beta_{0,2} &= 6 - 2N_{\mathrm{g}} - \frac{1}{2}N_{\mathrm{Hd}} = -1, \\ \beta_{0,1} &= 0 - \frac{10}{3}N_{\mathrm{g}} - \frac{1}{2}N_{\mathrm{Hd}} = -11. \end{aligned} \tag{5.2.6}$$

Inserting this into (5.1.7) and (5.1.10) yields for the same set of input parameters as given after (5.1.11): $m_{\mathrm{GUT}} \approx 2 \times 10^{16}\,\mathrm{GeV}$, $s_{\mathrm{w}}^2(M_{\mathrm{Z}}) \approx 0.230$, and $\alpha_{\mathrm{s}}(m_{\mathrm{GUT}}) \approx 1/24$. The resulting value for $s_{\mathrm{w}}^2(M_{\mathrm{Z}})$ is in good agreement with the experimental $\overline{\mathrm{MS}}$ value $(s_{\mathrm{w}}^2(M_{\mathrm{Z}}))_{\mathrm{exp}} = 0.23117 \pm 0.00016$ [PDG00]. In other words, the consistency condition (5.1.11), which reads here

$$s_{\mathrm{w}}^2(\mu) = \frac{1}{5} + \frac{\alpha(\mu)}{\alpha_{\mathrm{s}}(\mu)}\frac{7}{15}, \tag{5.2.7}$$

is fulfilled in good approximation. The unification of couplings is illustrated in Fig. 5.2. This figure shows the running of the gauge couplings starting from their experimental values at the electroweak scale. It is clearly seen, that the three gauge couplings do not unify in the SM case but they do unify in the MSSM.

Moreover, the unification scale is about 30 times higher than in the non-supersymmetric case, which enhances the proton lifetime so that is not in conflict with the present experimental limits. On the other hand, the prediction for the ratio m_{b}/m_{τ} is changed by less than 10%.

In SU(5) grand unification the fermion and gauge-boson fields (and their superpartners) are assigned to the same representations $\bar{\mathbf{5}}$ and $\mathbf{10}$ as in the conventional formalism. Since we have one more Higgs doublet in the MSSM we need one more five-dimensional representation which is chosen as $\bar{\mathbf{5}}$ in order to ensure anomaly cancellation.

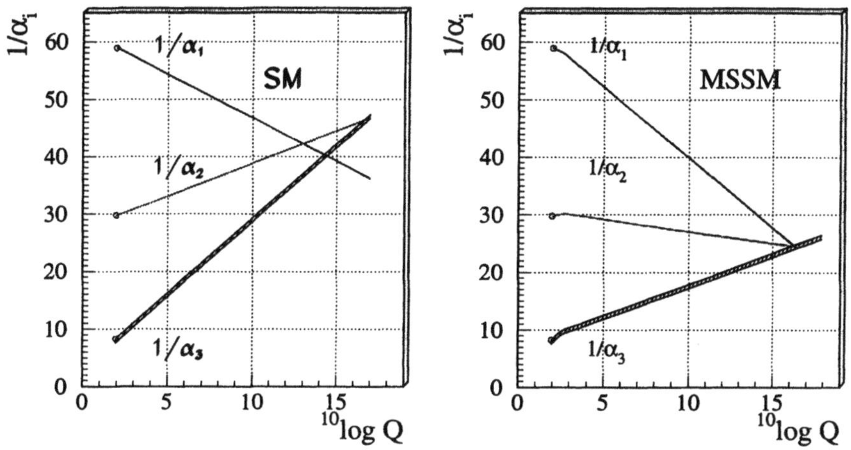

Fig. 5.2 Unification of couplings in the SM and the MSSM [dB00] (This plot is an update of Fig. 4.1 of [dB94])

With this short review of unified theories of the fundamental interactions and supersymmetry we conclude our discussion of gauge theories. It shows once again how broad the span of theoretical considerations on quantized gauge theories actually is. It results in a universal conception of the structure of matter and its interactions which includes today beyond the standard model grand unified and supersymmetric theories. All experiments, among them many with very high precision, from low energies up to the region of several hundred GeV are in excellent agreement with the standard model. Of course there are still many open problems. The most evident one is the huge number of parameters of the standard model and its extensions. Most of them are related to the mass matrices of the particles. A fundamental problem is the unification of the electroweak and strong interaction with gravity and its physical consequences in high energy physics and cosmology.

One of the most exciting aspects of present day physics is to further theoretically investigate these and other ideas, such as string theory, M theory, etc., and most important to look for new physics in order to prove or disprove these ideas by experiments.

References to Chapter 5

[Br83] L. Brink, O. Lindgren and B.E.W. Nilsson, *Nucl. Phys.* **B212** (1983) 401.

[Bu78] A.J. Buras, J. Ellis, M.K. Gaillard and D.V. Nanopoulos, *Nucl. Phys.* **B135** (1978) 66.

[Cl98a] B.T. Cleveland et al., *Astrophys. J* **496** (1998) 505;
Y. Fukada et al. (Kamiokande Coll.), *Phys. Rev. Lett.* **77** (1996) 1683;
W. Hampel et al. (GALLEX Coll.), *Phys. Lett.* **B447** (1999) 127;
J.N. Abdurashitov et al. (SAGE Coll.), *Phys. Rev.* **C60** (1999) 055801;
Y. Fukada et al. (Super-Kamiokande Coll.), *Phys. Rev. Lett.* **82** (1999) 2430.

[Co67] S. Coleman and J. Mandula, *Phys. Rev.* **159** (1967) 1251.

[Co93] A.G. Cohen, D.B. Kaplan and A.E. Nelson, *Annu. Rev. Nucl. Part. Sci.* **43** (1993) 27.

[dB94] W. de Boer, *Prog. Part. Nucl. Phys.* **33** (1994) 201.

[dB00] W. de Boer, private communication.

[De83] N.G. Deshpande, R.J. Johnson and E. Ma, *Phys. Lett.* **130B** (1983) 61.

[Fa77] P. Fayet and S. Ferrara, *Phys. Rept.* **C32** (1977) 249.

[Fr75] H. Fritzsch and P. Minkowski, *Ann. Phys.* **93** (1975) 193.

[Fu98] Y. Fukada et al. (Super-Kamiokande Coll.), *Phys. Rev. Lett.* **81** (1998) 1562.

[Ge74] H. Georgi and S.L. Glashow, *Phys. Rev. Lett.* **32** (1974) 438.

[Ge74a] H. Georgi, H.R. Quinn and S. Weinberg, *Phys. Rev. Lett.* **33** (1974) 451.

[Ge75] H. Georgi, *The state of the Art—Gauge Theories* in *Particles and Fields – 1974* AIP Conf. Proc. 23, ed. C.E. Carlson (New York, 1975) p. 575.

[Gi82a] L. Girardello and M.T. Grisaru, *Nucl. Phys.* **B194** (1982) 65.

[Gr79] M.T. Grisaru, M. Roček and W. Siegel, *Nucl. Phys.* **B159** (1979) 429.

[Gr82] M.T. Grisaru and W. Siegel, *Nucl. Phys.* **B201** (1982) 292.

[Ha75] R. Haag, J.T. Łopuszánski and M. Sohnius, *Nucl. Phys.* **B88** (1975) 257.

[Ha85] H.E. Haber and G.L. Kane, *Phys. Rept.* **117** (1985) 75.

[He99] S. Heinemeyer, W. Hollik and G. Weiglein, *Eur. Phys. J.* **C9** (1999) 343.

[Ho83] P.S. Howe, K.S. Stelle and P.C. West, *Phys. Lett.* **124B** (1983) 55.

[Ho83a] P.S. Howe, K.S. Stelle and P.K. Townsend, *Nucl. Phys.* **B214** (1983) 519 and **B236** (1984) 125.

[In82] T. Inami, N. Nishino and S. Watamura, *Phys. Lett.* **117B** (1982) 197.

[Ju83] J.W. Juer and D. Storey, *Nucl. Phys.* **B216** (1983) 185.

[Ka82] R.K. Kaul and P. Majumdar, *Nucl. Phys.* **B199** (1982) 36.

[La93] P. Langacker and N. Polonsky, *Phys. Rev.* **D47** (1993) 4028.

[Ma83] S. Mandelstam, *Nucl. Phys.* **B213** (1983) 149.

[Ma00a] W.A. Mann, *Athmospheric Neutrinos and the Oscillation Bonanza* in *Lepton and Photon Interactions at High Energies*, Proceedings of the XIX International Symposion on Lepton and Photon Interactions at High Energies, eds. J. Jaros and M. Peskin (Singapore, 2000) p. 229.

[Mo80] R.N. Mohapatra and R.E. Marshak, *Phys. Rev. Lett.* **44** (1980) 1316.

[Ni84] H.P. Nilles, *Phys. Rept.* **110** (1984) 1.

[PDG00] D.E. Groom et al. (Particle Data Group), *Eur. Phys. J.* **C15** (2000) 1.

[Sa67] A.D. Sakharov, *JETP Lett.* **5** (1967) 24.

[Sa74] A. Salam and J. Strathdee, *Phys. Lett.* **51B** (1974) 353.

[So81] M.F. Sohnius and P.C. West, *Phys. Lett.* **100B** (1981) 245.

[We60] S. Weinberg, *Phys. Rev.* **118** (1960) 838.

[We79a] S. Weinberg, *Phys. Rev. Lett.* **43** (1979) 1566.

[We74] J. Wess and B. Zumino, *Nucl. Phys.* **B70** (1974) 39.

[We85] P.C. West, *Supersymmetry and Finiteness* in *Shelter Island II, Proc. of the 1983 Shelter Island Conference on Quantum Field Theory and the Fundamental Problems of Physics*, eds. R. Jackiw et al. (Cambridge, Massachusetts, London, 1985) p. 127.

[Wi79b] F. Wilczek and A. Zee, *Phys. Rev. Lett.* **43** (1979) 1571.

A Appendix

A.1 Conventions

If not stated otherwise, we use the natural system of units, $c = 1$, $\hbar = 1$.

A.1.1 Minkowski space

A.1.1.1 Lorentz vectors and tensors

We use the *metric tensor*

$$g_{\mu\nu} = g^{\mu\nu} = \mathrm{diag}(1, -1, -1, -1). \tag{A.1.1}$$

Contravariant (V^μ) and covariant (V_μ) *Lorentz vectors* are formed as

$$V^\mu = (V^0, \mathbf{V}), \qquad V_\mu = (V^0, -\mathbf{V}), \tag{A.1.2}$$

where $\mathbf{V} = (V^1, V^2, V^3) = (V_x, V_y, V_z)$ denotes the usual vectors in space, in particular

$$x^\mu = (x^0, \mathbf{x}) = (t, x, y, z), \qquad x_\mu = (x^0, -\mathbf{x}) = (t, -x, -y, -z). \tag{A.1.3}$$

The *summation convention* is understood for repeated Lorentz (greek) or space (latin) indices unless explicitly stated otherwise,

$$\begin{aligned} VW &= V_\mu W^\mu = V^\mu W_\mu = g_{\mu\nu} V^\mu W^\nu = g^{\mu\nu} V_\mu W_\nu \\ &= V^0 W^0 - \mathbf{V}\mathbf{W} = V^0 W^0 - V^i W^i. \end{aligned} \tag{A.1.4}$$

Space–time derivatives are abbreviated as

$$\partial_\mu = \frac{\partial}{\partial x^\mu} = \left(\frac{\partial}{\partial t}, \nabla\right), \qquad \partial^\mu = \frac{\partial}{\partial x_\mu} = \left(\frac{\partial}{\partial t}, -\nabla\right) \tag{A.1.5}$$

with

$$\nabla = \left(\frac{\partial}{\partial x}, \frac{\partial}{\partial y}, \frac{\partial}{\partial z}\right) = (\partial_1, \partial_2, \partial_3) = (-\partial^1, -\partial^2, -\partial^3). \tag{A.1.6}$$

The four-momentum operator reads

$$p^\mu = \mathrm{i}\partial^\mu = (\mathrm{i}\partial^0, -\mathrm{i}\nabla), \tag{A.1.7}$$

and the d'Alembert operator is defined by

$$\Box = \partial^2 = \partial^\mu \partial_\mu = \partial_o^2 - \nabla^2. \tag{A.1.8}$$

The totally antisymmetric *Levi–Civita tensor* is defined by

$$\begin{aligned} \varepsilon^{\mu\nu\rho\sigma} &= -\varepsilon_{\mu\nu\rho\sigma} \\ &= \begin{cases} +1 & \text{if } \{\mu,\nu,\rho,\sigma\} \text{ is an even permutation of } \{0,1,2,3\} \\ -1 & \text{if } \{\mu,\nu,\rho,\sigma\} \text{ is an odd permutation of } \{0,1,2,3\} \\ 0 & \text{otherwise} \end{cases} \end{aligned} \tag{A.1.9}$$

and obeys the identities

$$\begin{aligned} \varepsilon^{\mu\nu\rho\sigma}\varepsilon^{\mu'\nu'\rho'\sigma'} &= -\begin{vmatrix} g^{\mu\mu'} & g^{\mu\nu'} & g^{\mu\rho'} & g^{\mu\sigma'} \\ g^{\nu\mu'} & g^{\nu\nu'} & g^{\nu\rho'} & g^{\nu\sigma'} \\ g^{\rho\mu'} & g^{\rho\nu'} & g^{\rho\rho'} & g^{\rho\sigma'} \\ g^{\sigma\mu'} & g^{\sigma\nu'} & g^{\sigma\rho'} & g^{\sigma\sigma'} \end{vmatrix}, \\ \varepsilon^{\mu\nu\rho\sigma}\varepsilon_{\mu}{}^{\nu'\rho'\sigma'} &= -\begin{vmatrix} g^{\nu\nu'} & g^{\nu\rho'} & g^{\nu\sigma'} \\ g^{\rho\nu'} & g^{\rho\rho'} & g^{\rho\sigma'} \\ g^{\sigma\nu'} & g^{\sigma\rho'} & g^{\sigma\sigma'} \end{vmatrix}, \\ \varepsilon^{\mu\nu\rho\sigma}\varepsilon_{\mu\nu}{}^{\rho'\sigma'} &= -2(g^{\rho\rho'}g^{\sigma\sigma'} - g^{\rho\sigma'}g^{\sigma\rho'}), \\ \varepsilon^{\mu\nu\rho\sigma}\varepsilon_{\mu\nu\rho}{}^{\sigma'} &= -6g^{\sigma\sigma'}, \qquad \varepsilon^{\mu\nu\rho\sigma}\varepsilon_{\mu\nu\rho\sigma} = -24. \end{aligned} \tag{A.1.10}$$

A.1.1.2 Pauli matrices

The Pauli matrices τ^a, $a = \{1,2,3\}$ (sometimes also denoted σ^a) read

$$\tau^1 = \begin{pmatrix} 0 & 1 \\ 1 & 0 \end{pmatrix}, \qquad \tau^2 = \begin{pmatrix} 0 & -\mathrm{i} \\ \mathrm{i} & 0 \end{pmatrix}, \qquad \tau^3 = \begin{pmatrix} 1 & 0 \\ 0 & -1 \end{pmatrix}. \tag{A.1.11}$$

They have the properties

$$\tau^a \tau^b = \delta^{ab} \mathbf{1} + \mathrm{i}\varepsilon^{abc}\tau^c, \qquad \sum_a \tau^a_{ij}\tau^a_{kl} = 2\delta_{il}\delta_{jk} - \delta_{ij}\delta_{kl}. \tag{A.1.12}$$

Here ε^{abc} is the totally antisymmetric *Levi–Civita tensor* in three dimensions ($\varepsilon^{123} = 1$), which obeys the identities

$$\varepsilon^{abc}\varepsilon^{a'b'c'} = \begin{vmatrix} \delta^{aa'} & \delta^{ab'} & \delta^{ac'} \\ \delta^{ba'} & \delta^{bb'} & \delta^{bc'} \\ \delta^{ca'} & \delta^{cb'} & \delta^{cc'} \end{vmatrix}, \qquad \varepsilon^{abc}\varepsilon^{ab'c'} = \begin{vmatrix} \delta^{bb'} & \delta^{bc'} \\ \delta^{cb'} & \delta^{cc'} \end{vmatrix},$$
$$\varepsilon^{abc}\varepsilon^{abc'} = 2\delta^{cc'}, \qquad \varepsilon^{abc}\varepsilon^{abc} = 6. \tag{A.1.13}$$

With (A.1.12) the following trace formulas can be derived

$$\mathrm{Tr}\{\tau^a\tau^b\} = 2\delta^{ab}, \qquad \mathrm{Tr}\{\tau^a\tau^b\tau^c\} = 2\mathrm{i}\varepsilon^{abc}. \tag{A.1.14}$$

A.1.1.3 Four-dimensional Dirac matrices

The Dirac matrices γ^μ are 4×4 matrices defined by the *Dirac algebra*

$$[\gamma^\mu, \gamma^\nu]_+ = 2g^{\mu\nu} \cdot \mathbf{1} \tag{A.1.15}$$

with γ^0 hermitian and γ^i antihermitian. The contraction of a Dirac matrix with a four-vector is conventionally abbreviated as

$$\gamma^\mu a_\mu = \not{a} \qquad \textit{(Feynman dagger)}. \tag{A.1.16}$$

The matrix

$$\gamma_5 = \gamma^5 = -\mathrm{i}\gamma_0\gamma_1\gamma_2\gamma_3 = \mathrm{i}\gamma^0\gamma^1\gamma^2\gamma^3 = -\frac{\mathrm{i}}{4!}\varepsilon_{\mu\nu\rho\sigma}\gamma^\mu\gamma^\nu\gamma^\rho\gamma^\sigma \tag{A.1.17}$$

has the properties

$$[\gamma_5, \gamma^\mu]_+ = 0, \qquad \gamma_5^2 = \mathbf{1}. \tag{A.1.18}$$

The commutator of two Dirac matrices

$$\sigma^{\mu\nu} = \frac{\mathrm{i}}{2}[\gamma^\mu, \gamma^\nu]_- = -\sigma^{\nu\mu} \tag{A.1.19}$$

satisfies

$$\gamma^\mu \gamma^\nu = g^{\mu\nu} - \mathrm{i}\sigma^{\mu\nu}, \qquad [\gamma_5, \sigma^{\mu\nu}]_- = 0, \qquad \gamma_5 \sigma^{\mu\nu} = \frac{\mathrm{i}}{2}\varepsilon^{\mu\nu\rho\sigma}\sigma_{\rho\sigma}. \tag{A.1.20}$$

The *Dirac spin matrix* is given by

$$\boldsymbol{\Sigma} = \gamma_5 \gamma^0 \boldsymbol{\gamma} \qquad \text{or} \quad \Sigma^i = \frac{1}{2}\varepsilon_{ijk}\sigma^{jk}. \tag{A.1.21}$$

Under hermitian conjugation the Dirac matrices behave as

$$\begin{aligned} \gamma^{\mu\dagger} &= \gamma^0 \gamma^\mu \gamma^0, & \gamma_5^\dagger &= -\gamma^0 \gamma_5 \gamma^0 = \gamma_5, \\ \sigma^{\mu\nu\dagger} &= \gamma^0 \sigma^{\mu\nu} \gamma^0, & (\gamma^\mu \gamma_5)^\dagger &= \gamma^0 (\gamma^\mu \gamma_5) \gamma^0. \end{aligned} \tag{A.1.22}$$

The *charge-conjugation matrix* C relates the Dirac matrices and their transpose

$$\begin{aligned} (\gamma^\mu)^{\mathrm{T}} &= -C\gamma^\mu C^{-1}, & (\gamma_5)^{\mathrm{T}} &= C\gamma_5 C^{-1}, \\ (\sigma^{\mu\nu})^{\mathrm{T}} &= -C\sigma^{\mu\nu} C^{-1}, & (\gamma^\mu \gamma_5)^{\mathrm{T}} &= C(\gamma^\mu \gamma_5) C^{-1}. \end{aligned} \tag{A.1.23}$$

In the usual representations, C is given by

$$C = \mathrm{i}\gamma^2 \gamma^0 = -C^{-1} = -C^\dagger = -C^{\mathrm{T}}. \tag{A.1.24}$$

As an example for an explicit representation of these matrices we list the *Weyl representation*:

$$
\begin{aligned}
\gamma^0 &= \begin{pmatrix} 0 & -\mathbf{1} \\ -\mathbf{1} & 0 \end{pmatrix}, \qquad \gamma^i = \begin{pmatrix} 0 & \tau^i \\ -\tau^i & 0 \end{pmatrix}, \\
\gamma_5 &= \begin{pmatrix} \mathbf{1} & 0 \\ 0 & -\mathbf{1} \end{pmatrix}, \qquad C = \begin{pmatrix} -\mathrm{i}\tau^2 & 0 \\ 0 & \mathrm{i}\tau^2 \end{pmatrix}, \\
\sigma^{0i} &= \begin{pmatrix} \mathrm{i}\tau^i & 0 \\ 0 & -\mathrm{i}\tau^i \end{pmatrix}, \qquad \sigma^{ij} = \varepsilon^{ijk} \begin{pmatrix} \tau^k & 0 \\ 0 & \tau^k \end{pmatrix}, \\
\Sigma^i &= \begin{pmatrix} \tau^i & 0 \\ 0 & \tau^i \end{pmatrix},
\end{aligned}
\tag{A.1.25}
$$

where τ^i are the Pauli matrices.

The matrices $\mathbf{1}$, γ^μ, $\sigma^{\mu\nu}$, $\gamma^\mu\gamma^5$, and γ^5 form a basis in the space of complex 4×4 matrices. Therefore, every 4×4 matrix Γ can be expanded (*Pauli identity*)

$$
\begin{aligned}
\Gamma = &\frac{1}{4}\mathbf{1}\,\mathrm{Tr}\{\Gamma\} + \frac{1}{4}\gamma_\mu\,\mathrm{Tr}\{\gamma^\mu\Gamma\} + \frac{1}{8}\sigma_{\mu\nu}\,\mathrm{Tr}\{\sigma^{\mu\nu}\Gamma\} \\
&- \frac{1}{4}\gamma_\mu\gamma_5\,\mathrm{Tr}\{\gamma^\mu\gamma_5\Gamma\} + \frac{1}{4}\gamma_5\,\mathrm{Tr}\{\gamma_5\Gamma\}.
\end{aligned}
\tag{A.1.26}
$$

The *Fierz identity* reads

$$
\begin{aligned}
(1 \otimes 1)_{\alpha\beta,\alpha'\beta'} = \frac{1}{4}\Big(&1 \otimes 1 + \gamma_\mu \otimes \gamma^\mu + \frac{1}{2}\sigma_{\mu\nu} \otimes \sigma^{\mu\nu} \\
&- \gamma_\mu\gamma_5 \otimes \gamma^\mu\gamma_5 + \gamma_5 \otimes \gamma_5\Big)_{\alpha\beta',\alpha'\beta}.
\end{aligned}
\tag{A.1.27}
$$

The contraction identities

$$
\begin{aligned}
\gamma^\lambda\gamma^{\mu_1}\ldots\gamma^{\mu_{2n+1}}\gamma_\lambda &= -2\gamma^{\mu_{2n+1}}\ldots\gamma^{\mu_1}, \\
\gamma^\lambda\gamma^{\mu_1}\ldots\gamma^{\mu_{2n}}\gamma_\lambda &= 2(\gamma^{\mu_{2n}}\gamma^{\mu_1}\ldots\gamma^{\mu_{2n-1}} + \gamma^{\mu_{2n-1}}\ldots\gamma^{\mu_1}\gamma^{\mu_{2n}})
\end{aligned}
\tag{A.1.28}
$$

are useful for practical calculations. We mention the special cases

$$
\begin{aligned}
&\gamma^\lambda\gamma_\lambda = 4, \qquad \gamma^\lambda\gamma^\mu\gamma_\lambda = -2\gamma^\mu, \\
&\gamma^\lambda\gamma^\mu\gamma^\nu\gamma_\lambda = 4g^{\mu\nu}, \qquad \gamma^\lambda\gamma^\mu\gamma^\nu\gamma^\rho\gamma_\lambda = -2\gamma^\rho\gamma^\nu\gamma^\mu, \\
&\gamma^\lambda\gamma^\mu\gamma^\nu\gamma^\rho\gamma^\sigma\gamma_\lambda = 2(\gamma^\sigma\gamma^\mu\gamma^\nu\gamma^\rho + \gamma^\rho\gamma^\nu\gamma^\mu\gamma^\sigma).
\end{aligned}
\tag{A.1.29}
$$

Finally, the following relations hold for *Dirac traces*

$$
\begin{aligned}
&\mathrm{Tr}\{\mathbf{1}\} = 4, \qquad \mathrm{Tr}\{\gamma^\mu\} = \mathrm{Tr}\{\sigma^{\mu\nu}\} = \mathrm{Tr}\{\gamma^\mu\gamma_5\} = \mathrm{Tr}\{\gamma_5\} = 0,\\
&\mathrm{Tr}\{\gamma^\mu\gamma^\nu\} = 4g^{\mu\nu}, \quad \mathrm{Tr}\{\gamma^\mu\gamma^\nu\gamma^\rho\gamma^\sigma\} = 4(g^{\mu\nu}g^{\rho\sigma} + g^{\mu\sigma}g^{\nu\rho} - g^{\mu\rho}g^{\nu\sigma}),\\
&\mathrm{Tr}\{\gamma^\mu\gamma^\nu\gamma_5\} = 0, \qquad \mathrm{Tr}\{\gamma^\mu\gamma^\nu\gamma^\rho\gamma^\sigma\gamma_5\} = -4\mathrm{i}\varepsilon^{\mu\nu\rho\sigma},\\
&\mathrm{Tr}\{\gamma^{\mu_1}\dots\gamma^{\mu_{2n+1}}\} = 0,\\
&\mathrm{Tr}\{\gamma^{\mu_1}\dots\gamma^{\mu_{2n}}\} = 4\sum \sigma g^{\mu_{i_1}\mu_{j_1}}\dots g^{\mu_{i_n}\mu_{j_n}}, \qquad \text{(A.1.30)}
\end{aligned}
$$

where the sum in the last equation runs over all different pairings of the indices $\{1,\dots,2n\}$ satisfying $1 = i_1 < i_2 < \dots < i_n$, $i_k < j_k$ and σ is the signature of the permutation $\{i_1, j_1, \dots, i_n, j_n\}$.

A.1.1.4 Dirac matrices in D dimensions

In D dimensions, the defining relation (A.1.15) still holds. However, since $\gamma^\lambda\gamma_\lambda = D$, the contraction identities (A.1.29) get extra terms,

$$
\begin{aligned}
\gamma^\lambda\gamma^{\mu_1}\dots\gamma^{\mu_{2n+1}}\gamma_\lambda &= -2\gamma^{\mu_{2n+1}}\dots\gamma^{\mu_1} + (4-D)\gamma^{\mu_1}\dots\gamma^{\mu_{2n+1}},\\
\gamma^\lambda\gamma^{\mu_1}\dots\gamma^{\mu_{2n}}\gamma_\lambda &= 2(\gamma^{\mu_{2n}}\gamma^{\mu_1}\dots\gamma^{\mu_{2n-1}} + \gamma^{\mu_{2n-1}}\dots\gamma^{\mu_1}\gamma^{\mu_{2n}})\\
&\quad + (D-4)\gamma^{\mu_1}\dots\gamma^{\mu_{2n}}. \qquad \text{(A.1.31)}
\end{aligned}
$$

The identity $\mathrm{Tr}\{\mathbf{1}\} = 4$ can still be used in dimensional regularization so that the trace identities (A.1.30) remain valid.

For the treatment of γ_5 in dimensional regularization see Sect. 2.7.1.5. Note that the Fierz identities (A.1.27) have to be modified in D dimensions.

A.1.1.5 Dirac spinors

The Dirac spinors u and v are functions of the on-shell momentum p, with $p_0 = E_p = \sqrt{\mathbf{p}^2 + M^2}$ and are labelled by the polarization index j_3. They are solutions of the *momentum-space Dirac equation*

$$(\not{p} - M)u(p, j_3) = 0, \qquad (\not{p} + M)v(p, j_3) = 0. \qquad \text{(A.1.32)}$$

The conjugate spinors

$$\bar{u} = u^\dagger\gamma^0, \qquad \bar{v} = v^\dagger\gamma^0 \qquad \text{(A.1.33)}$$

satisfy the conjugate Dirac equation

$$\bar{u}(p, j_3)(\not{p} - M) = 0, \qquad \bar{v}(p, j_3)(\not{p} + M) = 0. \tag{A.1.34}$$

The polarization of the spinors is fixed with respect to a space-like vector s orthogonal to p, i.e. $sp = 0$, and the spinors obey the eigenvalue equations

$$\begin{aligned} \gamma_5 \not{s}\, u\left(p, \pm\frac{1}{2}\right) &= \pm u\left(p, \pm\frac{1}{2}\right), \\ \gamma_5 \not{s}\, v\left(p, \pm\frac{1}{2}\right) &= \pm v\left(p, \pm\frac{1}{2}\right). \end{aligned} \tag{A.1.35}$$

In the rest system, $p = (M, 0)$, $s = (0, \mathbf{s})$, these equations correspond to

$$\begin{aligned} \mathbf{\Sigma s}\, u\left(p, \pm\frac{1}{2}\right) &= \pm u\left(p, \pm\frac{1}{2}\right), \\ -\mathbf{\Sigma s}\, v\left(p, \pm\frac{1}{2}\right) &= \pm v\left(p, \pm\frac{1}{2}\right). \end{aligned} \tag{A.1.36}$$

The spinors are *normalized* as

$$\begin{aligned} \sum_\alpha \bar{u}_\alpha(p, j_3')u_\alpha(p, j_3) &= 2M\delta_{j_3 j_3'}, \\ \sum_\alpha \bar{v}_\alpha(p, j_3')v_\alpha(p, j_3) &= -2M\delta_{j_3 j_3'} \end{aligned} \tag{A.1.37}$$

and obey

$$\sum_\alpha \bar{v}_\alpha(p, j_3')u_\alpha(p, j_3) = 0 = \sum_\alpha \bar{u}_\alpha(p, j_3')v_\alpha(p, j_3). \tag{A.1.38}$$

An equivalent normalization condition, which also applies for massless fermions, reads

$$\begin{aligned} \sum_\alpha u_\alpha^\dagger(p, j_3')u_\alpha(p, j_3) &= 2E\delta_{j_3 j_3'}, \\ \sum_\alpha v_\alpha^\dagger(p, j_3')v_\alpha(p, j_3) &= 2E\delta_{j_3 j_3'}. \end{aligned} \tag{A.1.39}$$

The *polarization sums* are given by

$$\sum_{j_3} u_\alpha(p, j_3)\bar{u}_\beta(p, j_3) = (\not{p} + M)_{\alpha\beta},$$
$$\sum_{j_3} v_\alpha(p, j_3)\bar{v}_\beta(p, j_3) = (\not{p} - M)_{\alpha\beta}. \tag{A.1.40}$$

The *Gordon identities* read

$$\bar{u}(p, j_3)\gamma^\mu u(p', j_3') = \frac{1}{2m}\bar{u}(p, j_3)[(p+p')^\mu + \mathrm{i}\sigma^{\mu\nu}(p+p')_\nu]u(p', j_3'),$$
$$\bar{u}(p, j_3)\gamma^\mu\gamma^5 u(p', j_3') =$$
$$\frac{1}{2m}\bar{u}(p, j_3)[(p-p')^\mu\gamma^5 + \mathrm{i}\sigma^{\mu\nu}(p+p')_\nu\gamma^5]u(p', j_3'), \tag{A.1.41}$$

in particular

$$\bar{u}(p, j_3)\gamma^\mu u(p, j_3') = \frac{p^\mu}{m}\bar{u}(p, j_3)u(p, j_3') = 2p^\mu\delta_{j_3 j_3'}. \tag{A.1.42}$$

A.1.1.6 Polarization vectors

The polarization vectors for vector-boson fields $\varepsilon^\mu(k, \lambda)$ depend on the on-shell momentum k and a polarization index λ. The polarization vectors are transverse

$$k^\mu \varepsilon_\mu(k, \lambda) = 0 \tag{A.1.43}$$

and normalized as

$$\varepsilon^*_\mu(k, \lambda)\varepsilon^\mu(k, \lambda') = -\delta_{\lambda\lambda'}. \tag{A.1.44}$$

In the massive case, $k^2 = M^2 \neq 0$, there are two (with respect to the three momentum) transverse and one longitudinal polarization vector, and the *polarization sum* reads

$$\sum_{\lambda=1}^{3} \varepsilon^*_\mu(k, \lambda)\varepsilon_\mu(k, \lambda) = -g_{\mu\nu} + \frac{k_\mu k_\nu}{M^2}. \tag{A.1.45}$$

In the massless case, $k^2 = 0$, there are only two physical (transverse) polarizations, and the polarization sum is gauge-dependent. In the gauge $n_\mu \varepsilon^\mu(k,\lambda) = 0$, where n_μ is an external four-vector, it reads

$$\sum_{\lambda=1}^{2} \varepsilon^*_\mu(k,\lambda)\varepsilon_\mu(k,\lambda) = -g_{\mu\nu} - \frac{k_\mu k_\nu}{(kn)^2} n^2 + \frac{n_\mu k_\nu + n_\nu k_\mu}{kn}. \qquad \text{(A.1.46)}$$

A.1.2 Euclidean space

In euclidean space the *metric tensor* is $\delta^{\mu\nu}$. The *euclidean coordinates* are related to the Minkowski coordinates as

$$x^4_{\mathrm{E}} = \mathrm{i}x^0, \qquad x^i_{\mathrm{E}} = x^i. \qquad \text{(A.1.47)}$$

The euclidean totally antisymmetric *Levi-Civita tensor* is defined by

$$\varepsilon_{\mathrm{E}}^{\mu\nu\rho\sigma} = \varepsilon_{\mathrm{E};\mu\nu\rho\sigma}, \qquad \varepsilon_{\mathrm{E};1234} = 1. \qquad \text{(A.1.48)}$$

It obeys identities like those in (A.1.10), e.g.

$$\varepsilon_{\mathrm{E};\mu\nu}{}^{\alpha\beta}\varepsilon_{\mathrm{E};\alpha\beta\rho\sigma} = 2(\delta_{\mu\rho}\delta_{\nu\sigma} - \delta_{\mu\sigma}\delta_{\nu\rho}). \qquad \text{(A.1.49)}$$

For the euclidean description of Dirac fields we use the conventions

$$\gamma^4_{\mathrm{E}} = \gamma^0, \qquad \gamma^i_{\mathrm{E}} = -\mathrm{i}\gamma^i, \quad i = 1,2,3. \qquad \text{(A.1.50)}$$

Therefore,

$$[\gamma^\mu_{\mathrm{E}}, \gamma^\nu_{\mathrm{E}}]_+ = 2\delta^{\mu\nu} \qquad \text{(A.1.51)}$$

and

$$\gamma^5_{\mathrm{E}} = -\gamma^1_{\mathrm{E}}\gamma^2_{\mathrm{E}}\gamma^3_{\mathrm{E}}\gamma^4_{\mathrm{E}} = \gamma^5. \qquad \text{(A.1.52)}$$

Moreover, the *euclidean Dirac matrices* are hermitian,

$$\gamma^{\mu\dagger}_{\mathrm{E}} = \gamma^\mu_{\mathrm{E}}, \qquad \gamma^{5\dagger}_{\mathrm{E}} = \gamma^5_{\mathrm{E}}. \qquad \text{(A.1.53)}$$

The *euclidean Dirac operator*

$$\not{\partial} = \gamma^\mu_{\mathrm{E}}\partial_\mu \qquad \text{(A.1.54)}$$

gets anti-hermitian and the Dirac Lagrangian turns to

$$\mathcal{L} = \overline{\psi}(\mathrm{i}\not{\partial} - m)\psi \mapsto -\mathcal{L}_{\mathrm{E}} = -\overline{\psi}(\not{\partial} + m)\psi. \qquad \text{(A.1.55)}$$

A.1.3 Field theory

A.1.3.1 Normalization of states and S matrix

One particle states are normalized as

$$\langle p, \sigma | p', \sigma' \rangle = 2E_p \, (2\pi)^3 \, \delta(\mathbf{p} - \mathbf{p}') \, \delta_{\sigma\sigma'}, \tag{A.1.56}$$

where $E_p = \sqrt{M^2 + \mathbf{p}^2}$ and σ and σ' denote polarizations.
The *invariant matrix element* $\mathcal{M}_{fi}$ is related to the S *matrix* as

$$\langle f | S | i \rangle = \langle f | i \rangle + \mathrm{i}(2\pi)^4 \delta^{(4)}(p_i - p_f) \mathcal{M}_{fi}, \tag{A.1.57}$$

where $|i\rangle$ denotes the initial state and $|f\rangle$ the final state. The *differential cross section* for the scattering of two particles with momenta p_1 and p_2 and masses M_1 and M_2 into n particles with momenta $q_1, \dots, q_n$ is obtained as

$$\begin{aligned} \mathrm{d}\sigma = {} & \frac{1}{4\sqrt{(p_1 p_2)^2 - M_1^2 M_2^2}} |\mathcal{M}_{fi}|^2 (2\pi)^4 \delta^{(4)} \left(p_1 + p_2 - \sum_{i=1}^{n} q_i \right) \\ & \times \frac{\mathrm{d}^3 q_1}{(2\pi)^3 2E_{q_1}} \cdots \frac{\mathrm{d}^3 q_n}{(2\pi)^3 2E_{q_n}}. \end{aligned} \tag{A.1.58}$$

A.1.3.2 Green functions and their generating functionals

In order to define our conventions we use generic charged fields ϕ with adjoints $\phi^\dagger$. General Green functions are denoted as

$$\begin{aligned} G^{\phi\ldots\phi^\dagger\ldots}(x, \dots, y, \dots) &= \langle T\phi(x) \cdots \phi^\dagger(y) \cdots \rangle \\ &= \frac{\delta}{\mathrm{i}\delta J_\phi(x) \cdots \mathrm{i}\delta J_{\phi^\dagger}(y) \cdots} T\{J_\phi, J_{\phi^\dagger}\} \Big|_{J_\phi(x) = J_{\phi^\dagger}(x) \equiv 0} \end{aligned}$$

$$=$$

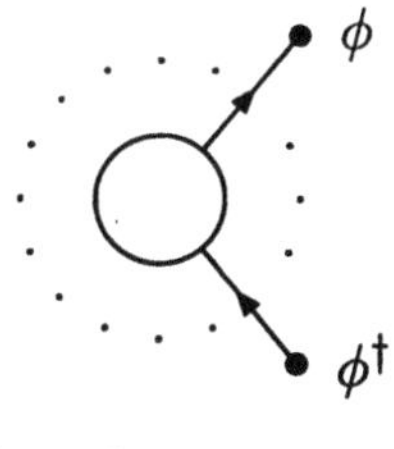

$$. \tag{A.1.59}$$

where $T\{J_\phi, J_{\phi^\dagger}\}$ is the *generating functional of Green functions* and J_ϕ and $J_{\phi^\dagger}$ are the sources corresponding to ϕ and $\phi^\dagger$, respectively. Because the field operator ϕ creates antiparticles and annihilates particles, the particles and thus the fields in $\langle T\phi(x)\cdots\phi^\dagger(y)\cdots\rangle$ have to be considered as outgoing [cf. (A.1.67)]. The order of the field indices equals the order of fields in the vacuum expectation value and the order of corresponding derivatives. This is crucial for anticommuting fields, where the exchange of two field indices yields a minus sign. In the diagrams, the arrows indicate the flow of the particles, which is opposite to the flow of antiparticles.

The transformation to momentum space is defined as

$$\begin{aligned} G^{\phi\ldots\phi^\dagger\ldots}(x,\ldots,y,\ldots) &= \int \frac{\mathrm{d}^4p}{(2\pi)^4}\cdots\int\frac{\mathrm{d}^4q}{(2\pi)^4} \\ &\quad\times \exp\left\{\mathrm{i}(px+\ldots+qy+\ldots)\right\}\,\tilde{G}^{\phi\ldots\phi^\dagger\ldots}(p,\ldots,q,\ldots), \end{aligned} \tag{A.1.60}$$

where the momenta p,q are incoming. In *momentum-space Green functions* we usually split off the momentum conservation δ-function

$$\begin{aligned} &\tilde{G}^{\phi\ldots\phi^\dagger\ldots}(p,\ldots,q,\ldots) \\ &\quad= (2\pi)^4\delta^{(4)}(p+\ldots+q+\ldots)G^{\phi\ldots\phi^\dagger\ldots}(p,\ldots,q,\ldots). \end{aligned} \tag{A.1.61}$$

The *generating functional for connected Green functions* is defined as

$$T_{\mathrm{c}}\{J_\phi, J_{\phi^\dagger}\} = \log T\{J_\phi, J_{\phi^\dagger}\}. \tag{A.1.62}$$

The corresponding Green functions are denoted as in (A.1.59) with an additional index "c".

Truncated Green functions are denoted as

$$\begin{aligned} G^{\phi\ldots\phi^\dagger\ldots}_{\mathrm{trunc}}(x,\ldots,y,\ldots) &= \langle T\phi(x),\ldots,\phi^\dagger(y)\ldots\rangle_{\mathrm{trunc}} \\ &= G^{\underline{\phi},\ldots,\underline{\phi}^\dagger\ldots}(x,\ldots,y,\ldots) = \langle T\underline{\phi}(x),\ldots,\underline{\phi}^\dagger(y)\ldots\rangle \end{aligned}$$

$=$ 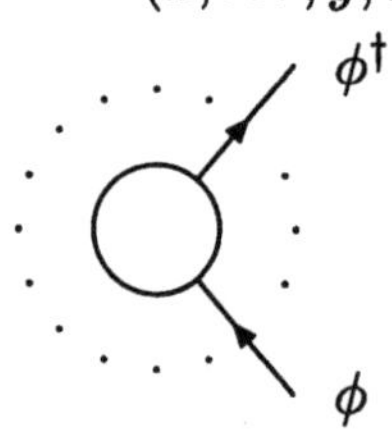

. (A.1.63)

Underlining is also used to indicate *truncation* of single lines. Because of the relation

$$G^{\cdots\phi\cdots}(x,\ldots)=\int \mathrm{d}^4y\, G^{\phi\phi'}(x,y)\, G^{\cdots\underline{\phi'}\cdots}(y,\ldots), \tag{A.1.64}$$

(ϕ' denotes the field that appears in the propagator together with ϕ; usually $\phi' = \phi^\dagger$) the field indices in truncated Green functions denote incoming particles or fields.

The *generating functional of vertex functions* is defined by

$$\phi(x)=\frac{\delta T_{\mathrm{c}}\{J_\phi,J_{\phi^\dagger}\}}{\mathrm{i}\delta J_\phi(x)},\qquad \phi^\dagger(x)=\frac{\delta T_{\mathrm{c}}\{J_\phi,J_{\phi^\dagger}\}}{\mathrm{i}\delta J_{\phi^\dagger}(x)}, \tag{A.1.65}$$

$$\mathrm{i}\Gamma\{\phi,\phi^\dagger\}=-\mathrm{i}\int \mathrm{d}^4x\,[J_\phi(x)\phi(x)+J_{\phi^\dagger}(x)\phi^\dagger(x)]+T_{\mathrm{c}}\{J_\phi,J_{\phi^\dagger}\},$$

and from this the vertex functions result as

$$\Gamma^{\phi\ldots\phi^\dagger\ldots}(x,\ldots,y,\ldots)=\frac{\delta\Gamma\{\phi,\phi^\dagger\}}{\delta\phi(x)\cdots\delta\phi^\dagger(y)\cdots}. \tag{A.1.66}$$

Note that in the vertex functions, as in truncated Green functions, the field arguments denote incoming quanta. The Fourier transformation of truncated functions or vertex functions is as in (A.1.60), i.e. all momenta are incoming.

For Grassmann fields the above formula hold as well. The order of labels corresponds to the order of left derivatives. Interchange of labels of the Green functions leads to signs.

From the truncated Green functions, the S-matrix elements are obtained via the *reduction formula*

$$\begin{aligned}&\langle\phi(k_1)\ldots\phi^+(k_2)\ldots|S|\phi(k_3)\ldots\phi^+(k_4)\ldots\rangle=\\&=R_\phi^{n/2}\langle T\underline{\phi}(k_1)\ldots\underline{\phi}^\dagger(k_2)\ldots\underline{\phi}^\dagger(k_3)\ldots\underline{\phi}(k_4)\ldots\rangle\end{aligned} \tag{A.1.67}$$

For Grassmann fields the above formula hold as well. The order of labels corresponds to the order of left derivatives. Interchange of labels of the Green functions leads to signs.

In particular, in our conventions the lowest-order two-point functions for fermions read

$$\begin{aligned} \Gamma^{\psi\bar{\psi}}_{\beta\alpha}(p,-p) &= -\Gamma^{\bar{\psi}\psi}_{\alpha\beta}(-p,p) = (\not{p} - m)_{\alpha\beta}, \\ G^{\psi\bar{\psi}}_{\alpha\beta}(-p,p) &= -G^{\bar{\psi}\psi}_{\beta\alpha}(p,-p) = \frac{\mathrm{i}(\not{p} + m)_{\alpha\beta}}{p^2 - m^2 - \mathrm{i}\varepsilon}. \end{aligned} \tag{A.1.68}$$

Note that $G^{\psi\bar{\psi}}$ has to be used for the truncation of a ψ line but $G^{\bar{\psi}\psi}$ for the truncation of a $\bar{\psi}$ line in (A.1.64).

The relation corresponding to (2.2.41) is

$$\delta^{(4)}(x-y) = \int \mathrm{d}^4 z \, \frac{\mathrm{i}\delta^2\Gamma\{\psi,\bar{\psi}\}}{\delta\bar{\psi}(y)\delta\psi(z)} \frac{\delta^2 T_{\mathrm{c}}\{J_\psi, J_{\bar{\psi}}\}}{\mathrm{i}\delta J_\psi(z)\mathrm{i}\delta J_{\bar{\psi}}(x)}, \tag{A.1.69}$$

and the one corresponding to (2.2.43) reads

$$\begin{aligned} \frac{\delta^2 T_{\mathrm{c}}\{J\}}{\mathrm{i}\delta J_\psi(x_1)\mathrm{i}\delta J_{\bar{\psi}}(x_2)} = &-\int \mathrm{d}^4 z_1 \, \mathrm{d}^4 z_2 \, \frac{\delta^2 T_{\mathrm{c}}\{J_\psi, J_{\bar{\psi}}\}}{\mathrm{i}\delta J_\psi(x_1)\mathrm{i}\delta J_{\bar{\psi}}(z_1)} \\ &\times \frac{\delta^2 T_{\mathrm{c}}\{J_\psi, J_{\bar{\psi}}\}}{\mathrm{i}\delta J_{\bar{\psi}}(x_2)\mathrm{i}\delta J_\psi(z_2)} \frac{\mathrm{i}\delta^2\Gamma\{\psi,\bar{\psi}\}}{\delta\bar{\psi}(z_1)\delta\psi(z_2)}. \end{aligned} \tag{A.1.70}$$

Acting on this relation with

$$\begin{aligned} \frac{\delta}{\mathrm{i}\delta J_{\bar{\psi}}(x)} &= \int \mathrm{d}^4 z \, \frac{\delta^2 T_{\mathrm{c}}\{J_\psi, J_{\bar{\psi}}\}}{\mathrm{i}\delta J_{\bar{\psi}}(x)\mathrm{i}\delta J_\psi(z)} \frac{\delta}{\delta\psi(z)}, \\ \frac{\delta}{\mathrm{i}\delta J_\psi(x)} &= \int \mathrm{d}^4 z \, \frac{\delta^2 T_{\mathrm{c}}\{J_\psi, J_{\bar{\psi}}\}}{\mathrm{i}\delta J_\psi(x)\mathrm{i}\delta J_{\bar{\psi}}(z)} \frac{\delta}{\delta\bar{\psi}(z)} \end{aligned} \tag{A.1.71}$$

gives the decomposition of the connected Green functions in terms of vertex functions as discussed in Sect. 2.2.2.2 for the case of commuting fields. The sign conventions of the vertex functions agree with those of the truncated Green functions as defined by (A.1.64) (with the order of fields given there).

The conventions for Faddeev–Popov ghosts are equivalent to those for the fermions.

A.1.3.3 Feynman rules

The Feynman rules for the calculation of Green functions or scattering amplitudes in momentum space read as follows:

1. Draw all possible topologically distinct diagrams (connected or disconnected but without vacuum subdiagrams) contributing to the process under consideration in the desired order of perturbation theory.

For each diagram the following steps have to be performed:

2. Associate external momenta to all external lines and L loop momenta to the internal loops. Determine the momenta of the internal lines so that four-momentum conservation holds at each vertex.

3. For each internal line attach a propagator,

$$= \frac{\mathrm{i}}{p^2 - M^2 + \mathrm{i}\varepsilon}, \qquad \text{for a scalar boson} \tag{A.1.72}$$

$$= \left(\frac{\mathrm{i}(\not{p} + M)}{p^2 - M^2 + \mathrm{i}\varepsilon}\right)_{\alpha\beta}, \qquad \text{for a fermion} \tag{A.1.73}$$

$$\begin{aligned} &= \frac{-\mathrm{i}g^{\mu\nu}}{p^2 - M^2 + \mathrm{i}\varepsilon} + \frac{\mathrm{i}(1-\xi)p^\mu p^\nu}{(p^2 - M^2 + \mathrm{i}\varepsilon)(p^2 - \xi M^2 + \mathrm{i}\varepsilon)} \\ &= -\mathrm{i}\frac{g^{\mu\nu} - p^\mu p^\nu/M^2}{p^2 - M^2 + \mathrm{i}\varepsilon} - \mathrm{i}\frac{p^\mu p^\nu/M^2}{p^2 - \xi M^2 + \mathrm{i}\varepsilon}. \end{aligned} \tag{A.1.74}$$

for a vector boson in a gauge with the kinetic Lagrangian

$$\mathcal{L} = -\frac{1}{4}(\partial_\mu V_\nu - \partial_\nu V_\mu)^2 - \frac{1}{2\xi}(\partial_\mu V^\mu)^2 + \frac{M^2}{2}V_\mu V^\mu. \tag{A.1.75}$$

4. For each *vertex* attach a factor derived from the relevant monomial of the interaction Lagrangian. It can be obtained by acting with functional derivatives with respect to the external fields corresponding to the considered vertex on $\int \mathrm{d}^4x\, \mathcal{L}_{\text{int}}$, transforming to momentum space and omitting the momentum conservation δ-function $(2\pi)^4\delta^{(4)}(\sum p)$ (cf. Sect. 2.4.2). In this process each field derivative $\partial_\mu\phi$ results in a factor $-\mathrm{i}p_\mu$, where p is the corresponding incoming momentum.

   The vertex factors for QED and a non-abelian gauge theory are given in (1.2.30) and (2.4.32), respectively. Further examples are given below.

5. Integrate over the L loop momenta q_μ using the measure $\int \mathrm{d}^4q/(2\pi)^4$, possibly after a regularization.

6. Multiply the contribution of each diagram by

   i. a factor -1 for each closed loop of anticommuting fields,

   ii. a global sign for external lines of anticommuting fields, coming from their permutation in the actual graph as compared to the argument of the considered Green function, (A simple rule for external fermion lines is given in item 5 of Sect. 1.2.1.7.)

   iii. a symmetry factor $1/S$, where S equals the number of possibilities of mapping the graph onto itself by permutation of lines and vertices.

7. Summation over the contribution of all diagrams yields truncated functions with no factor on the external lines. Contributions to one-particle irreducible functions are obtained from one-particle-irreducible diagrams. Non-truncated Green functions are obtained by putting propagators on the external lines. Connected Green functions are obtained by retaining only connected diagrams.

8. The scattering amplitude is obtained by multiplying with $-\mathrm{i}$, putting the external lines on their mass shell, i.e. $p_i^2 = M_i^2$, and multiplying with the appropriate wave-function factors, i.e. for incoming scalars, fermions, antifermions, and vector bosons:

$$1 \qquad u_\alpha(p, j_3) \qquad \bar{v}_\alpha(p, j_3) \qquad \varepsilon_\mu(k, j_3), \tag{A.1.76}$$

and for outgoing scalars, fermions, antifermions, and vector bosons:

$$1 \qquad \bar{u}_\alpha(p, j_3) \qquad v_\alpha(p, j_3) \qquad \varepsilon_\mu^*(k, j_3). \tag{A.1.77}$$

Finally we give some examples for vertices.

The vertex in the ϕ^4 theory,

$$\mathcal{L}_\mathrm{I} = -\frac{g}{4!}\phi^4,$$

reads

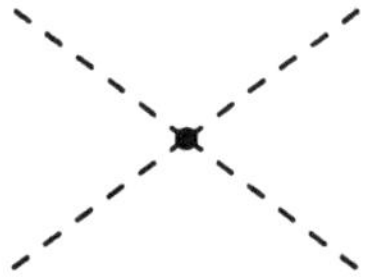

$$= -ig. \tag{A.1.78}$$

The vertex in the ϕ^3 theory,

$$\mathcal{L}_{\rm I} = -\frac{g}{3!}\phi^3,$$

is given by

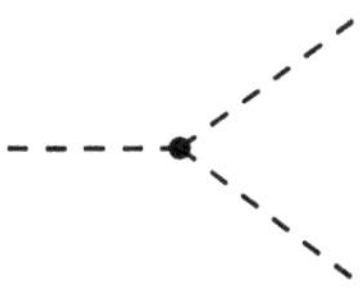

$$= -ig. \tag{A.1.79}$$

An example for a derivative coupling is given by

$$\mathcal{L}_{\rm I} = g\phi V^\mu \partial_\mu \phi, \tag{A.1.80}$$

which leads to the vertex factor

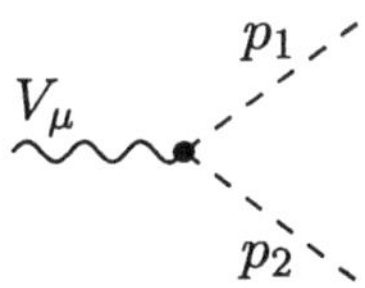

$$= g(p_1 + p_2)_\mu. \tag{A.1.81}$$

The appearance of the sum $p_1 + p_2$ in the Feynman rule results from the Bose symmetry of the two scalar fields in (A.1.80). If the two scalar fields would be different, only the momentum corresponding to the field on which the derivative acts would be present in the Feynman rule.

A.2 Feynman rules for the Electroweak Standard Model

In this appendix we list the Feynman rules of the SM in an arbitrary 't Hooft gauge. We write down generic Feynman rules for all propagators and vertices and give the possible actual insertions. The rules for the external lines are as in (A.1.76) and (A.1.77). For brevity we introduce the shorthand notation

$$c = c_{\mathrm{w}} = \cos\theta_{\mathrm{w}}, \qquad s = s_{\mathrm{w}} = \sin\theta_{\mathrm{w}}. \tag{A.2.1}$$

In the vertices all momenta and fields are considered as incoming.

The propagators read as follows for

gauge bosons $V = A$, Z, and W $(M_A = 0)$,

$$= \frac{-\mathrm{i}g_{\mu\nu}}{k^2 - M_V^2} + \frac{\mathrm{i}(1-\xi_V)k_\mu k_\nu}{(k^2 - M_V^2)(k^2 - \xi_V M_V^2)}, \tag{A.2.2}$$

Faddeev–Popov ghosts $G = u^A$, u^Z, and $u^\pm$ $(M_{u^A} = 0$, $M_{u^Z} = \sqrt{\xi_Z}M_{\mathrm{Z}}$, and $M_{u^\pm} = \sqrt{\xi_W}M_{\mathrm{W}})$,

$$= \frac{\mathrm{i}}{k^2 - M_G^2}, \tag{A.2.3}$$

scalar fields $S = \eta$, χ, and ϕ $(M_\chi = \sqrt{\xi_Z}M_{\mathrm{Z}}$ and $M_\phi = \sqrt{\xi_W}M_{\mathrm{W}})$,

$$= \frac{\mathrm{i}}{k^2 - M_S^2}, \tag{A.2.4}$$

and fermion fields $F = f_i$,

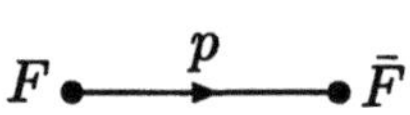

$$= \frac{\mathrm{i}(\not{p} + m_F)}{p^2 - m_F^2}. \tag{A.2.5}$$

The vertices are given by the following expressions:

VVVV vertex:

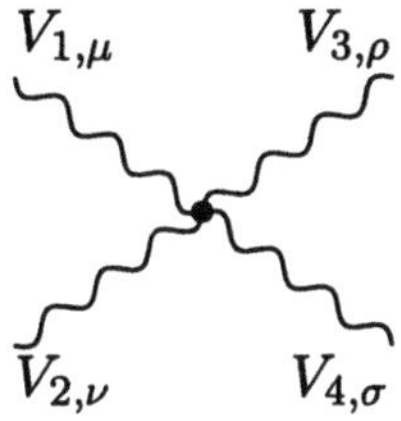

$$= \mathrm{i}e^2 C\Big[2g_{\mu\nu}g_{\sigma\rho} - g_{\nu\rho}g_{\mu\sigma} - g_{\rho\mu}g_{\nu\sigma}\Big] \qquad \text{(A.2.6)}$$

with the actual values of V_1, V_2, V_3, V_4, and C

$V_1V_2V_3V_4$	W^+W^-ZZ	W^+W^-AZ	W^+W^-AA
C	$-\frac{c^2}{s^2}$	$\frac{c}{s}$	-1

$V_1V_2V_3V_4$	$W^+W^+W^-W^-$
C	$\frac{1}{s^2}$

. (A.2.7)

VVV vertex:

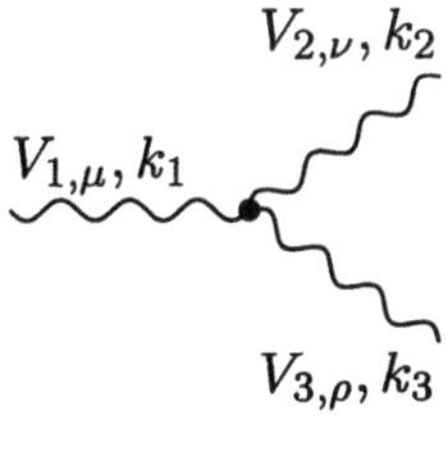

$$= \mathrm{i}eC\Big[g_{\mu\nu}(k_1 - k_2)_\rho + g_{\nu\rho}(k_2 - k_3)_\mu + g_{\rho\mu}(k_3 - k_1)_\nu\Big] \qquad \text{(A.2.8)}$$

with the actual values of V_1, V_2, V_3, and C

$V_1V_2V_3$	AW^+W^-	ZW^+W^-
C	1	$-\frac{c}{s}$

. (A.2.9)

$SSSS$ vertex:

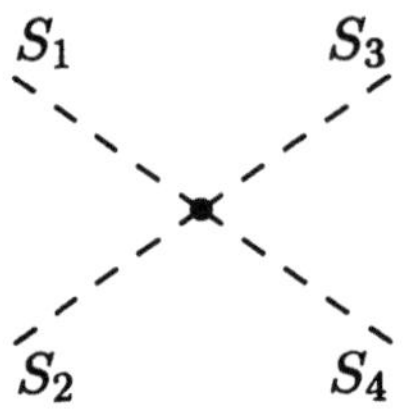

$$= \mathrm{i}e^2 C \tag{A.2.10}$$

with the actual values of S_1, S_2, S_3, S_4, and C

$S_1S_2S_3S_4$	$\eta\eta\eta\eta,\ \chi\chi\chi\chi$	$\eta\eta\chi\chi,\ \eta\eta\phi^+\phi^-,\ \chi\chi\phi^+\phi^-$	$\phi^+\phi^-\phi^+\phi^-$
C	$-\frac{3}{4s^2}\frac{M_H^2}{M_W^2}$	$-\frac{1}{4s^2}\frac{M_H^2}{M_W^2}$	$-\frac{1}{2s^2}\frac{M_H^2}{M_W^2}$

. (A.2.11)

SSS vertex:

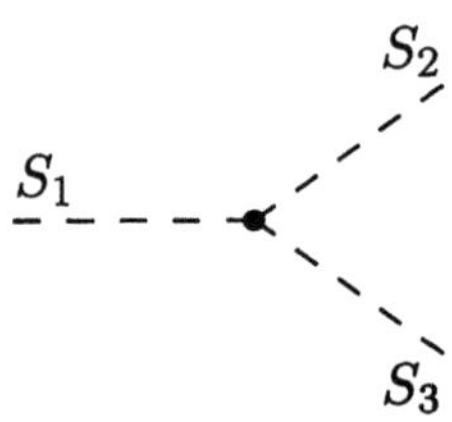

$$= \mathrm{i}eC \tag{A.2.12}$$

with the actual values of S_1, S_2, S_3, and C

$S_1S_2S_3$	$\eta\eta\eta$	$\eta\chi\chi,\ \eta\phi^+\phi^-$
C	$-\frac{3}{2s}\frac{M_H^2}{M_W}$	$-\frac{1}{2s}\frac{M_H^2}{M_W}$

. (A.2.13)

$VVSS$ vertex:

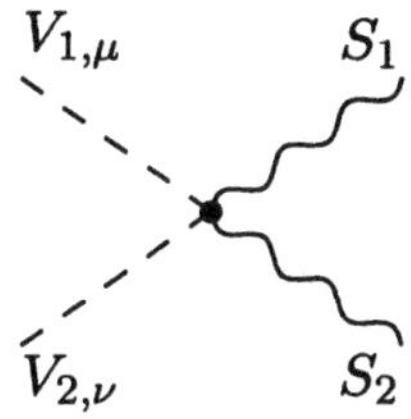

$$= \mathrm{i}e^2 g_{\mu\nu} C \tag{A.2.14}$$

with the actual values of V_1, V_2, S_1, S_2, and C

$V_1V_2S_1S_2$	$ZZ\eta\eta$ $ZZ\chi\chi$	$W^+W^-\eta\eta$ $W^+W^-\chi\chi$ $W^+W^-\phi^+\phi^-$	$AA\phi^+\phi^-$	$ZA\phi^+\phi^-$	$ZZ\phi^+\phi^-$
C	$\frac{1}{2c^2s^2}$	$\frac{1}{2s^2}$	2	$-\frac{c^2-s^2}{cs}$	$\frac{(c^2-s^2)^2}{2c^2s^2}$

$V_1V_2S_1S_2$	$W^\pm A\phi^\mp\eta$	$W^\pm A\phi^\mp\chi$	$W^\pm Z\phi^\mp\eta$	$W^\pm Z\phi^\mp\chi$
C	$-\frac{1}{2s}$	$\mp\frac{\mathrm{i}}{2s}$	$-\frac{1}{2c}$	$\mp\frac{\mathrm{i}}{2c}$

.(A.2.15)

VSS vertex:

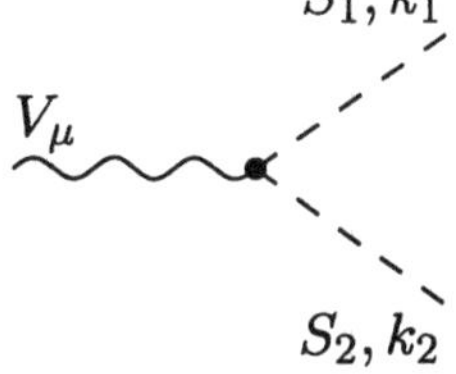

$$= \mathrm{i}eC(k_1 - k_2)_\mu \tag{A.2.16}$$

with the actual values of V, S_1, S_2, and C

VS_1S_2	$Z\chi\eta$	$A\phi^+\phi^-$	$Z\phi^+\phi^-$	$W^\pm\phi^\mp\eta$	$W^\pm\phi^\mp\chi$
C	$-\frac{\mathrm{i}}{2cs}$	-1	$\frac{c^2-s^2}{2cs}$	$\mp\frac{1}{2s}$	$-\frac{\mathrm{i}}{2s}$

. (A.2.17)

SVV vertex:

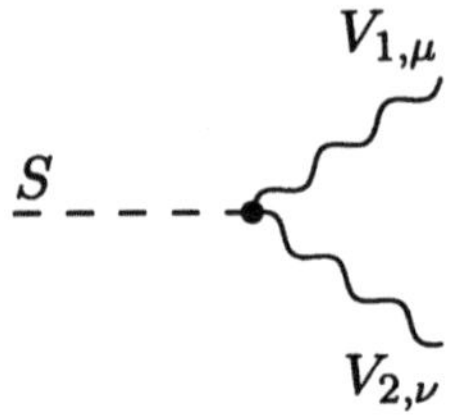

$$= \mathrm{i}e g_{\mu\nu} C \tag{A.2.18}$$

with the actual values of S, V_1, V_2, and C

SV_1V_2	ηZZ	ηW^+W^-	$\phi^\pm W^\mp A$	$\phi^\pm W^\mp Z$
C	$\frac{1}{c^2 s} M_\mathrm{W}$	$\frac{1}{s} M_\mathrm{W}$	$-M_\mathrm{W}$	$-\frac{s}{c} M_\mathrm{W}$

. (A.2.19)

$V\bar{F}F$ vertex:

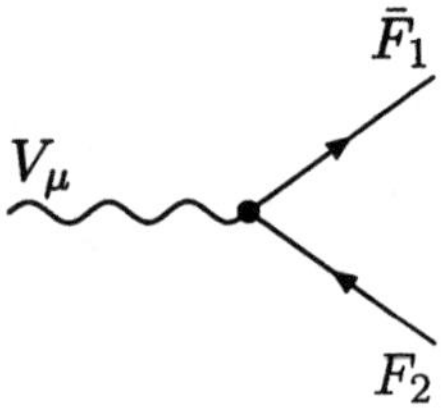

$$= \mathrm{i}e\gamma_\mu \left(C_\mathrm{L} \frac{1-\gamma_5}{2} + C_\mathrm{R} \frac{1+\gamma_5}{2} \right) \tag{A.2.20}$$

with the actual values of V, $\bar{F}_1$, F_2, C_R, and C_L

$V\bar{F}_1F_2$	$A\bar{f}_i f_j$	$Z\bar{f}_i f_j$	$W^+\bar{u}_i d_j$	$W^-\bar{d}_j u_i$	$W^+\bar{\nu}_i l_j$	$W^-\bar{l}_j \nu_i$
C_L	$-Q_f \delta_{ij}$	$g_f^- \delta_{ij}$	$\frac{1}{\sqrt{2}s} V_{ij}$	$\frac{1}{\sqrt{2}s} V_{ji}^\dagger$	$\frac{1}{\sqrt{2}s} \delta_{ij}$	$\frac{1}{\sqrt{2}s} \delta_{ij}$
C_R	$-Q_f \delta_{ij}$	$g_f^+ \delta_{ij}$	0	0	0	0

, (A.2.21)

where

$$g_f^+ = -\frac{s}{c} Q_f, \qquad g_f^- = \frac{I^3_{\mathrm{w},f} - s^2 Q_f}{sc}. \tag{A.2.22}$$

The *vector and axial-vector couplings* of the Z-boson are given by

$$
\begin{aligned}
v_f &= \frac{1}{2}(g_f^- + g_f^+) = \frac{I^3_{\mathrm{W},f} - 2s^2 Q_f}{2sc}, \\
a_f &= \frac{1}{2}(g_f^- - g_f^+) = \frac{I^3_{\mathrm{W},f}}{2sc}.
\end{aligned}
\tag{A.2.23}
$$

$S\bar{F}F$ vertex:

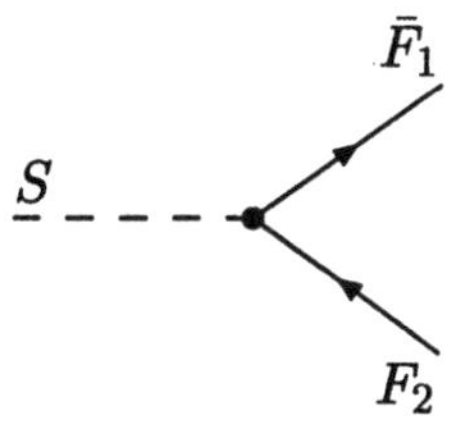

$$
= \mathrm{ie}\left(C_\mathrm{L}\frac{1-\gamma_5}{2} + C_\mathrm{R}\frac{1+\gamma_5}{2}\right) \tag{A.2.24}
$$

with the actual values of S, $\bar{F}_1$, F_2, C_R, and C_L

$S\bar{F}_1F_2$	$\eta\bar{f}_i f_j$	$\chi\bar{f}_i f_j$	$\phi^+\bar{u}_i d_j$	$\phi^-\bar{d}_j u_i$
C_L	$-\frac{1}{2s}\frac{m_{f,i}}{M_\mathrm{W}}\delta_{ij}$	$-\frac{\mathrm{i}}{2s}2I^3_{\mathrm{W},f}\frac{m_{f,i}}{M_\mathrm{W}}\delta_{ij}$	$+\frac{1}{\sqrt{2}s}\frac{m_{u,i}}{M_\mathrm{W}}V_{ij}$	$-\frac{1}{\sqrt{2}s}\frac{m_{d,j}}{M_\mathrm{W}}V^\dagger_{ji}$
C_R	$-\frac{1}{2s}\frac{m_{f,i}}{M_\mathrm{W}}\delta_{ij}$	$\frac{\mathrm{i}}{2s}2I^3_{\mathrm{W},f}\frac{m_{f,i}}{M_\mathrm{W}}\delta_{ij}$	$-\frac{1}{\sqrt{2}s}\frac{m_{d,j}}{M_\mathrm{W}}V_{ij}$	$+\frac{1}{\sqrt{2}s}\frac{m_{u,i}}{M_\mathrm{W}}V^\dagger_{ji}$

$S\bar{F}_1F_2$	$\phi^+\bar{\nu}_i l_j$	$\phi^-\bar{l}_j\nu_i$
C_L	0	$-\frac{1}{\sqrt{2}s}\frac{m_{l,j}}{M_\mathrm{W}}\delta_{ij}$
C_R	$-\frac{1}{\sqrt{2}s}\frac{m_{l,j}}{M_\mathrm{W}}\delta_{ij}$	0

. (A.2.25)

$V\bar{G}G$ vertex:

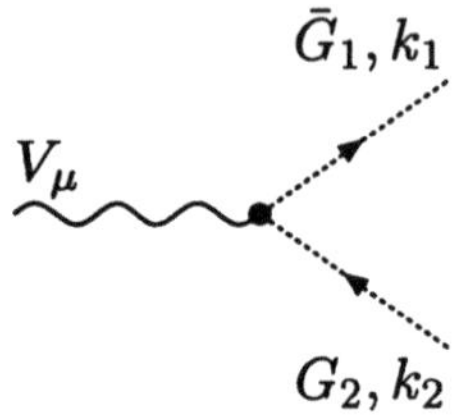

$$= \mathrm{i}ek_{1,\mu}C \tag{A.2.26}$$

with the actual values of V, $\bar{G}_1$, G_2, and C

$V\bar{G}_1G_2$	$A\bar{u}^{\pm}u^{\pm}, W^{\pm}\bar{u}^{A}u^{\mp}$ $W^{\mp}\bar{u}^{\mp}u^{A}$	$Z\bar{u}^{\pm}u^{\pm}, W^{\pm}\bar{u}^{Z}u^{\mp}$ $W^{\mp}\bar{u}^{\mp}u^{Z}$
C	± 1	$\mp\frac{c}{s}$

. (A.2.27)

$S\bar{G}G$ vertex:

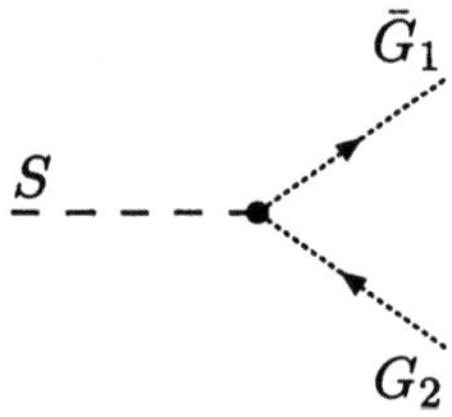

$$= \mathrm{i}eC\xi_1 \tag{A.2.28}$$

with $\xi_1 = \xi_W, \xi_Z$ for $G_1 = u^{\pm}, u^{Z}$, respectively, and the actual values of S, $\bar{G}_1$, G_2, and C

$S\bar{G}_1G_2$	$H\bar{u}^{Z}u^{Z}$	$H\bar{u}^{\pm}u^{\pm}$	$\chi\bar{u}^{\pm}u^{\pm}$
C	$-\frac{1}{2c^2s}M_{\mathrm{W}}$	$-\frac{1}{2s}M_{\mathrm{W}}$	$\mp\frac{\mathrm{i}}{2s}M_{\mathrm{W}}$

$S\bar{G}_1G_2$	$\phi^{\pm}\bar{u}^{\pm}u^{A}$	$\phi^{\pm}\bar{u}^{\pm}u^{Z}$	$\phi^{\pm}\bar{u}^{Z}u^{\mp}$
C	M_{W}	$-\frac{c^2-s^2}{2cs}M_{\mathrm{W}}$	$\frac{1}{2cs}M_{\mathrm{W}}$

. (A.2.29)

General references

[EP] Elementary particle physics:
D. Griffiths, *Introduction to Elementary Particle Physics*, (John Wiley & Sons, New York, etc., 1987);
E. Lohrmann, *Hochenergiephysik*, (Teubner, Stuttgart, 1992);
D.H. Perkins, *Introduction to High-Energy Physics*, (Addison-Wesley, Menlo Park, 1987);
W.B. Rolnick, *The Fundamental Particles and their Interactions*, (Addison-Wesley, Reading, Massachusetts, 1994).

[EW] Electroweak interaction:
E.D. Commins and P.H. Bucksbaum, *Weak Interactions of Leptons and Quarks*, (Cambridge University Press, Cambridge, 1983);
K. Grotz and H.V. Klapdor, *The Weak Interaction in Nuclear, Particle and Astrophysics*, (Hilger, Bristol, etc., 1990);
P. Renton, *Electroweak Interactions: an Introduction to the Physics of Quarks and Leptons*, (Cambridge University Press, Cambridge, 1990).

[FT] Field theory:
J.D. Bjorken and S. Drell, *Relativistic Quantum Mechanics/Relativistic Quantum Fields*, (McGraw–Hill, New York, 1964);
C. Itzykson and J.B. Zuber, *Quantum Field Theory*, (McGraw–Hill, New York, 1980);
C. Kaku, *Quantum Field Theory*, (Oxford University Press, New York, Oxford, 1993);
T. Kugo, *Eichtheorie*, (Springer, Berlin, etc., 1997);
M.E. Peskin and D.V. Schroeder, *An Introduction to Quantum Field Theory*, (Addison-Wesley, Reading, Massachusetts, 1995);
S. Pokorski, *Gauge Field Theories*, (Cambridge University Press, Cambridge, 2000);
S. Weinberg, *The Quantum Theory of Fields, Vol. 1 and 2*, (Cambridge University Press, Cambridge, 1995).

[GSW] Glashow–Salam–Weinberg theory:
E.S. Abers and B.W. Lee, *Phys. Rept.* **9** (1973) 1;
J.C. Taylor, *Gauge Theories of Weak interactions*, (Cambridge University Press, Cambridge, 1976);
D. Bardin and G. Passarino, *The Standard Model in the Making*, (Oxford University Press, Oxford, 1999).

[GT] Group theory:
H. Georgi, *Lie Algebras in Particle Physics: from Isospin to Unified Theories*, (Benjamin/Cummings, Reading, Massachusetts, 1982);
M. Hamermesh, *Group Theory and its Aplication to Physical Problems*, (Addison Wesley, Reading, Massachusetts, 1962);
D.B. Lichtenberg, *Unitary Symmetry and Elementary Particles*, (Academic Press, New York, 1978);
B.G. Wybourne, *Classical Groups for Physicists*, (John Wiley & Sons, New York, 1974).

[GUT] Grand Unified Theories:
P. Langacker, *Phys. Rept.* **72** (1981) 185;
G.G. Ross, *Grand Unified Theories*, (Benjamin-Cummings, Reading, Massachusetts, 1984).

[QED] Quantum Electrodynamics:
J.D. Bjorken and S. Drell, *Relativistic Quantum Mechanics/Relativistic Quantum Fields*, (McGraw-Hill, New York, 1964);
J.M. Jauch and F. Rohrlich, *The Theory of Photons and Electrons*, (Springer, Berlin, 1976);
T. Kinoshita, *Quantum Electrodynamics*, (World Scientific, Singapore, 1990).

[StM] Statistical Mechanics:
R.P. Feynman, *Statistical Mechanics: A Set of Lectures*, (W.A. Benjamin Inc., Reading, Massachusetts, 1972);
V.N. Popov, *Functional Integrals in Quantum Field Theory and Statistical Physics*, (D. Reidel Publishing Company, Dordrecht, etc., 1983);
J. Zinn-Justin, *Quantum Field Theory and Critical Phenomena*, (Oxford University Press, Oxford, 1989).

[SUSY] Supersymmetry:
P. Fayet and S. Ferrara, *Phys. Rept.* **32** (1977) 249;
H.E. Haber and G.L. Kane, *Phys. Rept.* **117** (1985) 75;
H.P. Nilles, *Phys. Rept.* **110** (1984) 1;
J. Wess and J. Bagger, *Supersymmetry and Supergravity*, (Princeton University Press, Princeton, New Jersey, 1992).

[LAT] Lattice gauge theory:
M. Creutz, *Quarks, Gluons and Lattices*, (Cambridge University Press, Cambridge, 1983);
I. Montvay and G. Münster, *Quantum Fields on a Lattice*, (Cambridge University Press, Cambridge, 1994);
H.J. Rothe, *Lattice Gauge Theories: an Introduction, 2nd ed.*, (World Scientific, Singapore, etc., 1997).

Index